Age of Information

At the forefront of cutting-edge technologies, this text provides a comprehensive treatment of a crucial network performance metric, ushering in new opportunities for rethinking the whole design of communication systems. A detailed exposition of the communication and network theoretic foundations of Age of Information (AoI) gives the reader a solid background, and a discussion of the implications for signal processing and control theory shed light on the important potential of recent research. The text includes extensive real-world applications of this vital metric, including caching, the Internet of Things (IoT), and energy-harvesting networks. The far-reaching applications of AoI include networked monitoring systems, cyber-physical systems such as the IoT, and information-oriented systems and data analytics applications ranging from the stock market to social networks. The future of this exciting subject in 5G communication systems and beyond makes this a vital resource for graduate students, researchers and professionals.

Nikolaos Pappas is an associate professor at the Department of Computer and Information Science, Linköping University. He serves as an editor for four IEEE journals.

Mohamed A. Abd-Elmagid is a graduate research assistant with the Department of Electrical and Computer Engineering at Virginia Tech.

Bo Zhou is a professor at the College of Electronic and Information Engineering, Nanjing University of Aeronautics and Astronautics, China. He received the best paper awards at IEEE GLOBECOM in 2018 and IFIP NTMS in 2019. He was recognized as an exemplary reviewer of IEEE Transactions on Communications in 2019 and 2020.

Walid Saad is a professor at the Department of Electrical and Computer Engineering at Virginia Tech, where he leads the Network Science, Wireless and Security laboratory. He is the author/coauthor of ten conference best paper awards and recipient of the 2015 IEEE ComSoc Fred W. Ellersick Prize. He is an IEEE fellow.

Harpreet S. Dhillon is a professor of Electrical and Computer Engineering at Virginia Tech. He is a Clarivate Analytics Highly Cited Researcher and a recipient of six best paper awards, including the IEEE Leonard G. Abraham Prize, the IEEE Heinrich Hertz Award, and the IEEE Communications Society Young Author Best Paper Award.

"As communication networks become increasingly integrated with sensing, computing, and storage platforms, the age of information is emerging as a versatile and tractable performance measure that captures the relevance of information for applications such as control and machine learning. Edited and contributed by leading experts in the field, this book is an excellent introduction to the topic and an essential reference for researchers in communications."

Osvaldo Simeone, King's College London

"With the inception of the fifth generation (5G) of wireless cellular systems and its projected use in various real-time applications there has been an increased interest in the low-latency performance of wireless systems. Beyond the sole focus of low latency, during the last decade the notion of Age of Information (AoI), information freshness, and its derivatives have sparked interest into analysis and design of communication systems with general timing requirements. This book provides a timely and fresh view on the area of AoI, featuring a highly competent Editorial team and excellent contributions from well-established researchers. Various important aspects are covered, ranging from analytical treatment and stochastic models for AoI, scheduling and transmission policies, interplay with machine learning techniques, as well as the very recent aspects of data economics and valuation. Beyond 5G and towards 6G systems, the relevance of timing performance in wireless communication in systems and standards will only continue to grow, rendering this book into a valuable and reliable reference for researchers and engineers."

Petar Popovski, Aalborg University, Denmark

Age of Information

Foundations and Applications

Edited by

NIKOLAOS PAPPAS
Linköping University, Sweden

MOHAMED A. ABD-ELMAGID
Virginia Tech

BO ZHOU
Nanjing University of Aeronautics and Astronautics

WALID SAAD
Virginia Tech

HARPREET S. DHILLON
Virginia Tech

CAMBRIDGE
UNIVERSITY PRESS

CAMBRIDGE
UNIVERSITY PRESS

Shaftesbury Road, Cambridge CB2 8EA, United Kingdom

One Liberty Plaza, 20th Floor, New York, NY 10006, USA

477 Williamstown Road, Port Melbourne, VIC 3207, Australia

314–321, 3rd Floor, Plot 3, Splendor Forum, Jasola District Centre, New Delhi – 110025, India

103 Penang Road, #05–06/07, Visioncrest Commercial, Singapore 238467

Cambridge University Press is part of Cambridge University Press & Assessment, a department of the University of Cambridge.

We share the University's mission to contribute to society through the pursuit of education, learning and research at the highest international levels of excellence.

www.cambridge.org
Information on this title: www.cambridge.org/9781108837873
DOI: 10.1017/9781108943321

First published 2023

A catalogue record for this publication is available from the British Library.

A Cataloging-in-Publication data record for this book is available from the Library of Congress.

ISBN 978-1-108-83787-3 Hardback

Contents

Contributors

Ahmed Arafa
Department of Electrical and Computer Engineering, University of North Carolina at Charlotte, United States

Sajjad Baghaee
Department of Electrical and Electronics Engineering, Middle East Technical University, Turkey

Bo Bai
Theory Lab (FKA Future Network Theory Lab), 2012 Labs, Huawei Technologies Co., Ltd., Hong Kong

John S. Baras
Institute for Systems Research, University of Maryland College Park, United States

Melih Bastopcu
Department of Electrical and Computer Engineering, University of Maryland, United States

Ahmed M. Bedewy
Department of ECE, The Ohio State University, Columbus, OH USA

Hasan Burhan Beytur
Department of Electrical and Computer Engineering, University of Texas at Austin, United States

Donald Richard Brown III
Worcester Polytechnic Institute, United States

Baturalp Buyukates
Department of Electrical and Computer Engineering, University of Maryland, United States

Elif Tugce Ceran
Department of Electrical and Electronic Engineering, Middle East Technical University, Turkey

Jaya Champati
School of Electrical Engineering and Computer Science, KTH Royal Institute of Technology, Sweden

Marian Codreanu
Department of Science and Technology (ITN), Linköping University, Sweden

Huaiyu Dai
Department of Electrical and Computer Engineering, North Carolina State University, United States

György Dán
Division of Network and Systems Engineering, School of Electrical Engineering and Computer Science, KTH Royal Institute of Technology, Sweden

Chenghao Deng
Department of Electronic Engineering, Tsinghua University, China

Anthony Ephremides
Electrical and Computer Engineering, Institute for Systems Research, University of Maryland, United States

Atilla Eryilmaz
Ohio State University, United States

Shahab Farazi
Worcester Polytechnic Institute, United States

Songtao Feng
School of Electrical Engineering and Computer Science, Pennsylvania State University, United States

James Gross
School of Electrical Engineering and Computer Science, KTH Royal Institute of Technology, Sweden

Deniz Gunduz
Department of Electrical and Electronic Engineering, Imperial College London, United Kingdom

András György
Deepmind, United Kingdom

Qing He
Division of Network and Systems Engineering, School of Electrical Engineering and Computer Science, KTH Royal Institute of Technology, Sweden

Jianwei Huang
School of Science and Engineering, Chinese University of Hong Kong, China

Yoshiaki Inoue
Department of Information and Communications Technology, Osaka University, Japan

Bo Ji
Virginia Tech, United States

Karl Henrik Johansson
Division of Decision and Control Systems, KTH Royal Institute of Technology, Sweden

Clement Kam
U.S. Naval Research Laboratory, United States

Sanjit K. Kaul
Indraprastha Institute of Information Technology Delhi, India

Onur Kaya
Department of Electrical and Electronics Engineering, Isik University, Turkey

Andrew G. Klein
Western Washington University College of Science and Engineering, United States

Sastry Kompella
Information Technology Division, Naval Research Laboratory, United States

Markus Leinonen
Centre for Wireless Communications – Radio Technologies, University of Oulu, Finland

Bin Li
Pennsylvania State University, United States

Juan Liu
School of Electrical Engineering and Computer Science, Ningbo University, China

Mohammad Moltafet
Centre for Wireless Communications – Radio Technologies, University of Oulu, Finland

Tasmeen Zaman Ornee
Department of Electrical and Computer Engineering, Auburn University, United States

H. Vincent Poor
Electrical Engineering Department, Princeton University, United States

Tony Q. S. Quek
Singapore University of Technology and Design, Singapore

Ness Shroff
Electrical and Computer Engineering Department, Ohio State University, United States

Touraj Soleymani
Division of Decision and Control Systems, KTH Royal Institute of Technology, Sweden

Yin Sun
Department of Electrical and Computer Engineering, Auburn University, United States

Tetsuya Takine
Department of Information and Communications Technology, Osaka University, Japan

Toshiyuki Tanaka
Department of Systems Science, Kyoto University, Japan

Sennur Ulukus
Department of Electrical and Computer Engineering, University of Maryland, United States

Elif Uysal
Middle East Technical University (METU), Turkey

Boyu Wang
Konux, China

Xijun Wang
School of Electronics and Communication Engineering, Sun Yat-sen University, China

Howard H. Yang
Zhejiang University/University of Illinois at Urbana-Champaign Institute, China

Jing Yang
School of Electrical Engineering and Computer Science, Pennsylvania State University, United States

Roy D. Yates
Rutgers University, United States

Di Yuan
Division of Computing Science, Department of Information Technology, Uppsala University, Sweden

Meng Zhang
Department of Electrical and Computer Engineering, Northwestern University, United States

Acknowledgments

The editors would like to express their gratitude to all the authors and reviewers who contributed to this book. In addition, the editors gratefully acknowledge the support of the Swedish Research Council (VR), the Excellence Center at Linköping-Lund in Information Technology (ELLIIT), the Center for Industrial Information Technology (CENIIT), the US National Science Foundation (Grants CNS-1739642 and CNS-1814477), and the US Office of Naval Research under MURI Grant N00014-19-1-2621.

1 The Probability Distribution of the Age of Information

Yoshiaki Inoue, Tetsuya Takine, and Toshiyuki Tanaka

The freshness of information is the most important factor in designing real-time monitoring systems. The theory of the Age of Information (AoI) provides an explicit way to incorporate this perspective into the system design. This chapter is aimed at introducing the basic concept of the AoI and explaining its mathematical aspects. We first present a general introduction to the AoI and its standard analytical method. We then proceed to advanced material regarding the characterization of distributional properties of the AoI. Some bibliographical notes are also provided at the end of this chapter.

1.1 A General Introduction to the Age of Information

The Age of Information is a performance metric quantifying *the information freshness* in real-time monitoring systems. Let us consider a situation that a time-varying information source is monitored remotely (Figure 1.1). A sensor is attached to the information source and it observes (samples) the current status with some frequency. Each time the sensor samples the information source, it generates an update containing the obtained sample and sends it to a remote server, where some computational task is performed to extract state-information from raw data. The extracted information is then transmitted to a monitor and updates the information being displayed. The AoI A_t at time t is defined as *the elapsed time from the generation time* η_t *of the update whose information is displayed on the monitor at time* t:

$$A_t := t - \eta_t, \quad t \geq 0. \tag{1.1}$$

We thus have $\eta_t = t - A_t$, that is, the information displayed at time t was reported by an update generated at time $t - A_t$ by the sensor. In this sense, the AoI A_t directly quantifies the freshness of the information being displayed: the smaller value the AoI A_t takes, the fresher the information is. Because "the sampling of the information source" and "the generation of an update" occur simultaneously, these words could be used interchangeably. In the rest of this chapter, however, we shall describe the behavior of the system, consistently focusing only on the generation of updates.

By definition, η_t ($t \geq 0$) is a piecewise constant function of t: η_t jumps upward when the displayed information is updated, while it does not change its value elsewhere. Therefore, we see from (1.1) that A_t is a piecewise linear function of t with downward jumps at update instants. In particular, the AoI plotted along the time axis

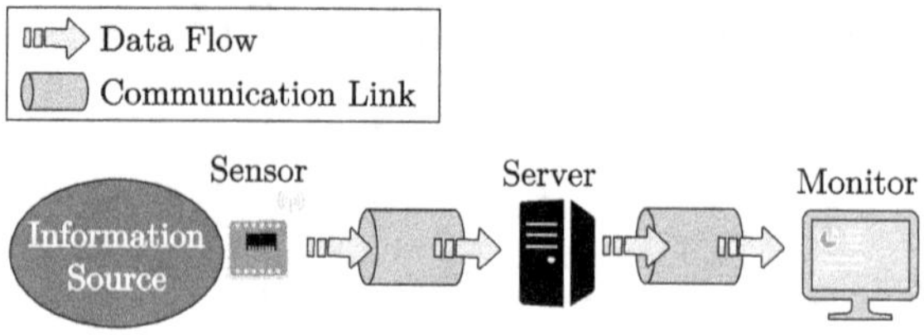

Figure 1.1 A remote monitoring system.

has a sawtooth graph as depicted in Figure 1.2, where $\beta_\ell^\dagger$ and $\alpha_\ell^\dagger$ ($\ell = 0, 1, \ldots$) represent the ℓth update time at the monitor and the generation time of that update at the sensor.

Note here that some updates generated by the sensor may *not* be displayed on the monitor forever because of the loss in communication links or subsequent updates' overtaking, where the latter is typically due to the management policy at the server (see Figure 1.1). Therefore, $(\alpha_\ell^\dagger)_{\ell=0,1,\ldots}$ and $(\beta_\ell^\dagger)_{\ell=0,1,\ldots}$ shown in Figure 1.2 are subsequences of $(\alpha_n)_{n=0,1,\ldots}$ of generation times of updates and $(\beta_n)_{n=0,1,\ldots}$ of their reception times, where $\beta_n = \infty$ if the update generated at time α_n is not displayed on the monitor forever. In this sense, we refer to $\alpha_\ell^\dagger$ and $\beta_\ell^\dagger$ as the generation and reception times of the ℓth *effective* update. A more formal discussion on overtaking of updates will be given in Section 1.3.1.

Let $G_\ell^\dagger$ and $D_\ell^\dagger$ denote the intergeneration time and the system delay of the ℓth effective update:

$$G_\ell^\dagger = \alpha_\ell^\dagger - \alpha_{\ell-1}^\dagger, \qquad D_\ell^\dagger = \beta_\ell^\dagger - \alpha_\ell^\dagger. \tag{1.2}$$

We observe from the definition of η_t that the AoI just after an update of the monitor equals the system delay experienced by the latest update:

$$A_{\beta_\ell^\dagger} = \beta_\ell^\dagger - \alpha_\ell^\dagger = D_\ell^\dagger.$$

On the other hand, the AoI just before an update of the monitor is called *the peak AoI*, as it corresponds to the peak of the sawtooth graph of the AoI process:

$$A_{\text{peak},\ell} := \lim_{t \to \beta_\ell^\dagger -} A_t = \beta_\ell^\dagger - \alpha_{\ell-1}^\dagger = D_\ell^\dagger + G_\ell^\dagger, \tag{1.3}$$

where the last equality follows from (1.2). We thus see that for each interval $[\beta_\ell^\dagger, \beta_{\ell+1}^\dagger)$ between the monitor's updates, the AoI process linearly increases from the system delay $D_\ell^\dagger$ to the peak AoI $A_{\text{peak},\ell+1}^\dagger$ with slope one (see Figure 1.2). This observation highlights the key difference between the AoI and the conventional delay metric: *The delay $D_\ell^\dagger$ represents only the information freshness immediately after an update* and it does not provide any information about the evolution of the information freshness between updates.

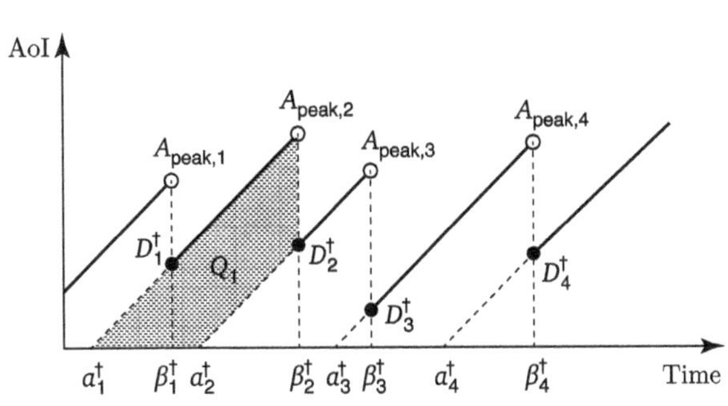

Figure 1.2 An example of the AoI process.

1.1.1 A Graphical Analysis of the Average AoI

Since the AoI $(A_t)_{t\geq 0}$ is a time-varying process, we need to consider some summary metric for the system performance, which is obtained by applying a functional to $(A_t)_{t\geq 0}$. The most commonly used summary metric is the time-averaged AoI defined as

$$m_A^\sharp := \lim_{T\to\infty} \frac{1}{T} \int_0^T A_t \mathrm{d}t. \tag{1.4}$$

The average AoI $m_A^\sharp$ can be analyzed with a graphical argument. Here, we provide only an informal argument to avoid technical complications; a more rigorous proof will be given in Section 1.3.

Observe that the area of the shaded trapezoid in Figure 1.2 is given by

$$Q_i = \frac{(A_{\text{peak},i+1})^2}{2} - \frac{(D_{i+1}^\dagger)^2}{2}.$$

By summing up Q_i for $i = 1, 2, \ldots$, we can calculate the area under the graph of the AoI, except for both ends. The boundary effects are negligible under suitable regularity conditions and we obtain

$$m_A^\sharp = \lim_{T\to\infty} \frac{M(T)}{T} \cdot \frac{1}{M(T)} \sum_{i=1}^{M(T)} Q_i = \lambda^\dagger \left\{ \frac{m_{(A_{\text{peak}})^2}}{2} - \frac{m_{(D^\dagger)^2}}{2} \right\}, \tag{1.5}$$

where $M(T) = \max\{n;\ \beta_n^\dagger \leq T\}$ denotes the total number of displayed information updates in time interval $(0, T]$, $\lambda^\dagger := \lim_{T\to\infty} M(T)/T$ denotes the average effective update rate, and

$$m_{(A_{\text{peak}})^2} := \lim_{N\to\infty} \frac{1}{N} \sum_{i=1}^N (A_{\text{peak},i})^2, \quad m_{(D^\dagger)^2} := \lim_{N\to\infty} \frac{1}{N} \sum_{i=1}^N (D_i^\dagger)^2.$$

If the system is represented as a stationary and ergodic stochastic process,[1] (1.5) is equivalent to the following relation for the mean AoI and the second moments of the peak AoI and system delay:

$$\mathrm{E}[A] = \lambda^{\dagger} \cdot \frac{\mathrm{E}[(A_{\mathrm{peak}})^2] - \mathrm{E}[(D^{\dagger})^2]}{2}, \tag{1.6}$$

where A, A_{peak}, and $D^{\dagger}$ denote generic random variables for stationary A_t, $A_{\mathrm{peak},\ell}$, and $D_{\ell}^{\dagger}$. Furthermore, using (1.3) and noting the stationarity, (1.6) is rewritten as

$$\mathrm{E}[A] = \lambda^{\dagger} \left(\frac{\mathrm{E}[(G^{\dagger})^2]}{2} + \mathrm{E}\left[G_{\ell}^{\dagger} D_{\ell}^{\dagger}\right] \right), \tag{1.7}$$

where $G^{\dagger}$ denotes a generic random variable for stationary $G_{\ell}^{\dagger}$. It is worth noting that the peak AoI is also given in terms of the system delay $D_{\ell}^{\dagger}$ and the inter-update time $J_{\ell}^{\dagger} := \beta_{\ell}^{\dagger} - \beta_{\ell-1}^{\dagger}$ by

$$A_{\mathrm{peak},\ell+1} = D_{\ell}^{\dagger} + J_{\ell+1}^{\dagger},$$

so that we have yet another equivalent formula for $\mathrm{E}[A]$:

$$\mathrm{E}[A] = \lambda^{\dagger} \left(\frac{\mathrm{E}[(J^{\dagger})^2]}{2} + \mathrm{E}\left[D_{\ell}^{\dagger} J_{\ell+1}^{\dagger}\right] \right), \tag{1.8}$$

where $J^{\dagger}$ denotes a generic random variable for stationary $J_{\ell}^{\dagger}$. While these three formulas (1.6), (1.7), and (1.8) are equivalent, the choice of which expression to use in the analysis often affects the degree of tractability. In the following subsection, we briefly demonstrate applications of these formulas to first-come first-served (FCFS) single-server queueing models.

1.1.2 Queueing Modeling of Monitoring Systems

As we have seen, the AoI process $(A_t)_{t\geq 0}$ is characterized in terms of effective generation times $(\alpha_{\ell}^{\dagger})_{\ell=0,1,\ldots}$ of updates by the sensor and their reception times $(\beta_{\ell}^{\dagger})_{\ell=0,1,\ldots}$ by the monitor. Because $\beta_{\ell}^{\dagger} \geq \alpha_{\ell}^{\dagger}$ always holds by their definitions, we can think of *a queueing system* defined by the sequences of *arrival times* $(\alpha_{\ell}^{\dagger})_{\ell=0,1,\ldots}$ and *departure times* $(\beta_{\ell}^{\dagger})_{\ell=0,1,\ldots}$.[2] To be more specific, consider a virtual *service system*, where updates enter it immediately after their generations and leave it when they are received by the monitor (see Figure 1.3). This service system can be considered as an abstraction of a series of components that intermediate between the information source and the monitor. For example, it may represent a communication network to transfer update

[1] Stationarity refers to the property that probability distributions representing the system dynamics are time-invariant. Also, ergodicity refers to the property that time-averages coincide with corresponding ensemble averages. A more detailed explanation will be given in Section 1.3.3.

[2] In the later sections, we will consider a modeling that includes arrival and departure times of noneffective updates as well as effective updates.

Figure 1.3 An abstraction of monitoring systems (cf. Figure 1.1).

packets, a server to extract status information from raw data, or a combination of them (see Figures 1.1 and 1.3).

The simplest model of the service system would be an FCFS single-server queue, that is, assuming that the service system consists of a first-in first-out (FIFO) buffer and a server. In what follows, we consider the mean AoI $\mathrm{E}[A]$ in two different FCFS single-server queues, where service times are assumed to follow an exponential distribution with mean $1/\mu$ ($\mu > 0$).

First, suppose that the intergeneration time $G^\dagger$ of effective updates is constant and equal to τ, that is, the D/M/1 queue in Kendall's notation. For the system stability, we assume $\tau > 1/\mu$. In this case, the formula in (1.7) is useful because we readily have $\lambda^\dagger = 1/\tau$, $\mathrm{E}[(G^\dagger)^2] = \tau^2$, and $\mathrm{E}[G^\dagger_\ell D^\dagger_\ell] = \tau \mathrm{E}[D^\dagger]$, yielding

$$\mathrm{E}[A] = \frac{\tau}{2} + \mathrm{E}[D^\dagger].$$

From the elementary queueing theory (Kleinrock 1975, p. 252), the system delay in the FCFS D/M/1 queue follows an exponential distribution with mean $1/\{\mu(1 - x^\star)\}$, where $x^\star$ is the unique solution of the following equation:

$$x = e^{-\tau\mu(1-x)}, \quad 0 < x < 1. \tag{1.9}$$

The mean AoI is thus given by

$$\mathrm{E}[A] = \frac{\tau}{2} + \frac{1}{\mu(1 - x^\star)}. \qquad \text{(D/M/1)}$$

Next, suppose that the sensor randomly generates effective updates according to a Poisson process with rate $\lambda^\dagger$, that is, the FCFS M/M/1 queue, where we assume $0 < \lambda^\dagger < \mu$ for stability. The intergeneration time $G^\dagger$ then follows an exponential distribution with mean $1/\lambda^\dagger$. In this case, the formula in (1.8) provides an easy way to calculate $\mathrm{E}[A]$. If an update does not depart the service system before the arrival of the next update (i.e., $D^\dagger_\ell \geq G^\dagger_{\ell+1}$), then the next service starts just after the departure, so that the next inter-departure time $J^\dagger_{\ell+1}$ equals a service time. If $D^\dagger_\ell < G^\dagger_{\ell+1}$, on the other hand, $J^\dagger_{\ell+1}$ equals the sum of the residual inter-arrival time (exponentially distributed with mean $1/\lambda^\dagger$ because of the memoryless property) and a service time. In both cases, $J^\dagger_{\ell+1}$ is conditionally independent of $D^\dagger_\ell$, given either $D^\dagger_\ell \geq G^\dagger_{\ell+1}$ or $D^\dagger_\ell < G^\dagger_{\ell+1}$. Using the fact that the system delay in the FCFS M/M/1 queue follows an exponential distribution with mean $1/(\mu - \lambda^\dagger)$ (Kleinrock 1975, p. 205), we obtain

$$
\begin{aligned}
\mathrm{E}\left[D_\ell^\dagger J_{\ell+1}^\dagger\right] &= \mathrm{E}\left[\mathbb{1}\{D_\ell^\dagger \geq G_{\ell+1}^\dagger\} D_\ell^\dagger J_{\ell+1}^\dagger\right] + \mathrm{E}\left[\mathbb{1}\{D_\ell^\dagger < G_{\ell+1}^\dagger\} D_\ell^\dagger J_{\ell+1}^\dagger\right] \\
&= \int_0^\infty x \cdot \frac{1}{\mu} \cdot (1 - e^{-\lambda^\dagger x}) \cdot (\mu - \lambda^\dagger) e^{-(\mu-\lambda^\dagger)x} \mathrm{d}x \\
&\qquad + \int_0^\infty x \cdot \left(\frac{1}{\lambda^\dagger} + \frac{1}{\mu}\right) \cdot e^{-\lambda^\dagger x} \cdot (\mu - \lambda^\dagger) e^{-(\mu-\lambda^\dagger)x} \mathrm{d}x \\
&= \frac{1}{\mu\lambda^\dagger}\left(1 - \rho + \frac{\rho}{1-\rho}\right),
\end{aligned}
\tag{1.10}
$$

and

$$
\begin{aligned}
\mathrm{E}\left[(J^\dagger)^2\right] &= \mathrm{E}\left[\mathbb{1}\{D_\ell^\dagger \geq G_{\ell+1}^\dagger\}(J^\dagger)^2\right] + \mathrm{E}\left[\mathbb{1}\{D_\ell^\dagger < G_{\ell+1}^\dagger\}(J^\dagger)^2\right] \\
&= \int_0^\infty \frac{2}{\mu^2} \cdot (1 - e^{-\lambda^\dagger x}) \cdot (\mu - \lambda^\dagger) e^{-(\mu-\lambda^\dagger)x} \mathrm{d}x \\
&\qquad + \int_0^\infty \left(\frac{2}{(\lambda^\dagger)^2} + \frac{2}{\mu\lambda^\dagger} + \frac{2}{\mu^2}\right) \cdot e^{-\lambda^\dagger x} (\mu - \lambda^\dagger) e^{-(\mu-\lambda^\dagger)x} \mathrm{d}x \\
&= \frac{2\rho^2}{\mu\lambda^\dagger} + \frac{2(1-\rho)}{(\lambda^\dagger)^2} \cdot (1 + \rho + \rho^2),
\end{aligned}
\tag{1.11}
$$

where $\rho := \lambda^\dagger/\mu$ denotes the traffic intensity. Therefore, we obtain from (1.8), (1.10), and (1.11),

$$
\mathrm{E}[A] = \frac{1}{\mu}\left(1 + \frac{1}{\rho} + \frac{\rho^2}{1-\rho}\right). \qquad \text{(M/M/1)}
$$

We conclude this section by presenting some numerical examples. We set the time unit so that $\mu = 1$ holds throughout. Figure 1.4 shows the mean AoI $\mathrm{E}[A]$ as a function of the generation rate $\lambda^\dagger$ (i.e., the rate at which effective updates are generated). We observe that $\mathrm{E}[A]$ forms a U-shaped curve with respect to $\lambda^\dagger$. This U-shaped curve of $\mathrm{E}[A]$ is understood to be due to *the trade-off between the generation interval $G^\dagger$ and the system delay $D^\dagger$*: while reducing the generation interval would be effective in keeping the information fresher, it would also increase the delay in the service system. To illustrate this fact, in Figure 1.4, we also plot the mean backward recurrence time $\mathrm{E}[(G^\dagger)^2]/(2\mathrm{E}[G^\dagger])$ of the generation process (i.e., the expected elapsed time since the last generation instant) and the mean system delay $\mathrm{E}[D]$. Note here that the sum of these terms $\mathrm{E}[(G^\dagger)^2]/(2\mathrm{E}[G^\dagger]) + \mathrm{E}[D]$ can be regarded as an approximation to the mean AoI $\mathrm{E}[A]$ ignoring the dependence between $G_\ell^\dagger$ and $D_\ell^\dagger$ (cf. (1.7)). We observe that the former term is dominant for small $\lambda^\dagger$, whereas the latter term becomes dominant for large values of $\lambda^\dagger$. The mean AoI $\mathrm{E}[A]$ is then minimized at a moderate value of $\lambda^\dagger$, where these effects on $\mathrm{E}[A]$ are appropriately balanced.

Figure 1.5 compares the mean AoI in the FCFS M/M/1 and D/M/1 queues. We observe that constant generation intervals are preferable to exponential generation intervals in terms of the mean AoI. Intuitively, the superiority of constant generation intervals can be understood as follows. Firstly, the constant generation intervals

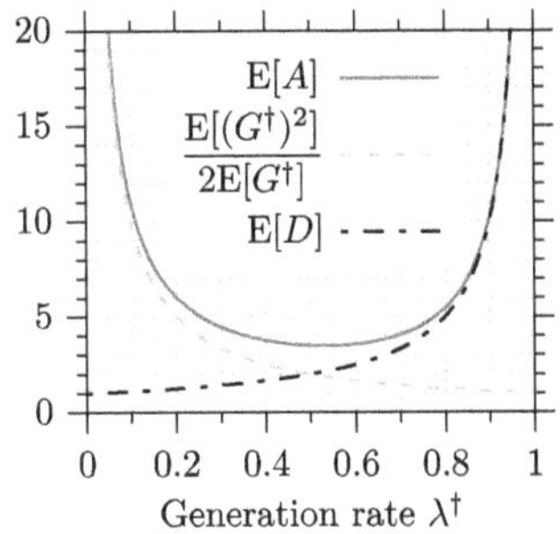

Figure 1.4 The mean AoI E[A] in the FCFS M/M/1 queue.

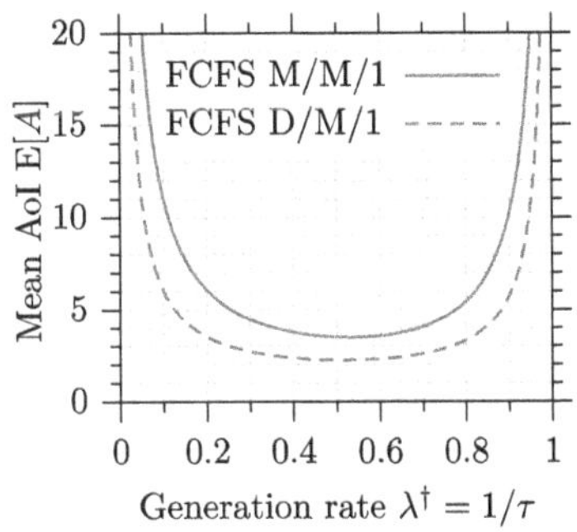

Figure 1.5 The mean AoI E[A] in the FCFS M/M/1 and D/M/1 queues.

minimize the mean backward recurrence time of the generation process, which can be verified with

$$\frac{\mathrm{E}[(G^{\dagger})^2]}{2\mathrm{E}[G^{\dagger}]} = \frac{\mathrm{Var}[G^{\dagger}] + (\mathrm{E}[G^{\dagger}])^2}{2\mathrm{E}[G^{\dagger}]} = \frac{1 + (\mathrm{Cv}[G^{\dagger}])^2}{2} \cdot \mathrm{E}[G^{\dagger}],$$

where $\mathrm{Cv}[Y] = \sqrt{\mathrm{Var}[Y]}/\mathrm{E}[Y]$ denotes the coefficient of variation of random variable Y. Because we have $\mathrm{Cv}[G^{\dagger}] \geq 0$ and the equality holds if and only if $G^{\dagger}$ is constant, the mean backward recurrence time is minimized by constant generation intervals. Secondly, the mean delay $1/\{\mu(1 - x^{\star})\}$ in the FCFS D/M/1 queue is smaller than the mean delay $1/\{\mu(1 - \rho)\}$ in the FCFS M/M/1 queue because $x^{\star}$ defined by (1.9) is smaller than ρ.

1.2 The Probability Distribution of the AoI

The rest of this chapter is devoted to discussions on the *probability distribution* of the AoI. As mentioned in the previous section, we need to use some summary metric of the AoI process $(A_t)_{t\geq 0}$ for performance evaluation. The *asymptotic frequency distribution (AFD)* of the AoI process is considered to be one of the most fundamental

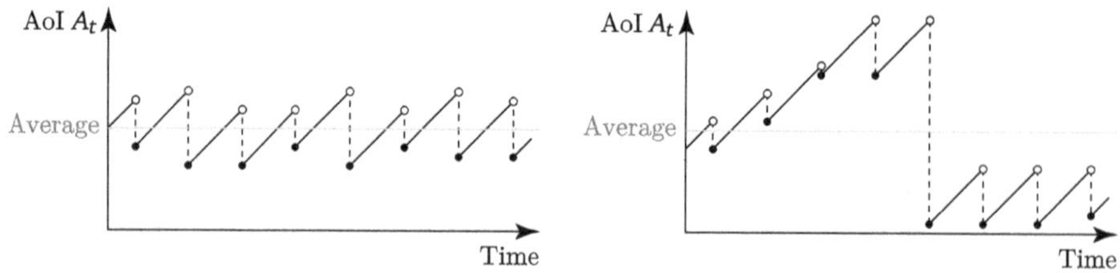

Figure 1.6 An example of different AoI processes with equal time-averages.

quantities among various kinds of summary metrics. The AFD $F_A^\sharp(x)$ of the AoI process is defined as

$$F_A^\sharp(x) = \lim_{T\to\infty} \frac{1}{T} \int_0^T \mathbb{1}\{A_t \le x\}\mathrm{d}t, \quad x \ge 0,$$

where $\mathbb{1}\{\cdot\}$ denotes an indicator function. The value of the AFD for fixed $x \ge 0$ thus represents the long-run fraction of time that the AoI does not exceed the threshold x. As in the previous section, the system is usually modeled as a stationary and ergodic stochastic process in theoretical studies on the AoI. Within such a framework, the AFD can be equated with the probability distribution of the stationary AoI A, which has the same distribution as A_t for all $t \ge 0$:

$$F_A^\sharp(x) = \Pr(A \le x) = \Pr(A_t \le x).$$

For the time being, we again focus on the stationary and ergodic system. We will provide a more detailed discussion on this point later in Section 1.3.

The probability distribution of the AoI (the AoI distribution in short) has several appealing properties in characterizing AoI performances. Firstly, although the time-averaged AoI (1.4) is the most widely used summary metric, it has a serious weakness as a performance metric: It cannot capture *how the information freshness fluctuates over time*. Figure 1.6 depicts an example of two AoI processes, which differ substantially in their fluctuations, but cannot be distinguished by the time-average alone. On the other hand, the AoI distribution contains much information about the fluctuation of the process. For example, a standard method to quantify the degree of fluctuation is to use the variance of the process

$$(\sigma_A^\sharp)^2 := \lim_{T\to\infty} \frac{1}{T} \int_0^T \left(A_t - m_A^\sharp\right)^2 \mathrm{d}t,$$

which can be readily computed from the probability distribution:

$$m_A^\sharp = \int_0^\infty x\mathrm{d}F_A(x),$$

$$(\sigma_A^\sharp)^2 = \int_0^\infty \left(x - m_A^\sharp\right)^2 \mathrm{d}F_A(x) = \int_0^\infty x^2\mathrm{d}F_A(x) - \left(\int_0^\infty x\mathrm{d}F_A(x)\right)^2.$$

Also, another common way to capture the variability is the use of a box-and-whisker diagram, which will be demonstrated in Section 1.4.

Secondly, it is often the case that the cost of stale information increases *nonlinearly* as time passes since its generation. The AoI value averaged with a nonlinear cost function $g(\cdot)$ is of interest in such a situation:

$$m^{\sharp}_{g(A)} := \lim_{T\to\infty} \frac{1}{T} \int_0^T g(A_t)\mathrm{d}t. \tag{1.12}$$

The AoI distribution $F_A(\cdot)$ provides a simple way to calculate the average nonlinear cost:

$$m^{\sharp}_{g(A)} = \int_0^\infty g(x)\mathrm{d}F_A(x),$$

that is, an analysis of the AoI process with any nonlinear cost functions reduces to that of the AoI distribution. An indicator function $g(y) = \mathbb{1}\{y > \theta\}$ with a threshold θ is a particular example of a nonlinear cost function, whose time-average is equal to the value of the complementary AoI distribution $\overline{F}_A(x) := 1 - F_A(x)$ evaluated at $x = \theta$. This form of the cost function is of interest for system reliability because making $\overline{F}_A(\theta) < \epsilon$ be satisfied for small ϵ guarantees that the AoI value is below the threshold θ for a fraction $(1 - \epsilon)$ of the time.

Thirdly, the AoI distribution is useful in characterizing monitoring errors when the dynamics of the information source is specified. Suppose that the information source is represented as a stochastic process $(X_t)_{t\geq 0}$ and that the monitor displays the latest state information $\hat{X}_t$ received:

$$\hat{X}_t = X_{t-A_t}.$$

Assuming that the AoI $(A_t)_{t\geq 0}$ and the monitored state $(X_t)_{t\geq 0}$ are independent,[3] the expected error measured with a penalty function $L(\cdot,\cdot)$ is given by

$$\mathrm{E}\left[L(X_t, \hat{X}_t)\right] = \mathrm{E}\left[L(X_t, X_{t-A_t})\right] = \int_0^\infty \mathrm{E}\left[L(X_t, X_{t-x})\right]\mathrm{d}F_A(x),$$

that is, it is represented in terms of the AoI distribution $F_A(\cdot)$, given the knowledge about the transition dynamics of X_t. This quantity further equals the time-average of the error $L(X_t, \hat{X}_t)$ if the monitored process $(X_t)_{t\geq 0}$ is also ergodic.

Finally, the AoI distribution will play an important role in developing statistical theory of the AoI, which would allow us to perform such tasks as parameter estimation, hypothesis testing, model selection, and so on the basis of a collection of observed AoI data, under situations where one does not have complete knowledge about the AoI-generating process of interest. The usefulness of the AoI distribution in this context as well is ascribed to the fact that a distribution is far more informative than a small number of statistics (summary metrics).

Since one cannot generally expect to explicitly write down the likelihood function for common models of the AoI process discussed in the literature, one would perform parameter estimation by first taking a set of statistics, evaluating their values on the basis of the observed AoI data, and then estimating the model parameters therefrom.

[3] This is usually the case if generation timings of updates are determined independently of the state X_t.

As an AoI model defines a mapping from its parameter space to the space of the statistics, parameter estimation amounts to evaluating the inverse of this mapping. It is, however, only possible if the forward mapping is one-to-one. Several existing AoI models, on the other hand, have more than one parameter, so that any single statistic, like the mean AoI, should be insufficient for parameter estimation, and one would therefore require at least as many statistics as the number of parameters in the model. Knowledge of the AoI distribution not only allows us to evaluate the forward mapping for a selected set of statistics, but furthermore would provide us with guidance on how to choose the statistics to be used in parameter estimation.

In model selection, in its simplest form, one takes two alternative AoI models, and decides which of the two models better explains the observed AoI data. For successful model selection, it is desirable that the ranges of the forward mappings associated with the two models are disjoint and well-separated. Such properties also depend on the choice of the statistics to be used, and knowledge of the AoI distributions plays an essential role here as well.

It should be noted that for a full-fledged statistical theory of the AoI one should go beyond the AoI distribution: in statistical procedures, such as obtaining confidence interval, performing hypothesis testing, and so on, one usually requires knowledge about distributions of the statistics evaluated on the basis of a finite-sized dataset from a prescribed AoI model. For example, in order to decide how reliable an estimate of the mean AoI from a finite dataset is, one would need to evaluate its variance, which in turn requires knowledge of the autocorrelation of the AoI process (Bhat & Rao 1987). In this regard, a full statistical theory of AoI is yet to be explored, and the AoI distribution may be recognized as a first step toward this direction.

1.3 A General Formula for the AoI Distribution

1.3.1 Model Description

We start by providing a formal description of the mathematical model to be considered. We suppose that updates are generated by the sensor with generation intervals $(G_n)_{n=1,2,\ldots}$. The sequence of generation times $(\alpha_n)_{n=0,1,\ldots}$ is then determined by the initial generation time α_0 and the recursion

$$\alpha_{n+1} = \alpha_n + G_{n+1}, \quad n = 0, 1, \ldots.$$

We refer to the update generated at time α_n as the nth update. Just after its generation, the nth update arrives at the service system (cf. Figure 1.3), which imposes a delay of length D_n. The information contained in the nth update is thus received by the monitor at time β_n, where

$$\beta_n = \alpha_n + D_n, \quad n = 0, 1, \ldots.$$

Without loss of generality, we set the time origin so that $\beta_0 \leq 0$.

The AoI process $(A_t)_{t\geq 0}$ is then constructed as follows. We first note that the AoI refers to the elapsed time of the *latest* status information displayed on the monitor.

That is, the AoI process is not affected by status updates overtaken by newer updates. More formally, let $\mathcal{I}$ denote the index set of updates that are not overtaken by other updates:

$$\mathcal{I} = \{n;\ \beta_n < \min\{\beta_{n+1}, \beta_{n+2}, \ldots\}\}. \tag{1.13}$$

Also, let $\mathcal{I}^{\mathrm{c}}$ denote the complement of $\mathcal{I}$:

$$\mathcal{I}^{\mathrm{c}} = \{0, 1, \ldots\} \setminus \mathcal{I}.$$

An update in $\mathcal{I}$ (resp. $\mathcal{I}^{\mathrm{c}}$) is said to be *effective* (resp. *noneffective*) in the sense that its information is newer (resp. older) than that displayed on the monitor just before its reception by the monitor update. As shown in what follows, the AoI process is completely characterized in terms of the generation times $(\alpha_n)_{n\in\mathcal{I}}$ and the reception times $(\beta_n)_{n\in\mathcal{I}}$ of effective updates only.

Recall that the AoI at time t is given by (1.1) in terms of η_t, which denotes the generation time of the latest information displayed on the monitor at time t. In the current setting, η_t is written as

$$\eta_t = \sup\{\alpha_n\,;\ n \in \{0, 1, \ldots\},\ \beta_n \le t\}.$$

For each noneffective update $i \in \mathcal{I}^{\mathrm{c}}$, there exists an integer $k \ge i+1$ such that $\beta_k \le \beta_i$ (cf. Eq. (1.13)), that is, $\beta_i \le t$ implies the existence of an update k with $\beta_k \le t$ and $\alpha_k \ge \alpha_i$. Therefore, the value of η_t is not affected by excluding all noneffective updates from consideration:

$$\eta_t = \sup\{\alpha_n\,;\ n \in \mathcal{I},\ \beta_n \le t\}. \tag{1.14}$$

We thus restrict our attention to the effective updates only. Recall that we use the superscript "†" to represent quantities of effective updates. Let $(\alpha_\ell^\dagger)_{\ell=0,1,\ldots}$ and $(\beta_\ell^\dagger)_{\ell=0,1,\ldots}$ denote the sequences of effective generation and reception times:

$$(\alpha_\ell^\dagger)_{\ell=0,1,\ldots} = (\alpha_n)_{n\in\mathcal{I}}, \quad (\beta_\ell^\dagger)_{\ell=0,1,\ldots} = (\beta_n)_{n\in\mathcal{I}}.$$

Also, we define the intergeneration time $G_\ell^\dagger$ ($\ell = 1, 2, \ldots$) and the system delay $D_\ell^\dagger$ ($\ell = 0, 1, \ldots$) of the ℓth effective update as in (1.2). Note here that while the effective system delay $D_\ell^\dagger$ equals the original system delay D_n for some $n \in \mathcal{I}$, the effective intergeneration time $G_\ell^\dagger$ is given by the sum of intergeneration times

$$G_\ell^\dagger = G_{n+1} + G_{n+2} + \cdots + G_{n+k+1}, \tag{1.15}$$

for some $n \in \mathcal{I}$ such that $n+1 \in \mathcal{I}^{\mathrm{c}}, n+2 \in \mathcal{I}^{\mathrm{c}}, \ldots, n+k \in \mathcal{I}^{\mathrm{c}}$, and $n+k+1 \in \mathcal{I}$.

From the construction of the sequence of effective updates, it is clear that

$$\ell < \ell' \Rightarrow \beta_\ell^\dagger < \beta_{\ell'}^\dagger,$$

that is, the effective updates enter and depart the service system in a first-in first-out (FIFO) manner. In other words, during the time interval $[\beta_\ell^\dagger, \beta_{\ell+1}^\dagger)$ between the consecutive receptions of the ℓth and $(\ell+1)$st effective updates, the ℓth update's information is the latest at the monitor. With this observation, (1.14) is considerably simplified as

$$\eta_t = \alpha_\ell^\dagger, \qquad t \in [\beta_\ell^\dagger, \beta_{\ell+1}^\dagger).$$

Therefore, the expression (1.1) for the AoI is rewritten as follows:

$$A_t = t - \alpha_\ell^\dagger, \qquad t \in [\beta_\ell^\dagger, \beta_{\ell+1}^\dagger),\ \ell = 0, 1, \ldots. \tag{1.16}$$

Our assumption $\beta_0 \le 0$ implies $\beta_0^\dagger \le 0$, so that the AoI A_t is well-defined for all $t \in [0, \beta_\infty^\dagger)$, where $\beta_\infty^\dagger := \lim_{\ell\to\infty} \beta_\ell^\dagger$. Note that we must have $\beta_\infty^\dagger = \infty$ in practical situations because otherwise there exists $T_{\text{sup}} < \infty$ such that the monitor will never be updated again after time $t = T_{\text{sup}}$.

Finally, recall that the ℓth peak AoI $A_{\text{peak},\ell}$ is defined as the AoI just before the reception of the ℓth update, and it is given by (1.3).

1.3.2 The Asymptotic Frequency Distribution (AFD) of the AoI

In this subsection, we present a sample-path analysis of the AoI process. Mathematical analysis in this subsection deals with a *deterministic* (i.e., not random) sequence of generation intervals $(G_n)_{n=1,2,\ldots}$, system delays $(D_n)_{n=0,1,\ldots}$, and a deterministic value of the initial generation time α_0. As we have seen in the previous subsection, these quantities completely determine the AoI process $(A_t)_{t\ge 0}$, the peak AoI process $(A_{\text{peak},\ell})_{\ell=1,2,\ldots}$, and the effective system delay process $(D_\ell^\dagger)_{\ell=0,1,\ldots}$. In the next subsection, we will turn our attention to a stochastic version of the AoI process, which is usually dealt with in the AoI literature.

The main purpose of this subsection is to derive a general relation satisfied by the AFDs of the (deterministic) AoI process $(A_t)_{t\ge 0}$, peak AoI process $(A_{\text{peak},\ell})_{\ell=1,2,\ldots}$, and effective system delay process $(D_\ell^\dagger)_{\ell=0,1,\ldots}$. As mentioned previously, the AFD of a process is defined for each $x \ge 0$ as the long-run fraction of time that the process does not exceed x. More specifically, the AFDs of the AoI, peak AoI, and the effective system delay are defined respectively as

$$F_A^\sharp(x) = \lim_{T\to\infty} \frac{1}{T} \int_0^T \mathbb{1}\{A_t \le x\} \mathrm{d}t, \qquad x \ge 0, \tag{1.17}$$

$$F_{A_{\text{peak}}}^\sharp(x) = \lim_{N\to\infty} \frac{1}{N} \sum_{\ell=1}^{N} \mathbb{1}\{A_{\text{peak},\ell} \le x\}, \qquad x \ge 0, \tag{1.18}$$

$$F_{D^\dagger}^\sharp(x) = \lim_{N\to\infty} \frac{1}{N} \sum_{\ell=0}^{N-1} \mathbb{1}\{D_\ell^\dagger \le x\}, \qquad x \ge 0, \tag{1.19}$$

provided that these limits exist.

In order to obtain nontrivial results from the analysis, we need to impose several basic assumptions: (i) finiteness and positivity of the effective generation rate, (ii) rate stability of the FIFO sequence of effective updates, and (iii) the existence of the AFDs of the peak AoI and system delay. To be more specific, let $\lambda^\dagger$ denote the mean effective generation rate:

$$\lambda^\dagger := \lim_{T\to\infty} \frac{1}{T} \sum_{\ell=1}^{\infty} \mathbb{1}\{\alpha_\ell^\dagger \le T\} = \lim_{N\to\infty} \left(\frac{1}{N} \sum_{\ell=1}^{N} G_\ell^\dagger \right)^{-1}, \tag{1.20}$$

provided the limits exist. We note that the second equality of this equation implies that the mean effective generation rate equals the reciprocal of the mean effective generation interval of updates; a formal proof of this intuitive relation can be given with a deterministic version of the elementary renewal theorem (El-Taha & Stidham Jr. 1999, Lemma 1.1).

The assumptions just mentioned are then formally stated as follows:

ASSUMPTION 1.1

(i) The effective generation rate satisfies $\lambda^\dagger \in (0, \infty)$.
(ii) The effective update rate equals the effective generation rate:

$$\lim_{T\to\infty} \frac{1}{T} \sum_{\ell=1}^{\infty} \mathbb{1}\{\beta_\ell^\dagger \le T\} = \lambda^\dagger. \tag{1.21}$$

(iii) The limits in (1.18) and (1.19) exist.

The key observation in our analysis is that for each interval $[\beta_\ell^\dagger, \beta_{\ell+1}^\dagger)$ between effective updates at the monitor, the contribution of the AoI process to its frequency distribution (for a fixed $x \ge 0$) is represented as

$$\begin{aligned}\int_{\beta_\ell^\dagger}^{\beta_{\ell+1}^\dagger} \mathbb{1}\{A_t \le x\}\mathrm{d}t &= \int_{\beta_\ell^\dagger}^{\beta_{\ell+1}^\dagger} \mathbb{1}\{t - \alpha_\ell^\dagger \le x\}\mathrm{d}t \\ &= \int_{\beta_\ell^\dagger-\alpha_\ell^\dagger}^{\beta_{\ell+1}^\dagger-\alpha_\ell^\dagger} \mathbb{1}\{u \le x\}\mathrm{d}u \\ &= \int_{D_\ell^\dagger}^{A_{\mathrm{peak},\ell+1}} \mathbb{1}\{u \le x\}\mathrm{d}u \\ &= \int_0^x \mathbb{1}\{A_{\mathrm{peak},\ell+1} > u\}\mathrm{d}u - \int_0^x \mathbb{1}\{D_\ell^\dagger > u\}\mathrm{d}u, \end{aligned} \tag{1.22}$$

where we have the first equality from (1.16), the second equality by letting $u = t - \alpha_\ell^\dagger$, the third equality from (1.2) and (1.3), and the last equality from the following identity:

$$\int_0^y \mathbb{1}\{u \le x\}\mathrm{d}u = \int_0^\infty \mathbb{1}\{u \le x\}\mathbb{1}\{u < y\}\mathrm{d}u = \int_0^x \mathbb{1}\{y > u\}\mathrm{d}u, \quad x \ge 0, y \ge 0.$$

Using $\mathbb{1}\{y > x\} = 1 - \mathbb{1}\{y \le x\}$, we further rewrite (1.22) as

$$\int_{\beta_\ell^\dagger}^{\beta_{\ell+1}^\dagger} \mathbb{1}\{A_t \le x\}\mathrm{d}t = \int_0^x \mathbb{1}\{D_\ell^\dagger \le u\}\mathrm{d}u - \int_0^x \mathbb{1}\{A_{\mathrm{peak},\ell+1} \le u\}\mathrm{d}u.$$

Therefore, we obtain the following relation by summing up both sides of the above equation for $\ell = 0, 1, \dots, N-1$:

$$\int_{\beta_0^\dagger}^{\beta_N^\dagger} \mathbb{1}\{A_t \le x\}\mathrm{d}t = \int_0^x \sum_{\ell=0}^{N-1} \mathbb{1}\{D_\ell^\dagger \le u\}\mathrm{d}u - \int_0^x \sum_{\ell=1}^{N} \mathbb{1}\{A_{\mathrm{peak},\ell} \le u\}\mathrm{d}u,$$

which is equivalent to that for $T > 0$,

$$\frac{1}{T}\int_{\beta_0^\dagger}^{\beta_N^\dagger} \mathbb{1}\{A_t \le x\}\mathrm{d}t$$
$$= \frac{N}{T}\left\{\int_0^x \frac{1}{N}\sum_{\ell=0}^{N-1} \mathbb{1}\{D_\ell^\dagger \le u\}\mathrm{d}u - \int_0^x \frac{1}{N}\sum_{\ell=1}^{N} \mathbb{1}\{A_{\mathrm{peak},\ell} \le u\}\mathrm{d}u\right\}. \tag{1.23}$$

Letting $T = \beta_N^\dagger - \beta_0^\dagger$, we then see that this equation relates the frequency distribution of the AoI to those of the effective system delay and the peak AoI for the finite time interval $[\beta_0^\dagger, \beta_N^\dagger)$, using the effective update rate N/T in that interval.

With Assumption 1.1, this relation is further extended to its limiting version, which is the main result of this subsection:

LEMMA 1.2 *Under Assumption 1.1, the AFD of the AoI $F_A^\sharp(x)$ is related to those of the peak AoI $F_{A_{\mathrm{peak}}}^\sharp(x)$ and effective system delay $F_{D^\dagger}^\sharp(x)$ as*

$$F_A^\sharp(x) = \lambda^\dagger \int_0^x \big(F_{D^\dagger}^\sharp(y) - F_{A_{\mathrm{peak}}}^\sharp(y)\big)\mathrm{d}y, \quad x \ge 0,$$

where $\lambda^\dagger$ denotes the mean generation rate defined as in (1.20).

Proof Let $M(t)$ denote the total number of effective updates at the monitor in a time interval $(\beta_0^\dagger, t]$:

$$M(t) = \sup\{\ell \in \{0, 1, \ldots\};\ \beta_\ell^\dagger \le t\} = \sum_{\ell=1}^{\infty} \mathbb{1}\{\beta_\ell^\dagger \le T\}.$$

For an interval $[0, T)$, the frequency distribution of the AoI is then written as

$$\frac{1}{T}\int_0^T \mathbb{1}\{A_t \le x\}\mathrm{d}t$$
$$= \frac{1}{T}\int_{\beta_0^\dagger}^{\beta_{M(T)}^\dagger} \mathbb{1}\{A_t \le x\}\mathrm{d}t - \frac{1}{T}\int_{\beta_0^\dagger}^{0} \mathbb{1}\{A_t \le x\}\mathrm{d}t + \frac{1}{T}\int_{\beta_{M(T)}^\dagger}^{T} \mathbb{1}\{A_t \le x\}\mathrm{d}t$$
$$= \frac{M(T)}{T}\cdot S_T(x) + \frac{\epsilon_T(x)}{T},$$

where $S_T(x)$ and $\epsilon_T(x)$ are defined as follows (cf. (1.23)):

$$S_T(x) = \int_0^x \frac{1}{M(T)}\sum_{\ell=0}^{M(T)-1} \mathbb{1}\{D_\ell^\dagger \le u\}\mathrm{d}u - \int_0^x \frac{1}{M(T)}\sum_{\ell=1}^{M(T)} \mathbb{1}\{A_{\mathrm{peak},\ell} \le u\}\mathrm{d}u, \tag{1.24}$$

$$\epsilon_T(x) = -\int_{\beta_0^\dagger}^{0} \mathbb{1}\{A_t \le x\}\mathrm{d}t + \int_{\beta_{M(T)}^\dagger}^{T} \mathbb{1}\{A_t \le x\}\mathrm{d}t.$$

To prove Lemma 1.2, it is then sufficient to show that

$$\lim_{T\to\infty} \frac{M(T)}{T} = \lambda^{\dagger}, \tag{1.25}$$

$$\lim_{T\to\infty} S_T(x) = \int_0^x \big(F^{\sharp}_{D^{\dagger}}(y) - F^{\sharp}_{A_{\text{peak}}}(y)\big)\mathrm{d}y, \tag{1.26}$$

$$\lim_{T\to\infty} \frac{\epsilon_T(x)}{T} = 0. \tag{1.27}$$

Note first that (1.25) immediately follows from Assumption 1.1 (ii). This further implies $M(T) \to \infty$ as $T \to \infty$, since we have $\lambda^{\dagger} > 0$ from Assumption 1.1 (i). We thus readily obtain (1.26) from (1.18), (1.19), and (1.24) using the dominated convergence theorem.

We then consider $\epsilon_T(x)$. By definition of $\epsilon_T(x)$ and $M(T)$, we have

$$\left|\frac{\epsilon_T(x)}{T}\right| \leq \left|\frac{\beta^{\dagger}_0}{T}\right| + \left|\frac{\beta_{M(T)+1} - \beta^{\dagger}_{M(T)}}{T}\right|.$$

The first term on the right-hand side converges to zero as $T \to \infty$ because $|\beta^{\dagger}_0| < \infty$. It thus suffices to consider the second term. Similarly to (1.20), we have from the deterministic version of the renewal theorem (El-Taha & Stidham Jr. 1999, Lemma 1.1)

$$\begin{aligned}
\left(\lim_{T\to\infty} \frac{M(T)}{T}\right)^{-1} &= \lim_{N\to\infty} \frac{1}{N}\sum_{\ell=1}^{N}(\beta^{\dagger}_{\ell} - \beta^{\dagger}_{\ell-1}) \\
&= \lim_{T\to\infty} \frac{1}{M(T)}(\beta^{\dagger}_{M(T)} - \beta^{\dagger}_1) \\
&= \lim_{T\to\infty} \frac{T}{M(T)} \cdot \frac{\beta^{\dagger}_{M(T)}}{T},
\end{aligned}$$

which together with (1.25) imply

$$\lim_{T\to\infty} \frac{\beta^{\dagger}_{M(T)}}{T} = 1.$$

We then have

$$\lim_{T\to\infty} \left|\frac{\beta_{M(T)+1} - \beta^{\dagger}_{M(T)}}{T}\right| = 0,$$

so that (1.27) holds. □

1.3.3 The Stationary AoI Distribution in Ergodic Systems

In the previous subsection, we have derived Lemma 1.2 assuming deterministic sequences of intergeneration times $(G_n)_{n=1,2,\ldots}$ and system delays $(D_n)_{0,1,\ldots}$, and a deterministic value of the initial generation time α_0. In this subsection, we investigate implications of this result to status update systems formulated as ergodic stochastic models. In the rest of this chapter, we follow a convention that for any nonnegative random variable Y, the cumulative distribution function (CDF) is denoted by $F_Y(x)$ ($x \geq 0$), the probability density function (if it exists) is denoted by $f_Y(x)$ ($x \geq 0$), and the Laplace–Stieltjes transform (LST) is denoted by $f_Y^*(s)$ ($s > 0$):

$$F_Y(x) := \Pr(Y \leq x), \ f_Y(x) = \frac{\mathrm{d}F_Y(x)}{\mathrm{d}x}, \ f_Y^*(s) = \mathrm{E}[e^{-sY}] = \int_0^\infty e^{-sx}\mathrm{d}F_Y(x).$$

Suppose that intergeneration times $(G_n)_{n=1,2,\ldots}$, system delays $(D_n)_{0,1,\ldots}$, and the initial generation time α_0 are given as random variables defined on a common probability space $(\Omega, \mathcal{F}, \Pr)$; G_n, D_n, and α_0 are then considered as functions $G_n(\cdot)$, $D_n(\cdot)$, and $\alpha_0(\cdot)$ from the sample space Ω to real values $\mathbb{R}$. From the discussions in Section 1.3.1, the effective intergeneration times $(G_\ell^\dagger(\omega))_{\ell=1,2,\ldots}$, the effective system delays $(D_\ell^\dagger(\omega))_{\ell=0,1,\ldots}$, the peak AoI values $(A_{\text{peak},\ell}(\omega))_{\ell=1,2,\ldots}$, and the AoI values $(A_t(\omega))_{t\geq 0}$ for each sample-path $\omega \in \Omega$ are given in terms of $(G_n(\omega))_{n=1,2,\ldots}$, $(D_n(\omega))_{n=0,1,\ldots}$, and $\alpha_0(\omega)$.

Also, we see that Lemma 1.2 holds for each $\omega \in \Omega$. For stationary and ergodic systems, we can rewrite Lemma 1.2 as a relation of stationary probability distributions. More specifically, we make the following assumptions:

ASSUMPTION 1.3

(i) The joint process $(G_\ell^\dagger, D_\ell^\dagger)_{\ell=1,2,\ldots}$ of effective intergeneration times and effective delays is stationary and ergodic.
(ii) The effective generation rate is constant $\lambda^\dagger(\omega) = \lambda^\dagger$ almost surely (a.s.) with some $\lambda^\dagger \in (0, \infty)$.
(iii) The rate stability (1.21) holds a.s.
(iv) The AoI process $(A_t)_{t\geq 0}$ is stationary.

Remark 1.1 Assumption 1.3 (i) and (ii) have a little redundancy because the ergodicity of $(G_\ell^\dagger)_{\ell=1,2,\ldots}$ implies that $\lambda^\dagger(\omega)$ is constant a.s.

Recall that we have $A_{\text{peak},\ell} = G_\ell^\dagger + D_\ell^\dagger$ as given in (1.3). The stationarity assumed in Assumption 1.3 thus implies the existence of generic random variables A, A_{peak}, and $D^\dagger$ with the same distributions as A_t, $A_{\text{peak},\ell}$, and $D_\ell^\dagger$, respectively:

$$\begin{aligned}
\Pr(A_t \leq x) &= F_A(x), & x &\geq 0, \ t \geq 0,\\
\Pr(A_{\text{peak},\ell} \leq x) &= F_{A_{\text{peak}}}(x), & x &\geq 0, \ \ell = 1, 2, \ldots,\\
\Pr(D_\ell^\dagger \leq x) &= F_{D^\dagger}(x), & x &\geq 0, \ \ell = 1, 2, \ldots.
\end{aligned}$$

For each sample-path $\omega \in \Omega$, let $F_A^\sharp(\omega, x)$, $F_{A_{\text{peak}}}^\sharp(\omega, x)$, and $F_{D^\dagger}^\sharp(\omega, x)$ denote the AFDs of the AoI $(A_t(\omega))_{t\geq 0}$, the peak AoI $(A_{\text{peak},\ell}(\omega))_{\ell=1,2,\ldots}$, and the effective system delay $(D_\ell^\dagger(\omega))_{\ell=0,1,\ldots}$, respectively (cf. (1.17), (1.18), and (1.19)).

The ergodicity of $(G_\ell^\dagger, D_\ell^\dagger)_{\ell=1,2,\ldots}$ (Assumption 1.3 (i)) implies that for $x \geq 0$,

$$F_{A_{\text{peak}}}^\sharp(\omega, x) = F_{A_{\text{peak}}}(x), \quad F_{D^\dagger}^\sharp(\omega, x) = F_{D^\dagger}(x), \quad \text{a.s.}$$

Therefore, we have, from Lemma 1.2,

$$F_A^\sharp(\omega, x) = \lambda^\dagger \int_0^x \big(F_{D^\dagger}(y) - F_{A_{\text{peak}}}(y)\big)\text{d}y, \quad \text{a.s,}$$

which obviously implies

$$\text{E}[F_A^\sharp(x)] = \lambda^\dagger \int_0^x \big(F_{D^\dagger}(y) - F_{A_{\text{peak}}}(y)\big)\text{d}y.$$

On the other hand, we have, from the dominated convergence theorem,

$$\text{E}[F_A^\sharp(x)] = \lim_{T\to\infty} \frac{1}{T} \int_0^T \text{E}[\mathbb{1}\{A_t \le x\}]\text{d}t = F_A(x), \quad x \ge 0.$$

We thus conclude as in the following theorem:

THEOREM 1.4 *Under Assumption 1.3, the CDF of the stationary AoI distribution is given by*

$$F_A(x) = \lambda^\dagger \int_0^x \big(F_{D^\dagger}(y) - F_{A_{\text{peak}}}(y)\big)\text{d}y, \quad x \ge 0.$$

Therefore, the AoI has an absolutely continuous distribution with density

$$f_A(x) = \lambda^\dagger \big(F_{D^\dagger}(x) - F_{A_{\text{peak}}}(x)\big), \quad x \ge 0. \tag{1.28}$$

Furthermore, the LST of the AoI is given by

$$f_A^*(s) = \lambda^\dagger \cdot \frac{f_{D^\dagger}^*(s) - f_{A_{\text{peak}}}^*(s)}{s}, \quad s > 0. \tag{1.29}$$

COROLLARY *The kth moment of the stationary AoI is given by*

$$\text{E}[A^k] = \lambda^\dagger \cdot \frac{\text{E}[(A_{\text{peak}})^{k+1}] - \text{E}[(D^\dagger)^{k+1}]}{k+1}, \tag{1.30}$$

provided that $\text{E}[(A_{\text{peak}})^{k+1}] < \infty$.

Notice that the previously presented formula (1.6) for the mean AoI $\text{E}[A]$ is reproduced by letting $k = 1$ in (1.30). In this sense, Theorem 1.4 can be regarded as a generalization of (1.6) to its distributional version. It is also worth noting that the expression (1.28) for the density function of the AoI distribution can be interpreted as *a level-crossing identity* (Brill & Posner 1977; Cohen 1977) for the AoI process. More specifically, for fixed $x \ge 0$, the density function $f_A(x)$ represents the mean number of upcrossings at level x of the AoI process per time unit because the AoI A_t linearly increases with slope one almost every t (cf. Figure 1.2). On the other hand, the right-hand side of (1.28) represents the mean number of downcrossings at level x per time unit, which is the product of the number of downcrossings (i.e., monitor updates) occurring per time unit and the probability of $\{A_{\text{peak},\ell} > x \ge D_\ell^\dagger\}$, which can be rewritten as

$$\Pr(A_{\text{peak},\ell} > x \geq D^\dagger_\ell) = \Pr(A_{\text{peak},\ell} > x) - \Pr(D^\dagger_\ell > x) = F_{D^\dagger}(x) - F_{A_{\text{peak}}}(x).$$

The formula (1.28) thus indicates that the mean numbers of upcrossings and downcrossings at level x occurring per time unit are equal.

Theorem 1.4 shows that the stationary distribution of the AoI A is given in terms of those of the peak AoI A_{peak} and the effective system delay $D^\dagger$. In many cases, characterizing the probability distributions of the peak AoI A_{peak} and the effective system delay $D^\dagger$ is quite easier than directly analyzing the AoI process. In the following section, we demonstrate how to apply Theorem 1.4 to AoI analysis, dealing with basic single-server queueing models as examples.

1.4 The AoI Distribution in FCFS and LCFS Single-Server Queues

In this section, we present an analysis of the AoI distribution for FCFS and last-come first-served (LCFS) single-server queues, which are described as follows. The sensor observes the state of a time-varying information source with independent and identically distributed (i.i.d.) generation intervals $(G_n)_{n=1,2,\ldots}$. The service system is represented as a single-server queue, where each update receives service with i.i.d. length of time. Let $(H_n)_{n=1,2,\ldots}$ denote the i.i.d. sequence of service times. We define G and H as generic random variables for intergeneration times and service times. We also define λ as the mean generation rate:

$$\lambda = \frac{1}{\mathrm{E}[G]}.$$

Let $\rho := \lambda\mathrm{E}[H]$ denote the traffic intensity. With Kendall's notation, this queueing model is denoted by GI/GI/1. We define $\widetilde{G}$ as a generic random variable for residual (equilibrium) intergeneration times, that is, the time to the next generation from a randomly chosen time instant. Similarly, we define $\widetilde{H}$ as a generic random variable for residual service times. The density functions and LSTs of $\widetilde{G}$ and $\widetilde{H}$ are given by

$$f_{\widetilde{G}}(x) = \frac{1 - F_G(x)}{\mathrm{E}[G]}, \qquad f^*_{\widetilde{G}}(s) = \frac{1 - f^*_G(s)}{s\mathrm{E}[G]}, \tag{1.31}$$

$$f_{\widetilde{H}}(x) = \frac{1 - F_H(x)}{\mathrm{E}[H]}, \qquad f^*_{\widetilde{H}}(s) = \frac{1 - f^*_H(s)}{s\mathrm{E}[H]}. \tag{1.32}$$

We consider two different service policies of the server: the FCFS and the preemptive LCFS service policies. Under the FCFS service policy, status updates are served in order of their arrivals, so that no overtaking of updates can occur. Under the preemptive LCFS service policy, on the other hand, the newest update is given priority: each update starts to receive service immediately after its arrival at the server, whereas its service is preempted when a newer update has arrived before the service completion. Recall that an update is said to be effective if it is not overtaken by other updates. It is readily seen that all updates are effective under the FCFS service policy, while there are noneffective updates under the preemptive LCFS service policy.

1.4.1 The Stationary AoI Distribution in FCFS Single-Server Queues

We first consider the FCFS case. Because all updates are effective in this case,

$$G_n^{\dagger} = G_n, \quad D_n^{\dagger} = D_n, \quad \lambda^{\dagger} = \lambda, \tag{1.33}$$

so that the peak AoI is given by (cf. (1.3))

$$A_{\text{peak},n} = D_n + G_n. \tag{1.34}$$

In order for Assumption 1.3 to be satisfied, we assume $\rho < 1$, which ensures the stability of the queueing system.

We note that D_n and G_n on the right-hand side of (1.34) are *dependent* in general: an update arriving after a long interval G_n tends to find less congested system than the time-average. For the FCFS queue, however, (1.34) can be rewritten in terms of independent random variables, using the well-known Lindley recursion for system delays $(D_n)_{n=0,1,\dots}$: because the waiting time of the nth update equals $D_{n-1} - G_n$ if $D_{n-1} > G_n$ and otherwise it equals zero, we have

$$D_n = \max(0, D_{n-1} - G_n) + H_n, \quad n = 1, 2, \dots.$$

(1.34) is then rewritten as

$$A_{\text{peak},n} = \max(G_n, D_{n-1}) + H_n, \quad n = 1, 2, \dots. \tag{1.35}$$

Observe that G_n, D_{n-1}, and H_n are independent. Because we have

$$\begin{aligned}\Pr(\max(G_n, D_{n-1}) \le x) &= \Pr(G_n \le x, D_{n-1} \le x) \\ &= \Pr(G_n \le x)\Pr(D_{n-1} \le x),\end{aligned}$$

the relation (1.35) implies

$$F_{A_{\text{peak}}}(x) = \int_0^x F_G(x-y)F_D(x-y)\mathrm{d}F_H(y), \quad x \ge 0,$$

where D denotes a generic random variable for the stationary D_n. Therefore, we obtain the following result from Theorem 1.4:

THEOREM 1.5 *In the stationary FCFS GI/GI/1 queue, the probability density function of the AoI distribution is given by*

$$f_A(x) = \lambda \left(F_D(x) - \int_0^x F_G(x-y)F_D(x-y)\mathrm{d}F_H(y) \right). \tag{1.36}$$

Noting that λ, $F_G(\cdot)$, and $F_H(\cdot)$ are model parameters, we see from Theorem 1.5 that the AoI distribution is given in terms of the stationary delay distribution $F_D(\cdot)$. We can find expressions for the delay distribution $F_D(\cdot)$ for standard FCFS queueing systems in textbooks on the queueing theory (Asmussen 2003; Kleinrock 1975).

We first introduce a general result, assuming that service times follow a *phase-type* distribution with representation $(\boldsymbol{\gamma}, \boldsymbol{S})$:

$$F_H(x) = 1 - \boldsymbol{\gamma}\exp[\boldsymbol{S}x]\boldsymbol{e}, \quad f_H(x) = \boldsymbol{\gamma}\exp[\boldsymbol{S}x](-\boldsymbol{S})\boldsymbol{e}, \tag{1.37}$$

where $\boldsymbol{e}$ denotes a column vector (with the same size as $\boldsymbol{S}$) whose elements are all equal to one. The queueing model considered is then denoted by GI/PH/1. We do not lose much generality with this assumption on service times because the set of phase-type distributions covers a fairly wide class of nonnegative probability distributions; in fact, it is known that the phase-type distributions form a dense subset in the set of all nonnegative probability distributions (Asmussen 2003, p. 84). Readers who are not familiar with phase-type distributions are advised to take a look at Appendix A.1, where we provide a brief introduction.

It is readily verified that for the phase-type service time distribution,

$$\mathrm{E}[H] = \gamma(-\boldsymbol{S})^{-1}\boldsymbol{e}, \tag{1.38}$$

and the residual service time distribution defined in (1.32) is also of phase-type with representation $(\widetilde{\gamma}, \boldsymbol{S})$, where

$$\widetilde{\gamma} := \frac{\gamma(-\boldsymbol{S})^{-1}}{\gamma(-\boldsymbol{S})^{-1}\boldsymbol{e}}. \tag{1.39}$$

LEMMA 1.6 (Asmussen 1992) *Consider an FCFS GI/PH/1 queue, which has a general inter-arrival time distribution with CDF $F_G(x)$ and the phase-type service time distribution with representation $(\gamma, \boldsymbol{S})$. The stationary system delay in this model follows a phase-type distribution with representation $(\gamma, \boldsymbol{Q})$:*

$$F_D(x) = 1 - \gamma \exp[\boldsymbol{Q}x]\boldsymbol{e}, \quad x \geq 0,$$

where $\boldsymbol{Q}$ is defined as

$$\boldsymbol{Q} = \boldsymbol{S} + (-\boldsymbol{S})\boldsymbol{e}\boldsymbol{\pi}_*, \tag{1.40}$$

with $\boldsymbol{\pi}_*$ defined as the limit $\boldsymbol{\pi}_* := \lim_{n\to\infty} \boldsymbol{\pi}_n$ of a sequence $(\boldsymbol{\pi}_n)_{n=0,1,\ldots}$ given by $\boldsymbol{\pi}_0 = \boldsymbol{0}$ and the following recursion:

$$\boldsymbol{\pi}_n = \gamma \int_0^\infty \exp[(\boldsymbol{S} + (-\boldsymbol{S})\boldsymbol{e}\boldsymbol{\pi}_{n-1})y]\mathrm{d}F_G(y), \quad n = 1, 2, \ldots. \tag{1.41}$$

Remark 1.2 (Asmussen 2003, p. 241) $(\boldsymbol{\pi}_n)_{n=0,1,\ldots}$ is an (elementwise) nondecreasing sequence of subprobability vectors (i.e., $\boldsymbol{\pi}_n\boldsymbol{e} < 1$), and its limit $\boldsymbol{\pi}_*$ is also a subprobability vector, provided that the stability condition $\rho < 1$ holds.

We hereafter focus on the Poisson and constant generation policies discussed in Section 1.1.2, that is, we consider the FCFS M/PH/1 and D/PH/1 queues:

$$F_G(x) = 1 - e^{-\lambda x}, \quad x \geq 0, \quad \text{(M/PH/1)} \tag{1.42}$$

$$F_G(x) = \mathbb{1}\{x \geq \tau\}, \quad x \geq 0, \quad \text{(D/PH/1)}, \tag{1.43}$$

where $\tau := 1/\lambda$. In the M/PH/1 queue, $\boldsymbol{\pi}_*$ is given explicitly by

$$\boldsymbol{\pi}_* = \rho\widetilde{\gamma}, \qquad \text{(M/PH/1)} \tag{1.44}$$

which can be verified with the following observation: substituting (1.42) into (1.41) and taking the limit $n \to \infty$, we have

$$\boldsymbol{\pi}_*(-\boldsymbol{S} + \lambda \boldsymbol{I} + \boldsymbol{S}\boldsymbol{e}\boldsymbol{\pi}_*) = \lambda\gamma, \tag{1.45}$$

which is equivalent to

$$\pi_*(-S) = \lambda\gamma, \tag{1.46}$$

because $\pi_*(-S)e = \lambda$ is obtained by post-multiplying both sides of (1.45) by e and rearranging terms. We thus obtain (1.44) from (1.38), (1.39), and (1.46). In the D/PH/1 queue, on the other hand, (1.41) is simplified as

$$\pi_n = \gamma \exp[(S + (-S)e\pi_{n-1})\tau], \qquad \text{(D/PH/1)}$$

so that π_* is easily computed by iterations.

Furthermore, the formula (1.36) for the AoI distribution is simplified in the M/PH/1 and D/PH/1 queues:

THEOREM 1.7

(i) In the stationary FCFS M/PH/1 queue, the density function and the CDF of the AoI are given by

$$f_A(x) = \rho\gamma_0 \exp[B_0 x](-B_0)e_0 + \gamma_1 \exp[B_1 x](-B_1)e_1, \tag{1.47}$$

$$F_A(x) = 1 - \rho\gamma_0 \exp[B_0 x]e_0 - \gamma_1 \exp[B_1 x]e_1, \tag{1.48}$$

where B_0 and B_1 are defined as

$$B_0 = \exp\left[\begin{pmatrix} Q & (-Q)e\tilde{\gamma} \\ 0 & S \end{pmatrix} x\right],$$

$$B_1 = \begin{pmatrix} Q - \lambda I & -(Q - \lambda I)e\gamma & 0 \\ 0 & S & (-S)e \\ 0 & 0 & -\lambda \end{pmatrix},$$

and γ_0 (resp. γ_1) denotes a row vector with the same size as B_0 (resp. B_1), which is expressed as

$$\gamma_0 = [\gamma \;\; 0], \qquad \gamma_1 = [\gamma - \pi_* \;\; 0].$$

Also, e_0 (resp. e_1) denotes a column vector with the same size as B_0 (resp. B_1) whose elements are all equal to one.

(ii) In the stationary FCFS D/PH/1 queue, the density function and the CDF of the AoI are given by

$$f_A(x) = \begin{cases} \dfrac{1 - \gamma \exp[Qx]e}{\tau}, & 0 \le x < \tau, \\[2ex] \dfrac{\gamma(I - \exp[Q\tau]) \exp[Q(x - \tau)]e}{\tau}, & x \ge \tau. \end{cases} \tag{1.49}$$

$$F_A(x) = \begin{cases} \dfrac{x - \gamma(-Q)^{-1}e + \gamma \exp[Qx](-Q)^{-1}e}{\tau}, & 0 \le x < \tau, \\[2ex] 1 - \dfrac{\gamma(I - \exp[Q\tau]) \exp[Q(x - \tau)](-Q)^{-1}e}{\tau}, & x \ge \tau. \end{cases} \tag{1.50}$$

The derivation of Theorem 1.7 is detailed in Appendix A.2.

Using Theorem 1.7, the probability distribution of the AoI in M/PH/1 and D/PH/1 queues is easily computed. We provide a few numerical examples using the following special class of phase-type distributions, which is uniquely identified by its mean $\mathrm{E}[H]$ and coefficient of variation $\mathrm{Cv}[H]$ (the standard deviation divided by the mean):

Mixed Erlang Distribution ($0 < \mathrm{Cv}[H] < 1$)

$$f_H(x) = p\mu \cdot \frac{e^{-\mu x}(\mu x)^{k-1}}{(k-1)!} + (1-p)\mu \cdot \frac{e^{-\mu x}(\mu x)^k}{k!},$$

where

$$k = \lfloor 1/(\mathrm{Cv}[H])^2 \rfloor,$$
$$p = \frac{k+1}{1+(\mathrm{Cv}[H])^2}\left((\mathrm{Cv}[H])^2 - \sqrt{\frac{1-k(\mathrm{Cv}[H])^2}{k+1}}\right),$$
$$\mu = \frac{pk+(1-p)(k+1)}{\mathrm{E}[H]}.$$

This distribution is also represented as a phase-type distribution by letting γ and $\boldsymbol{T}$ be a row vector and a matrix of size $(k+1)$ given by

$$\boldsymbol{\alpha} = (1\ 0\ \dots\ 0), \qquad \boldsymbol{T} = \begin{pmatrix} -\mu & \mu & 0 & \cdots & 0 & 0 & 0 \\ 0 & -\mu & \mu & \cdots & 0 & 0 & 0 \\ 0 & 0 & -\mu & \cdots & 0 & 0 & 0 \\ \vdots & \vdots & \vdots & \ddots & \vdots & \vdots & \vdots \\ 0 & 0 & 0 & \cdots & -\mu & \mu & 0 \\ 0 & 0 & 0 & \cdots & 0 & -\mu & (1-p)\mu \\ 0 & 0 & 0 & \cdots & 0 & 0 & -\mu \end{pmatrix}.$$

Exponential Distribution ($\mathrm{Cv}[H] = 1$)

$$f_H(x) = \mu e^{-\mu x}.$$

This is a phase-type distribution with $\gamma = 1$ and $\boldsymbol{S} = -\mu$, and we have $\mu = 1/\mathrm{E}[H]$. This case corresponds to the M/M/1 and D/M/1 queues.

Hyper-Exponential Distribution with Balanced Means ($\mathrm{Cv}[H] > 1$)

$$f_H(x) = p\mu_1 e^{-\mu_1 x} + (1-p)\mu_2 e^{-\mu_2 x},$$

where

$$p = \frac{1}{2}\left(1 + \sqrt{\frac{(\mathrm{Cv}[H])^2 - 1}{(\mathrm{Cv}[H])^2 + 1}}\right), \quad \mu_1 = \frac{2p}{\mathrm{E}[H]}, \quad \mu_2 = \frac{2(1-p)}{\mathrm{E}[H]}.$$

This is a phase-type distribution with

$$\gamma = (p\ \ 1-p), \qquad \boldsymbol{S} = \begin{pmatrix} -\mu_1 & 0 \\ 0 & -\mu_2 \end{pmatrix}.$$

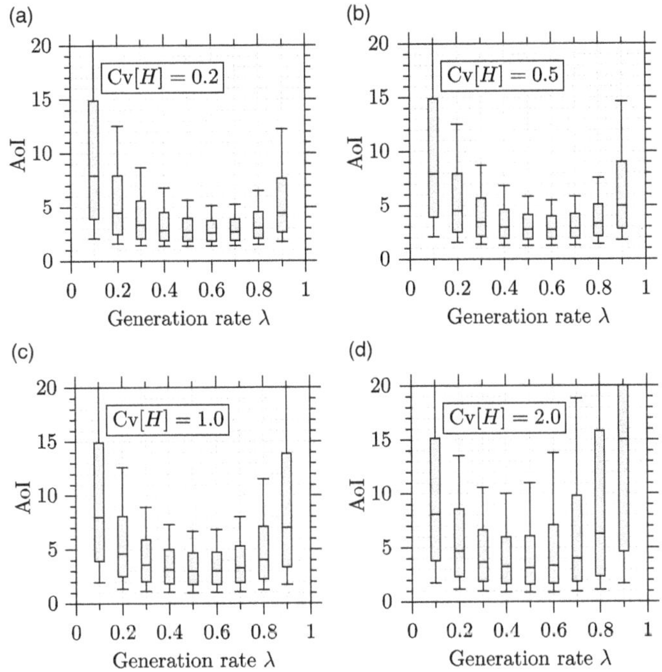

Figure 1.7 Boxplots of the AoI distribution in the FCFS M/PH/1 queue. The whiskers represent 10 and 90 percentiles.

We set $\mathrm{E}[H] = 1$ throughout. Figure 1.7 shows boxplots of the AoI distribution in the M/PH/1 queue for several values of the coefficient of variation $\mathrm{Cv}[H]$ of service times and the generation rate λ. As previously mentioned, the case of $\mathrm{Cv}[H] = 1.0$ refers to the M/M/1 queue, whose mean AoI was examined with Figure 1.4. We observe that using too small or too large generation rate λ leads to a significant increase in the variability of the AoI as well as the increase in the median value. We also observe that larger variability of service times leads to a significant increase in the AoI percentiles: its effect on the AoI distribution is prominent, particularly for large values of λ, because the system delay has a dominant impact on the AoI in that region, as discussed in Section 1.1. Figure 1.8 shows similar boxplots of the AoI distribution in the D/PH/1 queue. We see that the same observations as those for the M/M/1 queue apply regarding the impacts of the generation rate and the service-time variability on the AoI distribution. On the other hand, from Figures 1.7 and 1.8, we see that the variability of the AoI is reduced drastically by using constant generation intervals instead of exponential generation intervals. This again highlights the superiority of using constant generation intervals, which we observed at the end of Section 1.1.2.

1.4.2 The Stationary AoI Distribution in Preemptive LCFS Single-Server Queues

We next consider the case of preemptive LCFS service policy. In this case, an update immediately starts to receive service just after its generation. The update finishes the

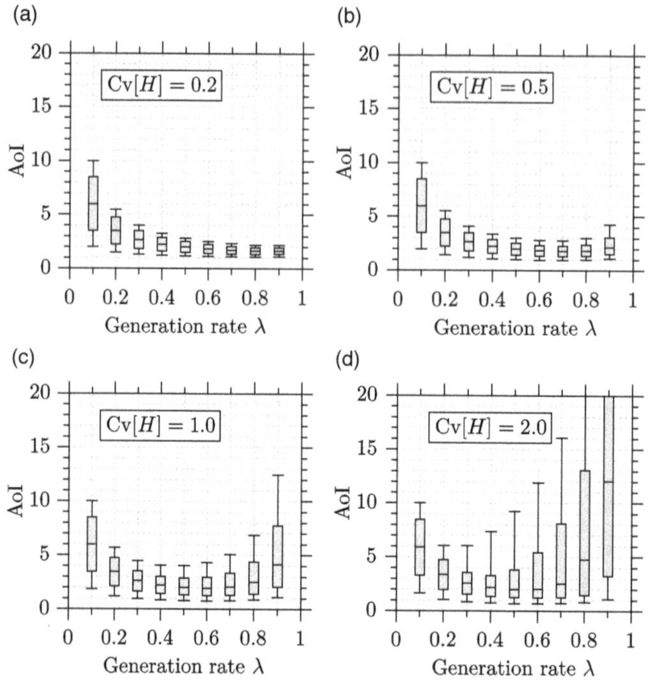

Figure 1.8 Boxplots of the AoI distribution in the FCFS D/PH/1 queue. The whiskers represent 10 and 90 percentiles.

service without being overtaken (i.e., becomes an effective update) if the next update is not generated until the end of the service. If the generation of the next update occurs before the service completion, on the other hand, the update is overtaken and becomes noneffective. Therefore, the probability that an update becomes noneffective is given by

$$\zeta := \Pr(G < H).$$

If $\zeta = 1$, then updates cannot finish receiving service with probability one, so that the AoI continues to increase all the time and never decreases a.s. If $\zeta = 0$, on the other hand, updates become effective with probability one, so that the model reduces to the FCFS case. We thus assume $0 < \zeta < 1$ hereafter. Also, we assume $\Pr(G = H) = 0$ for simplicity. Similarly to the FCFS case, we first derive a general result for the GI/GI/1 queue and then specialize it to the M/PH/1 and D/PH/1 queues.

We use the notation $[Y \mid \mathcal{E}]$ to represent a random variable Y conditioned on an event $\mathcal{E}$. From the previous discussion, we see that the effective system delay $D^\dagger$ has the same distribution as the conditional random variable $H_{<G} := [H \mid H < G]$:

$$D^\dagger \overset{\mathrm{d}}{=} H_{<G}, \tag{1.51}$$

where $\overset{\mathrm{d}}{=}$ stands for the equality in distribution. Recall that the effective intergeneration time $G_\ell^\dagger$ is given by the sum of intergeneration times of the form (1.15). In the current setting, we have that the terms on the right-hand side of (1.15) are characterized as

$$G_{n+1} \stackrel{\mathrm{d}}{=} G_{>H}, \quad G_{n+i} \stackrel{\mathrm{d}}{=} G_{<H}, \quad i = 2, 3, \ldots, k+1,$$

where we define $G_{>H} = [G \mid G > H]$ and $G_{<H} = [G \mid G < H]$. We can also verify that these intergeneration times $G_{n+1}, G_{n+2}, \ldots, G_{n+k+1}$ are independent. Furthermore, the number k of intermediate noneffective updates follows a geometric distribution with success probability $\Pr(G > H) = 1 - \zeta$; let K denote this discrete random variable:

$$\Pr(K = k) = \zeta^k (1 - \zeta), \quad k = 0, 1, \ldots. \tag{1.52}$$

The distribution of the effective intergeneration time is then characterized as

$$G_\ell^\dagger \stackrel{\mathrm{d}}{=} G_{>H} + \sum_{i=1}^{K} G_{<H}^{[i]},$$

where $(G_{<H}^{[i]})_{i=1,2,\ldots}$ denotes an i.i.d. sequence of random variables with the same distribution as $G_{<H}$. From (1.3) and (1.51), we then have that

$$A_{\mathrm{peak},\ell} \stackrel{\mathrm{d}}{=} G_{>H} + \sum_{i=1}^{K} G_{<H}^{[i]} + H_{<G}, \tag{1.53}$$

and that all random variables on the right-hand side of this equation are independent; note that $D_\ell^\dagger$ is dependent on *the next* intergeneration time G_{n+k+1} (as $D_\ell^\dagger < G_{n+k+1}$ should hold for this update to be effective), whereas it is independent of the preceding intergeneration times $G_{n+1}, G_{n+2} \ldots, G_{n+k}$.

As the expression (1.53) for the peak AoI contains the geometric random sum of i.i.d. random variables, it is easier to consider its LST instead of directly dealing with its CDF. More specifically, the LST of the geometric random sum is expressed as

$$\mathrm{E}\left[\exp\left[-s \sum_{i=1}^{K} G_{<H}^{[i]}\right]\right] = \sum_{k=0}^{\infty} \zeta^k (1 - \zeta) \{f_{G_{<H}}^*(s)\}^k = \frac{1 - \zeta}{1 - \zeta f_{G_{<H}}^*(s)}.$$

Therefore, the LST of the peak AoI is given by

$$f_{A_{\mathrm{peak}}}^*(s) = f_{G_{>H}}^*(s) \cdot \frac{1 - \zeta}{1 - \zeta f_{G_{<H}}^*(s)} \cdot f_{H_{<G}}^*(s).$$

Substituting this equation, $\lambda^\dagger = \lambda(1 - \zeta)$, and $f_{D^\dagger}^*(s) = f_{H_{<G}}^*(s)$ (cf. (1.51)) into (1.29), and rearranging terms using

$$f_G^*(s) = \zeta f_{G_{<H}}^*(s) + (1 - \zeta) f_{G_{>H}}^*(s),$$

we obtain the following result:

THEOREM 1.8 *In the preemptive LCFS GI/GI/1 queue, the LST of the stationary AoI distribution is given by*

$$f_A^*(s) = f_{\tilde{G}}^*(s) \cdot \frac{(1 - \zeta) f_{H_{<G}}^*(s)}{1 - \zeta f_{G_{<H}}^*(s)}, \tag{1.54}$$

where $\tilde{G}$ is defined as in (1.31).

Furthermore, we can interpret (1.54) as follows:

COROLLARY *In the preemptive LCFS GI/GI/1 queue, the stationary AoI has the same distribution as the sum of two independent random variables,*

$$A = \widetilde{G} + Z,$$

where Z is defined as

$$Z = \sum_{n=1}^{\widehat{K}} \min(G_n, H_n), \tag{1.55}$$

$$\widehat{K} = \min\{n = 1, 2, \ldots; H_n < G_n\}, \tag{1.56}$$

in terms of the i.i.d. sequences $(G_n)_{n=1,2,\ldots}$ and $(H_n)_{n=1,2,\ldots}$ of intergeneration and service times.

Remark 1.3 By definition, we have $\widehat{K} \stackrel{\text{d}}{=} K + 1$ (cf. (1.52)).

We then focus on the M/PH/1 and D/PH/1 queues, where the service time distribution is given by (1.37). We can readily verify from (1.31) that the residual intergeneration time $\widetilde{G}$ has simple characterizations in these cases:

$$f_{\widetilde{G}}(x) = \lambda e^{-\lambda x}, \quad x \geq 0, \qquad \text{(M/PH/1)} \tag{1.57}$$

$$f_{\widetilde{G}}(x) = \frac{1}{\tau} \cdot \mathbb{1}\{x \leq \tau\}, \quad x \geq 0, \qquad \text{(D/PH/1)}, \tag{1.58}$$

that is, $\widetilde{G}$ follows an exponential (resp. a uniform) distribution in the M/PH/1 (resp. the D/PH/1) queue. Therefore, it is sufficient to consider the distribution of Z defined in (1.55).

In the M/PH/1 queue, Z follows a phase-type distribution with representation $(\boldsymbol{\gamma}, \boldsymbol{S} - \lambda \boldsymbol{I} + \lambda \boldsymbol{e}\boldsymbol{\gamma})$, that is,

$$F_Z(x) = 1 - \boldsymbol{\gamma} \exp[(\boldsymbol{S} - \lambda \boldsymbol{I} + \lambda \boldsymbol{e}\boldsymbol{\gamma})x]\boldsymbol{e}. \quad \text{(M/PH/1)} \tag{1.59}$$

This fact can be verified with the following observation. Note that $\min(G_n, H_n)$ in the M/PH/1 queue is equal to the absorption time of a Markov chain with transition rate matrix $\boldsymbol{S} - \lambda \boldsymbol{I}$ and initial-state distribution $\boldsymbol{\gamma}$. This Markov chain transitions to its absorbing state according to a rate vector $(-\boldsymbol{S})\boldsymbol{e} + \lambda \boldsymbol{e}$, where the first (resp. second) term corresponds to the event $\min(G_n, H_n) = H_n$ (resp. $\min(G_n, H_n) = G_n$). Therefore, we can obtain a realization of Z with the following procedure: (i) start this Markov chain with the initial-state distribution $\boldsymbol{\gamma}$, (ii) whenever an absorption with the rate vector $\lambda \boldsymbol{e}$ occurs, restart the Markov chain with the initial-state distribution $\boldsymbol{\gamma}$, and (iii) terminate the process when an absorption with the rate vector $(-\boldsymbol{S})\boldsymbol{e}$ occurs. The whole process of this procedure is then represented as a Markov chain with initial-state vector $\boldsymbol{\gamma}$ and transition rate matrix $\boldsymbol{S} - \lambda \boldsymbol{I} + \lambda \boldsymbol{e}\boldsymbol{\gamma}$, which proves the preceding claim.

In the D/PH/1 queue, on the other hand, we have $G_{<H} = \tau$ and $H_{<G} \in [0, \tau)$ with probability one. Also, we have

$$\zeta = \boldsymbol{\gamma} \exp[\boldsymbol{S}\tau]\boldsymbol{e}. \quad \text{(D/PH/1)} \tag{1.60}$$

It is then readily verified that for $k = 0, 1, \ldots,$

$$F_Z(x) = 1 - \zeta^k \gamma \exp\big[\boldsymbol{S}(x - k\tau)\big]\boldsymbol{e}, \quad x \in [k\tau, (k+1)\tau). \quad \text{(D/PH/1)} \tag{1.61}$$

Therefore, we obtain the AoI distribution in the M/PH/1 and D/PH/1 queues from Corollary 1.4.2, (1.57), (1.58), (1.59), and (1.61):

THEOREM 1.9

(i) In the stationary preemptive LCFS M/PH/1 queue, the density function and the CDF of the AoI are given by

$$f_A(x) = \gamma \exp[\boldsymbol{C}x](-\boldsymbol{C})\boldsymbol{e}, \quad x \geq 0,$$
$$F_A(x) = 1 - \gamma \exp[\boldsymbol{C}x]\boldsymbol{e}, \quad x \geq 0,$$

where $\boldsymbol{C}$ is defined as

$$\boldsymbol{C} = \begin{pmatrix} \boldsymbol{S} - \lambda \boldsymbol{I} + \lambda \boldsymbol{e}\gamma & (-\boldsymbol{S})\boldsymbol{e} \\ \boldsymbol{0} & -\lambda \end{pmatrix}.$$

(ii) In the stationary preemptive LCFS D/PH/1 queue, the density function and the CDF of the AoI are given as follows: for $x \in [0, \tau)$,

$$f_A(x) = \frac{1 - \gamma \exp[\boldsymbol{S}x]\boldsymbol{e}}{\tau},$$
$$F_A(x) = \frac{x - \gamma(\boldsymbol{I} - \exp[\boldsymbol{S}x])(-\boldsymbol{S})^{-1}\boldsymbol{e}}{\tau},$$

and for $x \in [k\tau, (k+1)\tau)$ $(k = 1, 2, \ldots)$,

$$f_A(x) = \frac{\zeta^{k-1}(1-\zeta)}{\tau} \cdot \gamma \exp[\boldsymbol{S}(x - k\tau)]\boldsymbol{e},$$
$$F_A(x) = 1 - \frac{\zeta^{k-1}\gamma}{\tau}\Big(\zeta \boldsymbol{I} - \exp[\boldsymbol{S}\tau] + (1-\zeta)\exp\big[\boldsymbol{S}(x - k\tau)\big]\Big)(-\boldsymbol{S})^{-1}\boldsymbol{e},$$

where ζ is given by (1.60).

Finally, we present numerical examples employing the same class of service time distributions as those in the previous subsection, that is, the mixed Erlang and balanced hyper-exponential distributions with mean $\mathrm{E}[H] = 1$, which is uniquely identified by the coefficient of variation $\mathrm{Cv}[H]$. Figures 1.9 and 1.10 show comparisons of the FCFS and preemptive LCFS service policies in terms of the AoI distribution for M/PH/1 and D/PH/1 queues. As we previously observed in Figures 1.7 and 1.8, the large variability of service times has a significant negative impact on the AoI performance under the FCFS service policy. From Figures 1.9 and 1.10, however, we observe that the AoI performance is not negatively impacted under the preemptive LCFS service policy. Moreover, quite the contrary to the FCFS case, we find that *under the preemptive LCFS service policy, the AoI even tends to be smaller when service times have larger variability*. Intuitively, this phenomenon is due to the fact that the larger the variability

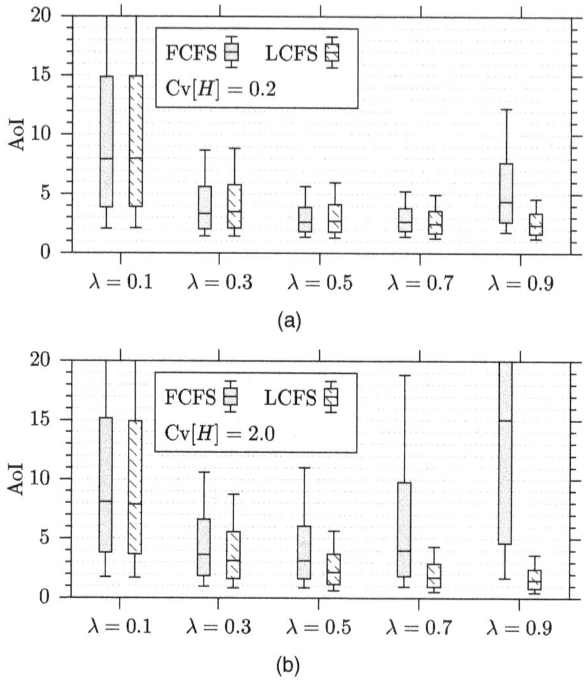

Figure 1.9 Comparison of the AoI distribution in FCFS and LCFS M/PH/1 queues.

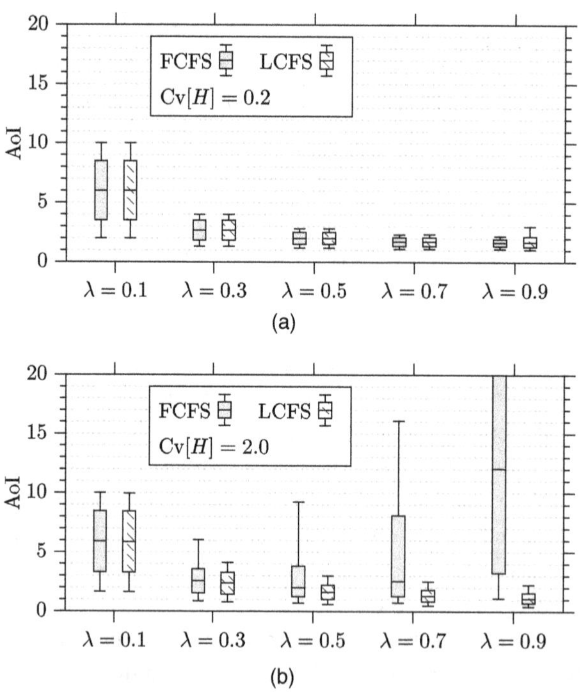

Figure 1.10 Comparison of the AoI distribution in FCFS and LCFS D/PH/1 queues.

of the service times, the more frequently service times take substantially smaller values than their mean. Updates with such small service times can update the monitor just after their generations under the preemptive LCFS service policy, which contributes to the timeliness of information updates.

While the preemptive LCFS service policy is quite efficient for highly variable service times, it should be noted that this policy *does not* always achieve superior performance compared to the FCFS service policy. For the case of $\mathrm{Cv}[H] = 0.2$ in Figure 1.9, we observe that the preemptive LCFS service policy has slightly worse AoI performance (in terms of the percentiles) than the FCFS service policy, except for large values of λ. This is because, when service times have small variability, the preemptive LCFS service policy may cause inefficient interruptions of service, resulting in a long update interval.

1.5 Bibliographical Notes

Section 1.1 The graphical analysis of the mean AoI and the queueing modeling of monitoring systems are due to the seminal work (Kaul, Yates, & Gruteser 2012b) on the AoI analysis, where the formulas for the mean AoI in the FCFS M/M/1 and D/M/1 queues are given. The expression (1.8) for the mean AoI focusing on the interdeparture time is given by Costa, Codreanu, and Ephremides (2016), where the authors also introduce the concept of the peak AoI A_{peak} as an alternative freshness metric.

Section 1.2 The nonlinear AoI cost function (1.12) is introduced by Sun, Uysal-Biyikoglu, Yates, Koksal, and Shroff (2017), where the optimal sampling policy minimizing the mean nonlinear AoI cost is discussed. Several expressions for the mean nonlinear AoI cost are also derived by Kosta, Pappas, Ephremides, and Angelakis (2020a) assuming specific cost functions. The effect of the AoI on the monitoring accuracy is first considered by Costa, Valentin, and Ephremides (2015) for a discrete-time two-state Markov chain, where the monitoring accuracy is defined as the probability that the actual and displayed states coincide. The monitoring accuracy for a continuous-time Markov chain is also analyzed by Inoue and Takine (2019). Sun, Polyanskiy, and Uysal (2020) show that if the information source is represented as a Wiener process and the AoI process is not affected by its state, then the mean AoI equals the mean squared error of the displayed information. Sun and Cyr (2018) also propose the use of the mutual information in quantifying the monitoring accuracy.

Sections 1.3 and 1.4 Most of the discussions in these sections are based on those given by Inoue, Masuyama, Takine, and Tanaka (2019), where the formula (1.4) for the AoI distribution is derived under a more general setting. In fact, this formula follows from the observation that the stochastic properties of the stationary AoI are characterized in terms of those of AoI values immediately before and after an update, so that the assumptions on underlying mechanisms (i.e., sampling intervals $(G_n)_{n=0,1,\ldots}$ and system delays $(D_n)_{n=0,1,\ldots}$) are not essential. Theorems 1.7 and 1.9 for the AoI distribution in queues with phase-type service time distributions are new, although most of those are specializations of more general results derived by Inoue et al. (2019). We have not included discussions on LCFS service policies without service preemption for the brevity of explanation. The analysis of the AoI distribution for non-preemptive LCFS queues can be found in the papers (Inoue et al. 2019) and Champati, Al-Zubaidy, and Gross (2019), which are extensions of earlier studies on the mean AoI analysis (Kaul, Yates, and Gruteser 2012a; Costa et al. 2016).

Further Studies on the AoI Distribution The general formula (1.4) serves as a starting point of the analysis of the distributional properties of the AoI. For instance, its applications to a single-server queue with packet deadlines (Inoue 2018) and infinite-server queues (Inoue 2020) are reported. While this formula is derived assuming continuous-time systems, a discrete-time analog of this formula is also considered by Kosta, Pappas, Ephremides, and Angelakis (2020b). Furthermore, Jiang and Miyoshi (2020) give a generalization of (1.4) to *the joint distribution* of the AoI values considering multiple source-monitor pairs.

Appendix

A.1 A Brief Summary of Phase-Type Distributions

Consider an absorbing Markov chain $(X_t)_{t\geq 0}$ with finite states. Generally, states in an absorbing Markov chain are classified into transient states and absorbing states, and the Markov chain eventually reaches an absorbing state after repeating transitions among transient states. The phase-type distribution is a class of probability distributions on $[0, \infty)$, which can be formulated as the absorption time T (i.e., the first passage time to a set of absorbing states) in an absorbing Markov chain. In the context of the phase-type distribution, transient states are called phases. In this appendix, the absorbing Markov chain $(X_t)_{t\geq 0}$ is assumed to have M ($M \geq 1$) phases and let $\mathcal{M} = \{1, 2, \ldots, M\}$.

In general, the absorbing Markov chain $(X_t)_{t\geq 0}$ is defined by the M-dimensional initial state vector $\boldsymbol{\alpha}$ and $M \times M$ matrix $\boldsymbol{Q}$, where the ith ($i \in \mathcal{M}$) element of $\boldsymbol{\alpha}$ is given by $\Pr(X_0 = i)$ and the (i,j)th ($i,j \in \mathcal{M}$, $i \neq j$) off-diagonal element $q_{i,j}$ of $\boldsymbol{Q}$ is given by the transition rate from state i to state j, that is,

$$q_{i,j} = \lim_{\Delta t \to 0} \frac{\Pr(X_{t+\Delta t} = j \mid X_t = i)}{\Delta t}, \quad i \neq j,$$

and the ith ($i \in \mathcal{M}$) diagonal element $q_{i,i}$ is given by

$$q_{i,i} = -\sum_{j \in \mathcal{M} \setminus \{i\}} q_{i,j}.$$

We assume that $\boldsymbol{Q} + (-\boldsymbol{Q})\boldsymbol{e}\boldsymbol{\alpha}$ is irreducible. For a given $(\boldsymbol{\alpha}, \boldsymbol{Q})$, the probability distribution function $F_T(x) = \Pr(T \leq x)$ of the absorption time T is given by

$$F_T(x) = 1 - \boldsymbol{\alpha} \exp[\boldsymbol{Q}x]\boldsymbol{e}, \quad x \geq 0,$$

where $\boldsymbol{e}$ denotes a column vector with an appropriate dimension, whose elements are all equal to one. Note here that $F_T(0) = \Pr(T = 0) = 1 - \boldsymbol{\alpha}\boldsymbol{e} \geq 0$. For simplicity,

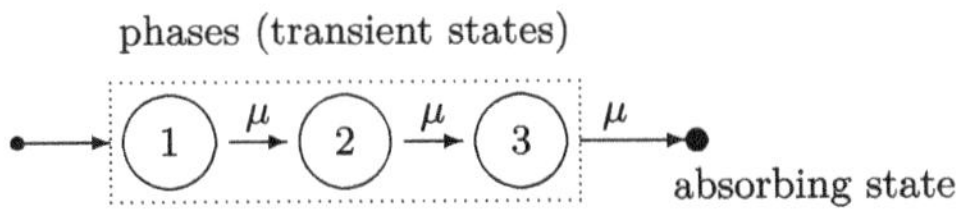

$$\alpha = (1\ 0\ 0), \quad Q = \begin{pmatrix} -\mu & \mu & 0 \\ 0 & -\mu & \mu \\ 0 & 0 & -\mu \end{pmatrix}$$

Figure A.1 A three-stage Erlang distribution.

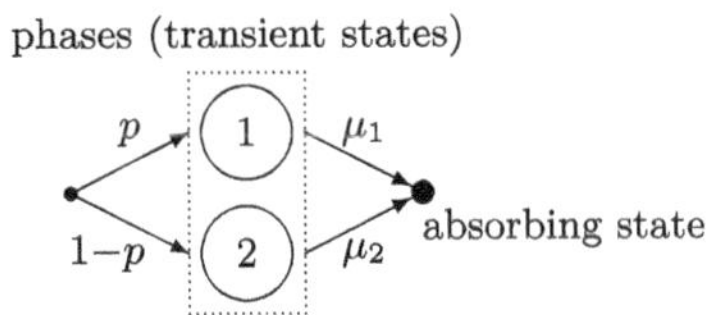

$$\alpha = (p\ 1{-}p), \quad Q = \begin{pmatrix} -\mu_1 & 0 \\ 0 & -\mu_2 \end{pmatrix}$$

Figure A.2 A two-stage hyper-exponential distribution.

we assume $F_T(0) = 0$, that is, $\boldsymbol{\alpha e} = 1$. The probability density function $f_T(x)$ is then given by

$$f_T(x) = \boldsymbol{\alpha} \exp[\boldsymbol{Q}x](-\boldsymbol{Q})\boldsymbol{e}, \quad x \geq 0,$$

and the nth ($n = 1, 2, \ldots$) moment $\mathrm{E}[T^n]$ is given by

$$\mathrm{E}[T^n] = \int_0^\infty x^n f(x)\mathrm{d}x = n!\boldsymbol{\alpha}\left[(-\boldsymbol{Q})^{-1}\right]^n \boldsymbol{e}.$$

Figures A.1 and A.2 show phase diagrams for a three-stage Erlang distribution and a two-stage hyper-exponential distribution.

We define $\widetilde{\boldsymbol{\alpha}}$ as a $1 \times M$ probability vector given by

$$\widetilde{\boldsymbol{\alpha}} = \frac{\boldsymbol{\alpha}(-\boldsymbol{Q})^{-1}}{\boldsymbol{\alpha}(-\boldsymbol{Q})^{-1}\boldsymbol{e}}.$$

Note that $\widetilde{\boldsymbol{\alpha}}$ satisfies $\widetilde{\boldsymbol{\alpha}}[\boldsymbol{Q} + (-\boldsymbol{Q})\boldsymbol{e\alpha}] = \boldsymbol{0}$. The probability distribution function $F_{\widetilde{T}}(x)$ of the equilibrium random variable $\widetilde{T}$ for T is then given by

$$F_{\widetilde{T}}(x) = 1 - \widetilde{\boldsymbol{\alpha}} \exp[\boldsymbol{Q}x]\boldsymbol{e}, \quad x \geq 0.$$

The phase-type distribution is close under the convolution, minimum, and maximum operations. Let T_i ($i = 1, 2$) denote an independent phase-type random variable with representation $(\boldsymbol{\alpha}_i, \boldsymbol{Q}_i)$. The probability distribution function of $T_1 + T_2$ is then given by

$$\Pr(T_1 + T_2 \leq x) = 1 - \boldsymbol{\alpha}_{\mathrm{sum}} \exp[\boldsymbol{Q}_{\mathrm{sum}}x]\boldsymbol{e}, \quad x \geq 0,$$

where

$$\boldsymbol{\alpha}_{\mathrm{sum}} = \begin{pmatrix} \boldsymbol{\alpha}_1 & \boldsymbol{0} \end{pmatrix}, \qquad \boldsymbol{Q}_{\mathrm{sum}} = \begin{pmatrix} \boldsymbol{Q}_1 & (-\boldsymbol{Q}_1)\boldsymbol{e\alpha}_2 \\ \boldsymbol{O} & \boldsymbol{Q}_2 \end{pmatrix}.$$

We can also obtain the phase-type distribution functions of $\min(T_1, T_2)$ and $\max(T_1, T_2)$ using Kronecker product/sum. In particular, if T_2 follows an exponential distribution with parameter λ (i.e., the phase-type distribution with representation $(1, -\lambda)$), we have

$$\Pr(\min(T_1, T_2) \le x) = 1 - \boldsymbol{\alpha}_1 \exp[(\boldsymbol{Q}_1 - \lambda \boldsymbol{I})x]\boldsymbol{e}, \quad x \ge 0,$$
$$\Pr(\max(T_1, T_2) \le x) = 1 - \boldsymbol{\alpha}_{\max} \exp[\boldsymbol{Q}_{\max} x]\boldsymbol{e}, \quad x \ge 0,$$

where

$$\boldsymbol{\alpha}_{\max} = \begin{pmatrix} \boldsymbol{\alpha}_1 & \mathbf{0} & 0 \end{pmatrix}, \qquad \boldsymbol{Q}_{\max} = \begin{pmatrix} \boldsymbol{Q}_1 - \lambda \boldsymbol{I} & \lambda \boldsymbol{I} & (-\boldsymbol{Q}_1)\boldsymbol{e} \\ \boldsymbol{O} & \boldsymbol{Q}_1 & \mathbf{0} \\ \mathbf{0} & \mathbf{0} & -\lambda \end{pmatrix}.$$

Another advantage of the phase-type distribution is computational feasibility, that is, the uniformization technique is applicable. For a phase-type distribution with representation $(\boldsymbol{\alpha}, \boldsymbol{Q})$, we define θ as

$$\theta = \max_{i \in \mathcal{M}} |q_{i,i}|.$$

It can be readily verified that

$$\exp[\boldsymbol{Q}x] = e^{-\theta x} \exp[(\boldsymbol{I} + \theta^{-1}\boldsymbol{Q})\theta x] = \sum_{n=0}^{\infty} e^{-\theta x} \frac{(\theta x)^n}{n!} \boldsymbol{P}^n,$$

where $\boldsymbol{P} = \boldsymbol{I} + \theta^{-1}\boldsymbol{Q}$. Note here that $\boldsymbol{P}$ can be regarded as a defective transition probability matrix because $\boldsymbol{I} + \theta^{-1}\boldsymbol{Q} \ge \boldsymbol{O}$ and $(\boldsymbol{I} + \theta^{-1}\boldsymbol{Q})\boldsymbol{e} \le \boldsymbol{e}$. The technique for expressing $\exp[\boldsymbol{Q}t]$ in terms of $\boldsymbol{P}$ is called uniformization.

The probability distribution of the phase-type random variable T with representation $(\boldsymbol{\alpha}, \boldsymbol{Q})$ may be computed as follows. For $x \ge 0$, let $\overline{F}_T(x) = \Pr(T > x) = \boldsymbol{\alpha} \exp[\boldsymbol{Q}x]\boldsymbol{e}$. The following procedure yields an approximation $\overline{F}_T^{\text{approx}}(x)$ to $\overline{F}_T(x)$, which satisfies

$$0 < \overline{F}_T(x) - \overline{F}_T^{\text{approx}}(x) \le \epsilon,$$

for a predefined error bound ϵ $(0 < \epsilon \ll 1)$.

1. Find a positive integer N^* such that

$$\sum_{n=0}^{N^*} e^{-\theta x} \frac{(\theta x)^n}{n!} \ge 1 - \epsilon.$$

2. Let

$$\boldsymbol{y}_{N^*} = \frac{\theta x}{N^*} \boldsymbol{P}\boldsymbol{e},$$

and compute

$$\boldsymbol{y}_n = \frac{\theta x}{n} \boldsymbol{P}(\boldsymbol{e} + \boldsymbol{y}_{n+1}),$$

recursively in descending order from $n = N^* - 1$ to 1.

3. Let $\overline{F}_T^{\text{approx}}(x) = \exp(-\theta x)\boldsymbol{\alpha}(\boldsymbol{e} + \boldsymbol{y}_1)$.

A.2 Derivation of Theorem 1.7

We first consider (i). Because the intergeneration time G and the system delay D are absolutely continuous in the M/PH/1 queue, we have that $\max(G, D)$ is also absolutely continuous with density:

$$\begin{aligned} f_{\max(G,D)}(x) &= f_G(x)F_D(x) + F_G(x)f_D(x) \\ &= \lambda e^{-\lambda x}F_D(x) + (1 - e^{-\lambda x})f_D(x) \\ &= f_D(x) + e^{-\lambda x}(\lambda F_D(x) - f_D(x)) \\ &= f_D(x) - \frac{\mathrm{d}}{\mathrm{d}x}\left[F_D(x)e^{-\lambda x}\right]. \end{aligned}$$

We then rewrite (1.36) as

$$\begin{aligned} f_A(x) &= \lambda \left\{ \int_0^x f_D(y)\mathrm{d}y - \int_0^x \left(f_D(y) - \frac{\mathrm{d}}{\mathrm{d}y}\left[F_D(x)e^{-\lambda y}\right] \right) F_H(x-y)\mathrm{d}y \right\} \\ &= \rho \int_0^x f_D(y) \cdot \frac{1 - F_H(x-y)}{\mathrm{E}[H]}\mathrm{d}y + \lambda \int_0^x \frac{\mathrm{d}}{\mathrm{d}y}\left[F_D(y)e^{-\lambda y}\right] F_H(x-y)\mathrm{d}y \\ &= \rho f_D * f_{\widetilde{H}}(x) + \lambda \int_0^x F_D(y)e^{-\lambda y} f_H(x-y)\mathrm{d}y \\ &= \rho f_D * f_{\widetilde{H}}(x) + \int_{y=0}^x \left(\int_{t=0}^y f_D(t)e^{-\lambda t} \cdot \lambda e^{-\lambda(y-t)}\mathrm{d}t \right) f_H(x-y)\mathrm{d}y \\ &= \rho f_D * f_{\widetilde{H}}(x) + (1-\rho) f_{D_{<G}} * f_G * f_H(x), \end{aligned} \tag{A.62}$$

where $D_{<G} := [D \mid D < G]$ denotes the system delay D conditioned to be smaller than an intergeneration time G. The density function of $D_{<G}$ is given by

$$\begin{aligned} f_{D_{<G}}(x) &= \frac{f_D(x)e^{-\lambda x}}{\Pr(D < G)} \\ &= \frac{f_D(x)e^{-\lambda x}}{1-\rho} \\ &= \frac{\boldsymbol{\gamma} \exp[(\boldsymbol{Q} - \lambda \boldsymbol{I})x](-\boldsymbol{Q})\boldsymbol{e}}{1-\rho} \\ &= \frac{\boldsymbol{\gamma}\{(-\boldsymbol{Q} - \lambda \boldsymbol{I})\}^{-1}(-\boldsymbol{Q}) \exp[(\boldsymbol{Q} - \lambda \boldsymbol{I})x]\{(-\boldsymbol{Q} - \lambda \boldsymbol{I})\}\boldsymbol{e}}{1-\rho} \\ &= \frac{(\boldsymbol{\gamma} - \boldsymbol{\pi}_*) \exp[(\boldsymbol{Q} - \lambda \boldsymbol{I})x]\{(-\boldsymbol{Q} - \lambda \boldsymbol{I})\}\boldsymbol{e}}{1-\rho}, \end{aligned} \tag{A.63}$$

where the second equality ($\Pr(D < G) = 1 - \rho$) follows because (i) $\Pr(D < G)$ equals the stationary probability p_0 that an arriving customer finds the system empty and (ii) due to the Poisson-arrivals-see-time-averages (PASTA) property, p_0 in the M/PH/1 queue equals the time-average probability that the system is empty, which is given by $1 - \rho$. Also, the third equality follows because $\boldsymbol{Q}$ and $\boldsymbol{Q} - \lambda \boldsymbol{I}$ commute and the last equality ($\boldsymbol{\gamma} - \boldsymbol{\pi}_* = \boldsymbol{\gamma}\{-(\boldsymbol{Q} - \lambda \boldsymbol{I})\}^{-1}(-\boldsymbol{Q})$) is verified, noting that $\boldsymbol{\pi}_*$ satisfies (cf. (1.40) and (1.45))

$$\boldsymbol{\pi}_* = \lambda \boldsymbol{\gamma}\{-(\boldsymbol{Q} - \lambda \boldsymbol{I})\}^{-1}.$$

We then readily obtain (1.47) from (A.62) and (A.63). Also, (1.48) is obtained by integrating (1.47) and using

$$\rho \boldsymbol{\gamma}_0 \boldsymbol{e}_0 + \boldsymbol{\gamma}_1 \boldsymbol{e}_1 = \rho + (1 - \boldsymbol{\pi}_* \boldsymbol{e}) = 1,$$

where the last equality follows from (1.44).

We next consider (ii). Recall that (1.36) has been derived using the expression (1.35) for the peak AoI instead of (1.34). For the case of deterministic intergeneration times, however, it is simpler to directly work on (1.34) because it reduces to $A_{\text{peak},n} = D_n + \tau$, so that we have

$$F_{A_{\text{peak}}}(x) = \begin{cases} 0 & 0 \le x < \tau, \\ F_D(x - \tau), & x \ge \tau. \end{cases}$$

It then follows from Theorem 1.4,

$$\begin{aligned} f_A(x) &= \begin{cases} \dfrac{F_D(x)}{\tau}, & 0 \le x < \tau, \\ \dfrac{F_D(x) - F_D(x - \tau)}{\tau}, & x \ge \tau. \end{cases} \\ &= \begin{cases} \dfrac{1 - \boldsymbol{\gamma} \exp[\boldsymbol{Q}x]\boldsymbol{e}}{\tau}, & 0 \le x < \tau, \\ \dfrac{\boldsymbol{\gamma} \exp[\boldsymbol{Q}(x - \tau)]\boldsymbol{e} - \boldsymbol{\gamma} \exp[\boldsymbol{Q}\tau] \exp[\boldsymbol{Q}(x - \tau)]\boldsymbol{e}}{\tau}, & x \ge \tau, \end{cases} \end{aligned}$$

which implies (1.49). Also, (1.50) is readily obtained by integrating (1.49).

References

Asmussen, S. (1992), "Phase-type representations in random walk and queueing problems," *Ann. Prob.* **20**, 772–789.

Asmussen, S. (2003), *Applied Probability and Queues*, 2nd ed., Springer.

Bhat, U. N., & Rao, S. S. (1987), "Statistical analysis of queueing systems," *Queueing Systems* **1**(3), 217–247.

Brill, P. H., & Posner, M. J. M. (1977), "Level crossings in point processes applied to queues: Single-server case," *Oper. Res.* **25**, 662–674.

Champati, J. P., Al-Zubaidy, H., & Gross, J. (2019), On the distribution of AoI for the GI/GI/1/1 and GI/GI/1/2* systems: Exact expressions and bounds, *in* "Proc. of IEEE INFOCOM 2019," pp. 37–45.

Cohen, J. W. (1977), "On up- and downcrossing," *J. Appl. Probab.* **14**, 405–410.

Costa, M., Codreanu, M., and Ephremides, A. (2016), "On the age of information in status update systems with packet management," *IEEE Trans. Inf. Theory* **62**(4), 1897–1910.

Costa, M., Valentin, S., & Ephremides, A. (2015), On the age of channel state information for a finite-state Markov model, *in* "Proc. of IEEE ICC 2015," pp. 4101–4106.

El-Taha, M., & Stidham Jr., S. (1999), *Sample-Path Analysis of Queueing Systems*, Kluwer Academic Publishers.

Inoue, Y. (2018), Analysis of the age of information with packet deadline and infinite buffer capacity, *in* "Proc. of IEEE ISIT 2018," pp. 2639–2643.

Inoue, Y. (2020), The probability distribution of the AoI in queues with infinitely many servers, *in* "Proc. of The 3rd Age of Information Workshop (AoI'20)," pp. 297–302.

Inoue, Y., Masuyama, H., Takine, T., & Tanaka, T. (2019), "A general formula for the stationary distribution of the age of information and its application to single-server queues," *IEEE Trans. Inf. Theory* **65**(12), 8305–8324.

Inoue, Y., & Takine, T., (2019), AoI perspective on the accuracy of monitoring systems for continuous-time Markovian sources, *in* "Proc. of The 2nd Age of Information Workshop (AoI '19)."

Jiang, Y., & Miyoshi, N. (2020), "Joint performance analysis of ages of information in a multi-source pushout server," *arXiv preprint arXiv:2006.03404.*

Kaul, S. K., Yates, R. D., & Gruteser, M. (2012a), Status updates through queues, *in* "Proc. of IEEE CISS 2012."

Kaul, S., Yates, R., & Gruteser, M. (2012b), Real-time status: How often should one update?, *in* "Proc. of IEEE INFOCOM 2012," pp. 2731–2735.

Kleinrock, L. (1975), *Queueing Systems, Volume I: Theory*, John Wiley & Sons.

Kosta, A., Pappas, N., Ephremides, A., & Angelakis, V. (2020a), "The cost of delay in status updates and their value: Non-linear ageing," *IEEE Trans. Commun.* **68**(8), 4905–4918.

Kosta, A., Pappas, N., Ephremides, A., & Angelakis, V. (2020b), Non-linear age of information in a discrete time queue: Stationary distribution and average performance analysis, *in* "Proc. of ICC 2020."

Sun, Y., & Cyr, B. (2018), Information aging through queues: A mutual information perspective, *in* "Proc. of SPAWC 2018."

Sun, Y., Polyanskiy, Y., & Uysal, E. (2020), "Sampling of the Wiener process for remote estimation over a channel with random delay," *IEEE Trans. Inf. Theory* **66**(2), 1118–1135.

Sun, Y., Uysal-Biyikoglu, E., Yates, R., Koksal, C. M., & Shroff, N. B. (2017), "Update or wait: How to keep your data fresh," *IEEE Trans. Inf. Theory* **63**(11), 7492–7508.

2 On the Distribution of AoI

Jaya Prakash Champati and James Gross

Initial research on the analysis and minimization of AoI for different queueing systems focused on the average AoI statistics. However, several applications of interest, for example in safety-critical industrial systems, pose stringent requirements on the freshness of packets exchanged over a network. This circumstance, which we further elaborate in what follows, together with related applications, motivates the research on studying the distribution of AoI.

Safety-critical control systems can be certified in two different ways (Leveson 2011). In the first case, *all* individual components of the system are subject to a stringent certification process. However, this requires one to quantify failure probabilities of the components under operational conditions, which is not always possible. An alternative approach is to employ so-called *safety layers* (Voss 2020), which is especially the case in distributed control systems that are closed over unreliable networks. Safety layers rest on the black channel principle, that is, no internal state of the underlying communication system can be taken into consideration at run-time. Instead, the safety is validated periodically by a simple message exchange between a pair of safety layers, which are essentially software entities residing on top of layer 7 of the ISO/OSI stack. Example implementations comprise PROFIsafe, SafetyNetP, as well as SafetyCAN, which operate according to the black channel principle (Voss 2020). In more detail, the entities exchange enumerated safety frames with message integrity codes generated from payload as well as shared secrets. If within a predefined safety deadline one of the safety layers involved cannot validate the integrity of the system through receiving a correctly encoded safety frame, then the corresponding safety layer (a) raises a safety exception, which can include switching the involved plant to a failsafe mode (i.e. stopping the machine), and (b) stops sending safety messages to its peers. As a consequence, safety can be guaranteed over an unreliable network at the price of (potentially frequent) events that shut down the plant, which do not reflect an immediate safety breach of the system (i.e. the trigger of a safety light barrier).

From this description it becomes clear why the freshness of the involved safety frames, and in particular the statistical distribution of the freshness is of interest. If we consider the messaging between two safety layers as an update process over a communication channel with random (unreliable) service and consider the stochastic process resembled by AoI of this system, then the tail mass of the AoI distribution for this system with respect to the given deadline is a direct measure of the availability of the control system. The lower the tail mass, the higher will be the availability of the system.

Thus, in a second step one might consider the optimization (minimization) of the tail mass, for instance, by tuning the frequency of safety messages. This emphasizes the relevance and potential applicability of AoI analysis with respect to its distribution.

This chapter is concerned with fundamental analysis, closed-form expressions, and bounds on the distribution of AoI. We first provide a general formula for the distribution of AoI for a single-source single-server system, and focus primarily on two non-preemptive service systems: (1) GI/GI/1/1, which has no queue, and (2) GI/GI/1/2*, which has a unit capacity queue and a new arrival always replaces a packet in the queue. The relevance of these systems in the context of AoI is due to the fact that in both systems every packet served is an *information update packet*, that is, every packet received at the destination has latest generation time. Furthermore, these systems potentially result in lower AoI statistics compared to that of the systems with larger capacity queue, because storing and transmitting old packets from the queue does not reduce AoI (Costa, Codreanu, & Ephremides 2016; Kosta, Pappas, Ephremides, & Angelakis 2019).

Most of the results of this chapter are from Champati, Al-Zubaidy, and Gross (2019a). A general formula for the distribution of AoI was also derived in Inoue, Masuyama, Takine, and Tanaka (2019), where the violation probability is characterized in terms of the distribution of peak AoI and the distribution of system delay. The distribution of peak AoI for M/M/1/1, M/M/1/2, and M/M/1/2* systems was derived in Costa, Codreanu, & Ephremides (2016) and that of PH/PH/1/1 and M/PH/1/2 systems[1] was derived in Akar, Dogan, and Atay (2020). Kesidis, Konstantopoulos, and Zazanis (2020) characterized the distribution of AoI for bufferless systems with arrival/service distributions with dependencies. The characterization of the distribution of AoI was used to compute average statistics of nonlinear function of AoI for discrete-time queueing systems in Kosta, Pappas, Ephremides, and Angelakis (2020). Distribution of AoI in a tandem network with single source was studied in Yates (2020), where the moment-generating function and stationary distribution were derived for Poisson arrivals and exponential service times under preemptive last-come-first-served policy, and in Champati, Al-Zubaidy, and Gross (2020), where periodic sampling rate is optimized to minimize the AoI violation probability.

Next, we present some notation that is used for the analysis in the sequel.

2.1 Notation, Definitions, and System Model

Consider a single source generating status updates or packets that are immediately dispatched to a single-server queueing system as shown in Figure 2.1. The inter-arrival time between the packets is denoted by the random variable Z with mean-arrival rate $\lambda = \frac{1}{\mathbb{E}[Z]}$. The arriving packets may be stored in a queue and are served by a server using some scheduling policy. We use the random variable X to denote the service time with mean-service rate $\mu = \frac{1}{\mathbb{E}[X]}$. We use packet n to refer to a packet that is nth

[1] PH stands for phase-type distribution.

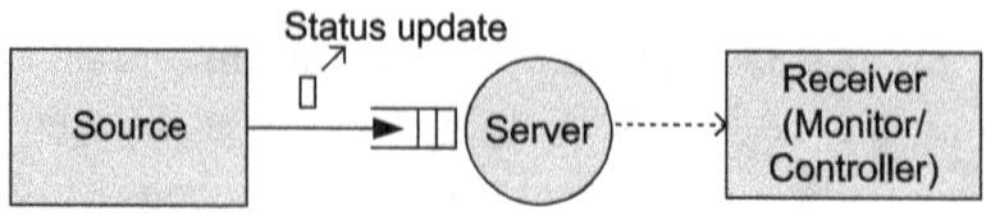

Figure 2.1 A single-source single-server system.

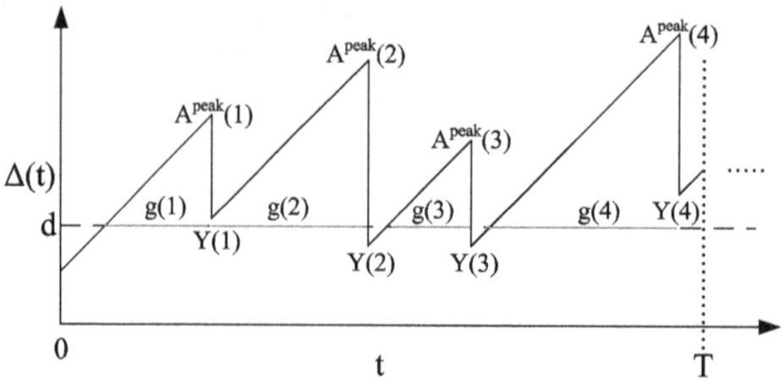

Figure 2.2 A sample path of AoI process.

in the sequence of departures at the server. Let $T_{\mathrm{D}}(n)$ denote the time instant of nth packet departure and $T_{\mathrm{A}}(n)$ denote the corresponding arrival instant.

The AoI metric, denoted by $\Delta(t)$, is defined as

$$\Delta(t) \triangleq t - \max\{T_{\mathrm{A}}(n) : T_{\mathrm{D}}(n) \leq t\}. \tag{2.1}$$

For a given age limit $d \geq 0$, we are interested in computing the steady-state violation probability or simply violation probability given by

$$\mathbb{P}(\Delta > d) = \lim_{t \to \infty} \mathbb{P}(\Delta(t) > d).$$

The AoI process increases linearly in time with a slope of 1 until the departure of information an update packet at the server and it drops to a value equal to the system delay of that packet. Let $\{A^{\mathrm{peak}}(k), k \geq 1\}$ denote the peak AoI process, where $A^{\mathrm{peak}}(k)$ denotes the kth peak of $\Delta(t)$ as shown in Figure 2.2. Let $M(t)$ denote the number of peaks in the interval $(0, t]$. Also, in Figure 2.2 we plot $g(k)$, which is defined as the time duration for which AoI is greater than an age limit d in the interval between $(k-1)$th peak and kth peak. As mentioned before, the characterization of the violation probability and the bounds presented in this chapter are obtained in terms of $g(k)$. Finally, we define $Y(k) = \Delta(T_{\mathrm{D}}(k))$, that is, $Y(k)$ is the system delay of an update that departed at time $T_{\mathrm{D}}(k)$.

Note that the AoI peaks occur only at the departure instants of packets, but the converse might not be true, as some packet departures might not result in a drop in the AoI. This may happen, for instance, in a GI/GI/1 queue under LCFS scheduling. If there is no new arrival during the service of a packet, the next packet from the queue does not reduce AoI upon its departure, as its arrival time would be older than that of the previous departure. As noted before, we refer to packets that reduce AoI upon their departure as information update packets, and use k to index them, as it uniquely identifies an information update packet that departs at kth AoI peak.

Table 2.1 List of symbols.

k	Index of an information update packet
$T_{\rm A}(k)$	Arrival time of packet k
$T_{\rm D}(k)$	Departure time of packet k
$\check{Z}_k$	Inter-arrival time between packet k and its *previous arrival*
$\hat{Z}_k$	Inter-arrival time between packet k and its *next arrival*
X_k	Service time of packet k
Y_k	System delay of packet k
ν	Expected departure rate of information update packets
I_k	Idle time of the server before starting service for packet k
W_k	Waiting time of packet k
$M(t)$	Number of AoI peaks in the interval $(0, t]$

We note that packet n and packet k may not refer to the same packet for $k = n$. We assume that the time average departure rate of information update packets, denoted by ν, is positive and finite. For stationary and ergodic AoI process, we have, almost surely, $\nu = 1/\mathbb{E}[T_{\rm D}(k) - T_{\rm D}(k-1)]$.

We study the GI/GI/1/1 and GI/GI/1/2* systems under non-preemptive scheduling. In both systems the inter-arrival times and the service times are i.i.d. As mentioned before, a packet being served always has arrival time later than that of the previous departure. Thus, AoI is reduced at each departure instant and all departures are information update packets. In these systems, packet n and packet k refer to the same packet for $n = k$, $T_{\rm D}(k) - T_{\rm D}(k-1)$ represents the inter-departure time, $M(t)$ is the number of departures until time t, and ν is simply the expected departure rate.

We use ω to denote a sample path of AoI, and Ω to denote the set of all sample paths. Let $\gamma(k)$ and $\Gamma(k)$ denote sample-path-wise lower and upper bounds for $g(k)$, that is,

$$\gamma(\omega, k) \leq g(\omega, k) \leq \Gamma(\omega, k),\ \forall k \text{ and } \forall \omega \in \Omega. \tag{2.2}$$

In the rest of the chapter, we explicitly drop ω if it is clear from the context.

The list of symbols used in this chapter are summarized in Table 2.1. We use $(x)^+$ for $\max(0, x)$, and $\mathbb{1}\{\cdot\}$ for the indicator function, where $\mathbb{1}\{E\}$ equals one if event E is true, and is zero, otherwise.

2.2 General Formula for the Distribution of AoI

Recall that AoI process increases linearly with slope one until the next information packet departure. Therefore, $A^{\rm peak}(k)$ can be determined from $\Delta(t)$ at departure of packet $(k-1)$ and the inter-departure time between kth and $(k-1)$th packets:

$$\begin{aligned} A^{\rm peak}(k) &= T_{\rm D}(k) - T_{\rm A}(k-1) \\ &= Y(k-1) + T_{\rm D}(k) - T_{\rm D}(k-1). \end{aligned} \tag{2.3}$$

Analyzing $g(k)$ is central to the results presented in this chapter. In the following lemma, we express $g(k)$ in terms of kth AoI peak and inter-departure time between kth and $(k-1)$th packet.

LEMMA 2.1 *Given $d \geq 0$, for any sample path of $\Delta(t)$,*

$$g(k) = \min\{(A^{peak}(k) - d)^+, T_D(k) - T_D(k-1)\}, \ \forall k. \tag{2.4}$$

Proof Consider the case where $A^{\text{peak}}(k) \leq d$. For this case $g(k)$ is zero, by definition, which is satisfied by (2.4) as $T_{\text{D}}(k) \geq T_{\text{D}}(k-1)$. For $A^{\text{peak}}(k) > d$ we further consider the following cases.

Case 1: $A^{\text{peak}}(k) > d$ and $A^{\text{peak}}(k) - d > T_{\text{D}}(k) - T_{\text{D}}(k-1)$. Using this in (2.3), we obtain $Y(k-1) > d$. This implies that $\Delta(t) > d$ during the entire interval $[T_{\text{D}}(k-1), T_{\text{D}}(k))$. Therefore, $g(k) = T_{\text{D}}(k) - T_{\text{D}}(k-1)$. This is the case for $g(2)$ in Figure 2.2.

Case 2: $A^{\text{peak}}(k) > d$ and $A^{\text{peak}}(k) - d \leq T_{\text{D}}(k) - T_{\text{D}}(k-1)$. Using this in (2.3), we obtain $Y(k-1) \leq d$. In this case the horizontal line, with y coordinate equal to d, intersects $\Delta(t)$ at some time $t' \in [T_{\text{D}}(k-1), T_{\text{D}}(k))$. Since $\Delta(t)$ increases linearly with slope one, by geometry we obtain $g(k) = T_{\text{D}}(k) - t' = A^{\text{peak}}(k) - d$.

From the preceding analysis, we conclude that $g(k)$ takes the minimum value of $(A^{\text{peak}}(k) - d)^+$ and $T_{\text{D}}(k) - T_{\text{D}}(k-1)$, and the lemma follows. □

Next, we characterize the violation probability in terms of $g(k)$ in the following theorem.

THEOREM 2.2 *If the AoI process is stationary and ergodic, the AoI violation probability, if it exists, is given by*

$$\mathbb{P}(\Delta > d) = \lim_{T \to \infty} \frac{1}{T} \sum_{k=1}^{M(T)} g(k), \ a.s., \tag{2.5}$$

where $g(k)$ is given in (2.4).

Proof Since $\Delta(t)$ is stationary and ergodic, by Birkhoff's ergodic theorem, we have

$$\mathbb{P}(\Delta > d) = \lim_{T \to \infty} \frac{1}{T} \int_0^T \mathbb{1}\{\Delta(\tau) > d\} d\tau, \text{ a.s.} \tag{2.6}$$

The RHS in (2.6) is the fraction of time $\Delta(t)$ is greater than d in a given sample path.

Consider a sample path of $\Delta(t)$ presented in Figure 2.2. Let $\delta(T)$ denote the duration for which $\Delta(t)$ is greater than d after the $M(T)$th peak and before time T. It is easy to see that

$$\int_0^T \mathbb{1}\{\Delta(\tau) > d\} d\tau = \sum_{k=1}^{M(T)} g(k) + \delta(T)$$

$$\Rightarrow \lim_{T \to \infty} \frac{1}{T} \int_0^T \mathbb{1}\{\Delta(\tau) > d\} d\tau = \lim_{T \to \infty} \frac{1}{T} \sum_{k=1}^{M(T)} g(k). \tag{2.7}$$

In the last step, we used $\frac{\delta(T)}{T}$ goes to zero as T goes to infinity owing to the assumption that the departure rate is positive and finite. The result follows by substituting (2.7) in (2.6). □

Theorem 2.2 is quite general in the sense that it holds for any scheduling policy (e.g., FCFS/LCFS, preemptive/non-preemptive), general service times (possibly correlated), and general inter-arrival times (possibly correlated), as long as it is ensured that the resulting AoI process is stationary and ergodic. Note that even if the AoI process is stationary and ergodic, the violation probability may not exist. For example, for a D/G/1 system using FCFS the violation probability does not exist if $d < \frac{1}{\lambda}$ (Champati, Al-Zubaidy, & Gross 2018).

The challenge in evaluating the infinite summation in the RHS of (2.5) is that the sequence $\{g(k), k \geq 1\}$ is not i.i.d., and we cannot directly use the Strong Law of Large Numbers (SLLN). However, we will later show that quantities involving $g(k)$ have the structural independence property, defined in Definition 2.3, which enables us to use SLLN.

DEFINITION 2.3 An infinite sequence of random variables $\{\Theta_n, n \geq 1\}$ is structurally independent and identically distributed (s.i.i.d.) if Θ_n are identically distributed and have the following structural independence: for $1 \leq m < \infty$, Θ_{i+jm} is independent of Θ_{i+km}, for all $1 \leq i \leq m, j \geq 0, k \geq 0$, and $j \neq k$.

In the results that follow we make use of the following lemma, which extends SLLN for s.i.i.d. random variables.

LEMMA 2.4 *For any sequence $\{\Theta_n, n \geq 1\}$ that is s.i.i.d. according to Definition 1, we have*

$$\lim_{N\to\infty} \frac{1}{N} \sum_{n=0}^{N} \Theta_n = \mathbb{E}[\Theta], \ a.s.,$$

where $\mathbb{E}[\Theta] = \mathbb{E}[\Theta_n]$ for all n.

Proof The proof is based on partitioning the sum into multiple terms, which themselves are infinite sums of i.i.d. random variables, and then apply SLLN for these summations.

$$\begin{aligned}
\lim_{N\to\infty} \frac{1}{N} \sum_{n=0}^{N} \Theta_n &= \lim_{N\to\infty} \frac{1}{N} \sum_{i=1}^{m} \sum_{j=1}^{\lfloor \frac{N-i+m}{m} \rfloor} \Theta_{i+(j-1)m} \\
&= \frac{1}{m} \sum_{i=1}^{m} \lim_{N\to\infty} \frac{\lfloor \frac{N-i+m}{m} \rfloor}{N/m} \sum_{j=1}^{\lfloor \frac{N-i+m}{m} \rfloor} \frac{\Theta_{i+(j-1)m}}{\lfloor \frac{N-i+m}{m} \rfloor} \\
&= \frac{1}{m} \sum_{i=1}^{m} \mathbb{E}[\Theta] = \mathbb{E}[\Theta], \ \text{a.s.}
\end{aligned}$$

In the third step, we have used SLLN as $\{\Theta_{i+(j-1)m}, j \geq 1\}$ are i.i.d. (Definition 2.3), and $\lfloor \frac{N-i+m}{m} \rfloor$ differs from $\frac{N}{m}$ by at most 1. □

THEOREM 2.5 *Given age limit* $d \geq 0$, $\lambda > 0$, $0 < \mathbb{E}[X] = \frac{1}{\mu} < \infty$, $\{g(k), k \geq 1\}$ *are s.i.i.d., and* $\{T_{\mathrm{D}}(k) - T_{\mathrm{D}}(k-1), k \geq 1\}$ *are s.i.i.d., then*

$$\lim_{T\to\infty} \frac{1}{T} \sum_{k=1}^{M(T)} g(k) = \nu\mathbb{E}[g(k)], \ a.s.$$

Proof We have

$$\lim_{T\to\infty} \frac{1}{T} \sum_{k=1}^{M(T)} g(k) = \lim_{T\to\infty} \frac{M(T)}{T} \cdot \frac{\sum_{k=1}^{M(T)} g(k)}{M(T)}. \tag{2.8}$$

Since $\lambda > 0$ and $\mathbb{E}[X] < \infty$, $M(T)$ approaches infinity, almost surely, as T approaches infinity, and we obtain

$$\lim_{T\to\infty} \frac{T}{M(T)} = \lim_{T\to\infty} \sum_{k=1}^{M(T)} (T_{\mathrm{D}}(k) - T_{\mathrm{D}}(k-1))/M(T).$$

Since $\{T_{\mathrm{D}}(k) - T_{\mathrm{D}}(k-1), k \geq 1\}$ are s.i.i.d., from Lemma 2.4 we have

$$\lim_{T\to\infty} \frac{T}{M(T)} = \mathbb{E}[T_{\mathrm{D}}(k) - T_{\mathrm{D}}(k-1)], \text{ a.s.} \tag{2.9}$$

Similarly, we invoke Lemma 2.4 for $\{g(k), k \geq 1\}$ and obtain

$$\lim_{M(T)\to\infty} \frac{\sum_{k=1}^{M(T)} g(k)}{M(T)} = \mathbb{E}[g(k)], \text{ a.s.} \tag{2.10}$$

The result follows by substituting (2.9) and (2.10) in (2.8). □

Theorem 2.5 can be seen as an extension of renewal reward theorem for s.i.i.d. renewals and rewards. Later, we use the theorem to derive exact expressions for the violation probability for the D/GI/1/1 and M/GI/1/1 systems.

2.2.1 An Alternative Formula

In this section, we derive an alternative general formula (due to Inoue et al. (2019)). It characterizes the distribution of AoI in terms of the distribution of the system delay and the distribution of the peak AoI and is presented in the following theorem.

THEOREM 2.6 *For* $\nu \in (0, \infty)$, *assume that* $\lim_{t\to\infty} M(t)/t = \nu$; *then for stationary and ergodic AoI process, the distribution of AoI, if it exists, is given by*

$$\mathbb{P}(\Delta \leq d) = \nu \int_0^d (\mathbb{P}(Y \leq u) - \mathbb{P}(A^{peak} \leq u))du,$$

where $\mathbb{P}(Y \leq u)$ *and* $\mathbb{P}(A^{peak} \leq u)$ *are the steady-state distributions of the system delay and peak AoI processes, respectively.*

Proof Let $h(k)$ denote the time duration for which AoI is *smaller* than the age limit d in the interval between $(k-1)$th peak and kth peak. Following the analysis in the proof of Theorem 2.2, we obtain

$$\begin{aligned}
\mathbb{P}(\Delta \le d) &= \lim_{T\to\infty} \frac{1}{T} \sum_{k=1}^{M(T)} h(k), \text{ a.s.} \\
&= \lim_{K\to\infty} \frac{\nu}{K} \sum_{k=1}^{K} h(k), \text{ a.s.}
\end{aligned} \tag{2.11}$$

We obtain the last equality in (2.11) by using $M(T)$ goes to infinity as T goes to infinity (since ν is positive and finite), $\lim_{t\to\infty} M(t)/t = \nu$, and replacing $M(T)$ with K. Also, from Birkhoff's ergodic theorem, we obtain

$$\mathbb{P}(Y \le u) = \lim_{K\to\infty} \frac{1}{K} \sum_{k=1}^{K} \mathbb{1}\{Y(k) \le u\}, \text{ a.s.} \tag{2.12}$$

$$\mathbb{P}(A^{\text{peak}} \le u) = \lim_{K\to\infty} \frac{1}{K} \sum_{k=1}^{K} \mathbb{1}\{A^{\text{peak}}(k) \le u\}, \text{ a.s.} \tag{2.13}$$

By the definition of $h(k)$, we have

$$\begin{aligned}
h(k) &= \int_{T_{\text{D}}(k-1)}^{T_{\text{D}}(k)} \mathbb{1}\{\Delta(t) \le d\} dt \\
&= \int_{T_{\text{D}}(k-1)}^{T_{\text{D}}(k)} \mathbb{1}\{Y(k-1) + t - T_{\text{D}}(k-1) \le d\} dt \\
&= \int_{Y(k-1)}^{A^{\text{peak}}(k)} \mathbb{1}\{u \le d\} du \\
&= \int_0^\infty \mathbb{1}\{u \le d\}\mathbb{1}\{Y(k-1) \le u\}\mathbb{1}\{A^{\text{peak}}(k) > u\} du \\
&= \int_0^d \mathbb{1}\{Y(k-1) \le u\}(1 - \mathbb{1}\{A^{\text{peak}}(k) \le u\}) du \\
&= \int_0^d (\mathbb{1}\{Y(k-1) \le u\} - \mathbb{1}\{A^{\text{peak}}(k) \le u\}) du.
\end{aligned} \tag{2.14}$$

In the third step we have used change of variable, and in the last step we have used $\mathbb{1}\{A^{\text{peak}}(k) \le u\} = \mathbb{1}\{Y(k-1) \le u\}\mathbb{1}\{A^{\text{peak}}(k) \le u\}$. We substitute (2.14) in (2.11), use bounded convergence theorem, and obtain the final result using (2.12) and (2.13). □

Using the preceding result, the Inoue et al. (2019) derived the Laplace-Stieltjes transform of the distribution of AoI and thereby the moments of AoI for several systems including FCFS GI/GI/1, non-preemptive LCFS, preemptive LCFS, and the -/-/1/2* system with either exponential service or exponential inter-arrival times (cf. Table II (Inoue et al. 2019)).

In the rest of the chapter, we primarily use the formula in Theorem 2.5. This formula does not necessarily provide computational ease in deriving exact expressions when compared with that of the formula presented in Theorem 2.6. Nevertheless, as we will see later, it does provide an approach to derive worst-case performance guarantees for the upper bounds for the AoI violation probability. Deriving these performance guarantees using the formula in Theorem 2.6 is not known to us.

2.3 GI/GI/1/1

In a GI/GI/1/1 system, packet k is served upon its arrival, which implies $T_{\mathrm{D}}(k) = T_{\mathrm{A}}(k)+X_k$. Further, the inter-departure time is given by $T_{\mathrm{D}}(k) - T_{\mathrm{D}}(k-1) = X_k + I_k$. We note that this relation is equally valid for the GI/GI/1/2* system. Therefore, for both systems

$$\nu = 1/(\mathbb{E}[X_k] + \mathbb{E}[I_k]). \tag{2.15}$$

In the following we compute $A^{\text{peak}}(k)$ for a GI/GI/1/1 system.

$$\begin{aligned} A^{\text{peak}}(k) &= T_{\mathrm{D}}(k) - T_{\mathrm{A}}(k-1) \\ &= T_{\mathrm{D}}(k) - T_{\mathrm{A}}(k) + T_{\mathrm{A}}(k) - T_{\mathrm{D}}(k-1) + T_{\mathrm{D}}(k-1) - T_{\mathrm{A}}(k-1) \\ &= X_k + I_k + X_{k-1}. \end{aligned} \tag{2.16}$$

The following lemma immediately follows from the preceding analysis and Lemma 2.1.

LEMMA 2.7 *In a GI/GI/1/1 system, given age limit d, for any sample path of $\Delta(t)$ the corresponding $g(k)$ is given by*

$$g(k) = \min\left\{(X_{k-1} + I_k + X_k - d)^+, X_k + I_k\right\}, \forall k. \tag{2.17}$$

We now provide a general expression for the violation probability in the following theorem.

THEOREM 2.8 *Consider a GI/GI/1/1 system; assuming the AoI process is stationary and ergodic, then for all $d \geq 0$, $\lambda > 0$, and $0 < \mathbb{E}[X] = \frac{1}{\mu} < \infty$, the violation probability, if it exists, is given by*

$$\mathbb{P}(\Delta > d) = \nu \mathbb{E}[g(k)], \ \textit{a.s.},$$

where $g(k)$ is given by (2.17) *and ν is given by* (2.15).

Proof We note that the inter-arrival times $\{T_{\mathrm{A}}(k) - T_{\mathrm{A}}(k-1), k \geq 1\}$ in a GI/GI/1/1 system are i.i.d. To see this, the duration $T_{\mathrm{A}}(k) - T_{\mathrm{A}}(k-1)$ equals the sum of inter-arrival times of all dropped packets and the packet k starting from packet $k-1$, and only depends on the inter-arrival time and the service time of packet $k-1$. Therefore, the start of service of a packet is a renewal instant. This implies I_k are i.i.d., which further implies that $T_{\mathrm{D}}(k) - T_{\mathrm{D}}(k-1)$ are i.i.d. From (2.17) we infer that $g(k)$ are identically distributed random variables, and $g(k+2)$ is independent of the random variables $\{g(n), 1 \leq n \leq k\}$ for all k. Therefore, the sequence $\{g(k), k \geq 1\}$ is s.i.i.d. The result then follows from Theorems 2.2 and 2.5. □

Note that to compute the violation probability, we must compute $\mathbb{E}[g(k)]$. In the derivations that follow, we first compute the distribution of $g(k)$ toward this purpose. The following lemma presents a simplified expression for the distribution of $g(k)$.

LEMMA 2.9 *For a GI/GI/1/1 system,*

$$\mathbb{P}(g(k) > y) = \int_0^d \mathbb{P}(X_k + I_k > y - x + d) f_X(x) dx + \int_d^\infty \mathbb{P}(X_k + I_k > y) f_X(x) dx.$$

Proof From (2.17), we have

$$\begin{aligned}
\mathbb{P}(g(k) > y) &= \mathbb{P}\left(\min\{(X_{k-1} + X_k + I_k - d)^+, X_k + I_k\} > y\right) \\
&= \mathbb{P}\left(\max\{0, X_{k-1} + X_k + I_k - d\} > y, X_k + I_k > y\right) \\
&= \mathbb{P}\left((y < 0, X_k + I_k > y) \cup (X_{k-1} + X_k + I_k - d > y, X_k + I_k > y)\right) \\
&= \mathbb{P}\left(X_{k-1} + X_k + I_k - d > y, X_k + I_k > y\right) \\
&= \int_0^\infty \mathbb{P}\left(X_k + I_k > y + d - x, X_k + I_k > y\right) f_X(x) dx \\
&= \int_0^d \mathbb{P}\left(X_k + I_k > y - x + d\right) f_X(x) dx \\
&\quad + \int_d^\infty \mathbb{P}(X_k + I_k > y) f_X(x) dx.
\end{aligned}$$

□

Zero-Wait Policy

In a single-source single-server queueing system using zero-wait policy, the source generates a packet only when there is a departure. It is easy to see that the statistics of the AoI process for this system will be same as that of GI/GI/1/1 when the input rate approaches infinity. Therefore, the following theorem immediately follows from Theorem 2.8 by substituting $I_k = 0$, as input rate is infinity.

THEOREM 2.10 *For the system using zero-wait policy, the violation probability is given by* $\nu\mathbb{E}[g(k)]$, *almost surely, where* $g(k) = \min\{(X_{k-1}+X_k-d)^+, X_k\}$ *and* $\nu = \mu$.

Since the AoI process is nonnegative, the expected AoI for zero-wait policy is given by

$$\mathbb{E}[\Delta(t)] = \int_0^\infty \nu\mathbb{E}[\min\{(X_{k-1} + X_k - y)^+, X_k\}] dy.$$

Next, we derive exact expressions for AoI violation probability for the D/GI/1/1 and M/GI/1/1 systems.

2.3.1 D/GI/1/1: Exact Expressions

In a D/GI/1/1 system, the inter-arrival time is deterministic and is equal to $\frac{1}{\lambda}$. Intuitively, in a D/GI/1/1 system, we only need to consider the rate region $\lambda \geq \frac{1}{d}$, as AoI cannot be less than $\frac{1}{\lambda}$ when the samples are generated at rate λ. The following lemma asserts this intuition.

LEMMA 2.11 *For the D/GI/1/1 system, given $d \geq 0$ and $\lambda > 0$, the AoI violation probability only exists for $d \geq \frac{1}{\lambda}$.*

Proof We prove that $\mathbb{P}(\Delta > d)$ does not exist when $d < \frac{1}{\lambda}$. Consider the event $\{\Delta(t) > d\}$ at time t. If $d < \frac{1}{\lambda}$, there will be time instances, say $\hat{t}$, for which there is no arrival in the interval $[\hat{t} - d, \hat{t})$. This implies that at $\hat{t}$ the receiver cannot have a packet with arrival time greater than $\hat{t} - d$. Therefore, the event $\{\Delta(\hat{t}) > d\}$ is true for all such $\hat{t}$. Let $\bar{t}$ denote any time instance $t \neq \hat{t}$, that is, at $\bar{t}$ there exists an arrival in the interval $[\bar{t} - d, \bar{t})$. Since $d < \frac{1}{\lambda}$, there can be only one arrival in this interval. Therefore, for this case the event $\{\Delta(\bar{t}) > d\}$ is true if either the server is busy, in which case the packet is dropped, or the departure time of this packet exceeds $\bar{t}$.

From the preceding analysis, we conclude that $\mathbb{P}(\Delta(t) > d)$ depends on the value of t. Specifically, we infer that $\limsup_{t\to\infty} \mathbb{P}(\Delta(t) > d) = 1$, because the event $\{\Delta(\hat{t}) > d\}$ is true for all $\hat{t}$, which occur infinitely often as t goes to infinity. Similarly, we infer that $\liminf_{t\to\infty} \mathbb{P}(\Delta(t) > d) < 1$, because the time instances $\bar{t}$ also occur infinitely often and at these time instances the occurrence of the event $\{\Delta(\bar{t}) > d\}$ is uncertain. Since the limit supremum and limit infimum are not equal, $\mathbb{P}(\Delta > d) = \lim_{t\to\infty} \mathbb{P}(\Delta(t) > d)$ does not exist for $d < \frac{1}{\lambda}$. □

We now present a closed form expression for the violation probability in the following theorem.

THEOREM 2.12 *For a D/GI/1/1 system, given $d \geq \frac{1}{\lambda}$, $\lambda > 0$, and $0 < \mathbb{E}[X] = \frac{1}{\mu} < \infty$, the violation probability is given by $\nu\mathbb{E}[g(k)]$, almost surely, where $g(k)$ is given by (2.17), $\nu = \lambda/\mathbb{E}[\lceil\lambda X_k\rceil]$, and $I_k = \lceil\lambda X_{k-1}\rceil/\lambda - X_{k-1}$.*

Proof Using the results from Lemma 2.7 and Theorem 2.8, it is sufficient to show that $I_k = \frac{\lceil\lambda X_{k-1}\rceil}{\lambda} - X_{k-1}$, which we argue to be true in the following. The time difference between the arrival of packet k and packet $(k-1)$ is given by $\frac{\lceil\lambda X_{k-1}\rceil}{\lambda}$. To see this, the service of packet $k-1$ starts upon its arrival, that is, at $T_A(k-1)$. During the service of packet $k-1$ the packets that arrived would be dropped and the packet that arrived immediately after $T_A(k-1) + X_{k-1}$ is served. The number of arrivals since $T_A(k-1)$ is given by $\lceil\lambda X_{k-1}\rceil$, and the time elapsed is $\frac{\lceil\lambda X_{k-1}\rceil}{\lambda}$. This implies that the idle time I_k is given by $\frac{\lceil\lambda X_{k-1}\rceil}{\lambda} - X_{k-1}$. □

In the following we compute the expression provided in Theorem 2.12 for exponential-service-time distribution.

THEOREM 2.13 *For a D/M/1/1 queue, given $d \geq \frac{1}{\lambda}$, $\lambda > 0$, and $0 < \mathbb{E}[X] = \frac{1}{\mu} < \infty$, the violation probability is given by $\nu\mathbb{E}[g(k)]$, almost surely, where $\nu = \lambda(1 - e^{-\mu/\lambda})$ and*

$$\mathbb{E}[g(k)] = \frac{e^{-\mu\frac{\lceil\lambda d\rceil}{\lambda}}}{\lambda(1 - e^{-\frac{\mu}{\lambda}})} + e^{-\mu\frac{\lfloor\lambda d\rfloor}{\lambda}}\left[\frac{\lceil\lambda d\rceil}{\lambda} - d + \frac{1}{\mu}\right] + \frac{e^{-\mu d}}{\mu}\left((e^{\frac{\mu}{\lambda}} - 1)\lfloor\lambda d\rfloor - 1\right).$$

Proof In the following we first derive $\mathbb{E}[\lceil \lambda X \rceil]$.

$$\begin{aligned}
\mathbb{E}[\lceil \lambda X \rceil] &= \int_0^\infty \lceil \lambda x \rceil \mu e^{-\mu x} dx \\
&= \sum_{m=1}^{\infty} m \int_{\frac{m-1}{\lambda}}^{\frac{m}{\lambda}} \mu e^{-\mu x} dx \\
&= (e^{\mu/\lambda} - 1) \sum_{m=1}^{\infty} m (e^{-\mu/\lambda})^m \\
&= (e^{\mu/\lambda} - 1) e^{-\mu/\lambda} / (1 - e^{-\mu/\lambda})^2 = 1/(1 - e^{-\mu/\lambda}).
\end{aligned}$$

In the following we compute $\mathbb{P}(g(k) > y)$. Recall that $I_k = \frac{\lceil \lambda X_{k-1} \rceil}{\lambda} - X_{k-1}$ (Theorem 2.8). Using this and Lemma 2.9, we obtain

$$\begin{aligned}
\mathbb{P}(g(k) > y) &= \int_0^d \mathbb{P}\left(X_k + \frac{\lceil \lambda d \rceil}{\lambda} - x > y + d - x\right) f_X(x) dx \\
&\quad + \int_d^\infty \mathbb{P}\left(X_k + \frac{\lceil \lambda d \rceil}{\lambda} - x > y\right) f_X(x) dx \\
&= \int_0^d \mathbb{P}\left(X_k > y + d - \frac{\lceil \lambda d \rceil}{\lambda}\right) f_X(x) dx \\
&\quad + \int_d^\infty \mathbb{P}\left(X_k > y + x - \frac{\lceil \lambda d \rceil}{\lambda}\right) f_X(x) dx \\
&= A + B.
\end{aligned}$$

We compute the terms A and B[2], and use $\mathbb{E}[g(k)] = \int_0^\infty \mathbb{P}(g(k) > y) dy$ to obtain the result. □

2.3.2 M/GI/1/1: Exact Expressions

For an M/GI/1/1 system, Najm, Yates, and Soljanin (2017) derived expressions for the expected AoI and the expected peak AoI. For this system we provide an expression for the violation probability of AoI.

THEOREM 2.14 *For an M/GI/1/1 system, $\lambda > 0$, and $0 < \mathbb{E}[X] = \frac{1}{\mu} < \infty$, the violation probability, if it exists, is given by $\nu \mathbb{E}[g(k)]$, almost surely, where $g(k)$ is given in* (2.17), $\frac{1}{\nu} = \frac{1}{\lambda} + \frac{1}{\mu}$, *and $I_k \sim Exp(\lambda)$.*

Proof The result follows from Theorem 2.8 and using the fact that in an M/G/1/1 system I_k and the inter-arrival times are identically distributed. □

For the special case of M/M/1/1, we have the following theorem.

[2] The computation of A and B is not shown here, as it involves lengthy expressions and can be referred from Champati, Al-Zubaidy, and Gross (2019b).

THEOREM 2.15 *For the M/M/1/1 system,* $\lambda > 0$, *and* $0 < \mathbb{E}[X] = \frac{1}{\mu} < \infty$, *the violation probability, if it exists, is given by* $\nu\mathbb{E}[g(k)]$, *almost surely, where* $\frac{1}{\nu} = \frac{1}{\lambda} + \frac{1}{\mu}$, *and*

$$\mathbb{E}[g(k)] = \begin{cases} \frac{\mu^2(e^{-\lambda d} - e^{-\mu d})}{\lambda(\mu-\lambda)^2} + e^{-\mu d}\left(\frac{1}{\lambda} + \frac{1}{\mu} - \frac{\lambda d}{\mu-\lambda}\right) & \lambda \neq \mu, \\ \frac{\mu e^{-\mu d}}{2}\left(d + \frac{2}{\mu}\right)^2 & \lambda = \mu. \end{cases}$$

Proof Since $X_k \sim Exp(\mu)$ and $I_k \sim Exp(\lambda)$, we have

$$\mathbb{P}(X_k + I_k > y) = \begin{cases} \frac{\mu e^{-\lambda(y-x+d)} - \lambda e^{-\mu(y-x+d)}}{\mu-\lambda} & \lambda \neq \mu, \\ (1+\mu y)e^{-\mu y} & \lambda = \mu. \end{cases} \tag{2.18}$$

In the following we compute the distribution of $g(k)$ by substituting (2.18) in $\mathbb{P}(g(k) > y)$ given in Lemma 2.9.

Case 1: $\mu \neq \lambda$. For this case, we have

$$\begin{aligned} \mathbb{P}(g(k) > y) &= \int_0^d \frac{\mu e^{-\lambda(y-x+d)} - \lambda e^{-\mu(y-x+d)}}{\mu - \lambda} f_X(x)dx \\ &\quad + \frac{\mu e^{-\lambda y} - \lambda e^{-\mu y}}{\mu - \lambda}\int_d^\infty f_X(x)dx \\ &= \frac{\mu e^{-\mu(y+d)}}{\mu - \lambda}\left[\frac{\mu(e^{d(\lambda-\mu)} - 1)}{\lambda - \mu} - \lambda d\right] + \frac{(\mu e^{-\lambda y} - \lambda e^{-\mu y})e^{-\mu d}}{\mu - \lambda}. \end{aligned}$$

Integrating the preceding expression over y, we obtain the desired result.

Case 2: $\mu = \lambda$. For this case, we have

$$\begin{aligned} \mathbb{P}(g(k) > y) &= \int_0^d (1 + \mu(y - x + d))e^{-\mu(y-x+d)} f_X(x)dx \\ &\quad + \int_d^\infty (1 + \mu y)e^{-\mu y} f_X(x)dx \\ &= \mu e^{-\mu(y+d)}\int_0^d (1 + \mu(y - x + d))dx + (1 + \mu y)e^{-\mu(y+d)} \\ &= \mu e^{-\mu(y+d)}\left[(1 + \mu(y + d))d - \frac{\mu d^2}{2}\right] + (1 + \mu y)e^{-\mu(y+d)} \\ &= \mu d e^{-\mu(y+d)}\left[1 + \mu\left(y + \frac{d}{2}\right)\right] + (1 + \mu y)e^{-\mu(y+d)} \\ &= (\mu d + 1)(1 + \mu y)e^{-\mu(y+d)} + \frac{\mu^2 d^2}{2}e^{-\mu(y+d)}. \end{aligned}$$

Therefore, integrating the preceding expression over y, we obtain

$$\begin{aligned} \mathbb{E}[g(x)] &= e^{-\mu d}\left[(\mu d + 1)\frac{2}{\mu} + \frac{\mu d^2}{2}\right] \\ &= \frac{\mu e^{-\mu d}}{2}\left(d + \frac{2}{\mu}\right)^2. \end{aligned}$$

□

In the following, we illustrate the computation of the distribution for the case $\lambda = \mu$, using the formula in Theorem 2.6. For this case, we have $\nu = \frac{\mu}{2}$. The system delay of packet k is equal to X_k, and therefore $Y_k = X_k$ for all k, and Y has the same distribution as the service time. Recall that $A_k^{\text{peak}} = X_{k-1} + I_k + X_k$. Since $I_k \sim Exp(\lambda)$ and $\lambda = \mu$, peak AoI has Erlang distribution with scale parameter 3 and rate parameter μ. We have

$$\begin{aligned}
\mathbb{P}(\Delta \le d) &= \nu \int_0^d (\mathbb{P}(Y \le x) - \mathbb{P}(A^{\text{peak}} \le x))dx \\
&= \frac{\mu}{2} \int_0^d \left(1 - e^{-\mu x} - (1 - e^{-\mu x} - xe^{-\mu x} - \frac{x^2 e^{-\mu x}}{2})\right) dx \\
&= \frac{\mu}{2} \int_0^d \left(xe^{-\mu x} + \frac{x^2 e^{-\mu x}}{2}\right) dx \\
&= 1 - \frac{\mu^2 e^{-\mu d}}{4} \left(d + \frac{2}{\mu}\right)^2.
\end{aligned}$$

In this case, the number of steps is less for computing the expression using the preceding formula because the distribution of peak AoI is readily available. However, in general, computing the distribution of peak AoI presents an additional step.

In the following theorem, we derive the violation probability for the system with zero-wait policy and exponentially distributed service times.

THEOREM 2.16 *For the system with zero-wait policy and exponentially distributed service times, the violation probability is given by*

$$\mathbb{P}(\Delta > d) = (1 + \mu d)e^{-\mu d}, \ a.s. \tag{2.19}$$

Proof The result can be obtained from Theorem 2.15 by utilizing the fact that the statistics of this system will be same as that for $M/M/1/1$ when λ approaches infinity. □

Interestingly, the distribution in (2.19) is gamma distribution with shape parameter 2 and scale parameter $\frac{1}{\mu}$. Further, the expected AoI in this case is $\frac{2}{\mu}$, a result reported in Kaul, Yates, and Gruteser (2012) and Costa et al. (2016).

2.4 Upper Bounds

As one can expect, $g(k)$ and $T_D(k) - T_D(k-1)$ depend on the idle time I_k and waiting time W_k in the queueing system. Therefore, computing $\mathbb{E}[g(k)]$ and ν is hard, in general, as the distributions of I_k and W_k become intractable for general inter-arrival time and service-time distributions. To this end, in the following theorem we present a result that is useful in deriving upper bounds for the violation probability and only requires the AoI process to be stationary.

THEOREM 2.17 *If the AoI process is stationary, then*

$$\mathbb{E}_\omega\left[\lim_{T\to\infty}\frac{1}{T}\sum_{k=1}^{M(T)}\gamma(\omega,k)\right]\le\mathbb{P}(\Delta>d)\le\mathbb{E}_\omega\left[\lim_{T\to\infty}\frac{1}{T}\sum_{k=1}^{M(T)}\Gamma(\omega,k)\right].$$

Proof Since $\Delta(t)$ is stationary, we have

$$\mathbb{P}(\Delta(t)>d)=\mathbb{E}_\omega[\mathbb{1}\{\Delta(\omega,t)>d\}],\ \forall t.$$

Therefore, for any t,

$$\begin{aligned}\mathbb{P}(\Delta(t)>d)&=\lim_{T\to\infty}\frac{1}{T}\int_0^T\mathbb{E}_\omega[\mathbb{1}\{\Delta(\omega,t)>d\}]dt\\&=\mathbb{E}_\omega\left[\lim_{T\to\infty}\frac{1}{T}\int_0^T\mathbb{1}\{\Delta(\omega,t)>d\}dt\right]\\&=\mathbb{E}_\omega\left[\lim_{T\to\infty}\frac{1}{T}\sum_{k=1}^{M(T)}g(\omega,k)\right].\end{aligned}\tag{2.20}$$

Second step is due to the fact that indicator function is nonnegative. The third step is due to the fact that (2.7) is true for any ω. The result follows from (2.20) and (2.2). □

In terms of applicability, Theorem 2.17 is more general than Theorem 2.2, as it does not require ergodicity of the AoI process. Following Theorem 2.17, we strive to obtain upper bounds for the violation probability for GI/GI/1/1 and GI/GI/1/2* systems by finding bounds for $g(k)$.

In the following we establish a lower bound for $g(k)$ that is applicable to any single-source single-server queueing system.

LEMMA 2.18 *For a single-source single-server queueing system, it is true that* $g(k)\ge\gamma^*(k)$, *for all* k, *where*

$$\gamma^*(k)=\min\{(X_k+X_{k-1}+I_k-d)^+,X_k+I_k\}.$$

Proof For a single-server system it is easy to see that the inter-departure time between information update packets is at least the service time of a packet and idle time before its service started, that is,

$$T_{\mathrm{D}}(k)-T_{\mathrm{D}}(k-1)\ge X_k+I_k.\tag{2.21}$$

From (2.3) we have

$$\begin{aligned}A^{\mathrm{peak}}(k)&=T_{\mathrm{D}}(k)-T_{\mathrm{A}}(k-1)\\&\ge T_{\mathrm{D}}(k)-(T_{\mathrm{D}}(k-1)-X_{k-1})\\&\ge X_k+I_k+X_{k-1}.\end{aligned}$$

The second step is due to the fact that a packet departure time is at least equal to its arrival time plus its service time. The last step is due to (2.21). □

We use the lower bound in Lemma 2.18 to analyze the performance of the upper bounds derived for the AoI violation probability for GI/GI/1/1 and GI/GI/1/2* systems. Nevertheless, this method is quite general and can be applied to other queueing systems.

2.4.1 Upper Bound for the GI/GI/1/1 System

In this section, we provide an upper bound for the violation probability for the GI/GI/1/1 system and also analyze its performance. To this end, we first provide an upper bound for $g(k)$ in the following lemma.

LEMMA 2.19 *For a GI/GI/1/1 system,* $g(k) \leq \Gamma_1(k)$ *for all* k*, where*

$$\Gamma_1(k) = \min\left\{(X_{k-1} + \check{Z}_k + X_k - d)^+, X_k + \check{Z}_k\right\}. \tag{2.22}$$

Proof Recall that $\check{Z}_k$ is the inter-arrival time between packet k and its previous arrival. Therefore, we have $I_k \leq \check{Z}_k$. The result follows from using this in (2.17). □

Remark 1: In an M/G/1/1 system, $\mathbb{E}[\Gamma_1(k)] = \mathbb{E}[g(k)]$, since for this system both I_k and $\check{Z}_k$ have the same distribution Exp(λ). Thus, $\mathbb{E}[\Gamma_1(k)]$ is a tight upper bound for $\mathbb{E}[g(k)]$ for the GI/GI/1/1 system.

The following theorem presents an upper bound Φ_1 for the violation probability.

THEOREM 2.20 *For a GI/GI/1/1 system, given* $d > 0$*, assuming that the AoI process is stationary, the violation probability is bounded as follows:*

$$\mathbb{P}(\Delta > d) = \nu\mathbb{E}[\gamma^*(k)] \leq \Phi_1,$$

where γ^* *is given by Lemma 2.18, and* $\Phi_1 = \hat{\nu}\mathbb{E}[\Gamma_1(k)]$*, for some* $\hat{\nu} \geq \nu$*, where* ν *is given in* (2.15).

Proof The equality follows from the fact that $\gamma^*(k)$ is equal to $g(k)$ given in (2.17) for the GI/GI/1/1 system. It is easy to see that $\Gamma_1(k)$ are s.i.i.d., and as noted in the proof of Theorem 2.8, $T_D(k) - T_D(k-1)$ are i.i.d. Therefore, from Theorem 2.5 we infer that

$$\lim_{T\to\infty} \frac{1}{T} \sum_{k=1}^{K(T)} \Gamma_1(k) = \nu\mathbb{E}[\Gamma_1(k)], \text{ a.s.}$$

Using the preceding equation in Theorem 2.17, we obtain $\mathbb{P}(\Delta > d) \leq \nu\mathbb{E}[\Gamma_1(k)]$. The result follows as $\hat{\nu} \geq \nu$. □

Here we define η that will be used in describing the worst-case performance of Φ_1.

$$\eta \triangleq \frac{1}{\lambda} + \frac{1}{\mu} - \frac{1}{\nu}. \tag{2.23}$$

In the following theorem we present a worst-case-performance guarantee for Φ_1.

THEOREM 2.21 *For a GI/GI/1/1 system, for a given $\hat{\nu} \geq \nu$, Φ_1 has the following worst-case-performance guarantee:*

$$\Phi_1 \leq \frac{\hat{\nu}}{\nu} \cdot \mathbb{P}(\Delta > d) + \hat{\nu}\eta.$$

Proof Noting that $I_k \leq \check{Z}_k$, we have

$$\begin{aligned}\Gamma_1(k) &= \min\left\{(X_{k-1} + \check{Z}_k + X_k - d)^+, X_k + \check{Z}_k\right\}\\ &\leq \min\left\{(X_{k-1}+I_k+X_k - d)^+, X_k+I_k\right\} + (\check{Z}_k - I_k)\\ &\leq g(k) + (\check{Z}_k - I_k).\end{aligned}$$

Therefore, using Theorem 2.20, we obtain

$$\begin{aligned}\Phi_1 &\leq \hat{\nu}(\mathbb{E}[g(k)] + \mathbb{E}[\check{Z}_k] - \mathbb{E}[I_k])\\ &= \frac{\hat{\nu}}{\nu} \cdot \mathbb{P}(\Delta(t) > d) + \hat{\nu}\left(\frac{1}{\lambda} + \frac{1}{\mu} - \frac{1}{\nu}\right).\end{aligned}$$

In the last step we have used Theorem 2.8 and (2.15). □

From Theorem 2.21, we infer that if $\hat{\nu} = \nu$, that is, the departure rate is given, then Φ_1 overestimates the violation probability by at most η. We note that $\frac{1}{\lambda} + \frac{1}{\mu} \geq \frac{1}{\nu}$, and the relation holds with equality for an M/GI/1/1 system. Further, ν increases sublinearly with λ in a GI/GI/1/1 system, in general. For example, $\nu = \lambda(1 - e^{-\mu/\lambda})$ for the D/M/1/1 system (Theorem 2.13). Therefore, for a fixed μ, η decreases with λ, in general. In other words, the derived upper bound is tighter at higher utilization. Finally, the worst-case guarantee in Theorem 2.21 is provided for any $d \geq 0$. Therefore, we expect that Φ_1 may not be tight for larger d values for which the violation probability takes smaller values.

We require to compute the value of expected idle time to obtain ν. When ν is not tractable, we propose to use $\hat{\nu} = \min\{\lambda, \mu\}$, a trivial upper bound on the departure rate. We note, however, that the conclusion about tightness of the upper bound at higher utilization may no longer be valid in this case.

Note that by replacing I_k and using an appropriate quantity one can obtain upper bounds using the formula in Theorem 2.6, but obtaining performance guarantees for those bounds is not straightforward and requires further study.

2.5 The GI/GI/1/2* System

The analysis of a GI/GI/1/2* system follows similar steps to the analysis we have presented for the GI/GI/1/1 system. We first obtain expressions for I_k and $A^{\text{peak}}(k)$, and use them to obtain $g(k)$.

In Figure 2.3, we present a possible sequence of arrivals (in blue) and departures (in red) in a GI/GI/1/2* system. Note that there are no arrivals during the service

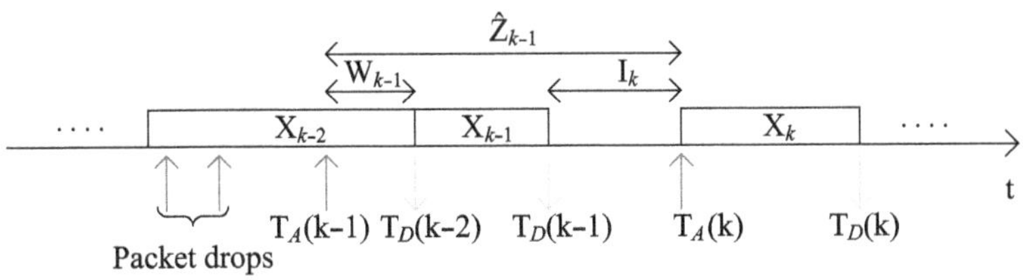

Figure 2.3 An example illustration of arrivals and departures in a GI/GI/1/2* system.

of packet $(k-1)$. This happens only when $\hat{Z}_{k-1} > W_{k-1} + X_{k-1}$ and in this case, $I_k = \hat{Z}_{k-1} - W_{k-1} - X_{k-1}$. If $\hat{Z}_{k-1} \leq W_{k-1} + X_{k-1}$, then $I_k = 0$. Therefore, we have

$$I_k = (\hat{Z}_{k-1} - X_{k-1} - W_{k-1})^+. \tag{2.24}$$

Recall that $A^{\text{peak}}(k) = T_{\text{D}}(k) - T_{\text{A}}(k-1)$. From Figure 2.3, it is easy to infer that $A^{\text{peak}}(k) = X_k + X_{k-1} + I_k + W_k$. The following lemma immediately follows from the preceding analysis and Lemma 2.1.

LEMMA 2.22 *Given $d \geq 0$, for any sample path of $\Delta(t)$ in a GI/GI/1/2* system, we have, for all k,*

$$g(k) = \min\left\{(X_k + X_{k-1} + I_k + W_{k-1} - d)^+,\, X_k + I_k\right\}. \tag{2.25}$$

Unlike the case of the GI/GI/1/1 system, for the GI/GI/1/2* system it is hard to derive a closed-form expression for the violation probability in terms of X_k, X_{k-1}, I_k, and W_{k-1}, because $g(k)$, given in (2.25), does not satisfy the s.i.i.d. property. Further, computing the violation probability requires the distributions of both I_k and W_{k-1}. While these quantities can be computed for exponential service or exponential inter-arrival times (cf. Inoue et al. 2019), they become intractable for general inter-arrival and service-time distributions. To this end we present upper bounds in the next section.

2.5.1 Upper Bound for the GI/GI/1/2* System

In this subsection we propose an upper bound for the violation probability and analyze its worst-case performance.

LEMMA 2.23 *For a GI/GI/1/2* system, $g(k) \leq \Gamma_2(k)$ for all k, where*

$$\Gamma_2(k) = \min\{(X_k + X_{k-1} + \hat{Z}_{k-1} - d)^+, X_k + (\hat{Z}_{k-1} - X_{k-1})^+\}.$$

Proof Noting the expression for $g(k)$ given in (2.25), it is sufficient to show that $I_k + W_{k-1} \leq \hat{Z}_{k-1}$ and $I_k \leq (\hat{Z}_{k-1} - X_{k-1})^+$. The latter inequality follows from (2.24). The former inequality is obviously true if there are no arrivals during the service of packet $(k-1)$; see Figure 2.3. If there is an arrival during the service of packet $(k-1)$, then $I_k = 0$. In this case $I_k + W_{k-1} = W_{k-1} < \hat{Z}_{k-1}$, since by definition there should be no arrival after packet $(k-1)$ arrived and before its service started. □

In the following theorem we present an upper bound Φ_2 for the violation probability.

THEOREM 2.24 *For a GI/GI/1/2* system, assuming that the AoI process is stationary, the violation probability is bounded by*

$$\nu\mathbb{E}[\gamma^*(k)] \leq \mathbb{P}(\Delta > d) \leq \Phi_2,$$

where $\Phi_2 = \hat{\nu}\mathbb{E}[\Gamma_2(k)]$*, for some* $\hat{\nu} \geq \nu$.

Proof The proof follows similar steps to the proof of Theorem 2.20 and is omitted. □

A worst-case-performance guarantee for Φ_2 is presented in the following theorem.

THEOREM 2.25 *For the GI/GI/1/2* system, for a given* $\hat{\nu} \geq \nu$, Φ_2 *has the following worst-case-performance guarantee:*

$$\Phi_2 \leq \frac{\hat{\nu}}{\nu} \cdot \mathbb{P}(\Delta > d) + \hat{\nu}\eta.$$

Proof It is easy to show that $\Gamma_2(k) \leq \gamma^*(k) + \hat{Z}_{k-1} - I_k$. The rest of the proof follows similar steps as the proof of Theorem 2.21 and is omitted. □

Thus, Φ_2 also overestimates the violation probability by at most η, if ν is given. Therefore, given ν and for a fixed average service, Φ_2 is tighter at higher utilization. Since it is hard to compute ν, in general, in the numerical section we compute Φ_2 using $\hat{\nu} = \min\{\lambda, \mu\}$.

Remark 2: For both GI/GI/1/1 and GI/GI/1/2* systems $\nu = 1/(\mathbb{E}[X_k] + \mathbb{E}[I_k])$, and $g(k)$ for GI/GI/1/1 given by (2.17) seems to be closely related to $g(k)$ for GI/GI/1/2* given by (2.25). Also, one can expect that the idle time in GI/GI/1/2* will be lower compared to that of GI/GI/1/1. However, for a given d, a comparison between the violation probabilities in these systems is nontrivial because of the waiting time in GI/GI/1/2* and higher idle time in GI/GI/1/1.

Remark 3: When the input rate approaches infinity, the inter-arrival time, waiting time, and idle time approach zero. Therefore, the upper bounds Φ_1, Φ_2, and the respective violation probabilities in GI/GI/1/1 and GI/GI/1/2* all converge to the violation probability in the system using zero-wait policy. Thus, both Φ_1 and Φ_2 are asymptotically tight.

2.6 Numerical Results

In this section, we validate the proposed upper bounds against the violation probability obtained through simulation for selected service-time and inter-arrival-time distributions. For all simulations we set $\mu = 1$, and thus the utilization increases with λ. We use $\lambda = .45$ and $d = 5$ as default values.

We first study the performance of Φ_1 in comparison with overestimation factor η, when ν is given. To this end we consider the D/M/1/1 system and compute Φ_1 by setting $\hat{\nu} = \nu = \lambda(1 - e^{-\mu/\lambda})$. In Figure 2.4, we plot Φ_1 against the exact value for the violation probability given in Theorem 2.13. Observe that the gap between Φ_1

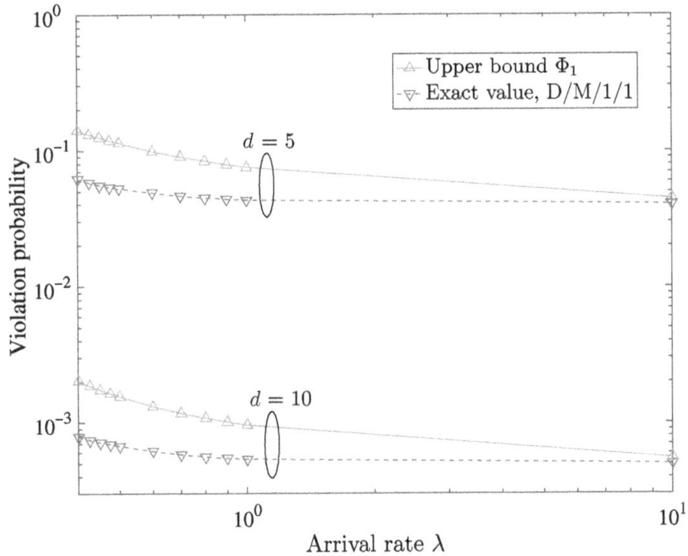

Figure 2.4 Performance of Φ_1 with varying λ, when ν is given, and $\mu = 1$.

and violation probability reduces as the arrival rate increases, confirming our initial conclusion that the bound is tighter at higher utilization. Furthermore, Φ_1 approaches the simulated violation probability asymptotically. For $d = 5$ and $\lambda = 0.4$, we compute η to be 0.28, while the actual gap is 0.08. For the same setting, but for $d = 10$, η remains the same while the actual gap is 0.0012. This suggests that the proposed upper bound is much lower than the worst-case-performance guarantee.

Next, we consider two example systems where exact expressions for the distribution of AoI are hard to compute. For both systems, we use $\hat{\nu} = \min(\lambda, \mu)$ to compute Φ_1 and Φ_2. In the first example system, we choose deterministic arrivals and Shifted-Exponential (SE) service times, that is, D/SE/1/1 and D/SE/1/2*. We set values of d and λ such that $d \geq \frac{1}{\lambda}$, $\mu = 1$ and shift parameter equal to 0.11. In Figures 2.5 and 2.6, we study the performance of the upper bounds, presented in Theorems 2.20 and 2.24, for varying arrival rate λ and varying age limit d, respectively. From Figure 2.5, we again observe that the upper bounds are tighter at higher utilization. For $\lambda > 1$ both upper bounds and the violation probabilities converge to 0.029. Interestingly, in contrast to D/SE/1/1 where the violation probability decreases with λ, D/SE/1/2* has minimum violation probability of 0.026 at around $\lambda = 0.6$. From Figure 2.6, we observe that both bounds are tighter at smaller d values. While the decay rate of Φ_1 matches with that of the simulated violation probability, Φ_2 becomes loose as d increases. We conjecture that this is due to the inequality $I_k + W_{k-1} \leq \hat{Z}_{k-1}$ that we use to obtain this bound.

In Figures 2.7 and 2.8, we present a comparison for deterministic service and Erlang distributed inter-arrival times, that is, Er/D/1/1 and Er/D/1/2*. We first note that for the parameter values chosen, Φ_1 and Φ_2 are equal in this case. From Figure 2.7, we observe that the bounds are not tight at larger arrival rate. This can be attributed to the

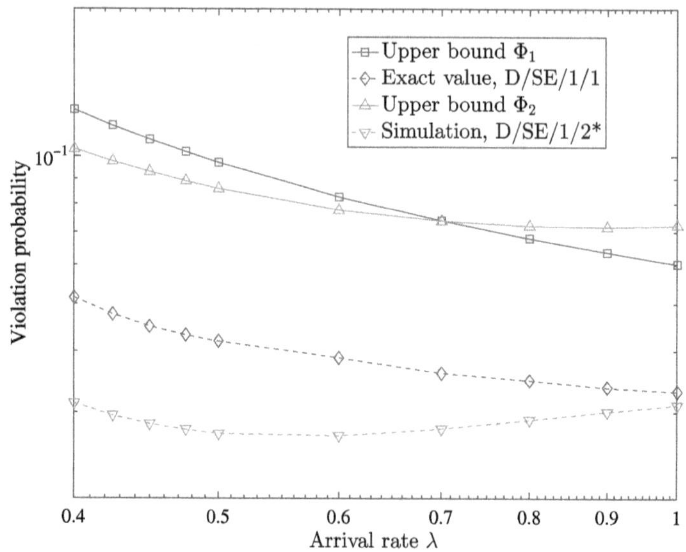

Figure 2.5 Performance of upper bounds with varying λ, $d = 5$, $\mu = 1$, and shift equal to 0.11.

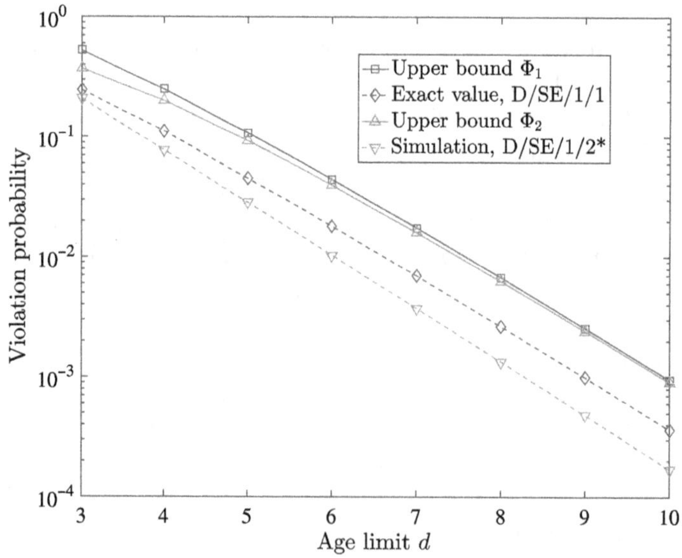

Figure 2.6 Performance of upper bounds with varying d, $\lambda = .45$, $\mu = 1$, and shift equal to 0.11.

use of $\hat{\nu} = \min(\lambda, \mu)$. From Figure 2.8, we observe that the decay rate of the bounds matches the decay rate of the violation probabilities. Finally, it is worth noting that, the violation probability in -/-/1/2* is lower than that in -/-/1/1 for the above example systems.

In conclusion, for the considered systems, the upper bounds are well within an order of magnitude from the violation probability. For most cases the decay rate of

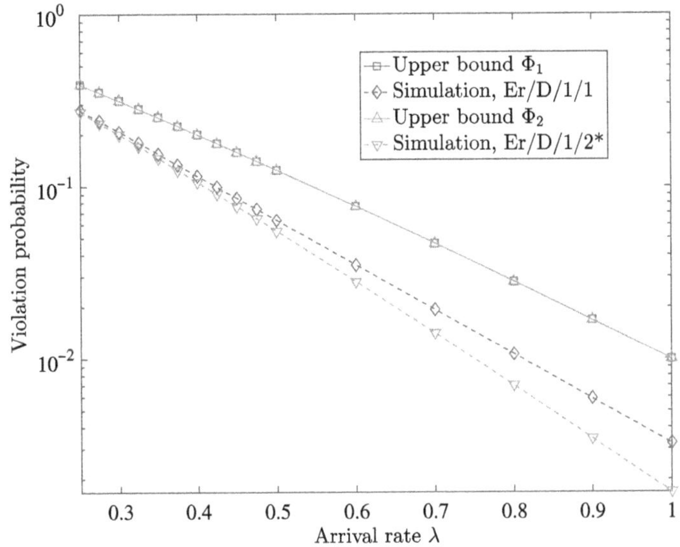

Figure 2.7 Performance of upper bounds with varying λ, Erlang shape parameter equal to 2, $d = 5$, and $\mu = 1$.

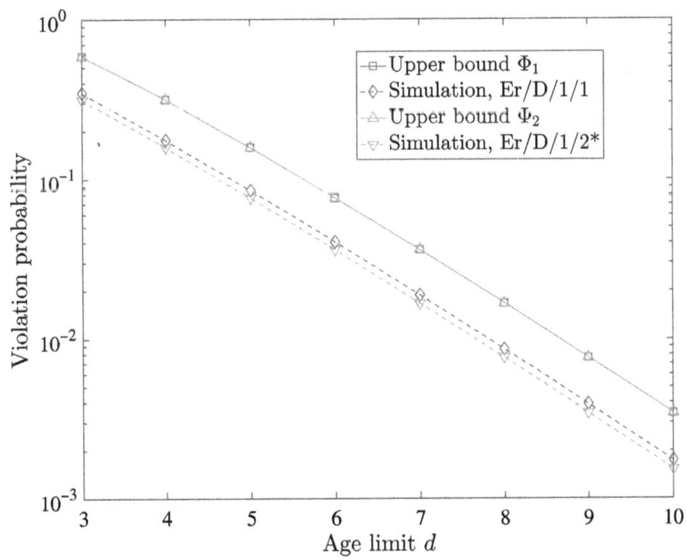

Figure 2.8 Performance of upper bounds with varying d, $\lambda = .45$, Erlang shape parameter equal to 2, and $\mu = 1$.

the proposed bounds follows the decay rate of the simulated violation probability as d increases. Also, the performance of these upper bounds can be improved further by finding nontrivial upper bounds for ν. Thus, we believe that the proposed upper bounds can be useful as first-hand metrics for measuring freshness of status updates in these systems.

References

Akar, N., Dogan, O., & Atay, E. U. (2020), "Finding the exact distribution of (peak) age of information for queues of PH/PH/1/1 and M/PH/1/2 type," *IEEE Transactions on Communications*, pp. 1–1.

Champati, J. P., Al-Zubaidy, H., & Gross, J. (2018), Statistical guarantee optimization for age of information for the D/G/1 queue, *in* "Proc. IEEE INFOCOM Workshop," pp. 130–135.

Champati, J. P., Al-Zubaidy, H., & Gross, J. (2019a), On the distribution of AoI for the GI/GI/1/1 and GI/GI/1/2* systems: Exact expressions and bounds, *in* "Proc. IEEE INFOCOM," pp. 37–45.

Champati, J. P., Al-Zubaidy, H., & Gross, J. (2019b), "On the distribution of AoI for the GI/GI/1/1 and GI/GI/1/2* systems: Exact expressions and bounds," *CoRR*, vol. **abs/1905.04068**.

Champati, J. P., Al-Zubaidy, H., & Gross, J. (2020), "Statistical guarantee optimization for AoI in single-hop and two-hop FCFS systems with periodic arrivals," *IEEE Transactions on Communications*, pp. 1–1.

Costa, M., Codreanu, M., & Ephremides, A. (2016), "On the age of information in status update systems with packet management," *IEEE Transactions on Information Theory* **62**(4), 1897–1910.

Inoue, Y., Masuyama, H., Takine, T., & Tanaka, T. (2019), "A general formula for the stationary distribution of the age of information and its application to single-server queues," *IEEE Transactions on Information Theory* **65**(12), 8305–8324.

Kaul, S., Yates, R., & Gruteser, M. (2012), Status updates through queues, *in* "Proc. Conference on Information Sciences and Systems (CISS)."

Kesidis, G., Konstantopoulos, T., & Zazanis, M. A. (2020), "The distribution of age-of-information performance measures for message processing systems," *Queueing Systems: Theory and Applications* **95**(3), 203–250.

Kosta, A., Pappas, N., Ephremides, A., & Angelakis, V. (2019), "Age of information performance of multiaccess strategies with packet management," *Journal of Communications and Networks* **21**(3), 244–255.

Kosta, A., Pappas, N., Ephremides, A., & Angelakis, V. (2020), Non-linear age of information in a discrete time queue: Stationary distribution and average performance analysis, *in* "Proc. IEEE ICC," pp. 1–6.

Leveson, N. (2011), *Engineering a Safer World – Systems Thinking Applied to Safety*, Massachusetts Institute of Technology Press.

Najm, E., Yates, R. D., & Soljanin, E. (2017), Status updates through M/G/1/1 queues with HARQ, *in* "Proc. IEEE ISIT," pp. 131–135.

Voss, W. (2020), "A comprehensible guide to industrial ethernet,"

Yates, R. D. (2020), "The age of information in networks: Moments, distributions, and sampling," *IEEE Transactions on Information Theory* **66**(9), 5712–5728.

3 Multisource Queueing Models

Sanjit K. Kaul and Roy D. Yates

3.1 Introduction

Often multiple sources send updates to their intended destinations, which desire timely updates, using shared constrained network and computational resources. In this chapter we model such settings using a queue theoretic abstraction that has updates from the sources arrive at a service facility with a single server. The server can process at most one update at a given time. New updates that arrive while the server is busy may be queued. The service facility may have the updates from the sources share a common queue or may have a separate queue for each source. The facility may apply packet management techniques like preemption in service or waiting and assigning different priorities to the sources. On completion of service, an update is delivered to a monitor. The monitor would like to have updates from sources that are as fresh as possible.

To exemplify, the different sources could be sensors on a smartwatch, for example, heart rate, location, temperature, and oximetry. A smartwatch application queues the updates generated by the sensors at the wireless interface of the watch where they await transmission to a phone paired with the watch. An update is in service while its transmission is being actively attempted by the interface. The application desires that updates available at its counterpart on the phone (*monitor*) are fresh.

Arrivals of updates from sources are modeled as independent renewal processes. The server takes a random amount of time to service an update. This service time is independent of the time taken to service other updates. Updates from different sources may have nonidentical service time distributions. The typical goal is to analyze the metrics of average age of information and peak age as a function of arrival rates of sources, the corresponding service rates, and for different packet management strategies.

Work on such analysis has been fairly limited. In [1] the authors analyzed the $M/M/1$ first-come-first-served (FCFS) queue. They used age sample functions and graphical methods to calculate the average age at the monitor. Arrivals from a source are modeled as a Poisson process. An update that arrives when the server is busy is queued. Updates are serviced in a FCFS manner. Service times of updates are independent and are exponentially distributed.

In addition, in [2], the authors considered M/M/1* and M/M/1/2* queues.[1] In M/M/1*, a new arrival preempts any update that is currently in service. In M/M/1/2*,

[1] We use the asterisk in Kendall notation to signify preemption.

we can have at most two updates in the queueing system (queue and server), one in waiting and the other in service. A new arrival can replace any update currently in waiting. However, an update in service can't be preempted. They propose and use a stochastic hybrid systems (SHS) approach to analyze the average age for the two queues. Unlike the graphical method that requires an appropriate choice of random variables to represent the age sample function and often convoluted calculations of joint moments, the SHS approach instead requires tracking of discrete states, for example, system occupancy, using a continuous time Markov chain (CTMC), and the continuous state, which is a vector of the relevant age processes. This makes the analysis less reliant on probabilistic arguments and less prone to error. In [3], they redid the M/M/1 analysis using SHS to rectify an error in analysis, spotted by [4], of the system in [1] that also trickled into [2].

In [4], in addition to finding the expression for the average age for a multisource M/M/1 FCFS system using a sample function-based graphical method, the authors provide approximations of the average age for general service times. In [5] the authors look at a variant of the M/M/1/2* system for when there are exactly two sources, in which at most one update from each source may wait in the queue. A new arrival replaces a packet of the same source, if any, in waiting. One would expect such source-aware preemption to be beneficial, in comparison to source-agnostic preemption considered in [2], to a source with a lower arrival rate, as the source's relatively infrequent arrivals are not preempted by the more frequent arrivals of the other. A SHS-based approach is used to calculate the average age of each source. It is empirically shown that the sum of the obtained average ages, over a range of selection of arrival rates, is smaller in comparison to when using M/M/1* and M/M/1/2*. In [6], the authors consider a variation in which the new arrival not only replaces any waiting packet of the same source but also goes to the head of the queue.

In [7] the authors analyze the peak age for an M/G/1 and an M/G/1/1 queueing system. Updates from multiple sources arrive at the service facility as independent Poisson processes. Updates from different sources may have different service time distributions. For the M/G/1/1 system, any update that arrives when the server is busy is discarded. Instead of discarding a new arrival to a busy server, [8] consider preemption of the update currently in service by the new arrival (M/G/1*). They analyze both the peak and average age.

Often sources have intrinsic relative priorities. Sources may also be assigned priorities to enable those with low arrival rates priority access to the service facility, enabling a smaller age of their updates at the monitor. Multiple sources with different priorities were looked at in the M/M/1* and M/M/1/2* settings in [9]. In both settings an update is preempted by a newer arrival of the same source or of a source of higher priority. They used an SSH-based approach to arrive at the average age of any source for each setting. In [10] the authors consider a variant in which each source can have exactly one update in waiting while the server is busy processing an update. If an update is preempted in service by a higher priority update, it is saved in its source's waiting room for resuming service later. They compare with when there is no waiting room and observe that not having a waiting room is better with regards to average age in certain settings.

In [11], the authors also consider sources with priorities. They calculate the peak age for when each source has its own queue. They analyze the cases when the queue can store exactly one packet, as in [10], and when the queues are of infinite size. They consider Poisson arrivals and general service times.

In [12] arrivals consist of a mix of ordinary and priority updates. A priority update can preempt any update that is currently in service. An ordinary update if preempted is queued for resuming service later. Ordinary updates are serviced in a FCFS manner and may have a service distribution different from that of priority updates.

In the rest of this chapter we will elaborate on the works [2, 3, 9]. These include an SHS-based AoI analysis of M/M/1*, M/M/1/2*, and M/M/1 FCFS systems. Also, we have sources with different priorities [9]. We start by detailing the SHS-based method for calculating the average age in Section 3.2. In Sections 3.3 and 3.4, we analyze, respectively, the queues M/M/1* and M/M/1/2*. Here we also use the M/M/1* system to highlight the flexibility in modeling provided by the SHS approach. In Section 3.5 we consider the M/M/1 FCFS queue. Section 3.6 considers M/M/1* and M/M/1/2* when the sources have different priorities. This is followed by a chapter summary in Section 3.7.

3.2 Stochastic Hybrid Systems for AoI

We now describe aspects of SHS for AoI that suffice to enable calculation of the average age of a source's updates at the monitor. Details on how SHS for AoI is derived from a more generic SHS model and the derivation of the key equations that must be solved to obtain average age are in [2].

The state of our system is hybrid, with discrete and continuous components. The discrete component evolves as a point process and may capture, among other things, the occupancy of queues and the source of the packet that is in waiting and/or in service. The continuous component consists of one or more age processes. These include (a) the age process that tracks the age of updates of the source of interest at the monitor and (b) the age processes that are associated with the updates in the system (in service or in waiting). At any time, an age process *associated* with an update in the system takes the value to which the age at the monitor would be reset, in case the update was received by the monitor then. Note that, as we will see later, these age processes *may not track the true ages* of the updates with which they are associated.

Formally, let $\mathcal{Q} = \{0, 1, \ldots, m\}$ be the set of all discrete states. At any time t, the discrete state is given by the Markov process $q(t)$. Let the continuous state be given by $\mathbf{x}(t) = [x_0(t) \quad \cdots \quad x_n(t)] \in \mathbb{R}^{n+1}$, $n \geqslant 1$, where $x_i(t)$, $0 \leqslant i \leqslant n+1$, are all age processes of interest over all states in $\mathcal{Q}$. We will use the convention of $x_0(t)$ being the age of updates at the monitor. Thus we are assuming that there are at most n updates in the system. Note that the descriptions that follow assume a finite n and m. However, later in Section 3.5 we will demonstrate how SSH-based analysis may be used when the number of discrete states and age processes is infinite.

Events like a new arrival or a completed service may cause the discrete state to transition. They may also lead to a reset of one or more age processes in $\mathbf{x}(t)$. For example, when a source's update completes service, we have one less update in the system (occupancy changes), and the age process that tracks the age at the monitor must be reset to the age process associated with the serviced update. Note that not all transitions may cause the discrete state to change. For example, in the case of a single source M/M/1* queue, if an arrival takes place when the server is busy, it will preempt the packet that is currently in service. The occupancy of the system, which is a valid choice of discrete state, doesn't change. However, the age process associated with the update that is currently in service is reset to zero. This is because the new arrival that preempts the update in service has age zero.

It is instructive to think of the ways in which an age process may be reset. It may be reset to 0 when a transition has it begin tracking the age of a new arrival. A transition may have an age process reset to another age process. This, for example, happens when an update finishes service and leaves the system. The age process that tracks the age at the monitor will be reset to the age process associated with the update that just finished service. Such age process resets can be captured by linear reset maps.

Lastly, note that in any discrete state, relevant age processes increase at rate 1. This simply captures the fact that the age sample function has a slope 1 in between resets. Here, by relevant we mean those age processes in $\mathbf{x}(t)$ that are associated with updates in the system that may eventually reset age at the monitor. To summarize, in addition to $\mathcal{Q}$ and $\mathbf{x}(t)$, we must enumerate all transitions in every discrete state $q \in \mathcal{Q}$. We must also clearly identify the relevant age processes in any discrete state q.

A transition l is defined by the tuple $a_l = (q_l, q'_l, \lambda^{(l)}, \mathbf{A}_l)$, where q_l is start state, q'_l is the end state, $\lambda^{(l)}$ is the infinitesimal rate with which the discrete state moves from state q_l to q'_l, and $\mathbf{A}_l$ is the linear reset map. Note that $q(t)$ is essentially a continuous-time Markov chain with the difference that we allow for self-transitions, that is, transitions in which $q_l = q'_l$.

The rate of change of the age processes in state q is given by $\dot{\mathbf{x}}(t) = \mathbf{b}_q \in \{0,1\}^{n+1}$. For an age process that is relevant in state q, the corresponding element in $\mathbf{b}_q$ is set to 1. Otherwise, we set it to 0. An age process may become irrelevant when the discrete state transitions to state q. The value such a process is reset to is unimportant. However, in the systems we will discuss later in this chapter, we will reset an irrelevant process to 0.

Let $\mathcal{L}$ be the set of all transitions. Define the sets $\mathcal{B} = \{\mathbf{b}_q : q \in \mathcal{Q}\}$ and $\mathcal{A} = \{a_l : l \in \mathcal{L}\}$. The following definition summarizes an AoI SHS.

DEFINITION 3.1 An age-of-information SHS $(\mathcal{Q}, \mathcal{B}, \mathcal{A})$ is an SHS in which the discrete state $q(t) \in \mathcal{Q}$ is a continuous-time Markov chain with transitions $l \in \mathcal{L}$ from state q_l to q'_l at rate $\lambda^{(l)}$ and the continuous state evolves according to $\dot{\mathbf{x}}(t) = \mathbf{b}_q \in \{0,1\}^{n+1}$ in each discrete state $q \in \mathcal{Q}$ and is subject to the linear transition reset map $\mathbf{x}' = \mathbf{x}\mathbf{A}_l$ in transition l.

Define

$$\mathcal{L}'_{\bar{q}} = \{l \in \mathcal{L} : q'_l = \bar{q}\}, \tag{3.1a}$$

$$\mathcal{L}_{\bar{q}} = \{l \in \mathcal{L} : q_l = \bar{q}\} \tag{3.1b}$$

as the respective sets of incoming and outgoing transitions for each state $\bar{q}$. We will assume that the continuous-time Markov chain $q(t)$ is ergodic as otherwise time-average age analysis makes little sense. Under this assumption, there must exist a unique stationary vector $\bar{\boldsymbol{\pi}} = [\bar{\pi}_0 \ \cdots \ \bar{\pi}_m]$ satisfying

$$\bar{\pi}_{\bar{q}} \sum_{l \in \mathcal{L}_{\bar{q}}} \lambda^{(l)} = \sum_{l \in \mathcal{L}'_{\bar{q}}} \lambda^{(l)} \bar{\pi}_{q_l}, \tag{3.2a}$$

$$\sum_{\bar{q} \in \mathcal{Q}} \bar{\pi}_{\bar{q}} = 1. \tag{3.2b}$$

Define $\mathbf{v}_{\bar{q}}(t) = \mathrm{E}\left[\mathbf{x}(t)\delta_{\bar{q},q(t)}\right]$, where $\delta_{\bar{q},q(t)}$ is the Kronecker delta function. When $q(t) = \bar{q}$, $\delta_{\bar{q},q(t)} = 1$. Otherwise, $\delta_{\bar{q},q(t)} = 0$. Observe that $\mathbf{v}_{\bar{q}}(t) = \mathrm{E}[\mathbf{x}(t)|q(t) = \bar{q}]$ $P[q(t) = \bar{q}]$, which is the conditional expectation of $\mathbf{x}(t)$, given that the discrete Markov chain is in state $\bar{q}$, weighed by the probability that the chain is in state $\bar{q}$. Define $\bar{\mathbf{v}}_{\bar{q}}$ as the limit[2] to which $\mathbf{v}_{\bar{q}}(t)$ converges as $t \to \infty$. The vector $\bar{\mathbf{v}}_{\bar{q}} = [\bar{v}_{\bar{q}0} \ \cdots \ \bar{v}_{\bar{q}n}]$ has the limiting values corresponding to each age process in $\mathbf{x}(t)$, given that the discrete state is $\bar{q}$. As a result, we have a total of $n + 1 \times m$ such values.

The following theorem[3] provides a system of linear equations in the $v_{\bar{q}k}$, $\bar{q} \in \mathcal{Q}$, and $0 \leqslant k \leqslant n$ that must be solved, and the conditions that must be satisfied, to obtain average age.

THEOREM 3.2 *If the discrete-state Markov chain $q(t)$ is ergodic with stationary distribution $\bar{\boldsymbol{\pi}}$ and we can find a nonnegative solution $\bar{\mathbf{v}} = [\bar{\mathbf{v}}_0 \ \cdots \bar{\mathbf{v}}_m]$ such that*

$$\bar{\mathbf{v}}_{\bar{q}} \sum_{l \in \mathcal{L}_{\bar{q}}} \lambda^{(l)} = \mathbf{b}_{\bar{q}} \bar{\pi}_{\bar{q}} + \sum_{l \in \mathcal{L}'_{\bar{q}}} \lambda^{(l)} \bar{\mathbf{v}}_{q_l} \mathbf{A}_l, \qquad \bar{q} \in \mathcal{Q}, \tag{3.3a}$$

then the average age of the AoI SHS is given by

$$\Delta = \sum_{\bar{q} \in \mathcal{Q}} \bar{v}_{\bar{q}0}. \tag{3.3b}$$

It is worth emphasizing here that ergodicity of the Markov chain doesn't guarantee the existence of nonnegative $\bar{\mathbf{v}}$.

In the sections that follow we will analyze various systems using the SHS method and in the process elucidate its workings. We have N sources that share the service facility. The service of an update is exponentially distributed with rate μ. Source $1 \leqslant i \leqslant N$ generates updates that arrive at the service facility as a rate λ_i Poisson process.

[2] The limit exists when the first-order differential equations [2, Equation (33)] that govern the evolution of $\mathbf{v}_{\bar{q}}(t)$, for any $\bar{q} \in \mathcal{Q}$, are stable.

[3] The theorem essentially restates [2, Theorem 4]. We skip the fact that the existence of nonnegative $\bar{\mathbf{v}}$ implies the stability of the differential equations [2, Equation (33)].

The corresponding load is $\rho_i = \lambda_i/\mu$. The total offered load $\rho = \sum_{i=1}^{N} \rho_i$ and the sum rate of arrivals is $\lambda = \sum_{i=1}^{N} \lambda_i$.

3.3 M/M/1*: LCFS with Preemption in Service

An update from any source that arrives at an idle server begins service immediately. In case an arrival sees a busy server, it preempts the update currently in service and begins service. The preempted update is discarded. Note that the preemption is source agnostic, that is, an arrival preempts any source's update that may be in service. The service distribution is exponential with rate μ.

The average age of source i for such a system was shown in [2, Theorem 2, part (a)] to be

$$\Delta_i = \frac{1}{\mu}(1+\rho)\frac{1}{\rho_i}. \tag{3.4}$$

Without loss of generality, we demonstrate how to calculate the age of source 1, whose updates arrive at the rate λ_1. Since arrivals from all the other sources that share the facility with source 1 are Poisson processes, we can model these other sources together as a Poisson process with rate λ_2. We have $\lambda = \lambda_1 + \lambda_2$.

Note that it suffices to have an age process associated with an update in service and another that tracks the age of updates of source 1 at the monitor. The age at the monitor is given by $x_0(t)$. Let $x_1(t)$ be the age process associated with an update in service. When a source 1 update completes service, $x_0(t)$ is reset to the age of that update. However, in case a source 2 update completes service, the age at the monitor should stay unaffected.

Next we look at how appropriate choices of age reset maps can help us come up with different selections of the discrete state set $\mathcal{Q}$. These give us AoI SHS models with varied complexity for the M/M/1* queueing system. We will start with a model that has the discrete state track the source of the update in service, which results in three discrete states and seven transitions. Next, we will have the state instead track busyness of the server, resulting in two discrete states and five transitions. In the third SSH model, we introduce the notion of fake updates, which results in just one discrete state and three transitions. Having described the models, we will go through the mechanics of using equations (3.2) and (3.3) to arrive at the average age (3.3b) for the M/M/1* system.

3.3.1 Markov Chain Tracks Source in Service

One way of ensuring updates of source 2 do not reset the age at the monitor is to have the discrete state track the source of an update, if any, in service. This results in the set of discrete states $\mathcal{Q} = \{0, 1, 2\}$. The Markov Chain is in state 0 when the server is idle. It is in state i when a source $i \in \{1, 2\}$ update is being serviced. The chain is shown in Figure 3.1. The corresponding table of transitions is Table 3.1. In the table, for each transition l we specify the start and end states, the rate $\lambda^{(l)}$, the linear reset map $\mathbf{A}_l$,

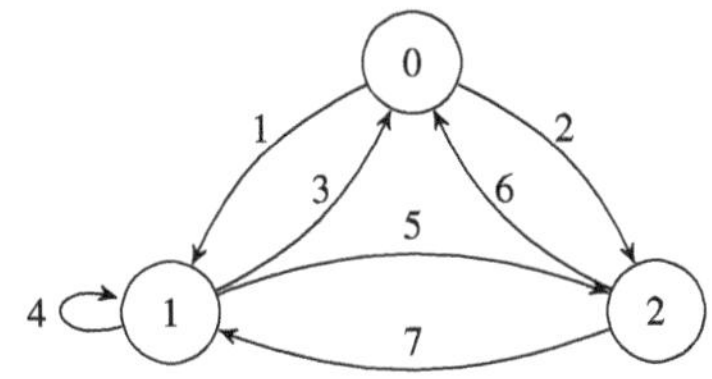

Figure 3.1 The SHS Markov chain for updates of source 1 in the two-source LCFS-S system. In state 0 the system is idle while in state $i \in \{1,2\}$ a source i update is in service. The transition rates and transition/reset maps for links $l = 1,\dots,7$ are shown in Table 3.1.

the age vector $\mathbf{x}' = \mathbf{x}\mathbf{A}_l$ that results from the reset associated with the transition, and $\mathbf{v}_{q_l}\mathbf{A}_l$, which helps in quickly writing down the equations (3.3).

When the chain is in state 2, the age process $x_1(t)$ tracks the age of an update from source 2 as it undergoes service. This age process, however, does not impact the age of source 1 updates at the monitor and is therefore set as irrelevant. Note that $x_0(t)$ is always relevant. We set $\mathbf{b}_2 = [1 \quad 0]$. As stated earlier for irrelevant age processes, we set $x_1(t) = 0$ in $\bar{q} = 2$. We must further ensure that the age processes are reset appropriately when the chain leaves state 2. In case the update completes service and the chain transitions into state 0, the age vector reset must leave both $x_0(t)$ and $x_1(t)$ unchanged. On the other hand, in case the update is preempted by a source 1 arrival, transitioning the chain into state 1, we must reset $x_1(t)$ to 0, which is the age of the new arrival. Also, note that in state 1, $x_1(t)$ is a relevant age process. That is, we must set $\mathbf{b}_1 = [1 \quad 1]$.

Observe that a source 2 arrival when $\bar{q} = 2$ changes neither the discrete state nor resets the age vector and hence can be ignored. This is why we don't show a self-transition of rate λ_2 into state 2. However, we can't ignore the self-transition into state 1 on arrival of a source 1 update. This is because the arrival results in the reset $x_1(t) = 0$.

Next we summarize the rate of change of the age processes in different discrete states followed by an explanation for all transitions. We have

$$\dot{\mathbf{x}}(t) = \mathbf{b}_q = \begin{cases} [1 \quad 0] & q = 0,2, \\ [1 \quad 1] & q = 1. \end{cases} \tag{3.5}$$

The transitions, see Figure 3.1 and Table 3.1, are

$l = 1$ A source 1 update arrives at an empty queue. With this arrival, the age at the monitor post reset $x_0' = x_0$ is unchanged because the arrival does not yield an age reduction until it departs. However, the age of the update in service post reset is $x_1' = 0$, because the arriving source 1 update is fresh and its age is zero at that instant.

$l = 2$ A source 2 update arrives at an empty queue. The age $x_0' = x_0$ is unchanged because the arrival does not change the age. However, $x_1' = 0$ because x_1 is irrelevant in state 2.

$l = 3$ A source 1 update completes service and is delivered to the monitor. In this transition, $x_0' = x_1$, since age at the monitor is reset to the age of the source

Table 3.1 Table of transitions for the Markov chain in Figure 3.1.

l	$q_l \to q'_l$	$\lambda^{(l)}$	$\mathbf{x}\mathbf{A}_l$	$\mathbf{A}_l$	$\mathbf{v}_{q_l}\mathbf{A}_l$
1	$0 \to 1$	λ_1	$[x_0\ 0]$	$\begin{bmatrix}1 & 0\\0 & 0\end{bmatrix}$	$[v_{00}\ 0]$
2	$0 \to 2$	λ_2	$[x_0\ 0]$	$\begin{bmatrix}1 & 0\\0 & 0\end{bmatrix}$	$[v_{00}\ 0]$
3	$1 \to 0$	μ	$[x_1\ 0]$	$\begin{bmatrix}0 & 0\\1 & 0\end{bmatrix}$	$[v_{11}\ 0]$
4	$1 \to 1$	λ_1	$[x_0\ 0]$	$\begin{bmatrix}1 & 0\\0 & 0\end{bmatrix}$	$[v_{10}\ 0]$
5	$1 \to 2$	λ_2	$[x_0\ 0]$	$\begin{bmatrix}1 & 0\\0 & 0\end{bmatrix}$	$[v_{10}\ 0]$
6	$2 \to 0$	μ	$[x_0\ 0]$	$\begin{bmatrix}1 & 0\\0 & 0\end{bmatrix}$	$[v_{20}\ 0]$
7	$2 \to 1$	λ_1	$[x_0\ 0]$	$\begin{bmatrix}1 & 0\\0 & 0\end{bmatrix}$	$[v_{20}\ 0]$

1 update that just completed service. Also note that $x'_1 = 0$ since x_1 becomes irrelevant when the system enters state 0.

$l = 4$ The source 1 update in service is preempted by a fresh source 1 update. The age x_0 remains unchanged while x_1 is reset to zero because the new update is fresh.

$l = 5$ The source 1 update in service is preempted by a source 2 update. The age $x'_0 = x_0$ is unchanged and $x'_1 = 0$ since x_1 becomes irrelevant in state 2.

$l = 6$ A source 2 update completes service. The source 1 age x_0 is unchanged. In the transition to state 0, x_1 remains irrelevant and is set to zero.

$l = 7$ The source 2 update in service is preempted by a fresh source 1 update. The age x_0 is unchanged while $x'_1 = 0$ because the new update is fresh.

To employ Theorem 3.2, we first use (3.2a) to show that the stationary probability vector $\bar{\boldsymbol{\pi}}$ satisfies $\bar{\boldsymbol{\pi}}\mathbf{D} = \bar{\boldsymbol{\pi}}\mathbf{Q}$ with

$$\mathbf{D} = \text{diag}[\lambda, \mu + \lambda, \mu + \lambda_1], \qquad \mathbf{Q} = \begin{bmatrix} 0 & \lambda_1 & \lambda_2 \\ \mu & \lambda_1 & \lambda_2 \\ \mu & \lambda_1 & 0 \end{bmatrix}. \tag{3.6}$$

Applying $\sum_{i=0}^{2} \bar{\pi}_i = 1$, the stationary probabilities are

$$\begin{bmatrix}\bar{\pi}_0 & \bar{\pi}_1 & \bar{\pi}_2\end{bmatrix} = (1+\rho)^{-1}\begin{bmatrix}1 & \rho_1 & \rho_2\end{bmatrix}. \tag{3.7}$$

Since $\mathbf{b}_0 = [1 \quad 0]$, evaluation of (3.3a) in Theorem 3.2 at $\bar{q} = 0$ yields

$$\lambda[\bar{v}_{00}\ \bar{v}_{01}] = [\bar{\pi}_0\ 0] + \mu[\bar{v}_{11}\ 0] + \mu[\bar{v}_{20}\ 0]. \tag{3.8}$$

We see from (3.8) that $\bar{v}_{01} = 0$. This is a consequence of x_1 being irrelevant in state $q = 0$. In particular, $x_1(t)\delta_{0,q(t)} = 0$ for all t because $x_1(t)$ is held at 0 when $q(t) = 0$

(by our convention for irrelevant variables) and $\delta_{0,q(t)} = 0$ when $q(t) \neq 0$. Thus, $v_{01}(t) = \mathrm{E}\left[x_1(t)\delta_{0,q(t)}\right] = 0$. Evaluating (3.3a) at $\bar{q} = 1$ and $\bar{q} = 2$ produces

$$(\mu + \lambda)[\bar{v}_{10}\ \bar{v}_{11}] = [\bar{\pi}_1\ \bar{\pi}_1] + \lambda_1[\bar{v}_{00}\ 0] + \lambda_1[\bar{v}_{10}\ 0] + \lambda_1[\bar{v}_{20}\ 0], \tag{3.9a}$$

$$(\mu + \lambda_1)[\bar{v}_{20}\ \bar{v}_{21}] = [\bar{\pi}_2\ 0] + \lambda_2[\bar{v}_{00}\ 0] + \lambda_2[\bar{v}_{10}\ 0]. \tag{3.9b}$$

In terms of the vectors

$$\bar{\boldsymbol{\pi}} = [\bar{\pi}_0\ \bar{\pi}_1\ \bar{\pi}_2], \tag{3.10}$$

$$\bar{\mathbf{v}} = [\bar{\mathbf{v}}_0\quad \bar{\mathbf{v}}_1\quad \bar{\mathbf{v}}_2] = [\bar{v}_{00}\ \bar{v}_{01}\ \bar{v}_{10}\ \bar{v}_{11}\ \bar{v}_{20}\ \bar{v}_{21}], \tag{3.11}$$

we have

$$\bar{\mathbf{v}}\mathbf{D} = \bar{\boldsymbol{\pi}}\mathbf{B} + \bar{\mathbf{v}}\mathbf{R} \tag{3.12}$$

where

$$\mathbf{D} = \mathrm{diag}[\lambda, \lambda, \mu + \lambda, \mu + \lambda, \mu + \lambda_1, \mu + \lambda_1], \tag{3.13}$$

$$\mathbf{B} = \begin{bmatrix} 1 & 0 & 0 & 0 & 0 & 0 \\ 0 & 0 & 1 & 1 & 0 & 0 \\ 0 & 0 & 0 & 0 & 1 & 0 \end{bmatrix}, \qquad \mathbf{R} = \begin{bmatrix} 0 & 0 & \lambda_1 & 0 & \lambda_2 & 0 \\ 0 & 0 & 0 & 0 & 0 & 0 \\ 0 & 0 & \lambda_1 & 0 & \lambda_2 & 0 \\ \mu & 0 & 0 & 0 & 0 & 0 \\ \mu & 0 & \lambda_1 & 0 & 0 & 0 \\ 0 & 0 & 0 & 0 & 0 & 0 \end{bmatrix}. \tag{3.14}$$

We observe that the columns and rows of **R** corresponding to the irrelevant variables $\bar{v}_{01}$ and $\bar{v}_{21}$ are zero. Gathering the relevant variables, we obtain

$$\frac{1}{\mu}[\bar{\pi}_0\quad \bar{\pi}_1\quad \bar{\pi}_1\quad \bar{\pi}_2] = [\bar{v}_{00}\quad \bar{v}_{10}\quad \bar{v}_{11}\quad \bar{v}_{20}] \begin{bmatrix} \rho & -\rho_1 & 0 & -\rho_2 \\ 0 & 1 + \rho_2 & 0 & -\rho_2 \\ -1 & 0 & 1 + \rho & 0 \\ -1 & -\rho_1 & 0 & 1 + \rho_1 \end{bmatrix}. \tag{3.15}$$

It follows from (3.7) and (3.15) that

$$\bar{v}_{00} = \frac{1}{\mu(1 + \rho)}\left[\frac{1 + \rho_2}{\rho_1} + \frac{1}{1 + \rho}\right], \tag{3.16a}$$

$$\bar{v}_{10} = \frac{1}{\mu(1 + \rho)}\left[1 + \rho + \frac{\rho_1}{1 + \rho}\right], \tag{3.16b}$$

$$\bar{v}_{20} = \frac{1}{\mu(1 + \rho)}\left[\frac{\rho_2(1 + \rho)}{\rho_1} + \frac{\rho_2}{1 + \rho}\right]. \tag{3.16c}$$

From (3.9), it can be seen that $\bar{v}_{11}$ is also nonnegative. Thus, Theorem 3.2 implies that the average age for source 1 is $\Delta = \sum_{q=0}^{2} \bar{v}_{q0}$. Applying (3.16) yields Equation (3.4) for source $i = 1$.

3.3.2 Markov Chain Tracks Server Busyness

In our choice of discrete states and transitions in Figure 3.1 and Table 3.1 we ensured that an update of source 2 doesn't reset the age at the monitor by assigning a separate discrete state to when a source 2 update is in service.

We could, however, ensure the preceding situation by a smart resetting of the age vector on arrival of a source 2 update and without the need for a separate discrete state for when a source 2 update is in service. The age process $x_0(t)$ will stay unaffected by a source 2 departure if on arrival of a source 2 update we set $x_1(t) = x_0(t)$ and treat $x_1(t)$ as a relevant age process while the update undergoes service. This resetting of $x_1(t)$ ensures that the age $x_0(t)$ at the monitor is unaffected when the source 2 packet finishes service and departs. We can simply set $x_0(t) = x_1(t)$ on it completing service as we do when a source 1 packet completes service. At this point, it is useful to observe that while $x_1(t)$ is associated with the update in service, it tracks the true age of the update only when it is of source 1. However, irrespective of the update in service, $x_1(t)$ is such that it resets $x_0(t)$ appropriately when the update finishes service.

Given the preceding reset method, we no longer need a separate state for a source 2 update that is in service. We only need to distinguish between a busy and an idle server. The Markov chain is shown in Figure 3.2, and the transitions are shown in Table 3.2. The chain tracks the busyness of the server. It is in state 0 when the server is idle and is in state 1 while servicing either a source 1 or a source 2 packet. We have $\mathcal{Q} = \{0, 1\}$.

A summary of the rate of change of the age processes and a description of all transitions is next. We have

$$\dot{\mathbf{x}}(t) = \mathbf{b}_q = \begin{cases} [1 \quad 0] & q = 0, \\ [1 \quad 1] & q = 1. \end{cases} \tag{3.17}$$

The transitions, see Figure 3.2 and Table 3.2, are

$l = 1$ A fresh source 1 update goes into service; $x_1' = 0$ because the update is fresh.

$l = 2$ A fresh source 2 update goes into service and $x_1' = x_0$. As explained earlier, this ensures that if this source 2 update completes service, it doesn't reduce the age of source 1 at the monitor.

$l = 3$ The update in service is delivered. The age x_0 is reset to $x_0' = x_1$. If this delivered update is from source 1, then $x_0' < x_0$. However, if this update is from source

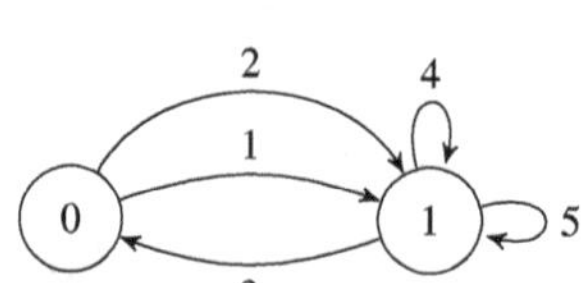

Figure 3.2 The simplified SHS Markov chain for updates of source 1 in the two-source LCFS-S system. In state 0 the system is idle, while in state 1 an update of either source 1 or 2 is in service. The transition rates and transition/reset maps for links $l = 1, \ldots, 5$ are shown in Table 3.2.

Table 3.2 Table of transitions for the Markov chain in Figure 3.2.

l	$q_l \to q'_l$	$\lambda^{(l)}$	$\mathbf{x}\mathbf{A}_l$	$\mathbf{A}_l$	$\mathbf{v}_{q_l}\mathbf{A}_l$
1	$0 \to 1$	λ_1	$[x_0\ 0]$	$\begin{bmatrix}1 & 0\\ 0 & 0\end{bmatrix}$	$[v_{00}\ 0]$
2	$0 \to 1$	λ_2	$[x_0\ x_0]$	$\begin{bmatrix}1 & 1\\ 0 & 0\end{bmatrix}$	$[v_{00}\ v_{00}]$
3	$1 \to 0$	μ	$[x_1\ 0]$	$\begin{bmatrix}0 & 0\\ 1 & 0\end{bmatrix}$	$[v_{11}\ 0]$
4	$1 \to 1$	λ_1	$[x_0\ 0]$	$\begin{bmatrix}1 & 0\\ 0 & 0\end{bmatrix}$	$[v_{10}\ 0]$
5	$1 \to 1$	λ_2	$[x_0\ x_0]$	$\begin{bmatrix}1 & 1\\ 0 & 0\end{bmatrix}$	$[v_{10}\ v_{10}]$

2, then $x'_0 = x_0$ and no age reduction occurs. Note that this age reduction was encoded in the prior transition that put this update in service.

$l = 4$ The update in service is replaced by a fresh source 1 update. This reset map is essentially the same as for transition $l = 1$.

$l = 5$ The update in service is replaced by a fresh source 2 update. This reset map is essentially the same as for transition $l = 2$.

The Markov chain for the discrete state has stationary probabilities:

$$\bar{\boldsymbol{\pi}} = [\bar{\pi}_0 \quad \bar{\pi}_1] = \left[\frac{1}{1+\rho} \quad \frac{\rho}{1+\rho}\right]. \tag{3.18}$$

In this system, $\mathbf{v}_0 = [v_{00} \quad v_{01}]$ and $\mathbf{v}_1 = [v_{10} \quad v_{11}]$. Evaluating (3.3a) at $\bar{q} = 0, 1$ produces

$$\lambda[\bar{v}_{00}\ \bar{v}_{01}] = [\bar{\pi}_0\ 0] + \mu[\bar{v}_{11}\ 0], \tag{3.19a}$$

$$(\mu + \lambda)[\bar{v}_{10}\ \bar{v}_{11}] = [\bar{\pi}_1\ \bar{\pi}_1] + \lambda_1[\bar{v}_{00}\ 0] + \lambda_2[\bar{v}_{00}\ \bar{v}_{00}] + \lambda_1[\bar{v}_{10}\ 0] + \lambda_2[\bar{v}_{10}\ \bar{v}_{10}]. \tag{3.19b}$$

As expected, we see from (3.19a) that $\bar{v}_{01} = 0$ because x_1 is irrelevant in state 0. Normalizing by the service rate μ, we obtain

$$\rho\bar{v}_{00} = \bar{\pi}_0/\mu + \bar{v}_{11}, \tag{3.20a}$$

$$(1 + \rho)\bar{v}_{10} = \bar{\pi}_1/\mu + \rho\bar{v}_{00} + \rho\bar{v}_{10}, \tag{3.20b}$$

$$(1 + \rho)\bar{v}_{11} = \bar{\pi}_1/\mu + \rho_2\bar{v}_{00} + \rho_2\bar{v}_{10}. \tag{3.20c}$$

Solving (3.20), it can be shown that $\bar{v}_{00}$, $\bar{v}_{10}$, and $\bar{v}_{11}$ are all nonnegative. Moreover, calculation of $\Delta = \bar{v}_{00} + \bar{v}_{10}$ yet again yields Equation (3.4) for source $i = 1$.

3.3.3 Introducing Fake Updates

Can we reduce the discrete state space further to just one state? It turns out that we can by introducing *fake* updates, which are updates that the server processes in lieu of

Table 3.3 Table of transitions for the Markov chain in Figure 3.2.

l	$q_l \to q'_l$	$\lambda^{(l)}$	$\mathbf{x}\mathbf{A}_l$	$\mathbf{A}_l$	$\mathbf{v}_{q_l}\mathbf{A}_l$
1	$0 \to 0$	λ_1	$[x_0 \; 0]$	$\begin{bmatrix} 1 & 0 \\ 0 & 0 \end{bmatrix}$	$[v_{00} \; 0]$
2	$0 \to 0$	λ_2	$[x_0 \; x_0]$	$\begin{bmatrix} 1 & 1 \\ 0 & 0 \end{bmatrix}$	$[v_{00} \; v_{00}]$
3	$0 \to 0$	μ	$[x_1 \; x_1]$	$\begin{bmatrix} 0 & 0 \\ 1 & 1 \end{bmatrix}$	$[v_{01} \; v_{01}]$

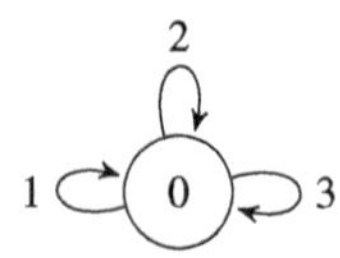

Figure 3.3 The simplified SHS Markov chain for updates of source 1 in the two-source LCFS-S system. The system is always busy serving either a real or a fake update. The transition rates and transition/reset maps for links $l = 1, 2, 3$ are shown in Table 3.3.

being idle. These belong to neither source 1 nor source 2. Such updates are serviced at a rate μ and can be preempted by updates of either source 1 or source 2. The age of a fake update when it begins processing is simply the age $x_0(t)$ at the monitor and therefore simply tracks the age at the monitor.

As before, we have the three transitions of rates λ_1, λ_2, and μ. A source 1 arrival has us set $x_1(t) = 0$, and on an arrival from source 2 we reset $x_1(t) = x_0(t)$. On completion of service of a fake update, as with any other update, we set $x_0(t) = x_1(t)$. The key is that a server keeps servicing fake updates until an update from source 1 or 2 arrives.

We obtain the Markov chain in Figure 3.3. The three transitions and the resets are listed in Table 3.3. Note that, as one would expect, both $x_0(t)$ and $x_1(t)$ are relevant in the only discrete state 0. We have $\mathbf{b}_0 = [1 \quad 1]$.

The transitions are

$l = 1$ A fresh source 1 update goes into service; $x'_1 = 0$ because the update is fresh.
$l = 2$ A fresh source 2 update goes into service and $x'_1 = x_1$. If the source 2 update does complete service, it doesn't reduce the age of the process of interest.
$l = 3$ The update in service is delivered. The age x_0 is reset to $x'_0 = x_1$, but x_1 is unchanged: $x'_1 = x_1$. This corresponds to creating a fake update with the same time stamp as the update that was just delivered.

The Markov chain for the discrete state has the trivial stationary probability $\pi_0 = 1$. In this system, $\mathbf{v}_0 = [v_{00} \quad v_{01}]$. Evaluating (3.3a) at $\bar{q} = 0$ produces

$$(\mu + \lambda)[\bar{v}_{00} \; \bar{v}_{01}] = [1 \; 1] + \lambda_1[v_{00} \; 0] + \lambda_2[v_{00} \; v_{00}] + \mu[v_{01} \quad v_{01}]. \tag{3.21}$$

Solving these two equations for v_{00} and v_{01}, the average age $\Delta = v_{00}$ yet again yields (3.4) for source $i = 1$.

3.4 M/M/1/2*: LCFS with Preemption in Waiting

In the M/M/1/2* queueing system, sources share a common waiting room that may be occupied by one update at any time. Therefore, we can have at most two updates in the system. An update in waiting is preempted by a new arrival. As in the case of M/M/1*, the preemption is source agnostic. However, once an update enters service, it can't be preempted and finishes service. An arrival to an empty system enters service immediately.

Updates from source i arrive as a rate λ_i Poisson process. As before, we will calculate the age Δ_1 of source 1 and we will treat all other sources as a Poisson process of rate λ_2. From [2, Theorem 2, part (b)], we know that the average age Δ_i of source i is given by

$$\Delta_i = \frac{1}{\mu}\left[\alpha_W(\rho) + \left(1 + \frac{\rho^2}{1+\rho}\right)\frac{1}{\rho_i}\right], \tag{3.22}$$

where

$$\alpha_W(\rho) = \frac{(1+\rho+\rho^2)^2 + 2\rho^3}{(1+\rho+\rho^2)(1+\rho)^2} \tag{3.23}$$

is a ratio of fourth-order polynomials. Direct calculation will verify that

$$0.837 < \alpha_W(\rho) < 1.09, \qquad \rho \geqslant 0. \tag{3.24}$$

Next, we will come up with the states and transitions of the Markov chain. The discrete state is simply the occupancy of the system and, as in the simpler M/M/1* SHS models, doesn't need to track the source of the update. We have $\mathcal{Q} = \{0, 1, 2\}$. Also, we require three age processes. As before, $x_0(t)$ tracks the age of updates from source 1 at the monitor, $x_1(t)$ is associated with the update that enters service, and $x_2(t)$ is associated with the update in waiting. The Markov chain is shown in Figure 3.4.

An arrival of an update to an empty system has the Markov chain transition from state 0 to state 1. The arrival leaves the age at the monitor unchanged. The age $x_1(t)$ is set to zero if the arrival is from source 1, since the arrival is fresh. In case the arrival is from source 2, we must set $x_1(t) = x_0(t)$. This ensures that on completion of service of this update, the age of source 1's updates at the monitor is unaffected. Irrespective of the source of update, $x_2(t)$ should be reset to 0 because the age process is irrelevant, as there is no packet in waiting.

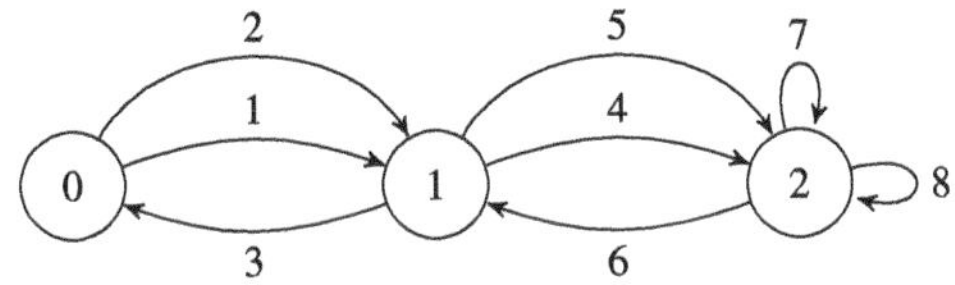

Figure 3.4 The simplified SHS Markov chain for updates of source 1 in the two-source M/M/1/2* system. The state i indicates the number of updates in the system. The transition rates and transition/reset maps for links $l = 1, \ldots, 8$ are shown in Table 3.4.

Table 3.4 Table of transitions for the Markov chain in Figure 3.4.

l	$q_l \to q'_l$	$\lambda^{(l)}$	$\mathbf{x}\mathbf{A}_l$	$\mathbf{v}_{q_l}\mathbf{A}_l$
1	$0 \to 1$	λ_1	$[x_0\ 0\ 0]$	$[v_{00}\ 0\ 0]$
2	$0 \to 1$	λ_2	$[x_0\ x_0\ 0]$	$[v_{00}\ v_{00}\ 0]$
3	$1 \to 0$	μ	$[x_1\ 0\ 0]$	$[v_{11}\ 0\ 0]$
4	$1 \to 2$	λ_1	$[x_0\ x_1\ 0]$	$[v_{10}\ v_{11}\ 0]$
5	$1 \to 2$	λ_2	$[x_0\ x_1\ x_1]$	$[v_{10}\ v_{11}\ v_{11}]$
6	$2 \to 1$	μ	$[x_1\ x_2\ 0]$	$[v_{21}\ v_{22}\ 0]$
7	$2 \to 2$	λ_1	$[x_0\ x_1\ 0]$	$[v_{20}\ v_{21}\ 0]$
8	$2 \to 2$	λ_2	$[x_0\ x_1\ x_1]$	$[v_{20}\ v_{21}\ v_{21}]$

If an update arrives in state 1, the system transitions to 2. Note that since we now have an update in waiting, $x_2(t)$ becomes relevant. However, it must be appropriately set on arrival of the update. If the update is from source 1, $x_2(t)$ is set to 0, as the update is fresh and, if it eventually completes service, we would want the age at the monitor to be reset to this update's age. Otherwise, in case the update is from source 2, we set $x_2(t)$ to $x_1(t)$. This ensures that in case the new source 2 arrival eventually enters and completes service, it does not change the age $x_0(t)$.

Last but not least, note that when a packet in waiting enters service, $x_1(t)$ is reset to $x_2(t)$. Also, when a packet completes service, $x_0(t)$ is reset to $x_1(t)$.

The rate of change of age processes in the different states is summarized by

$$\dot{\mathbf{x}}(t) = \mathbf{b}_q = \begin{cases} [1 \quad 0 \quad 0], & q = 0, \\ [1 \quad 1 \quad 0], & q = 1, \\ [1 \quad 1 \quad 1], & q = 2. \end{cases} \tag{3.25}$$

The list of all transitions is given in Table 3.4. For the Markov chain in Figure 3.4, it follows from (3.2) that the discrete state has stationary distribution

$$[\bar{\pi}_0\ \bar{\pi}_1\ \bar{\pi}_2] = C_\pi[1\ \rho\ \rho^2], \tag{3.26}$$

where $C_\pi = (1+\rho+\rho^2)^{-1}$ is the normalizing constant. From (3.3a) with $\bar{q} \in \mathcal{Q}$, we obtain

$$\lambda\bar{\mathbf{v}}_0 = [\bar{\pi}_0\ 0\ 0] + \mu[\bar{v}_{11}\ 0\ 0], \tag{3.27a}$$

$$(\lambda+\mu)\bar{\mathbf{v}}_1 = [\bar{\pi}_1\ \bar{\pi}_1\ 0] + \lambda_1[\bar{v}_{00}\ 0\ 0] + \lambda_2[\bar{v}_{00}\ \bar{v}_{00}\ 0] + \mu[\bar{v}_{21}\ \bar{v}_{22}\ 0], \tag{3.27b}$$

$$\begin{aligned}(\lambda+\mu)\bar{\mathbf{v}}_2 = {} & [\bar{\pi}_2\ \bar{\pi}_2\ \bar{\pi}_2] + \lambda_1[\bar{v}_{10}\ \bar{v}_{11}\ 0] + \lambda_2[\bar{v}_{10}\ \bar{v}_{11}\ \bar{v}_{11}] \\ & + \lambda_1[\bar{v}_{20}\ \bar{v}_{21}\ 0] + \lambda_2[\bar{v}_{20}\ \bar{v}_{21}\ \bar{v}_{21}]. \end{aligned} \tag{3.27c}$$

We see from (3.27a) that $\bar{v}_{01}$ and $\bar{v}_{02}$ are zero because $x_1(t)$ and $x_2(t)$ are irrelevant in state 0. Similarly, (3.27b) implies $\bar{v}_{12} = 0$ because $x_2(t)$ is irrelevant in state 1. Gathering the relevant variables and normalizing by the service rate μ, we obtain

$$\begin{aligned}
\rho\bar{v}_{00} &= \bar{\pi}_0/\mu + \bar{v}_{11}, && (3.28a)\\
(1+\rho)\bar{v}_{10} &= \bar{\pi}_1/\mu + \rho\bar{v}_{00} + \bar{v}_{21}, && (3.28b)\\
(1+\rho)\bar{v}_{11} &= \bar{\pi}_1/\mu + \rho_2\bar{v}_{00} + \bar{v}_{22}, && (3.28c)\\
\bar{v}_{20} &= \bar{\pi}_2/\mu + \rho\bar{v}_{10}, && (3.28d)\\
\bar{v}_{21} &= \bar{\pi}_2/\mu + \rho\bar{v}_{11}, && (3.28e)\\
(1+\rho)\bar{v}_{22} &= \bar{\pi}_2/\mu + \rho_2\bar{v}_{11} + \rho_2\bar{v}_{21}. && (3.28f)
\end{aligned}$$

We employ (3.28e) and (3.28f) to write

$$\bar{v}_{22} = \frac{1+\rho_2}{1+\rho}\bar{\pi}_2 + \rho_2\bar{v}_{11}. \tag{3.29}$$

We now apply (3.28e) and (3.29) to the other equations in (3.28), yielding

$$\begin{aligned}
\rho\bar{v}_{00} &= \bar{\pi}_0 + \bar{v}_{11}, && (3.30a)\\
\bar{v}_{10} &= \frac{1}{\mu(1+\rho)} + \bar{v}_{11}, && (3.30b)\\
(1+\rho)\bar{v}_{11} &= \bar{\pi}_1/\mu + \rho_2\bar{v}_{00} + \frac{1+\rho_2}{\mu(1+\rho)}\bar{\pi}_2 + \rho_2\bar{v}_{11}, && (3.30c)\\
\bar{v}_{20} &= \bar{\pi}_2/\mu + \rho\bar{v}_{10}. && (3.30d)
\end{aligned}$$

From (3.30), some algebra will show

$$\bar{v}_{11} = \frac{\rho}{\mu(1+\rho)}\frac{1}{\rho_1} - \frac{C_\pi(1+\rho+\rho^3)}{\mu(1+\rho)^2}. \tag{3.31}$$

To verify that $\bar{v}_{11}$ is nonnegative, we note that $\rho_1 \leqslant \rho$ and that for fixed ρ, $\bar{v}_{11}$ is minimized over all ρ_1 at $\rho_1 = \rho$. Some algebra will verify that $\bar{v}_{11} \geqslant 0$ when $\rho_1 = \rho$. It then follows from (3.28) and (3.30) that all components of $\bar{\mathbf{v}}$ are nonnegative. Moreover, it also follows from (3.3b) and (3.30) that the average age is

$$\begin{aligned}
\Delta &= \bar{v}_{00} + \bar{v}_{10} + \bar{v}_{20}\\
&= \frac{1}{\mu} + \frac{\pi_0 + \pi_2}{\rho} + \frac{1+\rho+\rho^2}{\rho}\bar{v}_{11}. && (3.32)
\end{aligned}$$

Equation (3.22) then follows from substitution of (3.26) and (3.31) in (3.32).

3.5 M/M/1 FCFS

In the M/M/1 FCFS system we have an infinite queue. Updates that arrive when the server is busy are queued. Queued updates are serviced in a first-come-first-served manner. An update that arrives to an empty system enters service. All arrivals complete service.

In the systems that we have looked at so far in this chapter, the old updates were preempted and discarded when new updates arrived. This ensured a finite discrete state space. The FCFS queue, on the other hand, can have an arbitrarily large discrete state space, as the state is the occupancy of the queue. Our approach involves carrying out the SSH-based analysis for the M/M/1/m queue in which new arrivals are discarded in

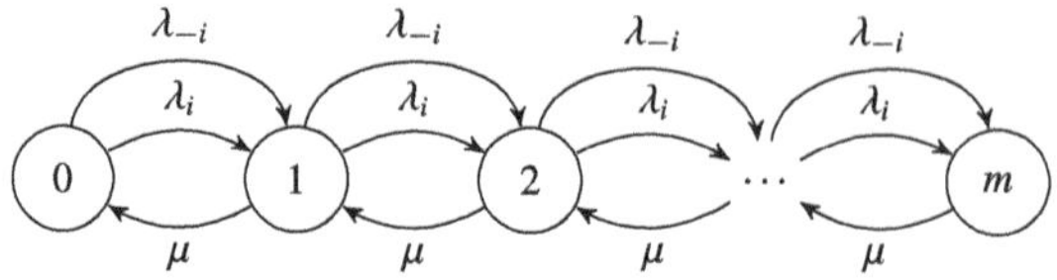

Figure 3.5 The SHS Markov chain for transitions between discrete states. We show the rates on the edges instead of the transition indices l.

case a total of m updates occupy the system. This ensures a finite discrete space and requires a finite number of age processes. To calculate the average age of source i for the M/M/1 system, we let $m \to \infty$.

Before we detail the approach, we summarize the age Δ_i of source i at the monitor, which was calculated in [3]. As before, $\rho = \sum_i \rho_i$. Also, $\rho_{-i} = \rho - \rho_i$. The age

$$\Delta_i = \frac{1}{\mu}\left[\frac{1-\rho}{(\rho - \rho_{-i}\mathcal{E}_i)(1-\rho\mathcal{E}_i)} + \frac{1}{1-\rho} + \frac{\rho_{-i}}{\rho_i}\right], \tag{3.33}$$

where

$$\mathcal{E}_i \equiv \frac{1+\rho-\sqrt{(1+\rho)^2 - 4\rho_{-i}}}{2\rho_{-i}}. \tag{3.34}$$

The discrete Markov state $q(t) \in \{0, 1, 2, \ldots\}$ tracks the number of updates in the system at time t. Figure 3.5 shows the evolution of the state for a corresponding blocking system that has a finite occupancy of m. In states $0 < k < m$, an arrival, which results from a rate λ transition, increases the state to $k + 1$. A departure on completion of service, which results from a rate μ transition, reduces the occupancy to $k - 1$. An empty system, $q(t) = 0$, can only see an arrival; while when the system is full and $q(t) = m$, all arrivals are blocked and discarded and only a departure may take place.

Let Δ_i^m be the average age of source i for the size m blocking system. The average age Δ_i of source i at the monitor can be obtained as

$$\Delta_i = \lim_{m\to\infty} \Delta_i^m. \tag{3.35}$$

For the blocking system, $\mathbf{x}(t) = [x_0(t)\ x_1(t)\ \ldots\ x_m(t)]$ is the age state. Here $x_0(t)$ is the age process of source i at the monitor. The age process $x_j(t)$, $1 \leqslant j \leqslant k$, is associated with the packet currently in the jth position in the queue, where $j = 1$ corresponds to the packet in service. As we will explain shortly, $x_j(t)$ is not necessarily the age of the update, since the update may not be of source i. Also, in any discrete state k we must track only the $k + 1$ age processes $x_0(t), x_1(t), \ldots, x_k(t)$. The rest are irrelevant. In state $q(t) = k$, we set $x_j(t) = 0$ for $j > k$. The age state evolves according to

$$\dot{\mathbf{x}}(t) = \mathbf{b}_k \equiv [\mathbf{1}_{k+1}\ \mathbf{0}_{m-k}]. \tag{3.36}$$

We use the shorthand notation $(\mathbf{x})_{j:l}$ and $(\mathbf{v}_q)_{j:l}$, respectively, to denote the vectors $[x_j \cdots x_l]$ and $[v_{q,j}\ \ldots\ v_{q,l}]$. Table 3.5 shows the transition reset maps for the transitions in the Markov chain in Figure 3.5. We also show the age state $\mathbf{x}_{q'}$ and the vector $\mathbf{v}_{q'}$ obtained on transitioning from state q to q'.

Table 3.5 SSH transitions and reset maps for the chain in Figure 3.5.

$q_l \to q'_l$	$\lambda^{(l)}$	$\mathbf{x}\mathbf{A}_l$	$\mathbf{A}_l$	$\mathbf{v}_{q_l}\mathbf{A}_l$
$1 \to 0$	μ	$\begin{bmatrix} x_1 & \mathbf{0} \end{bmatrix}$	$\mathbf{D}_0$	$\begin{bmatrix} v_{1,1} & \mathbf{0} \end{bmatrix}$
$0 \to 1$	λ_i	$\begin{bmatrix} x_0 & 0 & \mathbf{0} \end{bmatrix}$	Λ_1	$\begin{bmatrix} v_{0,0} & 0 & \mathbf{0} \end{bmatrix}$
$0 \to 1$	λ_{-i}	$\begin{bmatrix} x_0 & x_0 & \mathbf{0} \end{bmatrix}$	Γ_1	$\begin{bmatrix} v_{0,0} & v_{0,0} & \mathbf{0} \end{bmatrix}$
$2 \to 1$	μ	$\begin{bmatrix} x_1 & x_2 & \mathbf{0} \end{bmatrix}$	$\mathbf{D}_1$	$\begin{bmatrix} v_{21} & v_{22} & \mathbf{0} \end{bmatrix}$
$\vdots$	$\vdots$	$\vdots$	$\vdots$	$\vdots$
$k-1 \to k$	λ_i	$\begin{bmatrix} (\mathbf{x})_{0:k-1} & 0 & \mathbf{0} \end{bmatrix}$	Λ_k	$\begin{bmatrix} (\mathbf{v}_{k-1})_{0:k-1} & 0 & \mathbf{0} \end{bmatrix}$
$k-1 \to k$	λ_{-i}	$\begin{bmatrix} (\mathbf{x})_{0:k-1} & x_{k-1} & \mathbf{0} \end{bmatrix}$	Γ_k	$\begin{bmatrix} (\mathbf{v}_{k-1})_{0:k-1} & v_{k-1,k-1} & \mathbf{0} \end{bmatrix}$
$k+1 \to k$	μ	$\begin{bmatrix} (\mathbf{x})_{1:k+1} & \mathbf{0} \end{bmatrix}$	$\mathbf{D}_k$	$\begin{bmatrix} (\mathbf{v}_{k+1})_{1:k+1} & \mathbf{0} \end{bmatrix}$
$\vdots$	$\vdots$	$\vdots$	$\vdots$	$\vdots$
$m-1 \to m$	λ_i	$\begin{bmatrix} (\mathbf{x})_{0:m-1} & 0 & 0 \end{bmatrix}$	Λ_m	$\begin{bmatrix} (\mathbf{v}_{m-1})_{0:m-1} & 0 & 0 \end{bmatrix}$
$m-1 \to m$	λ_{-i}	$\begin{bmatrix} (\mathbf{x})_{0:m-1} & x_{m-1} & 0 \end{bmatrix}$	Γ_m	$\begin{bmatrix} (\mathbf{v}_{m-1})_{0:m-1} & v_{m-1,m-1} & 0 \end{bmatrix}$

Consider an arrival that transitions the system into state $0 < k \leqslant m$ from $k-1$ as a result of an arrival from source i. The age processes $x_0(t), x_1(t), \ldots, x_{k-1}(t)$ stay unaffected by the transition and continue to evolve as in $k-1$. That is, $x'_j = x_j$, $0 \leqslant j \leqslant k-1$. In state k we must also track $x_k(t)$. We set $x_k(t) = 0$, since the age of this new update from source i is 0. The transition reset map is given by the $(m+1) \times (m+1)$ matrix:

$$\Lambda_k = \left[\begin{array}{c|c} \underbrace{\begin{matrix} 1 & 0 & \cdots & 0 & 0 \\ 0 & 1 & \cdots & 0 & 0 \\ \vdots & \vdots & \ddots & \vdots & \vdots \\ 0 & 0 & \cdots & 1 & 0 \end{matrix}}_{k \times (k+1)} & \mathbf{0}_{m+1 \times m-k} \\ \mathbf{0} & \end{array}\right]. \tag{3.37}$$

If the arrival is of a source other than i, we must set $x_k(t)$ in a manner such that when this arrival completes service, it must not change the age $x_0(t)$ of i at the monitor. To do so, we set $x_k(t) = x_{k-1}(t)$. As before, $\dot{x}_k(t) = 1$. Note that in this case we are effectively setting $x_k(t)$ to the age of the freshest update of source i, if any, in the system. If there is no update of source i in the system, we are effectively setting $x_k(t)$ to $x_0(t)$, as in this case all relevant age processes $x_0(t), x_1(t), \ldots, x_k(t)$ will be tracking the age at the monitor. The transition reset map is

$$\Gamma_k = \left[\begin{array}{c|c} \underbrace{\begin{matrix} 1 & 0 & \cdots & 0 & 0 \\ 0 & 1 & \cdots & 0 & 0 \\ \vdots & \vdots & \ddots & \vdots & \vdots \\ 0 & 0 & \cdots & 1 & 1 \end{matrix}}_{k \times (k+1)} & \mathbf{0} \\ \mathbf{0} & \end{array}\right]. \tag{3.38}$$

Now consider a departure in state $k+1$. The SSH transitions to state k. The age process $x_1(t)$ in $k+1$ corresponds to the departure. On completion of service, this resets the process $x_0(t)$ at the monitor in state k. Once the packet departs, the one next in queue enters service. Thus the process $x_2(t)$ in state $k+1$ resets $x_1(t)$ in state k and so on. The reset map is given by

$$\mathbf{D}_k = \left[\begin{array}{c|c} \underbrace{\begin{matrix} 0 & 0 & \cdots & 0 \\ 1 & 0 & \cdots & 0 \\ 0 & 1 & \cdots & 0 \\ \vdots & \vdots & \ddots & \vdots \\ 0 & 0 & \cdots & 1 \end{matrix}}_{(k+2)\times(k+1)} & \mathbf{0} \\ \mathbf{0} & \end{array}\right]. \tag{3.39}$$

We can now write the equations given by (3.3a) in Theorem 3.2 for our blocking system.

$$\lambda \bar{\mathbf{v}}_0 = \mathbf{b}_0 \bar{\pi}_0 + \mu \bar{\mathbf{v}}_1 \mathbf{D}_0, \tag{3.40a}$$

$$\begin{aligned}(\lambda + \mu)\bar{\mathbf{v}}_k &= \mathbf{b}_k \bar{\pi}_k + \lambda_i \bar{\mathbf{v}}_{k-1} \Lambda_k \\ &\quad + \lambda_{-i} \bar{\mathbf{v}}_{k-1} \Gamma_k + \mu \bar{\mathbf{v}}_{k+1} \mathbf{D}_k, \quad 0 < k < m,\end{aligned} \tag{3.40b}$$

$$\mu \bar{\mathbf{v}}_m = \mathbf{b}_m \bar{\pi}_m + \lambda_i \bar{\mathbf{v}}_{m-1} \Lambda_m + \lambda_{-i} \bar{\mathbf{v}}_{m-1} \Gamma_m. \tag{3.40c}$$

We extract equations corresponding to the discrete states $\bar{q}$ and the relevant age processes $x_j(t)$ in the states. We have $k+1$ relevant age processes in state $0 < k < m$. For such a state k and $x_j(t)$, $0 \leqslant j \leqslant k$, we have

$$\begin{aligned}(\lambda + \mu)\bar{v}_{k,j} &= \bar{\pi}_k + \lambda \bar{v}_{k-1,j} \\ &\quad + \mu \bar{v}_{k+1,j+1}, \quad 0 \leqslant j \leqslant k-1,\end{aligned} \tag{3.41a}$$

$$(\lambda + \mu)\bar{v}_{k,k} = \bar{\pi}_k + \lambda_{-i} \bar{v}_{k-1,k-1} + \mu \bar{v}_{k+1,k+1}. \tag{3.41b}$$

In state $k = 0$, no departures take place and the only relevant age process is $x_0(t)$. We have

$$\lambda \bar{v}_{0,0} = \bar{\pi}_0 + \mu \bar{v}_{1,1}. \tag{3.42}$$

Lastly, in state $k = M$, arrivals are inconsequential while a departure may take place. All age processes $x_0(t), \ldots, x_m(t)$ are relevant. We have

$$\mu \bar{v}_{m,j} = \bar{\pi}_m + \lambda \bar{v}_{m-1,j}, \quad 0 \leqslant j \leqslant n-1, \tag{3.43a}$$

$$\mu \bar{v}_{m,m} = \bar{\pi}_m + \lambda_{-i} \bar{v}_{m-1,m-1}. \tag{3.43b}$$

The steady-state probability $\bar{\pi}_k$ of k updates in the blocking M/M/1/m system can be obtained by solving (3.2a)–(3.2b). The equations are

$$\lambda \bar{\pi}_0 = \mu \bar{\pi}_1, \tag{3.44a}$$

$$(\lambda + \mu)\bar{\pi}_k = \lambda \bar{\pi}_{k-1} + \mu \bar{\pi}_{k+1}, \; 0 < k < m, \tag{3.44b}$$

$$\mu \bar{\pi}_m = \lambda \bar{\pi}_{m-1}. \tag{3.44c}$$

Combined with the normalization constraint $\sum_{q=0}^{m} \bar{\pi}_q = 1$, (3.44) yields the M/M/1/m stationary distribution

$$\bar{\pi}_k = \left(\frac{1-\rho}{1-\rho^{m+1}}\right)\rho^k, \quad 0 \leqslant k \leqslant m. \tag{3.45}$$

From Theorem 3.2 we know that the average age of source i in the blocking system is

$$\Delta_i^m = \sum_{\bar{q}=0}^{m} \bar{v}_{\bar{q}0}. \tag{3.46}$$

The average age Δ_i, Equation (3.33), is obtained using Equation (3.35). Details regarding calculation of Δ_i^m and Δ_i can be found in [3]. We wrap up the M/M/1 FCFS analysis by observing that Δ_i approaches the age of a single user M/M/1 queue as $\rho_{-i} \to 0$. This is easy to see given the fact that $\lim_{\rho_{-i}\to 0} \mathcal{E}_i = 1/(1+\rho)$.

3.6 Priority Queueing

As before, our system consists of N independent sources that send their status updates to a monitor through a service facility consisting of a single server. However, now each source is assigned a priority by the service facility, and this determines the precedence of its updates with respect to the updates of the other sources. We will index the sources $1, 2, \ldots, N$ and will associate a higher priority with a smaller index. Thus, updates of source 1 have the highest priority.

We will analyze two kinds of service facilities. Firstly, in Section 3.6.1, we will consider a facility that has no waiting room and allows *preemption in service*, which is in essence M/M/1* with sources assigned priorities. Specifically, an update in service is preempted only by an arrival that is either from a source that has higher priority or from its own source. In either case, the fresh update begins service and the preempted update is discarded by the facility. An arrival that finds a higher-priority update in service is discarded.

Secondly, in Section 3.6.2, we will consider a facility whose waiting room can have at most one update waiting. An update that is in service cannot be preempted. However, an update in waiting can be preempted by an arrival of equal or higher priority. In either case, the waiting update is discarded and replaced by the new arrival. Also, an arrival that finds a higher-priority update in waiting is discarded. This system is M/M/1/2* with source priorities.

Let $\hat{\lambda}_i$ and $\check{\lambda}_i$ be, respectively, the sum-rate of arrivals of updates with a priority higher and lower than i. As before, λ is the total rate of arrivals. We have $\hat{\lambda}_i = \sum_{k=1}^{i-1} \lambda_k$, $\check{\lambda}_i = \sum_{k=i+1}^{N} \lambda_k$, and $\lambda = \hat{\lambda}_i + \lambda_i + \check{\lambda}_i$. The load offered by i is $\rho_i = \lambda_i/\mu$, and the total offered load is $\rho = \lambda/\mu$. Define $\hat{\rho}_i = \hat{\lambda}_i/\mu$ and $\check{\rho}_i = \check{\lambda}_i/\mu$.

3.6.1 M/M/1* with Source Priorities

Our status updating system has no waiting room. A status update (packet), if any, must be in service. Evidently, the server may be idle, or it may be busy servicing an update of source i, or of a higher-priority source, or of a source of lower priority than i. We will detail these states of the server and arrive at the two discrete Markov states and the continuous state $\mathbf{x}(t)$ that we will use in the SHS model for the system. As we did for M/M/1*, we will use fake updates. The two states are needed to distinguish between an update from a higher-priority source in service and an update from the source of interest, or of lower priority, in service. We want to calculate the average age Δ_i of source i.

This age was shown in [9] to be

$$\Delta_i = \frac{1 + \rho_i + 3\widehat{\rho}_i + 3\widehat{\rho}_i\rho_i + 3\widehat{\rho}_i^2 + \widehat{\rho}_i^2\rho_i + \widehat{\rho}_i^3}{\mu\rho_i(1 + \widehat{\rho}_i)}. \tag{3.47}$$

Our continuous-state vector is $\mathbf{x}(t) = \begin{bmatrix} x_0(t) & x_1(t) \end{bmatrix}$, where $x_0(t)$ is the age of i at the monitor and $x_1(t)$ is associated with the update in service. If an arrival of i takes place while an update of i is in service, the update in service is preempted by the new arrival. The preemption doesn't change the age $x_0(t)$ of i. However, source i now has a fresh update with age $x_1(t) = 0$ in service. An arrival from a lower-priority source doesn't affect the service, and that from a higher-priority source preempts the update of i and starts service instead.

Now consider when the update in service is of a higher-priority source. The completion of service of such an update does not change the age $x_0(t)$ of i. Therefore, we do not need to track the age of such an update. That is, $x_1(t)$ is *irrelevant* and we set $x_1(t) = 0$ and its rate of change to 0. Any arrivals of i that take place are discarded during such a service and don't influence the age process of i.

As far as source i is concerned, there is no difference between an idle server and one that is processing lower-priority updates. In either state, an arrival of i or that of a higher-priority source enters service immediately. In addition, the only difference between these and when there is an update of i in service is that on completion of service, an update of i resets the age $x_0(t)$ of i at the monitor to its own age.

All these three, an idle server, or one that is serving source i, or serving a lower-priority source, can be subsumed into one state using fake updates. Fake updates, as any other, are serviced at a rate μ. During their service, as for a real update of i, their age is tracked by $x_1(t)$. However, when they enter service they have age $x_0(t)$. Thus, they don't reset the age $x_0(t)$ at the monitor on completion of service. The server is processing a fake update when it is not serving an update from either i or a higher-priority source. Like any update of i, a fake update is preempted by an arrival of i or of a higher-priority source. Completion of service of a fake update by the server is followed by the beginning of service of another one.

We have just two discrete states: state $\widehat{1}$ in which a higher-priority update is being serviced, and state 1 in which either an update of source i or a fake update is being serviced. Updates from sources with a lower priority than i are discarded in both the states. Note that we must maintain a distinct discrete state $\widehat{1}$ for service of higher-priority updates because, unlike in 1, arrivals of i are discarded in $\widehat{1}$.

Table 3.6 Table of transitions for the chain in Figure 3.6.

l	$(q_l, q'_l, \lambda^{(l)})$	$\mathbf{x}\mathbf{A}_l$	$\mathbf{A}_l$	$\mathbf{v}_{q_l}\mathbf{A}_l$
1	$(1, 1, \lambda_i)$	$[x_0 \ \ 0]$	$\begin{bmatrix} 1 & 0 \\ 0 & 0 \end{bmatrix}$	$[v_{10} \ \ 0]$
2	$(1, 1, \mu)$	$[x_1 \ \ x_1]$	$\begin{bmatrix} 0 & 0 \\ 1 & 1 \end{bmatrix}$	$[v_{11} \ \ v_{11}]$
3	$(1, \hat{1}, \hat{\lambda}_i)$	$[x_0 \ \ 0]$	$\begin{bmatrix} 1 & 0 \\ 0 & 0 \end{bmatrix}$	$[v_{10} \ \ 0]$
4	$(\hat{1}, 1, \mu)$	$[x_0 \ \ x_0]$	$\begin{bmatrix} 1 & 1 \\ 0 & 0 \end{bmatrix}$	$[v_{\hat{1}0} \ \ v_{\hat{1}0}]$

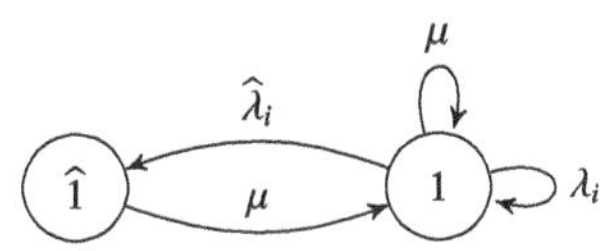

Figure 3.6 The SHS Markov chain for source i for the case when an update may be preempted in service. In state $\hat{1}$, a higher-priority update is in service. In 1, either an update of i or a fake update is in service.

The SHS Markov chain for source i is shown in Figure 3.6. The transitions and their reset maps are shown in Table 3.6. The rate of change of the age processes is given by

$$\dot{\mathbf{x}}(t) = \mathbf{b}_q = \begin{cases} [1 \ \ 1] & q = 1, \\ [1 \ \ 0] & q = \hat{1}. \end{cases} \tag{3.48}$$

The transitions are

$l = 1$: This transition occurs when an update of source i arrives while another update of the source, either fake or real, is in service. The new update preempts the current update and begins service. This sets $x_1(t) = 0$. The age $x_0(t)$ is unaffected.

$l = 2$: An update (fake or real) of source i finishes service. The age at the monitor $x_0(t)$ is reset to the age $x_1(t)$ of this update. The age $x_1(t)$ stays as is and is the age of the fake update that starts service on the completion of the previous service.

$l = 3$: An update (fake or real) of i is preempted by a higher-priority update. This doesn't reset $x_0(t)$ but makes $x_1(t)$ *irrelevant*. Hence, we set $x_1(t) = 0$.

$l = 4$: A higher-priority update completes service. This doesn't reset the age $x_0(t)$. Note that this transition into 1 marks the beginning of service of a fake update. So we set $x_1(t) = x_0(t)$.

We can now calculate the steady-state probabilities of the discrete states using Equations (3.2a)–(3.2b). They are $\bar{\pi}_1 = \mu/(\hat{\lambda}_i + \mu)$ and $\bar{\pi}_{\hat{1}} = \hat{\lambda}_i/(\hat{\lambda}_i + \mu)$.

Next, we calculate the $\bar{v}_{\bar{q}}$ for the two discrete states. Equation (3.3a) in Theorem 3.2 gives us the required system of equations. Substituting the rates of change of the continuous state given in (3.48), we obtain the following equations, one for each discrete state:

$$(\widehat{\lambda}_i + \lambda_i + \mu)[v_{10}\ v_{11}] = [1\ 1]\bar{\pi}_1 + [v_{10}\ 0]\lambda_i + [v_{11}\ v_{11}]\mu + [v_{\widehat{1}0}\ v_{\widehat{1}0}]\mu, \tag{3.49a}$$

$$\mu[v_{\widehat{1}0}\ v_{\widehat{1}1}] = [1\ 0]\bar{\pi}_{\widehat{1}} + [v_{10}\ 0]\widehat{\lambda}_i. \tag{3.49b}$$

Note that $v_{\widehat{1}1} = 0$. This is because $x_1(t)$ is irrelevant in state $\widehat{1}$. Solving for the rest gives us $v_{10} = (\mu + \lambda_i + \widehat{\lambda}_i)/(\mu\lambda_i)$, $v_{11} = (\mu + \widehat{\lambda}_i)/(\mu\lambda_i)$, and $v_{\widehat{1}0} = (\widehat{\lambda}_i)/(\mu(\widehat{\lambda}_i + \mu)) + (\widehat{\lambda}_i(\mu + \widehat{\lambda}_i + \lambda_i))/(\mu^2\lambda_i)$. These are all nonnegative. We use Theorem 3.2 to calculate the AoI $\Delta_i = v_{10} + v_{\widehat{1}0}$, which gives the expression given by (3.47). We end the analysis by making a few observations based on the obtained age.

Consider source 1. Since its arrivals can preempt any update in service, the source sees a $M/M/1$ queue with preemption in service. We have $\widehat{\rho}_1 = 0$. Substituting in (3.47), we get

$$\Delta_1 = \frac{1}{\mu}\left(1 + \frac{1}{\rho_1}\right),$$

which is in accordance with the age of a single source ($\rho = \rho_1$) in a M/M/1* queueing system.

Finally, consider source $i > 1$. From (3.47), observe that for given finite ρ_j, $1 \leqslant j < i$, Δ_i is nonincreasing in ρ_i. Therefore,

$$\lim_{\rho_i \to \infty} \Delta_i = \frac{1}{\mu}\left(1 + \widehat{\rho}_i + \frac{\widehat{\rho}_i}{1 + \widehat{\rho}_i}\right)$$

is the minimum age of information that source i may achieve for a given $\widehat{\rho}_i$. In the preceding limit and for $\widehat{\rho}_i \gg 1$, Δ_i increases linearly in $\widehat{\rho}_i$ and is $\approx \widehat{\rho}_i/\mu$.

3.6.2 M/M/1/2* With Source Priorities

An update of a source that is in waiting is preempted by an arrival of itself or that of a higher-priority source. However, an update in service cannot be preempted and always completes service. We show how the average age Δ_i is calculated. It was found in [9] to be

$$\Delta_i = v_{00} + v_{10} + v_{20} + v_{\widehat{2}0}, \text{ where} \tag{3.50a}$$

$$v_{00} = \frac{(\rho_i(1 + \widehat{\rho}_i)(1 + \rho) - \rho\rho_i(\rho - \widehat{\rho}_i))\bar{\pi}_0 + (1 + \widehat{\rho}_i)^2(1 + \widehat{\rho}_{i+1})}{\mu\rho\rho_i(1 + \widehat{\rho}_i)(1 + \rho) + \rho_i(1 + \widehat{\rho}_{i+1})(1 + \widehat{\rho}_i)^2}, \tag{3.50b}$$

$$v_{10} = \rho v_{00} - \frac{\bar{\pi}_0}{\mu} + \frac{1 + \widehat{\rho}_i}{\mu(1 + \rho)}, \tag{3.50c}$$

$$v_{20} = \frac{\rho(\rho - \widehat{\rho}_i)}{1 + \widehat{\rho}_i} v_{00} - \frac{(\rho - \widehat{\rho}_i) - (\rho - \widehat{\rho}_i)^2}{\mu(1 + \rho_i)^2}\bar{\pi}_0 + \frac{\rho - \widehat{\rho}_i}{\mu(1 + \rho)}, \tag{3.50d}$$

$$v_{\widehat{2}0} = \frac{\rho\widehat{\rho}_i(1 + \rho)}{1 + \widehat{\rho}_i} v_{00} + \frac{\rho\widehat{\rho}_i}{\mu(1 + \widehat{\rho}_i)}\left(\frac{\rho - \widehat{\rho}_i}{1 + \widehat{\rho}_i} - \frac{1 - \rho^2}{\rho}\right)\bar{\pi}_0 + \frac{\widehat{\rho}_i}{\mu}. \tag{3.50e}$$

There can be at most two updates in the system, one in service and the other in waiting. Thus the age vector $\mathbf{x}(t) = [x_0(t)\ x_1(t)\ x_2(t)]$, where $x_1(t)$ is associated with the update in service and $x_2(t)$ is associated with the update waiting for service.

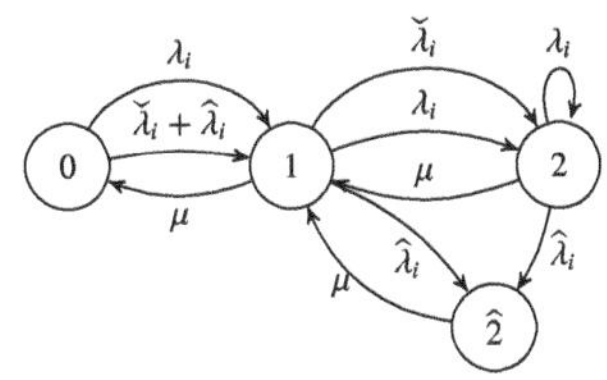

Figure 3.7 The SHS Markov chain for updates of source i for when preemption is allowed only in waiting.

An idle server can receive an arrival from any source. An arrival of source i updates the age $x_1(t) = 0$. Arrivals from other sources must not update the age of i on completion of service. With a careful resetting of $x_1(t)$ we don't need to track the source of the update. Specifically, when an arrival of a source other than i arrives to an idle server, we set $x_1(t) = x_0(t)$. This ensures that when the arrival completes service, it leaves the age $x_0(t)$ of source i unaffected. Also, since there is no update in waiting, the age $x_2(t)$ is irrelevant.

While there is an update in waiting of a source of higher priority than i, new arrivals of i will be discarded. The age $x_2(t)$ is irrelevant, as when such an update enters service, it does not update the age $x_1(t)$.

If the update in waiting is that of source i or of a lower-priority source, an arrival of i will preempt the update and instead occupy the waiting room. We do not need to distinguish between these two states if we set $x_2(t) = x_1(t)$ when a lower-priority update occupies the waiting room. On the other hand, when an update of i enters waiting, we set $x_2(t) = 0$.

The SHS Markov chain that models our system is shown in Figure 3.7. It has the discrete states $0, 1, 2, \widehat{2}$. The state 0 corresponds to an idle server. In state 1 there is exactly one update that is in service. In state 2, there is an update of source i or a lower-priority source in waiting. In state $\widehat{2}$, the update in waiting is of a higher-priority source.

The rate of change of the age vector in each of the discrete states is given by

$$\dot{\mathbf{x}}(t) = \mathbf{b}_q = \begin{cases} [1 \quad 0 \quad 0] & q = 0, \\ [1 \quad 1 \quad 0] & q = 1, \widehat{2}, \\ [1 \quad 1 \quad 1] & q = 2. \end{cases} \tag{3.51}$$

Table 3.7 lists the transitions and the reset maps. The transitions are

$l = 1$: This transition occurs when a source i update arrives to an empty system. It sets $x_1(t) = 0$. As no update is in waiting in state 1, $x_2(t)$ is *irrelevant*. The age $x_0(t)$ of source i is unaffected.

$l = 2$: An update of a source other than i arrives to an empty system. The discrete state transitions from 0 to 1. For reasons explained earlier, we set $x_1(t) = x_0(t)$.

$l = 3$: The only update in the system completes service. The age of the source at the monitor must be set to the age of the update. We set $x_0(t) = x_1(t)$. The ages $x_1(t)$ and $x_2(t)$ are irrelevant in state 0 and are set to 0.

Table 3.7 Table of transitions for the chain in Figure 3.7.

l	$(q_l, q'_l, \lambda^{(l)})$	$\mathbf{x}\mathbf{A}_l$	$\mathbf{A}_l$	$\mathbf{v}_{q_l}\mathbf{A}_l$
1	$(0,1,\lambda_i)$	$[x_0\ 0\ 0]$	$\begin{bmatrix}1&0&0\\0&0&0\\0&0&0\end{bmatrix}$	$[v_{00}\ 0\ 0]$
2	$(0,1,\check{\lambda}_i+\hat{\lambda}_i)$	$[x_0\ x_0\ 0]$	$\begin{bmatrix}1&1&0\\0&0&0\\0&0&0\end{bmatrix}$	$[v_{00}\ v_{00}\ 0]$
3	$(1,0,\mu)$	$[x_1\ 0\ 0]$	$\begin{bmatrix}0&0&0\\1&0&0\\0&0&0\end{bmatrix}$	$[v_{11}\ 0\ 0]$
4	$(1,2,\check{\lambda}_i)$	$[x_0\ x_1\ x_1]$	$\begin{bmatrix}1&0&0\\0&1&1\\0&0&0\end{bmatrix}$	$[v_{10}\ v_{11}\ v_{11}]$
5	$(1,2,\lambda_i)$	$[x_0\ x_1\ 0]$	$\begin{bmatrix}1&0&0\\0&1&0\\0&0&0\end{bmatrix}$	$[v_{10}\ v_{11}\ 0]$
6	$(2,1,\mu)$	$[x_1\ x_2\ 0]$	$\begin{bmatrix}0&0&0\\1&0&0\\0&1&0\end{bmatrix}$	$[v_{21}\ v_{22}\ 0]$
7	$(1,\hat{2},\hat{\lambda}_i)$	$[x_0\ x_1\ 0]$	$\begin{bmatrix}1&0&0\\0&1&0\\0&0&0\end{bmatrix}$	$[v_{10}\ v_{11}\ 0]$
8	$(\hat{2},1,\mu)$	$[x_1\ x_1\ 0]$	$\begin{bmatrix}0&0&0\\1&1&0\\0&0&0\end{bmatrix}$	$[v_{\hat{2}0}\ v_{\hat{2}1}\ 0]$
9	$(2,\hat{2},\hat{\lambda}_i)$	$[x_0\ x_1\ 0]$	$\begin{bmatrix}1&0&0\\0&1&0\\0&0&0\end{bmatrix}$	$[v_{20}\ v_{21}\ 0]$
10	$(2,2,\lambda_i)$	$[x_0\ x_1\ 0]$	$\begin{bmatrix}1&0&0\\0&1&0\\0&0&0\end{bmatrix}$	$[v_{20}\ v_{21}\ 0]$

$l=4$: An update of a source of priority lower than i arrives to a system in which an update is in service but none is waiting. While $x_0(t)$ and $x_1(t)$ stay unaffected by the transition, as explained earlier, we set $x_2(t) = x_1(t)$.

$=5$: Similar to the transition above. However, the update entering waiting is from source i. So we set $x_2(t) = 0$.

$l=6$: The update in service exits and that in waiting enters service. We set $x_0(t) = x_1(t)$, $x_1(t) = x_2(t)$, and $x_2(t) = 0$ (irrelevant).

$l=7$: An update of a source of priority higher than i arrives to a busy server. This transition doesn't impact $x_0(t)$ and $x_1(t)$. Also, in state $\hat{2}$, $x_2(t) = 0$ is irrelevant.

$l=8$: An update leaves service. We set $x_0(t) = x_1(t)$. It is replaced by an update of a source of priority higher than i and so $x_1(t)$ stays unaffected. In state 1, $x_2(t)$ is irrelevant.

$l = 9$: An update of source i or a lower-priority source is preempted in waiting by a higher-priority update. While this does not impact $x_0(t)$ and $x_1(t)$, we set $x_2(t) = 0$ because $x_2(t)$ is irrelevant in $\widehat{2}$.

$l = 10$: An update from i preempts the update in waiting, resulting in a fresh update of i in waiting. We set $x_2(t) = 0$.

The steady-state probabilities of the discrete states calculated using Equation (3.2) are

$$\bar{\pi}_0 = \frac{1}{1+\rho+\rho^2}, \qquad \bar{\pi}_1 = \rho\bar{\pi}_0,$$
$$\bar{\pi}_2 = \rho\frac{\rho-\widehat{\rho}_i}{1+\widehat{\rho}_i}\bar{\pi}_0, \qquad \bar{\pi}_{\widehat{2}} = \widehat{\rho}_i\rho\frac{1+\rho}{1+\widehat{\rho}_i}\bar{\pi}_0. \tag{3.52}$$

The system of equations in (3.3a) in Theorem 3.2, with substitutions from (3.51), can be written as

$$\rho v_{00} = \bar{\pi}_0/\mu + v_{11}, \tag{3.53a}$$
$$(1+\rho)v_{10} = \bar{\pi}_1/\mu + \rho v_{00} + v_{21} + v_{\widehat{2}0}, \tag{3.53b}$$
$$(1+\rho)v_{11} - v_{\widehat{2}1} = \bar{\pi}_1/\mu + (\rho-\rho_i)v_{00} + v_{22}, \tag{3.53c}$$
$$(1+\widehat{\rho}_i)v_{20} = \bar{\pi}_2/\mu + (\rho-\widehat{\rho}_i)v_{10}, \tag{3.53d}$$
$$(1+\widehat{\rho}_i)v_{21} = \bar{\pi}_2/\mu + (\rho-\widehat{\rho}_i)v_{11}, \tag{3.53e}$$
$$(1+\rho_i+\widehat{\rho}_i)v_{22} = \bar{\pi}_2/\mu + (\rho-\rho_i-\widehat{\rho}_i)v_{11}, \tag{3.53f}$$
$$v_{\widehat{2}0} = \bar{\pi}_{\widehat{2}}/\mu + \widehat{\rho}_i v_{10} + \widehat{\rho}_i v_{20}, \tag{3.53g}$$
$$v_{\widehat{2}1} = \bar{\pi}_{\widehat{2}}/\mu + \widehat{\rho}_i v_{11} + \widehat{\rho}_i v_{21}. \tag{3.53h}$$

The terms v_{01}, v_{02}, v_{12}, and $v_{\widehat{2}2}$ are all zero, as they correspond to irrelevant age processes. Theorem 3.2 can now be used to calculate the AoI of source i. Next, we make a few observations using Δ_i, given by Equations (3.50a)–(3.50e)

Consider a system that receives updates from a single source $i = 1$. The source sees an M/M/1 queue with preemption in waiting. We have $\rho_i = \rho$, $\widehat{\rho}_i = 0$, and $\widehat{\rho}_{i+1} = \widehat{\rho}_i + \rho_i = \rho$. Average age obtained using the equations is that of a single-source M/M/1/2* queue.

Further, observe that Δ_i is nondecreasing in ρ_i. For fixed rates of arrivals of the other sources, in the limit as $\rho_i \to \infty$, using Equations (3.50a)–(3.50e), we get $v_{00} = v_{10} = \bar{\pi}_0 = 0$, and the age converges to the limit $\frac{2+4\widehat{\rho}_i+4\widehat{\rho}_i^2+\widehat{\rho}_i^3}{\mu(1+\widehat{\rho}_i)^2}$. Note that this age is only a function of the sum-rate of updates offered by higher-priority sources and is not affected by lower-priority arrivals. This is because as $\rho_i \to \infty$, there is always an update in waiting ($\bar{\pi}_0, \bar{\pi}_1 \to 0$) and it is either of a higher-priority source or that of source i. The difference between this limiting age and the corresponding age for when preemption is allowed in service (Section 3.6.1) is $(1/\mu)(1/(1+\widehat{\rho}_i^2))$. However, note that this difference vanishes as $\widehat{\rho}_i$ becomes large. This is because at large $\widehat{\rho}_i$, in both systems, the limiting age of i is $\approx \widehat{\rho}_i/\mu$.

We end with a comparison, in Figure 3.8, of the average age achieved by three sources for M/M/1* and M/M/1/2*, when the sources have different priorities and

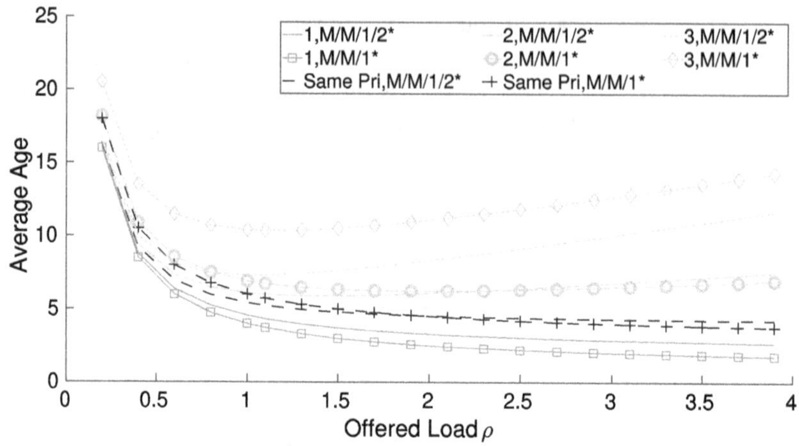

Figure 3.8 Average age of sources 1, 2, and 3 as a function of ρ. Each source offers a load of $\rho/3$. We show the age for when the sources have different priorities and when they have the same priority (Sections 3.3 and 3.4). Average age is shown for the M/M/1* and the M/M/1/2* queue. The age plots corresponding to when the sources have the same priority are denoted in the legend by Same Pri, M/M/1* and Same Pri, M/M/1/2*. The service rate is $\mu = 1$.

when they all have the same priority. The sources offer equal load. That is, $\rho_i = \rho/3$, for $i = 1, 2, 3$. Observe that source 1 benefits as a result of having the highest priority and its average age is reduced in comparison to when all sources have the same priority. However, the other sources suffer. Source 3, which has the least priority, does worse when preemption is allowed in service (M/M/1*) in comparison to when it is allowed only in waiting (M/M/1/2*). This is likely because more of its updates complete service in the latter.

3.7 Summary

In this chapter, we summarized works on multiple sources sharing a service facility with a single server. We provided a simplified explanation of the SHS for AoI approach to calculate the average age of updates of any source at the monitor. We demonstrated the approach for queueing systems including FCFS, M/M/1*, and M/M/1/2*, and the latter two with and without source priorities.

References

[1] R. D. Yates and S. Kaul, "Real-time status updating: Multiple sources," in *2012 IEEE International Symposium on Information Theory Proceedings*, 2012, pp. 2666–2670.

[2] R. D. Yates and S. K. Kaul, "The age of information: Real-time status updating by multiple sources," *IEEE Transactions on Information Theory*, vol. 65, no. 3, pp. 1807–1827, 2019.

[3] S. K. Kaul and R. D. Yates, "Timely updates by multiple sources: The M/M/1 queue revisited," in *2020 54th Annual Conference on Information Sciences and Systems (CISS)*, 2020, pp. 1–6.

[4] M. Moltafet, M. Leinonen, and M. Codreanu, "On the age of information in multi-source queueing models," *IEEE Transactions on Communications*, vol. 68, no. 8, pp. 5003–5017, 2020.

[5] M. Moltafet, M. Leinonen, and M. Codreanu, "Average age of information for a multi-source m/m/1 queueing model with packet management," in *2020 IEEE International Symposium on Information Theory (ISIT)*, 2020, pp. 1765–1769.

[6] M. Moltafet, M. Leinonen, and M. Codreanu, "Average age of information in a multi-source m/m/1 queueing model with lcfs prioritized packet management," in *IEEE INFOCOM 2020 - IEEE Conference on Computer Communications Workshops (INFOCOM WKSHPS)*, 2020, pp. 303–308.

[7] L. Huang and E. Modiano, "Optimizing age-of-information in a multi-class queueing system," in *2015 IEEE International Symposium on Information Theory (ISIT)*, 2015, pp. 1681–1685.

[8] E. Najm and E. Telatar, "Status updates in a multi-stream m/g/1/1 preemptive queue," in *IEEE INFOCOM 2018 – IEEE Conference on Computer Communications Workshops (INFOCOM WKSHPS)*, 2018, pp. 124–129.

[9] S. K. Kaul and R. D. Yates, "Age of information: Updates with priority," in *2018 IEEE International Symposium on Information Theory (ISIT)*, 2018, pp. 2644–2648.

[10] A. Maatouk, M. Assaad, and A. Ephremides, "Age of information with prioritized streams: When to buffer preempted packets?" in *2019 IEEE International Symposium on Information Theory (ISIT)*, 2019, pp. 325–329.

[11] J. Xu and N. Gautam, "Peak age of information in priority queueing systems," *IEEE Transactions on Information Theory*, vol. 67, no. 1, pp. 373–390, 2021.

[12] E. Najm, R. Nasser, and E. Telatar, "Content based status updates," *IEEE Transactions on Information Theory*, vol. 66, no. 6, pp. 3846–3863, 2020.

4 Controlling the Age of Information: Buffer Size, Deadlines, and Packet Management

Sastry Kompella and Clement Kam

The *Age of Information (AoI)* metric, which is a measure of the freshness of a continually updating piece of information as observed at a remote monitor, has been studied for a variety of different update monitoring systems. In this chapter, we introduce three network control mechanisms for controlling the age, namely, buffer size, packet deadlines and packet management. In the case of packet deadlines, we analyze an update monitoring system for the cases of a fixed deadline and a random exponential deadline and derive closed-form expressions for the average age. We also derive a closed-form expression for the optimal average deadline for the random exponential case.

4.1 Introduction

We consider applications in which the goal is to continually communicate the most recently updated state of some time-varying process to a monitor. For example, a device regularly transmits packets containing some status (e.g., sensor data, list of neighboring nodes) to a network manager such that the observed status at the network manager stays relatively fresh at all times. In this chapter, we focus on the metric of *status age* or the *age of information* for a system in which updates randomly pass through a queue. We define *age* at the time of observation as the current (observation) time minus the time at which the observed state was generated, and it directly applies to this objective of achieving timely updating in a way that traditional metrics (e.g., delay, throughput) do not [1–4].

Research on the age metric has focused on optimizing the performance of systems that are modeled by different types of queues, with various arrival/departure processes, number of servers, and queue capacities. In particular, it was shown in [2] that deterministic arrival and departure processes achieve a lower average age than memoryless processes. In [5, 6], it was shown that the average age decreases as the number of servers increases. There has been other work that suggested that the age decreases as the queue capacity decreases or when packets in the queue are replaced with newer packets [7, 8, 9]. Specifically, it was shown in [7] that the age with a system capacity of one or two can be much lower than that of a system with infinite capacity, and the ability to replace packets in the buffer when newer packets arrive does even better. Other related works on age of information include analysis of the average (peak) age for an M/M/1 queue with packet errors [10], the average age for a system with gamma-distributed service time [11], the average age in a multi-access channel [12], and the stationary distribution of the age and peak age in first-come, first-served queues [13]. The optimal sampling strategy for minimizing the age in a delay system was studied in [14]. It has also been shown that a preemptive last-generated, first-served queueing policy optimizes the age, throughput, and delay in a single-hop, multi-server system [15], and it optimizes the age in a multi-hop system [16].

Aside from the various queue models and policies, in this chapter, we would like to uncover and understand other mechanisms for optimizing the age for different queues. One mechanism to study is the size of the buffer, which has only been studied for 0, 1, and infinity. It was somewhat counterintuitive to see in [7] that a smaller buffer size may not always be better, since for some smaller values of packet arrival rate λ, the average age is lower for a buffer size of 1 than 0, but for larger λ, it is better to have a buffer size of 0. Thus, it would be interesting to investigate other buffer sizes and see which is optimal under different conditions.

Another mechanism we are interested in studying is the use of a packet deadline to discard packets in the buffer. In this chapter, we study the age metric when imposing a deadline on data packets that are waiting in a queue, such that they are dropped from the system when the deadline expires. Intuitively, a deadline that is too short would have more packets expiring, leading to less frequent updates at the monitor and a larger average age. However, a deadline that is too long would not discard packets that grow very stale in the queue, resulting in the inefficient use of server resources on old packets, leading to an increase in the average age. Our prior work [17] studies the impact of this type of packet deadline for a buffer size of 1, and a properly chosen deadline was shown to improve the average age when compared with not using a deadline. Further study into the age with a packet deadline for various buffer sizes as well as random deadlines, packet control in the server, and *packet replacement* are of interest in this chapter.

4.2 System Model

We study a system in which a source transmits packets to a monitor through an M/M/1/K queue, where the last entry in the Kendall notation describes a total capacity of $K-1$ packets in the queue and one packet in service. As an example, a plot of the age of information for an M/M/1/K system for $K=2$ is shown in Figure 4.1, where transmissions occur at times $t_1, t_2, \ldots$, and receptions at the monitor occur at times $t'_1, t'_2, \ldots$. Typically, an arriving packet that encounters a full-capacity system never enters the system and is dropped. We refer to the time between packet generations as the inter-arrival time $X_i, i = 2, 3, \ldots$, which is equal to $t_i - t_{i-1}$. The inter-arrival times are modeled as random; consequently, the source does not have control over the exact times at which it can transmit updates. Here, the X_i are i.i.d. exponential random variables with rate λ.

We call the time spent in the server by packet k the service time $S_k, k = 1, 2, \ldots$, which is equal to $t'_k - t_k$. The service time S_k is modeled as exponential with rate μ,

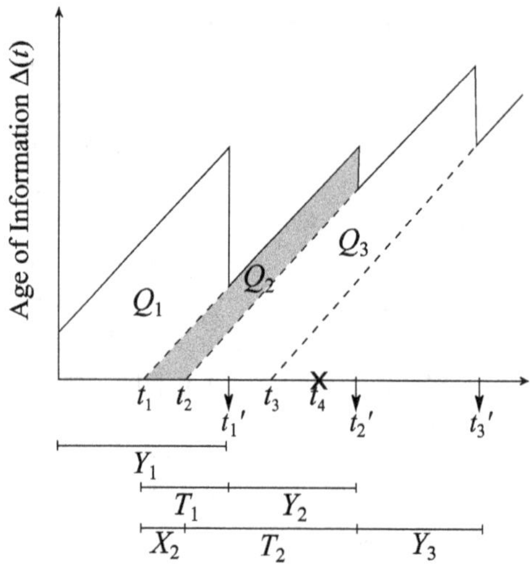

Figure 4.1 Age of information for an M/M/1/2 system.

and all the S_k are i.i.d. and independent of the X_i. The total time spent in the system from arrival to service is given by $T_k, k = 1, 2, \ldots$, where $T_k = W_k + S_k$, with W_k being the time spent waiting in the queue. We also define the *inter-departure time* Y_k as the time between the instants of complete service for the $k-1$st packet served and the kth packet served. This will be useful in the computation of the average age.

The *age of information* at time t is defined as $\Delta(t) = t - u(t)$ [2], where $u(t)$ is the time stamp of the most recent information at the receiver as of time t. Given this definition, we can see that the age increases linearly with t but is reset to a smaller value with each packet received that contains newer information, resulting in the saw-tooth pattern shown in Figure 4.1. Assuming that the age process is ergodic, the *time average age* in an observation interval $(0, \tau)$ can be calculated using time averaging as follows:

$$\Delta_\tau = \frac{1}{\tau} \int_0^\tau \Delta(t) dt, \tag{4.1}$$

with the integration providing the area under the curve $\Delta(t)$. It can be noticed from Figure 4.1 that the area can be divided into a series of trapezoids identified by $Q_k,\ k = 1, 2, ..., I(\tau)$, where $I(\tau)$ is the index of the most recently received update. Summing the areas under all $Q_k,\ k = 1, 2, ..., I(\tau)$, we can write Δ_τ as follows:

$$\begin{aligned}\Delta_\tau &= \frac{1}{\tau}\Big(Q_1 + \sum_{k=2}^{I(\tau)} Q_k + \check{Q}\Big)\\ &= \frac{Q_1 + \check{Q}}{\tau} + \frac{I(\tau)-1}{\tau}\frac{1}{I(\tau)-1}\sum_{k=2}^{I(\tau)} Q_k,\end{aligned}$$

where $\check{Q}$ is the partial area that needs to be taken into consideration in the case that $\tau > t'_{I(\tau)}$. In order to compute the *average age* Δ, we extend τ to ∞, that is,

$$\Delta = \lim_{\tau \to \infty} \Delta_\tau .$$

Noting that as $\tau \to \infty$, the ratio $\frac{I(\tau)}{\tau}$ converges to the average rate of transmitted packets given by λ_e, and given the ergodicity of Q_k, the average age can be rewritten as

$$\begin{aligned}\Delta &= \lambda_e E[Q_k] \\ &= \lambda_e \Big(\frac{1}{2} E[(T_{k-1} + Y_k)^2] - \frac{1}{2} E[T_k^2] \Big) \\ &= \lambda_e \Big(\frac{1}{2} E[Y_k^2] + E[T_{k-1} Y_k] \Big). \end{aligned} \tag{4.2}$$

The second equality is based on the fact that the area of Q_k can be computed by taking the area of the bigger triangle $(T_{k-1} + Y_k)$ and subtracting the area of the smaller triangle with sides T_k, while the third equality is based on the fact that T_{k-1} and T_k are identically distributed for $k \geq 3$.

4.3 Effect of Buffer Size

We first consider the impact of the buffer size on the age of information. This was studied for a buffer size of 0 and 1 in [8], which can be analyzed using a graphical argument to arrive at the expression (4.2). Solving for the various terms in (4.2), the average age for the M/M/1/1 and M/M/1/2 were shown to be

$$\Delta_{M/M/1/1} = \frac{1}{\lambda} + \frac{2}{\mu} - \frac{1}{\lambda + \mu} = \frac{1}{\mu} \Big(\frac{1}{\rho} + 2 - \frac{1}{\rho + 1} \Big) \tag{4.3}$$

$$\begin{aligned}\Delta_{M/M/1/2} &= \frac{1}{\lambda} + \frac{3}{\mu} - \frac{2(\lambda + \mu)}{\lambda^2 + \lambda\mu + \mu^2} \\ &= \frac{1}{\mu} \Big(\frac{1}{\rho} + 3 - \frac{2(\rho + 1)}{\rho^2 + \rho + 1} \Big). \end{aligned} \tag{4.4}$$

From (4.3) and (4.4), we determine here the values of $\rho = \lambda/\mu$ for which the age $\Delta_{M/M/1/1}$ is less than $\Delta_{M/M/1/2}$ (and vice versa):

$$\begin{aligned}\Delta_{M/M/1/1} < \Delta_{M/M/1/2} \iff \qquad 2 - \frac{1}{\rho + 1} &< 3 - \frac{2(\rho + 1)}{\rho^2 + \rho + 1} \\ \frac{2(\rho + 1)}{\rho^2 + \rho + 1} - \frac{1}{\rho + 1} &< 1 \\ 0 &< \rho(\rho^2 + \rho - 1). \end{aligned}$$

For $\rho > 0$, the inequality is true when $\rho > (-1 + \sqrt{5})/2 \approx 0.618$. That is, when $\rho > 0.618$, M/M/1/1 achieves a lower age than M/M/1/2, since packets are arriving frequently enough relative to the service rate. However, when $\rho < 0.618$, M/M/1/2 achieves a lower age since packets do not arrive frequently enough, and it helps to

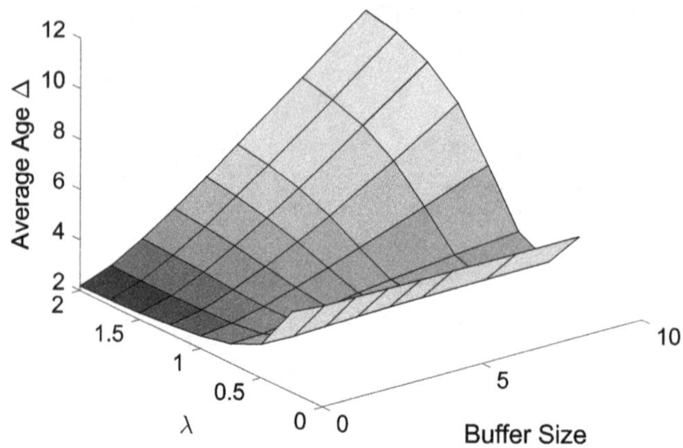

Figure 4.2 Average age for M/M/1/K and various arrival rates λ, $\mu = 1$.

have an update packet stored in the buffer to be transmitted. From this analysis of the average age of the M/M/1/1 and M/M/1/2, we have demonstrated that the relationship between age and buffer size is not simple, in that while having a smaller buffer reduces the waiting time, this does not necessarily achieve a lower age.

Given the complexity of solving for the average age for the M/M/1/1 and M/M/1/2 systems and the certain complexity of solving for the average age of the general M/M/1/K system, we use simulation to find the age for an M/M/1/K system, for various values of K. The results are plotted in Figure 4.2 for $\mu = 1$. We again see that for lower packet arrival rate λ, increasing the buffer size actually leads to a slight decrease in the average age; but for larger λ, larger buffer sizes have a more detrimental impact on the average age.

If we fix the buffer size to be 0 (i.e., M/M/1/1), the average age is strictly decreasing in λ since there is no waiting in the buffer. For buffer sizes greater than 0, the average age initially decreases in λ, but eventually starts to increase since at some point packets are arriving frequently enough that it is not necessary to hold any in the buffer. This effect is more significant for larger buffer sizes; but even for the M/M/1/2 case, there is a point at which the average age starts increasing. We can determine that point as follows:

$$\frac{\partial \Delta_{M/M/1/2}}{\partial \rho} = -\frac{1}{\rho} - \frac{2}{\rho^2+\rho+1} + \frac{2(\rho+1)(2\rho+1)}{(\rho^2+\rho+1)^2)} = 0$$
$$\rho^4 + 2\rho^3 - 3\rho^2 - 2\rho - 1 = 0.$$

For nonnegative and noncomplex values of ρ, we can see that the last equality is true for $\rho \approx 1.427$. In our simulations, the minimum among $\lambda = 0.25, 0.5, \ldots, 1.75, 2$ occurs at $\lambda = 1.5$. As the buffer size increases, the value of λ at which the minimum age occurs decreases, since this prevents the buffer from filling up with older packets.

The optimum average ages for $\lambda = 0.25$, 0.5, 1, and 1.5 and the optimum buffer size are provided in Table 4.1. For $\lambda = 0.25$, a buffer size of 10 is optimal, but for

Table 4.1 Optimum average age, effect of buffer size.

λ	Minimum Age	Optimum Buffer Size	Improvement vs. M/M/1/1	% improvement vs. M/M/1/2
0.25	5.0698	10	2.49%	0.51%
0.5	3.2850	1	1.28%	–
1	2.4997	0	–	6.19%
1.5	2.2705	0	–	13.13%
2	2.1628	0	–	18.01%

$\lambda = 0.5$, the optimum buffer size quickly goes to 1, and then to 0 for larger values of λ. (Recall, M/M/1/1 is better than M/M/1/2 for $\rho > 0.618$.) Over all values of λ and buffer sizes, the minimum age is achieved for a buffer size of 0 and $\lambda \to \infty$, which achieves an average age of $2/\mu$ (can be derived from (4.3)).

4.4 Effect of Packet Deadlines

There are numerous studies on systems in which packets are under a deadline constraint. Most of those works focus on the problem of scheduling packets to minimize the number of packets that expire before delivery (e.g., [18, 19]). In the context of wireless sensor networks (WSNs), deadlines have been used to limit delay and energy consumption [20]; but the use of deadlines has not been studied from the perspective of controlling the age of information.

In this section, we study a system in which a source transmits update packets to a monitor through an M/M/1/K system for $K = 2$. This system has a total capacity of one packet in the buffer and one packet in service. Additionally, we consider that the packet waiting in the queue is subject to a deadline, such that if it waits in the queue for a time period longer than the deadline, it is dropped from the system and never enters service.[1] We choose an M/M/1/2 system for tractability and for the fact that small buffer size achieves lower average age [8]. If a packet enters the server before its deadline expires, it is guaranteed to be served and is never dropped. The case where packets in service can expire will be considered in Section 4.5.2. A plot of the age with deadline D is shown in Figure 4.3, where arrivals that are actually served occur at times $t_1, t_2, \ldots$, with their receptions at the monitor occurring at times $t'_1, t'_2, \ldots$, respectively. Arrivals occurring at $t_*, t_{**}, \ldots$ are discarded. A packet can be discarded due to a deadline (time of deadline expiration denoted as t^d_*) or due to a full buffer (indicated by a strikethrough). The packet arrival process is Poisson with arrival rate λ.

[1] In this section, we do not consider the capability of replacing the packet in the buffer with a newly arriving packet, as studied in [7]. Our goal here is to assess the effect of deadlines on models that have been studied before, not to optimize the age over all strategies. The *packet replacement* capability is studied in Sections 4.5.3 and 4.5.4.

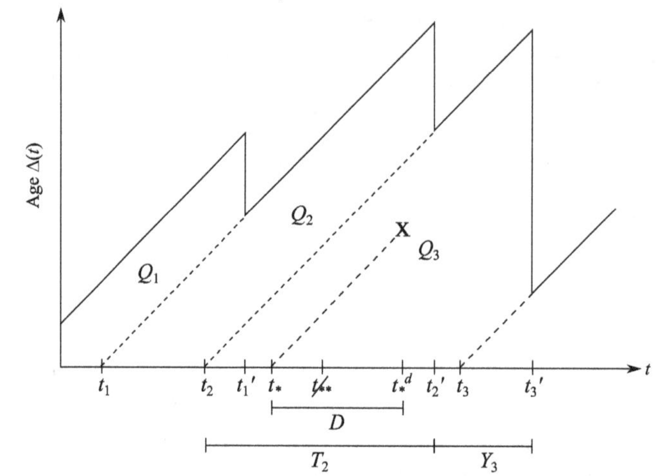

Figure 4.3 Age of information for an M/M/1/2 system with deadlines.

Recall that the service time S_k is the time packet k spends in the server and is modeled as exponential with rate μ, and all the S_k are i.i.d. and independent of the inter-arrival times. The total time spent in the system from arrival to service is given by $T_k = t'_k - t_k, k = 1, 2, \ldots$, and $T_k = W_k + S_k$, with W_k being the time spent waiting in the queue. Also recall that the *interdeparture time* Y_k was defined as the time between the instants of completed service for the k-1st packet served and the kth packet served. This will be useful in the computation of the average age.

We derive the average age for two cases: deterministic deadline and random (exponential) deadline. For the deterministic case where the deadline is fixed, we derive a closed-form expression for the average age as given in (4.20). For the random case where the deadline is randomly generated for each packet, the closed-form expression for the average age as given in (4.32). It is interesting to note that the average age performance of the random deadline is similar to that of the deterministic deadline; but when minimizing the average age over the deadline, the deterministic deadline achieves a lower minimum average age. We also prove that there exists an optimal deadline in $(0, \infty)$ for $0 < \lambda < \infty$. For the random case, we also solve for the optimal average deadline that minimizes the average age. Numerical results for the average age in both cases demonstrate that incorporating a deadline can further improve the age performance of the system. The analysis of the average age is challenging even for simple queueing systems like the M/M/1/2. When we add the deadline requirement, the analysis is further complicated, since we must account for packets that may exit the system without being served, which is the focus of the rest of this section.

4.4.1 Deterministic Deadlines

We begin our analysis for the system in which the deadline is a constant D for all packets; that is, packets that enter the buffer expire after D time units if they do not

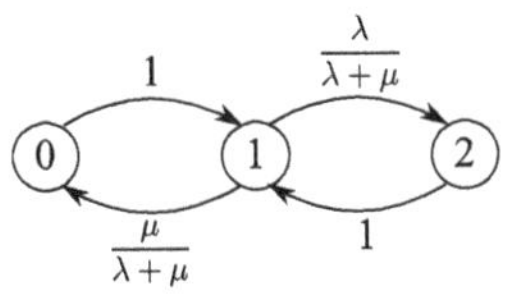

Figure 4.4 Embedded Markov chain for number of packets in the system.

make it into the server. We will discuss the case of random deadlines in the next section.

To compute the average age for the M/M/1/2 system with deadline, we again apply the graphical argument as described in Section 4.2. We compute the area under the sawtooth curve using the sum of the trapezoids Q_k, $k \geq 1$ in Figure 4.3, where each trapezoid is associated with a unique packet that is successfully transmitted. Following the approach described in Section 4.2, we compute the average area of trapezoid k by taking the area of the large isosceles triangle $(^1/_2)(T_{k-1}+Y_k)^2$ and subtracting the area of the smaller isosceles triangle $(^1/_2)T_k^2$. The average age is given by $\Delta = \lambda_e E[Q_k]$, where λ_e is the effective arrival rate of packets that eventually complete service. Since T_{k-1} and T_k are identically distributed, the average age can be expressed as

$$\Delta_{M/M/1/2_D} = \lambda_e \left(\frac{1}{2} E[Y_k^2] + E[T_{k-1} Y_k] \right). \tag{4.5}$$

We derive the terms λ_e, $E[Y_k^2]$, and $E[T_{k-1} Y_k]$ in the following subsections.

Equilibrium Distribution of System State

Before deriving the terms in the average age expression, we first need to find the equilibrium distribution of the number of packets in the system, denoted as p_0, p_1, and p_2. If $N(t)$ is the number of packets in the system at time t, then $N(t)$ is a semi-Markov process with an irreducible, positive-recurrent, embedded Markov chain, shown in Figure 4.4, and the times between state transitions have continuous probability distributions with finite mean.

The transition probabilities in Figure 4.4 are derived as follows. Since states 0 and 2 can only directly transition to 1, those transition probabilities are equal to 1. Since the inter-arrival times and service times are memoryless, the transition from 1 to 0 occurs when a service time is less than an inter-arrival time, which has probability $\frac{\mu}{\lambda+\mu}$. Likewise, the transition from 1 to 2 occurs when an inter-arrival time is less than a service time, which has probability $\frac{\lambda}{\lambda+\mu}$.

Using the balance equations, we can show that the equilibrium distribution of the embedded Markov chain is given by

$$\pi_0 = \frac{\mu}{2(\lambda+\mu)}, \quad \pi_1 = \frac{1}{2}, \quad \pi_2 = \frac{\lambda}{2(\lambda+\mu)}.$$

We then derive the expected time spent in each state between state transitions. We denote the times spent in states 0, 1, and 2 between state transitions as the random variables V_0, V_1, V_2, respectively. For V_0, the time spent is simply an inter-arrival time, so $E[V_0] = 1/\lambda$. For V_1, the average time spent is given by

$$\begin{aligned} E[V_1] &= \Pr(X < S)E[X|X < S] + \Pr(S < X)E[S|S < X] \\ &= \frac{\lambda}{\lambda+\mu}\frac{1}{\lambda+\mu} + \frac{\mu}{\lambda+\mu}\frac{1}{\lambda+\mu} \\ &= \frac{1}{\lambda+\mu}. \end{aligned}$$

Finally, $E[V_2]$ is the expectation of the minimum between an exponential service time and D. (If the residual service of the packet in the server is longer than the deadline D, the packet in the queue is dropped.) This is given by

$$E[V_2] = \frac{1}{\mu}(1 - e^{-\mu D}). \tag{4.6}$$

Finally, the equilibrium probability of the semi-Markov process [21] being in state i can be computed as

$$p_i = \frac{\pi_i E[V_i]}{\sum_{j=0}^{2} \pi_j E[V_j]}.$$

We then have

$$\begin{aligned} p_0 &= \frac{\mu^2}{\mu^2 + \lambda\mu + \lambda^2(1 - e^{-\mu D})} \\ p_1 &= \frac{\lambda\mu}{\mu^2 + \lambda\mu + \lambda^2(1 - e^{-\mu D})} \\ p_2 &= \frac{\lambda^2(1 - e^{-\mu D})}{\mu^2 + \lambda\mu + \lambda^2(1 - e^{-\mu D})}. \end{aligned}$$

Effective Arrival Rate λ_e

To derive the effective arrival rate λ_e, we compute the probability that a packet is neither dropped due to deadline nor blocked by a full system. This is simply the probability that an arriving packet does not experience a residual service time greater than D nor does it see a full system:

$$\Pr(\text{not blocked or dropped}) = 1 - (p_1 e^{-\mu D} + p_2). \tag{4.7}$$

Thus, the effective arrival rate is

$$\begin{aligned} \lambda_e &= \lambda(1 - (p_1 e^{-\mu D} + p_2)) \\ &= \lambda\Big(\frac{\mu^2 + \lambda\mu(1 - e^{-\mu D})}{\mu^2 + \lambda\mu + \lambda^2(1 - e^{-\mu D})}\Big). \end{aligned} \tag{4.8}$$

Second Moment of the Interdeparture Time $E[Y_k^2]$

To compute the second moment of the inter-departure time, we condition on whether a packet $k-1$ departing the server leaves behind an empty system. We denote this event ψ and its complement $\bar{\psi}$. Packet $k-1$ enters a system in state 1, and from there, it can leave behind an empty queue if it completes service before the next packet arrives; or the next packet arrives before service and its deadline expires, in which case the system is back in state 1, which is identical to the state in which it started due to the Markovianness of the embedded system. (The event ψ does not occur if the next packet arrives and packet $k-1$ is served before the deadline expires.) The probability of ψ is thus given by

$$\Pr(\psi) = \frac{\mu}{\lambda + \mu} + \frac{\lambda(1 - e^{-\mu D})}{\mu + \lambda} \Pr(\psi),$$

which simplifies to

$$\Pr(\psi) = \frac{\mu}{\mu + \lambda(1 - e^{-\mu D})}, \tag{4.9}$$

and the probability of its complement is

$$\Pr(\bar{\psi}) = 1 - \Pr(\psi).$$

The inter-departure time conditioned on ψ is the sum of a residual inter-arrival time and a service time. Taking the convolution of the two exponential random variables as in [8], we obtain

$$E[Y_k^2|\psi] = \frac{2(\lambda^2 + \lambda\mu + \mu^2)}{\lambda^2\mu^2}.$$

The inter-departure time conditioned on $\bar{\psi}$ is simply a service time, so $E[Y_k^2|\bar{\psi}] = \frac{2}{\mu^2}$. Finally, to get $E[Y_k^2]$ we substitute the conditional statistics in the following expression:

$$\begin{aligned} E[Y_k^2] &= E[Y_k^2|\psi]\Pr(\psi) + E[Y_k^2|\bar{\psi}]\Pr(\bar{\psi}) \\ &= \frac{2(\mu(\lambda^2 + \lambda\mu + \mu^2) + \lambda^3(1 - e^{-\mu D}))}{\lambda^2\mu^2(\mu + \lambda(1 - e^{-\mu D}))}. \end{aligned} \tag{4.10}$$

Computing $E[T_{k-1}Y_k]$

Next, we need to compute the quantity $E[T_{k-1}Y_k]$. Again, we condition on the events ψ and $\bar{\psi}$. In each case, the system time of packet $k-1$ is conditionally independent of the inter-departure time for packet k, since the event ψ or $\bar{\psi}$ determines whether Y_k is a residual inter-arrival time plus a service time or just a service time, independent of the just completed system time T_{k-1}.

To compute $E[T_{k-1}]$, we first consider the waiting time for served packet $k-1$. For the case where packet $k-1$ encounters a busy system, its waiting time is the residual service time of the packet in service, which must be less than the deadline for packet $k-1$ to not be dropped. The expected waiting time in this case is

$$\begin{aligned} E[S|S < D] &= \frac{1}{1 - e^{-\mu D}}\left(\frac{1}{\mu} - (D + \frac{1}{\mu})e^{-\mu D}\right) \\ &= \frac{1}{\mu} - \frac{De^{-\mu D}}{1 - e^{-\mu D}}. \end{aligned} \tag{4.11}$$

Using this, we compute the expected waiting time for a served packet $k-1$ by conditioning on whether the previous packet $k-2$ left the system empty or not:

$$\begin{aligned} E[W_{k-1}] &= 0 \cdot \Pr(\psi) + E[S'_{k-2}|S'_{k-2} < D](1-\Pr(\psi)) \\ &= \left(\frac{1}{\mu} - \frac{De^{-\mu D}}{1-e^{-\mu D}}\right)\frac{\lambda(1-e^{-\mu D})}{\mu+\lambda(1-e^{-\mu D})}, \end{aligned} \tag{4.12}$$

where S'_{k-2} is the residual service time for packet $k-2$ at the time of packet $k-1$'s arrival.

To compute $E[S_{k-1}|\psi]$, we condition on the event that there are l packets dropped while packet $k-1$ is in service when it leaves the server idle. Let Z_l be the sum of the inter-arrival times for the l dropped packets. Before computing $E[S_{k-1}|\psi, l \text{ dropped}]$, we first compute the probability of a packet with service time s leaving the system idle and l packets being dropped during service. For $s \le lD$, $\Pr(\psi, l \text{ dropped}|s) = 0$. For $s > lD$,

$$\begin{aligned} \Pr(\psi, l \text{ dropped}|s) &= \Pr(Z_l + lD < s < Z_l + lD + X_{l+1}) \\ &= \int_0^{s-lD} e^{-\lambda(s-lD-z)} f_{Z_l}(z)\mathrm{d}z \\ &= \int_0^{s-lD} e^{-\lambda(s-lD-z)} \frac{\lambda^l z^{l-1} e^{-\lambda z}}{(l-1)!}\mathrm{d}z \\ &= \frac{(\lambda(s-lD))^l e^{-\lambda(s-lD)}}{l(l-1)!}. \end{aligned} \tag{4.13}$$

Therefore we have

$$\Pr(\psi, l \text{ dropped}|s) = \begin{cases} \frac{(\lambda(s-lD))^l e^{-\lambda(s-lD)}}{l!} & s > lD \\ 0 & s \le lD \end{cases}. \tag{4.14}$$

By using repeated integration by parts, we compute

$$\begin{aligned} \Pr(\psi, l \text{ dropped}) &= \int_0^{\infty} \Pr(\psi, l \text{ dropped}|s)\, f_{S_{k-1}}(s)\mathrm{d}s \\ &= \frac{\lambda^l \mu e^{\lambda l D}}{l!} \int_{lD}^{\infty} (s-lD)^l e^{-(\lambda+\mu)s}\mathrm{d}s \\ &\ \ \vdots \\ &= \frac{\mu}{\lambda+\mu}\left(\frac{\lambda}{\lambda+\mu}e^{\mu D}\right)^l. \end{aligned} \tag{4.15}$$

We use (4.14) and (4.15) to get the following conditional service time distribution:

$$\begin{aligned} f_{S_{k-1}}(s|\psi, l \text{ dropped}) &= \frac{\Pr(\psi, l \text{ dropped}|s)\, f_{S_{k-1}}(s)}{\Pr(\psi, l \text{ dropped})} \\ &= \frac{(\lambda+\mu)^{l+1}(s-lD)^l e^{-(\lambda+\mu)s}}{l! e^{-(\lambda+\mu)lD}}. \end{aligned}$$

Again using repeated integration by parts, we compute

$$
\begin{aligned}
E[S_{k-1}|\psi, l \text{ dropped}] &= \int_0^\infty s f_{S_{k-1}}(s|\psi, l \text{ dropped}) \mathrm{d}s \\
&= \frac{(\lambda+\mu)^{l+1}}{l! e^{-(\lambda+\mu)lD}} \int_{lD}^\infty s(s-lD)^l e^{-(\lambda+\mu)s} \mathrm{d}s \\
&\vdots \\
&= lD + \frac{l+1}{\lambda+\mu}.
\end{aligned} \tag{4.16}
$$

Finally, we can compute the expected service time, given that the packet leaves the system idle (let $\eta = \frac{\lambda}{\lambda+\mu} e^{-\mu D}$):

$$
\begin{aligned}
E[S_{k-1}|\psi] &= \sum_{l=0}^{\infty} \Pr(l \text{ dropped}|\psi) E[S_{k-1}|\psi, l \text{ dropped}] \\
&= \sum_{l=0}^{\infty} \frac{\Pr(l \text{ dropped}, \psi)}{\Pr(\psi)} E[S_{k-1}|\psi, l \text{ dropped}] \\
&= \frac{1}{\Pr(\psi)} \sum_{l=0}^{\infty} \frac{\mu}{\lambda+\mu} \Big(\frac{\lambda}{\lambda+\mu} e^{-\mu D}\Big)^l \Big(lD + \frac{l+1}{\lambda+\mu}\Big) \\
&= \frac{\mu}{(\lambda+\mu)\Pr(\psi)} \sum_{l=0}^{\infty} \Big(\frac{1}{\lambda+\mu} + l\Big(D + \frac{1}{\lambda+\mu}\Big)\Big)\eta^l \\
&= \frac{\mu}{(\lambda+\mu)\Pr(\psi)} \Big(\frac{1}{\lambda+\mu}\Big(\frac{1}{1-\eta}\Big) + \Big(D + \frac{1}{\lambda+\mu}\Big)\Big(\frac{\eta}{(1-\eta)^2}\Big).
\end{aligned}
$$

Substituting back for η and $\Pr(\psi)$, and after some algebra, we have

$$
E[S_{k-1}|\psi] = \frac{1+\lambda D e^{-\mu D}}{\mu + \lambda(1-e^{-\mu D})}. \tag{4.17}
$$

As we have mentioned, the system time for packet $k-1$ and the inter-departure time for packet k are conditionally independent, given ψ and $\bar{\psi}$. We now use (4.9), (4.12), and (4.17) to evaluate the following expression:

$$
\begin{aligned}
E[T_{k-1}Y_k] &= E[T_{k-1}Y_k|\psi]\Pr(\psi) + E[T_{k-1}Y_k|\bar{\psi}]\Pr(\bar{\psi}) \\
&= \Big(\frac{1}{\lambda} + \frac{1}{\mu}\Big) E[T_{k-1}|\psi]\Pr(\psi) + \frac{1}{\mu} E[T_{k-1}|\bar{\psi}]\Pr(\bar{\psi}) \\
&= \Big(\frac{1}{\lambda} + \frac{1}{\mu}\Big)(E[W_{k-1}] + E[S_{k-1}|\psi])\Pr(\psi) \\
&\quad + \frac{1}{\mu}(E[W_{k-1}] + E[S_{k-1}|\bar{\psi}])\Pr(\bar{\psi}) \\
&= \frac{1}{\lambda}(E[W_{k-1}] + E[S_{k-1}|\psi])\Pr(\psi) \\
&\quad + \frac{1}{\mu}(E[W_{k-1}] + E[S_{k-1}])
\end{aligned} \tag{4.18}
$$

$$\vdots$$

$$= \frac{\mu^2 + \lambda\mu + \lambda^2(2 - (2 + \mu D)e^{-\mu D})}{\lambda\mu^2(\mu + \lambda(1 - e^{-\mu D}))}. \tag{4.19}$$

Expression for the Average Age

We have now derived the terms necessary to compute our main result. By substituting the expressions (4.8), (4.10), and (4.19) into (4.5), we obtain the closed-form expression for the average age for the deterministic deadline case:

$$\Delta_{M/M/1/2_D} = \frac{\mu^3 + 2\lambda\mu^2 + 2\lambda^2\mu + \lambda^3(3 - (3 + \mu D)e^{-\mu D})}{\lambda\mu(\mu^2 + \lambda\mu + \lambda^2(1 - e^{-\mu D}))}. \tag{4.20}$$

Substituting the terms of the system utilization $\rho = \lambda/\mu$ and normalized deadline $D' = \mu D$, the average age expression can be rewritten as follows:

$$\Delta_{M/M/1/2_D} = \frac{1}{\mu}\left(\frac{1 + 2\rho + 2\rho^2 + \rho^3(3 - (3 + D')e^{-D'})}{(\rho + \rho^2 + \rho^3(1 - e^{-D'}))}\right). \tag{4.21}$$

From this expression, we see that if we use the normalized terms ρ and D', the average age is independent of μ except for the $1/\mu$ scale factor. Having obtained the average age expression, we consider the average age for extreme values of the deadline. If we let $D' = 0$ in (4.21), we obtain

$$\Delta_{M/M/1/2_D}|_{D'=0} = \frac{1 + 2\rho + 2\rho^2}{\mu(\rho + \rho^2)}, \tag{4.22}$$

which, through straightforward algebraic manipulation, can be shown to be equal to the average age of an M/M/1/1 queue ((21) in [8]). If we let D' go to infinity in (4.21), we obtain

$$\lim_{D' \to \infty} \Delta_{M/M/1/2_D} = \frac{1 + 2\rho + 2\rho^2 + 3\rho^3}{\mu(\rho + \rho^2 + \rho^3)}, \tag{4.23}$$

which can also be easily shown to be equal to the average age of an M/M/1/2 queue ((45) in [8]). This confirms our intuition about how the system behaves when the deadline is zero (packets entering the buffer immediately expire) and when it is infinity (packets entering the buffer never expire). The following theorem is an intuitive result about the optimal deadline as the rate of update arrivals goes to infinity.

THEOREM 4.1 *In the case of a deterministic deadline, as ρ goes to infinity, the deadline that minimizes the average age is zero.*

Proof If we take the limit of the average age expression as ρ goes to infinity, we arrive at the following expression:

$$\lim_{\rho \to \infty} \Delta_{M/M/1/2_D} = \frac{1}{\mu}\left(3 + \frac{D'}{1 - e^{D'}}\right). \tag{4.24}$$

If we take the derivative with respect to D', we have

$$\frac{\partial}{\partial D'} \lim_{\rho \to \infty} \Delta_{M/M/1/2_D} = \frac{1 + (D' - 1)e^{D'}}{\mu(1 - e^{D'})^2}. \tag{4.25}$$

The denominator is greater than 0 for all D'. The numerator is equal to zero at $D' = 0$, and its derivative ($= D'e^{D'}$) is greater than zero for $D' > 0$. Therefore, the numerator, and thus equation (4.25), is greater than or equal to 0 for $D' > 0$, and (4.24) is an increasing function for $D' > 0$. Therefore, for the case when ρ approaches infinity, the minimum average age occurs for $D' = 0$. □

The next theorem states that for $0 < \lambda < \infty$, the optimal deadline is never zero or infinity.

THEOREM 4.2 *For $0 < \rho < \infty$, there exists a deterministic deadline $0 < D' < \infty$ that minimizes the average age.*

Proof We need to show that (4.21) is less than (4.22) and (4.23) for some value of $0 < D' < \infty$. We start with the $D' = 0$ case, where we require (4.21) be less than (4.22):

$$\frac{1 + 2\rho + 2\rho^2 + \rho^3(3 - (3 + D')e^{-D'})}{\mu(\rho + \rho^2 + \rho^3(1 - e^{-D'}))} < \frac{1 + 2\rho + 2\rho^2}{\mu(\rho + \rho^2)}.$$

After some algebra, we can obtain the following condition:

$$(1 - e^{-D'})(1 - \rho(1 + \rho)) + D'\rho(1 + \rho)e^{-D'} > 0.$$

The $D'\rho(1 + \rho)e^{D'}$ term is greater than zero for all $\rho > 0$ and $D' > 0$. For the first term to be positive, we consider two cases of ρ: (1) if $\rho(1 + \rho) < 1$, we choose D' such that $e^{-D'} < 1$, and (2) if $\rho(1 + \rho) > 1$, we choose D' such that $e^{-D'} > 1$. This ensures that for any value of $\rho > 0$, we can find a value of $0 < D' < \infty$ such that (4.21) is less than (4.22).

For the $D' \to \infty$ case, we require (4.21) be less than (4.23):

$$\frac{1 + 2\rho + 2\rho^2 + \rho^3(3 - (3 + D')e^{-D'})}{\mu(\rho + \rho^2 + \rho^3(1 - e^{-D'}))} < \frac{1 + 2\rho + 2\rho^2 + 3\rho^3}{\mu(\rho + \rho^2 + \rho^3)}.$$

After some algebra, we can obtain the following condition:

$$[\rho(1 + \rho) - 1 + D'\rho(1 + \rho + \rho^2)]e^{-D'} > 0.$$

The exponential $e^{-D'}$ is greater than zero for $0 < D' < \infty$. If we choose $D' > \frac{1-\rho(1+\rho)}{\mu\rho(1+\rho+\rho^2)}$, the condition is satisfied, and (4.21) is less than (4.23).

We have shown that for any value of $0 < \rho < \infty$, we can find a $0 < D' < \infty$ which is less than both (4.22) and (4.23). Therefore, the optimal deadline is neither $D' = 0$ nor $D' \to \infty$.

□

4.4.2 Average Age of M/M/1/2 with Random Exponential Deadlines

In this section, we consider the case of random exponential deadlines [17], in which a packet arriving in the buffer is independently given a randomly generated deadline according to the pdf $f_D(x) = \gamma e^{-\gamma x}$, where γ is the average value of D. One motivation for considering exponential deadlines is in solving for the optimal deadline to minimize the age. In the case of deterministic deadlines, solving for the optimal deadline is quite involved. It turns out that when we consider deadlines modeled as exponentials, the analysis for optimizing over the deadline is comparatively straightforward. The analysis in the random case can provide some insight into the deterministic deadline case. Studying a random deadline is also useful for understanding a system in which different packets are given different deadlines, which can be modeled by considering a distribution over them.

We define D_k as the deadline associated with the kth served packet. We note that not all served packets enter into a busy system and thus are not given a deadline. Additionally, some packets arrive and have their deadlines expire, and thus are not served and are not part of the D_k sequence.

Equilibrium Distribution of System State

As with the deterministic deadline case, we start by finding the equilibrium distribution of the number of packets in the system. The only difference in the case of random deadlines is the average time spent in state 2 per visit, which is the time until a service occurs or a deadline expires. This is given by averaging (4.6) over the pdf of the exponential deadline:

$$\begin{aligned} E[V_2] &= \int_0^\infty \frac{1}{\mu}\left(1 - e^{-\mu x}\right)\gamma e^{-\gamma x} dx \\ &= \frac{1}{\mu + \gamma}. \end{aligned}$$

Using the same approach as in the deterministic case, the resulting distribution of time spent in each state is given by

$$\begin{aligned} p_0 &= \frac{\mu(\mu+\gamma)}{(\lambda+\mu)(\mu+\gamma)+\lambda^2} \\ p_1 &= \frac{\lambda(\mu+\gamma)}{(\lambda+\mu)(\mu+\gamma)+\lambda^2} \\ p_2 &= \frac{\lambda^2}{(\lambda+\mu)(\mu+\gamma)+\lambda^2}. \end{aligned}$$

Effective Arrival Rate λ_e

The probability that a packet is not blocked or dropped due to its deadline is given in (4.7). The effective arrival rate is given by averaging that expression over the pdf of the exponential deadline, multiplied by λ [17]:

$$\begin{aligned} \lambda_e &= \lambda \int_0^\infty \Pr(\text{packet not dropped or blocked}|D_k = x) f_{D_k}(x) dx \\ &= \frac{\lambda\mu(\lambda+\mu+\gamma)}{(\lambda+\mu)(\mu+\gamma)+\lambda^2}. \end{aligned} \tag{4.26}$$

Second Moment of Interdeparture Time $E[Y_k^2]$

To compute the second moment of the inter-departure time $E[Y_k^2]$, we use the same approach as in the deterministic case of conditioning on whether a packet leaves a system idle or busy. We first compute the probability that a packet is left idle, in which we again consider the cases in which a packet is served before the next arrival and in which a packet arrives and is dropped due to a deadline:

$$\Pr(\psi) = \frac{\mu}{\lambda + \mu} + \frac{\lambda\gamma}{(\mu + \lambda)(\mu + \gamma)} \Pr(\psi),$$

which simplifies to

$$= \frac{\mu + \gamma}{\lambda + \mu + \gamma}. \tag{4.27}$$

The inter-departure time is the time between packet $k - 1$ and k's service, which does not include any of packet k's waiting time, and thus does not depend on D_k. As a result, $E[Y_k^2|\psi]$ and $E[Y_k^2|\bar{\psi}]$ are the same as in the deterministic deadline case. The second moment of the inter-departure time is given by

$$E[Y_k^2] = E[Y_k^2|\psi]\Pr(\psi) + E[Y_k^2|\bar{\psi}]\Pr(\bar{\psi})$$
$$= \frac{2((\mu + \gamma)(\lambda^2 + \lambda\mu + \mu^2) + \lambda^3}{\lambda^2\mu^2(\lambda + \mu + \gamma)}. \tag{4.28}$$

Computing $E[T_{k-1}Y_k]$

Following the approach in the deterministic case, we find $E[T_{k-1}Y_k]$ by conditioning on the system being idle or busy, in which case the system time T_{k-1} is conditionally independent of the inter-departure time Y_k. To compute the waiting time portion of the system time for the $k - 1$st packet transmitted, we first need the pdf of D_{k-1}, which is the pdf of a deadline D given that its associated packet entered a busy system and is eventually served. The condition for being served is equivalent to packet $k - 2$'s residual service being less than D. The conditional cdf is computed as follows [17]:

$$\Pr(D_{k-1} < x) = \Pr(D < x|S'_{k-2} < D)$$
$$= 1 - e^{-(\mu+\gamma)x} - \frac{\mu + \gamma}{\mu} e^{-\gamma x}(1 - e^{-\mu x}).$$

We then obtain the pdf by taking the derivative of the cdf:

$$f_{D_{k-1}}(x) = \frac{\gamma(\mu + \gamma)}{\mu} e^{-\gamma x}(1 - e^{-\mu x}).$$

We are now ready to compute the expected waiting time for packet $k - 1$ by conditioning on whether packet $k - 2$ leaves behind a busy system:

$$E[W_{k-1}] = (1 - \Pr(\psi)) \int_0^\infty E[S'_{k-2}|S'_{k-2} < x] f_{D_{k-1}}(x)\mathrm{d}x$$
$$= \frac{\lambda}{(\mu + \gamma)(\lambda + \mu + \gamma)}, \tag{4.29}$$

where $E[S'_{k-2}|S'_{k-2} < x]$ is derived as in (4.11).

The other part of the system time is the service time. Following the analysis of the expected service time in the deterministic deadline case, we first condition on the number of packets dropped during the service of a packet. The expected service time of packet $k-1$, given that it leaves the server idle and there were l packets dropped during its service, and the l packets have deadlines of $D_1, \ldots, D_l$ (denoted as $\mathbf{D}_1^l = \{D_1, \ldots, D_l\}$), is given by

$$E[S_{k-1}|\psi, l \text{ dropped}, \mathbf{D}_1^l] = \sum_{i=1}^{l} D_i + \frac{l+1}{\lambda+\mu}.$$

This result follows from the deterministic deadline case (4.16) with $\sum_{i=1}^{l} D_i$ substituted for the lD term. We also need the probability of l packets being dropped during a service time where the system is left idle, conditioned on all of the deadlines $\mathbf{D}_1^l$:

$$\Pr(l \text{ dropped}, \psi|\mathbf{D}_1^l) = \frac{\mu}{\lambda+\mu}\left(\frac{\lambda}{\lambda+\mu}\right)^l \prod_{i=1}^{l} e^{-\mu D_i},$$

where $S^{(k)}$ (or $S^{(l+1)}$) refers to the residual service time after the $k-1$st (or lth) packet has been dropped. These arise after invoking the memoryless property of the exponential service time.

Now we can compute the expected service time given that the system is left idle:

$$\begin{aligned} E[S_{k-1}|\psi] &= \sum_{l=0}^{\infty} \int \Pr(l \text{ dropped}|\psi, \mathbf{D}_1^l = \mathbf{x}) E[S_{k-1}|\psi, l \text{ dropped}, \mathbf{D}_1^l = \mathbf{x}] \\ &\quad \times f_{\mathbf{D}_1^l}(\mathbf{x})\mathrm{d}\mathbf{x} \\ &= \sum_{l=0}^{\infty} \frac{\mu(\lambda+\mu+\gamma)}{(\lambda+\mu)(\mu+\gamma)}\left(\frac{\lambda}{\lambda+\mu}\right)^l \left[\frac{l\gamma^l}{(\mu+\gamma)^{l+1}} + \frac{l+1}{\lambda+\mu}\left(\frac{\gamma}{\mu+\gamma}\right)^l\right]. \end{aligned}$$

We let $\alpha = \frac{\lambda\gamma}{(\lambda+\mu)(\mu+\gamma)}$, and use $\sum_{l=0}^{\infty} l\alpha^l = \alpha/(1-\alpha)^2$ and $\sum_{l=0}^{\infty}(l+1)\alpha^l = 1/(1-\alpha)^2$ to simplify the expression:

$$\begin{aligned} E[S_{k-1}|\psi] &= \frac{\mu(\lambda+\mu+\gamma)}{(\lambda+\mu)(\mu+\gamma)}\left(\frac{1}{\mu+\gamma}\frac{\alpha}{(1-\alpha)^2} + \frac{1}{\lambda+\mu}\frac{1}{(1-\alpha)^2}\right) \\ &= \frac{\lambda\gamma + (\mu+\gamma)^2}{\mu(\mu+\gamma)(\lambda+\mu+\gamma)}. \end{aligned} \tag{4.30}$$

Now that we have obtained $E[W_{k-1}]$ and $E[S_{k-1}|\psi]$, we can compute $E[T_{k-1}Y_k]$ by substituting (4.27), (4.29), and (4.30) into (4.18) and obtain

$$E[T_{k-1}Y_k] = \frac{\mu+\gamma}{\lambda\mu(\lambda+\mu+\gamma)} + \frac{\lambda\mu + (\mu+\gamma)(\lambda+\mu+\gamma)}{\mu^2(\mu+\gamma)(\lambda+\mu+\gamma)}. \tag{4.31}$$

Average Age

Finally, we can compute our second main result in this chapter. By substituting (4.26), (4.28), and (4.31) into (4.5), and after some algebra, we obtain the closed-form expression for the average age for the random deadline case:

$$\Delta_{M/M/1/2_{D,\text{ random}}} = \frac{2}{\mu} + \frac{\mu(\mu+\gamma)^2 + \lambda^3}{\lambda(\lambda+\mu)(\mu+\gamma)^2 + \lambda^3(\mu+\gamma)}. \tag{4.32}$$

In terms of $\rho = \lambda/\mu$ and $\gamma' = \gamma/\mu$, the average age is

$$\Delta_{M/M/1/2_{D,\text{ random}}} = \frac{1}{\mu}\left(2 + \frac{(1+\gamma')^2 + \rho^3}{\rho(1+\rho)(1+\gamma')^2 + \rho^3(1+\gamma')}\right). \tag{4.33}$$

As in the deterministic case, we see that if we use the normalized terms ρ and γ', the average age is independent of μ except for the $1/\mu$ scale factor. It is straightforward to show that if we take the limit of (4.33) as $\gamma' \to \infty$, we get the average age of an M/M/1/1, just as in the deterministic case when $D' = 0$. Likewise, it is straightforward to show that if we set $\gamma' = 0$, we get the average age of an M/M/1/1, just as in the deterministic case as $D' \to \infty$.

Using the expression in (4.33), we can find the average deadline that minimizes the average age. We take the derivative of (4.33) with respect to the deadline parameter γ' and obtain

$$\begin{aligned}\frac{\partial\Delta}{\partial\gamma'} &= \frac{2(1+\gamma')}{\mu(\rho(1+\rho)(1+\gamma')^2 + \rho^3(1+\gamma'))} \\ &\quad - \frac{((1+\gamma')^2+\rho^3)(2\rho(1+\rho)(1+\gamma')+\rho^3)}{\mu(\rho(1+\rho)(1+\gamma')^2+\rho^3(1+\gamma'))^2}.\end{aligned}$$

We solve for the roots using the quadratic formula and obtain

$$\gamma' = \rho\left(1 + \rho \pm \sqrt{(1+\rho)^2 + \rho}\right) - 1.$$

Since $1 + \rho$ is less than $\sqrt{(1+\rho)^2 + \rho}$, the root using the minus sign is negative. Therefore we focus on the root using the plus sign, and we take the inverse of the root, which for nonnegative values of the root corresponds to the average deadline that minimizes the average age:

$$\gamma'^{-1} = \left[\rho\left(1 + \rho + \sqrt{(1+\rho)^2 + \rho}\right) - 1\right]^{-1}. \tag{4.34}$$

The value of ρ where this root crosses over from negative to positive occurs when $\rho\left(1 + \rho + \sqrt{(1+\rho)^2 + \rho}\right) - 1 = 0$. The only real root is at $\rho \approx 0.3532$, so no deadline should be imposed for smaller values of ρ. This differs from the deterministic case, in which there always exists a deadline in the interval $(0, \infty)$ that minimizes the age for $0 < \rho < \infty$.

4.4.3 Numerical Results

We numerically evaluate the average age for the deterministic case ($\Delta_{M/M/1/2_D}$) and for the random case ($\Delta_{M/M/1/2_{D,\text{ random}}}$) for various normalized arrival rates ρ and plot the ages versus the normalized deadline (or average deadline in the random case) in

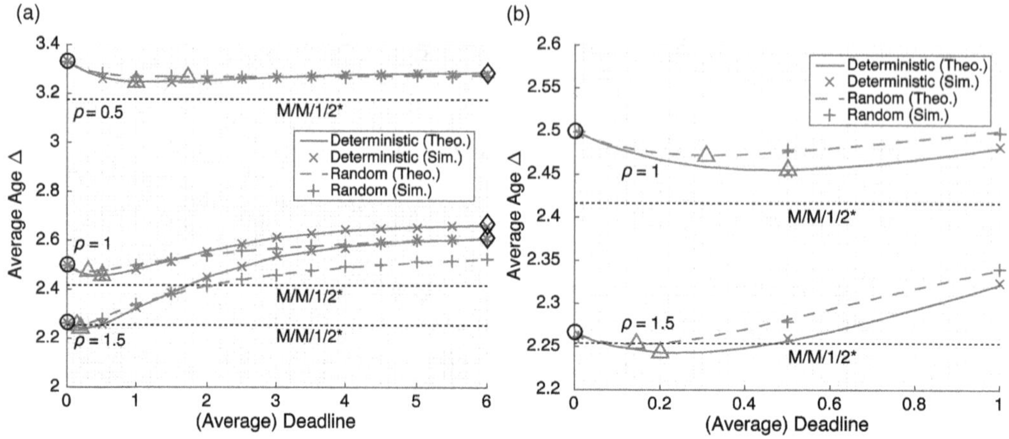

Figure 4.5 Average age vs. deadline for $\mu = 1$, $\rho = 0.5$, 1, and 1.5. Minima are indicated by "Δ," M/M/1/1 age indicated by "$\circ$," M/M/1/2 age indicated by "$\diamond$."

Figure 4.5(a) for $\mu = 1$. As noted in the expressions for (4.21) and (4.33), the service rate μ only impacts the vertical scaling of the age, so we only plot for $\mu = 1$. We also plot the age calculated from simulations of 10^4 samples averaged over 100 runs, and the plot points are displayed as "$\times$" in the deterministic case and "+" in the random case. The simulation agrees closely with theory. The minimum for each case of ρ is marked with a "Δ." In Figure 4.5(a), we observe that for small values of ρ, the average age appears to decrease more initially before only slightly increasing toward an asymptote. Since packets arrive infrequently, there is less of a need to drop the packet in queue since it will not be replaced frequently. For larger values of ρ, as the deadline increases, the average age is initially decreasing but quickly starts to increase and asymptotically approach a limit. As ρ continues to increase, the age only slightly decreases before increasing more, since a packet dropped from the queue is likely to be replaced immediately, keeping packets fresh. These results show that the deadline has a larger impact for higher ρ.

As shown in Sections 4.4.1 and 4.4.2, when the deadline is set to 0, the system is equivalent to an M/M/1/1 system since no packet can wait in the queue. Additionally, for a deadline equal to infinity, the system is equivalent to an M/M/1/2 system. We have also plotted the values of the age for these systems in Figure 4.5(a) ("$\circ$" for M/M/1/1, "$\diamond$" for M/M/1/2). It is shown in [8] that for large values of ρ, the M/M/1/1 has a higher average age than the M/M/1/2, but for large values of ρ, the M/M/1/1 has the lower average age. Again, the intuition is that for lower arrival rates, the M/M/1/1 will have to wait longer after a packet departure for another packet to send, thus increasing the age. On the other hand, for higher arrival rates, the M/M/1/1 will typically be filled with a fresh packet shortly after a departure while the M/M/1/2 will have a packet in waiting that is typically older. A packet deadline can be viewed as a way to transition between the behavior of an M/M/1/1 and an M/M/1/2, at worst getting the best of either approach, but in actuality improving upon them in many cases.

We also include in Figures 4.5(a) and 4.5(b) (detailed plot) a comparison to the M/M/1/2* system in [8], indicated by the horizontal dashed line (it is not a function of the deadline since no deadline is used). The M/M/1/2* is an M/M/1/2 system in which packets waiting in the buffer can be replaced by newer arriving packets, thus improving the average age. This *packet replacement* capability was shown to yield a lower average age than the M/M/1/1 and M/M/1/2, and from the plots we see that it also yields a lower age than the M/M/1/2 with deadline for $\rho \leq 1$. For the case where $\rho = 1.5$, we can see in Figure 4.5(b) that there is a value of the deadline for which the system yields a lower age than the M/M/1/2*. This suggests that for a high enough arrival rate, removing packets from the buffer with a properly chosen deadline is better than keeping a packet in the buffer even if it is refreshed with each new arrival. The high arrival rate ensures a healthy rate of update packets, while the deadline prevents packets that are older than the deadline from occupying the server.

Comparing the random case versus the deterministic case, we note that the minimum age for the deterministic case appears to be smaller than that of the random case. This indicates that limiting the maximum possible deadline, which is true of the deterministic case, is important for getting good age performance. We also observe that for smaller ρ, the average ages for the deterministic and random cases are similar, since the deadline affects fewer packets. For larger ρ and smaller values of the average deadline (close to the optimal), the exponential deadline results in a larger age. The intuition is as follows: a significant number of packets in the random exponential case have deadlines larger than in the deterministic case, which is detrimental with respect to age, while the rest of the packets are smaller than the average, but do not do much to mitigate that effect. For larger values of the average deadline, the exponential deadline results in a smaller age because having a deadline larger than the deterministic case less than half of the time does not hurt the age significantly, since the probability of such a stale packet going into service before the deadline expires is not much greater. However, having a smaller deadline more than half of the time has a more significant impact on lowering the age.

In Figure 4.6, we plot the optimal deadline as a function of ρ for the deterministic and the random cases. The optimal deadline for the deterministic case was computed by repeated numerical evaluation of the average age expression (4.21) for various D' and plotting the D' that achieves the minimum age. In the random case, the optimal deadline was computed directly from (4.34). In both cases, the optimal deadline decreases for increasing ρ, since more frequent arrivals means that packets can be dropped and replaced more frequently. The optimal deadline in the deterministic case is smaller than in the random case for smaller ρ, and it is larger than in the random case for larger ρ. This is because dropping packets increases the age more when the update packets arrive infrequently, so it is better to use a higher average deadline in the random case to reduce dropped packets. When update packets arrive more frequently, it is better to choose a smaller average deadline to drop packets more aggressively, reducing the probability of packets with long deadlines entering the system. In the random case, values of $\rho < 0.3532$ do not have an optimal deadline, which confirms our theoretical result. In such cases, no deadline should be imposed. In the deterministic

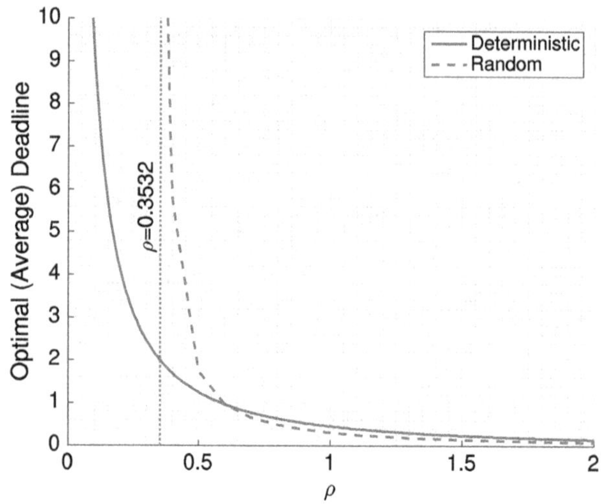

Figure 4.6 Optimal (average) deadline vs. ρ, $\mu = 1$.

case, there are no values of ρ that do not have an optimal deadline, which confirms Theorem 4.2.

4.5 Effect of Packet Management

In the previous section, we have mathematically analyzed the M/M/1/K queueing system with deadlines, for $K = 2$, by solving for the terms in (4.2). Numerical results showed that a properly chosen deadline can improve the age compared to the M/M/1/1 and M/M/1/2 without a deadline. But, what about for other values of $K > 2$? How do larger buffer sizes impact the age with deadlines? This is important to know, especially for systems where the user may not have a choice in selecting the buffer size. Additionally, the mathematical analysis in the previous section did not allow the deadline to affect the packets that are already in service. Furthermore, we have not yet considered the impact of the ability to replace an older packet in the buffer upon arrival of a new packet, the so-called *packet replacement* capability on average age.

In this section, we conduct a simulation study on the aforementioned control mechanisms and their impact on the average age in a single server queue, as including these aspects quickly renders the analysis intractable. We use MATLAB to simulate a queueing system with varying buffer sizes, deadline policies, and packet replacement policies. Simulations are again run for an average of 10^4 samples, and results are averaged over 100 runs.

4.5.1 Packet Control in Buffer Only

We start with simulating the M/M/1/K system with a deadline, for $0 \le K \le 30$, where a packet that has not been served may get dropped from the system if the deadline expires. We first consider the case where the deadline affects packets in the buffer

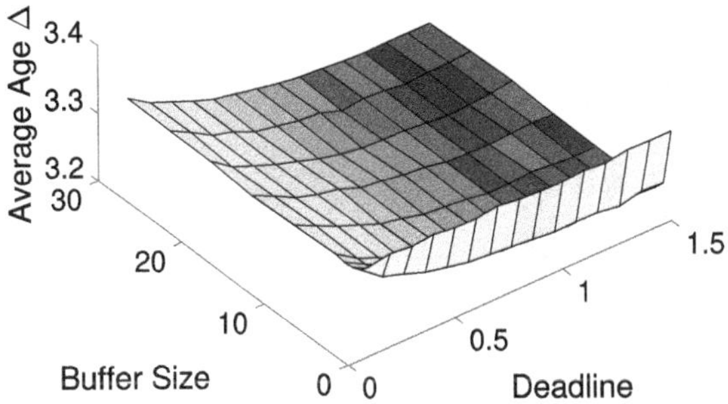

Figure 4.7 Average age for M/M/1/K with deadline, packet control in buffer only, $\lambda = 0.5$, $\mu = 1$.

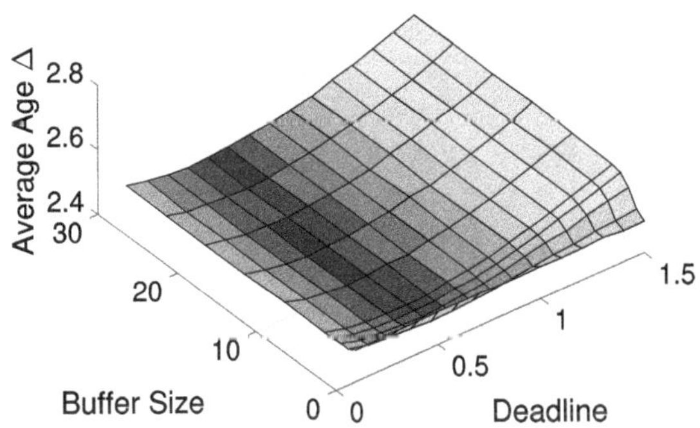

Figure 4.8 Average age for M/M/1/K with deadline, packet control in buffer only, $\lambda = 1$, $\mu = 1$.

only, that is, packets that make it to the server before the deadline are never dropped. Results are displayed in Figures 4.7–4.10 for $\lambda = 0.5$, 1, 1.5, 2, and $\mu = 1$.

For smaller λ (= 0.5, Figure 4.7), we observe that increasing the deadline reduces the age, and that larger buffer sizes seem to do better, since more packets are stored and the deadline prevents them from getting too stale in the buffer. For $\lambda = 1$ (Figure 4.8), the age appears to decrease and then increase with the deadline, and a similar phenomenon is observed with the buffer size. If the deadline is too small, then packets are removed too quickly, which leads to fewer updates. If a deadline is too large, packets can get too stale and it would be better to drop them and wait for another packet to arrive. For a properly chosen deadline, the buffer size should also be carefully chosen for this value of λ. The minimum average age is lower than that of the $\lambda = 0.5$ case by 24% (2.4454 vs. 3.2258) and is achieved with a buffer size of 24 and a deadline of 0.5. For a larger $\lambda = 1.5$ (Figure 4.9), a lower age is achieved with a smaller deadline, since packets are generated frequently enough to be trimmed via deadline, so, in effect, fresh packets are sent at a sufficiently high rate. Of the values simulated, the optimum occurs here for a buffer size of 14 and a deadline of 0.3. The minimum

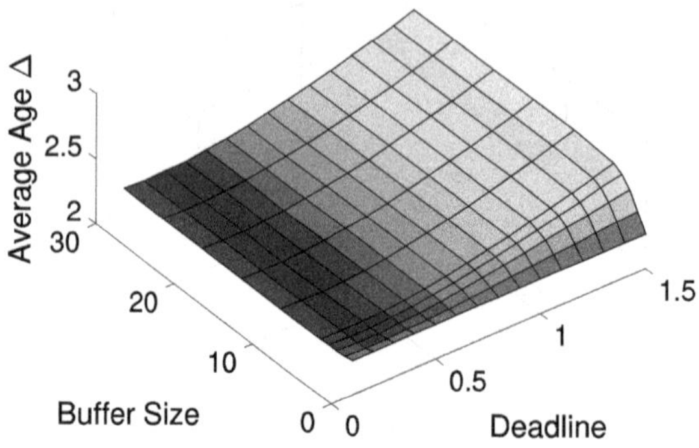

Figure 4.9 Average age for M/M/1/K with deadline, packet control in buffer only, $\lambda = 1.5$, $\mu = 1$.

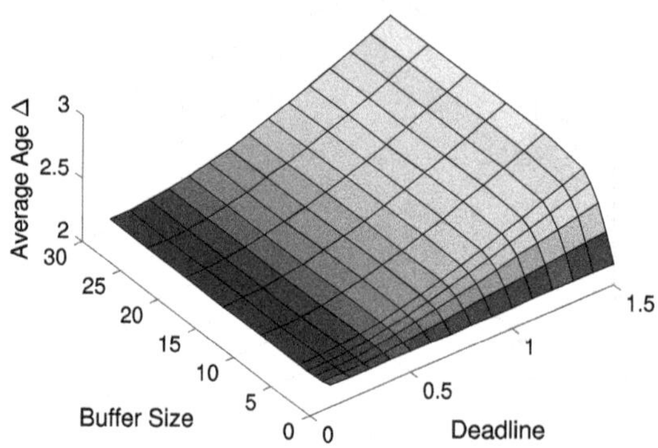

Figure 4.10 Average age for M/M/1/K with deadline, packet control in buffer only, $\lambda = 2$, $\mu = 1$.

average age is lower than that of the $\lambda = 0.5$ case by 31% (2.2369 vs. 3.2258). Finally, for $\lambda = 2$ (Figure 4.10), the optimum buffer size is still 14 but the optimum deadline is reduced to 0.1. The minimum average age is lower than that of the $\lambda = 0.5$ case by 33% (2.1486 vs. 3.2258).

We have provided the optimal ages for the four values of λ in Table 4.2. The observation is as follows: as λ increases, the minimum age decreases, and it is achieved by reducing the deadline, since more arrivals require more aggressively trimming the packets in queue. The buffer size does not have as clear a trend. It is noted that the % improvement appears small compared to the case with no deadline, but such an improvement may be critical for a real-time system. In addition, we will see in the next section that the deadline has a significantly greater effect when it can affect the packet in the server.

4.5.2 Packet Control in Buffer and Server

We now study a case of using a deadline, where both packets in the server and the buffer can be dropped if the deadline expires. The simulation results are provided in

Table 4.2 Optimum average age, packet control in buffer only.

λ	Minimum Age	Optimum Buffer Size	Optimum Deadline	% Improvement vs. No Deadline
0.5	3.2258	19	1.2	1.80%
1	2.4454	24	0.5	2.17%
1.5	2.2369	14	0.3	1.48%
2	2.1489	14	0.1	0.90%

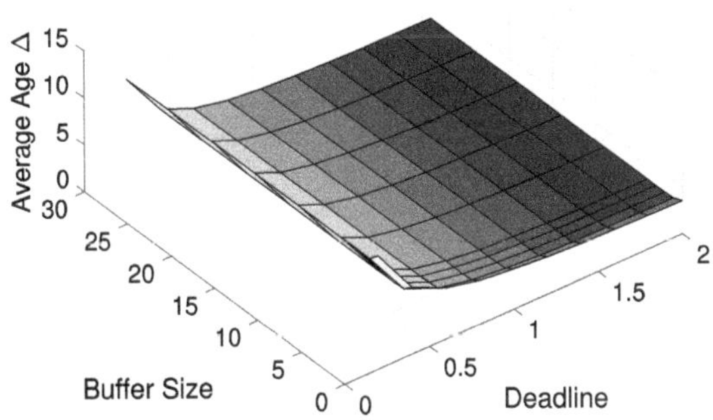

Figure 4.11 Average age for M/M/1/K with deadline, packet control in buffer and server, $\lambda = 0.5$, $\mu = 1$.

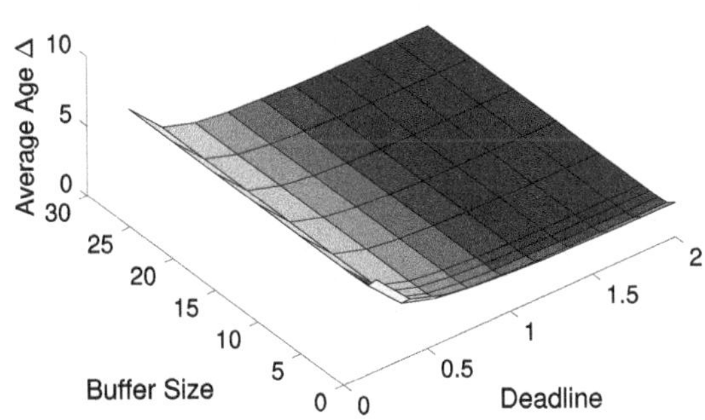

Figure 4.12 Average age for M/M/1/K with deadline, packet control in buffer and server, $\lambda = 1$, $\mu = 1$.

Figures 4.11–4.13. For all λ, the age is relatively large at smaller deadlines because the packets in the server are now also subject to a deadline, so most packets are dropped. As the deadline starts increasing, the age starts to decrease. The age and the optimum deadline and buffer size are provided in Table 4.3. We observe that the optimum deadline decreases as λ increases. As in the case with packet control in the buffer only, the

Table 4.3 Optimum average age, packet control in buffer and server.

λ	Minimum Age	Optimum Buffer Size	Optimum Deadline	% Improvement vs. Buffer Only
0.5	3.0305	24	2	6.05%
1	1.9034	14	1.6	22.16%
1.5	1.4820	14	1.2	33.75%

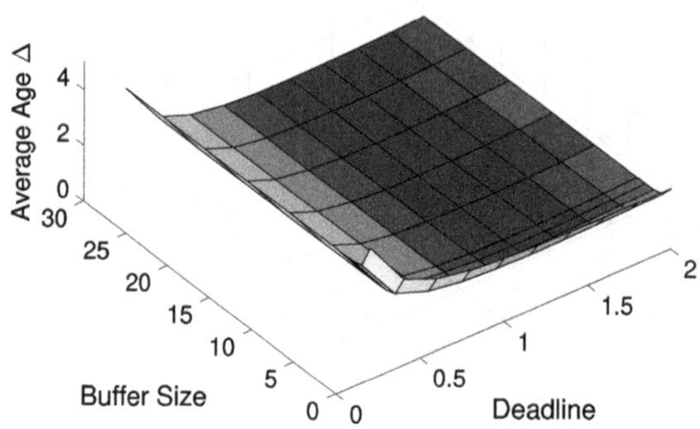

Figure 4.13 Average age for M/M/1/K with deadline, packet control in buffer and server, $\lambda = 1.5$, $\mu = 1$.

buffer size does not show as clear a trend. In this case, the average age is less sensitive to the buffer size once it is sufficiently large (≥ 1). Overall, we observe that the ability to drop packets in the server using a deadline can improve the age by as much as 33% when compared to only dropping packets in the buffer.

4.5.3 Packet Replacement in Buffer Only

We now consider a greater level of packet control in the queue, in which we have the ability to replace an old packet in the buffer upon arrival of a new packet. In this case, no new packets are blocked from entering a full system. This was analyzed in [7] (denoted as M/M/1/2*) and was shown to achieve a lower age than the M/M/1/1 and M/M/1/2 without this packet replacement capability. If we always choose to replace a packet in the buffer with a newly arriving packet, having a buffer size larger than one has no impact since any arriving packet that sees a packet in the buffer will replace it. Therefore, our focus here is on the M/M/1/2 system with packet replacement. Other types of queue management, such as a last-come, first-served discipline or replacing packets after more than one packet arrives in the buffer, are outside the scope of this paper.

We first consider the impact of a deadline on an M/M/1/2 system with packet replacement, where a packet can only be dropped in the buffer if its deadline expires. We plot the results of our simulations in Figure 4.14 for $\lambda = 0.5$, 1, 1.5, 2, and $\mu = 1$.

Table 4.4 Optimum average age with packet replacement, packet control in buffer only.

λ	Minimum Age	Optimum Deadline	% Improvement vs. No Deadline	% Improvement vs. No Packet Replacement, over All Deadlines
0.5	3.1652	3	0.29%	2.08%
1	2.4145	2	0.09%	1.39%
1.5	2.227	0.8	1.2%	0.53%
2	2.1441	0.3	2.47%	0.36%

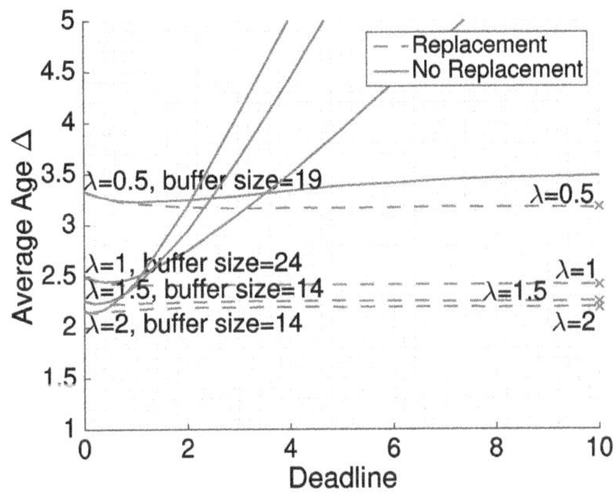

Figure 4.14 Average age for M/M/1/2* and M/M/1/K with deadline.

We see that the deadline has a very small impact on the average age. This suggests that the packet replacement capability prevents the deadline from having much of an impact on the average age. Of the λ plotted, only for $\lambda = 2$ does the age noticeably (albeit slightly) decrease as the deadline decreases. The minimum age for the M/M/1/2 with packet replacement is shown in Table 4.4, and the percent improvement over the case where there is no deadline is shown to be as much as 2%.

We are interested in studying whether the performance of a system with packet replacement policy can be achieved with a deadline only. Figure 4.14 includes the results for M/M/1/K with a deadline in the buffer for $\lambda = 0.5, 1, 1.5, 2$ and we choose the buffer size that minimizes the age according to Table 4.2. Table 4.4 shows the percent improvement of the M/M/1/2 with packet replacement over these M/M/1/K without packet replacement results, minimized over the deadline. If the buffer size and deadline are chosen properly, the age for the M/M/1/K without packet replacement can be close to that of the case with packet replacement, particularly for higher λ.

4.5.4 Packet Replacement in Buffer and Server

Now we consider the case where in addition to packets in the buffer, packets in the server can be dropped due an expiring deadline. The results for our simulations are provided in Figure 4.15, where we see that the deadline has a significant effect on

Table 4.5 Optimum average age with packet replacement, packet control in buffer and server.

λ	Minimum Age	Optimum Deadline	% Improvement vs. Buffer Only
0.5	3.022	3	4.52%
1	1.9125	1.6	20.79%
1.5	1.4968	1.2	32.79%
2	1.2667	1	40.92%

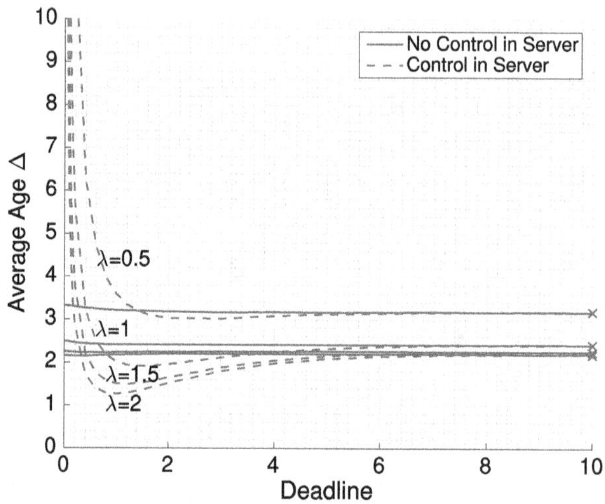

Figure 4.15 Average age for M/M/1/2* and M/M/1/2 with deadline and packet control in server.

the average age when it affects the packet in the server. The minimum age values are provided in Table 4.5, where the age improvement is up to 40% ($\lambda = 2$) compared to the buffer only case. However, the age is highly sensitive to the deadline in the vicinity of the minimum age, since the age approaches infinity as the deadline approaches 0.

4.6 Final Remarks

In this chapter, we introduce the idea of using buffer size, packet deadlines, and packet management as mechanisms for improving the performance of real-time monitoring systems, as measured by the Age of Information metric. We have explored the impact of different buffer sizes, derived closed-form expressions of the average age using deterministic deadlines and random exponential deadlines, solved for the optimal average deadline for the random case, and explored the idea of packet replacement in the buffer and the server. Our numerical evaluation confirms that the age approaches the M/M/1/1 and M/M/1/2 ages as the deadline approaches 0 and ∞, respectively, but there is also an optimal deadline that yields an even lower age. Our work demonstrates that the use of a packet deadline can add a new dimension to optimizing the age performance, providing the freshest information in real-time applications. We followed the theoretical work on the use of packet deadlines with a simulation-based study on a broader range of control mechanisms and their effect on age, both independently and jointly. We notice that the best performance for the case without packet replacement can be very similar to that of the case with packet replacement, if the buffer size and deadline are chosen optimally. When the deadline can impact the packet in the server, the improvement is again more significant as in the case without packet replacement, but the age is highly sensitive to the deadline in the vicinity of the optimal age. Choosing a slightly lower deadline can lead to a significant increase in the average age.

References

[1] S. Kaul, M. Gruteser, V. Rai, and J. Kenney, "Minimizing age of information in vehicular networks," in *IEEE Conference on Sensor, Mesh and Ad Hoc Communications and Networks (SECON)*, June 2011, pp. 350–358.

[2] S. Kaul, R. Yates, and M. Gruteser, "Real-time status: How often should one update?" in *Proc. IEEE INFOCOM*, Orlando, FL, March 2012, pp. 2731–2735.

[3] S. Kaul, R. Yates, and M. Gruteser, "Status updates through queues," in *Conference on Information Sciences and Systems (CISS)*, Princeton, NJ, March 2012, pp. 1–6.

[4] R. D. Yates and S. Kaul, "Real-time status updating: Multiple sources," in *Proc. IEEE International Symposium on Information Theory (ISIT)*, Cambridge, MA, July 2012, pp. 2666–2670.

[5] C. Kam, S. Kompella, and A. Ephremides, "Age of information under random updates," in *Proc. IEEE International Symposium on Information Theory (ISIT)*, Istanbul, Turkey, July 2013, pp. 66–70.

[6] C. Kam, S. Kompella, and A. Ephremides, "Effect of message transmission diversity on status age," in *2014 IEEE International Symposium on Information Theory (ISIT)*, June 2014, pp. 2411–2415.

[7] M. Costa, M. Codreanu, and A. Ephremides, "Age of information with packet management," in *IEEE International Symposium on Information Theory (ISIT)*, June 2014, pp. 1583–1587.

[8] M. Costa, M. Codreanu, and A. Ephremides, "On the age of information in status update systems with packet management," *IEEE Transactions on Information Theory*, vol. 62, no. 4, pp. 1897–1910, April 2016.

[9] N. Pappas, J. Gunnarsson, L. Kratz, M. Kountouris, and V. Angelakis, "Age of information of multiple sources with queue management," in *2015 IEEE International Conference on Communications (ICC)*, June 2015, pp. 5935–5940.

[10] K. Chen and L. Huang, "Age-of-information in the presence of error," in *2016 IEEE International Symposium on Information Theory (ISIT)*, July 2016, pp. 2579–2583.

[11] E. Najm and R. Nasser, "Age of information: The gamma awakening," in *2016 IEEE International Symposium on Information Theory (ISIT)*, July 2016, pp. 2574–2578.

[12] R. D. Yates and S. K. Kaul, "Status updates over unreliable multiaccess channels," in *2017 IEEE International Symposium on Information Theory (ISIT)*, June 2017, pp. 331–335.

[13] Y. Inoue, H. Masuyama, T. Takine, and T. Tanaka, "The stationary distribution of the age of information in fcfs single-server queues," in *2017 IEEE International Symposium on Information Theory (ISIT)*, June 2017, pp. 571–575.

[14] Y. Sun, E. Uysal-Biyikoglu, R. D. Yates, C. E. Koksal, and N. B. Shroff, "Update or wait: How to keep your data fresh," *IEEE Transactions on Information Theory*, vol. 63, no. 11, pp. 7492–7508, November 2017.

[15] A. M. Bedewy, Y. Sun, and N. B. Shroff, "Optimizing data freshness, throughput, and delay in multi-server information-update systems," in *2016 IEEE International Symposium on Information Theory (ISIT)*, July 2016, pp. 2569–2573.

[16] A. M. Bedewy, Y. Sun, and N. B. Shro, "Age-optimal information updates in multihop networks," in *2017 IEEE International Symposium on Information Theory (ISIT)*, June 2017, pp. 576–580.

[17] C. Kam, S. Kompella, G. D. Nguyen, and A. Ephremides, "On the age of information with packet deadlines," *IEEE Transactions on Information Theory*, vol. 64, no. 9, pp. 6419–6428, September 2018.

[18] C. L. Liu and J. W. Layland, "Scheduling algorithms for multiprogramming in a hard-real-time environment," *J. ACM*, vol. 20, no. 1, pp. 46–61, January 1973.

[19] P. P. Bhattacharya and A. Ephremides, "Optimal scheduling with strict deadlines," *IEEE Transactions on Automatic Control*, vol. 34, no. 7, pp. 721–728, July 1989.

[20] C. Chen and J. Ma, "Designing energy-efficient wireless sensor networks with mobile sinks," in *In WSW'06 at Sensys'06*, 2006.

[21] S. M. Ross, *Stochastic Processes.* John Wiley & Sons, Inc., 1996.

5 Timely Status Updating via Packet Management in Multisource Systems

Mohammad Moltafet, Markus Leinonen, and Marian Codreanu

5.1 Introduction

In many Internet of things applications and cyber-physical control systems, freshness of the status information at receivers is a critical factor. Recently, the Age of Information (AoI) was proposed as a destination-centric metric to measure the information freshness in status update systems [1–3]. A status update packet contains the measured value of a monitored process and a time stamp representing the time when the sample was generated. Due to wireless channel access, channel errors, fading, and so on, communicating a status update packet through the network experiences a random delay. If at a time instant t, the most recently received status update packet contains the time stamp $U(t)$, AoI is defined as the random process $\Delta(t) = t - U(t)$. Thus, the AoI measures for each sensor the time elapsed since the last received status update packet was generated at the sensor. The average AoI is the most commonly used metric to evaluate the AoI [1–23].

The work [2] is the seminal queueing theoretic work on the AoI in which the authors derived the average AoI for a single-source first-come first-served (FCFS) M/M/1 queueing model. In [24], the authors proposed peak AoI as an alternative metric to evaluate the information freshness. The work [12] was the first to investigate the average AoI in a multisource setup. The authors of [12] derived the average AoI for a multisource FCFS M/M/1 queueing model. The authors of [25] considered a multisource M/G/1 queueing system and optimized the arrival rates of each source to minimize the peak AoI. The authors of [13] derived an exact expression for the average AoI for a multisource FCFS M/M/1 queueing model and an approximate expression for the average AoI for a multisource FCFS M/G/1 queueing model having a general service time distribution.

The preceding works address the FCFS policy under the infinite queue size. However, it has been shown that the AoI can be significantly decreased if there is a possibility to apply *packet management* in the system (either in the queue or server) [1, 3, 9, 11, 14–18, 20, 26]. To this end, the average AoI for a last-come first-served (LCFS) M/M/1 queueing model with preemption was analyzed in [3]. The average AoI for different packet management policies in a single-source M/M/1 queueing model were derived in [9]. The authors of [20] derived a closed-form expression for the average AoI of a single-source M/G/1/1 preemptive queueing model (where the last entry in the Kendall notation shows the total capacity of the queueing system; 1 indicates that there is one packet under service, whereas the queue holds zero packets). The work [11] considered a single-source LCFS queueing model where the packets arrive according to a Poisson process and the service time follows a gamma distribution. They derived the average AoI for two packet management policies, LCFS with and without preemption. The closed-form expressions for the average AoI and average peak AoI in a multisource M/G/1/1 preemptive queueing model were derived in [14]. In [26], the authors considered a slotted time status update system and derived the average AoI with and without packet management for various access methods.

In [1], the authors gave an in-depth introduction on a powerful technique, *stochastic hybrid systems* (SHS), that can be used to evaluate the AoI in different continuous-time queueing systems. They considered a multisource queueing model in which the packets of different sources are generated according to the Poisson process and served according to an exponentially distributed service time. The authors derived the average AoI for two packet management policies: (1) LCFS with preemption under service (LCFS-S), and (2) LCFS with preemption only in waiting (LCFS-W). Under the LCFS-S policy, a new arriving packet preempts any packet that is currently under service (regardless of the source index). Under the LCFS-W policy, a new arriving packet replaces any older packet waiting in the queue (regardless of the source index); however, the new packet has to wait for any packet under service to finish.

Since its establishment as an efficient tool for the AoI analysis [1], the SHS technique has recently been applied to derive the average AoI for various queueing models and packet management policies [15–19, 21–23, 27]. The authors of [15] studied a multisource M/M/1 queueing model in which sources have different priorities and proposed two packet management policies: (1) there is no waiting room and an update under service is preempted on arrival of an equal or higher priority update, and (2) there is a waiting room for at most one update and preemption is allowed in waiting but not in service. In [16], the author considered a single-source M/M/1 status update system in which the updates follow a route through a series of network nodes where each node is an LCFS queue that supports preemption in service. In [17], the author considered a single-source LCFS queueing model with multiple servers with preemption in service. The authors of [18] considered a multisource LCFS queueing model with multiple servers that employ preemption in service. In [19], the authors derived the average AoI for a multisource FCFS M/M/1 queueing model with an infinite queue size. In [27], the authors studied moments and the moment-generating function of the AoI.

In this chapter, we consider a status update system in which two independent sources generate packets according to the Poisson process and the packets are served according to an exponentially distributed service time. To emphasize the importance of minimizing the average AoI of each individual source and enhance the fairness between different sources in the system, we propose three different *source-aware* packet management policies.

In Policy 1, the queue can contain at most two waiting packets at the same time (in addition to the packet under service), one packet of source 1 and one packet of source 2. When the server is busy and a new packet arrives, the possible packet of the same source waiting in the queue (not being served) is replaced by the fresh packet. In Policy 2, the system (i.e., the waiting queue and the server) can contain at most two packets, one from each source. When the server is busy at an arrival of a packet, the possible packet of the same source either waiting in the queue or being served is replaced by the fresh packet. Policy 3 is similar to Policy 2 but it does not permit preemption in service; that is, while a packet is under service, all new arrivals from the same source are blocked and cleared.

We derive the average AoI for each source under the proposed packet management policies using the SHS technique. By numerical experiments, we investigate the effectiveness of the proposed packet management policies in terms of the sum average AoI and fairness between different sources. The results show that our proposed policies provide better fairness than that of the existing policies. In addition, Policy 2 outperforms the existing policies in terms of the sum average AoI.

5.2 System Model

We consider a status update system consisting of two independent sources,[1] one server, and one sink, as depicted in Figures 5.1 and 5.2. Each source observes a random process at random time instants. The sink is interested in timely information about the status of these random processes. Status updates are transmitted as packets, containing the measured value of the monitored process and a time stamp representing the time when the sample was generated. We assume that the packets of sources 1 and 2 are generated according to the Poisson process with rates λ_1 and λ_2, respectively, and the packets are served according to an exponentially distributed service time with mean $1/\mu$. Let $\rho_1 = \lambda_1/\mu$ and $\rho_2 = \lambda_2/\mu$ be the load of source 1 and 2, respectively. Since packets of the sources are generated according to the Poisson process and the sources are independent, the packet generation in the system follows the Poisson process with rate $\lambda = \lambda_1 + \lambda_2$. The overall load in the system is $\rho = \rho_1 + \rho_2 = \lambda/\mu$.

In the next subsections, we first explain each packet management policy, and then, give a formal definition of AoI.

5.2.1 Packet Management Policies

The structure of the queueing system for all considered policies is illustrated in Figures 5.1 and 5.2. In all policies, when the system is empty, any arriving packet immediately enters the server. However, the policies differ in how they handle the arriving packets when the server is busy.

In Policy 1 (see Figure 5.1), when the server is busy and a new packet arrives, the possible packet of the same source waiting in the queue (not being served) is replaced by the fresh packet.

[1] We consider two sources for simplicity of presentation; the same methodology as used in this chapter can be applied for more than two sources. However, the complexity of the calculations increases exponentially with the number of sources.

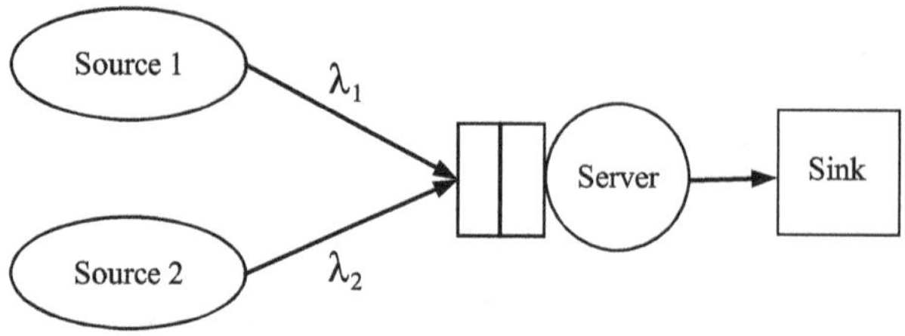

Figure 5.1 Policy 1: The queue can contain at most two waiting packets at the same time (in addition to the packet under service), one packet of source 1 and one packet of source 2; when the server is busy and a new packet arrives, the possible packet of the same source waiting in the queue (not being served) is replaced by the fresh packet.

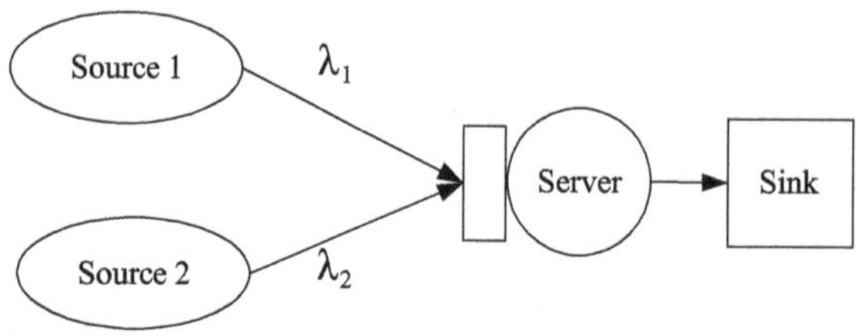

Figure 5.2 Policies 2 and 3: The system (i.e., the waiting queue and the server) can contain at most two packets, one from each source. In Policy 2, when the server is busy and a new packet arrives, the possible packet of the same source either waiting in the queue or being served is replaced by the fresh packet. Policy 3 is similar to Policy 2, but it does not permit preemption in service.

In Policy 2 (see Figure 5.2), when the server is busy and a new packet arrives, the possible packet of the same source either waiting in the queue or being served (called *self-preemption*) is replaced by the fresh packet.

Policy 3 is similar to Policy 2, but it does not permit preemption in service (see Figure 5.2). While a packet is under service, all new arrivals from the same source are blocked and cleared. However, the packet waiting in the queue is replaced upon the arrival of a newer one from the same source. It is also interesting to remark that this policy is also similar to Policy 1, but it has a one-unit-shorter waiting queue.

5.2.2 AoI Definition

For each source, the AoI at the destination is defined as the time elapsed since the last successfully received packet was generated. Formal definition of the AoI is given next.

Let $t_{c,i}$ denote the time instant at which the ith status update packet of source c was generated, and $t'_{c,i}$ denote the time instant at which this packet arrives at the sink. At a time instant τ, the index of the most recently received packet of source c is given by $N_c(\tau) = \max\{i' | t'_{c,i'} \leq \tau\}$, and the time stamp of the most recently received packet of source c is $U_c(\tau) = t_{c,N_c(\tau)}$. The AoI of source c at the destination is defined as the random process $\Delta_c(t) = t - U_c(t)$. Let $(0, \tau)$ denote an observation interval. Accordingly, the time average AoI of the source c at the sink, denoted as $\Delta_{\tau,c}$, is defined as

$$\Delta_{\tau,c} = \frac{1}{\tau}\int_0^{\tau} \Delta_c(t)\mathrm{d}t.$$

The average AoI of source c, denoted by Δ_c, is defined as

$$\Delta_c = \lim_{\tau\to\infty} \Delta_{\tau,c}. \tag{5.1}$$

5.3 A Brief Introduction to the SHS Technique

In the following, we briefly present the main idea behind the SHS technique, which is the key tool for our AoI analysis in Section 5.4. We refer the reader to [1] for more details.

The SHS technique models a queueing system through the states $(q(t), \mathbf{x}(t))$, where $q(t) \in \mathcal{Q} = \{0, 1, \ldots, m\}$ is a continuous-time finite-state Markov chain that describes the occupancy of the system, and $\mathbf{x}(t) = [x_0(t)\, x_1(t) \cdots x_n(t)] \in \mathbb{R}^{1\times(n+1)}$ is a continuous process that describes the evolution of age-related processes at the sink. Following the approach in [1], we label the source of interest as source 1 and employ the continuous process $\mathbf{x}(t)$ to track the age of source 1 status updates at the sink.

The Markov chain $q(t)$ can be presented as a graph $(\mathcal{Q}, \mathcal{L})$ where each discrete state $q(t) \in \mathcal{Q}$ is a node of the chain and a (directed) link $l \in \mathcal{L}$ from node q_l to node q'_l indicates a transition from state $q_l \in \mathcal{Q}$ to state $q'_l \in \mathcal{Q}$.

A transition occurs when a packet arrives or departs in the system. Since the time elapsed between departures and arrivals is exponentially distributed according to the M/M/1 queueing model, transition $l \in \mathcal{L}$ from state q_l to state q'_l occurs with the exponential rate $\lambda^{(l)}\delta_{q_l,q(t)}$[2], where the Kronecker delta function $\delta_{q_l,q(t)}$ ensures that the transition l occurs only when the discrete state $q(t)$ is equal to q_l. When a transition l occurs, the discrete state q_l changes to state q'_l, and the continuous state $\mathbf{x}$ is reset to $\mathbf{x}'$ according to a binary transition reset map matrix $\mathbf{A}_l \in \mathbb{B}^{(n+1)\times(n+1)}$ as $\mathbf{x}' = \mathbf{x}\mathbf{A}_l$. In addition, at each state $q(t) = q \in \mathcal{Q}$, the continuous state $\mathbf{x}$ evolves as a piecewise linear function through the differential equation $\dot{\mathbf{x}}(t) \triangleq \frac{\partial \mathbf{x}(t)}{\partial t} = \mathbf{b}_q$, where $\mathbf{b}_q = [b_{q,0}\, b_{q,1} \cdots b_{q,n}] \in \mathbb{B}^{1\times(n+1)}$ is a binary vector with elements $b_{q,j} \in \{0, 1\}, \forall j \in \{0, \ldots, n\}, q \in \mathcal{Q}$. If the age process $x_j(t)$ increases at a unit rate, we have $b_{q,j} = 1$; otherwise, $b_{q,j} = 0$.

Note that unlike in a typical continuous-time Markov chain, a transition from a state to itself (i.e., a self-transition) is possible in $q(t) \in \mathcal{Q}$. In the case of a self-transition, a reset of the continuous state $\mathbf{x}$ takes place, but the discrete state remains the same. In addition, for a given pair of states $s, s' \in \mathcal{Q}$, there may be multiple transitions l and l' so that the discrete state changes from s to s', but the transition reset maps $\mathbf{A}_l$ and $\mathbf{A}_{l'}$ are different (for more details, see [1, Section III]).

[2] In our system model, $\lambda^{(l)}$ can represent three quantities: arrival rate of source 1 (λ_1), arrival rate of source 2 (λ_2), and the service rate (μ).

To calculate the average AoI using the SHS technique, the state probabilities of the Markov chain and the correlation vector between the discrete state $q(t)$ and the continuous state $\mathbf{x}(t)$ need to be calculated. Let $\pi_q(t)$ denote the probability of being in state q of the Markov chain and $\mathbf{v}_q(t) = [v_{q0}(t) \cdots v_{qn}(t)] \in \mathbb{R}^{1\times(n+1)}$ denote the correlation vector between the discrete state $q(t)$ and the continuous state $\mathbf{x}(t)$. Accordingly, we have

$$\pi_q(t) = \Pr(q(t) = q) = \mathbb{E}[\delta_{q,q(t)}], \quad \forall q \in \mathcal{Q}, \tag{5.2}$$

$$\mathbf{v}_q(t) = [v_{q0}(t) \cdots v_{qn}(t)] = \mathbb{E}[\mathbf{x}(t)\delta_{q,q(t)}], \quad \forall q \in \mathcal{Q}. \tag{5.3}$$

Let $\mathcal{L}'_q$ denote the set of incoming transitions and $\mathcal{L}_q$ denote the set of outgoing transitions for state q, defined as

$$\mathcal{L}'_q = \{l \in \mathcal{L} : q'_l = q\}, \quad \forall q \in \mathcal{Q},$$
$$\mathcal{L}_q = \{l \in \mathcal{L} : q_l = q\}, \quad \forall q \in \mathcal{Q}.$$

Following the ergodicity assumption of the Markov chain $q(t)$ in the AoI analysis [1, 27, 28], the state probability vector $\boldsymbol{\pi}(t) = [\pi_0(t) \cdots \pi_m(t)]$ converges uniquely to the stationary vector $\bar{\boldsymbol{\pi}} = [\bar{\pi}_0 \cdots \bar{\pi}_m]$ satisfying [1]

$$\bar{\pi}_q \textstyle\sum_{l\in\mathcal{L}_q} \lambda^{(l)} = \sum_{l\in\mathcal{L}'_q} \lambda^{(l)}\bar{\pi}_{q_l}, \quad \forall q \in \mathcal{Q}, \tag{5.4}$$

$$\textstyle\sum_{q\in\mathcal{Q}} \bar{\pi}_q = 1. \tag{5.5}$$

Further, it has been shown in [1, Theorem 4] that under the ergodicity assumption of the Markov chain $q(t)$ with stationary distribution $\bar{\boldsymbol{\pi}} \succ 0$, the existence of a nonnegative solution $\bar{\mathbf{v}}_q = [\bar{v}_{q0} \cdots \bar{v}_{qn}], \forall q \in \mathcal{Q}$, for the following system of linear equations,

$$\bar{\mathbf{v}}_q \textstyle\sum_{l\in\mathcal{L}_q} \lambda^{(l)} = \mathbf{b}_q\bar{\pi}_q + \sum_{l\in\mathcal{L}'_q} \lambda^{(l)}\bar{\mathbf{v}}_{q_l}\mathbf{A}_l, \quad \forall q \in \mathcal{Q}, \tag{5.6}$$

implies that the correlation vector $\mathbf{v}_q(t)$ converges to $\bar{\mathbf{v}}_q = [\bar{v}_{q0} \cdots \bar{v}_{qn}], \forall q \in \mathcal{Q}$ as $t \to \infty$. Finally, the average AoI of source 1 is calculated by [1, Theorem 4]

$$\Delta_1 = \textstyle\sum_{q\in\mathcal{Q}} \bar{v}_{q0}. \tag{5.7}$$

As (5.7) implies, the main challenge in calculating the average AoI of a source using the SHS technique reduces to deriving the first elements of each correlation vector $\bar{\mathbf{v}}_q$, that is, $\bar{v}_{q0}$, $\forall q \in \mathcal{Q}$. Note that these quantities are, in general, different for each particular queueing model.

5.4 Average AoI Analysis Using the SHS Technique

In this section, we use the SHS technique to calculate the average AoI in (5.1) of each source under the considered packet management policies described in Section 5.2.1. Recall from (5.7) that the characterization of the average AoI in each of our queueing setups is accomplished by deriving the quantities $\bar{v}_{q0}$, $\forall q \in \mathcal{Q}$. The next three sections are devoted to elaborate derivations of these quantities.

Table 5.1 SHS Markov chain states for Policy 1.

State	Source index of the second packet in the queue	Source index of the first packet in the queue	Server
0	-	-	I
1	-	-	B
2	-	1	B
3	-	2	B
4	2	1	B
5	1	2	B

5.4.1 Average AoI under Policy 1

In Policy 1, the state space of the Markov chain is $\mathcal{Q} = \{0, 1, \ldots, 5\}$, with each state presented in Table 5.1. For example, $q = 0$ indicates that the server is idle, which is shown by I; $q = 1$ indicates that a packet is under service, that is, the queue is empty and the server is busy, which is shown by B; and $q = 5$ indicates that the server is busy, the first packet in the queue (i.e., the packet that is at the head of the queue as depicted in Figure 5.1) is a source 2 packet, and the second packet in the queue is a source 1 packet.

The continuous process is $\mathbf{x}(t) = [x_0(t)\, x_1(t)\, x_2(t)\, x_3(t)]$, where $x_0(t)$ is the current AoI of source 1 at time instant t, $\Delta_1(t)$; $x_1(t)$ encodes what $\Delta_1(t)$ would become if the packet that is under service is delivered to the sink at time instant t; $x_2(t)$ encodes what $\Delta_1(t)$ would become if the first packet in the queue is delivered to the sink at time instant t; and $x_3(t)$ encodes what $\Delta_1(t)$ would become if the second packet in the queue is delivered to the sink at time instant t.

Recall that our goal is to find $\bar{v}_{q0}, \forall q \in \mathcal{Q}$, to calculate the average AoI of source 1 in (5.7). To this end, we need to solve the system of linear equations (5.6) with variables $\bar{\mathbf{v}}_q, \forall q \in \mathcal{Q}$. To form the system of linear equations (5.6) for each state $\forall q \in \mathcal{Q}$, we need to determine $\mathbf{b}_q$, $\bar{\pi}_q$, and $\bar{\mathbf{v}}_{q_l}\mathbf{A}_l$ for each incoming transition $l \in \mathcal{L}'_q$. Next, we derive these for Policy 1.

Determining the Value of $\bar{\mathbf{v}}_{q_l}\mathbf{A}_l$ for Incoming Transitions for Each State $q \in \mathcal{Q}$

The Markov chain for the discrete state $q(t)$ with the incoming and outgoing transitions for each state $q \in \mathcal{Q}$ is shown in Figure 5.3. The transitions between the discrete states $q_l \to q'_l$, $\forall l \in \mathcal{L}$, and their effects on the continuous state $\mathbf{x}(t)$ are summarized in Table 5.2. In the following, we explain the transitions presented in Table 5.2:

- $l = 1$: A source 1 packet arrives at an empty system. With this arrival/transition, the AoI of source 1 does not change, that is, $x'_0 = x_0$. This is because the arrival of source 1 packet does not yield an age reduction until it is delivered to the sink. Since the arriving source 1 packet is fresh and its age is zero, we have $x'_1 = 0$. Since with this arrival the queue is still empty, x_2 and x_3 become irrelevant to the AoI of source 1, and thus, $x'_2 = x_2$ and $x'_3 = x_3$. Note that if the system moves into a new state where x_j is irrelevant, we set $x'_j = x_j, j \in \{1, 2, 3\}$. An interpretation of this

Table 5.2 Table of transitions for the Markov chain of Policy 1 in Figure 5.3.

l	$q_l \to q'_l$	$\lambda^{(l)}$	$\mathbf{A}_l$	$\mathbf{x}\mathbf{A}_l$	$\mathbf{v}_{q_l}\mathbf{A}_l$
1	$0 \to 1$	λ_1	$[x_0\ 0\ x_2\ x_3]$	$\begin{bmatrix}1&0&0&0\\0&0&0&0\\0&0&1&0\\0&0&0&1\end{bmatrix}$	$[v_{00}\ 0\ v_{02}\ v_{03}]$
2	$0 \to 1$	λ_2	$[x_0\ x_0\ x_2\ x_3]$	$\begin{bmatrix}1&1&0&0\\0&0&0&0\\0&0&1&0\\0&0&0&1\end{bmatrix}$	$[v_{00}\ v_{00}\ v_{02}\ v_{03}]$
3	$1 \to 0$	μ	$[x_1\ x_1\ x_2\ x_3]$	$\begin{bmatrix}0&0&0&0\\1&1&0&0\\0&0&1&0\\0&0&0&1\end{bmatrix}$	$[v_{11}\ v_{11}\ v_{12}\ v_{13}]$
4	$1 \to 2$	λ_1	$[x_0\ x_1\ 0\ x_3]$	$\begin{bmatrix}1&0&0&0\\0&1&0&0\\0&0&0&0\\0&0&0&1\end{bmatrix}$	$[v_{10}\ v_{11}\ 0\ v_{13}]$
5	$1 \to 3$	λ_2	$[x_0\ x_1\ x_1\ x_3]$	$\begin{bmatrix}1&0&0&0\\0&1&1&0\\0&0&0&0\\0&0&0&1\end{bmatrix}$	$[v_{10}\ v_{11}\ v_{11}\ v_{13}]$
6	$2 \to 1$	μ	$[x_1\ x_2\ x_2\ x_3]$	$\begin{bmatrix}0&0&0&0\\1&0&0&0\\0&1&1&0\\0&0&0&1\end{bmatrix}$	$[v_{21}\ v_{22}\ v_{22}\ v_{23}]$
7	$3 \to 1$	μ	$[x_1\ x_1\ x_2\ x_3]$	$\begin{bmatrix}0&0&0&0\\1&1&0&0\\0&0&1&0\\0&0&0&1\end{bmatrix}$	$[v_{31}\ v_{31}\ v_{32}\ v_{33}]$
8	$2 \to 2$	λ_1	$[x_0\ x_1\ 0\ x_3]$	$\begin{bmatrix}1&0&0&0\\0&1&0&0\\0&0&0&0\\0&0&0&1\end{bmatrix}$	$[v_{20}\ v_{21}\ 0\ v_{23}]$
9	$2 \to 4$	λ_2	$[x_0\ x_1\ x_2\ x_2]$	$\begin{bmatrix}1&0&0&0\\0&1&0&0\\0&0&1&0\\0&0&0&1\end{bmatrix}$	$[v_{20}\ v_{21}\ v_{22}\ v_{22}]$
10	$3 \to 5$	λ_1	$[x_0\ x_1\ x_1\ 0]$	$\begin{bmatrix}1&0&0&0\\0&1&1&0\\0&0&0&0\\0&0&0&0\end{bmatrix}$	$[v_{30}\ v_{31}\ v_{31}\ 0]$
11	$4 \to 4$	λ_1	$[x_0\ x_1\ 0\ 0]$	$\begin{bmatrix}1&0&0&0\\0&1&0&0\\0&0&0&0\\0&0&0&0\end{bmatrix}$	$[v_{40}\ v_{41}\ 0\ 0]$
12	$5 \to 5$	λ_1	$[x_0\ x_1\ x_1\ 0]$	$\begin{bmatrix}1&0&0&0\\0&1&1&0\\0&0&0&0\\0&0&0&0\end{bmatrix}$	$[v_{50}\ v_{51}\ v_{51}\ 0]$
13	$4 \to 3$	μ	$[x_1\ x_2\ x_2\ x_3]$	$\begin{bmatrix}0&0&0&0\\1&0&0&0\\0&1&1&0\\0&0&0&1\end{bmatrix}$	$[v_{41}\ v_{42}\ v_{42}\ v_{43}]$
14	$5 \to 2$	μ	$[x_1\ x_1\ x_3\ x_3]$	$\begin{bmatrix}0&0&0&0\\1&1&0&0\\0&0&0&0\\0&0&1&1\end{bmatrix}$	$[v_{51}\ v_{51}\ v_{53}\ v_{53}]$

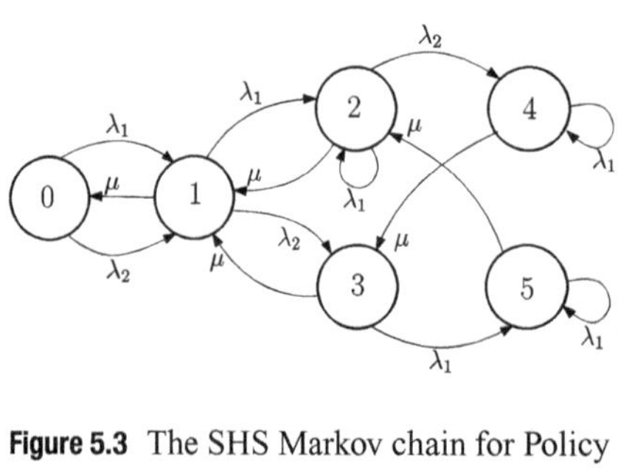

Figure 5.3 The SHS Markov chain for Policy 1

assignment is that x_j has not changed in the transition to the new state. Finally, we have

$$\mathbf{x}' = [x_0\ x_1\ x_2\ x_3]\mathbf{A}_1 = [x_0\ 0\ x_2\ x_3]. \tag{5.8}$$

According to (5.8), it can be shown that the binary matrix $\mathbf{A}_1$ is given by

$$\mathbf{A}_1 = \begin{bmatrix} 1 & 0 & 0 & 0 \\ 0 & 0 & 0 & 0 \\ 0 & 0 & 1 & 0 \\ 0 & 0 & 0 & 1 \end{bmatrix}. \tag{5.9}$$

Then, by using (5.9), $\mathbf{v}_0\mathbf{A}_1$ is calculated as

$$\mathbf{v}_0\mathbf{A}_1 = [v_{00}\ v_{01}\ v_{02}\ v_{03}]\mathbf{A}_1 = [v_{00}\ 0\ v_{02}\ v_{03}]. \tag{5.10}$$

It can be seen from (5.8)–(5.10) that when we have $\mathbf{x}'$ for a transition $l \in \mathcal{L}$, it is easy to calculate $\mathbf{v}_{q_l}\mathbf{A}_l$. Thus, for the rest of the transitions, we just explain the calculation of $\mathbf{x}'$ and present the final expressions of $\mathbf{A}_l$ and $\mathbf{v}_{q_l}\mathbf{A}_l$.

- $l = 2$: A source 2 packet arrives at an empty system. We have $x_0' = x_0$, because this arrival does not change the AoI at the sink. Since the arriving packet is a source 2 packet, its delivery does not change the AoI of source 1; thus, we have $x_1' = x_0$. Moreover, since the queue is empty, x_2 and x_3 become irrelevant, and we have $x_2' = x_2$ and $x_3' = x_3$.
- $l = 3$: A packet is under service and it completes service and is delivered to the sink. With this transition, the AoI at the sink is reset to the age of the packet that just completed service, and thus, $x_0' = x_1$. Since the system enters state $q = 0$, we have $x_1' = x_1$, $x_2' = x_2$, and $x_3' = x_3$.
- $l = 4$: A packet is under service and a source 1 packet arrives. In this transition, we have $x_0' = x_0$ because there is no departure. The delivery of the packet under service reduces the AoI to x_1 and thus, $x_1' = x_1$. Since the arriving source 1 packet is fresh and its age is zero we have $x_2' = 0$. Since there is only one packet in the queue, x_3 becomes irrelevant, and we have $x_3' = x_3$.
- $l = 5$: A packet is under service and a source 2 packet arrives. In this transition, we have $x_0' = x_0$ because there is no departure. The delivery of the packet under service reduces the AoI to x_1 and thus, $x_1' = x_1$. Since the arriving source 2 packet, its delivery does not change the AoI of source 1, and thus we have $x_2' = x_1$. Since there is only one packet in the queue, x_3 becomes irrelevant, and we have $x_3' = x_3$.
- $l = 6$: A source 1 packet is in the queue, a packet is under service, and it completes service and is delivered to the sink. With this transition, the AoI at the sink is reset to the age of the packet that just completed service, and thus, $x_0' = x_1$. Since the source 1 packet in the queue goes to the server, we have $x_1' = x_2$. In addition, since with this departure the queue becomes empty, we have $x_2' = x_2$ and $x_3' = x_3$.
- $l = 7$: A source 2 packet is in the queue, a packet is under service, and it completes service and is delivered to the sink. With this transition, the AoI at the sink is reset to the age of the packet that just completed service, and thus, $x_0' = x_1$. Since the

source 2 packet in the queue goes to the server and its delivery does not change the AoI of source 1, we have $x_1' = x_1$. In addition, since with this departure the queue becomes empty, we have $x_2' = x_2$ and $x_3' = x_3$.

- $l = 8$: A packet is under service, a source 1 packet is in the queue, and a source 1 packet arrives. According to Policy 1, the source 1 packet in the queue is replaced by the fresh source 1 packet. In this transition, we have $x_0' = x_0$ because there is no departure. The delivery of the packet under service reduces the AoI to x_1, and thus, $x_1' = x_1$. Since the arriving source 1 packet is fresh and its age is zero, we have $x_2 = 0$. Since there is only one packet in the queue, we have $x_3' = x_3$.
- $l = 9$: A packet is under service, a source 1 packet is in the queue, and a source 2 packet arrives. In this transition, $x_0' = x_0$ because there is no departure. The delivery of the packet under service reduces the AoI to x_1, and thus, $x_1' = x_1$. The delivery of the first packet in the queue reduces the AoI to x_2, and thus, $x_2' = x_2$. Since the second packet in the queue is a source 2 packet, its delivery does not change the AoI of source 1, and thus we have $x_3' = x_2$.
- $l = 10$: A packet is under service, a source 2 packet is in the queue, and a source 1 packet arrives. In this transition, $x_0' = x_0$ because there is no departure. The delivery of the packet under service reduces the AoI to x_1, and thus, $x_1' = x_1$. Since the first packet in the queue is a source 2 packet, its delivery does not change the AoI of source 1, and thus we have $x_1' = x_1$. Since the arriving source 1 packet is fresh and its age is zero, we have $x_3 = 0$.
- $l = 11$: A packet is under service, the first packet in the queue is a source 1 packet, the second packet in the queue is a source 2 packet, and a source 1 packet arrives. According to Policy 1, the source 1 packet in the queue is replaced by the fresh source 1 packet. In this transition, we have $x_0' = x_0$ because there is no departure. The delivery of the packet under service reduces the AoI to x_1, thus, $x_1' = x_1$. Since the arriving source 1 packet is fresh and its age is zero, we have $x_2' = 0$. Since the second packet in the queue is a source 2 packet, its delivery does not change the AoI of source 1, and thus we have $x_3' = 0$. The reset maps of transition $l = 12$ can be derived similarly.
- $l = 13$: The first packet in the queue is a source 1 packet, the second packet in the queue is a source 2 packet, and the packet under service completes service and is delivered to the sink. With this transition, the AoI at the sink is reset to the age of the source 1 packet that just completed service, and thus, $x_0' = x_1$. Since the first packet in the queue goes to the server, we have $x_1' = x_2$. In addition, since with this departure the queue holds the source 2 packet and its delivery does not change the AoI of source 1, we have $x_2' = x_2$ and $x_3' = x_3$. The reset maps of transition $l = 14$ can be derived similarly.

Having defined the sets of incoming and outgoing transitions, and the value of $\bar{\mathbf{v}}_{q_l}\mathbf{A}_l$ for each incoming transition for each state $q \in \mathcal{Q}$, the remaining task is to derive $\mathbf{b}_q$, $\forall q \in \mathcal{Q}$, and the stationary probability vector $\bar{\boldsymbol{\pi}}$. This is carried out next.

Calculation of $\mathbf{b}_q$ and $\bar{\pi}_q$ for Each State $q \in \mathcal{Q}$

The evolution of $\mathbf{x}(t)$ at each discrete state $q(t) = q$ is determined by the differential equation $\dot{\mathbf{x}} = \mathbf{b}_q$, as described in Section 5.3. Since as long as the discrete state $q(t)$ is unchanged, the age of each element $x_j(t), j \in \{0, \ldots, 3\}$, increases at a unit rate with time, and thus we have $\mathbf{b}_q = \mathbf{1}$, where $\mathbf{1}$ is the row vector $[1 \cdots 1] \in \mathbb{R}^{1\times(n+1)}$.

To calculate the stationary probability vector $\bar{\pi}$, we use (5.4) and (5.5). Using (5.4) and the transitions between the different states presented in Table 5.2, it can be shown that the stationary probability vector $\bar{\pi}$ satisfies $\bar{\pi}\mathbf{D} = \bar{\pi}\mathbf{Q}$, where the diagonal matrix $\mathbf{D} \in \mathbb{R}^{(n+1)\times(n+1)}$ and matrix $\mathbf{Q} \in \mathbb{R}^{(n+1)\times(n+1)}$ are given as

$$\mathbf{D} = \mathrm{diag}[\lambda, \lambda+\mu, \lambda+\mu, \lambda_1+\mu, \lambda_1+\mu, \lambda_1+\mu],$$

$$\mathbf{Q} = \begin{bmatrix} \mu & \lambda & 0 & 0 & 0 & 0 \\ 0 & 0 & \lambda_1 & \lambda_2 & \lambda_2 & 0 \\ 0 & \mu & \lambda_1 & 0 & 0 & 0 \\ 0 & \mu & 0 & 0 & 0 & \lambda_1 \\ 0 & 0 & 0 & \mu & \lambda_1 & 0 \\ 0 & 0 & \mu & 0 & 0 & \lambda_1 \end{bmatrix},$$

where $\mathrm{diag}[a_1, a_2, \ldots, a_n]$ denotes a diagonal matrix with elements $a_1, a_2, \ldots, a_n$ on its main diagonal. Using the above $\bar{\pi}\mathbf{D} = \bar{\pi}\mathbf{Q}$ and $\sum_{q\in\mathcal{Q}} \bar{\pi}_q = 1$ in (5.5), the stationary probabilities are given as

$$\bar{\pi} = \frac{1}{\rho^2 + \rho(2\rho_1\rho_2 + 1) + 1}\left[1 \;\; \rho \;\; \rho_1\rho \;\; \rho_2\rho \;\; \rho_1\rho_2\rho \;\; \rho_1\rho_2\rho\right]. \tag{5.11}$$

Average AoI Calculation

By substituting (5.11) into (5.6) and solving the corresponding system of linear equations, the values of $\bar{v}_{q0}$, $\forall q \in \mathcal{Q}$, are calculated as

$$\bar{v}_{00} = \frac{3\rho_1^4 + \rho_1^3(5\rho_2+9) + \rho_1^2(2\rho_2^2+11\rho_2+10) + \rho_1(4\rho_2^2+6\rho_2+5) + \rho_2^2+\rho_2+1}{\mu\rho_1(1+\rho_1)^2\big((\rho+1)^2-\rho_2\big)\big(\rho^2+\rho(2\rho_1\rho_2+1)+1\big)}, \tag{5.12}$$

$$\bar{v}_{10} = \frac{\rho_2^5 + 3\rho_2^4 + 4\rho_2^3 + 3\rho_2^2 + \rho_2 + \sum_{k=1}^{7}\rho_1^k\gamma_{1,k}}{\mu\rho_1(1+\rho)(1+\rho_2)(1+\rho_1)^2\big((\rho+1)^2-\rho_2\big)\big(\rho^2+\rho(2\rho_1\rho_2+1)+1\big)}, \tag{5.13}$$

where

$$\gamma_{1,1} = 4\rho_2^5 + 16\rho_2^4 + 26\rho_2^3 + 22\rho_2^2 + 9\rho_2 + 1,$$
$$\gamma_{1,2} = 2\rho_2^5 + 25\rho_2^4 + 64\rho_2^3 + 70\rho_2^2 + 35\rho_2 + 6,$$
$$\gamma_{1,3} = 11\rho_2^4 + 62\rho_2^3 + 107\rho_2^2 + 72\rho_2 + 16,$$
$$\gamma_{1,4} = \rho_2^4 + 23\rho_2^3 + 75\rho_2^2 + 77\rho_2 + 23,$$
$$\gamma_{1,5} = 3\rho_2^3 + 23\rho_2^2 + 41\rho_2 + 18,$$
$$\gamma_{1,6} = 3\rho_2^2 + 10\rho_2 + 7,$$

$$\gamma_{1,7} = \rho_2^2 + 10\rho_2 + 1.$$

$$\bar{v}_{20} = \frac{\rho_2^6 + 4\rho_2^5 + 7\rho_2^4 + 7\rho_2^3 + 4\rho_2^2 + \rho_2 + \sum_{k=1}^{7} \rho_1^k \gamma_{2,k}}{\mu(1+\rho)(1+\rho_1)^2(1+\rho_2)^2\big((\rho+1)^2 - \rho_2\big)\big(\rho^2 + \rho(2\rho_1\rho_2 + 1) + 1\big)}, \tag{5.14}$$

where

$$\begin{aligned}
\gamma_{2,1} &= 5\rho_2^6 + 23\rho_2^5 + 46\rho_2^4 + 51\rho_2^3 + 32\rho_2^2 + 10\rho_2 + 1,\\
\gamma_{2,2} &= 4\rho_2^6 + 36\rho_2^5 + 108\rho_2^4 + 156\rho_2^3 + 119\rho_2^2 + 46\rho_2 + 7,\\
\gamma_{2,3} &= \rho_2^6 + 2\rho_2^5 + 100\rho_2^4 + 213\rho_2^3 + 222\rho_2^2 + 111\rho_2 + 21,\\
\gamma_{2,4} &= 4\rho_2^5 + 40\rho_2^4 + 134\rho_2^3 + 202\rho_2^2 + 138 + 33,\\
\gamma_{2,5} &= 6\rho_2^4 + 39\rho_2^3 + 89\rho_2^2 + 87\rho_2 + 28,\\
\gamma_{2,6} &= 4\rho_2^3 + 18\rho_2^2 + 26\rho_2 + 12,\\
\gamma_{2,7} &= \rho_2^2 + 3\rho_2 + 2.
\end{aligned}$$

$$\bar{v}_{30} = \frac{\rho_2^6 + 3\rho_2^5 + 4\rho_2^4 + 3\rho_2^3 + \rho_2^2 + \sum_{k=1}^{7} \rho_1^k \gamma_{3,k}}{\mu\rho_1(1+\rho)(1+\rho_2)(1+\rho_1)^2\big((\rho+1)^2 - \rho_2\big)\big(\rho^2 + \rho(2\rho_1\rho_2 + 1) + 1\big)}, \tag{5.15}$$

where

$$\begin{aligned}
\gamma_{3,1} &= 5\rho_2^6 + 19\rho_2^5 + 30\rho_2^4 + 25\rho_2^3 + 10\rho_2^2 + \rho_2,\\
\gamma_{3,2} &= 5\rho_2^6 + 37\rho_2^5 + 84\rho_2^4 + 87\rho_2^3 + 41\rho_2^2 + 6\rho_2,\\
\gamma_{3,3} &= 2\rho_2^6 + 27\rho_2^5 + 101\rho_2^4 + 149\rho_2^3 + 91\rho_2^2 + 18\rho_2,\\
\gamma_{3,4} &= 8\rho_2^5 + 55\rho_2^4 + 126\rho_2^3 + 110\rho_2^2 + 30\rho_2,\\
\gamma_{3,5} &= 12\rho_2^4 + 51\rho_2^3 + 69\rho_2^2 + 27\rho_2,\\
\gamma_{3,6} &= 8\rho_2^3 + 20\rho_2^2 + 12\rho_2,\\
\gamma_{3,7} &= 2\rho_2^2 + 2\rho_2.
\end{aligned}$$

$$\bar{v}_{40} = \frac{\rho_2^7 + 4\rho_2^6 + 7\rho_2^5 + 7\rho_2^4 + 4\rho_2^3 + \rho_2^2 + \sum_{k=1}^{7} \rho_1^k \gamma_{4,k}}{\mu(1+\rho)(1+\rho_2)^2(1+\rho_1)^2\big((\rho+1)^2 - \rho_2\big)\big(\rho^2 + \rho(2\rho_1\rho_2 + 1) + 1\big)}, \tag{5.16}$$

where

$$\begin{aligned}
\gamma_{4,1} &= 6\rho_2^7 + 27\rho_2^6 + 53\rho_2^5 + 58\rho_2^4 + 36\rho_2^3 + 11\rho_2^2 + \rho_2,\\
\gamma_{4,2} &= 6\rho_2^7 + 48\rho_2^6 + 137\rho_2^5 + 193\rho_2^4 + 145\rho_2^3 + 55\rho_2^2 + 8\rho_2,\\
\gamma_{4,3} &= 2\rho_2^7 + 32\rho_2^6 + 143\rho_2^5 + 286\rho_2^4 + 287\rho_2^3 + 140\rho_2^2 + 26\rho_2,\\
\gamma_{4,4} &= 8\rho_2^6 + 67\rho_2^5 + 201\rho_2^4 + 281\rho_2^3 + 183\rho_2^2 + 43\rho_2,\\
\gamma_{4,5} &= 12\rho_2^5 + 67\rho_2^4 + 137\rho_2^3 + 123\rho_2^2 + 38\rho_2,
\end{aligned}$$

$$\gamma_{4,6} = 8\rho_2^4 + 31\rho_2^3 + 40\rho_2^2 + 14\rho_2,$$
$$\gamma_{4,7} = 2\rho_2^3 + 5\rho_2^2 + 3\rho_2.$$

$$\bar{v}_{50} = \frac{\rho_2^6 + 3\rho_2^5 + 4\rho_2^4 + 3\rho_2^3 + \rho_2^2 + \sum_{k=1}^{7} \rho_1^k \gamma_{5,k}}{\mu(1+\rho)(1+\rho_2)(1+\rho_1)^2\big((\rho+1)^2 - \rho_2\big)\big(\rho^2 + \rho(2\rho_1\rho_2 + 1) + 1\big)}, \quad (5.17)$$

where

$$\gamma_{5,1} = 6\rho_2^6 + 22\rho_2^5 + 34\rho_2^4 + 28\rho_2^3 + 11\rho_2^2 + \rho_2,$$
$$\gamma_{5,2} = 7\rho_2^6 + 47\rho_2^5 + 103\rho_2^4 + 105\rho_2^3 + 49\rho_2^2 + 7\rho_2,$$
$$\gamma_{5,3} = 3\rho_2^6 + 38\rho_2^5 + 133\rho_2^4 + 190\rho_2^3 + 115\rho_2^2 + 23\rho_2,$$
$$\gamma_{5,4} = 12\rho_2^5 + 78\rho_2^4 + 170\rho_2^3 + 145\rho_2^2 + 40\rho_2,$$
$$\gamma_{5,5} = 18\rho_2^4 + 73\rho_2^3 + 95\rho_2^2 + 37\rho_2,$$
$$\gamma_{5,6} = 12\rho_2^3 + 29\rho_2^2 + 17\rho_2,$$
$$\gamma_{5,7} = 3\rho_2^2 + 3\rho_2.$$

Finally, substituting the values of $\bar{v}_{q0}$, $\forall q \in \mathcal{Q}$, into (5.7) results in the average AoI of source 1 under the FCFS-PM policy, which is given as

$$\Delta_1 = \frac{\sum_{k=0}^{7} \rho_1^k \eta_k}{\mu\rho_1 (1+\rho_1)^2 \sum_{j=0}^{4} \rho_1^j \xi_j},$$

where

$$\eta_0 = \rho_2^4 + 2\rho_2^3 + 3\rho_2^2 + 2\rho_2 + 1,$$
$$\eta_1 = 7\rho_2^4 + 15\rho_2^3 + 21\rho_2^2 + 14\rho_2 + 6,$$
$$\eta_2 = 17\rho_2^4 + 46\rho_2^3 + 64\rho_2^2 + 42\rho_2 + 16,$$
$$\eta_3 = 15\rho_2^4 + 73\rho_2^3 + 118\rho_2^2 + 78\rho_2 + 26,$$
$$\eta_4 = 5\rho_2^4 + 52\rho_2^3 + 124\rho_2^2 + 102\rho_2 + 30,$$
$$\eta_5 = 15\rho_2^3 + 66\rho_2^2 + 79\rho_2 + 24,$$
$$\eta_6 = 15\rho_2^2 + 31\rho_2 + 11,$$
$$\eta_7 = 5\rho_2 + 2,$$
$$\xi_0 = \rho_2^4 + 2\rho_2^3 + 3\rho_2^2 + 2\rho_2 + 1,$$
$$\xi_1 = 2\rho_2^4 + 6\rho_2^3 + 9\rho_2^2 + 7\rho_2 + 3,$$
$$\xi_2 = 6\rho_2^3 + 12\rho_2^2 + 10\rho_2 + 4,$$
$$\xi_3 = 6\rho_2^2 + 8\rho_2 + 3,$$
$$\xi_4 = 2\rho_2 + 1.$$

Finally, substituting the values of $\bar{v}_{q0}$, $\forall q \in \mathcal{Q}$, into (5.7) results in the average AoI of source 1 under Policy 1, which is given as

$$\Delta_1 = \frac{\sum_{k=0}^{7} \rho_1^k \eta_k}{\mu\rho_1 (1+\rho_1)^2 \sum_{j=0}^{4} \rho_1^j \xi_j},$$

where

$$\eta_0 = \rho_2^4 + 2\rho_2^3 + 3\rho_2^2 + 2\rho_2 + 1, \qquad \eta_1 = 7\rho_2^4 + 15\rho_2^3 + 21\rho_2^2 + 14\rho_2 + 6,$$
$$\eta_2 = 17\rho_2^4 + 46\rho_2^3 + 64\rho_2^2 + 42\rho_2 + 16, \quad \eta_3 = 15\rho_2^4 + 73\rho_2^3 + 118\rho_2^2 + 78\rho_2 + 26,$$
$$\eta_4 = 5\rho_2^4 + 52\rho_2^3 + 124\rho_2^2 + 102\rho_2 + 30, \quad \eta_5 = 15\rho_2^3 + 66\rho_2^2 + 79\rho_2 + 24,$$
$$\eta_6 = 15\rho_2^2 + 31\rho_2 + 11, \qquad \eta_7 = 5\rho_2 + 2,$$
$$\xi_0 = \rho_2^4 + 2\rho_2^3 + 3\rho_2^2 + 2\rho_2 + 1, \qquad \xi_1 = 2\rho_2^4 + 6\rho_2^3 + 9\rho_2^2 + 7\rho_2 + 3,$$
$$\xi_2 = 6\rho_2^3 + 12\rho_2^2 + 10\rho_2 + 4, \qquad \xi_3 = 6\rho_2^2 + 8\rho_2 + 3,$$
$$\xi_4 = 2\rho_2 + 1.$$

Note that the expression is *exact*; it characterizes the average AoI in the considered queueing model in *closed form*.

5.4.2 Average AoI under Policy 2

Recall from Section 5.2 that the main difference of Policy 2 compared to Policy 1 treated above is that *the system* can contain only two packets, one packet of source 1 and one packet of source 2. Accordingly, for Policy 2, the state space of the Markov chain is $\mathcal{Q} = \{0, 1, 2, 3, 4\}$, where $q = 0$ indicates that the server is idle, that is, the system is empty; $q = 1$ indicates that a source 1 packet is under service and the queue is empty; $q = 2$ indicates that a source 2 packet is under service and the queue is empty; $q = 3$ indicates that a source 1 packet is under service, and a source 2 packet is in the queue; and $q = 4$ indicates that a source 2 packet is under service, and a source 1 packet is in the queue.

The continuous process is $\mathbf{x}(t) = [x_0(t)\, x_1(t)\, x_2(t)]$, where $x_0(t)$ is the current AoI of source 1 at time instant t, $\Delta_1(t)$; $x_1(t)$ encodes what $\Delta_1(t)$ would become if the packet that is under service is delivered to the sink at time instant t; $x_2(t)$ encodes what $\Delta_1(t)$ would become if the packet in the queue is delivered to the sink at time instant t. Next, we will determine the required quantities to form the system of linear equations in (5.6) under Policy 2.

Determining the Value of $\bar{\mathbf{v}}_{q_l}\mathbf{A}_l$ for Incoming Transitions for Each State $q \in \mathcal{Q}$

The Markov chain for the discrete state $q(t)$ is shown in Figure 5.4. The transitions between the discrete states $q_l \rightarrow q_l'$, $\forall l \in \mathcal{L}$, and their effects on the continuous state

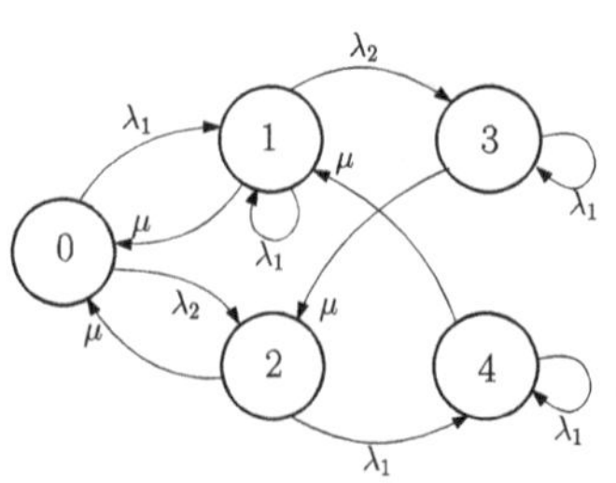

Figure 5.4 The SHS Markov chain for Policy 2.

Table 5.3 Table of transitions for the Markov chain of Policy 2 in Figure 5.4.

l	$q_l \to q'_l$	$\lambda^{(l)}$	$\mathbf{x}\mathbf{A}_l$	$\mathbf{A}_l$	$\mathbf{v}_{q_l}\mathbf{A}_l$
1	$0 \to 1$	λ_1	$[x_0\ 0\ x_2]$	$\begin{bmatrix}1&0&0\\0&0&0\\0&0&1\end{bmatrix}$	$[v_{00}\ 0\ v_{02}]$
2	$0 \to 2$	λ_2	$[x_0\ x_0\ x_2]$	$\begin{bmatrix}1&1&0\\0&0&0\\0&0&1\end{bmatrix}$	$[v_{00}\ v_{00}\ v_{02}]$
3	$1 \to 1$	λ_1	$[x_0\ 0\ x_2]$	$\begin{bmatrix}1&0&0\\0&0&0\\0&0&1\end{bmatrix}$	$[v_{10}\ 0\ v_{12}]$
4	$1 \to 3$	λ_2	$[x_0\ x_1\ x_1]$	$\begin{bmatrix}1&0&0\\0&1&1\\0&0&0\end{bmatrix}$	$[v_{10}\ v_{11}\ v_{11}]$
5	$2 \to 4$	λ_1	$[x_0\ x_0\ 0]$	$\begin{bmatrix}1&1&0\\0&0&0\\0&0&0\end{bmatrix}$	$[v_{20}\ v_{20}\ 0]$
6	$3 \to 3$	λ_1	$[x_0\ 0\ 0]$	$\begin{bmatrix}1&0&0\\0&0&0\\0&0&0\end{bmatrix}$	$[v_{30}\ 0\ 0]$
7	$4 \to 4$	λ_1	$[x_0\ x_0\ 0]$	$\begin{bmatrix}1&1&0\\0&0&0\\0&0&0\end{bmatrix}$	$[v_{40}\ v_{40}\ 0]$
8	$1 \to 0$	μ	$[x_1\ x_1\ x_2]$	$\begin{bmatrix}0&0&0\\1&1&0\\0&0&1\end{bmatrix}$	$[v_{11}\ v_{11}\ v_{12}]$
9	$2 \to 0$	μ	$[x_0\ x_1\ x_2]$	$\begin{bmatrix}1&0&0\\0&1&0\\0&0&1\end{bmatrix}$	$[v_{20}\ v_{21}\ v_{22}]$
10	$3 \to 2$	μ	$[x_1\ x_1\ x_2]$	$\begin{bmatrix}0&0&0\\1&1&0\\0&0&1\end{bmatrix}$	$[v_{31}\ v_{31}\ v_{32}]$
11	$4 \to 1$	μ	$[x_0\ x_2\ x_2]$	$\begin{bmatrix}1&0&0\\0&0&0\\0&1&1\end{bmatrix}$	$[v_{40}\ v_{42}\ v_{42}]$

$\mathbf{x}(t)$ are summarized in Table 5.3. In the following, we explain the transitions presented in Tablc 5.3:

- $l = 1$: A source 1 packet arrives at an empty system. With this transition we have $x'_0 = x_0$ because there is no departure. Since with this arrival the queue is still empty, x_2 becomes irrelevant to the AoI of source 1, and thus, $x'_2 = x_2$.
- $l = 2$: A source 2 packet arrives at an empty system. We have $x'_0 = x_0$, because this arrival does not change the AoI at the sink. Since the arriving packet is a source 2 packet, its delivery does not change the AoI of source 1, and thus we have $x'_1 = x_0$. Moreover, since the queue is empty, x_2 becomes irrelevant, and thus, we have $x'_2 = x_2$.
- $l = 3$: A source 1 packet is under service and a source 1 packet arrives. According to the self-preemptive service of Policy 2, the source 1 packet that is under service is preempted by the arriving source 1 packet. In this transition, we have $x'_0 = x_0$ because there is no departure. Since the arrived source 1 packet that entered the server through the preemption is fresh and its age is zero, we have $x'_1 = 0$. Since the queue is empty, x_2 becomes irrelevant, and thus, we have $x'_2 = x_2$.
- $l = 4$: A source 1 packet is under service and a source 2 packet arrives. In this transition, we have $x'_0 = x_0$ because there is no departure. The delivery of the packet under service reduces the AoI to x_1, and thus, we have $x'_1 = x_1$. Since the packet in

the queue is a source 2 packet, its delivery does not change the AoI of source 1, and thus we have $x_2' = x_1$. The reset map of transition $l = 6$ can be derived similarly.

- $l = 5$: A source 2 packet is under service and a source 1 packet arrives. In this transition, we have $x_0' = x_0$ because there is no departure. Since the packet under service is a source 2 packet, its delivery does not change the AoI of source 1, and thus we have $x_1' = x_0$. Since the arriving source 1 packet is fresh and its age is zero, we have $x_2 = 0$.
- $l = 6$: A source 1 packet is under service, the packet in the queue is a source 2 packet, and a source 1 packet arrives. According to the self-preemptive policy, the source 1 packet that is under service is preempted by the arriving source 1 packet. In this transition, we have $x_0' = x_0$ because there is no departure. Since the arrived source 1 packet that entered the server through the preemption is fresh and its age is zero, we have $x_1' = 0$. Since the packet in the queue is a source 2 packet, its delivery does not change the AoI of source 1, and thus we have $x_2' = 0$. The reset maps of transition $l = 7$ can be derived similarly.
- $l = 8$: A source 1 packet is under service and it completes service and is delivered to the sink. With this transition, the AoI at the sink is reset to the age of the source 1 packet that just completed service, and thus, $x_0' = x_1$. Since the system enters state $q = 0$, we have $x_1' = x_1$ and $x_2' = x_2$. The reset map of transition $l = 9$ can be derived similarly.
- $l = 10$: The packet in the queue is a source 2 packet, and the source 1 packet in the server completes service and is delivered to the sink. With this transition, the AoI at the sink is reset to the age of the source 1 packet that just completed service, that is, $x_0' = x_1$. Since the packet that goes to the server is a source 2 packet, its delivery does not change the AoI of source 1, and thus we have $x_1' = x_1$. In addition, since with this transition the queue becomes empty, we have $x_2' = x_2$. The reset map of transition $l = 11$ can be derived similarly.

Calculation of $\mathbf{b}_q$ and $\bar{\pi}_q$ for Each State $q \in \mathcal{Q}$

Similarly as in Section 5.4.1, as long as the discrete state $q(t)$ is unchanged, the age of each element $x_j(t), j \in \{0, \ldots, 2\}$, increases at a unit rate with time. Thus, we have $\mathbf{b}_q = \mathbf{1}$. Next, we calculate the stationary probability vector $\bar{\boldsymbol{\pi}}$. Using (5.4) and the transition rates among the different states presented in Table 5.3, it can be shown that the stationary probability vector $\bar{\boldsymbol{\pi}}$ satisfies $\bar{\boldsymbol{\pi}}\mathbf{D} = \bar{\boldsymbol{\pi}}\mathbf{Q}$ with

$$\mathbf{D} = \mathrm{diag}[\lambda, \lambda + \mu, \lambda_1 + \mu, \lambda_1 + \mu, \lambda_1 + \mu],$$

$$\mathbf{Q} = \begin{bmatrix} 0 & \lambda_1 & \lambda_2 & 0 & 0 \\ \mu & \lambda_1 & 0 & \lambda_2 & 0 \\ \mu & 0 & 0 & 0 & \lambda_1 \\ 0 & 0 & \mu & \lambda_1 & 0 \\ 0 & \mu & 0 & 0 & \lambda_1 \end{bmatrix}.$$

Using the above $\bar{\boldsymbol{\pi}}\mathbf{D} = \bar{\boldsymbol{\pi}}\mathbf{Q}$ and $\sum_{q \in \mathcal{Q}} \bar{\pi}_q = 1$ in (5.5), the stationary probabilities are given as

$$\bar{\pi} = \frac{1}{2\rho_1\rho_2 + \rho + 1}[1 \ \rho_1 \ \rho_2 \ \rho_1\rho_2 \ \rho_1\rho_2]. \tag{5.18}$$

Average AoI Calculation

By substituting (5.18) into (5.6) and solving the corresponding system of linear equations, the values of $\bar{v}_{q0}$, $\forall q \in \mathcal{Q}$, are calculated as

$$\begin{aligned}
\bar{v}_{00} &= \frac{\rho_1^2(2\rho+5) + (4\rho_1+1)(\rho_2+1)}{\mu\rho_1(1+\rho_1)^2(1+\rho)(1+\rho+2\rho_1\rho_2)}, \\
\bar{v}_{10} &= \frac{(1+\rho_2)(\rho_1^3+4\rho_1^2+1) + \rho_1(5\rho_2+4)}{\mu(1+\rho_2)(1+\rho_1)^2(1+\rho+2\rho_1\rho_2)}, \\
\bar{v}_{20} &= \frac{\rho_2\big(\rho_1^2(2\rho+6) + (4\rho_1+1)(\rho_2+1)\big)}{\mu\rho_1(1+\rho_1)^2(1+\rho)(1+\rho+2\rho_1\rho_2)}, \\
\bar{v}_{30} &= \frac{\rho_2\big((1+\rho_2)(2\rho_1^3+6\rho_1^2+1) + \rho_1(6\rho_2+5)\big)}{\mu(1+\rho_2)(1+\rho_1)^2(1+\rho+2\rho_1\rho_2)}, \\
\bar{v}_{40} &= \frac{\rho_2\big(\rho_1^2(\rho_1^2+5\rho_1+\rho_1\rho_2+4\rho_2+9) + (5\rho_1+1)(1+\rho_2)\big)}{\mu(1+\rho_1)^2(1+\rho)(1+\rho+2\rho_1\rho_2)}.
\end{aligned} \tag{5.19}$$

Finally, substituting the values of $\bar{v}_{q0}$, $\forall q \in \mathcal{Q}$, in (5.19) into (5.7) results in the average AoI of source 1 under Policy 2, which is given as

$$\Delta_1 = \frac{(\rho_2+1)^2 + \sum_{k=1}^{5} \rho_1^k \tilde{\eta}_k}{\mu\rho_1(1+\rho_1)^2\left(\rho_1^2(2\rho_2+1) + (\rho_2+1)^2(2\rho_1+1)\right)},$$

where

$$\begin{aligned}
\tilde{\eta}_1 &= 6\rho_2^2 + 11\rho_2 + 5, & \tilde{\eta}_2 &= 13\rho_2^2 + 24\rho_2 + 10, \\
\tilde{\eta}_3 &= 10\rho_2^2 + 27\rho_2 + 10, & \tilde{\eta}_4 &= 3\rho_2^2 + 14\rho_2 + 5, \\
\tilde{\eta}_5 &= 3\rho_2 + 1.
\end{aligned}$$

5.4.3 Average AoI under Policy 3

The main difference of Policy 3 compared to Policy 2 is that it does not permit preemption in service. The Markov chain and the continuous process of Policy 3 are the same as those for Policy 2. Thus, the stationary probability vector $\bar{\pi}$ of Policy 3 is given in (5.18). The transitions between the discrete states $q_l \rightarrow q'_l$, $\forall l \in \mathcal{L}$, and their effects on the continuous state $\mathbf{x}(t)$ are summarized in Table 5.4. The reset maps of transitions $l \in \{1, 2, 4, 5, 7, 8, 9, 10, 11\}$ are the same as those for Policy 2. Thus, we only explain transitions $l = 3$ and $l = 6$ (see Table 5.4).

- $l = 3$: A source 1 packet is under service and a source 1 packet arrives. According to Policy 3, the arrived source 1 packet is blocked and cleared. In this transition, we have $x'_0 = x_0$ because there is no departure. The delivery of the packet under service reduces the AoI to x_1, and thus, we have $x'_1 = x_1$. Since the queue is empty, x_2 becomes irrelevant, and thus, we have $x'_2 = x_2$.

Table 5.4 Table of transitions for the Markov chain of Policy 3.

l	$q_l \to q'_l$	$\lambda^{(l)}$	$\mathbf{x}\mathbf{A}_l$	$\mathbf{A}_l$	$\mathbf{v}_{q_l}\mathbf{A}_l$
1	$0 \to 1$	λ_1	$[x_0\ 0\ x_2]$	$\begin{bmatrix}1&0&0\\0&0&0\\0&0&1\end{bmatrix}$	$[v_{00}\ 0\ v_{02}]$
2	$0 \to 2$	λ_2	$[x_0\ x_0\ x_2]$	$\begin{bmatrix}1&1&0\\0&0&0\\0&0&1\end{bmatrix}$	$[v_{00}\ v_{00}\ v_{02}]$
3	$1 \to 1$	λ_1	$[x_0\ x_1\ x_2]$	$\begin{bmatrix}1&0&0\\0&1&0\\0&0&1\end{bmatrix}$	$[v_{10}\ v_{11}\ v_{12}]$
4	$1 \to 3$	λ_2	$[x_0\ x_1\ x_1]$	$\begin{bmatrix}1&0&0\\0&1&1\\0&0&0\end{bmatrix}$	$[v_{10}\ v_{11}\ v_{11}]$
5	$2 \to 4$	λ_1	$[x_0\ x_0\ 0]$	$\begin{bmatrix}1&1&0\\0&0&0\\0&0&0\end{bmatrix}$	$[v_{20}\ v_{20}\ 0]$
6	$3 \to 3$	λ_1	$[x_0\ x_1\ x_1]$	$\begin{bmatrix}1&0&0\\0&1&1\\0&0&0\end{bmatrix}$	$[v_{30}\ v_{31}\ v_{31}]$
7	$4 \to 4$	λ_1	$[x_0\ x_0\ 0]$	$\begin{bmatrix}1&1&0\\0&0&0\\0&0&0\end{bmatrix}$	$[v_{40}\ v_{40}\ 0]$
8	$1 \to 0$	μ	$[x_1\ x_1\ x_2]$	$\begin{bmatrix}0&0&0\\1&1&0\\0&0&1\end{bmatrix}$	$[v_{11}\ v_{11}\ v_{12}]$
9	$2 \to 0$	μ	$[x_0\ x_1\ x_2]$	$\begin{bmatrix}1&0&0\\0&1&0\\0&0&1\end{bmatrix}$	$[v_{20}\ v_{21}\ v_{22}]$
10	$3 \to 2$	μ	$[x_1\ x_1\ x_2]$	$\begin{bmatrix}0&0&0\\1&1&0\\0&0&1\end{bmatrix}$	$[v_{31}\ v_{31}\ v_{32}]$
11	$4 \to 1$	μ	$[x_0\ x_2\ x_2]$	$\begin{bmatrix}1&0&0\\0&0&0\\0&1&1\end{bmatrix}$	$[v_{40}\ v_{42}\ v_{42}]$

- $l = 6$: A source 1 packet is under service, the packet in the queue is a source 2 packet, and a source 1 packet arrives. The arrived source 1 packet is blocked and cleared. In this transition, we have $x'_0 = x_0$ because there is no departure. The delivery of the packet under service reduces the AoI to x_1, and thus, we have $x'_1 = x_1$. Since the packet in the queue is a source 2 packet, its delivery does not change the AoI of source 1, and thus we have $x'_2 = x_1$.

Having the stationary probability vector $\bar{\pi}$ (given in (5.18)) and the table of transitions (Table 5.4), we can form the system of linear equations (5.6). By solving the system of linear equations, the values of $\bar{v}_{q0}$, $\forall q \in \mathcal{Q}$, are calculated as

$$\bar{v}_{00} = \frac{\rho_1^3 + \rho_1^2((\rho_2+2)^2 - 1) + (\rho_2+1)^2(3\rho_1+1)}{\mu\rho_1(1+\rho_1)(1+\rho_2)(1+\rho)(1+\rho+2\rho_1\rho_2)}, \tag{5.20}$$

$$\bar{v}_{10} = \frac{(1+\rho_2)(2\rho_1^2+1) + \rho_1(4\rho_2+3)}{\mu(1+\rho_2)(1+\rho_1)(1+\rho+2\rho_1\rho_2)},$$

$$\bar{v}_{20} = \frac{\rho_2\big(\rho_1^3(\rho_2+2) + \rho_1^2(\rho_2^2+5\rho_2+4) + (3\rho_1+1)(\rho_2+1)^2\big)}{\mu\rho_1(1+\rho_1)(1+\rho_2)(1+\rho)(1+\rho+2\rho_1\rho_2)},$$

$$\bar{v}_{30} = \frac{\rho_2\big((\rho_2+1)(3\rho_1^2+1) + \rho_1(5\rho_2+4)\big)}{\mu(1+\rho_1)(1+\rho_2)(1+\rho+2\rho_1\rho_2)},$$

$$\bar{v}_{40} = \frac{\rho_2\big(\rho_1^3(2\rho_2+3) + 2\rho_1^2((\rho_2+2)^2-1) + (4\rho_1+1)(\rho_2+1)^2\big)}{\mu(1+\rho_1)(1+\rho_2)(1+\rho)(1+\rho+2\rho_1\rho_2)}.$$

Finally, substituting the values of $\bar{v}_{q0}$, $\forall q \in \mathcal{Q}$, in (5.20) into (5.7) results in the average AoI of source 1 under Policy 3, which is given as

$$\Delta_1 = \frac{(\rho_2+1)^3 + \sum_{k=1}^{4} \rho_1^k \hat{\eta}_k}{\mu\rho_1(1+\rho_1)(1+\rho_2)\left(\rho_1^2(2\rho_2+1)+(\rho_2+1)^2(2\rho_1+1)\right)},$$

where

$$\hat{\eta}_1 = 5\rho_2^3 + 14\rho_2^2 + 13\rho_2 + 4, \qquad \hat{\eta}_2 = 10\rho_2^3 + 28\rho_2^2 + 25\rho_2 + 7,$$
$$\hat{\eta}_3 = 5\rho_2^3 + 22\rho_2^2 + 23\rho_2 + 6, \qquad \hat{\eta}_4 = 5\rho_2^2 + 8\rho_2 + 2.$$

5.5 Numerical Results

In this section, we show the effectiveness of the proposed packet management policies in terms of the sum average AoI and fairness between the different sources in the system. Moreover, we compare our policies against the following existing policies: the source-agnostic packet management policies LCFS-S and LCFS-W proposed in [1], and the priority-based packet management policies proposed in [15], which we term PP-NW and PP-WW. Under the LCFS-S policy, a new arriving packet preempts any packet that is currently under service (regardless of the source index). Under the LCFS-W policy, a new arriving packet replaces any older packet waiting in the queue (regardless of the source index); however, the new packet has to wait for any packet under service to finish. Under the PP-NW policy, there is no waiting room and an update under service is preempted on arrival of an equal or higher priority update. Under the PP-WW policy, there is a waiting room for at most one update and preemption is allowed in waiting but not in service. Without loss of generality, for the PP-NW and PP-WW policies, we assume that source 2 has higher priority than source 1; for the opposite case, the results are symmetric.

Average AoI

Figure 5.5 depicts the contours of achievable average AoI pairs (Δ_1, Δ_2) for fixed values of system load $\rho = \rho_1 + \rho_2$ under different packet management policies with normalized service rate $\mu = 1$; in Figure 5.5(a), $\rho = 1$ and in Figure 5.5(b), $\rho = 6$. This figure shows that under an appropriate packet management policy in the system (either in the queue or server), by increasing the load of the system the average AoI decreases. Besides that, it shows that Policy 2 provides the lowest average AoI as compared to the other policies.

Sum Average AoI

Figure 5.6 depicts sum average AoI as a function of ρ_1 under different packet management policies with $\mu = 1$; in Figure 5.6(a), $\rho = \rho_1 + \rho_2 = 1$ and in Figure 5.6(b), $\rho = \rho_1 + \rho_2 = 6$. This figure shows that Policy 2 provides the lowest average AoI for all values of ρ_1 as compared to the other policies. In addition, we can observe that among Policy 1, Policy 3, PP-NW, PP-WW, LCFS-S, and LCFS-W policies, the policy that achieves the lowest value of the sum average AoI depends on the system

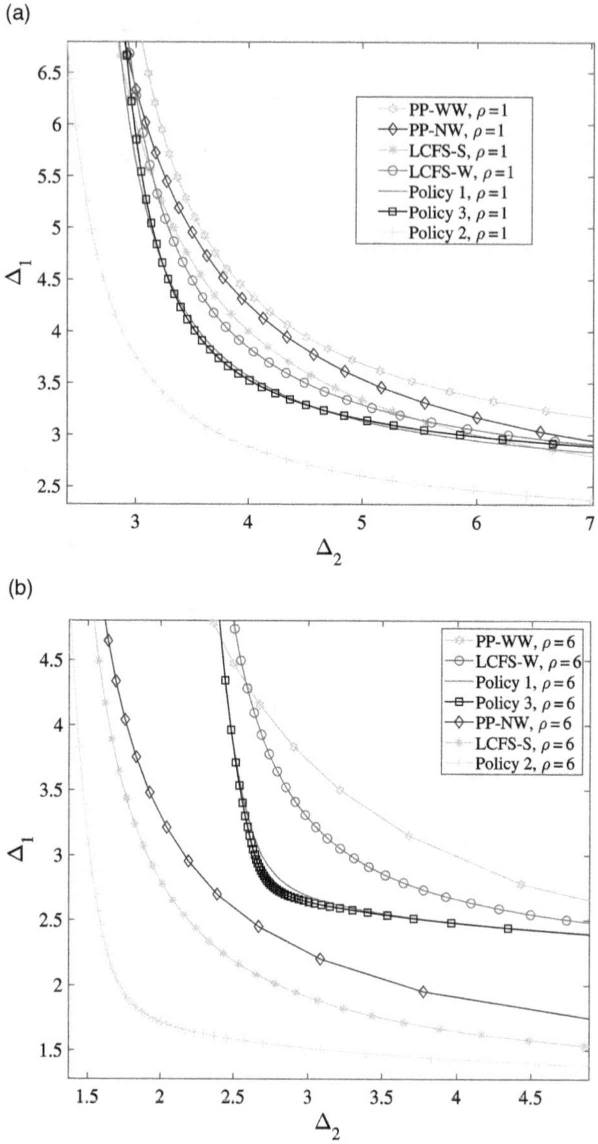

Figure 5.5 The average AoI of sources 1 and 2 under different packet management policies for $\mu = 1$ with (a) $\rho = \rho_1 + \rho_2 = 1$, and (b) $\rho = \rho_1 + \rho_2 = 6$

parameters. Moreover, we can observe that under the PP-NW and PP-WW policies the minimum value of sum average AoI is achieved for a high value of ρ_1. This is because when priority is with source 2, a high value of ρ_1 is needed to compensate for the priority. In addition, we can see that for a high value of total load, that is, $\rho = 6$, the range of values of ρ_1 for which PP-NW and PP-WW policies operate well becomes narrow.

Fairness

For some applications, besides the sum average AoI, the individual average AoI of each source is critical. In this case, fairness between different sources becomes

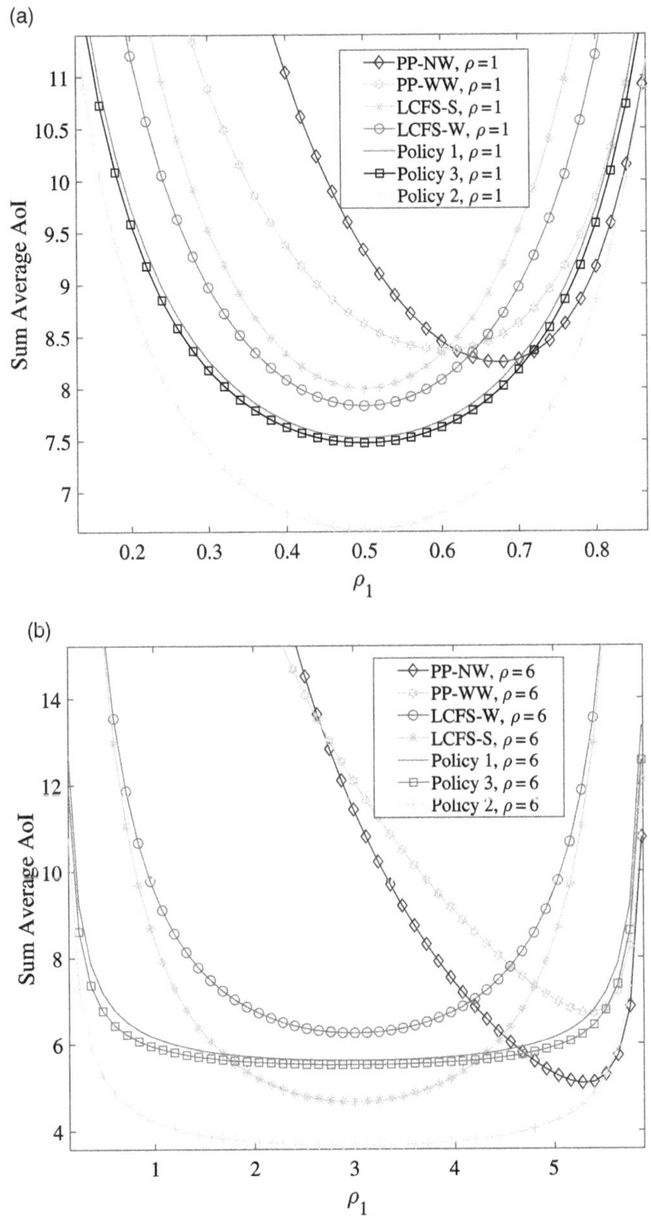

Figure 5.6 Sum average AoI as a function of ρ_1 under different packet management policies for $\mu = 1$ with (a) $\rho = \rho_1 + \rho_2 = 1$, and (b) $\rho = \rho_1 + \rho_2 = 6$

important to be taken into account. To compare the fairness between different sources under the different packet management policies, we use the Jain's fairness index [29]. For the average AoI of sources 1 and 2, the Jain's fairness index $J(\Delta_1, \Delta_2)$ is defined as [30, Section 3] [29, Definition 1]

$$J(\Delta_1, \Delta_2) = \frac{(\Delta_1 + \Delta_2)^2}{2(\Delta_1^2 + \Delta_2^2)}. \tag{5.21}$$

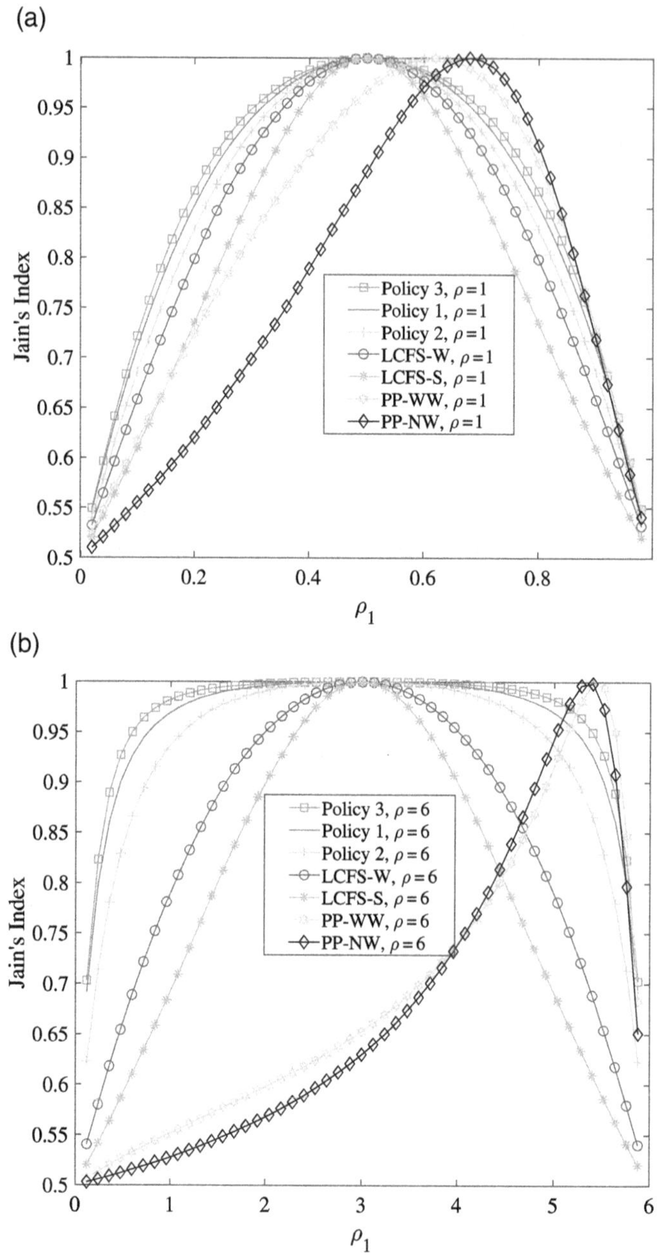

Figure 5.7 The Jain's fairness index for the average AoI of sources 1 and 2 as a function of ρ_1 under different packet management policies for $\mu = 1$ with (a) $\rho = \rho_1 + \rho_2 = 1$, and (b) $\rho = \rho_1 + \rho_2 = 6$

The Jain's index $J(\Delta_1, \Delta_2)$ is continuous and lies in $[0.5, 1]$, where $J(\Delta_1, \Delta_2) = 1$ indicates the fairest situation in the system.

Figure 5.7 depicts the Jain's fairness index for the average AoI of sources 1 and 2 as a function of ρ_1 under different packet management policies with $\mu = 1$; in

Figure 5.7(a), $\rho = \rho_1 + \rho_2 = 1$ and in Figure 5.7(b), $\rho = \rho_1 + \rho_2 = 6$. As can be seen, among Policy 1, Policy 2, Policy 3, LCFS-S, and LCFS-W policies, the LCFS-S policy provides the lowest fairness in the system. This is because the packets of a source with a lower packet arrival rate are most of the time preempted by the packets of the other source having a higher packet arrival rate. In addition, we observe that the proposed source-aware policies (i.e., Policy 1, Policy 2, and Policy 3) in general provide better fairness than that of the other policies and Policy 3 provides the fairest situation in the system. This is because under these policies, a packet in the queue or server can be preempted *only* by a packet with the same source index. Similarly as in Figure 5.6(b), for the high load case, the range of values of ρ_1 for which PP-NW and PP-WW policies provide a good fairness becomes narrow.

5.6 Conclusions

We considered an M/M/1 status update system consisting of two independent sources, one server, and one sink. We proposed three source-aware packet management policies where, differently from the existing works, a packet in the system can be preempted only by a packet with the same source index. We derived the average AoI for each source under the proposed packet management policies using the SHS technique. The numerical results showed that Policy 2 results in a lower sum average AoI in the system compared to the existing policies. In addition, the experiments showed that in general the proposed source-aware policies result in higher fairness in the system than that of the existing policies and, in particular, Policy 3 provides the fairest situation in the system.

References

[1] R. D. Yates and S. K. Kaul, "The age of information: Real-time status updating by multiple sources," *IEEE Transactions on Information Theory*, vol. 65, no. 3, pp. 1807–1827, Mar. 2019.

[2] S. Kaul, R. Yates, and M. Gruteser, "Real-time status: How often should one update?" in *Proceedings of the International Conference on Computer Communications (INFOCOM)*, Orlando, FL, Mar. 25–30, 2012, pp. 2731–2735.

[3] S. K. Kaul, R. D. Yates, and M. Gruteser, "Status updates through queues," in *Proceedings of the Conference on Information Sciences and Systems*, Princeton, NJ, Mar. 21–23, 2012, pp. 1–6.

[4] A. Kosta, N. Pappas, and V. Angelakis, "Age of information: A new concept, metric, and tool," *Foundations and Trends in Networking*, vol. 12, no. 3, pp. 162–259, 2017.

[5] M. Moltafet, M. Leinonen, and M. Codreanu, "Worst case age of information in wireless sensor networks: A multi-access channel," *IEEE Wireless Communications Letters*, vol. 9, no. 3, pp. 321–325, Mar. 2020.

[6] M. Moltafet, M. Leinonen, M. Codreanu, and N. Pappas, "Power minimization in wireless sensor networks with constrained AoI using stochastic optimization," in *Proceedings of*

the Annual Asilomar Conference on Signals, Systems and Computers, Pacific Grove, CA, Nov. 3–6, 2019, pp. 406–410.

[7] ——, "Power minimization for AoI of information constrained dynamic control in wireless sensor networks," *IEEE Transactions on Communications*, vol. 70, no. 1, pp. 419–432, 2022.

[8] S. Kaul, M. Gruteser, V. Rai, and J. Kenney, "Minimizing age of information in vehicular networks," in *Proceedings of the Communications Society Conference on Sensor, Mesh and Ad Hoc Communications and Networks*, Salt Lake City, UT, Jun. 27–30, 2011, pp. 350–358.

[9] M. Costa, M. Codreanu, and A. Ephremides, "On the age of information in status update systems with packet management," *IEEE Transactions on Information Theory*, vol. 62, no. 4, pp. 1897–1910, Apr. 2016.

[10] Y. Inoue, H. Masuyama, T. Takine, and T. Tanaka, "The stationary distribution of the age of information in FCFS single-server queues," in *Proceedings of the IEEE International Symposium on Information Theory*, Aachen, Germany, Jun. 25–30, 2017, pp. 571–575.

[11] E. Najm and R. Nasser, "Age of information: The gamma awakening," in *Proceedings of the IEEE International Symposium on Information Theory*, Barcelona, Spain, Jul. 10–16, 2016, pp. 2574–2578.

[12] R. D. Yates and S. Kaul, "Real-time status updating: Multiple sources," in *Proceedings of the IEEE International Symposium on Information Theory*, Cambridge, MA, Jul. 1–6, 2012, pp. 2666–2670.

[13] M. Moltafet, M. Leinonen, and M. Codreanu, "On the age of information in multi-source queueing models," *IEEE Transactions on Communications*, vol. 68, no. 8, pp. 5003–5017, May 2020.

[14] E. Najm and E. Telatar, "Status updates in a multi-stream M/G/1/1 preemptive queue," in *Proceedings of the International Conference on Computer Communications (INFOCOM)*, Honolulu, HI, Apr. 15–19, 2018, pp. 124–129.

[15] S. K. Kaul and R. D. Yates, "Age of information: Updates with priority," in *Proceedings of the IEEE International Symposium on Information Theory*, Vail, CO, Jun. 17–22, 2018, pp. 2644–2648.

[16] R. D. Yates, "Age of information in a network of preemptive servers," in *Proceedings of the International Conference on Computer Communications (INFOCOM)*, Honolulu, HI, Apr. 15–19, 2018, pp. 118–123.

[17] ——, "Status updates through networks of parallel servers," in *Proceedings of the IEEE International Symposium on Information Theory*, Vail, CO, Jun. 17–22, 2018, pp. 2281–2285.

[18] A. Javani, M. Zorgui, and Z. Wang, "Age of information in multiple sensing," in *Proceedings of the IEEE Global Telecommunication Conference*, Waikoloa, HI, Dec. 9–13, 2019.

[19] S. K. Kaul and R. D. Yates, "Timely updates by multiple sources: The M/M/1 queue revisited," in *Proceedings of the Conference on Information Sciences and Systems*, Princeton, NJ, Mar. 18–20, 2020.

[20] E. Najm, R. Yates, and E. Soljanin, "Status updates through M/G/1/1 queues with HARQ," in *Proceedings of the IEEE International Symposium on Information Theory*, Aachen, Germany, Jun. 25–30, 2017, pp. 131–135.

[21] M. Moltafet, M. Leinonen, and M. Codreanu, "Average age of information for a multi-source M/M/1 queueing model with packet management and self-preemption in service,"

in *Proceedings of the International Symposium on Modeling and Optimization in Mobile, Ad Hoc and Wireless Networks*, Volos, Greece, Jun. 15–19, 2020, pp. 1–5.

[22] ——, "Average age of information in a multi-source M/M/1 queueing model with lcfs prioritized packet management," in *Proceedings of the International Conference on Computer Communications (INFOCOM)*, Toronto, Canada, Jul. 6–9, 2020, pp. 1–5.

[23] ——, "Average age of information for a multi-source M/M/1 queueing model with packet management," in *Proceedings of the IEEE International Symposium on Information Theory*, Los Angeles, CA, Jun. 21–26, 2020, pp. 1–5.

[24] M. Costa, M. Codreanu, and A. Ephremides, "Age of information with packet management," in *Proceedings of the IEEE International Symposium on Information Theory*, Honolulu, HI, Jun. 20–23, 2014, pp. 1583–1587.

[25] L. Huang and E. Modiano, "Optimizing age-of-information in a multi-class queueing system," in *Proceedings of the IEEE International Symposium on Information Theory*, Hong Kong, China, Jun. 14–19, 2015, pp. 1681–1685.

[26] A. Kosta, N. Pappas, A. Ephremides, and V. Angelakis, "Age of information performance of multiaccess strategies with packet management," *Journal of Communications and Networks*, vol. 21, no. 3, pp. 244–255, 2019.

[27] R. D. Yates, "The age of information in networks: Moments, distributions, and sampling," *IEEE Transactions on Information Theory*, vol. 66, no. 9, pp. 5712–5728, May 2020.

[28] A. Maatouk, M. Assaad, and A. Ephremides, "On the age of information in a CSMA environment," *IEEE/ACM Transactions on Networking*, vol. 28, no. 2, pp. 818–831, Feb. 2020.

[29] A. B. Sediq, R. H. Gohary, R. Schoenen, and H. Yanikomeroglu, "Optimal tradeoff between sum-rate efficiency and Jain's fairness index in resource allocation," *IEEE Transactions on Wireless Communications*, vol. 12, no. 7, pp. 3496–3509, Jul. 2013.

[30] R. Jain, D. Chiu, and W. Hawe, "A quantitative measure of fairness and discrimination for resource allocation in shared systems," *Digital Equipment Corporation, DEC-TR-301, Tech. Rep., 1984*, available: www1.cse.wustl.edu/ jain/papers/ftp/fairness.pdf.

6 Age of Information in Source Coding

Melih Bastopcu, Baturalp Buyukates, and Sennur Ulukus

Throughput and delay are well-known metrics to assess the performance of communication networks. Throughput measures the amount of transmitted data in a certain time duration, and delay quantifies the time it takes to transmit a certain amount of data. Recently, with the proliferation of applications that require the delivery of real-time status information, such as Internet of Things (IoT) networks, autonomous vehicle systems, and social media networks, the concept of *timeliness* of information has emerged as a desirable feature, as in all of these systems, information is most valuable when it is most fresh.

The Age of Information metric has been proposed to assess timeliness of information in such systems. Unlike throughput and delay, age of information measures the time elapsed since the most recent update packet at the receiver was generated at the source. To obtain a good age performance, update packets need to be delivered regularly with low delay since the age captures not only the packet delay but also the intergeneration time of update packets. That is, good age performance corresponds to neither delay minimization nor throughput maximization alone, but a certain combined optimization of both.

This chapter focuses on age of information in the context of source coding, poses the timely source coding problem, and characterizes the age-optimal prefix-free (uniquely and instantaneously decodable) source codes that enable the delivery of timely status update packets over a communication network.

This chapter starts with the introduction of the age of information concept. Then, we present the timely source coding problem along with clarifying its distinction from the traditional source coding problem. Throughout the chapter, we discuss the timeliness of various encoding schemes including encoding all realizations and selective encoding policies, and find the corresponding age-optimal codes. We conclude this chapter with main takeaways and future research directions.

6.1 Introduction

In a typical network to study age of information, there is a source node that acquires time-stamped status updates regarding a phenomenon of interest. These time-sensitive status updates are transmitted to a receiver that wants to track the source process in a timely manner. The age of information at the receiver, or simply the age, is the time elapsed since the most recent update packet at the receiver was generated at the source. In a way, the age metric tracks the delay from the receiver's perspective. That is, instantaneous age at time t, $\Delta(t)$ is given by

$$\Delta(t) = t - u(t), \tag{6.1}$$

where $u(t)$ is the time-stamp of the most recent update packet at the receiver. Age increases linearly in time and drops to the packet delay of the received update upon packet delivery.

Age of information has been studied in the context of queueing networks [1–10], scheduling and optimization [11–31], energy harvesting [32–42], reinforcement learning [43–47] problems, and so on. The focus of this chapter is timely source coding, which will be described next.

6.1.1 Timely Source Coding Problem

In a status updating system, to represent the acquired information, a source code needs to be used. The traditional source coding problem [48] aims to minimize the expected codeword length. To this end, in traditional source coding, shorter codewords are assigned to more probable realizations of the data source, and longer codewords are reserved for less probable realizations [48]. In a communication setting where transmission of one bit of information takes a unit time, codeword lengths determine the transmission times. Thus, designing codewords to minimize the average codeword length minimizes the average transmission time, that is, expected packet delay.

When the status updates are time-sensitive, we need to minimize the age of information to maintain timeliness rather than minimize the average codeword length. We will show that age of information depends not only on the first moment of the codeword lengths but also on the second moment of the codeword lengths. Thus, minimizing age of information is different from minimizing average codeword lengths, and consequently, the timely source coding problem is different from the traditional source coding problem. We will show that well-known optimum traditional source codes are suboptimum for timely source coding.

Summary of Related Works

There are only a handful of works that study the timely source coding problem [49–54]. Reference [49] studies the timely source coding problem for a time-slotted system where the updates are generated at the source at each time unit. There is no buffer for the updates, and thus only the ones that arrive when the channel is empty are transmitted. Reference [49] finds real-valued codeword lengths by using Shannon codes based on a modified version of the given probability mass function (pmf) and shows that these codeword lengths are age-optimal up to a constant gap. As an extension, reference [49] proposes the idea of randomly skipping the transmission of some status updates even if the channel is free to achieve a lower average age at the information receiver.

Reference [50] considers a time-slotted system where the updates generated by the source arrive at the transmitter at each time slot. In this work, the encoder groups every B updates into a single group and encodes the entire group into variable-length bit strings. Status updates which arrive when the channel is busy are buffered such that the entire source message stream is reconstructed in a lossless manner, that is, updates are not dropped and discarded. This work proposes a block coding scheme that minimizes an approximated average age expression. Reference [51] further improves the average age performance of [50] and shows that a smaller average age at the receiver can be achieved by allowing different block lengths based on the backlog of the symbols at the encoder.

Reference [52] minimizes the average peak age of information for randomly arriving source symbols in a time-slotted system. In this work, status updates that find the transmitter busy are placed in a First In First Out (FIFO) queue such that the entire source message stream is reconstructed in a lossless manner. When the queue is empty and there are no update arrivals, the transmitter sends an empty status update. Reference [52] considers two different models for the transmission of the empty symbol. In an idealized model, a special signaling is available for the empty status update, and thus the empty status update is not a part of the encoding policy. If there is no special signal available for the empty symbol, however, the empty status update is encoded along with the updates generated at the source.

Reference [53] studies a timely source coding problem for randomly arriving status updates in a continuous time system. Status update packets that find the transmitter busy are not encoded and lost. In order to achieve a lower average age at the receiver, this work considers selective encoding mechanisms where the most probable k status updates are always encoded. For the remaining $n - k$ least probable status updates, reference [53] considers three different scenarios: they are never transmitted; they are transmitted randomly with certain probability; they are mapped into a designated empty symbol. In [53], for all these encoding schemes, real-valued age-optimal codeword lengths are determined, and for a given pmf, optimal k values are characterized numerically. This work shows that selective encoding methods achieve a lower average age than the greedy approach of encoding all possible realizations.

Reference [54] considers the problem of generating *partial updates* which have smaller information content and also smaller transmission times compared to the original updates. This work studies the generation of age-optimal partial updates together with the corresponding age-optimal real-valued codeword lengths, subject to a constraint on the information fidelity between the original and the partial updates.

Categorization

We categorize the existing literature on the timely source coding problem according to four main aspects. The first aspect is the nature of update arrivals. Update arrivals can be deterministic, that is, one arrival at each time slot, stochastic, or generated at will, that is, the updates are generated upon a request by the transmitter. The second aspect is the nature of codeword lengths. Codeword lengths can be integer or real-valued. The third aspect is whether the system has no update drops, that is, each arriving update

Table 6.1 Summary of the existing literature on age of information with source coding.

arrival profile	codeword type	update drops	age metric	reference
deterministic	real-valued	drops	average age	[49]
deterministic	integer-valued	no drops	average age	[50, 51]
stochastic	integer-valued	no drops	peak age	[52]
stochastic	real-valued	drops	average age	[53]
generation at will	real-valued	no drops	average age	[54]

packet is encoded and transmitted, or has update drops, that is, not every arriving update packet is encoded and transmitted. The fourth aspect is the targeted age metric, which can be average age of information, or peak age of information.

6.1.2 Chapter Outline and Focus

Over the next two sections, we will discuss two main source coding schemes. In the first source coding scheme, which is detailed in Section 6.2, the transmitter encodes all possible realizations such that an update packet is transmitted for each realization at the source side. In the second type of source coding scheme, which is detailed in Section 6.3, the transmitter performs a selective encoding scheme such that it only encodes and transmits a subset of all possible realizations. We implement this selection based on the realization probabilities, as well as randomly, and analyze the use of a designated empty symbol to indicate a skipped transmission. For each of these different encoding schemes, we characterize the age-optimal codes that minimize the average age of information at the receiver. A main takeaway of this chapter is emphasized below.

> **Greedy Encoding is Not Always Optimal**
> To minimize the average age, the optimum encoding scheme is not to encode and transmit every single realization. Selective encoding schemes that encode only a subset of the realizations keeps the codeword lengths short enough, without decreasing the effective arrival rate significantly, to achieve a better age performance.

In the next section, we first consider a greedy approach, which encodes every realization.

6.2 Encoding All Realizations

In this section, we discuss the results reported in [53, 55, 56]. In these works, an information source generates independent and identically distributed (i.i.d.) status update packets from the set $\mathcal{X} = \{x_1, x_2, \ldots, x_n\}$ with a known pmf $P_X(x_i)$ for $i \in \{1, \ldots, n\}$.

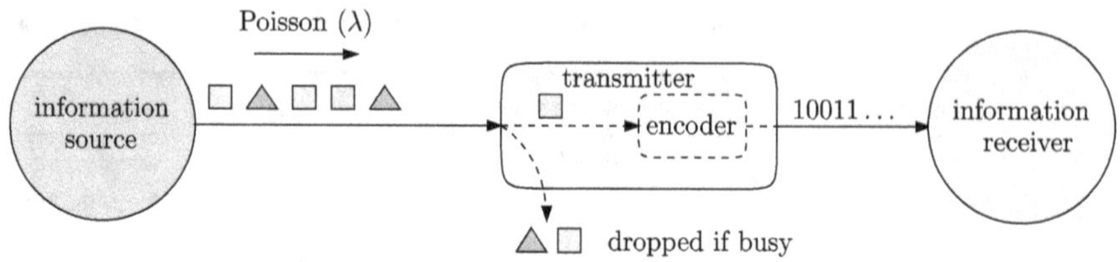

Figure 6.1 Update arrivals that find the transmitter idle are encoded and sent to the receiver. Update arrivals that find the transmitter busy are dropped.

Without loss of generality, we assume that the probabilities of the status updates are in a nonincreasing order, that is, $P_X(x_m) \geq P_X(x_{m+1})$ for all m. The status updates arrive at the transmitter following a Poisson process with rate λ. The updates that arrive when the transmitter is busy are dropped and lost. Thus, in these works, only the status updates that arrive when the transmitter is idle are sent to the information receiver (see Figure 6.1). We denote these updates as *successful* status updates.

After receiving a status update x_i, the transmitter assigns codeword $c(x_i)$ with length $\ell(x_i)$ to update x_i. Then, the first and the second moments of the codeword lengths are given by

$$\mathbb{E}[L] = \sum_{i=1}^{n} P_X(x_i)\ell(x_i), \tag{6.2}$$

$$\mathbb{E}[L^2] = \sum_{i=1}^{n} P_X(x_i)\ell(x_i)^2. \tag{6.3}$$

We assume that the channel between the transmitter and the receiver is error free and the transmitter can only send one bit at a unit time. Thus, the transmission time for status update x_i is equal to $\ell(x_i)$. Unlike the previous works in the age of information literature where the transmission times are based on a given distribution, in [53, 55, 56], the aim is to design the age-optimal codeword lengths, and therefore the transmission times, through source coding schemes.

As we have Poisson arrivals, and the service times are general, which are based on the codeword lengths, the communication system we consider is a $M/G/1/1$ queueing system. Performing a graphical age analysis using Figure 6.2, we find the long-term average age at the receiver, Δ, as [8]

$$\Delta = \frac{\mathbb{E}[Y^2]}{2\mathbb{E}[Y]} + \mathbb{E}[S], \tag{6.4}$$

where random variable Y denotes the time between two successful arrivals at the transmitter and is given by $Y = S + W$, where S denotes the service times and W represents the overall waiting time until the next successful update arrival.

As the service times of the status updates are equal to the codeword lengths, we have $\mathbb{E}[S] = \mathbb{E}[L]$ and $\mathbb{E}[S^2] = \mathbb{E}[L^2]$. Due to the memoryless property of the exponential random variable, the waiting time for the next successful status update packet W is exponentially distributed with rate λ where $\mathbb{E}[W] = \frac{1}{\lambda}$ and $\mathbb{E}[W^2] = \frac{2}{\lambda^2}$. We note that

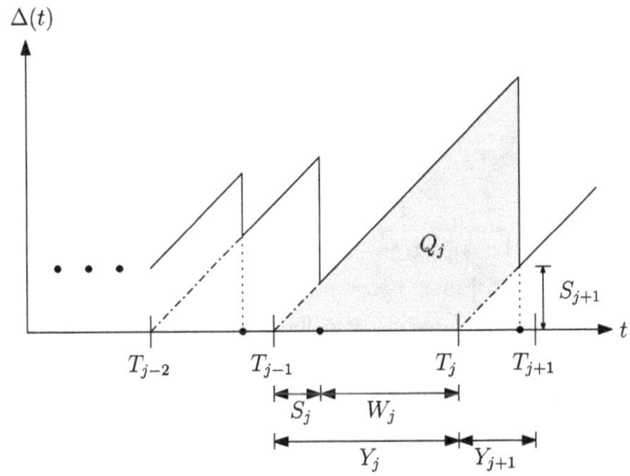

Figure 6.2 Sample age evolution $\Delta(t)$ at the receiver. Successful updates are indexed by j. The jth successful update arrives at the server node at T_{j-1}. Update cycle at the server node is the time in between two successive arrivals and is equal to $Y_j = S_j + W_j = T_j - T_{j-1}$.

$$\mathbb{E}[Y] = \mathbb{E}[L] + \mathbb{E}[W], \tag{6.5}$$

$$\mathbb{E}[Y^2] = \mathbb{E}[L^2] + 2\mathbb{E}[W]\mathbb{E}[L] + \mathbb{E}[W^2]. \tag{6.6}$$

By inserting (6.5) and (6.6) into (6.4), we obtain the average age as

$$\Delta = \frac{\mathbb{E}[L^2] + \frac{2}{\lambda}\mathbb{E}[L] + \frac{2}{\lambda^2}}{2\left(\mathbb{E}[L] + \frac{1}{\lambda}\right)} + \mathbb{E}[L]. \tag{6.7}$$

The average age expression in (6.7) depends on the first and the second moments of the codeword lengths as well as the status update arrival rate λ. Our aim is to find the age-optimal real-valued codeword lengths that minimize the average age given in (6.7). To this end, we formulate the following optimization problem:

$$\begin{aligned} \min_{\{\ell(x_i)\}} \quad & \frac{\mathbb{E}[L^2] + \frac{2}{\lambda}\mathbb{E}[L] + \frac{2}{\lambda^2}}{2\left(\mathbb{E}[L] + \frac{1}{\lambda}\right)} + \mathbb{E}[L] \\ \text{s.t.} \quad & \sum_{i=1}^{n} 2^{-\ell(x_i)} \leq 1 \\ & \ell(x_i) \in \mathbb{R}^+, \quad i \in \{1, \ldots, n\}. \end{aligned} \tag{6.8}$$

The objective function in (6.8) is equal to the average age found in (6.7); the first constraint is the Kraft inequality, which is needed for the feasibility of a uniquely decodable code; and the second constraint is the feasibility of the codeword lengths, that is, each codeword length must be nonnegative.

We note that age-minimizing source codes are different from traditional source codes that minimize the expected codeword length, $\mathbb{E}[L]$. To be precise, the traditional source coding problem with real-valued codeword lengths is given by

$$\begin{aligned} \min_{\{\ell(x_i)\}} \quad & \mathbb{E}[L] \\ \text{s.t.} \quad & \sum_{i=1}^{n} 2^{-\ell(x_i)} \leq 1 \\ & \ell(x_i) \in \mathbb{R}^+, \quad i \in \{1, \ldots, n\}. \end{aligned} \tag{6.9}$$

As the objective function in the optimization problem in (6.8) depends on both $\mathbb{E}[L]$ and $\mathbb{E}[L^2]$, its solution is not necessarily at the point that minimizes $\mathbb{E}[L]$ alone. Thus, the timely source coding problem in (6.8) is different from the traditional source coding problem in (6.9).

Next, we solve the optimization problem in (6.8). Similar to [21] and [35], we define the following intermediate problem parameterized by θ,

$$\begin{aligned} p(\theta) := \min_{\{\ell(x_i)\}} \quad & \frac{1}{2}\mathbb{E}[L^2] + \mathbb{E}[L]^2 + (2a - \theta)\mathbb{E}[L] + a^2 - \theta a \\ \text{s.t.} \quad & \sum_{i=1}^{n} 2^{-\ell(x_i)} \leq 1 \\ & \ell(x_i) \in \mathbb{R}^+, \quad i \in \{1, \ldots, n\}, \end{aligned} \tag{6.10}$$

where $a = \frac{1}{\lambda}$. One can show that $p(\theta)$ in (6.10) decreases with θ and the optimal solution is obtained when $p(\theta^*) = 0$. The optimal age for the problem in (6.8) is equal to θ^*, that is, $\Delta^* = \theta^*$ [57].

Next, we solve (6.10). We define the Lagrangian [58] function for (6.10) as

$$\mathcal{L} = \frac{1}{2}\mathbb{E}[L^2] + \mathbb{E}[L]^2 + (2a - \theta)\mathbb{E}[L] + a^2 - \theta a + \beta\left(\sum_{i=1}^{k} 2^{-\ell(x_i)} - 1\right), \tag{6.11}$$

where $\beta \geq 0$.

We note the fact that the optimal codeword lengths must satisfy the Kraft inequality with an equality, that is, $\sum_{i=1}^{n} 2^{-\ell(x_i)} = 1$. By taking the derivative of the Lagrangian in (6.11) with respect to $\ell(x_i)$ and equating to 0, we obtain the unique solution for $\ell(x_i)$ for $i \in \{1, 2, \ldots, n\}$ as

$$\ell(x_i) = -\frac{\log\left(\frac{P_X(x_i)}{\beta(\log 2)^2} W\left(\frac{\beta(\log 2)^2}{P_X(x_i)} 2^{\frac{-\theta+2\beta\log 2+2a}{3}}\right)\right)}{\log 2}, \tag{6.12}$$

where $W(\cdot)$ denotes the principal branch of the Lambert W function [59].

We note that there are two unknowns in (6.12), θ and β. We also note that θ and β must satisfy two equalities, $p(\theta) = 0$, and the Kraft inequality with equality, that is, $\sum_{i=1}^{n} 2^{-\ell(x_i)} = 1$. Thus, we solve for the two unknowns θ and β using these two equalities. Alternatively, we can use the following iterative procedure to solve for θ and β. Starting from an arbitrary (θ, β) pair, if $p(\theta) > 0$ (or $p(\theta) < 0$), we increase (or respectively decrease) θ in the next iteration, as $p(\theta)$ is a decreasing function of θ. Then, we update β such that the Kraft inequality holds with equality, that is, $\sum_{i=1}^{n} 2^{-\ell(x_i)} = 1$. We repeat these steps until $p(\theta) = 0$ and $\sum_{i=1}^{n} 2^{-\ell(x_i)} = 1$.

With the solution method just described, we can find the age-optimal real-valued codeword lengths for any given pmf. As an example, we consider the following Zipf(n, s) pmf,

$$P_X(x_i) = \frac{i^{-s}}{\sum_{j=1}^{n} j^{-s}}, \quad 1 \leq i \leq n. \tag{6.13}$$

We note that when $s = 0$, Zipf distribution becomes the uniform distribution. When s gets larger, the Zipf distribution becomes more polarized. In the following numerical example, we take the update arrival rate $\lambda = 1$, the number of status updates $n = 10$, and $s = 0, 1, 2$. We compare the codeword lengths of the age-optimal code obtained in (6.12) by solving the optimization problem in (6.8) with the codeword lengths obtained by solving the optimization problem in (6.9). The codeword lengths that solve (6.9) are given by $\ell(x_i) = -\log_2(P_X(x_i))$, and we call them Shannon* codes here.[1]

We observe in Figure 6.3(a) that when $s = 0$, that is, when the distribution is uniform, the age-optimal codeword lengths are equal to Shannon* codeword lengths. This result is also observed in [49]. When the distribution is more polarized, that is, when $s = 1$ in Figure 6.3(b), and $s = 2$ in Figure 6.3(c), we observe that codeword lengths of the age-optimal code become significantly different from Shannon* codeword lengths, which shows that the age minimization problem is different from the classical source coding problem where the aim is to minimize the average codeword length.

In the next section, we explore the idea of selective encoding where the transmitter does not encode all status updates even if it is idle. We will see that such a strategy achieves a smaller average age compared to the strategy where all realizations are encoded.

6.3 Selective Encoding Schemes

In Section 6.2, we presented a setting where all update realizations are encoded. Unlike this encoding structure, in this section, we discuss selective encoding schemes where only a subset of the update realizations is encoded and transmitted. That is, remaining realizations are not encoded even if the channel is free upon their arrival. Specifically, we present the highest k selective encoding, randomized selective encoding, and the highest k selective encoding with empty symbol schemes proposed in [53, 55, 56].

6.3.1 Highest k Selective Encoding Scheme

In this encoding scheme, the transmitter only encodes the most probable k realizations, that is, only the realizations from the set $\mathcal{X}_k = \{x_1, \ldots, x_k\}$ that have the highest probabilities among the possible n realizations, where $k \in \{1, \ldots, n\}$. The transmitter

[1] Note that the solution of (6.9) for integer-valued codeword lengths, i.e., $\ell(x_i) \in \mathcal{Z}^+$, is the Huffman code [48]. Shannon code is a well-known suboptimal source code with integer-valued codeword lengths $\ell(x_i) = \lceil -\log(P_X(x_i)) \rceil$ [48]. Codeword length produced by Shannon code is a feasible solution for (6.9). We use Shannon* to denote optimal real-valued codeword lengths for (6.9).

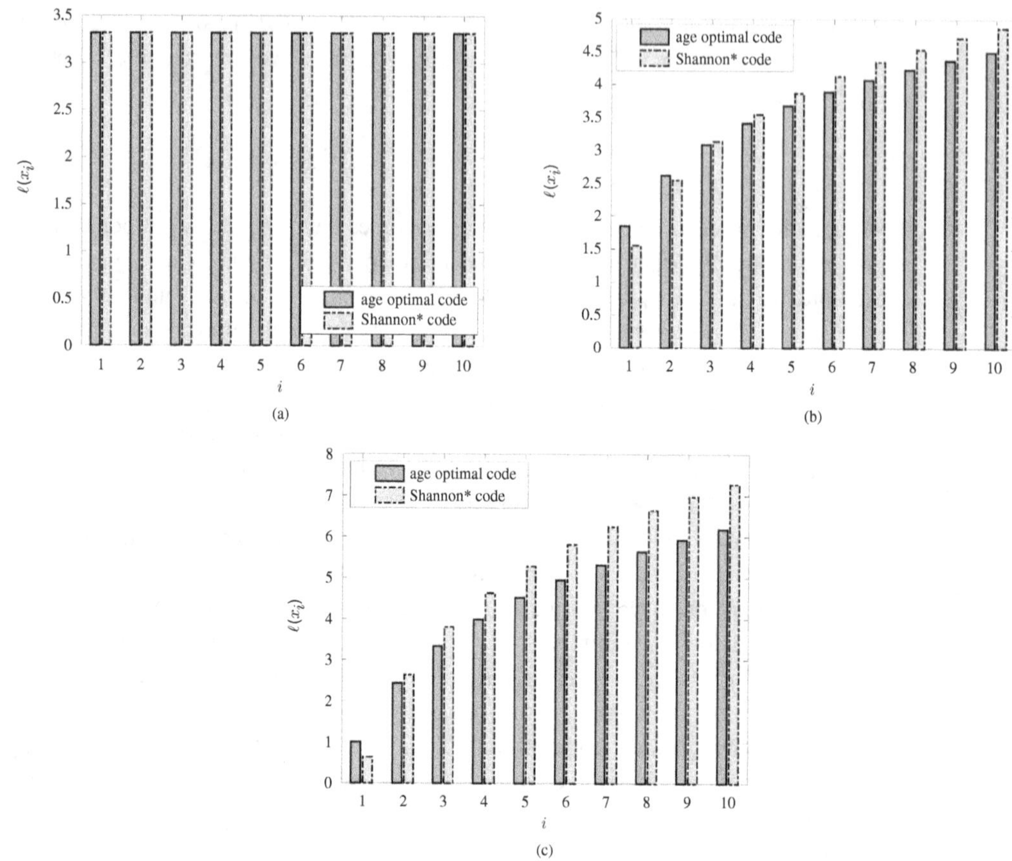

Figure 6.3 Codeword lengths under Shannon* code and the age-optimal code for $\lambda = 1$ and the pmf in (6.13) with the parameters $n = 10$, (a) $s = 0$, (b) $s = 1$, and (c) $s = 2$.

drops and never encodes the remaining nonselected $n - k$ realizations. This selective encoding scheme is illustrated in Figure 6.4. Here, we note that the transmitter drops a realization from the selected k realizations if it is busy upon arrival. Thus, in the context of this encoding scheme, we term update realizations from the set $\mathcal{X}_k$ that arrive when the transmitter is idle as successful arrivals. Upon a successful arrival, the transmitter encodes that update realization using the following conditional pmf:

$$P_{X_k}(x_i) = \begin{cases} \frac{P_X(x_i)}{q_k}, & i = 1, 2, \ldots, k \\ 0, & i = k+1, k+2, \ldots, n, \end{cases} \tag{6.14}$$

where

$$q_k \triangleq \sum_{\ell=1}^{k} P_X(x_\ell). \tag{6.15}$$

We note that the realization x_i has codeword $c(x_i)$ with length $\ell(x_i)$, for $i \in \{1, \ldots, k\}$.

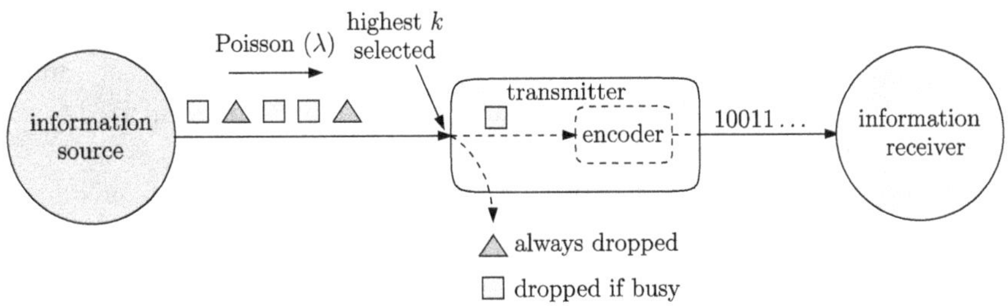

Figure 6.4 Selected update realizations (shown with a square) that find the transmitter idle are encoded and sent to the receiver. Nonselected realizations (shown with a triangle) are always dropped.

Under the highest k selective encoding scheme, the overall waiting time W is equal to

$$W = \sum_{\ell=1}^{M} Z_\ell, \tag{6.16}$$

where Z_ℓs denote the residual time that the transmitter waits for the next arrival upon completion of a service. We note that Z_ℓs are i.i.d. exponential random variables with rate λ from the memoryless property of the inter-arrival times. Here, M is a geometric random variable with parameter q_k and denotes the number of arrivals until an arrival from the set $\mathcal{X}_k$ occurs at the transmitter node. W is also an exponential random variable with rate λq_k [60, Prob. 9.4.1]. Then, using (6.5) and (6.6) along with (6.16) in (6.4), we find

$$\Delta = \frac{\mathbb{E}[L^2] + \frac{2}{q_k\lambda}\mathbb{E}[L] + \frac{2}{(q_k\lambda)^2}}{2\left(\mathbb{E}[L] + \frac{1}{q_k\lambda}\right)} + \mathbb{E}[L], \tag{6.17}$$

where the first and the second moments of the codeword lengths are given by $\mathbb{E}[L] = \sum_{i=1}^{k} P_{X_k}(x_i)\ell(x_i)$ and $\mathbb{E}[L^2] = \sum_{i=1}^{k} P_{X_k}(x_i)\ell(x_i)^2$.

We observe that the average age expression in the case of the highest k selective encoding scheme given in (6.17) depends on the first and second moments of the codeword lengths, the given pmf, selected k, and the arrival rate λ.

To find the real-valued prefix-free age-optimal codeword lengths, for a given pmf and k, we formulate the following optimization problem:

$$\begin{aligned} \min_{\{\ell(x_i)\}} \quad & \frac{\mathbb{E}[L^2] + 2a\mathbb{E}[L] + 2a^2}{2(\mathbb{E}[L] + a)} + \mathbb{E}[L] \\ \text{s.t.} \quad & \sum_{i=1}^{k} 2^{-\ell(x_i)} \le 1 \\ & \ell(x_i) \in \mathbb{R}^+, \quad i \in \{1, \ldots, k\}, \end{aligned} \tag{6.18}$$

where the objective function is the average age expression in (6.17) with $a = \frac{1}{\lambda q_k}$, the first constraint is the Kraft inequality, and the second constraint is the nonnegativity of the codeword lengths. We define an intermediate problem similar to the problem in (6.10) as detailed in Section 6.2 and find the age-optimal codeword lengths through this intermediate parameterized problem for a given pmf and fixed k (see the derivation details in [53]).

We note that a smaller k value results in shorter transmission times, that is, shorter codeword lengths, but larger waiting times in between successful update realizations. This is because for a smaller k most of the update realizations are not encoded and the transmitter waits longer to receive a realization from the set $\mathcal{X}_k$. On the other hand, when k is a large number, waiting times get shorter at the expense of increased codeword lengths. Thus, to get the minimum age performance out of the highest k selective encoding scheme, parameter k needs to be tuned carefully for a given pmf to balance these two opposing trends.

Having found the age-optimal codeword lengths for a fixed k value, we determine the age minimizing k value numerically. Next, we present numerical results for Zipf(n, s) distribution with the pmf in (6.13) for $n = 100$, $s = 0.4$.

In Figure 6.5 we show the average age as a function of k for infrequent arrival profiles, that is, for $\lambda = 0.3, 0.5, 1$ where the age-optimal k value for each case is indicated with an arrow. We see that as the arrival rate increases, age decreases as expected. Here, the key observation is that the optimal k is not equal to n. That is, to get the best age performance the transmitter needs to implement a selective encoding as opposed to encoding all realizations. We see that, under these arrival rates, age-optimal k values are not close to 1, which shows that the transmitter chooses to encode more updates as opposed to idly waiting for the next arrival when the updates are infrequent. The optimal k values are not close to n either, as this selection increases the codeword lengths of low probability updates, which in turn dominates the age performance.

In Figure 6.6, we present a similar result but for larger values of the arrival rate λ. The key takeaway from this figure is that the age-optimal k value approaches 1 when updates become more frequent. In fact, when $\lambda = 10$, the age-optimal k is equal to 1, that is, the transmitter only encodes the most probable realization and discards the remaining $n-1$ realizations. This indicates that when the arrivals are frequent enough, it is preferable not to encode more realizations to keep the codeword lengths, that is, transmission times, short.

One other observation from Figures 6.5 and 6.6 is that for large values of λ, for example, $\lambda = 10$, age is an increasing function of k since in this case codeword lengths dominate the age performance. On the other hand, for smaller λ values, for example, $\lambda = 0.3, 0.5, 1, 2$, we see that age is first a decreasing function of k because, as k increases, the waiting time in between updates decreases, which is critical since inter-arrivals are less frequent. For larger k values, however, codeword lengths start to dominate the performance, which in turn results in an increasing age with k.

In Figure 6.7, the performance of the age-optimal codes is compared with that of the well-known codes that minimize the average codeword length such as Huffman code and Shannon* code [48]. In these numerical results, we use the pmf in (6.13) with $n = 10$ and $s = 0, 3, 4$ for $\lambda = 1$. When $s = 0$, this distribution becomes a uniform distribution, in which case Shannon* code and the age-optimal code are identical as shown in Figure 6.7(a). In this figure, we see that when k is a power of 2, Huffman code becomes the same as the Shannon* and the age-optimal code. In Figures 6.7(b) and (c), we consider $s=3$ and $s=4$ cases, respectively, and observe that the age-optimal code outperforms Shannon* code and Huffman code. The key takeaway from these numerical results is that the age-optimal code significantly improves the age performance when we have a polarized distribution, that is, when s is higher.

We note that a similar k out of n type idea appears in the context of multicasting updates to monitor nodes in [61–65]. In these works, the source node transmits each update packet until the earliest k of the total n monitor nodes receive that particular packet. The main result of these works is that sending update packets until (earliest) k out of n monitors receive them as opposed to waiting for all monitors to receive them achieves a smaller average age. Analogously, in the context of source coding, sending updates for (most probable) k out of n realizations as opposed to sending updates for all realizations achieves a smaller average age as shown in Figures 6.5 and 6.6.

We note that even though so far we have selected the most probable k realizations for encoding, the same analysis holds true for any k-subset of the total n realizations. In other words, average age in an encoding scheme in which the transmitter encodes a predetermined k-subset of the realizations is still given by (6.17) and the corresponding age-optimal codeword lengths are found through the optimization problem in (6.18). In what follows, we discuss the optimality of the highest k selection among all $\binom{n}{k}$ selections for encoding.

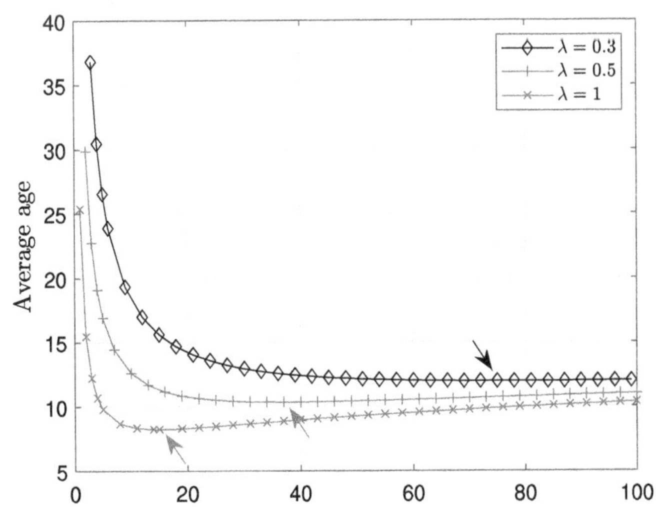

Figure 6.5 Average age under the age-optimal codeword lengths as a function of k for $\lambda \in \{0.3, 0.5, 1\}$ for the pmf provided in (6.13) with the parameters $n = 100$, $s = 0.4$ when the highest k selective encoding scheme is implemented. Age-minimizing k values are indicated with an arrow.

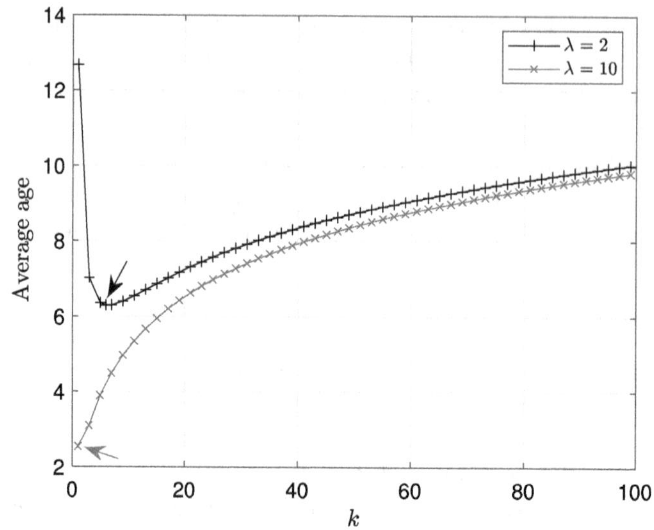

Figure 6.6 Average age under the age-optimal codeword lengths as a function of k for $\lambda \in \{2, 10\}$ for the pmf provided in (6.13) with the parameters $n = 100$, $s = 0.4$ when the highest k selective encoding scheme is implemented. Age-minimizing k values are indicated with an arrow.

Optimality of the Highest k Selection

The age expression in the highest k selective encoding scheme provided in (6.17) depends on the effective arrival rate of the encoded updates and codeword lengths, which depends on the given pmf of X. In this section, we present results from [53] that take a careful look at the selection of k realizations for encoding and establish optimality regimes for the highest k selection. Let λ_e denote the effective arrival rate $\lambda_e = \lambda \sum_{x \in \mathcal{X}_s} P_X(x)$, where $\mathcal{X}_s$ is the set of arbitrarily selected k updates for encoding. From Figure 6.5 we know that when arrivals are infrequent, the average age is mainly dominated by the effective arrival rate. In this case, choosing the most probable realizations for encoding may be preferable to achieve a higher effective arrival rate. On the other hand, when arrivals are frequent, codeword lengths dominate the age performance, which hints that under such frequent arrival profiles selecting most probable realizations for encoding may not be optimal.

To illustrate this structure, we find the age-optimal selections for given pmfs for $n = 10$ and $k = 5$ and present the results in Table 6.2. We use the Zipf distribution given in (6.13) with parameters $n = 10$, $s = 0.2$ and the following pmf with $n = 10$:

$$P_X(x_i) = \begin{cases} 2^{-i}, & i = 1, \ldots, n-1 \\ 2^{-n+1}, & i = n. \end{cases} \tag{6.19}$$

In both pmfs, realizations are in a decreasing order such that we have $P_X(x_i) \geq P_X(x_j)$ if $i \leq j$. We observe that for $\lambda = 0.1$ for the first pmf and for $\lambda = 0.5$ in the second pmf, choosing the realizations with the highest probabilities, that is, encoding realizations $\{1, 2, 3, 4, 5\}$, is optimal in line with the preceding discussion on the

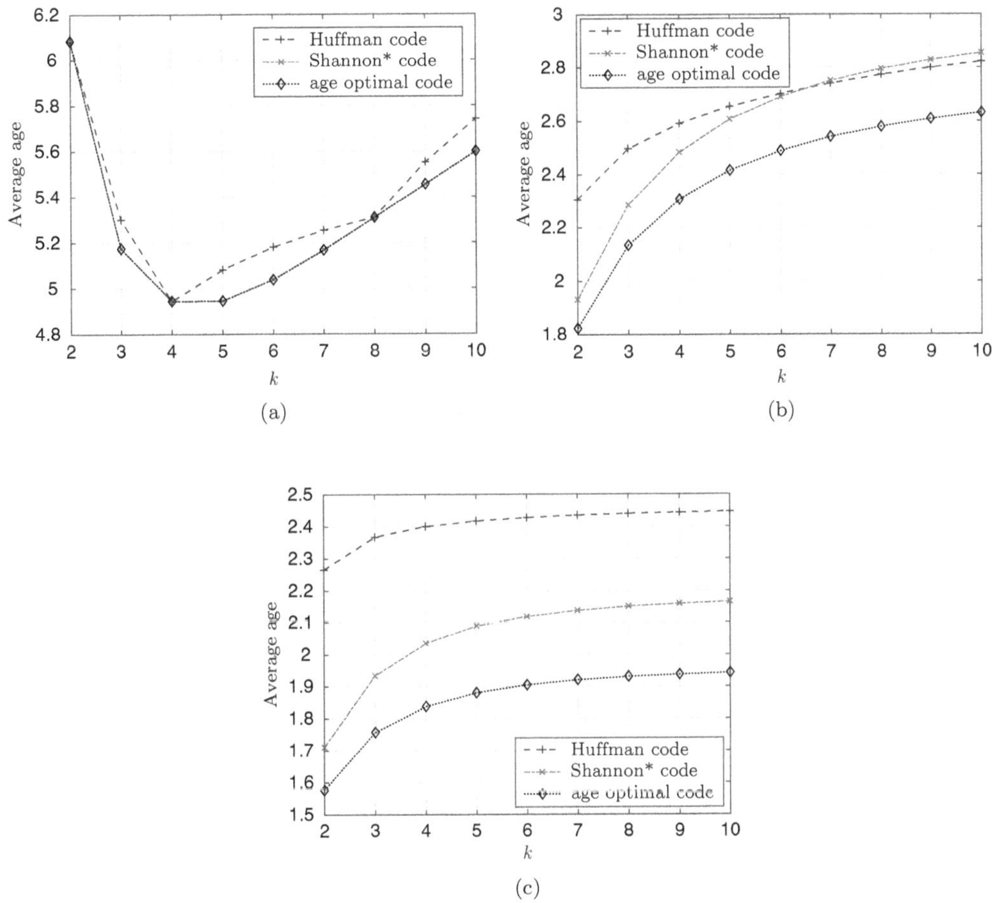

Figure 6.7 The average age under Huffman code, Shannon* code, and the age-optimal code for $\lambda = 1$ and the pmf in (6.13) with the parameters $n = 10$, (a) $s = 0$, (b) $s = 3$, and (c) $s = 4$. k varies from 2 to n.

optimality of highest k under low arrival profiles since this selection maximizes the effective arrival rate. However, when the arrival rate is high, for example, $\lambda = 1$ for the first pmf and $\lambda = 2$ for the second pmf, the optimal policy for both pmfs is to encode the most probable realization along with the $k - 1$ least probable realizations, that is, encoding realizations $\{1, 7, 8, 9, 10\}$, since this selection keeps the codeword lengths at a reasonable level for the optimum age performance. From these, we observe that the optimal selection for a fixed k aims to increase the effective arrival rate while keeping the moments of the codeword lengths at a suitable level, which is observed under intermediate arrival rates, for example, $\lambda = 0.5$ for the first pmf and $\lambda = 1$ in the second pmf, where the optimal selection is to encode the most probable two realizations along with the least probable $k - 2 = 3$ realizations, that is, encoding realizations $\{1, 2, 8, 9, 10\}$.

Table 6.2 Age-optimal update selection for fixed $k = 5$ with different arrival rates, λ.

pmf	λ	optimal selection	λ_e	optimal age
The pmf in (6.19) for $n = 10$	0.1	{1,2,3,4,5}	0.0969	12.292
	0.5	{1,2,8,9,10}	0.3789	3.867
	1	{1,7,8,9,10}	0.5156	2.4229
Zipf($n = 10, s = 0.2$)	0.5	{1,2,3,4,5}	0.3898	5.154
	1	{1,2,8,9,10}	0.6269	3.929
	2	{1,7,8,9,10}	1.01	3.304

Thus, from the preceding analysis, we observe that the highest k selection is not necessarily age-optimal for a given pmf and arrival rate. The results in [53] suggest that the highest k selection is age-optimal when the arrival rate is low, but a detailed theoretical analysis on when the highest k encoding scheme is optimal is needed. In addition, this selective encoding scheme can be further refined to reflect the importance of each realization since there can be models in which an unlikely realization may carry important information. In such a scenario, neglecting this realization could cause undesired information loss at the receiver. To mitigate this, selective encoding can be performed by considering both the importance of each realization and the realization probabilities.

6.3.2 Randomized Selective Encoding Scheme

So far, we discussed the highest k encoding scheme in which the encoding selection is deterministic and is based on the realization probabilities alone. In this section, we present results from [66] and [53] that consider randomized transmission decisions to enable the transmitter to skip the transmission of an update even if the channel is free.

The idea of randomized encoding is first proposed in [66], where the transmitter encodes each realization x_i with probability $\alpha(x_i)$ and sends a designated empty symbol $\emptyset$ to indicate a skipped transmission. We note that here the randomization enables the transmission of certain less probable realizations, unlike the highest k selective encoding scheme in which less probable realizations are never transmitted.

To illustrate the randomized encoding idea, we present the following example from reference [66], where the source generates $\mathcal{X} = \{x_1, x_2, \ldots, x_{64}\}$ with the following pmf:

$$P_X(x_i) = \begin{cases} \frac{1}{4}, & \text{if } i \in \{1, \ldots, 3\}, \\ \frac{1}{244}, & \text{if } i \in \{4, \ldots, 64\}. \end{cases} \tag{6.20}$$

Here, under the randomized encoding, the codeword lengths are given by

$$L(\alpha) = \begin{cases} \ell(x_i), & \text{w.p. } P_X(x_i)\alpha(x_i) \\ \ell(\emptyset), & \text{w.p. } 1 - \mathbb{E}[\alpha(X)]. \end{cases} \tag{6.21}$$

We note that an extreme case of the randomization is selecting $\alpha(x)$ equal to 1 for certain realizations and 0 for others. For example, let us consider a randomized encoding scheme in which realizations x_1, x_2, and x_3 are always encoded and transmitted whereas the remaining realizations x_i for $i \geq 4$ are discarded, and thus the empty symbol $\emptyset$ is transmitted in the case of these realizations. As a result of this randomization, each status update from the set $\{x_1, x_2, x_3, \emptyset\}$ is transmitted with probability $\frac{1}{4}$. We note that, under this α, this encoding scheme essentially becomes a highest k encoding scheme, for $k = 3$, with an empty symbol, which will be discussed in Section 6.3.3 in detail.

As a natural extension of the highest k encoding scheme to the randomized encoding setting, reference [53] proposes a randomized encoding scheme in which the most probable k realizations are always encoded and each of the remaining $n-k$ realizations is transmitted with probability α. That is, with probability $1 - \alpha$ these least probable $n - k$ realizations are discarded even if the channel is free upon their arrival. We note that unlike the randomized encoding scheme of [66], this scheme does not utilize an empty symbol. Thus, under this scheme codeword lengths become

$$L(\alpha) = \ell(x_i), \quad \text{w.p. } P_X(x_i)\alpha(x_i), \tag{6.22}$$

with $\alpha(x_i) = 1$ for the most probable k realizations given by the set $\mathcal{X}_k$ and $\alpha(x_i) = \alpha$ for the remaining realizations in set $\mathcal{X} \setminus \mathcal{X}_k$. With this randomization, previously discarded least likely realizations under the highest k selective encoding scheme are sometimes transmitted to the receiver node. In other words, this randomized selective encoding policy strikes a balance between encoding every single realization and the highest k selective encoding scheme discussed so far. Thus, the transmitter needs to generate codewords for all n realizations. Here, the transmitter assigns codeword $c(x_i)$ with length $\ell(x_i)$ to realization x_i for $i \in \{1, 2, \ldots, n\}$ using the following conditional probabilities,

$$P_{X_\alpha}(x_i) = \begin{cases} \frac{P_X(x_i)}{q_{k,\alpha}}, & i = 1, 2, \ldots, k \\ \alpha \frac{P_X(x_i)}{q_{k,\alpha}}, & i = k+1, k+2, \ldots, n, \end{cases} \tag{6.23}$$

where

$$q_{k,\alpha} \triangleq \sum_{\ell=1}^{k} P_X(x_\ell) + \alpha \sum_{\ell=k+1}^{n} P_X(x_\ell). \tag{6.24}$$

We find the average age experienced by the receiver node under this randomized selective encoding scheme as

$$\Delta_\alpha = \frac{\mathbb{E}[L^2] + \frac{2}{q_{k,\alpha}\lambda}\mathbb{E}[L] + \frac{2}{(q_{k,\alpha}\lambda)^2}}{2\left(\mathbb{E}[L] + \frac{1}{q_{k,\alpha}\lambda}\right)} + \mathbb{E}[L], \tag{6.25}$$

where $\mathbb{E}[L] = \sum_{i=1}^{n} P_{X_\alpha}(x_i)\ell(x_i)$, and $\mathbb{E}[L^2] = \sum_{i=1}^{n} P_{X_\alpha}(x_i)\ell(x_i)^2$. We note that (6.25) follows similarly to (6.17) by replacing q_k with $q_{k,\alpha}$.

Next, we formulate the following optimization problem to determine the age-optimal codewords for this encoding scheme,

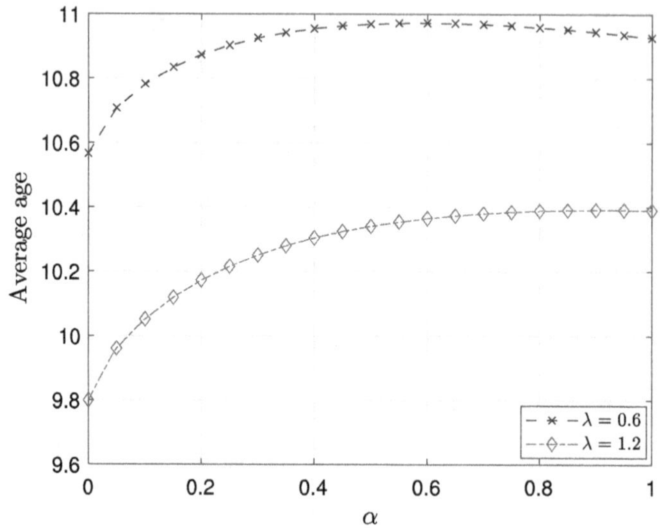

Figure 6.8 Average age under the age-optimal codeword lengths for different α values with the pmf provided in (6.13) with $n = 100$, $s = 0.2$ for $k = 70$ and $\lambda = 0.6, 1.2$ when the randomized highest k selective encoding is implemented.

$$
\begin{aligned}
\min_{\{\ell(x_i),\alpha\}} \quad & \frac{\mathbb{E}[L^2] + 2\bar{a}\mathbb{E}[L] + 2\bar{a}^2}{2(\mathbb{E}[L] + \bar{a})} + \mathbb{E}[L] \\
\text{s.t.} \quad & \sum_{i=1}^{n} 2^{-\ell(x_i)} \le 1 \\
& \ell(x_i) \in \mathbb{R}^+, \quad i \in \{1, \ldots, n\},
\end{aligned} \tag{6.26}
$$

where the objective function is the average age, Δ_α, in (6.25) with $\bar{a} = \frac{1}{\lambda q_{k,\alpha}}$ and the first constraint is the Kraft inequality, and the second constraint denotes the feasibility region of the codeword lengths.

We solve this optimization problem by introducing an intermediate problem as in Section 6.2 and characterize the age-optimal code for the randomized selective encoding scheme (see [53] for derivation details).

In Figure 6.8, we give numerical results for the randomized highest k selective encoding scheme with Zipf distribution in (6.13) with parameters $n = 100$, $s = 0.2$. Here, we observe that the age under the randomized encoding scheme is higher than the plain highest k selective encoding scheme, that is, $\alpha = 0$ case. At the expense of this increase, however, previously discarded least probable $n - k$ realizations can be received under this encoding scheme. We observe that when the arrival rate is high, for example, $\alpha = 1.2$, average age monotonically increases with α, whereas when the arrival rate is smaller, for example, $\alpha = 0.6$, average age initially increases with α and then decreases since waiting times decrease as opposed to increasing codeword lengths as α increases. One interesting observation here is that when α is larger than 0.3, it is preferable to select $\alpha = 1$, that is, encoding every realization.

6.3.3 Highest k Selective Encoding Scheme with Empty Symbol

The idea of sending empty status update is initially proposed in [49] where a special symbol is transmitted to indicate a skipped transmission. Based on this idea, reference [52] considers a system where the transmitter sends a special symbol when the FIFO buffer is empty. In [52], two different models are considered to transmit the empty symbol. In the first model, a separate empty buffer signal is available to inform the decoder when the transmitter buffer is empty. In other words, the empty symbol is not encoded like the other status updates generated by the source. In the second model, the special empty buffer signal is no longer available, and therefore the empty status update needs to be encoded along with the other status updates. If the transmitter mostly stays idle, then assigning a shorter codeword to the empty status update might be desirable, as the transmitter usually sends the empty status update. On the other hand, if the transmitter is busy most of the time, then assigning a longer codeword to the empty symbol might be desirable. We note that the codeword length of the empty status update affects the codeword lengths of the other status updates. The aim of [52] is to find the optimal transmission probability for the empty symbol to minimize peak average age of information at the receiver. Finally, we emphasize that the purpose of sending the empty symbol in [52] is different form the purpose in [66]. In [66], the empty symbol indicates that the transmission of a status update is purposely missed. However, in [52], the empty symbol indicates that there is no status update in the buffer.

Under the highest k selective encoding scheme presented in Section 6.3.1, the receiver node is not notified when one of the remaining $n-k$ least probable status updates is realized. In other words, when there is no transmission, the receiver cannot distinguish whether there is no update arrival at the transmitter or one of the least probable realizations has occurred. In order to notify the receiver when one of the remaining $n-k$ realizations has occurred, we present a modified selective encoding policy denoted as *highest k selective encoding with empty symbol*. With this encoding scheme, the most probable k status updates are always encoded, whereas in the case of the remaining $n-k$ least probable updates, a designated empty symbol denoted by x_e is transmitted. Thus, when the empty symbol is transmitted, the receiver knows that one of the remaining $n-k$ least probable status updates is realized at the source but it does not know which status update is realized specifically. When the empty status update x_e is received, the receiver may not reset its age, as the empty symbol is not a regular status update. On the other hand, as the receiver has some *partial* information about the status update generated at the source, that is, the receiver knows that one of the $n-k$ least probable status updates is realized, it may choose to reset its age. In this section, we consider both of these cases, that is, when the empty symbol does not reset the age and when the empty symbol resets the age.

As a result of this encoding scheme, the transmitter encodes status updates from the set $\mathcal{X}_k' = \mathcal{X}_k \cup \{x_e\}$ using the binary alphabet based on the pmf given by $\{P_X(x_1), P_X(x_2), \ldots, P_X(x_k), P_X(x_e)\}$, where $P_X(x_e) = 1 - q_k$. That is, the transmitter assigns codeword $c(x_i)$ with length $\ell(x_i)$ to realization x_i for $i \in \{1, \ldots, k, e\}$.

When the Empty Symbol Does Not Reset the Age

In this case, age at the receiver is not updated when the empty status update is received. As the empty status update x_e does not reset the age and increases the codeword lengths of the other status updates, sending empty status updates increases the average age at the receiver. Upon delivery of a successful status update, the total waiting time for the next successful status update is equal to

$$W = (M-1)\ell(x_e) + \sum_{\ell=1}^{M} Z_\ell, \tag{6.27}$$

where M denotes the total number of update arrivals until the first status update from the set $\mathcal{X}_k$ is realized at the transmitter. That is, there are $M-1$ transmissions of empty status updates between two successful updates from the set $\mathcal{X}_k$. Thus, M is a geometric random variable with parameter q_k, which is independent from the arrival and service processes.

As the service time of a successful update is equal to its codeword length, the first two moments of the service times are equal to

$$\mathbb{E}[S] = \mathbb{E}[L|X_k' \neq x_e] = \sum_{i=1}^{k} P_{X_k}(x_i)\ell(x_i), \tag{6.28}$$

$$\mathbb{E}[S^2] = \mathbb{E}[L^2|X_k' \neq x_e] = \sum_{i=1}^{k} P_{X_k}(x_i)\ell(x_i)^2, \tag{6.29}$$

where $P_{X_k}(x_i)$ is defined in (6.14). By using the independence of M and the arrival process, the first and the second moments of the waiting time become

$$\mathbb{E}[W] = \ell(x_e)\left(\frac{1}{q_k} - 1\right) + \frac{1}{\lambda q_k}, \tag{6.30}$$

$$\mathbb{E}[W^2] = \frac{(2-q_k)(1-q_k)}{q_k^2}\ell(x_e)^2 + \frac{4(1-q_k)}{\lambda q_k^2}\ell(x_e) + \frac{2}{(\lambda q_k)^2}. \tag{6.31}$$

The facts that $\mathbb{E}[M] = \frac{1}{q_k}$, $\mathbb{E}[M^2] = \frac{2-q_k}{q_k^2}$, and Z is exponential with rate λ are used to derive (6.30) and (6.31). By substituting (6.28)–(6.31) in (6.4), we obtain the average age expression Δ_e as

$$\Delta_e = \frac{\mathbb{E}[L^2|X_k' \neq x_e] + 2\mathbb{E}[W]\mathbb{E}[L|X_k' \neq x_e] + \mathbb{E}[W^2]}{2\left(\mathbb{E}[L|X_k' \neq x_e] + \mathbb{E}[W]\right)} + \mathbb{E}[L|X_k' \neq x_e]. \tag{6.32}$$

Then, we formulate the following optimization problem:

$$\begin{aligned} \min_{\{\ell(x_i),\ell(x_e)\}} \quad & \frac{\mathbb{E}[L^2|X_k' \neq x_e] + 2\mathbb{E}[W]\mathbb{E}[L|X_k' \neq x_e] + \mathbb{E}[W^2]}{2\left(\mathbb{E}[L|X_k' \neq x_e] + \mathbb{E}[W]\right)} + \mathbb{E}[L|X_k' \neq x_e] \\ \text{s.t.} \quad & 2^{-\ell(x_e)} + \sum_{i=1}^{k} 2^{-\ell(x_i)} \leq 1 \\ & \ell(x_i) \in \mathbb{R}^+, \quad i \in \{1,\ldots,k,e\}, \end{aligned} \tag{6.33}$$

where the objective function in (6.33) is equal to the average age Δ_e in (6.32). The optimization problem in (6.33) is not convex. However, for a fixed and given codeword length for the empty symbol $\ell(x_e)$, it becomes a convex optimization problem. Therefore, we first solve the problem in (6.33) for a given $\ell(x_e)$, then we find the optimal $\ell(x_e)$ that minimizes the average age in (6.32) numerically. For a fixed $\ell(x_e)$, the optimization problem in (6.33) becomes

$$\begin{aligned} \min_{\{\ell(x_i)\}} \quad & \frac{\mathbb{E}[L^2|X'_k \neq x_e] + 2\mathbb{E}[W]\mathbb{E}[L|X'_k \neq x_e] + \mathbb{E}[W^2]}{2\left(\mathbb{E}[L|X'_k \neq x_e] + \mathbb{E}[W]\right)} + \mathbb{E}[L|X'_k \neq x_e] \\ \text{s.t.} \quad & \sum_{i=1}^{k} 2^{-\ell(x_i)} \leq 1 - 2^{-c} \\ & \ell(x_i) \in \mathbb{R}^+, \quad i \in \{1, \ldots, k\}, \end{aligned} \tag{6.34}$$

where $\ell(x_e) = c$. We note that the Kraft inequality must be satisfied with equality, that is, $\sum_{i=1}^{k} 2^{-\ell(x_i)} = 1 - 2^{-c}$.

We solve the optimization problem in (6.34) by introducing an intermediate problem as in Section 6.2 and find the age-optimal codeword lengths for the highest k selective encoding scheme with empty symbol when the empty symbol does not reset the age (see [53] for detailed derivation). To find the overall optimal solution, we vary $\ell(x_e)$ over all possible values and choose the one that yields the least average age for a given arbitrary pmf.

As a numerical example, we consider the pmf in (6.19) for $n = 10$ and take $\lambda = 5$. When k is small, the probability of sending the empty symbol increases, and thus choosing a small codeword length for the empty symbol is desirable. We observe in Figure 6.9 that choosing $\ell(x_e) = 2$ when $k = 2$ and $\ell(x_e) = 3$ when $k = 4$ is optimal. However, when k is close to n, the probability of sending the empty symbol is small. That is why choosing a longer codeword length for the empty symbol is desirable for higher values of k. We observe in Figure 6.9 that choosing $\ell(x_e) = 5$ when $k = 6$ and $\ell(x_e) = 7$ when $k = 8$ is optimal. We note that sending an empty symbol incurs additional cost for the system as it increases the overall waiting time for the next successful status update and also increases the codeword lengths of the most probable k updates. We notice this effect especially when the value of k is small, that is, when we send the empty symbol more frequently. We observe in Figure 6.9 that the minimum achieved age increases significantly when we send the empty symbol for $k = 2$, whereas the effect of the empty symbol is negligible when $k = 8$.

Next, we explore the age-optimal solution for the highest k selective encoding scheme with an empty symbol when the empty symbol resets the age.

When the Empty Symbol Resets the Age

Here, the empty symbol resets the age as it carries some information about the update realized at the source. That is, when the empty symbol is transmitted, the receiver knows that one of the $n - k$ least probable realizations has occurred.

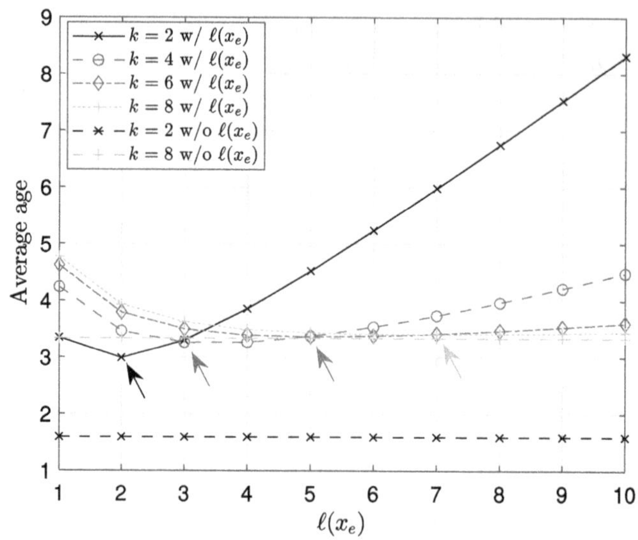

Figure 6.9 Average age under the age-optimal codeword lengths as a function of $\ell(x_e)$ for the pmf in (6.19) with $n = 10$ and $\lambda = 5$ when the empty symbol does not reset the age. The age-optimal $\ell(x_e)$ values are indicated with an arrow. We also provide the optimal age without sending the empty symbol for $k = 2$ and $k = 8$.

As all status updates from the set $\mathcal{X}'_k$, that is, the most probable k updates and the empty symbol x_e, are able to reset the age, average age Δ_e is equal to the age expression in (6.7) with

$$\mathbb{E}[L] = \sum_{i=1}^{k} P_X(x_i)\ell(x_i) + P_X(x_e)\ell(x_e), \tag{6.35}$$

$$\mathbb{E}[L^2] = \sum_{i=1}^{k} P_X(x_i)\ell(x_i)^2 + P_X(x_e)\ell(x_e)^2. \tag{6.36}$$

Next, we formulate the following optimization problem to minimize average age at the receiver:

$$\begin{aligned} \min_{\{\ell(x_i),\ell(x_e)\}} \quad & \frac{\mathbb{E}[L^2] + 2\frac{1}{\lambda}\mathbb{E}[L] + \frac{2}{\lambda^2}}{2\left(\mathbb{E}[L] + \frac{1}{\lambda}\right)} + \mathbb{E}[L] \\ \text{s.t.} \quad & 2^{-\ell(x_e)} + \sum_{i=1}^{k} 2^{-\ell(x_i)} \leq 1 \\ & \ell(x_i) \in \mathbb{R}^+, \quad i \in \{1, \ldots, k, e\}, \end{aligned} \tag{6.37}$$

where the objective function in (6.37) is equal to the average age in (6.7). For a given k, we follow the same solution method in Section 6.2 to solve the problem in (6.37).

Finally, we note that the value of k affects the optimal codeword selection for the empty symbol $\ell(x_e)$. When k is close to n, the probability of the empty symbol

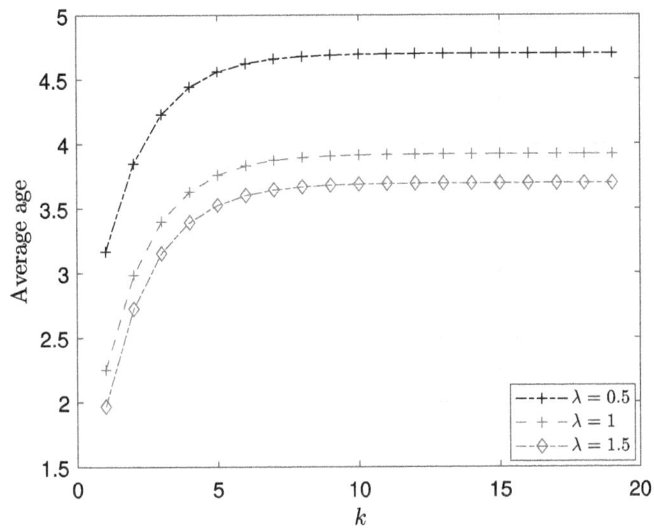

Figure 6.10 Average age under the age-optimal codeword lengths for varying k with the pmf in (6.19) for $n = 20$ and $\lambda = 0.5, 1, 1.5$ when the empty symbol resets the age.

becomes small, which leads to a higher codeword length for the empty symbol. However, when k is small, the probability of the empty symbol increases, which causes a shorter codeword length selection for the empty symbol.

As a numerical example, we use the pmf in (6.19) for $n = 20$ and vary $\lambda = 0.5, 1, 1.5$. We observe in Figure 6.10 that the minimum age is achieved when $k = 1$ since the waiting time does not depend on k and the codeword lengths increase with the higher values of k. With the selection of $k = 1$, only the most probable status update is transmitted and the remaining status updates are mapped into the empty symbol. We note that even though the selection of $k = 1$ is optimal in terms of the average age, this selection causes significant information loss at the receiver, which is not captured by the age metric alone.

6.4 Conclusion and Outlook

In this chapter, we explore age of information in the context of the timely source coding problem. In most of the existing literature, service (transmission) times are based on a given distribution. In the timely source coding problem, by using source coding schemes, we design the transmission times of the status updates. We observe that the average age minimization problem is different from the traditional source coding problem, as the average age depends on both the first and the second moments of the codeword lengths. For the age minimization problem, we first consider a greedy source coding scheme, where all realizations are encoded. For this source coding scheme, we find the age-optimal real-valued codeword lengths. Then, we explore the highest k selective encoding scheme, where instead of encoding all realizations, we encode only the most probable k realizations and discard the remaining least probable $n-k$ realizations. In order to further inform the receiver regarding the remaining $n-k$ realizations, we consider the randomized selective encoding scheme where the remaining least probable $n-k$ realizations are transmitted with a certain probability. We also consider the encoding scheme where the remaining least probable $n-k$ realizations are mapped into an empty status update. For this source coding scheme, we consider two different scenarios where the empty symbol resets the age and does not reset the age. For all these selective encoding schemes, we first determine the average age expressions and then, for a given pmf, characterize the age-optimal k value, and find the corresponding age-optimal codeword lengths. Through numerical results, we show that selective encoding schemes achieve lower average age than encoding all realizations.

There are several open research directions that can be addressed in the area of timely source coding. Here, we present two of these directions.

1. *Generalization of the selective encoding scheme.* In this chapter, we only consider selective encoding schemes in which only the most probable k realizations are always encoded. We note that there are $\binom{n}{k}$ possible selections for the encoded k realizations. Thus, for a given update arrival rate, the optimal selection of k out of n realizations and finding the corresponding age-optimal integer-valued codeword lengths remain as open problems.
2. *Minimization of average age along with the information loss.* We note that under the highest selective encoding scheme, the remaining $n-k$ realizations are not transmitted, which causes information loss at the receiver that is not captured by the age metric alone. Thus, the timely source coding problem can be reformulated along with a distortion constraint to measure the information loss at the receiver [54].

References

[1] S. K. Kaul, R. D. Yates, and M. Gruteser, "Real-time status: How often should one update?" in *IEEE Infocom*, March 2012.
[2] M. Costa, M. Codrenau, and A. Ephremides, "Age of information with packet management," in *IEEE ISIT*, June 2014.
[3] A. M. Bedewy, Y. Sun, and N. B. Shroff, "Optimizing data freshness, throughput, and delay in multi-server information-update systems," in *IEEE ISIT*, July 2016.
[4] Q. He, D. Yuan, and A. Ephremides, "Optimizing freshness of information: On minimum age link scheduling in wireless systems," in *IEEE WiOpt*, May 2016.
[5] C. Kam, S. Kompella, G. D. Nguyen, W. J. E., and A. Ephremides, "Age of information with a packet deadline," in *IEEE ISIT*, July 2016.
[6] Y. Sun, E. Uysal-Biyikoglu, R. D. Yates, C. E. Koksal, and N. B. Shroff, "Update or wait: How to keep your data fresh," *IEEE Transactions on Information Theory*, vol. 63, no. 11, pp. 7492–7508, November 2017.
[7] E. Najm and E. Telatar, "Status updates in a multi-stream M/G/1/1 preemptive queue," in *IEEE Infocom*, April 2018.

[8] E. Najm, R. D. Yates, and E. Soljanin, "Status updates through M/G/1/1 queues with HARQ," in *IEEE ISIT*, June 2017.

[9] A. Soysal and S. Ulukus, "Age of information in G/G/1/1 systems," in *Asilomar Conference*, November 2019.

[10] A. Soysal and S. Ulukus, "Age of information in G/G/1/1 systems: Age expressions, bounds, special cases, and optimization," May 2019, available on arXiv: 1905.13743.

[11] R. D. Yates, P. Ciblat, A. Yener, and M. Wigger, "Age-optimal constrained cache updating," in *IEEE ISIT*, June 2017.

[12] H. Tang, P. Ciblat, J. Wang, M. Wigger, and R. D. Yates, "Age of information aware cache updating with file- and age-dependent update durations," September 2019, available on arXiv: 1909.05930.

[13] S. Nath, J. Wu, and J. Yang, "Optimizing age-of-information and energy efficiency tradeoff for mobile pushing notifications," in *IEEE SPAWC*, July 2017.

[14] Y. Hsu, "Age of information: Whittle index for scheduling stochastic arrivals," in *IEEE ISIT*, June 2018.

[15] I. Kadota, A. Sinha, E. Uysal-Biyikoglu, R. Singh, and E. Modiano, "Scheduling policies for minimizing age of information in broadcast wireless networks," *IEEE/ACM Transactions on Networking*, vol. 26, no. 6, pp. 2637–2650, December 2018.

[16] B. Buyukates, A. Soysal, and S. Ulukus, "Age of information scaling in large networks," in *IEEE ICC*, May 2019.

[17] B. Buyukates, A. Soysal, and S. Ulukus, "Age of information scaling in large networks with hierarchical cooperation," in *IEEE Globecom*, December 2019.

[18] J. Gong, Q. Kuang, X. Chen, and X. Ma, "Reducing age-of-information for computation-intensive messages via packet replacement," January 2019, available on arXiv: 1901.04654.

[19] B. Buyukates and S. Ulukus, "Timely distributed computation with stragglers," *IEEE Transactions on Communications*, vol. 68, no. 9, pp. 5273–5282, September 2020.

[20] A. Arafa, K. Banawan, K. G. Seddik, and H. V. Poor, "On timely channel coding with hybrid ARQ," in *IEEE Globecom*, December 2019.

[21] Y. Sun, Y. Polyanskiy, and E. Uysal-Biyikoglu, "Remote estimation of the Wiener process over a channel with random delay," in *IEEE ISIT*, June 2017.

[22] Y. Sun and B. Cyr, "Information aging through queues: A mutual information perspective," in *IEEE SPAWC*, June 2018.

[23] J. Chakravorty and A. Mahajan, "Remote estimation over a packet-drop channel with Markovian state," July 2018, available on arXiv:1807.09706.

[24] M. Bastopcu and S. Ulukus, "Age of information for updates with distortion," in *IEEE ITW*, August 2019.

[25] M. Bastopcu and S. Ulukus, "Age of information for updates with distortion: Constant and age-dependent distortion constraints," December 2019, available on arXiv:1912.13493.

[26] M. Bastopcu and S. Ulukus, "Who should Google Scholar update more often?" in *Infocom Workshop on Age of Information*, July 2020.

[27] D. Ramirez, E. Erkip, and H. V. Poor, "Age of information with finite horizon and partial updates," October 2019, available on arXiv:1910.00963.

[28] P. Zou, O. Ozel, and S. Subramaniam, "Trading off computation with transmission in status update systems," in *IEEE PIMRC*, September 2019.

[29] A. Kosta, N. Pappas, A. Ephremides, and V. Angelakis, "Age and value of information: Non-linear age case," in *IEEE ISIT*, June 2017.

[30] M. Bastopcu and S. Ulukus, "Age of information with soft updates," in *Allerton Conference*, October 2018.

[31] M. Bastopcu and S. Ulukus, "Minimizing age of information with soft updates," *Journal of Communications and Networks*, vol. 21, no. 3, pp. 233–243, June 2019.

[32] A. Arafa and S. Ulukus, "Age minimization in energy harvesting communications: Energy-controlled delays," in *Asilomar Conference*, October 2017.

[33] A. Arafa and S. Ulukus, "Age-minimal transmission in energy harvesting two-hop networks," in *IEEE Globecom*, December 2017.

[34] X. Wu, J. Yang, and J. Wu, "Optimal status update for age of information minimization with an energy harvesting source," *IEEE Transactions on Green Communications and Networking*, vol. 2, no. 1, pp. 193–204, March 2018.

[35] A. Arafa, J. Yang, and S. Ulukus, "Age-minimal online policies for energy harvesting sensors with random battery recharges," in *IEEE ICC*, May 2018.

[36] A. Arafa, J. Yang, S. Ulukus, and H. V. Poor, "Age-minimal online policies for energy harvesting sensors with incremental battery recharges," in *UCSD ITA*, February 2018.

[37] A. Arafa, J. Yang, S. Ulukus, and H. V. Poor, "Age-minimal transmission for energy harvesting sensors with finite batteries: Online policies," *IEEE Transactions on Information Theory*, vol. 66, no. 1, pp. 534–556, January 2020.

[38] A. Arafa, J. Yang, S. Ulukus, and H. V. Poor, "Online timely status updates with erasures for energy harvesting sensors," in *Allerton Conference*, October 2018.

[39] A. Arafa, J. Yang, S. Ulukus, and H. V. Poor, "Using erasure feedback for online timely updating with an energy harvesting sensor," in *IEEE ISIT*, July 2019.

[40] S. Farazi, A. G. Klein, and D. R. Brown III, "Average age of information for status update systems with an energy harvesting server," in *IEEE Infocom*, April 2018.

[41] S. Leng and A. Yener, "Age of information minimization for an energy harvesting cognitive radio," *IEEE Transactions on Cognitive Communications and Networking*, vol. 5, no. 2, pp. 427–439, May 2019.

[42] Z. Chen, N. Pappas, E. Bjornson, and E. G. Larsson, "Age of information in a multiple access channel with heterogeneous traffic and an energy harvesting node," in *IEEE INFOCOM*, March 2019, available on arXiv: 1903.05066.

[43] M. A. Abd-Elmagid and H. S. Dhillon, "Average peak age-of-information minimization in UAV-assisted IoT networks," *IEEE Transactions on Vehicular Technology*, vol. 68, no. 2, pp. 2003–2008, February 2019.

[44] J. Liu, X. Wang, and H. Dai, "Age-optimal trajectory planning for UAV-assisted data collection," in *IEEE Infocom*, April 2018.

[45] E. T. Ceran, D. Gunduz, and A. Gyorgy, "A reinforcement learning approach to age of information in multi-user networks," in *IEEE PIMRC*, September 2018.

[46] H. B. Beytur and E. Uysal-Biyikoglu, "Age minimization of multiple flows using reinforcement learning," in *IEEE ICNC*, February 2019.

[47] M. A. Abd-Elmagid, H. S. Dhillon, and N. Pappas, "A reinforcement learning framework for optimizing age-of-information in RF-powered communication systems," August 2019, available on arXiv: 1908.06367.

[48] T. M. Cover and J. A. Thomas, *Elements of Information Theory*. Wiley Press, 2012.

[49] P. Mayekar, P. Parag, and H. Tyagi, "Optimal source codes for timely updates," *IEEE Transactions on Information Theory*, vol. 66, no. 6, pp. 3714–3731, March 2020.

[50] J. Zhong and R. D. Yates, "Timeliness in lossless block coding," in *IEEE DCC*, March 2016.

[51] J. Zhong, R. D. Yates, and E. Soljanin, "Backlog-adaptive compression: Age of information," in *IEEE ISIT*, June 2017.

[52] J. Zhong, R. D. Yates, and E. Soljanin, "Timely lossless source coding for randomly arriving symbols," in *IEEE ITW*, November 2018.

[53] M. Bastopcu, B. Buyukates, and S. Ulukus, "Selective encoding policies for maximizing information freshness," April 2020, available on arXiv:2004.06091.

[54] M. Bastopcu and S. Ulukus, "Partial updates: Losing information for freshness," in *IEEE ISIT*, June 2020.

[55] M. Bastopcu, B. Buyukates, and S. Ulukus, "Optimal selective encoding for timely updates," in *CISS*, March 2020.

[56] B. Buyukates, M. Bastopcu, and S. Ulukus, "Optimal selective encoding for timely updates with empty symbol," in *IEEE ISIT*, June 2020.

[57] W. Dinkelbach, "On nonlinear fractional programming," *Management Science*, vol. 13, no. 7, pp. 435–607, March 1967.

[58] S. P. Boyd and L. Vandenberghe, *Convex Optimization.* Cambridge University Press, 2004.

[59] R. M. Corless, G. H. Gonnet, D. E. G. Hare, D. J. Jeffrey, and D. E. Knuth, "On the Lambert W function," *Advances in Computational Mathematics*, vol. 5, no. 1, pp. 329–359, December 1996.

[60] R. D. Yates and D. J. Goodman, *Probability and Stochastic Processes.* Wiley, 2014.

[61] J. Zhong, E. Soljanin, and R. D. Yates, "Status updates through multicast networks," in *Allerton Conference*, October 2017.

[62] J. Zhong, R. D. Yates, and E. Soljanin, "Multicast with prioritized delivery: How fresh is your data?" in *IEEE SPAWC*, June 2018.

[63] B. Buyukates, A. Soysal, and S. Ulukus, "Age of information in two-hop multicast networks," in *Asilomar Conference*, October 2018.

[64] B. Buyukates, A. Soysal, and S. Ulukus, "Age of information in multihop multicast networks," *Journal of Communications and Networks*, vol. 21, no. 3, pp. 256–267, July 2019.

[65] B. Buyukates, A. Soysal, and S. Ulukus, "Age of information in multicast networks with multiple update streams," in *Asilomar Conference*, November 2019.

[66] P. Mayekar, P. Parag, and H. Tyagi, "Optimal lossless source codes for timely updates," in *IEEE ISIT*, June 2018.

7 Sampling and Scheduling for Minimizing Age of Information of Multiple Sources

Ahmed M. Bedewy, Yin Sun, Sastry Kompella, and Ness B. Shroff

This work has been supported in part by ONR grants N00014-17-1-2417 and N00014-15-1-2166, Army Research Office grants W911NF-14-1-0368 and MURI W911NF-12-1-0385, National Science Foundation grants CNS-1446582, CNS-1421576, CNS-1518829, and CCF-1813050, and a grant from the Defense Thrust Reduction Agency HDTRA1-14-1-0058.

7.1 Abstract

In this chapter, we consider a joint sampling and scheduling problem for optimizing data freshness in multisource systems. Data freshness is measured by a nondecreasing penalty function of *age of information*, where all sources have the same age-penalty function. Sources take turns to generate update samples, and forward them to their destinations one by one through a shared channel with random delay. There is a scheduler that chooses the update order of the sources, and a sampler that determines when a source should generate a new sample in its turn. We aim to find the optimal scheduler–sampler pairs that minimize the total-average age-penalty (Ta-AP).

We start the chapter by providing a brief explanation of the sampling problem in the light of single-source networks, as well as some useful insights and applications on age of information and its penalty functions. Then, we move on to the multi-source networks, where the problem becomes more challenging. We provide a detailed explanation of the model and the solution in this case. Finally, we conclude this chapter by providing an open question in this area and its inherent challenges.

7.2 Introduction

7.2.1 Sampling Problem in Single-Source Networks

In many applications, a controller observes one or more continuous-time processes and takes proper actions depending on these processes' states. However, continuous observation and processing of these continuous-time processes could be costly and not always available. For example, in autonomous vehicles, a control unit may observe multiple processes simultaneously, as shown in Figure 7.1 where sources transmit information to a control unit via a wireless channel. Continuous observation

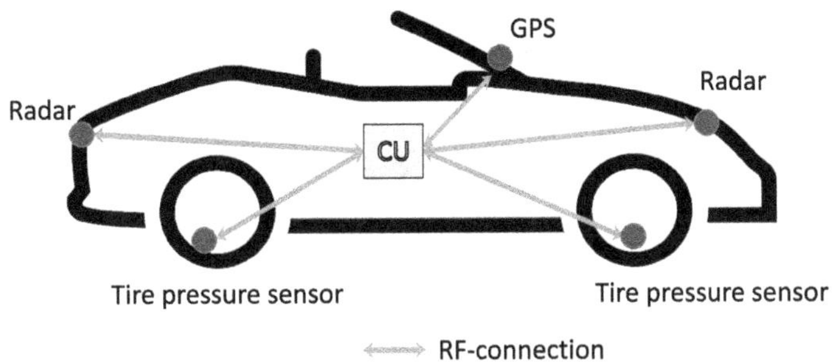

Figure 7.1 A control unit (CU) observes multiple processes, for example, tire pressure, location, orientation, speed, and so on, in an autonomous vehicle.

of these processes is difficult because of the limited resources in terms of available bands and energy. One way to overcome this difficulty is to deal with samples from these observed processes. As a result, controlling the sampling times is very important to save precious system resources such as energy, channel use, computation resources, and so on. Also, with the influence of uncertain factors, the channel delay between the controller and the observed processes varies with time. Hence, it is necessary to introduce a flexible sampling control. One important factor in this control system is the design of inter-sampling times. In particular, as the inter-sampling time increases, information at the destination about the observed processes becomes stale. This pushes us to ask the following question: Is it true that increasing the sampling rate (i.e., decreasing the inter-sampling times) always improves information freshness? We will show that, contrary to conventional wisdom, this is not always the case. To measure data freshness, Age of Information (AoI) metric is used, which is defined as the time elapsed since the most recently received sample was generated. Next, we discuss the sampling problem in the presence of a single source.

7.2.2 Sampling Problem in Single-Source Networks

In this section, we will focus on a simple setting, which includes a single source transmitting sensed information to a receiver. Our goal will be to understand the sampling phenomena in the single-source setting first and then use the insights to extend the analysis in Section 7.3 to the multisource setting. In particular, consider a single-source information update system as shown in Figure 7.2, where the source (e.g. a sensor) observes a time-varying process. There is a sampler that determines when the source should generate samples from the observed process. This is known as the "generate-at-will" model [1–3] (i.e., samples can be generated at any time). The generated samples are, thereafter, sent to the destination via a channel with random delay. The channel is modeled as a single-server First-Come, First-Served (FCFS) queue with i.i.d. service times. We use S_i and D_i to denote the generation time and the delivery time of the ith generated sample, respectively. We suppose that the sampler has full knowledge

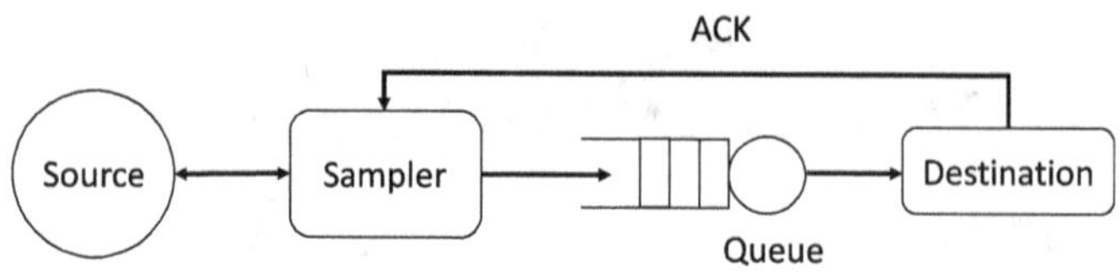

Figure 7.2 A single-source information update system.

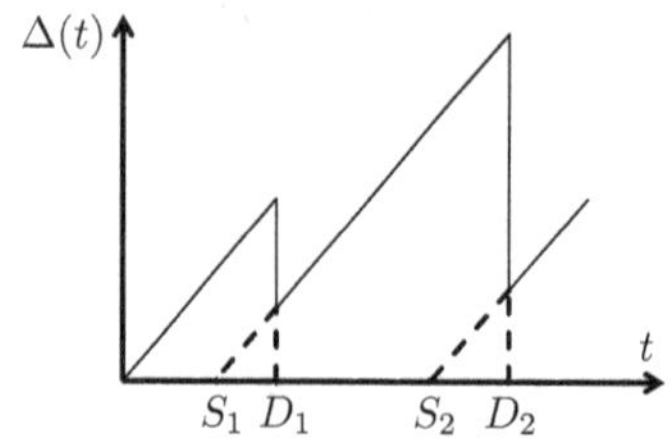

Figure 7.3 Sample path of the age process $\Delta(t)$.

of the idle/busy state of the server via acknowledgments (ACKs) from the destination with zero delay. Our target is to design the sampling times $(S_1, S_2, \ldots)$ such that age of information at the destination (or a nondecreasing penalty function of it) is minimized. Age of information is defined as follows [4–7]: At time t, if the freshest sample at the destination was generated at time $U(t) = \max\{S_i : D_i \leq t\}$, then the age $\Delta(t)$ is defined as

$$\Delta(t) = t - U(t). \tag{7.1}$$

As shown in Figure 7.3, the age increases linearly with t but is reset to a smaller value with the delivery of a fresher sample. Next, we show that this sampling problem is not trivial and the solution may counter the common wisdom.

7.2.3 Counterintuitive Phenomenon of the Optimal Sampler

One may think that, as long as the sampler knows the idle/busy state of the server in real time, then an intuitive solution is the zero-wait sampler, in which a sample is generated as soon as the server becomes idle (i.e., $S_{i+1} = D_i$). Clearly, this zero-wait sampler is throughput and delay optimal. However, surprisingly, this policy does not always minimize age of information. The following example reveals the reasons behind this counterintuitive phenomenon.

Example [1] Consider an information update system as shown in Figure 7.2. Suppose that the sample transmission times are i.i.d. across the samples and are either 0 or 2 with probability 0.5.[1] For simplicity, consider the following realization of the sample transmission times:

[1] The 0 transmission time here is just chosen for the simplicity of the illustration. Indeed, it represents the transmission times that are extremely small.

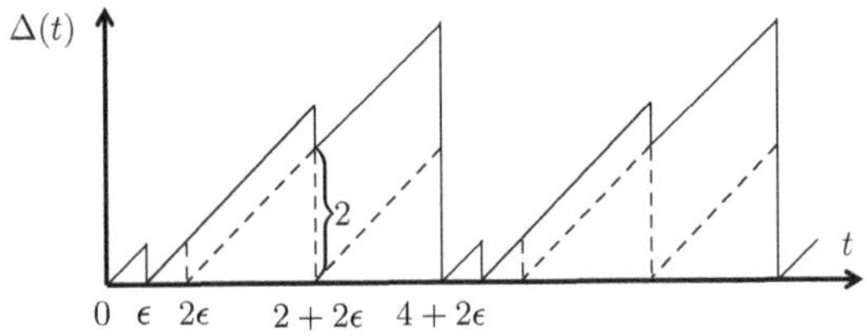

Figure 7.4 Evolution of the age $\Delta(t)$ under the ϵ-wait policy in the example [1].

$$0, 0, 2, 2, 0, 0, 2, 2, \ldots.$$

Now, if the zero-wait sampler is followed, then Sample 1 is generated at time 0 and delivered at time 0. After the delivery of Sample 1 at time 0, Sample 2 is generated, which occurs at time 0 as well. Hence, Samples 1 and 2 are both generated at the same time (time 0). As a result, Sample 2 does not bring any new information to the destination after the delivery of Sample 1. In other words, under zero-wait sampler, having zero transmission times has not been exploited well, but rather has resulted in wasted system resources. (Some samples with stale information are generated, and hence are wasted.) This issue is repeated as follows: Whenever a sample has a zero transmission time, the next generated sample does not carry new information, and hence, is wasted. This raises an important question: Can we do better?

The answer is yes. For the sake of comparison, let us consider an ϵ-wait policy, in which the sampler waits for ϵ seconds after each sample with a zero transmission time, and does not wait otherwise. Note that this policy is a causal policy, as it just needs to know the transmission time of the last delivered sample to specify the generation time of the next one. Figure 7.4 illustrates the age evolution under the ϵ-wait policy for the preceding realization of the sample transmission times. From Figure 7.4, we can compute the time-average age of the ϵ-wait policy, which is given by

$$(\epsilon^2/2 + \epsilon^2/2 + 2\epsilon + 4^2/2)/(4 + 2\epsilon) = (\epsilon^2 + 2\epsilon + 8)/(4 + 2\epsilon) \text{ seconds.}$$

If we set the waiting time $\epsilon = .5$, then the time-average age of the ϵ-wait policy is 1.85 seconds. Observe that, if we set $\epsilon = 0$, the ϵ-wait policy reduces to the zero-wait one, whose time-average age is 2 seconds. Hence, the zero-wait sampler is not age-optimal in this case, and we can do better. This makes the problem nontrivial, which needs to be attacked carefully. Next, we provide some motivation for using age of information as a metric and the benefit of using penalty functions.

7.2.4 Data Freshness in Real-Time Applications

In applications such as networked monitoring and control systems, wireless sensor networks, and autonomous vehicles, the destination node must receive timely status updates so that it can make accurate decisions. For example, real-time knowledge about the location, orientation, and speed of motor vehicles is crucial to avoid collisions and reduce traffic congestion. In addition, fresh information about stock price and interest rate is of ultimate importance in developing efficient business plans in the stock market. In light of this, age of information has emerged to provide a mathematical framework for modeling the data freshness at a particular destination. For this framework to be complete, it must capture the variation of the stale information's harmful impact from one application to another. Here comes the importance of using penalty functions of the age. Next, we discuss some examples of the used age-penalty functions and their applications.

Examples of Age-Penalty Functions and Their Applications

The impact of stale information depends on how fast the information source varies with time. For instance, the location of a motor vehicle is considered an information source that may vary quickly with time. In particular, a moving car with a speed of 65 mph will traverse almost 29 meters for 1 second. Hence, stale information has a dramatic serious impact on this situation. Meanwhile, the engine temperature in a vehicle, for example, is one of the information sources that may vary slowly with time, and hence a reading that was taken a few minutes ago is still valid for observing the engine's health. This illustrates how the value of the fresh information varies from one application to another. Unfortunately, age alone cannot capture such variation. Thus, we desperately need an age-penalty function in such applications.

In this chapter, we use $g : [0, \infty) \rightarrow \mathbb{R}$ to denote the nondecreasing age-penalty function in use. Note that $g(\cdot)$ does not have to be convex or continuous. As mentioned, $g(\cdot)$ is used to represent the level of dissatisfaction of data staleness in different applications based on their demands. Here are some examples of $g(\cdot)$:

- A stair-shape function $g(x) = \lfloor x \rfloor$ can be used to characterize the dissatisfaction for data staleness when the information of interest is checked periodically.
- An exponential function $g(x) = e^x$ can be utilized in online learning and control applications in which the demand for updating data increases quickly with age.
- An indicator function $g(x) = \mathbb{1}(x > q)$ can be used to indicate the dissatisfaction of the violation of an age threshold q.

Besides the preceding examples, a recent survey [8] showed that, under certain conditions, information freshness metrics expressed in terms of auto-correlation functions, the estimation error of signal values, and mutual information are monotonic functions of the age. To make it clearer, let us consider the following example.

Auto-correlation Function of Signals Consider the model that is illustrated in Figure 7.2. Let $X_t \in \mathbb{R}$ represent the process that is being observed by the source. The transmitted samples from the sampler are used for estimating the process X_t at the destination. At time t, the most recent received sample is generated at time $t - \Delta(t)$, where $\Delta(t)$ is the age of the most recent received sample. If the samples at t and $t - \Delta(t)$ are correlated, then the accuracy of the estimation will depend on the value of $\Delta(t)$. In other words, the smaller the value of $\Delta(t)$, the more accurate the estimate. Hence, the auto-correlation function $\mathbb{E}[X_t X_{t-\Delta(t)}]$ can be used to evaluate the freshness of the sample $X_{t-\Delta(t)}$ [9]. For some stationary processes, $|\mathbb{E}[X_t X_{t-\Delta(t)}]|$ becomes a nonnegative, nonincreasing function of the age $\Delta(t)$, which can be considered an age-penalty function.

7.3 Multisource Sampling and Scheduling Problem

In real-life applications, a controller may need observations from multiple sources in order to make accurate decisions. In autonomous vehicles, for example, many electronic control units (ECUs) are connected to one or more sensors and actuators via a controller area network (CAN) bus [10, 11]. As vehicles and commercial trucks get smarter, the number of needed sensors increases and can reach up to 200 sensors per vehicle [12]. With such a large number of sensors, some of them may send their information via a shared channel. Hence, the single-source model just considered cannot be fully applied in such applications. In other words, besides the sampler, a scheduler is needed to handle the transmission order of the sources. This leads to the need for designing a sampler and a scheduler that can jointly optimize the data freshness. This makes the optimization problem more challenging. In this section, we present in detail the sampling problem in multisource networks [13, 14]. We start by providing some useful notations and definitions.

7.3.1 Notations and Definitions

We use $\mathbb{N}^+$ to represent the set of nonnegative integers, $\mathbb{R}^+$ is the set of nonnegative real numbers, $\mathbb{R}$ is the set of real numbers, and $\mathbb{R}^n$ is the set of n-dimensional real Euclidean space. We use t^- to denote the time instant just before t. Let $\mathbf{x} = (x_1, x_2, \ldots, x_n)$ and $\mathbf{y} = (y_1, y_2, \ldots, y_n)$ be two vectors in $\mathbb{R}^n$; then we denote $\mathbf{x} \leq \mathbf{y}$ if $x_i \leq y_i$ for $i = 1, 2, \ldots, n$. Also, we use $x_{[i]}$ to denote the ith largest component of vector $\mathbf{x}$. A set $U \subseteq \mathbb{R}^n$ is called upper if $\mathbf{y} \in U$ whenever $\mathbf{y} \geq \mathbf{x}$ and $\mathbf{x} \in U$. We will need the following definitions:

DEFINITION 7.1 **Univariate Stochastic Ordering:** [15] Let X and Y be two random variables. Then, X is said to be stochastically smaller than Y (denoted as $X \leq_{\text{st}} Y$), if

$$\mathbb{P}\{X > x\} \leq \mathbb{P}\{Y > x\}, \quad \forall x \in \mathbb{R}.$$

DEFINITION 7.2 **Multivariate Stochastic Ordering:** [15] Let $\mathbf{X}$ and $\mathbf{Y}$ be two random vectors. Then, $\mathbf{X}$ is said to be stochastically smaller than $\mathbf{Y}$ (denoted as $\mathbf{X} \leq_{\text{st}} \mathbf{Y}$), if

$$\mathbb{P}\{\mathbf{X} \in U\} \leq \mathbb{P}\{\mathbf{Y} \in U\}, \quad \text{for all upper sets} \quad U \subseteq \mathbb{R}^n.$$

DEFINITION 7.3 **Stochastic Ordering of Stochastic Processes:** [15] Let $\{X(t), t \in [0, \infty)\}$ and $\{Y(t), t \in [0, \infty)\}$ be two stochastic processes. Then, $\{X(t), t \in [0, \infty)\}$ is said to be stochastically smaller than $\{Y(t), t \in [0, \infty)\}$ (denoted by $\{X(t), t \in$

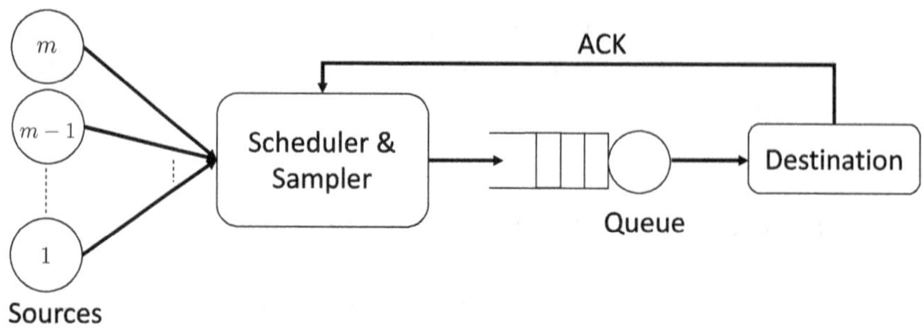

Figure 7.5 System model.

$[0, \infty)\} \leq_{\text{st}} \{Y(t), t \in [0, \infty)\})$, if, for all choices of an integer n and $t_1 < t_2 < \ldots < t_n$ in $[0, \infty)$, it holds that

$$(X(t_1), X(t_2), \ldots, X(t_n)) \leq_{\text{st}} (Y(t_1), Y(t_2), \ldots, Y(t_n)), \tag{7.2}$$

where the multivariate stochastic ordering in (7.2) was defined in Definition 7.2.

7.3.2 Multisource Network Model

We consider a status update system with m sources as shown in Figure 7.5, where each source observes a time-varying process. Sources take turns to generate samples, and forward the samples to their destinations one by one through a shared error-free channel with random delay. Hence, a decision maker consists of a scheduler that chooses the update order of the sources, and a sampler that determines when a source should generate a new sample in its turn.

We use S_i to denote the generation time of the ith generated sample from all sources, called sample i. Moreover, we use r_i to represent the source index from which sample i is generated. The channel is modeled as an FCFS queue with random *i.i.d.* service time Y_i, where Y_i represents the service time of sample i, $Y_i \in \mathcal{Y}$, and $\mathcal{Y} \subset \mathbb{R}^+$ is a finite and bounded set. We also assume that $0 < \mathbb{E}[Y_i] < \infty$ for all i. We suppose that the decision maker knows the idle/busy state of the server through acknowledgments (ACKs) from the destination with zero delay. If a sample is generated while the server is busy, it needs to wait in the queue until its transmission opportunity, and becomes stale while waiting. Hence, there is no loss of optimality to avoid generating a new sample during the busy periods. As a result, a sample is served immediately once it is generated. Let D_i denote the delivery time of sample i, where $D_i = S_i + Y_i$. After the delivery of sample i at time D_i, the decision maker may insert a waiting time Z_i before generating a new sample (hence, $S_{i+1} = D_i + Z_i$),[2] where $Z_i \in \mathcal{Z}$, and $\mathcal{Z} \subset \mathbb{R}^+$ is a finite and bounded set.[3]

At any time t, the most recently delivered sample from source l is generated at time

$$U_l(t) = \max\{S_i : r_i = l, D_i \leq t\}. \tag{7.3}$$

[2] We assume that $D_0 = 0$. Thus, $S_1 = Z_0$.
[3] We assume that $0 \in \mathcal{Z}$.

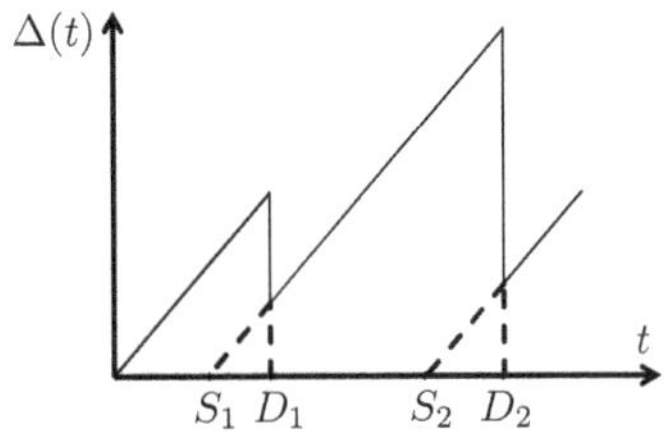

Figure 7.6 The age $\Delta_l(t)$ of source l, where we suppose that the first and third samples are generated from source l, i.e., $r_1 = r_3 = l$.

Hence, the age of source l is defined as

$$\Delta_l(t) = t - U_l(t), \tag{7.4}$$

which is plotted in Figure 7.6. We suppose that the age $\Delta_l(t)$ is right-continuous. Moreover, we assume that the initial age values $\Delta_l(0^-)$ at time $t = 0^-$ for all l are known to the system. The age process for source l is given by $\{\Delta_l(t), t \geq 0\}$. For each source l, we consider an age-penalty function $g(\Delta_l(t))$ of the age $\Delta_l(t)$. The function $g : [0, \infty) \to \mathbb{R}$ is nondecreasing and is not necessarily convex or continuous. We suppose that $\mathbb{E}[| \int_a^{a+x} g(\tau)d\tau |] < \infty$ whenever $x < \infty$.

7.3.3 Decision Policies in Multisource Networks

A decision policy, denoted by d, controls the following: (i) the scheduler, denoted by π, that determines the update order of the sources $\pi \triangleq (r_1, r_2, \ldots)$, (ii) the sampler, denoted by f, that controls the sampling times $f \triangleq (S_1, S_2, \ldots)$, or equivalently, the sequence of waiting times $f \triangleq (Z_0, Z_1, \ldots)$. Hence, $d = (\pi, f)$ implies that a decision policy d employs the scheduler π and the sampler f. Let $\mathcal{D}$ denote the set of causal decision policies in which decisions are made based on the history and current information of the system. Observe that $\mathcal{D}$ consists of Π and $\mathcal{F}$, where Π and $\mathcal{F}$ are the sets of causal schedulers and samplers, respectively.

A decision policy acts as follows: After each delivery, the decision maker determines the source to be served and then imposes a waiting time before the generation of the new sample. Next, we present our optimization problem.

7.3.4 Problem Formulation and Challenges

In this chapter, we aim to minimize the total-average age-penalty per unit time (Ta-AP). Consider the time interval $[0, D_n]$, the Ta-AP is defined for any decision policy $d = (\pi, f)$ as

$$\Delta_{\text{avg}}(\pi, f) = \limsup_{n \to \infty} \frac{\mathbb{E}\left[\sum_{l=1}^{m} \int_0^{D_n} g\left(\Delta_l(t)\right) dt\right]}{\mathbb{E}\left[D_n\right]}. \tag{7.5}$$

Hence, our optimization problem can be cast as

$$\bar{\Delta}_{\text{avg-opt}} \triangleq \min_{\pi \in \Pi, f \in \mathcal{F}} \Delta_{\text{avg}}(\pi, f), \tag{7.6}$$

where $\bar{\Delta}_{\text{avg-opt}}$ is the optimum objective values of Problem (7.6).

Due to the large decision policy space, the optimization problem is quite challenging. In other words, we need to seek the optimal decision policy that controls both the scheduler and sampler to minimize the Ta-AP. Moreover, the possible correlation between the optimal actions of the sampler and the scheduler makes this challenge more difficult. Thus, we have to find a way to tackle this challenge. To that end, we develop an important separation principle that helps us to bypass this difficulty, which is presented next.

7.4 Optimal Sampling and Scheduling Design

7.4.1 Separation Principle and Optimal Scheduler

We show that our optimization problem in (7.6) has an important separation principle: Given the sampling times, the Maximum Age First (MAF) scheduler provides the best age performance compared to any other scheduler. What then remains to be addressed is the question of finding the best sampler that solves Problem (7.6), given that the scheduler is fixed to the MAF. We start by defining the MAF scheduler as follows:

DEFINITION 7.4 ([16–20]) Maximum Age First (MAF) scheduler: In this scheduler, the source with the maximum age is served first among all sources. Ties are broken arbitrarily.

For simplicity, let π_{MAF} represent the MAF scheduler. The age performance of π_{MAF} scheduler is characterized in the following proposition:

PROPOSITION 1 *For all $f \in \mathcal{F}$*

$$\Delta_{\text{avg}}(\pi_{MAF}, f) = \min_{\pi \in \Pi} \Delta_{\text{avg}}(\pi, f). \tag{7.7}$$

That is, the MAF scheduler minimizes the Ta-AP in (7.5) *among all schedulers in* Π.

Proof We use a sample-path method to prove Proposition 1 as follows: Given any sampler that controls the sampling times, the scheduler only controls from which source a sample is generated. We couple the policies such that the sample delivery times are fixed under all decision policies. In the MAF scheduler, a source with maximum age becomes the source with minimum age among the m sources after each delivery. Under any arbitrary scheduler, a sample can be generated from any source, which is not necessarily the one with the maximum age, and the chosen source becomes the one with minimum age among the m sources after the delivery. Since the age-penalty function $g(\cdot)$ is nondecreasing, the MAF scheduler provides a better age performance compared to any other scheduler. For details, see Appendix 7.7.1. □

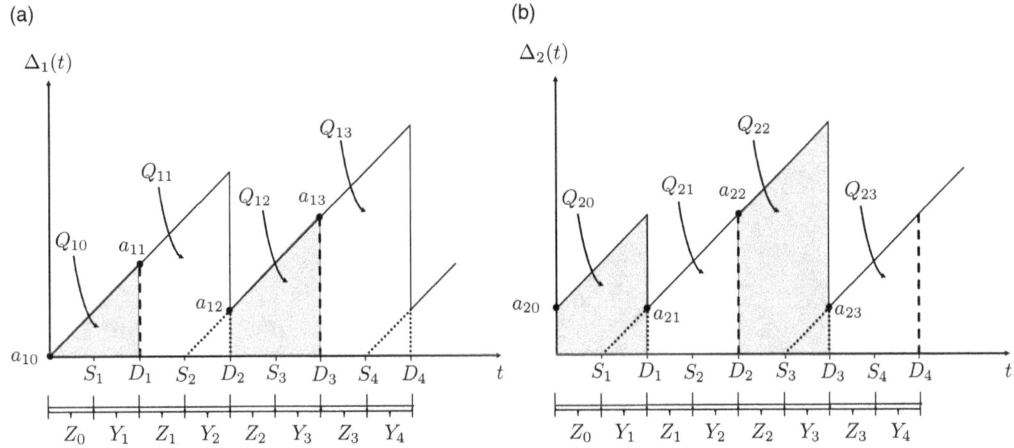

Figure 7.7 The age processes evolution of the MAF scheduler in a two-sources information update system. Source 2 has a higher initial age than Source 1. Thus, Source 2 starts service and Sample 1 is generated from Source 2, which is delivered at time D_1. Then, Source 1 is served and Sample 2 is generated from Source 1, which is delivered at time D_2. The same operation is repeated over time.

Proposition 1 concludes the separation principle that the optimal sampler can be optimized separately, given that the scheduling policy is fixed to the MAF scheduler. Hence, the optimization problem (7.6) can be rewritten as

$$\bar{\Delta}_{\text{avg-opt}} \triangleq \min_{f \in \mathcal{F}} \Delta_{\text{avg}}(\pi_{\text{MAF}}, f). \tag{7.8}$$

By fixing the scheduling policy to the MAF scheduler, the evolution of the age processes of the sources is as follows: The sampler may impose a waiting time Z_i before generating sample $i+1$ at time $S_{i+1} = D_i + Z_i$ from the source with the maximum age at time $t = D_i$. Sample $i+1$ is delivered at time $D_{i+1} = S_{i+1} + Y_{i+1}$ and the age of the source with maximum age drops to the minimum age with the value of Y_{i+1}, while the age processes of other sources increase linearly with time without change. This operation is repeated with time, and the age processes evolve accordingly. An example of age processes evolution is shown in Figure 7.7. Next, we shift our focus to obtain the optimal sampler for Problem (7.8).

7.4.2 Optimal Sampler in Multisource Networks

Now, we fix the scheduling policy to the MAF scheduler and seek for the optimal sampler for minimizing the Ta-AP. A naive solution for the optimal sampler would be the zero-wait policy, that is, $Z_i = 0$ for all i. However, we showed with a counterexample in Section 7.2.3 that this is not necessarily true. Thus, we need to be careful in obtaining this optimal sampler. Since solving Problem (7.8) in the current form is challenging, we reformulate it as an equivalent semi-Markov decision problem (SMDP). Next, we discuss this reformulation in detail.

Reformulation of the Optimal Sampling Problem

We start by analyzing the Ta-AP when the scheduling policy is fixed to the MAF scheduler. We decompose the area under each curve $g(\Delta_l(t))$ into a sum of disjoint geometric parts. Observing Figure 7.7,[4] this area in the time interval $[0, D_n]$, where $D_n = \sum_{i=0}^{n-1} Z_i + Y_{i+1}$, can be seen as the concatenation of the areas Q_{li}, $0 \le i \le n-1$. Thus,

$$\int_0^{D_n} g(\Delta_l(t))dt = \sum_{i=0}^{n-1} Q_{li}, \tag{7.9}$$

where

$$Q_{li} = \int_{D_i}^{D_{i+1}} g(\Delta_l(t))dt = \int_{D_i}^{D_i+Z_i+Y_{i+1}} g(\Delta_l(t))dt. \tag{7.10}$$

Let a_{li} denote the age value of source l at time D_i, that is, $a_{li} = \Delta_l(D_i)$.[5] Hence, for $t \in [D_i, D_{i+1})$, we have

$$\Delta_l(t) = t - U_l(t) = t - (D_i - a_{li}), \tag{7.11}$$

where $(D_i - a_{li})$ represents the generation time of the last delivered sample from source l before time D_{i+1}. By performing change of variables in (7.10), we get

$$Q_{li} = \int_{a_{li}}^{a_{li}+Z_i+Y_{i+1}} g(\tau)d\tau. \tag{7.12}$$

Hence, the Ta-AP can be rewritten as

$$\limsup_{n\to\infty} \frac{\sum_{i=0}^{n-1} \mathbb{E}\left[\sum_{l=1}^{m} \int_{a_{li}}^{a_{li}+Z_i+Y_{i+1}} g(\tau)d\tau\right]}{\sum_{i=0}^{n-1} \mathbb{E}\left[Z_i + Y_{i+1}\right]}. \tag{7.13}$$

Using this, the optimal sampling problem for minimizing the Ta-AP, given that the scheduling policy is fixed to the MAF scheduler, can be cast as

$$\bar{\Delta}_{\text{avg-opt}} \triangleq \min_{f\in\mathcal{F}} \limsup_{n\to\infty} \frac{\sum_{i=0}^{n-1} \mathbb{E}\left[\sum_{l=1}^{m} \int_{a_{li}}^{a_{li}+Z_i+Y_{i+1}} g(\tau)d\tau\right]}{\sum_{i=0}^{n-1} \mathbb{E}\left[Z_i + Y_{i+1}\right]}. \tag{7.14}$$

Since $|\int_{a_{li}}^{a_{li}+Z_i+Y_{i+1}} g(\tau)d\tau| < \infty$ for all $Z_i \in \mathcal{Z}$ and $Y_i \in \mathcal{Y}$, and $\mathbb{E}[Y_i] > 0$ for all i, $\bar{\Delta}_{\text{avg-opt}}$ is bounded. Note that Problem (7.14) is hard to solve in the current form. Therefore, we reformulate it. We consider the following optimization problem with a parameter $\beta \ge 0$:

$$\Theta(\beta) \triangleq \min_{f\in\mathcal{F}} \limsup_{n\to\infty} \frac{1}{n} \sum_{i=0}^{n-1} \mathbb{E}\left[\sum_{l=1}^{m} \int_{a_{li}}^{a_{li}+Z_i+Y_{i+1}} g(\tau)d\tau - \beta(Z_i + Y_{i+1})\right], \tag{7.15}$$

[4] Observe that a special age-penalty function is depicted in Figure 7.8, where we choose $g(x) = x$ for simplicity.

[5] Since the age process is right-continuous, if sample i is delivered from source l, then $\Delta_l(D_i)$ is the age value of source l just after the delivery time D_i.

where $\Theta(\beta)$ is the optimal value of (7.15). The following lemma is motivated by [21]. It explains the relationship between Problems (7.14) and (7.15).

LEMMA 7.5 *The following assertions are true:*

(i) $\bar{\Delta}_{avg\text{-}opt} \lesseqqgtr \beta$ *if and only if* $\Theta(\beta) \lesseqqgtr 0$.
(ii) *If* $\Theta(\beta) = 0$, *then the optimal sampling policies that solve* (7.14) *and* (7.15) *are identical.*

Proof See Appendix 7.7.2. □

As a result of Lemma 7.5, the solution to (7.14) can be obtained by solving (7.15) in a multilayer manner: In the inner layer, we optimize Z_i for a given β. Then, in the outer layer, we seek a $\beta = \bar{\Delta}_{\text{avg-opt}} \geq 0$ such that $\Theta(\bar{\Delta}_{\text{avg-opt}}) = 0$. Lemma 7.5 helps us to utilize the DP technique to obtain the optimal sampler. Note that without Lemma 7.5, it would be quite difficult to use the DP technique to solve (7.14) optimally.

Existence of an Optimal Stationary Deterministic Policy

We resort to the methodology proposed in [22]. When $\beta = \bar{\Delta}_{\text{avg-opt}}$, Problem (7.15) is equivalent to an average cost per stage problem. According to [22], we describe the components of this problem in detail in what follows.

- **States:** At stage i,[6] the system state is specified by

$$\mathbf{s}(i) = (a_{[1]i}, \ldots, a_{[m]i}), \tag{7.16}$$

  where $a_{[l]i}$ is the lth largest age of the sources at stage i, that is, it is the lth largest component of the vector $(a_{1i}, \ldots, a_{mi})$. We use $\mathcal{S}$ to denote the state-space including all possible states. Notice that $\mathcal{S}$ is finite and bounded because $\mathcal{Z}$ and $\mathcal{Y}$ are finite and bounded.
- **Control action:** At stage i, the action that is taken by the sampler is $Z_i \in \mathcal{Z}$.
- **Random disturbance:** In our model, the random disturbance occurring at stage i is Y_{i+1}, which is independent of the system state and the control action.
- **Transition probabilities:** If the control $Z_i = z$ is applied at stage i and the service time of sample $i+1$ is $Y_{i+1} = y$, then the evolution of the system state from $\mathbf{s}(i)$ to $\mathbf{s}(i+1)$ is as follows:

$$\begin{aligned} a_{[m]i+1} &= y, \\ a_{[l]i+1} &= a_{[l+1]i} + z + y, \ l = 1, \ldots, m-1. \end{aligned} \tag{7.17}$$

  We let $\mathbb{P}_{\mathbf{ss}'}(z)$ denote the transition probabilities

$$\mathbb{P}_{\mathbf{ss}'}(z) = \mathbb{P}(\mathbf{s}(i+1) = \mathbf{s}' | \mathbf{s}(i) = \mathbf{s}, Z_i = z), \ \mathbf{s}, \mathbf{s}' \in \mathcal{S}. \tag{7.18}$$

6 From here forward, we assume that the duration of stage i is $[D_i, D_{i+1})$.

When $\mathbf{s} = (a_{[1]}, \ldots, a_{[m]})$ and $\mathbf{s}' = (a'_{[1]}, \ldots, a'_{[m]})$, the law of the transition probability is given by

$$\mathbb{P}_{\mathbf{ss}'}(z) = \begin{cases} \mathbb{P}(Y_{i+1} = y) & \text{if } a'_{[m]} = y \text{ and} \\ & a'_{[l]} = a_{[l+1]} + z + y \text{ for } l \neq m; \\ 0 & \text{else.} \end{cases} \tag{7.19}$$

- **Cost function:** Each time the system is in stage i and control Z_i is applied, we incur a cost,

$$C(\mathbf{s}(i), Z_i, Y_{i+1}) = \sum_{l=1}^{m} \int_{a_{[l]i}}^{a_{[l]i}+Z_i+Y_{i+1}} g(\tau)d\tau - \bar{\Delta}_{\text{avg-opt}}(Z_i + Y_{i+1}). \tag{7.20}$$

To simplify notation, we use the expected cost $C(\mathbf{s}(i), Z_i)$ as the cost per stage, that is,

$$C(\mathbf{s}(i), Z_i) = \mathbb{E}_{Y_{i+1}} \left[C(\mathbf{s}(i), Z_i, Y_{i+1}) \right], \tag{7.21}$$

where $\mathbb{E}_{Y_{i+1}}$ is the expectation with respect to Y_{i+1}, which is independent of $\mathbf{s}(i)$ and Z_i. It is important to note that there exists $c \in \mathbb{R}^+$ such that $|C(\mathbf{s}(i), Z_i)| \leq c$ for all $\mathbf{s}(i) \in \mathcal{S}$ and $Z_i \in \mathcal{Z}$. This is because $\mathcal{Z}$, $\mathcal{Y}$, $\mathcal{S}$, and $\bar{\Delta}_{\text{avg-opt}}$ are bounded.

In general, the average cost per stage under a sampling policy $f \in \mathcal{F}$ is given by

$$\limsup_{n \to \infty} \frac{1}{n} \mathbb{E} \left[\sum_{i=0}^{n-1} C(\mathbf{s}(i), Z_i) \right]. \tag{7.22}$$

We say that a sampling policy $f \in \mathcal{F}$ is *average-optimal* if it minimizes the average cost per stage in (7.22). Our objective is to find the average-optimal sampling policy. A policy f is called a stationary randomized policy if it assigns a probability distribution $q_{\mathcal{Z}}(\mathbf{s}(i))$ over the control set based on the state $\mathbf{s}(i)$ such that it chooses the control Z_i randomly according to this distribution; while a stationary deterministic policy chooses an action with certainty such that $Z_i = Z_j$ whenever $\mathbf{s}(i) = \mathbf{s}(j)$ for any i, j. According to [22], there may not exist a stationary deterministic policy that is average-optimal. However, in the next proposition, we are able to show that there is a stationary deterministic policy that is average-optimal.

PROPOSITION 2 *There exist a scalar λ and a function h that satisfy the following Bellman's equation:*

$$\lambda + h(\mathbf{s}) = \min_{z \in \mathcal{Z}} \left(C(\mathbf{s}, z) + \sum_{\mathbf{s}' \in \mathcal{S}} \mathbb{P}_{\mathbf{ss}'}(z) h(\mathbf{s}') \right), \tag{7.23}$$

where λ is the optimal average cost per stage that is independent of the initial state $\mathbf{s}(0)$ and satisfies

$$\lambda = \lim_{\alpha \to 1} (1 - \alpha) J_\alpha(\mathbf{s}), \forall \mathbf{s} \in \mathcal{S}, \tag{7.24}$$

and $h(\mathbf{s})$ is the relative cost function that, for any state $\mathbf{o}$, satisfies

$$h(\mathbf{s}) = \lim_{\alpha \to 1} (J_\alpha(\mathbf{s}) - J_\alpha(\mathbf{o})), \forall \mathbf{s} \in \mathcal{S}, \tag{7.25}$$

where $J_\alpha(\mathbf{s})$ is the optimal total expected α-discounted cost function, which is defined by

$$J_\alpha(\mathbf{s}) = \min_{f\in\mathcal{F}} \limsup_{n\to\infty} \mathbb{E}\left[\sum_{i=0}^{n-1} \alpha^i C(\mathbf{s}(i), Z_i)\right], \mathbf{s}(0) = \mathbf{s} \in \mathcal{S}, \tag{7.26}$$

where $0 < \alpha < 1$ is the discount factor. Furthermore, there exists a stationary deterministic policy that attains the minimum in (7.23) for each $\mathbf{s} \in \mathcal{S}$ and is average-optimal.

Proof According to [22, Proposition 4.2.1 and Proposition 4.2.6], it is enough to show that for every two states $\mathbf{s}$ and $\mathbf{s}'$, there exists a stationary deterministic policy f such that for some k, we have $\mathbb{P}\left[\mathbf{s}(k) = \mathbf{s}'|\mathbf{s}(0) = \mathbf{s}, f\right] > 0$, that is, we have a communicating Markov decision process (MDP). For details, see Appendix 7.7.3. □

We can deduce from Proposition 2 that the optimal waiting time is a fixed function of the state $\mathbf{s}$. One possible way to solve this SMDP is by using the relative value iteration (RVI) algorithm. Next, we present this algorithm and reveal a useful structure of the optimal sampler that helps in simplifying this algorithm.

Simplified Relative Value Iteration Using Optimal Sampler Structure

The RVI algorithm [23, Section 9.5.3], [24, Page 171] can be used to solve Bellman's equation (7.23). Starting with an arbitrary state $\mathbf{o}$, a single iteration for the RVI algorithm is given as follows:

$$\begin{aligned} Q_{n+1}(\mathbf{s}, z) &= C(\mathbf{s}, z) + \sum_{\mathbf{s}'\in\mathcal{S}} \mathbb{P}_{\mathbf{s}\mathbf{s}'}(z) h_n(\mathbf{s}'), \\ J_{n+1}(\mathbf{s}) &= \min_{z\in\mathcal{Z}}(Q_{n+1}(\mathbf{s}, z)), \\ h_{n+1}(\mathbf{s}) &= J_{n+1}(\mathbf{s}) - J_{n+1}(\mathbf{o}), \end{aligned} \tag{7.27}$$

where $Q_{n+1}(\mathbf{s}, z)$, $J_n(\mathbf{s})$, and $h_n(\mathbf{s})$ denote the state action value function, value function, and relative value function for iteration n, respectively. In the beginning, we set $J_0(\mathbf{s}) = 0$ for all $\mathbf{s} \in \mathcal{S}$, and then we repeat the iteration of the RVI algorithm as described before.[7]

The complexity of the RVI algorithm is high due to many sources (i.e., the curse of dimensionality [25]). Thus, we need to simplify the RVI algorithm. To that end, we show that the optimal sampler has a threshold property that can reduce the complexity of the RVI algorithm. Define $z_{\mathbf{s}}^\star$ as the optimal waiting time for state $\mathbf{s}$, and Y as a random variable that has the same distribution as Y_i. The threshold property in the optimal sampler is manifested in the following proposition:

[7] According to [23, 24], a sufficient condition for the convergence of the RVI algorithm is the aperiodicity of the transition matrices of stationary deterministic optimal policies. In our case, these transition matrices depend on the service times. This condition can always be achieved by applying the aperiodicity transformation as explained in [23, Section 8.5.4], which is a simple transformation. However, this does not always need to be done.

Algorithm 1 RVI algorithm with reduced complexity

given $l = 0$, sufficiently large u, tolerance $\epsilon_1 > 0$, tolerance $\epsilon_2 > 0$;
while $u - l > \epsilon_1$ **do**
    $\beta = \frac{l+u}{2}$;
    $J(\mathbf{s}) = 0$, $h(\mathbf{s}) = 0$, $h_{\text{last}}(\mathbf{s}) = 0$ for all states $\mathbf{s} \in \mathcal{S}$;
    while $\max_{\mathbf{s} \in \mathcal{S}} |h(\mathbf{s}) - h_{last}(\mathbf{s})| > \epsilon_2$ **do**
        for *each* $\mathbf{s} \in \mathcal{S}$ **do**
            if $\mathbb{E}_Y \left[\sum_{l=1}^{m} g(a_{[l]} + Y)\right] \geq \beta$ **then**
                $z_s^{\star} = 0$;
            else
                $z_s^{\star} = \operatorname{argmin}_{z \in \mathcal{Z}} C(\mathbf{s}, z) + \sum_{\mathbf{s}' \in \mathcal{S}} \mathbb{P}_{\mathbf{ss}'}(z) h(\mathbf{s}')$;
            end
            $J(\mathbf{s}) = C(\mathbf{s}, z_s^{\star}) + \sum_{\mathbf{s}' \in \mathcal{S}} \mathbb{P}_{\mathbf{ss}'}(z_s^{\star}) h(\mathbf{s}')$;
        end
        $h_{\text{last}}(\mathbf{s}) = h(\mathbf{s})$;
        $h(\mathbf{s}) = J(\mathbf{s}) - J(\mathbf{o})$;
    end
    if $J(\mathbf{o}) \geq 0$ **then**
        $u = \beta$;
    else
        $l = \beta$;
    end
end

PROPOSITION 3 *If the state* $\mathbf{s} = (a_{[1]}, \ldots, a_{[m]})$ *satisfies* $\mathbb{E}_Y \left[\sum_{l=1}^{m} g(a_{[l]} + Y)\right] \geq \bar{\Delta}_{avg\text{-}opt}$, *then* $z_s^{\star} = 0$.

Proof See Appendix 7.7.4. □

We can exploit the threshold test in Proposition 3 to reduce the complexity of the RVI algorithm as follows: The optimal waiting time for any state $\mathbf{s}$ that satisfies $\mathbb{E}_Y \left[\sum_{l=1}^{m} g(a_{[l]} + Y)\right] \geq \bar{\Delta}_{\text{avg-opt}}$ is zero. Thus, we need to solve (7.27) only for the states that fail this threshold test. As a result, we reduce the number of computations required along the system state space, which reduces the complexity of the RVI algorithm in return. Note that $\bar{\Delta}_{\text{avg-opt}}$ can be obtained using the bisection method or any other one-dimensional search method. Combining this with the result of Proposition 3 and the RVI algorithm, we propose the "RVI with reduced complexity (RVI-RC) sampler" in Algorithm 1. According to [23, 24], $J(\mathbf{o})$ in Algorithm 1 converges to the optimal average cost per stage. Moreover, in the outer layer of Algorithm 1, bisection is employed to obtain $\bar{\Delta}_{\text{avg-opt}}$, where β converges to $\bar{\Delta}_{\text{avg-opt}}$. Finally, the value of u in Algorithm 1 can be initialized to the value of the Ta-AP of the zero-wait sampler (as the Ta-AP of the zero-wait sampler provides an upper bound on the optimal Ta-AP), which can be easily calculated.

7.4.3 Jointly Optimal Scheduler and Sampler

Thanks to the separation principle, we are able to design the scheduler and sampler separately, which are presented in Sections 7.4.1 and 7.4.2, respectively. We here conclude our presented results so far, where an optimal solution for Problem (7.6) is manifested in the following theorem:

THEOREM 7.6 *The MAF scheduler and the RVI-RC sampler form an optimal solution for Problem* (7.6).

Proof The theorem follows directly from Proposition 1, Proposition 2, and Proposition 3. □

7.5 Final Remarks and Open Questions

7.5.1 Low-Complexity Sampler Design via Bellman's Equation Approximation

In this section, we devise a low-complexity sampler via an approximate analysis for Bellman's equation in (7.23) whose solution is the RVI-RC sampler. During our discussion here and in what follows, we fix the scheduler to the MAF. The obtained low-complexity sampler in this section will be shown to have near-optimal age performance in our numerical results in Section 7.5.3. For a given state $\mathbf{s}$, we denote the next state given z and y by $\mathbf{s}'(z, y)$. We can observe that the transition probability in (7.19) depends only on the distribution of the service time Y, which is independent of the system state and the control action. Hence, the second term in Bellman's equation in (7.23) can be rewritten as

$$\sum_{\mathbf{s}'\in\mathcal{S}} \mathbb{P}_{\mathbf{ss}'}(z)h(\mathbf{s}'(z,y)) = \sum_{y\in\mathcal{Y}} \mathbb{P}(Y=y)h(\mathbf{s}'(z,y)). \tag{7.28}$$

As a result, Bellman's equation in (7.23) can be rewritten as

$$\lambda = \min_{z}\left(C(\mathbf{s},z)+\sum_{y\in\mathcal{Y}} \mathbb{P}(Y=y)(h(\mathbf{s}'(z,y))-h(\mathbf{s}))\right). \tag{7.29}$$

Although $h(\mathbf{s})$ is discrete, we can interpolate the value of $h(\mathbf{s})$ between the discrete values so that it is differentiable by following the same approach in [26] and [27]. Let $\mathbf{s} = (a_{[1]}, \ldots, a_{[m]})$; then using the first-order Taylor approximation around a state $\mathbf{v} = (a^v_{[1]}, \ldots, a^v_{[m]})$ (some fixed state), we get

$$h(\mathbf{s}) \approx h(\mathbf{v}) + \sum_{l=1}^{m}(a_{[l]} - a^v_{[l]})\frac{\partial h(\mathbf{v})}{\partial a_{[l]}}. \tag{7.30}$$

Again, we use the first-order Taylor approximation around the state $\mathbf{v}$, together with the state evolution in (7.17), to get

$$h(\mathbf{s}'(z,y)) \approx h(\mathbf{v}) + (y - a^v_{[m]})\frac{\partial h(\mathbf{v})}{\partial a_{[m]}} + \sum_{l=1}^{m-1}(a_{[l+1]} - a^v_{[l]} + z + y)\frac{\partial h(\mathbf{v})}{\partial a_{[l]}}. \tag{7.31}$$

From (7.30) and (7.31), we get

$$h(\mathbf{s}'(z,y)) - h(\mathbf{s}) \approx (y - a_{[m]})\frac{\partial h(\mathbf{v})}{\partial a_{[m]}} + \sum_{l=1}^{m-1}(a_{[l+1]} - a_{[l]} + z + y)\frac{\partial h(\mathbf{v})}{\partial a_{[l]}}. \tag{7.32}$$

This implies that

$$\begin{aligned}\sum_{y\in\mathcal{Y}} \mathbb{P}(Y=y)(h(\mathbf{s}'(z,y)) - h(\mathbf{s})) \approx (\mathbb{E}[Y] - a_{[m]})\frac{\partial h(\mathbf{v})}{\partial a_{[m]}} \\ + \sum_{l=1}^{m-1}(a_{[l+1]} - a_{[l]} + z + \mathbb{E}[Y])\frac{\partial h(\mathbf{v})}{\partial a_{[l]}}.\end{aligned} \tag{7.33}$$

Using (7.29) with (7.33), we can get the following approximated Bellman's equation:

$$\lambda \approx \min_{z} \left(C(\mathbf{s},z) + (\mathbb{E}[Y] - a_{[m]})\frac{\partial h(\mathbf{v})}{\partial a_{[m]}} + \sum_{l=1}^{m-1}(a_{[l+1]} - a_{[l]} + z + \mathbb{E}[Y])\frac{\partial h(\mathbf{v})}{\partial a_{[l]}} \right). \tag{7.34}$$

By following the same steps as in Appendix 7.7.4 to get the optimal z that minimizes the objective function in (7.34), we get the following condition: The optimal z, for a given state $\mathbf{s}$, must satisfy

$$\mathbb{E}_Y\left[\sum_{l=1}^{m} g(a_{[l]} + t + Y)\right] - \bar{\Delta}_{\text{avg-opt}} + \sum_{l=1}^{m-1}\frac{\partial h(\mathbf{v})}{\partial a_{[l]}} \geq 0 \tag{7.35}$$

for all $t > z$, and

$$\mathbb{E}_Y\left[\sum_{l=1}^{m} g(a_{[l]} + t + Y)\right] - \bar{\Delta}_{\text{avg-opt}} + \sum_{l=1}^{m-1}\frac{\partial h(\mathbf{v})}{\partial a_{[l]}} \leq 0 \tag{7.36}$$

for all $t < z$. The smallest z that satisfies (7.35)–(7.36) is

$$\hat{z}_s^\star = \inf\left\{t \geq 0 : \mathbb{E}_Y\left[\sum_{l=1}^{m} g(a_{[l]} + t + Y)\right] \geq \bar{\Delta}_{\text{avg-opt}} - \sum_{l=1}^{m-1}\frac{\partial h(\mathbf{v})}{\partial a_{[l]}}\right\}, \tag{7.37}$$

where $\hat{z}_s^*$ is the optimal solution of the approximated Bellman's equation for state $\mathbf{s}$. Note that the term $\sum_{i=1}^{m-1}\frac{\partial h(\mathbf{v})}{\partial a_{[i]}}$ is constant. Hence, (7.37) can be rewritten as

$$\hat{z}_s^\star = \inf\left\{t \geq 0 : \mathbb{E}_Y\left[\sum_{l=1}^{m} g(a_{[l]} + t + Y)\right] \geq T\right\}. \tag{7.38}$$

This simple threshold sampler can approximate the optimal sampler for the original Bellman's equation in (7.23). The optimal threshold (T) in (7.38) can be obtained using a golden-section method [28]. Moreover, for a given state $\mathbf{s}$ and the threshold (T), (7.38) can be solved using the bisection method or any other one-dimensional search method. As we have mentioned, we will show in our numerical results in Section 7.5.3 that, when the scheduling policy is fixed to the MAF scheduler, the performance of this approximated sampler is almost the same as that of the RVI-RC sampler.

7.5.2 Linear Age-Penalty Function and Useful Insights

In this section, we consider the special case when the age-penalty function is linear, that is, $g(x) = b_1x + b_2$ for $b_1 > 0$. Such a simplification in the penalty function will further simplify the obtained samplers.

Simplification in the Optimal Sampler

Define $A_s = \sum_{l=1}^{m} a_{[l]}$ as the sum of the age values of state $\mathbf{s}$. The threshold test in Proposition 3 is simplified as follows:

PROPOSITION 4 *If the state* $\mathbf{s} = (a_{[1]}, \ldots, a_{[m]})$ *satisfies* $A_s \geq \frac{\bar{\Delta}_{avg\text{-}opt} - mb_1\mathbb{E}[Y] - mb_2}{b_1}$, *then we have* $z_s^\star = 0$.

Proof The proposition follows directly by substituting $g(x) = b_1x + b_2$ into the threshold test in Proposition 3. □

Hence, the change in Algorithm 1 is to replace the threshold test in Step 7 by $A_s \geq (\bar{\Delta}_{\text{avg-opt}} - mb_1\mathbb{E}[Y] - mb_2)/b_1$. This threshold test is easier to check than that in Proposition 3. This further simplifies the RVI-RC algorithm.

Optimality of the Zero-Wait Sampler

It is obvious that the zero-wait sampler is throughput and delay optimal. However, as explained in Section 7.2.3 with a counterexample, the zero-wait sampler does not necessarily minimize the average age. For a special case when $g(x) = b_1x + b_2$, we provide a sufficient condition for the optimality of the zero-wait sampler for minimizing the Ta-AP. Let $y_{\text{inf}} = \inf\{y \in \mathcal{Y} : \mathbb{P}[Y = y] > 0\}$, that is, y_{inf} is the smallest possible transmission time in $\mathcal{Y}$. As a result of Proposition 4, the sufficient condition for the optimality of the zero-wait sampler is manifested in the following theorem:

THEOREM 7.7 *If*

$$y_{inf} \geq \frac{(m-1)\mathbb{E}[Y]^2 + \mathbb{E}[Y^2]}{(m+1)\mathbb{E}[Y]}, \tag{7.39}$$

then the zero-wait sampler is optimal for Problem (7.15).

Proof See Appendix 7.7.5 □

From Theorem 7.7, it immediately follows that:

COROLLARY *If the transmission times are positive and constant (i.e.,* $Y_i = const > 0$ *for all i), then the zero-wait sampler is optimal for Problem* (7.15).

Proof The corollary follows directly from Theorem 7.7 by showing that (7.39) always holds in this case. □

Corollary 7.5.2 suggests that designing the optimal schedulers could be enough if the transmission times are deterministic, where a source generates a sample whenever this source is scheduled for transmission. However, if there is a variation in the transmission times, these schedulers alone may not be optimal anymore, and we need to optimize the sampling times as well. In practice, such random transmission times occur in many applications, such as autonomous vehicles. In particular, as we have mentioned, there are many ECUs in a vehicle that are connected to one or more sensors and actuators via a CAN bus [10, 11]. These ECUs are given different priority, based on their criticality level (e.g., ECUs in the powertrain have a higher priority compared to those connected to infotainment systems). Since high-priority samples usually have hard deadlines, the transmissions of low-priority samples are interrupted whenever the higher-priority ones are transmitted. Therefore, update samples with lower priority see a time-varying bandwidth and hence encounter a random transmission time.

Low-Complexity Water-Filling Sampler

We here investigate how the designed low-complexity sampler in Section 7.5.1 can be further simplified when $g(x) = b_1 x + b_2$. In particular, we will show that the threshold sampler in Section 7.5.1 reduces to the water-filling sampler in this case.

Substituting $g(x) = b_1 x + b_2$ into (7.37), where the equality holds in this case, we get the following condition: The optimal z in this case, for a given state $\mathbf{s}$, must satisfy

$$b_1 A_s - \bar{\Delta}_{\text{avg-opt}} + m b_1 z + m b_1 \mathbb{E}[Y] + m b_2 + \sum_{l=1}^{m-1} \frac{\partial h(\mathbf{v})}{\partial a_{[l]}} = 0, \tag{7.40}$$

where A_s is the sum of the age values of state $\mathbf{s}$. Rearranging (7.40), we get

$$\hat{z}_s^* = \left[\frac{\bar{\Delta}_{\text{avg-opt}} - m b_1 \mathbb{E}[Y] - \sum_{l=1}^{m-1} \frac{\partial h(\mathbf{v})}{\partial a_{[l]}} - m b_2}{m b_1} - \frac{A_s}{m} \right]^+. \tag{7.41}$$

By observing that the term $\sum_{i=1}^{m-1} \frac{\partial h(\mathbf{v})}{\partial a_{[i]}}$ is constant, (7.41) can be rewritten as

$$\hat{z}_s^* = \left[T - \frac{A_s}{m} \right]^+. \tag{7.42}$$

The solution in (7.42) is in the form of the water-filling solution as we compare a fixed threshold (T) with the average age of a state $\mathbf{s}$. The solution in (7.42) suggests that this simple water-filling sampler can approximate the optimal solution of the original Bellman's equation in (7.23) when $g(x) = b_1 x + b_2$ for some $b_1 > 0$ and b_2. Similar to the general case, the optimal threshold (T) in (7.42) can be obtained using a golden-section method. We will also evaluate the performance of these water-filling samplers in the next section and show that its performance is almost the same as RVI-RC sampler.

7.5.3 Numerical Results

We consider an information update system with $m = 3$ sources. We use "RAND" to represent a random scheduler, where sources are chosen to be served with equal probability. By "constant-wait" we refer to the sampler that imposes a constant waiting time after each delivery with $Z_i = 0.3\mathbb{E}[Y], \forall i$. Moreover, we use "threshold" and "water-filling" to denote the proposed samplers in (7.38) and (7.42), respectively.

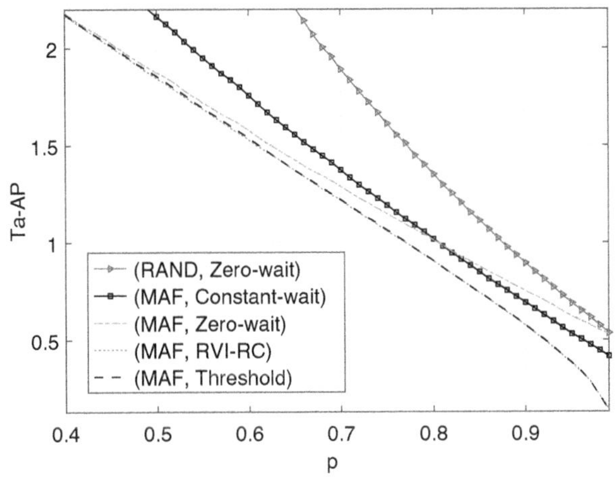

Figure 7.8 Ta-AP versus transmission probability p for an update system with $m = 3$ sources, where $g(x) = e^{0.1x} - 1$

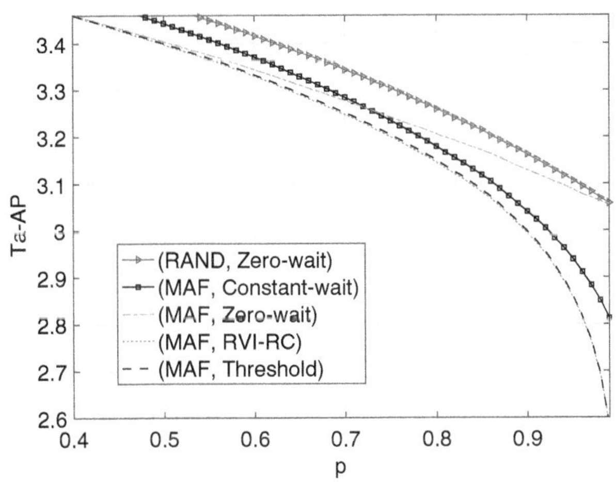

Figure 7.9 Ta-AP versus transmission probability p for an update system with $m = 3$ sources, where $g(x) = x^{0.1}$.

We now evaluate the performance of our proposed solutions for minimizing the Ta-AP. We set the transmission times to be either 0 or 3 with probability p and $1 - p$, respectively. Figures 7.8, 7.9, and 7.10 illustrate the Ta-AP versus the transmission probability p, where we set the age-penalty function $g(x)$ to be $e^{0.1x} - 1$, $x^{0.1}$, and x, respectively. The range of the probability p is [0.4; 0.99] in Figures 7.8, 7.9, and 7.10. When $p = 1$, $\mathbb{E}[Y] = \mathbb{E}[Y^2] = 0$ and hence the Ta-AP of the zero-wait sampler (for any scheduler) at $p = 1$ is undefined. Therefore, the point $p = 1$ is not plotted in Figures 7.8, 7.9, and 7.10. For the zero-wait sampler, we find that the MAF scheduler provides a lower Ta-AP than that of the RAND scheduler. This agrees with Proposition 1. Moreover, when the scheduling policy is fixed to the MAF scheduler, we find that the

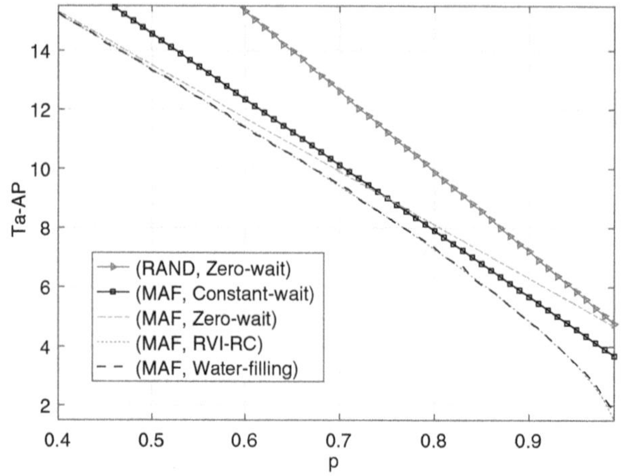

Figure 7.10 Ta-AP versus transmission probability p for an update system with $m = 3$ sources, where $g(x) = x$

Ta-AP resulting from the RVI-RC sampler is lower than those resulting from the zero-wait sampler and the constant-wait sampler. This observation suggests the following: (i) The zero-wait sampler does not necessarily minimize the Ta-AP; (ii) optimizing the scheduling policy only is not enough to minimize the Ta-AP, but we have to optimize both the scheduling policy and the sampling policy together to minimize the Ta-AP. In addition, as we can observe, the Ta-AP resulting from the threshold sampler in Figures 7.8 and 7.9, and the water-filling sampler in Figure 7.10 almost coincides with the Ta-AP resulting from the RVI-RC sampler.

We then set the transmission times to be either 0 or $Y_{\max}$ with probability 0.9 and 0.1, respectively. We vary the maximum transmission time $Y_{\max}$ and plot the Ta-AP in Figures 7.11, 7.12, and 7.13, where $g(x)$ is set to be $e^{0.1x} - 1$, $x^{0.1}$, and x, respectively. The scheduling policy is fixed to the MAF scheduler in all plotted curves. We can observe in all figures that the Ta-AP resulting from the RVI-RC sampler is lower than those resulting from the zero-wait sampler and the constant-wait sampler, and the gap between them increases as the variability (variance) of the transmission times increases. This suggests that when the transmission times have a big variation, we have to optimize the scheduler and the sampler together to minimize the Ta-AP. Finally, as we can observe, the Ta-AP of the threshold sampler in Figures 7.11 and 7.12, and the water-filling sampler in Figure 7.13 almost coincides with that of the RVI-RC sampler.

7.5.4 Open Questions

So far, we have considered only homogeneous age-penalty functions, that is, all sources have the same penalty on the age. This technical assumption allowed us to develop the separation principle. Using this principle, we were able to decouple the scheduler decisions and the sampler decisions, which allowed us to design each separately from the other. However, in practice, it is quite possible that different sources

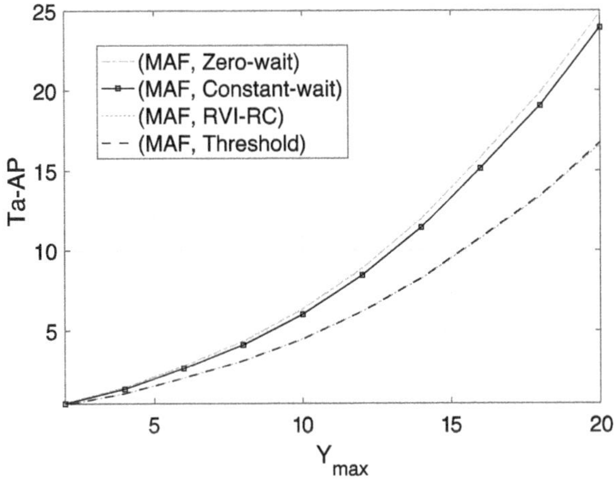

Figure 7.11 Ta-AP versus the maximum service time $Y_{\max}$ for an update system with $m = 3$ sources, where $g(x) = e^{0.1x} - 1$

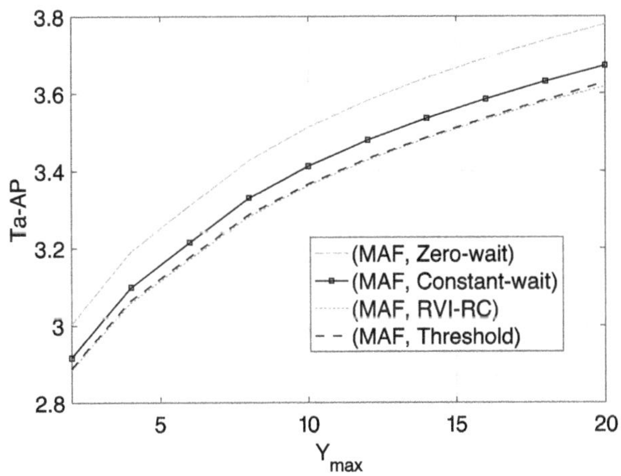

Figure 7.12 Ta-AP versus the maximum service time $Y_{\max}$ for an update system with $m = 3$ sources, where $g(x) = x^{0.1}$

may incur different penalties on the age, for example, for certain types of information may be quite sensitive to the age (e.g., in autonomous driving applications), while others may be far more age-tolerant (e.g., measuring the humidity in a region). Hence, for various practical problems, this condition may be violated, in which case the separation principle itself may not hold any longer. This is because of the variation in the sources' order, according to their age-penalty values, with time. This results in a strong correlation between the scheduler and sampler actions. Hence, our technique here may not be suitable for this case. Indeed, other tools and techniques, which could be more sophisticated, are needed to answer such a challenging question. We raised this question to the reader to convey that this research area is still at an early stage and

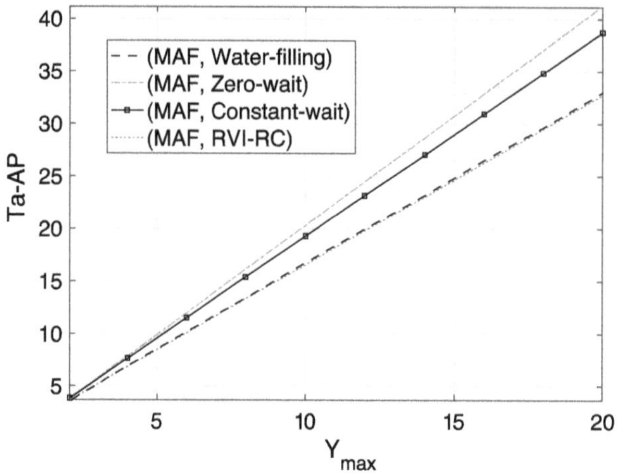

Figure 7.13 Ta-AP versus the maximum service time $Y_{\max}$ for an update system with $m = 3$ sources, where $g(x) = x$

expected to grow considerably in the next few years because various research challenges have to be addressed and solved. The aforementioned challenge is one of them that needs to be addressed.

7.6 Summary

In this chapter, we have investigated how controlling the sampling times can further improve the data freshness in information update systems. We started by providing a brief explanation of the sampling problem in single-source networks, supported by an example that shows that the optimal sampler is not trivial and has a counterintuitive phenomenon. Moreover, we provided real-time applications on age of information and its penalty functions.

Later on, we shifted our focus to study the optimal sampling problem in multisource networks. It turned out that the optimal sampling problem in multisource networks is more challenging. This is because the sources are communicating their generated samples to the destination via a shared channel, and hence the decision policy does not only control a sampler, but also a scheduler. Our target was to study the problem of finding the optimal decision policy that controls the sampling times, the sampler, and the transmission order of the sources, the scheduler, to minimize the Ta-AP in multisource networks. We showed that the MAF scheduler and the RVI-RC sampler, which results from reducing the computation complexity of the RVI algorithm, are jointly optimal for minimizing the Ta-AP. In addition, we devised a low-complexity threshold sampler via an approximate analysis of Bellman's equation. Finally, we considered a special case when the age-penalty function is linear and obtained a sufficient condition for the optimality of the zero-wait sampler in this case. We also showed that the approximated threshold sampler is further simplified to a simple water-filling sampler in the special case of linear age-penalty function. The numerical results showed that the performance of these approximated samplers is almost the same as that of the RVI-RC sampler.

7.7 Appendix

7.7.1 Proof of Proposition 1

Let the vector $\boldsymbol{\Delta}_\pi(t) = (\Delta_{[1],\pi}(t), \ldots, \Delta_{[m],\pi}(t))$ denote the system state at time t of the scheduler π, where $\Delta_{[l],\pi}(t)$ is the lth largest age of the sources at time t under the scheduler π. Let $\{\boldsymbol{\Delta}_\pi(t), t \geq 0\}$ denote the state process of the scheduler π. For notational simplicity, let P represent the MAF scheduler. Throughout the proof, we assume that $\boldsymbol{\Delta}_\pi(0^-) = \boldsymbol{\Delta}_P(0^-)$ for all π and the sampler is fixed to an arbitrarily chosen one. The key step in the proof of Proposition 1 is the following lemma, where we compare the scheduler P with any arbitrary scheduler π.

LEMMA 7.8 *Suppose that $\boldsymbol{\Delta}_\pi(0^-) = \boldsymbol{\Delta}_P(0^-)$ for all scheduler π and the sampler is fixed; then we have*

$$\{\boldsymbol{\Delta}_P(t), t \geq 0\} \leq_{st} \{\boldsymbol{\Delta}_\pi(t), t \geq 0\}. \tag{7.43}$$

We use a coupling and forward induction to prove Lemma 7.8. For any scheduler π, suppose that the stochastic processes $\widetilde{\boldsymbol{\Delta}}_P(t)$ and $\widetilde{\boldsymbol{\Delta}}_\pi(t)$ have the same stochastic laws as $\boldsymbol{\Delta}_P(t)$ and $\boldsymbol{\Delta}_\pi(t)$. The state processes $\widetilde{\boldsymbol{\Delta}}_P(t)$ and $\widetilde{\boldsymbol{\Delta}}_\pi(t)$ are coupled such that the service times of the samples are equal under both scheduling policies, that is, Y_i are the same under both scheduling policies. Such a coupling is valid since the service time distribution is fixed under all policies. Since the sampler is fixed, such a coupling implies that the sample generation and delivery times are the same under both schedulers. According to Theorem 6.B.30 of [15], if we can show

$$\mathbb{P}\left[\widetilde{\boldsymbol{\Delta}}_P(t) \leq \widetilde{\boldsymbol{\Delta}}_\pi(t), t \geq 0\right] = 1, \tag{7.44}$$

then (7.43) is proven. To ease the notational burden, we will omit the tildes on the coupled versions in this proof and just use $\boldsymbol{\Delta}_P(t)$ and $\boldsymbol{\Delta}_\pi(t)$. Next, we compare scheduler P and scheduler π on a sample path and prove (7.43) using the following lemma:

LEMMA 7.9 (Inductive Comparison) *Suppose that a sample with generation time S is delivered under the scheduler P and the scheduler π at the same time t. The system state of the scheduler P is $\boldsymbol{\Delta}_P$ before the sample delivery, which becomes $\boldsymbol{\Delta}'_P$ after the sample delivery. The system state of the scheduler π is $\boldsymbol{\Delta}_\pi$ before the sample delivery, which becomes $\boldsymbol{\Delta}'_\pi$ after the sample delivery. If*

$$\Delta_{[i],P} \leq \Delta_{[i],\pi}, i = 1, \ldots, m, \tag{7.45}$$

then

$$\Delta'_{[i],P} \leq \Delta'_{[i],\pi}, i = 1, \ldots, m. \tag{7.46}$$

Proof Since only one source can be scheduled at a time and the scheduler P is the MAF one, the sample with generation time S must be generated from the source with maximum age $\Delta_{[1],P}$, call it source l^*. In other words, the age of source l^* is reduced from the maximum age $\Delta_{[1],P}$ to the minimum age $\Delta'_{[m],P} = t - S$, and the age of the other $(m-1)$ sources remain unchanged. Hence,

$$\begin{aligned} \Delta'_{[i],P} &= \Delta_{[i+1],P}, i = 1, \ldots, m-1, \\ \Delta'_{[m],P} &= t - S. \end{aligned} \tag{7.47}$$

In the scheduler π, this sample can be generated from any source. Thus, for all cases of scheduler π, it must hold that

$$\Delta'_{[i],\pi} \geq \Delta_{[i+1],\pi}, i = 1, \ldots, m-1. \tag{7.48}$$

By combining (7.45), (7.47), and (7.48), we have

$$\Delta'_{[i],\pi} \geq \Delta_{[i+1],\pi} \geq \Delta_{[i+1],P} = \Delta'_{[i],P}, i = 1, \ldots, m-1. \tag{7.49}$$

In addition, since the same sample is also delivered under the scheduler π, the source from which this sample is generated under policy π will have the minimum age after the delivery, that is, we have

$$\Delta'_{[m],\pi} = t - S = \Delta'_{[m],P}. \tag{7.50}$$

By this, (7.46) is proven. □

Proof of Lemma 7.8 Using the coupling between the system state processes, and for any given sample path of the service times, we consider two cases:

Case 1: When there is no sample delivery, the age of each source grows linearly with a slope 1.

Case 2: When a sample is delivered, the ages of the sources evolve according to Lemma 7.9.

By induction over time, we obtain

$$\Delta_{[i],P}(t) \leq \Delta_{[i],\pi}(t), i = 1, \ldots, m, t \geq 0. \tag{7.51}$$

Hence, (7.44) follows which implies (7.43) by Theorem 6.B.30 of [15]. This completes the proof. □

Proof of Proposition 1 Since the Ta-AP for any scheduling policy π is the expectation of nondecreasing functional of the process $\{\mathbf{\Delta}_\pi(t), t \geq 0\}$, (7.43) implies (7.7) using the properties of stochastic ordering [15]. This completes the proof. □

7.7.2 Proof of Lemma 7.5

Part (i) is proven in two steps:

Step 1: We will prove that $\bar{\Delta}_{\text{avg-opt}} \leq \beta$ if and only if $\Theta(\beta) \leq 0$. If $\bar{\Delta}_{\text{avg-opt}} \leq \beta$, there exists a sampling policy $f = (Z_0, Z_1, \ldots) \in \mathcal{F}$ that is feasible for (7.14) and (7.15), which satisfies

$$\limsup_{n\to\infty} \frac{\sum_{i=0}^{n-1} \mathbb{E}\left[\sum_{l=1}^{m} \int_{a_{li}}^{a_{li}+Z_i+Y_{i+1}} g(\tau)d\tau\right]}{\sum_{i=0}^{n-1} \mathbb{E}[Z_i + Y_{i+1}]} \leq \beta. \tag{7.52}$$

Hence,

$$\limsup_{n\to\infty} \frac{\frac{1}{n}\sum_{i=0}^{n-1}\mathbb{E}\left[\sum_{l=1}^{m} \int_{a_{li}}^{a_{li}+Z_i+Y_{i+1}} g(\tau)d\tau - \beta(Z_i+Y_{i+1})\right]}{\frac{1}{n}\sum_{i=0}^{n-1} \mathbb{E}[Z_i+Y_{i+1}]} \leq 0. \tag{7.53}$$

Since Z_i and Y_i are bounded and positive and $\mathbb{E}[Y_i] > 0$ for all i, we have $0 < \liminf_{n\to\infty} \frac{1}{n}\sum_{i=0}^{n-1} \mathbb{E}[Z_i + Y_{i+1}] \leq \limsup_{n\to\infty} \frac{1}{n}\sum_{i=0}^{n-1} \mathbb{E}[Z_i + Y_{i+1}] \leq q$ for some $q \in \mathbb{R}^+$. By this, we get

$$\limsup_{n\to\infty} \frac{1}{n}\sum_{i=0}^{n-1}\mathbb{E}\left[\sum_{l=1}^{m} \int_{a_{li}}^{a_{li}+Z_i+Y_{i+1}} g(\tau)d\tau - \beta(Z_i+Y_{i+1})\right] \leq 0. \tag{7.54}$$

Therefore, $\Theta(\beta) \leq 0$.

In the reverse direction, if $\Theta(\beta) \leq 0$, then there exists a sampling policy $f = (Z_0, Z_1, \ldots) \in \mathcal{F}$ that is feasible for (7.14) and (7.15), which satisfies (7.54). Since we have $0 < \liminf_{n\to\infty} \frac{1}{n}\sum_{i=0}^{n-1} \mathbb{E}[Z_i+Y_{i+1}] \leq \limsup_{n\to\infty} \frac{1}{n}\sum_{i=0}^{n-1} \mathbb{E}[Z_i+Y_{i+1}] \leq q$, we can divide (7.54) by $\liminf_{n\to\infty} \frac{1}{n} \sum_{i=0}^{n-1} \mathbb{E}[Z_i + Y_{i+1}]$ to get (7.53), which implies (7.52). Hence, $\bar{\Delta}_{\text{avg-opt}} \leq \beta$. By this, we have proven that $\bar{\Delta}_{\text{avg-opt}} \leq \beta$ if and only if $\Theta(\beta) \leq 0$.

Step 2: We need to prove that $\bar{\Delta}_{\text{avg-opt}} < \beta$ if and only if $\Theta(\beta) < 0$. This statement can be proven by using the arguments in Step 1, in which "$\leq$" should be replaced by "$<$". Finally, from the statement of Step 1, it immediately follows that $\bar{\Delta}_{\text{avg-opt}} > \beta$ if and only if $\Theta(\beta) > 0$. This completes part (i).

Part (ii): We first show that each optimal solution to (7.14) is an optimal solution to (7.15). By the claim of part (i), $\Theta(\beta) = 0$ is equivalent to $\bar{\Delta}_{\text{avg-opt}} = \beta$. Suppose that policy $f = (Z_0, Z_1, \ldots) \in \mathcal{F}$ is an optimal solution to (7.14). Then, $\Delta_{\text{avg}(\pi_{\text{MAF}},f)} = \bar{\Delta}_{\text{avg-opt}} = \beta$. Applying this in the arguments of (7.52)–(7.54), we can show that policy f satisfies

$$\limsup_{n\to\infty} \frac{1}{n}\sum_{i=0}^{n-1}\mathbb{E}\left[\sum_{l=1}^{m} \int_{a_{li}}^{a_{li}+Z_i+Y_{i+1}} g(\tau)d\tau - \beta(Z_i+Y_{i+1})\right] = 0. \tag{7.55}$$

This and $\Theta(\beta) = 0$ imply that policy f is an optimal solution to (7.15).

Similarly, we can prove that each optimal solution to (7.15) is an optimal solution to (7.14). By this, part (ii) is proven. □

7.7.3 Proof of Proposition 2

According to [22, Proposition 4.2.1 and Proposition 4.2.6], it is enough to show that for every two states $\mathbf{s}$ and $\mathbf{s}'$, there exists a stationary deterministic policy f such that for some k, we have

$$\mathbb{P}\left[\mathbf{s}(k) = \mathbf{s}' | \mathbf{s}(0) = \mathbf{s}, f\right] > 0. \tag{7.56}$$

From the state evolution equation (7.17), we can observe that any state in $\mathcal{S}$ can be represented in terms of the waiting and service times. This implies (7.56). To clarify this, let us consider a system with three sources. Assume that the elements of state $\mathbf{s}'$ are as follows:

$$\begin{aligned} a'_{[1]} &= y_3 + z_2 + y_2 + z_1 + y_1, \\ a'_{[2]} &= y_3 + z_2 + y_2, \\ a'_{[3]} &= y_3, \end{aligned} \tag{7.57}$$

where y_i and z_i are any arbitrary elements in $\mathcal{Y}$ and $\mathcal{Z}$, respectively. Then, we will show that from any arbitrary state $\mathbf{s} = (a_{[1]}, a_{[2]}, a_{[3]})$, a sequence of service and waiting times can be followed to reach state $\mathbf{s}'$. If we have $Z_0 = z_1$, $Y_1 = y_1$, $Z_1 = z_1$, $Y_2 = y_2$, $Z_2 = z_2$, and $Y_3 = y_3$, then according to (7.17), we have in the first stage

$$\begin{aligned} a_{[1]1} &= a_{[2]} + z_1 + y_1, \\ a_{[2]1} &= a_{[3]} + z_1 + y_1, \\ a_{[3]1} &= y_1, \end{aligned} \tag{7.58}$$

and in the second stage, we have

$$\begin{aligned} a_{[1]2} &= a_{[3]} + z_1 + y_2 + z_1 + y_1, \\ a_{[2]2} &= y_2 + z_1 + y_1, \\ a_{[3]2} &= y_2, \end{aligned} \tag{7.59}$$

and in the third stage, we have

$$\begin{aligned} a_{[1]3} &= y_3 + z_2 + y_2 + z_1 + y_1 = a'_{[1]}, \\ a_{[2]3} &= y_3 + z_2 + y_2 = a'_{[2]}, \\ a_{[3]3} &= y_3 = a'_{[3]}. \end{aligned} \tag{7.60}$$

Hence, a stationary deterministic policy f can be designed to reach state $\mathbf{s}'$ from state $\mathbf{s}$ in three stages, if the aforementioned sequence of service times occurs. This implies that

$$\mathbb{P}\left[\mathbf{s}(3) = \mathbf{s}' | \mathbf{s}(0) = \mathbf{s}, f\right] = \prod_{i=1}^{3} \mathbb{P}(Y_i = y_i) > 0, \tag{7.61}$$

where we have used that Y_i are *i.i.d.*[8] The previous argument can be generalized to any number of sources. In particular, a forward induction over m can be used to show the result, where (7.56) trivially holds for $m = 1$, and the previous argument can be used to show that (7.56) holds for any general m. This completes the proof. □

[8] We assume that all elements in $\mathcal{Y}$ have a strictly positive probability, where the elements with zero probability can be removed without affecting the proof.

7.7.4 Proof of Proposition 3

We prove Proposition 3 into two steps:

Step 1: We first show that $h(\mathbf{s})$ is nondecreasing in $\mathbf{s}$. To do so, we show that $J_\alpha(\mathbf{s})$, defined in (7.26), is nondecreasing in $\mathbf{s}$, which together with (7.25) imply that $h(\mathbf{s})$ is nondecreasing in $\mathbf{s}$.

Given an initial state $\mathbf{s}(0)$, the total expected discounted cost under a sampling policy $f \in \mathcal{F}$ is given by

$$J_\alpha(\mathbf{s}(0); f) = \limsup_{n\to\infty} \mathbb{E}\left[\sum_{i=0}^{n-1} \alpha^i C(\mathbf{s}(i), Z_i)\right], \tag{7.62}$$

where $0 < \alpha < 1$ is the discount factor. The optimal total expected α-discounted cost function is defined by

$$J_\alpha(\mathbf{s}) = \min_{f\in\mathcal{F}} J_\alpha(\mathbf{s}; f), \ \mathbf{s} \in \mathcal{S}. \tag{7.63}$$

A policy is said to be α-optimal if it minimizes the total expected α-discounted cost. The discounted cost optimality equation of $J_\alpha(\mathbf{s})$ is discussed in what follows.

PROPOSITION 5 *The optimal total expected α-discounted cost $J_\alpha(\mathbf{s})$ satisfies*

$$J_\alpha(\mathbf{s}) = \min_{z\in\mathcal{Z}} C(\mathbf{s}, z) + \alpha \sum_{\mathbf{s}'\in\mathcal{S}} \mathbb{P}_{\mathbf{s}\mathbf{s}'}(z) J_\alpha(\mathbf{s}'). \tag{7.64}$$

Moreover, a stationary deterministic policy that attains the minimum in Eq. (7.64) for each $\mathbf{s} \in \mathcal{S}$ will be an α-optimal policy. Also, let $J_{\alpha,0}(\mathbf{s}) = 0$ for all $\mathbf{s}$ and any $n \geq 0$,

$$J_{\alpha,n+1}(\mathbf{s}) - \min_{z\in\mathcal{Z}} C(\mathbf{s}, z) + \alpha \sum_{\mathbf{s}'\in\mathcal{S}} \mathbb{P}_{\mathbf{s}\mathbf{s}'}(z) J_{\alpha,n}(\mathbf{s}'). \tag{7.65}$$

Then, we have $J_{\alpha,n}(\mathbf{s}) \to J_\alpha(\mathbf{s})$ as $n \to \infty$ for every $\mathbf{s}$, and α.

Proof Since we have bounded cost per stage, the proposition follows directly from [22, Proposition 1.2.2 and Proposition 1.2.3] and [29]. □

Next, we use the optimality equation (7.64) and the value iteration in (7.65) to prove that $J_\alpha(\mathbf{s})$ is nondecreasing in $\mathbf{s}$.

LEMMA 7.10 *The optimal total expected α-discounted cost function $J_\alpha(\mathbf{s})$ is nondecreasing in $\mathbf{s}$.*

Proof We use induction on n in equation (7.65) to prove Lemma 7.10. Obviously, the result holds for $J_{\alpha,0}(\mathbf{s})$.

Now, assume that $J_{\alpha,n}(\mathbf{s})$ is nondecreasing in $\mathbf{s}$. We need to show that for any two states $\mathbf{s}_1$ and $\mathbf{s}_2$ with $\mathbf{s}_1 \leq \mathbf{s}_2$, we have $J_{\alpha,n+1}(\mathbf{s}_1) \leq J_{\alpha,n+1}(\mathbf{s}_2)$. First, we note that, since the age-penalty function $g(\cdot)$ is nondecreasing, the expected cost per stage $C(\mathbf{s}, z)$ is nondecreasing in $\mathbf{s}$, that is, we have

$$C(\mathbf{s}_1, z) \leq C(\mathbf{s}_2, z). \tag{7.66}$$

From the state evolution equation (7.17) and the transition probability equation (7.19), the second term of the right-hand side (RHS) of (7.65) can be rewritten as

$$\sum_{\mathbf{s}'\in\mathcal{S}} \mathbb{P}_{\mathbf{s}\mathbf{s}'}(z)J_{\alpha,n}(\mathbf{s}') = \sum_{y\in\mathcal{Y}} \mathbb{P}(Y=y)J_{\alpha,n}(\mathbf{s}'(z,y)), \tag{7.67}$$

where $\mathbf{s}'(z,y)$ is the next state from state $\mathbf{s}$ given the values of z and y. Also, according to the state evolution equation (7.17), if the next states of $\mathbf{s}_1$ and $\mathbf{s}_2$ for given values of z and y are $\mathbf{s}_1'(z,y)$ and $\mathbf{s}_2'(z,y)$, respectively, then we have $\mathbf{s}_1'(z,y) \leq \mathbf{s}_2'(z,y)$. This implies that

$$\sum_{y\in\mathcal{Y}} \mathbb{P}(Y=y)J_{\alpha,n}(\mathbf{s}_1'(z,y)) \leq \sum_{y\in\mathcal{Y}} \mathbb{P}(Y=y)J_{\alpha,n}(\mathbf{s}_2'(z,y)), \tag{7.68}$$

where we have used the induction assumption that $J_{\alpha,n}(\mathbf{s})$ is nondecreasing in $\mathbf{s}$. Using (7.66), (7.68), and the fact that the minimum operator in (7.65) retains the nondecreasing property, we conclude that

$$J_{\alpha,n+1}(\mathbf{s}_1) \leq J_{\alpha,n+1}(\mathbf{s}_2). \tag{7.69}$$

This completes the proof. □

Step 2: We use Step 1 to prove Proposition 3. From Step 1, we have that $h(\mathbf{s})$ is nondecreasing in $\mathbf{s}$. Similar to Step 1, this implies that the second term of the right-hand side (RHS) of (7.23) ($\sum_{\mathbf{s}'\in\mathcal{S}} \mathbb{P}_{\mathbf{s}\mathbf{s}'}(z)h(\mathbf{s}')$) is nondecreasing in $\mathbf{s}'$. Moreover, from the state evolution (7.17), we can notice that, for any state $\mathbf{s}$, the next state $\mathbf{s}'$ is increasing in z. This argument implies that the second term of the right-hand side (RHS) of (7.23) ($\sum_{\mathbf{s}'\in\mathcal{S}} \mathbb{P}_{\mathbf{s}\mathbf{s}'}(z)h(\mathbf{s}')$) is increasing in z. Thus, the value of $z \in \mathcal{Z}$ that achieves the minimum value of this term is zero. If, for a given state $\mathbf{s}$, the value of $z \in \mathcal{Z}$ that achieves the minimum value of the cost function $C(\mathbf{s}, z)$ is zero, then $z = 0$ solves the RHS of (7.23). In the sequel, we obtain the condition on $\mathbf{s}$ under which $z = 0$ minimizes the cost function $C(\mathbf{s}, z)$.

Now, we focus on the cost function $C(\mathbf{s}, z)$. In order to obtain the optimal z that minimizes this cost function, we need to obtain the one-sided derivative of it. The one-sided derivative of a function q in the direction of ω at z is given by

$$\delta q(z;\omega) \triangleq \lim_{\epsilon\to 0^+} \frac{q(z+\epsilon\omega) - q(z)}{\epsilon}. \tag{7.70}$$

Let $r(\mathbf{s}, z, Y) = \sum_{l=1}^{m} \int_{a_{[l]}}^{a_{[l]}+z+Y} g(\tau)d\tau$. Since $r(\mathbf{s}, z, Y)$ is the sum of integration of a nondecreasing function $g(\cdot)$, it is easy to show that $r(\mathbf{s}, z, Y)$ is convex. According to [1, Lemma 4], the function $q(z) = \mathbb{E}_Y[r(\mathbf{s}, z, Y)]$ is convex as well. Hence, the one-sided derivative $\delta q(z;\omega)$ of $q(z)$ exists [30, p. 709]. Moreover, since $z \to r(\mathbf{s}, z, Y)$ is convex, the function $\epsilon \to [r(\mathbf{s}, z+\epsilon\omega, Y) - r(\mathbf{s}, z, Y)]/\epsilon$ is nondecreasing and bounded from above on $(0, \theta]$ for some $\theta > 0$ [31, Proposition 1.1.2(i)]. Using the monotone convergence theorem [32, Theorem 1.5.6], we can interchange the limit and integral operators in $\delta q(z;\omega)$ such that

$$
\begin{aligned}
\delta q(z;\omega) &= \lim_{\epsilon\to 0^+} \frac{1}{\epsilon}\mathbb{E}_Y[r(\mathbf{s}, z+\epsilon\omega, Y) - r(\mathbf{s}, z, Y)] \\
&= \mathbb{E}_Y\left[\lim_{\epsilon\to 0^+}\frac{1}{\epsilon}\{r(\mathbf{s}, z+\epsilon\omega, Y) - r(\mathbf{s}, z, Y)\}\right] \\
&= \mathbb{E}_Y\left[\lim_{t\to z^+}\sum_{l=1}^{m} g(a_{[l]}+t+Y)w\mathbb{1}_{\{\omega>0\}} + \lim_{t\to z^-}\sum_{l=1}^{m} g(a_{[l]}+t+Y)w\mathbb{1}_{\{\omega<0\}}\right] \\
&= \lim_{t\to z^+}\mathbb{E}_Y\left[\sum_{l=1}^{m} g(a_{[l]}+t+Y)w\mathbb{1}_{\{\omega>0\}}\right] + \lim_{t\to z^-}\mathbb{E}_Y\left[\sum_{l=1}^{m} g(a_{[l]}+t+Y)w\mathbb{1}_{\{\omega<0\}}\right],
\end{aligned}
$$

where $\mathbb{1}_E$ is the indicator function of event E. According to [30, p. 710] and the convexity of $q(z)$, z is optimal to the cost function $C(\mathbf{s}, z)$ if and only if

$$\delta q(z;\omega) - \bar{\Delta}_{\text{avg-opt}}\omega \geq 0, \ \forall \omega \in \mathbb{R}. \tag{7.71}$$

As ω in (7.71) is an arbitrary real number, considering $\omega = 1$, (7.71) becomes

$$\lim_{t\to z^+}\mathbb{E}_Y\left[\sum_{l=1}^{m} g(a_{[l]}+t+Y)\right] - \bar{\Delta}_{\text{avg-opt}} \geq 0. \tag{7.72}$$

Likewise, considering $\omega = -1$, (7.71) implies

$$\lim_{t\to z^-}\mathbb{E}_Y\left[\sum_{l=1}^{m} g(a_{[l]}+t+Y)\right] - \bar{\Delta}_{\text{avg-opt}} \leq 0. \tag{7.73}$$

Since $g(\cdot)$ is nondecreasing, we get from (7.71)–(7.73) that z must satisfy

$$\mathbb{E}_Y\left[\sum_{l-1}^{m} g(a_{[l]}+t+Y)\right] - \bar{\Delta}_{\text{avg-opt}} \geq 0, \text{ if } t > z, \tag{7.74}$$

$$\mathbb{E}_Y\left[\sum_{l=1}^{m} g(a_{[l]}+t+Y)\right] - \bar{\Delta}_{\text{avg-opt}} \leq 0, \text{ if } t < z. \tag{7.75}$$

Subsequently, the smallest z that satisfies (7.74)–(7.75) is

$$z = \inf\left\{t \geq 0 : \mathbb{E}_Y\left[\sum_{l=1}^{m} g(a_{[l]}+t+Y)\right] \geq \bar{\Delta}_{\text{avg-opt}}\right\}. \tag{7.76}$$

According to (7.76), since $g(\cdot)$ is nondecreasing, if $\mathbb{E}_Y\left[\sum_{l=1}^{m} g(a_{[l]}+Y)\right] \geq \bar{\Delta}_{\text{avg-opt}}$, then $z = 0$ minimizes $C(\mathbf{s}, z)$. This completes the proof. □

7.7.5 Proof of Theorem 7.7

We use the threshold test $A_s \geq (\bar{\Delta}_{\text{avg-opt}} - mb_1\mathbb{E}[Y] - mb_2)/b_1$, in Proposition 4, to prove Theorem 7.7. We will show that the condition in (7.39) implies that $A_s \geq (\bar{\Delta}_{\text{avg-opt}} - mb_1\mathbb{E}[Y] - mb_2)/b_1$ holds for all states $\mathbf{s} \in \mathcal{S}$, and hence the zero-wait sampler is optimal under this condition. From the state evolution (7.17), we can deduce that for any state $\mathbf{s} \in \mathcal{S}$, we have

$$a_{[l]} \geq (m - l + 1)y_{\text{inf}}, \ \forall l = 1, \ldots, m. \tag{7.77}$$

This implies

$$A_s \geq \sum_{l=1}^{m} l y_{\text{inf}} = \frac{m(m+1)}{2} y_{\text{inf}}, \ \forall \mathbf{s} \in \mathcal{S}. \tag{7.78}$$

Moreover, it is easy to show that the Ta-AP of the zero-wait sampler, when the scheduling policy is fixed to the MAF scheduler and $g(x) = b_1 x + b_s$, is given by

$$\bar{\Delta}_0 = \frac{\frac{m(m+1)}{2} b_1 \mathbb{E}[Y]^2 + \frac{m}{2} b_1 \mathbb{E}[Y^2] + m b_2 \mathbb{E}[Y]}{\mathbb{E}[Y]}. \tag{7.79}$$

Since $\bar{\Delta}_0 \geq \bar{\Delta}_{\text{avg-opt}}$, we have

$$\frac{\bar{\Delta}_0 - m b_1 \mathbb{E}[Y] - m b_2}{b_1} \geq \frac{\bar{\Delta}_{\text{avg-opt}} - m b_1 \mathbb{E}[Y] - m b_2}{b_1}. \tag{7.80}$$

Hence, if the following condition holds,

$$\frac{m(m+1)}{2} y_{\text{inf}} \geq \frac{\frac{m(m+1)}{2} b_1 \mathbb{E}[Y]^2 + \frac{m}{2} b_1 \mathbb{E}[Y^2] + m b_2 \mathbb{E}[Y]}{b_1 \mathbb{E}[Y]} - m\mathbb{E}[Y] - \frac{m b_2}{b_1}, \tag{7.81}$$

which is equivalent to

$$y_{\text{inf}} \geq \frac{(m-1)\mathbb{E}[Y]^2 + \mathbb{E}[Y^2]}{(m+1)\mathbb{E}[Y]}, \tag{7.82}$$

then we have $A_s \geq (\bar{\Delta}_{\text{avg-opt}} - m\mathbb{E}[Y])$ for all states $\mathbf{s} \in \mathcal{S}$. This implies that the zero-wait sampler is optimal under this condition. This completes the proof. □

References

[1] Y. Sun, E. Uysal-Biyikoglu, R. D. Yates, C. E. Koksal, and N. B. Shroff, "Update or wait: How to keep your data fresh," *IEEE Trans. Inf. Theory*, vol. 63, no. 11, pp. 7492–7508, Nov. 2017.

[2] B. T. Bacinoglu, E. T. Ceran, and E. Uysal-Biyikoglu, "Age of information under energy replenishment constraints," in *Proc. ITA*, Feb. 2015.

[3] R. D. Yates, "Lazy is timely: Status updates by an energy harvesting source," in *Proc. IEEE ISIT*, 2015.

[4] B. Adelberg, H. Garcia-Molina, and B. Kao, "Applying update streams in a soft real-time database system," in *ACM SIGMOD Record*, vol. 24, no. 2, 1995, pp. 245–256.

[5] J. Cho and H. Garcia-Molina, "Synchronizing a database to improve freshness," in *ACM SIGMOD Record*, vol. 29, no. 2, 2000, pp. 117–128.

[6] L. Golab, T. Johnson, and V. Shkapenyuk, "Scheduling updates in a real-time stream warehouse," in *Proc. IEEE 25th Int'l Conf. Data Eng. (ICDE)*, Mar. 2009, pp. 1207–1210.

[7] S. Kaul, R. D. Yates, and M. Gruteser, "Real-time status: How often should one update?" in *Proc. IEEE INFOCOM*, 2012, pp. 2731–2735.

[8] Y. Sun and B. Cyr, "Sampling for data freshness optimization: Non-linear age functions," *Journal of Communications and Networks – Special Issue on the Age of Information*, vol. 21, no. 3, pp. 204–219, Jun. 2019.

[9] A. Kosta, N. Pappas, A. Ephremides, and V. Angelakis, "Age and value of information: Non-linear age case," in *ISIT*. IEEE, 2017, pp. 326–330.

[10] L. Ran, W. Junfeng, W. Haiying, and L. Gechen, "Design method of CAN BUS network communication structure for electric vehicle," in *International Forum on Strategic Technology 2010*. IEEE, 2010, pp. 326–329.

[11] K. H. Johansson, M. Törngren, and L. Nielsen, "Vehicle applications of controller area network," in *Handbook of Networked and Embedded Control Systems.*. Springer, 2005, pp. 741–765.

[12] Automotive Sensors and Electronics Expo 2017. [Online]. Available: www.automotivesensors2017.com/

[13] A. M. Bedewy, Y. Sun, S. Kompella, and N. B. Shroff, "Age-optimal sampling and transmission scheduling in multi-source systems," in *Proc. MobiHoc*, ser. Mobihoc '19. Association for Computing Machinery, 2019, pp. 121–130. [Online]. Available: https://doi.org/10.1145/3323679.3326510

[14] A. M. Bedewy, Y. Sun, S. Kompella, and N. B. Shroff, "Optimal sampling and scheduling for timely status updates in multi-source networks," *IEEE Trans. Inf. Theory*, vol. 67, pp. 4019–4034, 2021.

[15] M. Shaked and J. G. Shanthikumar, *Stochastic Orders*. Springer Science & Business Media, 2007.

[16] R. Li, A. Eryilmaz, and B. Li, "Throughput-optimal wireless scheduling with regulated inter-service times," in *Proc. IEEE INFOCOM*, 2013, pp. 2616–2624.

[17] Y. Hsu, E. Modiano, and L. Duan, "Scheduling algorithms for minimizing age of information in wireless broadcast networks with random arrivals," *arXiv preprint arXiv:1712.07419*, 2017.

[18] I. Kadota, E. Uysal-Biyikoglu, R. Singh, and E. Modiano, "Minimizing the age of information in broadcast wireless networks," in *54th Annual Allerton Conference on Communication, Control, and Computing (Allerton), 2016*. IEEE, 2016, pp. 844–851.

[19] I. Kadota, A. Sinha, E. Uysal-Biyikoglu, R. Singh, and E. Modiano, "Scheduling policies for minimizing age of information in broadcast wireless networks," *IEEE/ACM Trans. Netw.*, vol. 26, no. 6, pp. 2637–2650, Dec. 2018.

[20] Y. Sun, E. Uysal-Biyikoglu, and S. Kompella, "Age-optimal updates of multiple information flows," in *IEEE INFOCOM 2018 – IEEE Conference on Computer Communications Workshops (INFOCOM WKSHPS)*, 2018, pp. 136–141.

[21] Y. Sun, Y. Polyanskiy, and E. Uysal, "Sampling of the Wiener process for remote estimation over a channel with random delay," *IEEE Trans. Inf. Theory*, vol. 66, no. 2, pp. 1118–1135, 2019.

[22] D. P. Bertsekas, *Dynamic Programming and Optimal Control*, 2nd ed. Athena Scientific, 2001, vol. 2.

[23] M. L. Puterman, *Markov Decision Processes: Discrete Stochastic Dynamic Programming*. Wiley Series in Probability and Statistics, 2005.

[24] L. P. Kaelbling and L. P. Kaelbling, *Recent Advances in Reinforcement Learning*. Springer, 1996.

[25] W. B. Powell, *Approximate Dynamic Programming: Solving the Curses of Dimensionality*. John Wiley & Sons, 2007, vol. 703.

[26] I. Bettesh and S. Shamai, "Optimal power and rate control for minimal average delay: The single-user case," vol. 52, no. 9, pp. 4115–4141, *IEEE Trans. Inf. Theory*, 2006.

[27] R. Wang and V. K. Lau, "Delay-optimal two-hop cooperative relay communications via approximate mdp and distributive stochastic learning," *IEEE Trans. Inf. Theory*, 2013.

[28] W. H. Press, S. A. Teukolsky, W. T. Vetterling, and B. P. Flannery, "Golden section search in one dimension," *Numerical Recipes in C: The Art of Scientific Computing*, p. 2, 1992.

[29] L. I. Sennott, "Average cost optimal stationary policies in infinite state Markov decision processes with unbounded costs," *Oper. Res.*, vol. 37, no. 4, pp. 626–633, 1989.

[30] D. P. Bertsekas, *Nonlinear Programming*, 2nd ed. Athena Scientific, 1999.

[31] D. Butnariu and A. N. Iusem, *Totally Convex Functions for Fixed Points Computation and Infinite Dimensional Optimization*. Kluwer Academic Publisher, 2000.

[32] R. Durrett, *Probability: Theory and Examples*. Cambridge University Press, 2010.

8 Age-Efficient Scheduling in Communication Networks

Bin Li, Bo Ji, and Atilla Eryilmaz

Age of Information (AoI) is used to characterize the freshness of information and is critical for information monitoring, tracking, and control, which is typically required in many network applications, such as autonomous vehicles, virtual/augmented reality, and Internet of Things (IoT). Both inter-arrival times and delays of packets affect AoI performance, and thus traditional delay-efficient algorithms do not necessarily exhibit low AoI performance. This calls for "age-efficient" algorithm design in communication networks, which forms the focus of this chapter. In particular, we first discuss the recent advances in the age-efficient algorithm design for three different types of common network traffic: (i) elastic traffic (cf. Section 8.1): packets are allowed to be delivered without any specific deadline constraints; (ii) inelastic traffic (cf. Section 8.2): packets will be dropped if they are not delivered within a specific deadline; (iii) heterogeneous traffic (cf. Section 8.3): different packets may have different size. To facilitate our discussions, we explicitly consider the discrete-time model and emphasize the difference between age-efficient and delay-efficient algorithm design paradigms. Then, we examine "fresh" scheduling design for remote estimation with the goal of optimally balancing the trade-off between the estimation accuracy and the communication cost (cf. Section 8.4).

8.1 Age-Efficient Scheduling for Elastic Traffic

In this section, we focus on the age-efficient scheduling design for elastic traffic, where packets are allowed to be delivered without any specific deadline constraints. We first consider a simple Geo/Geo/1 queue to illustrate the difference between the AoI metrics and the traditional delay metric (i.e., average delay). Then, we consider the age-efficient scheduling algorithm design in wireless networks.

8.1.1 Delay and AoI Analysis in the Geo/Geo/1 Queue

We consider an infinite-buffer single-server queue. In each time slot, a packet arrives at the queue with probability $\lambda \in (0, 1]$, and a packet at the front of the queue is served with probability $\mu \in (0, 1]$. In such a queue, both inter-arrival and interservice times are geometrically distributed, so the queue is called a Geo/Geo/1 queue.

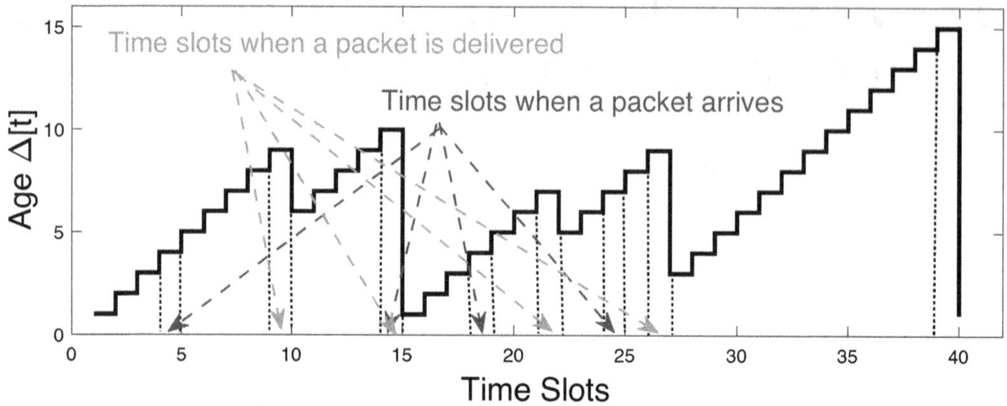

Figure 8.1 The evolution of the AoI

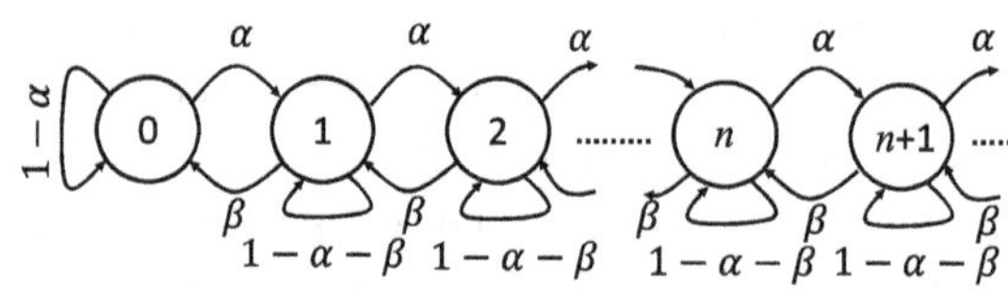

Figure 8.2 The Markov chain of a Geo/Geo/1 queue.

Let $\Delta[t]$ and $H[t]$ be the AoI associated with the receiver of the single-server queue and the delay of the Head-of-Line (HoL) packet at the queue in time slot t, respectively, where $\Delta[t]$ is the time since the generated time of the latest packet received by the receiver. Noting that the HoL delay measures the duration from the time that the HoL packet was generated to the current time and that at most one packet can be served in each time slot, we have $\Delta[t+1] = H[t] + 1$ whenever a packet is served in time slot t and $\Delta[t+1] = \Delta[t] + 1$ otherwise, that is,

$$\Delta[t+1] = \begin{cases} H[t] + 1, & \text{if a packet is served in time slot } t; \\ \Delta[t] + 1, & \text{otherwise.} \end{cases} \tag{8.1}$$

Figure 8.1 shows one sample path of the AoI.

The number of packets or the queue length in the Geo/Geo/1 queue evolves as a Markov chain, which is shown in Figure 8.2. Here, $\alpha \triangleq \Pr\{\text{one arrival, no departure}\}$ is the probability that the queue length increases by one and $\beta \triangleq \Pr\{\text{no arrival, one departure}\}$ is the probability that the queue length decreases by one. From this Markov Chain, we can calculate the average delay, average peak AoI, and average AoI, as shown in the following theorem.

THEOREM 8.1 *If $\lambda < \mu$, then average delay $\overline{D}$, average peak age $\overline{\Delta}^{(p)}$, and average age $\overline{\Delta}^{(avg)}$ are expressed as follows:*

$$\overline{D} = \frac{1-\lambda}{\mu-\lambda}, \tag{8.2}$$

$$\overline{\Delta}^{(p)} = \frac{\mu - \lambda^2}{\lambda(\mu - \lambda)}, \tag{8.3}$$

$$\overline{\Delta}^{(avg)} = \frac{\lambda^3 + \mu^3 - \lambda^3\mu - \lambda^2\mu}{\lambda\mu^2(\mu - \lambda)}. \tag{8.4}$$

Proof Let π_n denote the steady-state distribution of the Markov Chain. According to the local balance equation and the normalizing condition (i.e., $\sum_{n=0}^{\infty} \pi_n = 1$), we can obtain π_n:

$$\pi_n = \gamma^n(1 - \gamma), \quad n = 1, 2, 3, \ldots, \tag{8.5}$$

where $\gamma \triangleq \alpha/\beta = \lambda(1 - \mu)/(\mu(1 - \lambda))$. Since $\lambda < \mu$, we have $\gamma < 1$ and thus the mean queue length can be calculated as follows:

$$\mathbb{E}[Q] = \sum_{n=0}^{\infty} n\pi_n = (1 - \gamma)\sum_{n=0}^{\infty} n\gamma^n = \frac{\gamma}{1 - \gamma} = \frac{\lambda(1 - \mu)}{\mu - \lambda}. \tag{8.6}$$

The average delay of an incoming packet consists of the average queueing delay and its average service time, where the average queueing delay is the total average service time of all existing packets in the system. According to the BASTA (Bernoulli Arrivals See Time Averages) property (Srikant & Ying 2013, Section 3.4.3), the average delay $\overline{D}$ is expressed as follows:

$$\overline{D} = \mathbb{E}[Q]\mathbb{E}[S] + \mathbb{E}[S] = \frac{\lambda(1 - \mu)}{\mu - \lambda}\frac{1}{\mu} + \frac{1}{\mu} = \frac{1 - \lambda}{\mu - \lambda}, \tag{8.7}$$

where S is the service time that is identically and independently geometrically distributed with mean $1/\mu$, and we use the fact that all packets experience identical and independent service time distribution.

Since the average peak age is the sum of the average delay and the average inter-arrival time in a First-Come-First-Serve queue, the average peak age can be calculated as follows:

$$\overline{\Delta}^{(p)} = \overline{D} + \mathbb{E}[I_A] = \frac{1 - \lambda}{\mu - \lambda} + \frac{1}{\lambda} = \frac{\mu - \lambda^2}{\lambda(\mu - \lambda)}, \tag{8.8}$$

where I_A is the inter-arrival time that is i.i.d. geometrically distributed with mean $1/\lambda$. The calculation of the average age $\overline{\Delta}^{(avg)}$ is much more involved and is available in Talak, Karaman, and Modiano (2019, Theorem 5). □

In the light traffic regime (i.e., $\lambda \downarrow 0$), the average delay is close to $1/\mu$, which is the average service time of a packet and thus is the minimum delay experienced by a packet. In contrast, both average peak age and average age blow up to infinity, since the inter-arrival time is extremely large, and thus, it takes a really long time to receive a packet update. On the other hand, in the heavy traffic regime (i.e., $\lambda \uparrow \mu$), all these three metrics are dominated by the queueing delay and thus increase to infinity.

We verify these theoretical results via simulations with $\mu = 0.5$. We plot both theoretical and simulated results with varying arrival rate λ. From Figure 8.3, we can observe that simulated values exactly match their corresponding theoretical results,

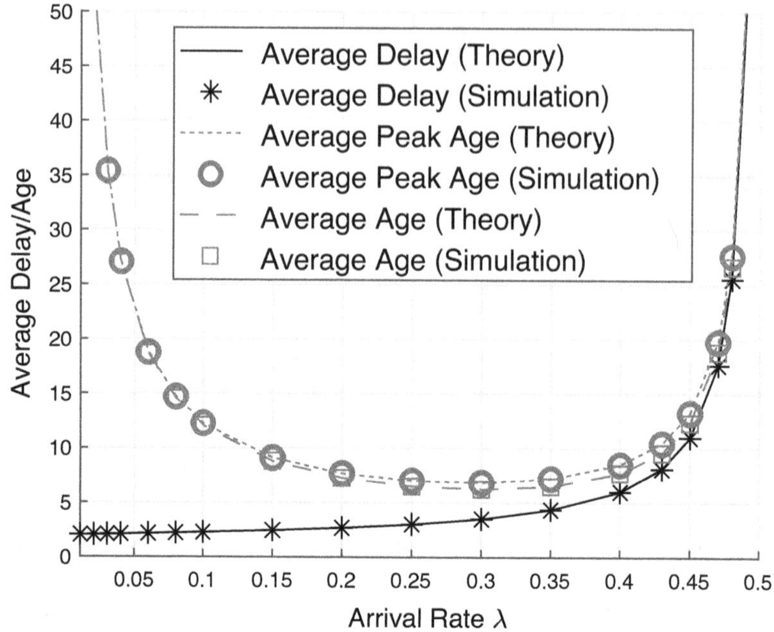

Figure 8.3 Average delay/age in the Geo/Geo/1 queue

which validates the correctness of these theoretical results. In addition, we can observe that in the light traffic regime, the average delay is quite small, while both the average peak age and average age are rather large. In contrast, all three metrics are large in the heavy traffic regime.

From the AoI analysis of the Geo/Geo/1 queue, we learned that the AoI performance is poor in both low-rate and high-rate regimes. This implies that an age-efficient scheduling design should take into account not only the queueing congestion as in the traditional delay-efficient scheduling algorithms but also the arrival rates. We will discuss the age-efficient scheduling design in the next subsection.

8.1.2 Age-Efficient Scheduling Algorithms

In this section, we consider the age-efficient scheduling design for elastic traffic in wireless networks. We first introduce the network model and then develop an age-efficient scheduling algorithm. Finally, we demonstrate the efficiency of the proposed scheduling algorithm via various simulations.

System Model: We consider a wireless network with N links, where each link represents a pair of transmitter and receiver within the transmission range of each other. Due to the wireless interference, the success or failure of transmission over a link depends on whether one of its interfering links is also active in the same time slot, commonly called the *link-based conflict model*. Let $S_n[t] = 1$ denote that link n is scheduled in time slot t and $S_n[t] = 0$ otherwise. We call $\mathbf{S}[t] \triangleq (S_n[t])_{n=1}^{N}$ the *feasible* *schedule* denoting the set of links that can be active simultaneously in time slot t. Let $\mathcal{S}$ be the collection of all feasible schedules. Figure 8.4 shows a network with four wireless links, where the neighboring links (e.g., link 1 and link 2) interfere with each other and thus cannot be scheduled at the same time. In such a network, either one link or two links can be scheduled at the same time, and none of the three links can be simultaneously active. Here, $(0,0,0,0)$, $(1,0,0,0)$, $(1,0,1,0)$, $(0,1,0,1)$ are feasible schedules, while $(1,1,1,0)$ and $(1,1,0,1)$ are not feasible.

We use $A_n[t]$ to denote the number of packets arriving at link n in time slot t that is independently distributed over links, and independently and identically distributed (i.i.d.) over time with finite mean $\lambda_n > 0$. We assume that each link n independently experiences i.i.d. ON-OFF channel fading over time with $C_n[t] = 1$ denoting that the channel is available for link n in time slot t. We maintain a queue at each link and assume that packets are served in the order that they arrive, that is, First-Come-First-Served (FCFS). Let $Q_n[t]$ be the queue length at link n in time slot t, denoting the number of packets waiting for transmission, and thus its evolution can be described as follows:

$$Q_n[t+1] = \max\{Q_n[t] + A_n[t] - C_n[t]S_n[t], 0\}, \tag{8.9}$$

where the maximum operation is due to the fact that the queue length is non-negative.

Let $\Delta_n(t)$ and $H_n(t)$ be the age associated with the receiver of link n and the delay of the Head-of-Line (HoL) packet at link n in time slot t, respectively. Noting that the HoL delay measures the duration from the time that the packet was generated to the current time and that at most one packet can transmit over a link in each time slot, we have $\Delta_n[t+1] = H_n[t] + 1$ whenever a packet is successfully delivered over link n in time slot t and $\Delta_n[t+1] = \Delta_n[t] + 1$ otherwise, that is,

$$\Delta_n[t+1] = \begin{cases} \Delta_n[t] + 1, & \text{if } S_n[t]C_n[t] = 0; \\ H_n[t] + 1, & \text{if } S_n[t]C_n[t] = 1. \end{cases} \tag{8.10}$$

We call the system *stable* if the average queue lengths of all links are finite, that is, $\limsup_{T\to\infty} \frac{1}{T}\sum_{t=0}^{T-1} \mathbb{E}[Q_n[t]] < \infty, \forall n = 1,2,\ldots,N$. The *capacity region* Λ is defined as the maximal set of the achievable link rates in packets per slot that can be supported by the network under stability for a given interference model. For example, in the wireless network shown in Figure 8.4, the capacity region is

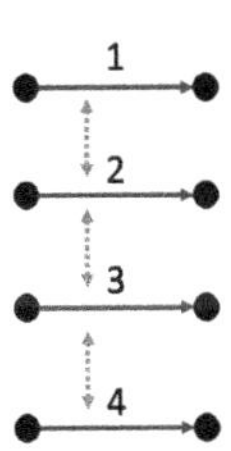

Figure 8.4 A wireless network with four links: a line denotes a transmitter–receiver pair and a dash line means that there is an interference between two links.

$\Lambda = \{(\lambda_n)_{n=1}^4 : \lambda_1 + \lambda_2 \leq 1, \lambda_2 + \lambda_3 \leq 1, \lambda_3 + \lambda_4 \leq 1\}$. We say that a scheduler is *throughput-optimal* if it achieves the network stability for any arrival rate vector $\boldsymbol{\lambda} \triangleq (\lambda_n)_{n=1}^N$ that lies strictly inside the capacity region Λ. We are interested in developing throughput-optimal algorithms that achieve low AoI performance (i.e., low average peak age and average age).

Algorithm Design: As we mentioned in the last section, the age-efficient scheduling design should take both queueing congestion and arrival rate into account. On the one hand, the smaller the arrival rate of a link, the larger the AoI of this link, and thus the scheduling priority should be given to the link with a low arrival rate. On the other hand, the queueing delay dominates the AoI performance in the heavy-traffic regime, and thus the scheduling priority should be given to the congested links (i.e., links with large HoL delays). These two observations naturally motivate the following scheduling algorithm (see also Joo & Eryilmaz (2018)).

(Scaled HoL-Delay-based MaxWeight (SHD-MW) Algorithm). In each time slot t, given the channel state $(C_n[t])_{n=1}^N$, select a feasible schedule $\mathbf{S}^*[t] \in \mathcal{S}$ such that

$$\mathbf{S}^*[t] \in \underset{\mathbf{S}[t]\in\mathcal{S}}{\arg\max} \sum_{n=1}^{N} \frac{1}{\lambda_n} H_n[t] C_n[t] S_n[t]. \tag{8.11}$$

In the SHD-MW Algorithm, we require the knowledge of the arrival rate of each link, which is usually not available. However, we can use the estimate $\widetilde{\lambda}_n[t] = (1 - 1/t)\widetilde{\lambda}[t-1] + A_n[t]/t$ to replace λ_n. According to the Law of Large Numbers, $\widetilde{\lambda}[t] \to \lambda_n$ as $t \to \infty$.

It has been well known that the HoL-Delay-based scheduling algorithm possesses not only the throughput optimality (i.e., achieving stability for any arrival rate vector within the capacity region) but also good delay performance (see McKeown, Mekkittikul, Anantharam, and Walrand (1999)). Therefore, the SHD-MW Algorithm is expected to be throughput-optimal and delay-efficient. Due to the fact that the mean peak age of each link is the sum of its average delay and average inter-arrival time when packets are served in FCFS, the SHD-MW Algorithm exhibits excellent AoI performance, which is demonstrated via simulations.

Simulation Results: We evaluate the performance of the SHD-MW Algorithm and compare it with the following existing policies:

(i) Queue-Length-based MaxWeight (Q-MW) Algorithm (Tassiulas & Ephremides 1990): Select a feasible schedule satisfying

$$\mathbf{S}^{(\text{Q-MW})}[t] \in \underset{\mathbf{S}[t]\in\mathcal{S}}{\arg\max} \sum_{n=1}^{N} Q_n[t] C_n[t] S_n[t].$$

(ii) HoL-Delay-based MaxWeight (HD-MW) Algorithm (McKeown et al. 1999): Select a feasible schedule satisfying

$$\mathbf{S}^{(\text{HD-MW})}[t] \in \arg\max_{\mathbf{S}[t]\in\mathcal{S}} \sum_{n=1}^{N} H_n[t]C_n[t]S_n[t].$$

(iii) Age-based MaxWeight (A-MW) Algorithm (Sun, Kadota, Talak, & Modiano 2019): Select a feasible schedule satisfying

$$\mathbf{S}^{(\text{A-MW})}[t] \in \arg\max_{\mathbf{S}[t]\in\mathcal{S}} \sum_{n=1}^{N} h(\Delta_n[t])C_n[t]S_n[t],$$

where $h(x)$ takes a linear form (i.e., $h(x) = x$) or quadratic form (i.e., $h(x) = x^2 + x$).

(iv) Largest-Age-Drop-based MaxWeight (LAD-MW) algorithm (Sun et al. 2019): Select a feasible schedule satisfying

$$\mathbf{S}^{(\text{LAD-MW})}[t] \in \arg\max_{\mathbf{S}[t]\in\mathcal{S}} \sum_{n=1}^{N} (\Delta_n[t] - H_n[t])C_n[t]S_n[t].$$

(v) Round-Robin (RR) Algorithm: in the case of fully connected networks (i.e., at most one link can be scheduled in each time slot), it serves links in turn by simply skipping transmission when a channel is OFF. However, it is unclear how to define an appropriate version of the RR Algorithm in the general wireless setup.

We consider a fully connected ON-OFF fading network with $N = 5$ links and a 3×3 switch. While algorithms were described for wireless networks, it also applies to other communication networking applications where interference-type constraints exist, such as a switch. Since switches are widely used in data centers and the Internet, we demonstrate our algorithm in that context as well. Packets arrive at link n in each time slot with probability λ'_n. Whenever arrivals happen at link n, the number of packets has the following distribution: it is equal to M_n with probability $(\eta_n - 1)/(M_n - 1)$ and 1 otherwise, where $M_n \geq 2$ is some parameter that measures the burstiness of the arrivals and η_n is the average number of arriving packets at link n when arrivals happen. In the fully connected network, the probability vector of channel being ON is $\mathbf{p} = [0.1, 0.1, 0.8, 0.8, 0.8]$, $\boldsymbol{\eta} = [2, 5, 5, 10, 15]$, $\mathbf{M} = [5, 10, 10, 20, 20]$, and we consider the following specific arrival rate vector: $\boldsymbol{\lambda}' = \theta \times 0.3821 \times [0.05, 0.02, 0.16, 0.08, 0.053]$, where $\theta \in (0, 1)$. It can be verified that $\boldsymbol{\lambda}' \otimes \boldsymbol{\eta}$ is within the capacity region according to Tassiulas and Ephremides (1993, Lemma 1), where $\mathbf{x} \otimes \mathbf{y}$ denotes the component-wise product of the vector $\mathbf{x}$ and $\mathbf{y}$. In the case of the 3×3 switch, $\eta = [5, 2, 2; 3, 10, 5; 5, 4, 5]$, and $\boldsymbol{\lambda}' = \theta \times [0.14, 0.1, 0.05; 0.05, 0.06, 0.05; 0.03, 0.05, 0.13]$, where $\theta \in (0, 1)$. Again, we can verify that $\boldsymbol{\lambda}' \otimes \boldsymbol{\eta}$ is within the capacity region.

Delay Performance: Figure 8.5 shows the average delay under our proposed SHD-MW Algorithm and other existing algorithms in both fully connected networks and 3×3 switch. It can be observed from Figure 8.5 that HD-MW and our SHD-MW have similar delay performance and uniformly perform better than other algorithms. In addition, SHD-MW can stabilize the system for any $\theta \in (0, 1)$ in the preceding network setups, which validate the throughput optimality of our proposed algorithm.

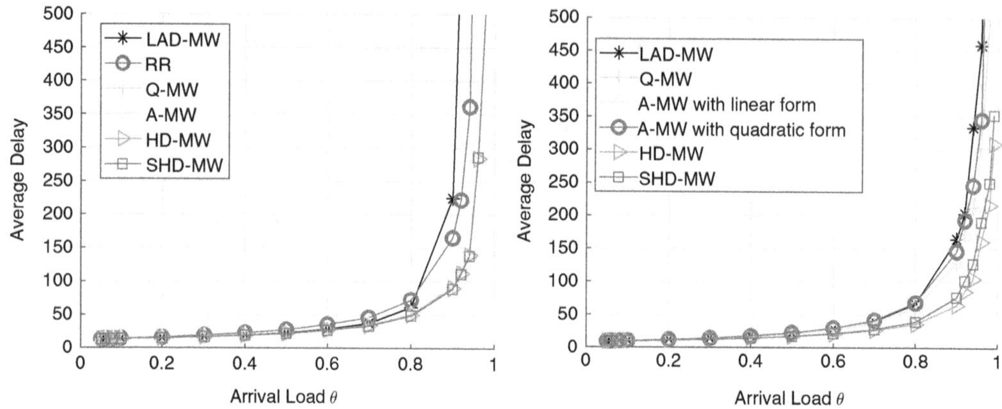

(a)Fully connected network (b) 3 × 3 switch

Figure 8.5 Average delay performance

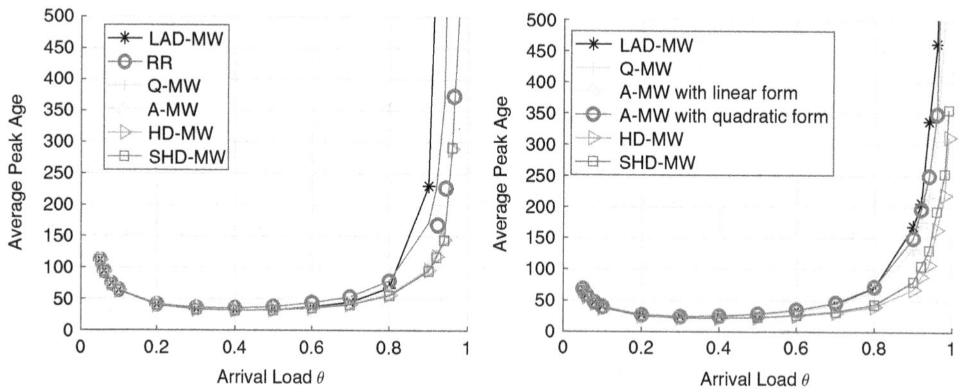

(a) Fully connected network (b) 3 × 3 switch

Figure 8.6 Average peak age performance

However, RR and LAD-MW suffer from relatively poor delay performance, especially in the heavy-traffic regime, since they do not prioritize congested links.

AoI Performance: We can observe from Figure 8.6 and Figure 8.7 that the SHD-MW performs uniformly better than all other algorithms except the HD-MW algorithm in terms of both average peak age and average age performance. Indeed, the HD-MW slightly outperforms the SHD-MW algorithm in terms of average peak age, especially in 3 × 3 switch in the heavy-traffic regime, as shown in Figure 8.6b. This makes sense since the average peak age is the sum of the average HoL delay and the average inter-arrival time, and HD-MW tries to minimize the HoL delay. However, HD-MW Algorithm performs slightly worse than A-MW and SHD-MW in terms of average age performance, since it does not incorporate the inter-arrival time into the scheduling design.

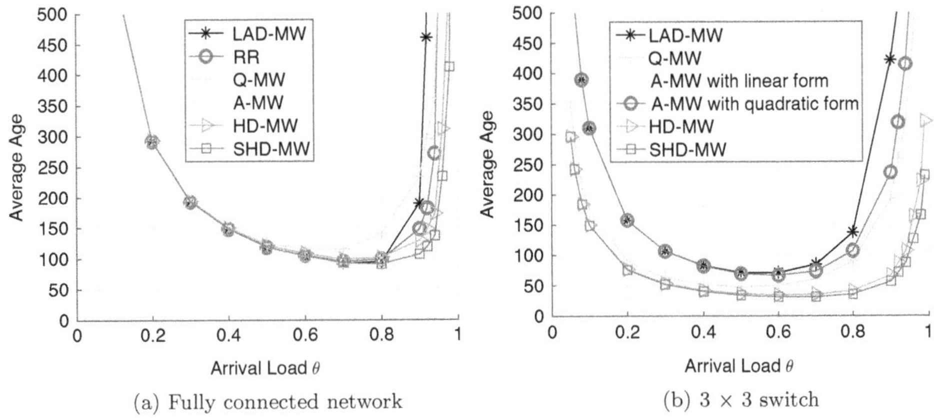

Figure 8.7 Average age performance

8.2 Age-Efficient Scheduling for Inelastic Traffic

In this section, we consider the case with inelastic traffic, where packets will be dropped if they are not delivered within a specific deadline. We first consider a simple single-server queue with packet deadlines and understand the impact of system parameters on the AoI performance. Then, we study the age-efficient scheduling design for inelastic traffic.

8.2.1 AoI Analysis in the Single-Server Queue with Strict Deadlines

Different from the previous Geo/Geo/1 queueing model, here each arriving packet has a strict deadline of $T \in \{1, 2, 3, \ldots\}$, that is, it will be dropped if it cannot be delivered within T time slots. We divide time into frames with a size of T time slots. We assume that packets arrive at the beginning of each time frame according to Bernoulli distribution with the mean of $\lambda \in (0, 1]$. In each time slot within a frame, a packet is served with the probability of $\mu \in (0, 1]$. In such a queue, the age is equal to the number of time slots that have been passed since the packet was last delivered, that is,

$$\Delta[(k+1)T] = \begin{cases} \Delta[kT] + T & \text{if a packet cannot be delivered in frame } k; \\ T & \text{otherwise.} \end{cases} \tag{8.12}$$

Figure 8.8 shows the age dynamics of the queue with deadlines. Hence, the age forms a Markov chain as illustrated in Figure 8.9, where $q \triangleq \lambda(1 - (1 - \mu)^T)$ denotes the probability that a packet is successfully delivered within a frame.

THEOREM 8.2 *The average peak age and average age are equal and can be expressed as follows.*

$$\overline{\Delta}^{(p)} = \overline{\Delta}^{(avg)} = \frac{T}{\lambda\left(1 - (1 - \mu)^T\right)}. \tag{8.13}$$

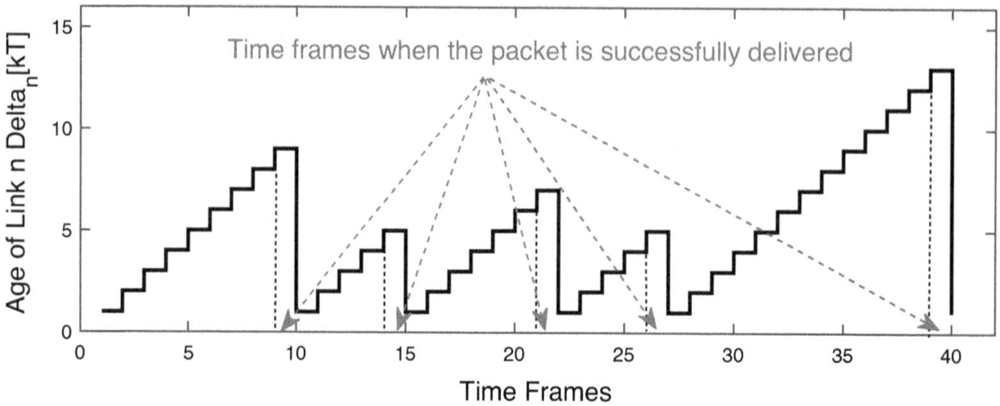

Figure 8.8 The evolution of the AoI of link n.

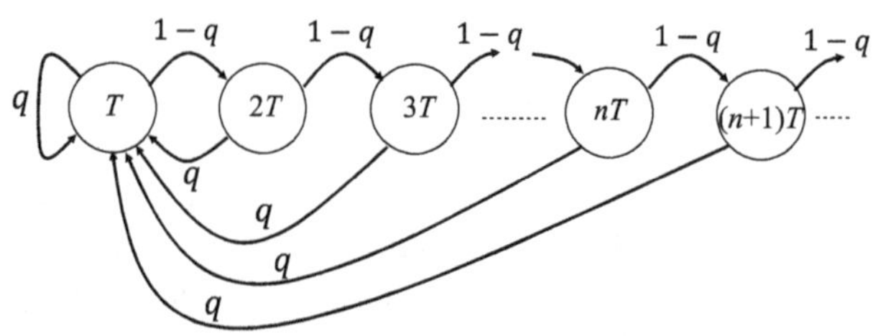

Figure 8.9 The Markov chain of a single-server queue with deadlines.

Proof Let π_n denote the probability of the AoI being equal to nT in the steady state in the Markov chain shown in Figure 8.9. According to the global balance equations, we have

$$\pi_1 = q \sum_{n=1}^{\infty} \pi_n = q \tag{8.14}$$

$$\pi_{n-1}(1-q) = \pi_n, \quad \forall n = 2, 3, 4, \ldots, \tag{8.15}$$

where we use the normalizing condition $\sum_{n=1}^{\infty} \pi_n = 1$ in the first equation. Hence, we obtain π_n as follows:

$$\pi_n = q(1-q)^{n-1}, \quad \forall n = 1, 2, 3, \ldots. \tag{8.16}$$

Therefore, the average age can be calculated as follows:

$$\overline{\Delta^{(avg)}} = \sum_{n=1}^{\infty} \pi_n \cdot nT = T \sum_{n=1}^{\infty} nq(1-q)^n = \frac{T}{q} = \frac{T}{\lambda\left(1-(1-\mu)^T\right)}. \tag{8.17}$$

Since the peak age happens whenever there is a successful delivery (i.e., age drops), the average peak age again can be expressed as

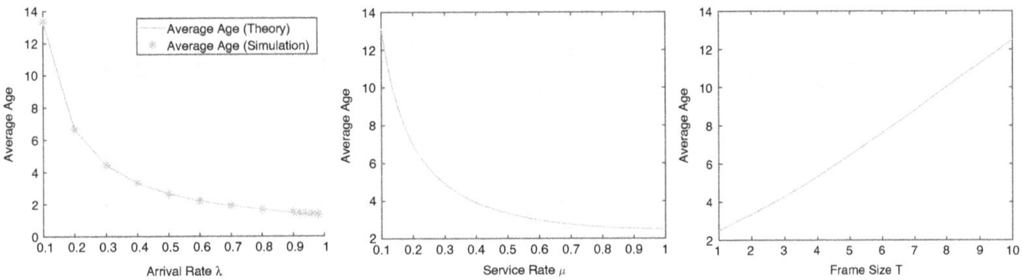

(a) Validation of the average age (b) Average age vs. service rate (c) Average age vs. frame size

Figure 8.10 Average age in the single-server queue with deadlines.

$$\overline{\Delta}^{(p)} = \sum_{n=1}^{\infty} \pi_n \cdot nT, \tag{8.18}$$

which is equal to average age $\overline{\Delta}^{(avg)}$. □

Unlike the previous Geo/Geo/1 queue, there is no constraint on the arrival rate λ. The larger the arrival rate λ (or the service rate μ), the higher the successful packet delivery probability q and thus the better the AoI performance. On the other hand, the larger the time frame size T, both average peak age and average age increase despite the high successful packet delivery probability.

We validate our theoretical results via simulations with $\mu = 0.5$ and $T = 2$. We plot both theoretical average age and its simulated values with respect to the arrival rate λ in Figure 8.10a. We can observe from Figure 8.10a that the simulation result matches the theoretical result quite well, which confirms the correctness of the theoretical average age. In addition, we can see from Figure 8.10b and Figure 8.10c that the average age indeed decreases with respect to the service rate μ and increases with respect to the time frame size T.

From the simple analysis of the single-server queue with deadlines, we know that the successful delivery probability increases with the increase of the frame size T, which, however, deteriorates AoI performance. In practice, the frame size T is typically fixed. In the next section, we consider the age-efficient scheduling design for wireless networks with inelastic traffic.

8.2.2 Age-Efficient Deadline-Constrained Scheduling

In this section, we focus on the age-efficient scheduling design for inelastic traffic. We first present the network model and then propose an age-efficient scheduling algorithm. Finally, we validate the efficiency of the proposed algorithm via various simulations.

System Model: We consider a wireless network consisting of N links. Here, each link models a pair of transmitter and receiver that are within the transmission range of each other. Each packet has a strict deadline of T time slots, that is, it will be dropped if it cannot be delivered within T time slots. We divide time into frames, each of which has consecutive T time slots. We assume that packets arrive at each link at the beginning of each time frame. Let $A_n[kT]$ denote the number of packets arriving at link $n \in \{1, 2, \cdots, N\}$ in frame $k \in \{0, 1, 2, \cdots\}$, which are independently distributed over links and i.i.d. over time with mean λ_n. All remaining packets are dropped at the end of a frame. Each link has a maximum allowable drop rate $\rho_n\lambda_n$, where $\rho_n \in (0, 1)$ is the maximum fraction of packets that can be dropped at link n on average.

Each link n experiences an i.i.d. ON-OFF channel fading, where the channel state is constant during a frame and can vary over frames. We use $C_n[kT]$ to capture channel fading of link n in frame k, that is, $C_n[kT] = 1$ if the channel of link n is ON within frame k and $C_n[kT] = 0$ otherwise. Due to the wireless interference, only a subset of links can be scheduled in each time slot, called feasible schedule denoted by $\mathbf{S}[t] = (S_n[t])_{n=1}^N$, where $S_n[t] = 1$ if user n is scheduled in slot t and $S_n[t] = 0$, otherwise. We use $\mathcal{S}$ to denote the set of all feasible schedules.

Let $\Delta_n[kT]$ be the age associated with the receiver of link n in frame k, which measures the number of frames elapsed up to frame k since the most recent successful packet delivery for link n. In other words, $\Delta_n[kT]$ increases by one until link n has at least one packet delivery in frame k and then it drops to one at the beginning of the next frame.

$$\Delta_n[(k+1)T] = \begin{cases} \Delta_n[kT] + T, & \text{if } C_n[kT] \sum_{t=kT}^{(k+1)T-1} S_n[t] = 0; \\ T, & \text{if } C_n[kT] \sum_{t=kT}^{(k+1)T-1} S_n[t] > 0. \end{cases} \tag{8.19}$$

Note that the age is similar to the concept of *Time-Since-Last-Service (TSLS)* considered in Li, Li, and Eryilmaz (2014) under our system settings. Yet, the TSLS drops to zero whenever there is a packet delivery.

We provide an example in Figure 8.11 to show an impact of real-time scheduling design on the AoI performance. As shown in Figure 8.11(a), the age of the user updates at the beginning of each frame. The delivery of packet a, b, and e results in the decrease of the age. If we only focus on the time window shown in the figure, Figure 8.11(a) shows a 60% delivery ratio (3 packet deliveries over 5 packet arrivals) on average with an average age of 2.27 frames. While in Figure 8.11(b), with the same packet arrival pattern and delivery ratio, the schedule of delivery of packet d instead of e leads to a drop of average age to 2 frames, indicating a better performance on the AoI performance.

In Kadota, Uysal-Biyikoglu, Singh, and Modiano (2016), the authors proposed a greedy policy (AoI-Greedy) for scheduling deadline-constrained traffic in a symmetric wireless network. However, they only focused on the AoI minimization without providing any guarantees on timely throughput. In this subsection, we are interested in developing a scheduling algorithm that ensures that the average drop rate of each link should not be greater than its maximum allowable drop rate while guaranteeing the average age performance.

Algorithm Design: We use the virtual queue technique to guarantee the desired drop rate requirement. In particular, each link n maintains a virtual queue to keep track of

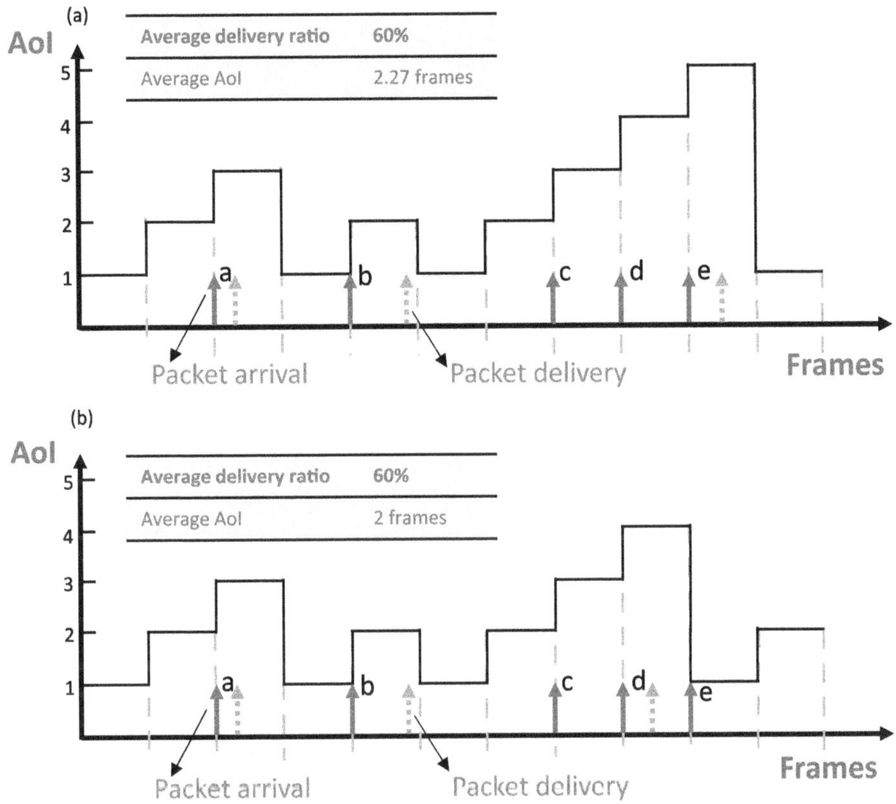

Figure 8.11 The impact of scheduling design on the AoI performance.

the number of dropped packets due to deadline expiry in each frame k denoted by $Q_n[kT]$. In Lu, Ji, and Li (2018), we develop a parametric class of maximum-weight type scheduling algorithms, where the weight of each link n consists of its own virtual queue-length $Q_n[kT]$ and age $\Delta_n[kT]$.

(Age-based Real-Time [AoI-RT] Algorithm). In each time frame k, given the channel state $(C_n[kT])_{n=1}^{N}$, select a schedule $\{\mathbf{S}^*[t]\}_{t=kT}^{(k+1)T-1}$ such that

$$\{\mathbf{S}^*[t]\}_{t=kT}^{(k+1)T-1} \in \arg\max_{\mathbf{S}[t]\in\mathcal{S}} \sum_{n=1}^{N} \Big(Q_n[kT] + \zeta\beta_n\Delta_n[kT] \Big) \times \min\left\{ \sum_{t=kT}^{(k+1)T-1} C_n[kT]S_n[t], A_n[kT] \right\},$$

where $\zeta \geq 0$ is some control parameter, and β_n is a parameter related to the user preference with the AoI performance.

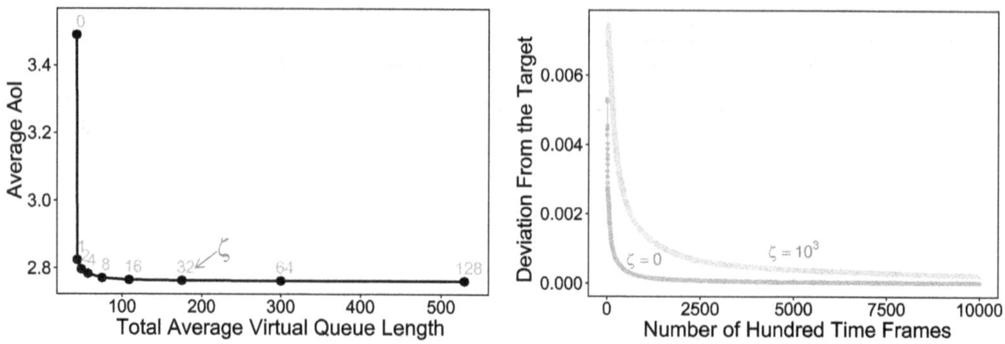

(a) Average age vs. average virtual queue length (b) Convergence speed

Figure 8.12 Performance of AoI-RT Algorithm.

It can be shown that the proposed algorithm is throughput-optimal in the sense that it can support any throughput that can be supported by any other algorithms. Moreover, we can derive an upper bound on a considered AoI metric, demonstrating that the network-wide data freshness is guaranteed and can be tuned under the proposed algorithm. However, the improvement of AoI performance is at the cost of slowing down the convergence of the timely throughput.

Simulation Results: Figure 8.12a shows the trade-off between the convergence speed of timely throughput and the AoI performance. We can observe from the first figure in Figure 8.12a that as parameter ζ increases, the value of AoI metric decreases (i.e., data freshness improves); while the total average virtual-queue length increases (i.e., the convergence of timely throughput slows down) as shown in the second figure in Figure 8.12b. In fact, when $\zeta = 0$, the timely throughput has the fastest convergence speed. However, if the data freshness is more important to applications (as is often the case), we could trade the convergence speed of timely throughput for data freshness by increasing the parameter ζ.

8.3 Age-Efficient Scheduling for Heterogeneous Traffic

In this section, we consider heterogeneous packet sizes, which can be used to model heterogeneous traffic in real-world applications, such as environmental sensing and monitoring, where the packet size depends on the actual measurements. We conduct a systematic and comparative study based on simulations to investigate the impact of scheduling policies on the AoI performance in a single-server queue. Here, scheduling policies are adopted to decide which packet in the queue will be served when the server becomes idle, so they are usually referred to as queueing disciplines.

System Model: We consider a single-server queue with infinite buffer space. To begin with, we make the following simplifying assumptions. Packets arrive at this queue according to an *i.i.d.* Bernoulli process with parameter λ. Packet sizes follow a geometric distribution with mean $1/\mu$, and one unit of data can be served in each time slot.[1] Such a single-server queue is called a Geo/Geo/1 queue as both the inter-arrival and departure times follow a geometric distribution.

Prior studies reveal that the AoI depends on both the inter-arrival time and delay of the packets (Kaul, Yates, & Gruteser 2012b). Although it is well known that scheduling policies play an important role in reducing the packet delay in a single-server queue (Harchol-Balter 2013), it remains largely unknown how exactly scheduling policies impact the AoI performance. To that end, we aim to holistically study the impact of various aspects of scheduling policies on the AoI performance in a single-server queue and suggest guidelines that are potentially useful for the design of AoI-efficient scheduling policies in more general settings. While a similar study for the continuous-time queueing model has been conducted in Liu, Huang, Li, and Ji (2020) and Liu, Huang, Li, and Ji (2021), we will focus on the discrete-time model.

While a lot of research effort has been made toward the analysis of the AoI performance, most of these studies have been focused on scheduling policies based on the packet arrival times, such as First-Come-First-Served (FCFS) and Last-Come-First-Served (LCFS), assuming that the packet-size information is unavailable (see, e.g., Bedewy, Sun, & Shroff 2016; Costa, Codreanu, & Ephremides 2016; Kaul, Yates, & Gruteser 2012a; Kam, Kompella, Nguyen, & Ephremides 2015). In some applications, such as environmental sensing and monitoring, however, the packet-size information can be obtained or fairly well estimated. It has been shown that scheduling policies that leverage the packet-size information can substantially reduce the delay, especially when the system load is high or when the packet-size variability is large (Harchol-Balter 2013). This motivates us to investigate the AoI performance of size-based policies.

Common Scheduling Policies: Following Harchol-Balter (2013), we first give the definitions of several common scheduling policies that can be divided into four types: depending on whether they are size-based or not (i.e., whether a scheduling policy uses the packet-size information for making scheduling decisions or not); depending on whether they are preemptive[2] or not (i.e., whether a policy allows a packet to be stopped partway through its execution and then to be restarted at a later time without losing intermediate work).

The first type consists of policies that are non-preemptive and blind to the packet size:

- *First-Come-First-Served (FCFS)*: When the server frees up, it chooses to serve the packet that arrived first if any.
- *Last-Come-First-Served (LCFS)*: When the server frees up, it chooses to serve the packet that arrived last if any.

[1] Equivalently, we can make the following assumption: packets are of unit size; one packet is served with probability μ and no packet is served otherwise. We adopt the current assumption of heterogeneous packet sizes, as we will consider scheduling policies that potentially make decisions based on the packet sizes.

[2] Here, we do not consider the cost of preemption.

- *Random-Order-Service (RANDOM)*: When the server frees up, it randomly chooses one packet to serve if any.

The second type consists of policies that are non-preemptive and make scheduling decisions based on the packet size:

- *Shortest-Job-First (SJF)*: When the server frees up, it chooses to serve the packet with the smallest size if any. When there are multiple packets of the same smallest size, the tie-breaking rule randomly chooses one of them.

The third type consists of policies that are preemptive and blind to the packet size:

- *Time-Since-Last-Service (TSLS)* (Li, Li, & Eryilmaz 2014): When the server frees up, it chooses to serve the packet with the largest Time-Since-Last-Service (TSLS). The TSLS of a packet is set to zero when the packet arrives, increases by one if the packet is not served, and drops to zero once the packet is served. The TSLS policy is preemptive and mimics the Round Robin policy in the discrete-time system.
- *Preemptive Last-Come-First-Served (LCFS_P)*: This is the preemptive version of LCFS. Specifically, a preemption happens when a new packet arrives.

The fourth type consists of policies that are preemptive and make scheduling decisions based on the packet size:

- *Preemptive Shortest-Job-First (SJF_P)*: This is the preemptive version of the SJF policy. Specifically, a preemption happens when there is a new packet that has the smallest size.
- *(Preemptive) Shortest-Remaining-Processing-Time (SRPT)*: When the server frees up, it chooses to serve the packet with the smallest remaining size. When there are multiple packets of the same smallest remaining size, the tie-breaking rule randomly chooses one of them. In addition, a preemption happens only when a new packet arrives, and its size is smaller than the remaining size of the packet in service.

The preceding common scheduling policies are summarized in Table 8.1.

Simulation Results: To begin with, we consider a Geo/Geo/1 queue with $\lambda = 0.4$ and $\mu = 0.5$ (i.e., the mean packet size is equal to $1/\mu = 2$). By default, each data

Table 8.1 Taxonomy of some common scheduling policies.

	Non-preemptive	Preemptive
Non-size-based	FCFS, LCFS, RANDOM	TSLS, LCFS_P
Size-based	SJF	SJF_P, SRPT

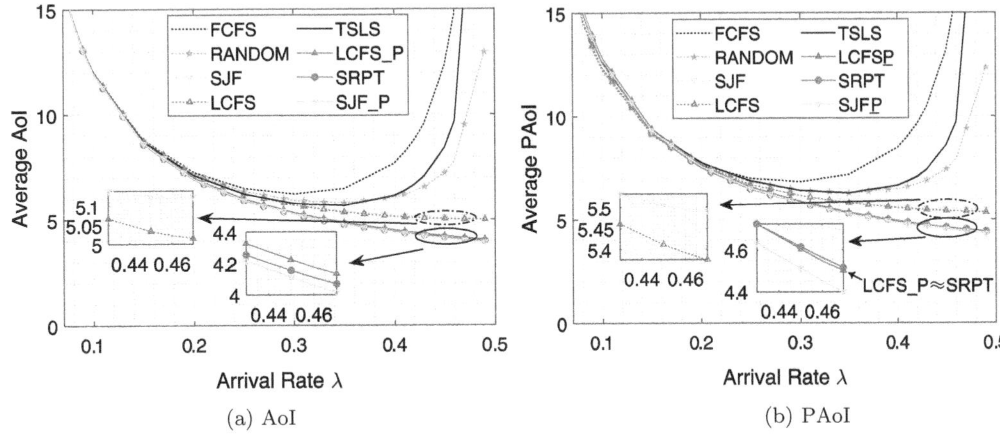

Figure 8.13 Comparisons of the average AoI and average PAoI performance of several common scheduling policies for a Geo/Geo/1 queue with varying arrival rate λ.

point in the plots is an average of 100 simulation runs. Each simulation run consists of 10^5 packets. In Figure 8.13, we present the simulation results of the average AoI and average PAoI performance with varying arrival rate λ under the scheduling policies we introduced previously.

We observe that for the non-preemptive case, SJF, a size-based policy, has a similar average AoI performance to LCFS but performs much better than FCFS and RANDOM, especially when the arrival rate is large. An explanation for LCFS and SJF having a similar AoI performance is the following. In the setting we consider, there is a high chance that the newest packet has the smallest size (i.e., one unit). In this case, SJF and LCFS would choose two packets of the same smallest size. They choose packets of different size only when the newest packet does not have the smallest size, which happens less often. This is why their AoI performance is similar. We also observe that LCFS performs very slightly better than SJF. The reason is that in the case that SJF and LCFS choose two packets of the same smallest size, the packet chosen by SJF is usually older due to its random tie-breaking rule.

Similarly, for the preemptive case, SJF_P and SRPT, two size-based policies, have an average AoI performance similar to LCFS_P but perform much better than TSLS, especially when the arrival rate is large. SJF_P and SRPT perform slightly better than LCFS_P because they make preemption decisions based on packet size. Moreover, we observe that preemptive policies generally have a better average AoI performance than non-preemptive ones, especially when the arrival rate is large.

Note that TSLS is also a preemptive policy. However, it performs much worse than the other preemptive policies. The reason is that TSLS mimics the Round Robin policy in an FCFS fashion. When the arrival rate becomes larger, its behavior is closer to that of FCFS, and thus, its performance becomes worse.

Similar observations can be made for the average PAoI performance.

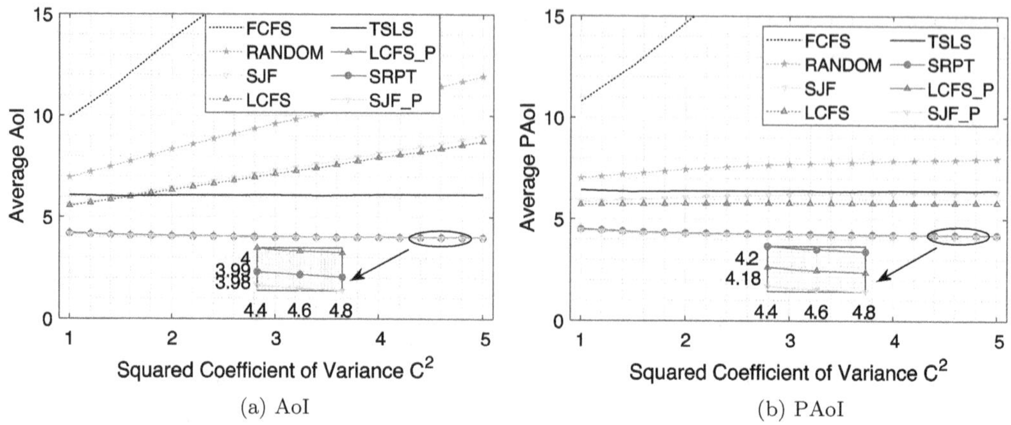

Figure 8.14 Comparisons of the average AoI and average PAoI performance of several common scheduling policies for a Geo/G/1 queue with varying squared coefficient of variance C^2

Next, we consider a Geo/G/1 queue, where the packet size follows a Zipf distribution[3] (Newman 2005). Let S be the packet size. We use $\mathbb{E}[S]$ and Var(S) to denote the mean and the variance of the packet size S. Then, we define the squared coefficient of variation of the packet size as $C^2 \triangleq \text{Var}(S)/\mathbb{E}[S]^2$, that is, the variance normalized by the square of the mean (Harchol-Balter 2013). A larger C^2 implies a larger variability of the packet size. While the mean packet arrival rate is fixed at $\lambda = 0.4$ and the mean packet size is fixed at $\mathbb{E}[S] = 2$, we change the value of the variance and thus the value of C^2 by choosing different values of the parameters of Zipf distribution. We present the simulation results in Figure 8.14.

We observe that the average AoI performance of preemptive policies becomes more superior as the packet-size variability (i.e., C^2) increases. Moreover, the average AoI performance of preemptive policies is only slightly impacted when the packet-size variability changes, while that of non-preemptive policies varies significantly. On the other hand, the PAoI exhibits a very different behavior for non-preemptive policies (except for FCFS). Specifically, their PAoI performance is insensitive to the packet-size variability: it either hardly changes (under LCFS and SJF) or increases very slightly (under RANDOM) as the packet-size variability increases. To the best of our knowledge, this is the first time such a counterintuitive observation about the PAoI is reported. An explanation for this observation is the following.

First, we explain why the PAoI under FCFS is still sensitive to the packet-size variability. Note that a key difference between FCFS and other non-preemptive policies is that under FCFS, every packet will lead to an AoI drop and thus corresponds to an AoI peak. When a large packet is in service, it will block all the following packets waiting in the queue, which results in a large delay for all such packets and thus a large

[3] The Zipf distribution comes from Zipf's law, which predicts that out of a population of N elements, the normalized frequency of elements of rank k is $f(k;s,N) = \frac{1/k^s}{\sum_{n=1}^{N}(1/n^s)}$, where s is the value of the exponent characterizing the distribution.

AoI peak corresponding to these packets. In contrast, under RANDOM, LCFS, and SJF the impact of such a blocking issue is minimal for the packets that lead to an AoI drop. Next, we explain why the PAoI has a different behavior than the AoI under RANDOM, LCFS, and SJF. We first consider LCFS. In the setting we consider, there is a high chance that the newest packet has the smallest size (i.e., one unit). Serving such unit-size packets leads to a small AoI peak (i.e., one). When the newest packet has a large size, the corresponding AoI peak would also be large. However, this happens less often. Therefore, the AoI trajectory would consist of a smaller percentage of large AoI peaks with many small AoI peaks in between. As the packet-size variability increases, there will be fewer but larger AoI peaks. In such cases, while the AoI is sensitive to the large AoI peaks (which comes from the large packet-size variability), the PAoI is much less sensitive. To illustrate this, we provide the following example. Consider a large AoI peak of 10, followed by 10 immediate small AoI peaks of 1. In this case, the average AoI is equal to $\frac{(1+2+\cdots+10)+10\times 1}{10+10\times 1} = 3.25$, and the average PAoI is equal to $\frac{10+10\times 1}{1+10} \approx 1.82$. Now, consider a larger AoI peak of 100, followed by 100 immediate small AoI peaks of 1. Then, the average AoI is equal to $\frac{(1+2+\cdots+100)+100\times 1}{100+100\times 1} = 25.75$, and the average PAoI is equal to $\frac{100+100\times 1}{1+100} \approx 1.98$. This example shows that a larger packet-size variability results in a larger average AoI but only minimally affects the average PAoI. A similar explanation also applies to SJF and RANDOM. SJF has a slightly worse PAoI performance due to its random tie-breaking rule. RANDOM has an even worse PAoI performance, as it randomly chooses a packet among all the packets waiting in the queue, and thus, the impact of the packet-size variability becomes more significant.

We have demonstrated that size-based policies achieve a better average AoI and average PAoI performance than non-size-based policies. However, the gain compared to the best non-size-based policies (e.g., LCFS_P) appears to be marginal. On the other hand, it is quite intuitive that when the packet-size information is unavailable, one should prioritize more recent updates. This is because while all the packets have the same mean packet size, the most recent packet arrives the last, leading to the smallest AoI once delivered.

Note that size-based policies do not utilize the arrival-time information, which also plays an important role in reducing the AoI. The packet-size information is "orthogonal" to the arrival-time information, both of which could significantly impact the AoI performance. Therefore, it is quite natural to further consider AoI-based policies that use both the size and arrival-time information of packets.

Age-Based Policies: In the following, we propose three AoI-based scheduling policies, which leverage both the packet-size and arrival-time information to reduce the AoI. We begin with the definitions of three AoI-based policies that *attempt to optimize the AoI in a specific time slot in the future* from three different perspectives:

- *AoI-Drop-Earliest (ADE)*: When the server frees up, it chooses to serve a packet such that once it is delivered, the AoI drops as soon as possible.

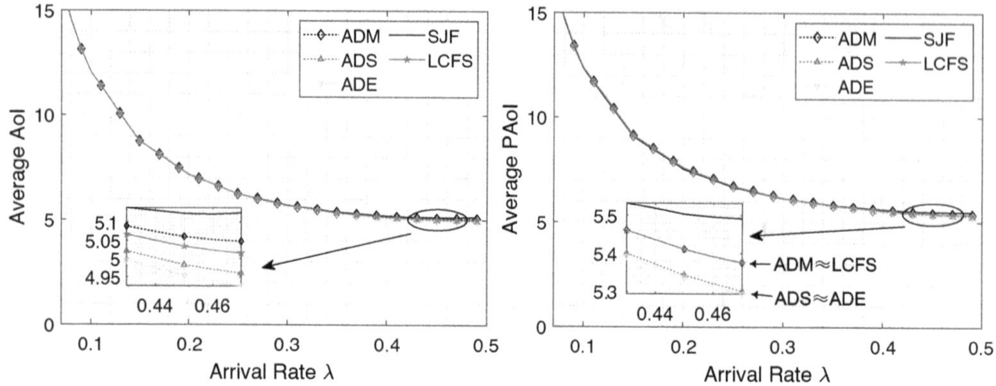

(a) AoI (b) PAoI

Figure 8.15 Comparisons of the average AoI and PAoI performance of non-preemptive policies for a Geo/Geo/1 queue with varying arrival rate λ: AoI-based policies vs. non-AoI-based policies.

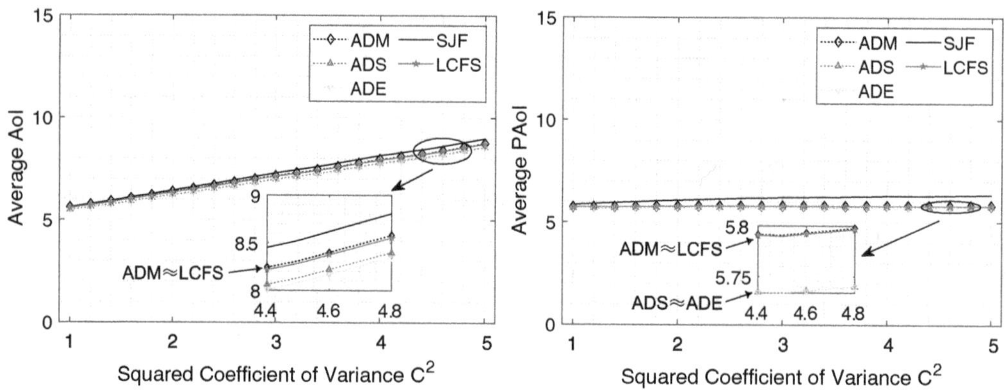

(a) AoI (b) PAoI

Figure 8.16 Comparisons of the average AoI and PAoI performance of non-preemptive policies for a Geo/G/1 queue with varying squared coefficient of variance C^2: AoI-based policies vs. non-AoI-based policies.

- *AoI-Drop-to-Smallest (ADS)*: When the server frees up, it chooses to serve a packet such that once it is delivered, the AoI drops to a value as small as possible.
- *AoI-Drop-Most (ADM)*: When the server frees up, it chooses to serve a packet such that once it is delivered, the AoI drops as much as possible.

If all packets waiting in the queue are obsolete (i.e., serving such packets does not lead to an AoI drop), then the preceding policies choose to serve a packet with the smallest size.

In Figures 8.15 and 8.16, we present the simulation results of the average AoI and average PAoI performance of the preceding three AoI-based policies in comparison with two non-AoI-based policies (LCFS and SJF), with varying arrival rate λ and with varying squared coefficient of variance C^2, respectively. While ADE performs slightly better than the other policies, all the considered policies have fairly close AoI performance.

We also consider the preemptive versions of the AoI-based policies (denoted by ADE_P, ADS_P, and ADM_P) and compare them with non-AoI-based preemptive policies (LCFS_P, SJF_P, and SRPT). In Figures 8.17 and 8.18, we present the simulation results of the average AoI and average PAoI performance of the considered policies, with varying arrival rate λ and with varying squared coefficient of variance C^2, respectively. The observations are similar to the non-preemptive case.

Remarks: In addition to the well-known guideline that one should prioritize new packets (as in LCFS), we suggest the following guidelines that will likely be useful

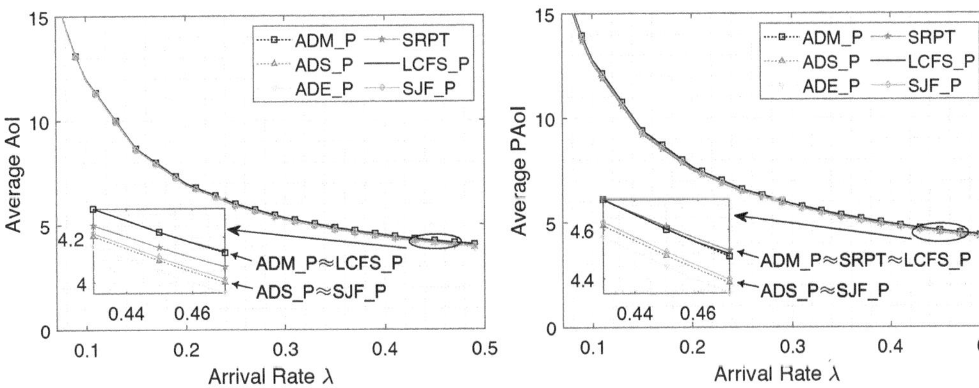

(a) AoI (b) PAoI

Figure 8.17 Comparisons of the average AoI and PAoI performance of preemptive policies for a Geo/Geo/1 queue with varying arrival rate λ: AoI-based policies vs. non-AoI-based policies.

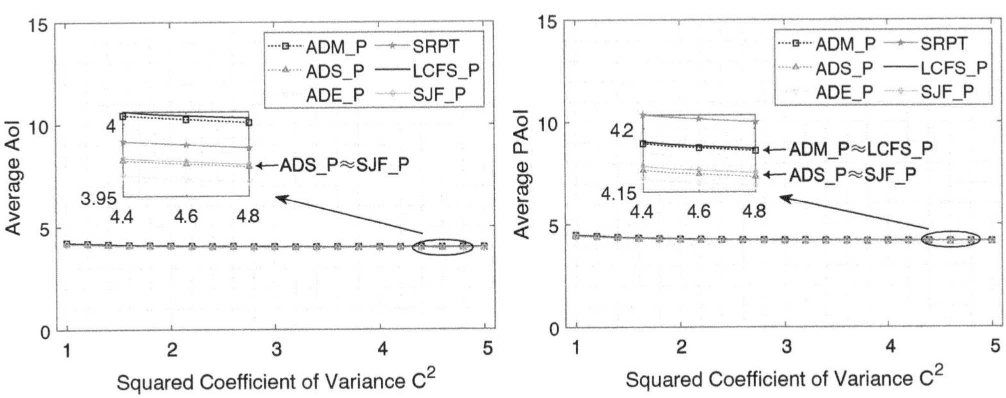

(a) AoI (b) PAoI

Figure 8.18 Comparisons of the average AoI and PAoI performance of preemptive policies for a Geo/G/1 queue with varying squared coefficient of variance C^2: AoI-based policies vs. non-AoI-based policies.

for designing age-efficient scheduling policies. (i) Service preemption should be employed when it is allowed. (ii) When the packet-size information is available, one should prioritize packets with a small size. (iii) The AoI can be further reduced if both the packet-size and arrival-time information can be used in the design of scheduling policies (such as age-based policies). (iv) Compared to LCFS and LCFS_P, the gain of the size-based and age-based policies is rather marginal. Therefore, it is good enough to use LCFS and LCFS_P when the packet-size information is unavailable or it incurs a high computational cost to make scheduling decisions based on such packet-size information.

8.4 Fresh Scheduling for Remote Estimation

In our discussion of the previous sections, we have predominantly focused on the timely transfer of dynamically generated and enqueued content over various circumstances and under various metrics, especially AoI, or *freshness* of updates at the receiver. A common characteristic of these investigations has been that the generation of content was not controllable by our designs, but was governed by external phenomena beyond our control. Also, the significance of a packet or file was directly measured by age metrics (e.g., mean delay, HOL-delay, mean AoI, peak-AoI). Both of these characteristics fail to hold in an important class of scenarios that are exemplified by the remote estimation/tracking setting that we will overview in this section.

Our discussion will reveal several interesting insights in maintaining *freshness* of useful information for dynamically evolving sources. In particular, we will discuss the design of both *transmitter-driven* and *receiver-driven* update mechanisms that aim to optimally balance the trade-off between estimation accuracy and the communication cost of updates in single-sensor (cf. Section 8.4.1) and multi-sensor (cf. Section 8.4.2) environments.

8.4.1 Single-Source Remote Estimation

The basic remote estimation problem is concerned with closely tracking the evolving state of a remote agent (such as a sensor or any other device with a dynamically changing state) with updates. This scenario is depicted in Figure 8.19 whereby $X[t]$ is the state of the source in slot[4] t and $\hat{X}[t]$ is its estimate at the receiver at the same time. The *source* S and the *receiver* are separated by a *transmitter* (which, in turn may be further separated into a *sampler* that decides when to sample the state, and a *queue* into which the sampled states are enqueued for transmission) and a *communication channel* between the transmitter and the receiver.

The broad purpose of remote estimation is to develop sampling and transmission strategies to keep the difference between $X[t]$ and $\hat{X}[t]$ small (in terms of an appropriate measure) while also accounting for the impact and costs of the communication channel that must be utilized in the process. Therefore, this problem naturally concerns

[4] In the spirit of the rest of the chapter, we keep the discrete-time evolution of the system dynamics here. However, the discussion can be extended to the evolution and update in continuous time.

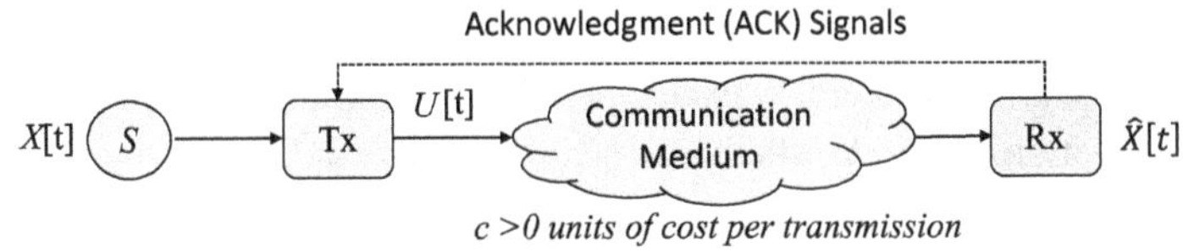

Figure 8.19 Basic remote estimation setting for a single source.

the transfer of fresh updates of the source state that is just enough to keep the estimation error low for the associated impact and cost of communication. As such, it is necessary to discuss the possible impact of the communication channel in this process.

• *Impact of the Communication Channel:* The communication channel between the transmitter and the receiver can have diverse effects on the remote sensing process. Transmissions may be erased, distorted, delayed, and so on based on the channel conditions and the choice of communication protocol. Each of these different effects may call for substantially different strategies to manage them. In our subsequent discussion, we will focus on a *cost-based* channel model by assigning a flexible *cost* value $c > 0$ to each transmission. Within this model, by choosing this cost value appropriately, we can increase or decrease the relative cost of a successful transmission under different channel impacts within a common setting. This allows us to focus on the design and analysis of policies that can optimize the trade-off between the estimation accuracy and communication cost. Next, we make a distinction between transmitter and receiver-driven update rules.

• *Transmitter vs. Receiver-Driven Update Policies:* One important question in the remote estimation of a dynamic source is whether the update decision is made by the transmitter or the receiver. These two entities have different information on the state of the source, namely, the transmitter can be expected to know the state $X[t]$ while the receiver only has an estimate $\hat{X}[t]$. Therefore, a transmitter-driven policy can make its decisions based on the accurate knowledge of $X[t]$, while a receiver-driven policy must work with other information, such as the time-since-last-update, for its update decision. In the multi-sensor setting in Section 8.4.2, we discuss the trade-off between these two paradigms in more detail.

In this basic setting, it is clear that the transmitter-driven update rule is better than its receiver-driven counterpart since its information is more accurate without any extra cost. The following example from Yun, Joo, and Eryilmaz (2018) summarizes useful insights on the optimal transmitter and receiver-driven design approaches for a randomly evolving source.

EXAMPLE 1 (Optimal Remote Estimation of a Random Walk (Yun et al. 2018)) Suppose we aim to remotely track a state $X[t]$ that evolves as a random walk according to $X[t+1] = X[t] + W[t]$, where $\{W[t]\}_t$ are i.i.d. random variables with

$$W[t] = \begin{cases} 1 & \text{with probability } \theta, \\ 0 & \text{with probability } 1-2\theta, \\ -1 & \text{with probability } \theta, \end{cases}$$

for some $\theta \in (0, \frac{1}{2}]$. Thus, the θ parameter determines how dynamic the state of $X[t]$ is within this basic structure. In each slot t, the transmitter-driven policy uses $X[t]$ and $\hat{X}[t]$ information to decide whether to send an update at the time by choosing $U[t] \in \{0, 1\}$.[5] Every update causes a fixed cost of c units, as modeled previously Supposing that $X[0] = \hat{X}[0] = 0$, and defining the error in the estimation as $\varepsilon[t] = |X[t] - \hat{X}[t]|$ at time t, the objective of the design is to

$$\text{minimize} \quad \lim_{T\to\infty} \frac{1}{T} \sum_{t=0}^{T-1} \mathbb{E}\left[f\left(\varepsilon[t]\right) + c \cdot U[t]\right], \tag{8.20}$$

where $f(\cdot)$ is a nonnegative-valued function with $f(0) = 0$ that measures the cost associated with the error. While this function can, in general, be an arbitrary non-decreasing function of its input, a very common case is when $f(\varepsilon) = \varepsilon^2$, which corresponds to *mean-squared-error* (MSE) measure of the estimation. As such, the objective (8.20) aims to optimize the trade-off between the estimation accuracy and the communication cost of updates. Investigation in Yun et al. (2018) proves that the optimal transmitter-driven policy that solves this problem is *threshold-type*, whereby the update is made whenever $[t] \geq \gamma$ for some γ that depends on the source dynamics θ, the estimation error measure $f(\cdot)$, and the update cost. In particular, when $f(\varepsilon) = \varepsilon^2$, the optimal threshold is given by $\approx \sqrt[4]{12\theta}$. If the parameters θ and are unknown to the controller, the paper also develops sequential learning strategies to find the optimal threshold level adaptively. Figure 8.20 illustrates the average cost (8.20) with MSE measure of a simple random walk, that is, $\theta = 1/2$, for the *optimal receiver-driven policy*, which performs updates periodically with the optimal choice of period T^*, and the *optimal transmitter-driven policy* with threshold-level γ^* selected as described previously. This reveals that as the communication cost c increases, the improvement from transmission-driven updates grows since it avoids unnecessary transmissions. The figure also includes another heuristic transmitter-driven policy, called *one-step-ahead* policy, which shows that even partial use of source state information can substantially improve over the optimal receiver-driven strategy.

This basic example, and many other works that we will partially overview in what follows, reveals several interesting new elements in maintaining fresh information at the receiver. In particular, we see that the need for an update is determined by the *level of error* in the estimation, and not by the *age or the delay of the last update*. The error level, in turn, depends on the dynamics of the source (i.e., how predictable it is) as well as the cost of communication needed for the update.

In the next section, we will see that these effects raise further new considerations when we expand the setting to a multisource environment, as needed in future networks. We note that, in the single-source setting, there have been numerous other works (e.g., Chakravorty & Mahajan 2015; Gao, Akyol, & Başar 2015; Sun, Polyanskiy, & Uysal-Biyikoglu 2017; Yun et al. 2018; Ornee & Sun 2019; Chakravorty & Mahajan 2019; Arafa, Banawan, Seddik, & Poor 2020) with a range of source

[5] The transmitter can know $[t]$ since it knows the estimation strategy of the receiver.

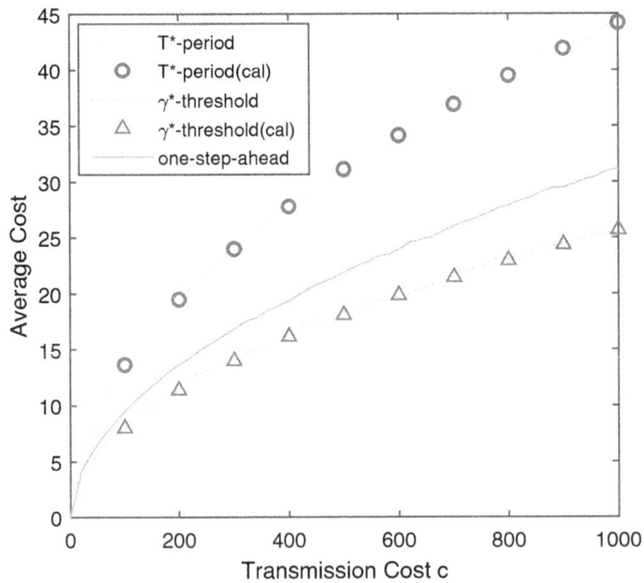

Figure 8.20 Average cost comparison with $\theta = 1/2$ of the three policies. Both the optimal transmitter-driven γ^*-threshold policy and the suboptimal transmitter driven one-step-ahead policy performs increasingly better than the optimal receiver-driven T^*-threshold policy as increases.

dynamics (such as Ornstein–Uhlenbeck processes, Wiener processes, Gauss–Markov processes, first-order auto-regressive models, etc.); varying communication channel models (such as random service time, noisy channels subject to erasures or errors, cost-based models); and various objectives (such as MSE minimization under update frequency bound, age-based nonlinear cost minimization under sampling frequency constraints, etc.). While the specific results differ in these works due to differences in the dynamics, costs, and constraints, the threshold-based design that is achieved in Example 1 remains a frequent characteristic of optimal designs for remote estimation.

8.4.2 Multisource Remote Estimation

In this section, we move from the single-source to a distributed multisource remote estimation setting, where the dynamically evolving states $[t] = (X_i[t])_{i=1}^n$ of n sources are accessible by k distributed transmitters that must share a wireless communication medium to update the common receiver, which estimates the state with $[t] = (i[t])_{i=1}^n$ as depicted in Figure 8.21.

Similar to the single-source setting, each transmitter k must make its update decision by selecting $U_k[t] \in \{0, 1\}$. Also, as before, the update policy can be transmitter-driven (with access to the true states of the connected sources) or receiver-driven (without access to the true source states). However, in this distributed setting, a new challenge emerges that concerns the cost characteristics of the multiuser

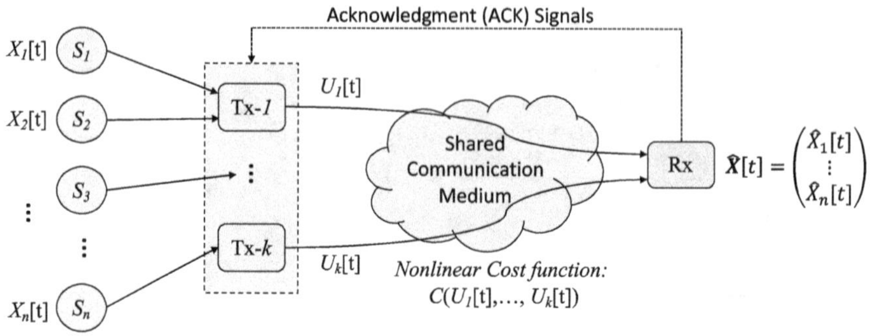

Figure 8.21 Multisource remote estimation setting.

communication medium. In what follows, we discuss the diverse nature that this communication cost can take under different circumstances.

• *Emergence of Nonlinear Cost Structures of the Multiuser Communication:* Recall that in the single-source setting, we assumed that the cost of each update communication was fixed to be $c > 0$. However, the cost of communication in the multiuser case can differ drastically depending on the medium access strategy, the coding policy, the interference effects, and protocol overhead, just to name a few. It is easy to see that, if $k = n$ and the cost function becomes a linear function of $\sum_{i=1}^{k} U_k[t]$, then the multiuser problem can be decomposed to k independent single-user problems. However, such a cost structure presumes completely orthogonal channels between each transmitter and the receiver, which requires increased coordination between transmitters. More often, the cost function will take a nonlinear structure, in which case interesting multiuser remote estimation problems emerge. In the rest of this section, we discuss the cases of distributed and collocated sources separately since they tend to exhibit *superlinear* versus *sublinear* cost functions, respectively, which raises different research questions and solutions.

Remote Estimation for Distributed Sources

In this setting, we consider the case of physically distributed sources so that the number of transmitters k in the generic setting is large. As the number k of transmitters in the wireless network increases, the multiuser uplink communication gets increasingly congested and distributed. Therefore, as the total number $\sum_{i=1}^{k} U_i[t]$ of updates increases in a slot, the communication will be subject to superlinear costs due to increased interference effects and coordination needs. A particular instance of such superlinear growth is illustrated in Figure 8.22 whereby we plot the energy consumption of k distributed transmitters making simultaneous updates over a shared random access channel. We see that for varying values of the attempt probabilities q of each transmitter, the cost function grows at various superlinear growth rates. This superlinear cost structure raises an interesting trade-off between transmitter-driven (TD) and receiver-driven (RD) update paradigms, which did not exist in the single-source setting, as discussed in the next example.

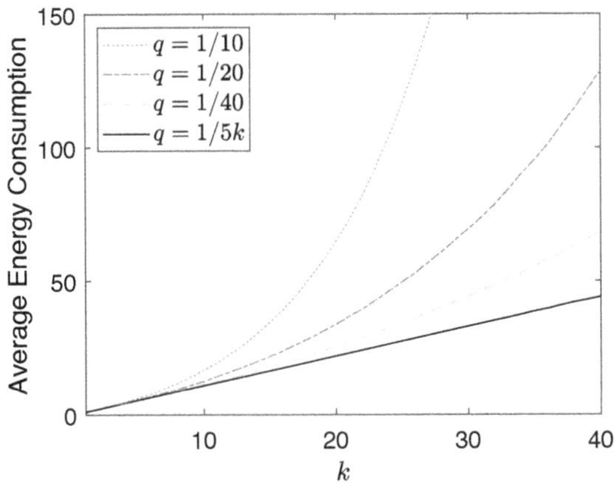

Figure 8.22 Average energy consumed for successful update with k simultaneous transmitters, when each transmitter attempts with probability q.

EXAMPLE 2 (TD vs. RD Updates for Distributed Sources (Kang, Eryilmaz, & Joo 2021)) Example 1 showed that the optimal transmitter-driven update policy always outperforms the optimal receiver-driven policy for a single source. This is no longer the case for distributed sources. In particular, the randomized timing of the threshold-based transmitter-driven design, such as the γ^*-threshold policy of Example 1, results in a random number of updates at a given slot. Since the communication cost function takes a superlinear form in this distributed setting, the mean communication cost (due to Jensen's Inequality) will be boosted. In contrast, a receiver-driven policy can schedule the update timings at the outset so that a deterministic number of transmissions are made at a given slot, thereby keeping the mean communication cost steady. This communication cost gain of receiver-driven design can compensate for some of its disadvantages from its knowledge of the source state information.

In Figure 8.23, we illustrate these trade-offs by assuming that the transmitter-driven communication cost function takes the form $C([t]) = (\sum_i U_i[t])^{1+\epsilon_s}$ and the receiver-driven communication cost function takes the form $C([t]) = (\sum_i U_i[t])^{\epsilon_r}$, with $\epsilon_s \geq \epsilon_r \geq 0$. For $c = 50$ random walk parameter $\theta = 0.3$, and $\epsilon_s = \epsilon_r = 2$, Figure 8.23a shows the average MSE and communication cost performances of both the optimized TD and the RD policies as the number of transmitters k grows. We can see that the TD policy outperforms RD for small enough number of transmitters until a *crossing point* is reached. Beyond this crossing point number of users, the RD policy starts outperforming since its advantage of centralized scheduling dominates advantage from the individual state information that is available to TD policy. Figure 8.23b illustrates how this crossing point changes as a function of ϵ_s and ϵ_r parameters of the TD and RD communication cost functions.

Remote Estimation for collocated Sources

In this setting, we envision the remote estimation of many collocated sources that share common transmitter(s) for their updates. In contrast to the superlinear cost structures

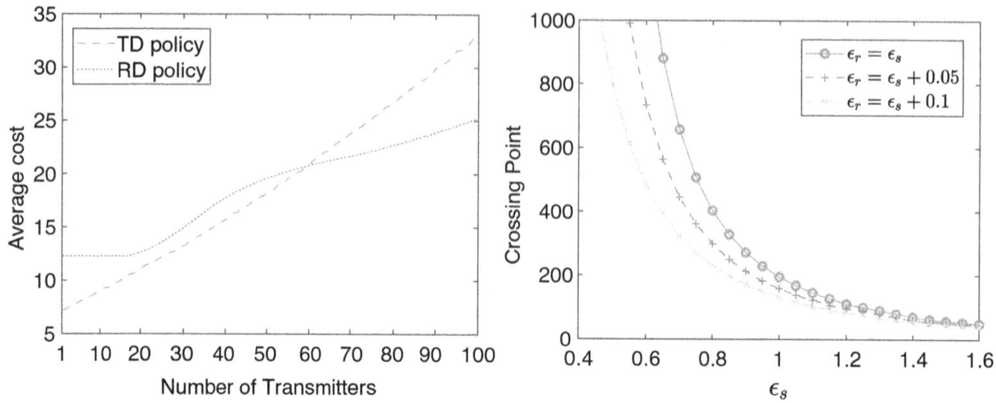

(a) Average cost when $\epsilon_s = \epsilon_r = 2$

(b) Crossing point with varying ϵ_s and ϵ_r

Figure 8.23 Performance comparison between TD and RD policies.

when the number of transmitters k increases, the communication cost function typically takes a sublinear form as the number of sources n increases while k remains small. In this new regime, more and more sensors are monitored by a common transmitter, which allows the transmitter to perform more-efficient strategies to encode and time the updates in its transmissions, thereby reducing the cost per unit update. Theoretically, this is supported by the information-theoretic foundations that encoding more information into the transmission increases the coding rate (since the codeblock length increases). Practically, this is supported by the fact that sending multiple updates at once allows the use of less overhead in each transmission (for synchronization, probing, channel estimation, etc.). Accordingly, in this regime it is sensible to assume $C([t])$ to be a sublinear function of $\sum_{i=1}^{k} U_i[t]$ such as $C([t]) = (\sum_i U_i[t])^{\alpha}$ for $\alpha \in [0, 1]$. In this setting, interesting new dynamics emerge that call for the design of new policies for performing updates.

In particular, these designs need to optimize the trade-off between two forces. On the one hand, sending updates without delay is desirable to keep the error low, but this calls for sending a small number of updates at a time, hence operating in the inefficient transmission regime. On the other hand, delaying updates is desirable so that multiple updates can be made at a time, hence operating in the efficient transmission regime; but this causes higher errors in the estimation of the source state. Good update policies must balance these two forces in order to maintain tolerable estimation error levels with low communication costs. More work is needed in this direction.

We would like to end the section by noting that the remote estimation problem for multiple sources has been attracting increasing attention in the literature, and continues to be an active area of research. Numerous interesting works have attacked the question of optimal update strategies for multisource settings (e.g., Nar & Başar 2014; Li, Quevedo, Lau, & Shi 2014; Leong, Dey, & Quevedo 2016; Han, Wu, Zhang, & Shi 2017; Ding, Li, Dey, & Shi 2017; Vasconcelos, Nayyar, & Mitra 2017; Gatsis, Ribeiro, & Pappas 2018; Wu, Ding, Cheng, & Shi 2020) under various source models, objectives, and/or the dynamics.

8.5 Conclusions and Future Works

In this chapter, we focused on the age-efficient scheduling design that is essential for information monitoring, tracking, and control in communication networks. In particular, we first considered the age-efficient scheduling algorithm design in the presence of elastic traffic, inelastic traffic, and heterogeneous traffic, respectively. Then, we discussed the fresh scheduling design for remote estimation. All these discussions cover some of the most recent advances in the age-efficient algorithm design in communication networks and motivate several interesting research efforts. Some of the possible future research directions include the following:

- In the presence of elastic traffic, age-efficient algorithm design is as hard as its delay-efficient counterpart. While there are delay-optimal algorithms in some special regimes (e.g., heavy-traffic regime or large-system regime), it is interesting to study age-optimal algorithms in these special regimes.
- With regard to the inelastic traffic, the existing age-efficient algorithm achieves a trade-off between the AoI and the algorithm convergence speed, which may not be an optimal trade-off. An interesting question is whether there are any algorithms that can achieve a provably better or even optimal trade-off.
- For the heterogeneous traffic, we discussed the impacts of various scheduling design on the AoI performance via simulations. Can we analytically quantify the AoI performance in such a case? In addition, how do we develop an analytical framework for age-efficient algorithm design for heterogeneous traffic?
- For distributed remote estimation, there are numerous interesting research directions to pursue whereby the source dynamics, the communication channel imperfections and limitations, and the objectives can differ to accommodate various network and user characteristics in future IoT and cyber-physical systems.

References

Arafa, A., Banawan, K., Seddik, K. G., & Poor, H. V. (2020), "Timely estimation using coded quantized samples," *arXiv preprint arXiv:2004.12982*.

Bedewy, A. M., Sun, Y., & Shroff, N. B. (2016), Optimizing data freshness, throughput, and delay in multi-server information-update systems, *in* "2016 IEEE International Symposium on Information Theory (ISIT)," IEEE, pp. 2569–2573.

Chakravorty, J., & Mahajan, A. (2015), Distortion-transmission trade-off in real-time transmission of Gauss-Markov sources, *in* "IEEE International Symposium on Information Theory (ISIT)," pp. 1387–1391.

Chakravorty, J., & Mahajan, A. (2019), "Remote estimation over a packet-drop channel with Markovian state," *IEEE Transactions on Automatic Control* **65**(5), 2016–2031.

Costa, M., Codreanu, M., & Ephremides, A. (2016), "On the age of information in status update systems with packet management," *IEEE Transactions on Information Theory* **62**(4), 1897–1910.

Ding, K., Li, Y., Dey, S., & Shi, L. (2017), "Multi-sensor transmission management for remote state estimation under coordination," *IFAC-PapersOnLine* **50**(1), 3829–3834.

Gao, X., Akyol, E., & Başar, T. (2015), Optimal estimation with limited measurements and noisy communication, *in* "2015 54th IEEE Conference on Decision and Control (CDC)," pp. 1775–1780.

Gatsis, K., Ribeiro, A., & Pappas, G. J. (2018), "Random access design for wireless control systems," *Automatica* **91**, 1–9.

Han, D., Wu, J., Zhang, H., & Shi, L. (2017), "Optimal sensor scheduling for multiple linear dynamical systems," *Automatica* **75**, 260–270.

Harchol-Balter, M. (2013), *Performance Modeling and Design of Computer Systems: Queueing Theory in Action*, Cambridge University Press.

Joo, C., & Eryilmaz, A. (2018), "Wireless scheduling for information freshness and synchrony: Drift-based design and heavy-traffic analysis," *IEEE/ACM Transactions on Networking* **26**(6), 2556–2568.

Kadota, I., Uysal-Biyikoglu, E., Singh, R., & Modiano, E. (2016), Minimizing the age of information in broadcast wireless networks, *in* "2016 54th Annual Allerton Conference on Communication, Control, and Computing (Allerton)," IEEE, pp. 844–851.

Kam, C., Kompella, S., Nguyen, G. D., & Ephremides, A. (2015), "Effect of message transmission path diversity on status age," *IEEE Transactions on Information Theory* **62**(3), 1360–1374.

Kang, S., Eryilmaz, A., & Joo, C. (2021), Comparison of decentralized and centralized update paradigms for remote tracking of distributed dynamic sources, *in* "Proceedings of Infocom," pp. 1–10.

Kaul, S. K., Yates, R. D., & Gruteser, M. (2012a), Status updates through queues, *in* "2012 46th Annual Conference on Information Sciences and Systems (CISS)," IEEE, pp. 1–6.

Kaul, S., Yates, R., & Gruteser, M. (2012b), Real-time status: How often should one update?, *in* "2012 Proceedings IEEE INFOCOM," IEEE, pp. 2731–2735.

Leong, A. S., Dey, S., & Quevedo, D. E. (2016), "Sensor scheduling in variance based event triggered estimation with packet drops," *IEEE Transactions on Automatic Control* **62**(4), 1880–1895.

Li, B., Li, R., & Eryilmaz, A. (2014), "Throughput-optimal scheduling design with regular service guarantees in wireless networks," *IEEE/ACM Transactions on Networking* **23**(5), 1542–1552.

Li, Y., Quevedo, D. E., Lau, V., & Shi, L. (2014), Multi-sensor transmission power scheduling for remote state estimation under sinr model, *in* "53rd IEEE Conference on Decision and Control," pp. 1055–1060.

Liu, Z., Huang, L., Li, B., & Ji, B. (2020), Anti-aging scheduling in single-server queues: A systematic and comparative study, *in* "IEEE INFOCOM 2020 - IEEE Conference on Computer Communications Workshops (INFOCOM WKSHPS)," pp. 309–316.

Liu, Z., Huang, L., Li, B., & Ji, B. (2021), "Anti-aging scheduling in single-server queues: A systematic and comparative study," *Journal of Communications and Networks* **23**(2), 91–105.

Lu, N., Ji, B., & Li, B. (2018), Age-based scheduling: Improving data freshness for wireless real-time traffic, *in* "Proceedings of the Eighteenth ACM International Symposium on Mobile Ad Hoc Networking and Computing," pp. 191–200.

McKeown, N., Mekkittikul, A., Anantharam, V., & Walrand, J. (1999), "Achieving 100% throughput in an input-queued switch," *IEEE Transactions on Communications* **47**(8), 1260–1267.

Nar, K., & Başar, T. (2014), Sampling multidimensional Wiener processes, *in* "53rd IEEE Conference on Decision and Control," pp. 3426–3431.

Newman, M. E. (2005), "Power laws, Pareto distributions and Zipf's law," *Contemporary Physics* **46**(5), 323–351.

Ornee, T. Z., & Sun, Y. (2019), "Sampling for remote estimation through queues: Age of information and beyond," *arXiv preprint arXiv:1902.03552.*

Srikant, R., & Ying, L. (2013), *Communication Networks: An Optimization, Control, and Stochastic Networks Perspective*, Cambridge University Press.

Sun, Y., Kadota, I., Talak, R., & Modiano, E. (2019), "Age of information: A new metric for information freshness," *Synthesis Lectures on Communication Networks* **12**(2), 1–224.

Sun, Y., Polyanskiy, Y., & Uysal-Biyikoglu, E. (2017), Remote estimation of the Wiener process over a channel with random delay, *in* "IEEE International Symposium on Information Theory (ISIT)," pp. 321–325.

Talak, R., Karaman, S., & Modiano, E. (2019), "Optimizing information freshness in wireless networks under general interference constraints," *IEEE/ACM Transactions on Networking* **28**(1), 15–28.

Tassiulas, L., & Ephremides, A. (1990), Stability properties of constrained queueing systems and scheduling policies for maximum throughput in multihop radio networks, *in* "29th IEEE Conference on Decision and Control," IEEE, pp. 2130–2132.

Tassiulas, L., & Ephremides, A. (1993), "Dynamic server allocation to parallel queues with randomly varying connectivity," *IEEE Transactions on Information Theory* **39**(2), 466–478.

Vasconcelos, M. M., Nayyar, A., & Mitra, U. (2017), Optimal sensor scheduling strategies in networked estimation, *in* "2017 IEEE 56th Annual Conference on Decision and Control (CDC)," pp. 5378–5384.

Wu, S., Ding, K., Cheng, P., & Shi, L. (2020), "Optimal scheduling of multiple sensors over lossy and bandwidth limited channels," *IEEE Transactions on Control of Network Systems*.

Yun, J., Joo, C., & Eryilmaz, A. (2018), Optimal real-time monitoring of an information source under communication costs, *in* "IEEE Conference on Decision and Control (CDC)," pp. 4767–4772.

9 Age-Driven Transmission Scheduling in Wireless Networks

Qing He, Di Yuan, György Dán, and Anthony Ephremides

9.1 Introduction

Optimal scheduling of data transmission is a fundamental problem class in wireless communications. The main challenge for this problem type is the coupling relation between links, namely, simultaneous transmissions of multiple links on a shared channel generate interference at the receivers. There are two main modeling approaches for studying optimal scheduling. The first approach takes a probabilistic viewpoint, with assumptions on the randomness of the data arrival processes at the transmitters along with those of the channel. The solution is then typically a scheduling policy. The second approach addresses scenarios where the amount of data residing at the transmitters and the channel information are assumed to be known. As a result, the solution is a schedule that specifies when each transmitter shall transmit, and for how long.

In this chapter, we adopt the second approach for system modeling, that is, the task is to compute a schedule for emptying the buffers of the transmitters. For this modeling approach, the most common objective function prior to the notion of AoI has been the schedule length, that is, the time it takes before all data are delivered. This objective function makes sense from the resource management standpoint. Namely, the minimum-length schedule is also the one that utilizes the resource, here a common channel, most efficiently. To address the arrival of new data, one can adopt rolling horizon scheduling, that is, the computed schedule is executed partially, after which a new scheduling optimization is performed to account for the remaining data as well as the new data arrivals.

AoI gives rise to new types of optimization problems for transmission scheduling in wireless networks. First, using AoI as the objective function, the minimum-length schedule is not necessarily optimal. Second, as will be clear later, optimizing AoI has a large impact on the mathematical formulation and candidate optimization approaches. Moreover, optimizing AoI with additional system considerations, such as age correlation, leads to new dimensions in the optimization process.

We structure the chapter as follows. We present and discuss the overall system model in Section 9.2. In the same section, we discuss minimum-length scheduling. Next, we address the scheduling problem for AoI minimization in Section 9.3. We then extend the problem by introducing the aspect of energy consumption in Section 9.4. In Section 9.5, we study another problem extension that addresses correlated AoI.

We discuss the main insights gained from the solutions to these problems, and present topics warranting further investigation in Section 9.7. Finally, in Section 9.8 we provide further reading notes.

9.2 Preliminaries

9.2.1 System Model

We consider a system containing N information sources. Each source generates data packets that carry information samples with time stamps. Each source is associated with a transmitter (TX) for transmitting its data packets to its intended receiver (RX). We consider a single-hop system, hence the network is composed of the N TX-RX wireless links, where the nth link with TX_n and RX_n is for source $n, n = 1, \ldots, N$. The number of data packets of source n is denoted by K_n. Due to the one-to-one mapping between information sources and transmission links, we use the same index. The case with multiple sources having the same destination receiver can be transformed to our scenario, by splitting the receiver into multiple identical logical receivers.

We consider that time is slotted, and we denote by t_j the time instant when slot j ends. Thus, time slot j spans the time interval $[t_{j-1}, t_j]$, with the convention that t_0 is the starting time of the scheduling horizon. A link, if scheduled in a time slot, transmits one data packet to its RX. We use U_{ni} to refer to the ith packet of source n, and we denote by τ_{ni} the time stamp of the packet, with $\tau_{n,i-1} < \tau_{ni}, i = 2, \ldots, K_n$. The K_n packets of link n are transmitted following the first-come-first-serve (FCFS) service discipline. Thus, one can view the data packets of a link being held in a queue, and the task is to construct a schedule for emptying these queues. Note that, from a pure AoI standpoint, it would be best to send the packet having the latest time stamp only, discarding the other packets in the queue; this is referred to as packet management [1, 2]. Nonetheless, we assume that all packets carry information that are of value to the intended receiver to build knowledge of the process monitored by the source over time. Thus, all data packets of a source are to be delivered. In Figure 9.1, we provide an illustration of key components of the system model [3].

In any time slot, concurrent transmissions on multiple links are subject to interference constraints. We use the term group to refer to a subset of the N links that can transmit successfully simultaneously. Such a set is also known as a compatible link set [4]. The criterion for compatible link sets depends on the interference model. For a wireless network with a shared channel, two interference models are widely used [5]. In the first model, any two links can be active simultaneously if and only if they are sufficiently spatially separated from each other. This model is called the protocol model; the naming originates from the fact that the model resembles collision avoidance protocols. In the second model, referred to as the physical model, a successful transmission requires that the signal-to-interference-and-noise ratio (SINR) meets a specified threshold at the receiver. More details of the two interference models are provided in the Section 9.2.2. In this chapter, we do not restrict ourselves to

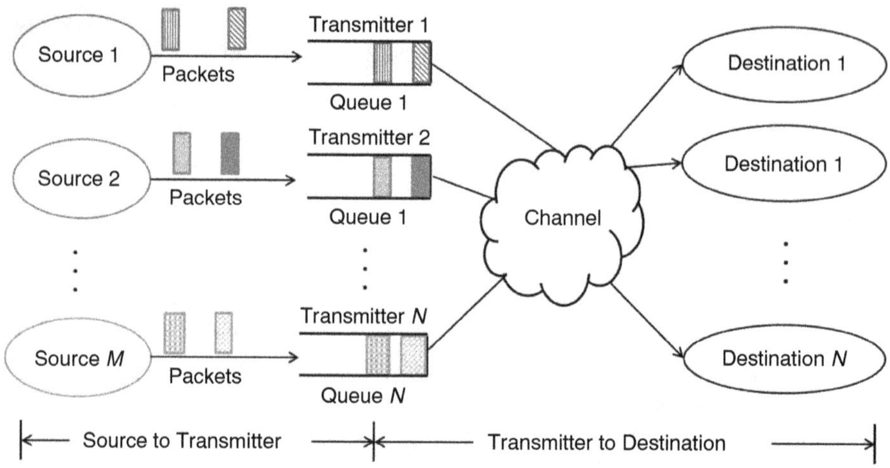

Figure 9.1 System scenario illustrated.

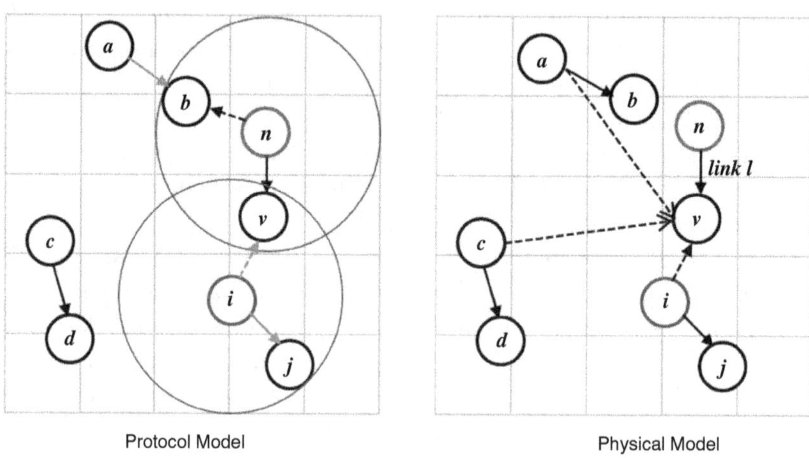

Figure 9.2 Interference models.

a specific interference model. Rather, we consider a given set of candidate groups $\mathcal{C}$, where each element $c \in \mathcal{C}$ is a group, that is, $c \subseteq \{1, \ldots, N\}$.

9.2.2 Interference Models

Due to the broadcast nature of the wireless medium, each receiver hears not only its desired signal but also all the other transmitted signals, which constitute interference. The capacity of the channel depends on interference, which in its turn is determined by the composition of concurrently active links. Given a set of co-channel links, to estimate the data rate and throughput, two interference models have been used, as illustrated in Figure 9.2.

The first model is referred to as the protocol model, where we use "node" to denote the transmitter and receiver of a link. For a set of nodes $\mathcal{V}$ on a Euclidean plane, each node $n \in \mathcal{V}$ is associated with a disk of radius $r(n)$, which depends on the transmission power level P_n of node n and the path-loss exponent. The signal from node n is assumed to be "heard" only within this range (as shown by the red circle in Figure 9.2). This results in a graph $\mathcal{G} = (\mathcal{V}, \mathcal{E})$, where set $\mathcal{E}$ is obtained by adding edges between n and v, $\forall n, v \in \mathcal{V}$, if $d(n, v) \leq r(n)$, where $d(n, v)$ denotes the Euclidean distance between nodes n and v. Interference in this so-called, "conflict graph" is modeled through independence constraints: if node n transmits to v, no adjacent node of v, for example, node i, should transmit concurrently.

The second model is referred to as the physical model, which utilizes the signal-to-interference-and-noise ratio (SINR) to measure channel quality. The model is based on practical transceiver designs of communication systems that treat interference as noise. Hence the SINR at the receiver of the lth link, that is, node v in the physical model shown in Figure 9.2, is quantified by (9.1):

$$\text{SINR}_l \triangleq \frac{G_{ll}P_l}{\sum\limits_{m \neq l} G_{ml}P_m + \sigma_l^2}. \tag{9.1}$$

In (9.1), σ_l^2 is the noise variance of the lth link and P_m is the transmission power of the mth link. G_{ml} is the channel gain between the transmitter of the mth link and the receiver of the lth link, incorporating the effects of path loss, shadowing, and fading. Under the physical model, aka the SINR model, a transmission that is successful means that the SINR value at the intended receiver exceeds a threshold so that the transmitted signal can be decoded with an acceptable bit error probability.

Comparing to the protocol model, the physical model takes into consideration the effects of interference and power on transmission rate, and hence is more realistic on reflecting of the wireless channel. The key practical issue associated with the physical model is its computational complexity, mainly for the reason that the SINR calculation is a non-convex function with respect to the transmission powers. Therefore, solving an optimization problem with the physical model or designing effective algorithms for it is more challenging, especially for large networks.

9.2.3 Minimum-Length Scheduling

If the performance is not AoI-driven, the most common objective is to find a minimum-length schedule [6] in terms of the number of time slots before all data packets are transmitted. For many scheduling problems in domains other than wireless communications, the schedule length is also known as the makespan. As one group is scheduled in each time slot, the makespan objective becomes equivalent to minimizing the usage of groups. This immediately gives us the following insight. Namely, for minimum-length scheduling, the order in which the groups appear in the schedule has no significance, as long as the total number of used groups is minimal. Thus, the schedule is oblivious to the time stamps of the data packets.

For a mathematical formulation, we first define $\mathcal{C}_n \triangleq \{c \in \mathcal{C} : n \in c\}$ to be the subset of groups containing link n. Then, the use of groups can be represented by nonnegative integer variables z_c, denoting the number of occurrences of group c in the schedule. The problem can then be formulated as the following integer linear optimization model. The constraint of the model states that the number of groups in the schedule containing link n has to be at least the number of packets to be delivered on the link.

$$\min \sum_{c \in \mathcal{C}} z_c \tag{9.2a}$$

$$\text{subject to} \sum_{c \in \mathcal{C}_n} z_c \geq K_n \quad n = 1, \ldots, N, \tag{9.2b}$$

$$z_c \geq 0, \text{integer.} \tag{9.2c}$$

The preceding integer formulation resembles the set cover problem [7]. By considering the subsets in the input of a set cover problem as the groups for transmission scheduling, the two problems become equivalent and hence the scheduling problem (with given groups) is NP-hard.

Accounting for group formation for any of the two interference models adds more to the problem complexity. For mathematical formulation, one option is to introduce additional optimization variables and constraints to characterize the compatible sets used for each time slot. A better modeling approach, which also enables an effective solution algorithm, is to consider (9.2) and use a second module to generate promising candidate groups. Embedding the second module into solving the linear programming (LP) version of (9.2) yields the column generation algorithm that has been very successful in deriving near-optimal solutions time-efficiently [8].

9.3 Link Scheduling with AoI Minimization

9.3.1 Problem Statement

We now turn our attention to optimal scheduling for AoI minimization. To this end, we define an AoI variable a_{nj} for each source, representing the AoI of source n at the end of time slot j. The initial AoI of source n is denoted by a_{n0}.

$$a_{nj} = \begin{cases} a_{n,j-1} + 1 & \text{if no packet is received by } \mathrm{RX}_n \text{ in time slot } j, \\ t_j - \tau_{ni} & \text{if } U_{ni} \text{ is received by } \mathrm{RX}_n \text{ in time slot } j. \end{cases} \tag{9.3}$$

By Equation (9.3), if the ith packet of source n is received in a time slot, the AoI, by definition, is set to be the difference of the end of the time slot and the packet's time stamp τ_{ni}; otherwise, the AoI of a source n increases by one time slot. For our cost function, we can adopt the average AoI, or the peak AoI. We use the peak AoI, which has the advantage of addressing also fairness.

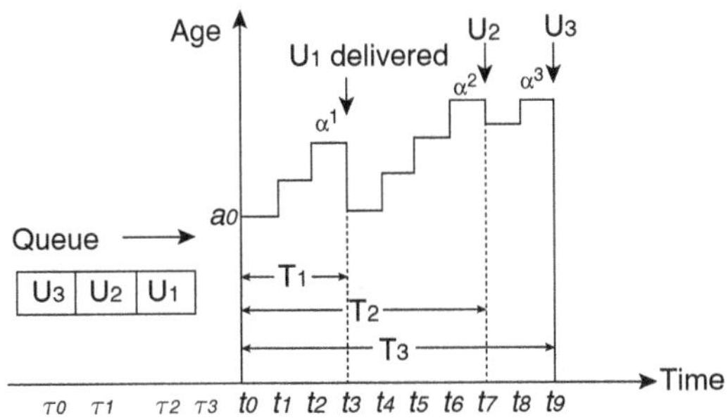

Figure 9.3 AoI evolution illustrated.

Recall that time slot j is formed by interval $[t_{j-1}, t_j]$. By (9.3), if a packet is delivered in a time slot, the arrival time at the receiver is the end of the slot.[1] Moreover, as the AoI is monotonically increasing when no packet is received, it is convenient to introduce α_n^i as the AoI of source n, attained immediately before U_{ni} is received. This quantity is defined in what follows. Note that for any valid schedule, there is a unique time point j defining the value of α_n^i. In Figure 9.3, we illustrate the key AoI entities for a single source.

$$\alpha_n^i = a_{nj} \text{ if } U_{ni} \text{ is received by } \mathrm{RX}_n \text{ in time slot } j. \tag{9.4}$$

Using the definitions and assumptions introduced, we now formalize the problem of optimal transmission scheduling for AoI minimization.

DEFINITION 1 Minimum Peak AoI Scheduling (MPAS): *We consider a set of information sources* $\mathcal{N} \triangleq \{1, \ldots, N\}$, *and the corresponding links with the candidate transmission groups forming set* $\mathcal{C}$. *We are given the set of data packets* $\mathcal{K}_n \triangleq \{1, \ldots, K_n\}$ *of source* $n \in \mathcal{N}$, *and the time stamps* τ_{ni} *of packet* U_{ni}, $i = \mathcal{K}_n, n \in \mathcal{N}$, *with* $\tau_{n,i-1} < \tau_{n,i}, i = 2, \ldots, K_n$. *For any source* n, *the initial AoI is* $a_{n0}, n \in \mathcal{N}$. *The starting schedule time is* t_0. *We construct a transmission schedule with one group* $c \in \mathcal{C}$ *per time slot, to transmit all* K_n *packets of each source* n *using FCFS, with the objective function of minimizing the maximum* $\alpha_n^i, i \in \mathcal{K}_n, n \in \mathcal{N}$.

By the definition just given as well as the illustration in Figure 9.3, the AoI of a source after all its packets are transmitted (which could occur before the end of the schedule) is not present in the problem statement and thus in effect is treated as zero. This is seemingly counterintuitive, as the AoI increases unless receiving new information. The assumption is not unreasonable as it may look, however. First, in some application context, information sources may dynamically enter and exit the system. When a source exits the system, its age becomes irrelevant and reasonably zero. The task of MPAS is to minimize AoI for the known packets, without assuming

[1] This particular choice does not affect the problem structure or the algorithms for solving the problem. One can also use the starting or middle point of the interval.

there will always be additional arrivals. Second, recall that MPAS corresponds to one scheduling cycle in a rolling horizon strategy. In this context, a new MPAS scheduling cycle will arise to account for new packet arrivals. The next cycle can, in fact, start when all packets for at least one source in the current problem input are delivered. Hence even if the age is treated as zero in the current schedule after a source is emptied, the cycle-by-cycle scheduling process shall still keep track of the age increase, which, in fact, will define the initial age for each cycle. Hence there will not be any zero AoI in scheduling with rolling horizon.

9.3.2 Structural Results

MPAS is inherently a combinatorial optimization problem, for which complexity is a fundamental aspect. We have commented on the NP-hardness of the minimum-length scheduling problem. However, one cannot use this fact as a proof for the complexity of MPAS, which has a different objective. In the following, we provide a theorem formalizing the complexity of MPAS, and then provide insights into how to derive a proof.

THEOREM 1 *MPAS is NP-hard.*

Instead of giving a fully detailed proof, it is more instructive to discuss the underlying rationale. Theorem 1 can be proven by using a reduction from the well known 3-satisfiability (3-SAT) that in fact was the first known NP-complete problem [9]. The idea is to transform 3-SAT into a specific instance of MPAS, such that solving the MPAS instance will also solve 3-SAT[2].

3-SAT is a decision problem from Boolean logic. Let $\chi_1, \dots, \chi_m, \dots, \chi_M$ be M Boolean variables. A literal is either a variable χ_m or its negation $\bar{\chi}_m$. The literals χ_m and $\bar{\chi}_m$ are true if and only if the variable χ_m is assigned to be true and false, respectively. A clause is a disjunction of exactly three literals, for example, $(\chi_1 \vee \bar{\chi}_5 \vee \chi_6)$. A clause is satisfied if and only if at least one of its literals is true. The problem is to determine if there is an assignment of Boolean values to the variables, such that all clauses hold true.

To make problem transformation, suppose we define one source for each Boolean variable and one for each clause. As before, each of these sources is associated with a link. All sources have a single data packet, and the initial AoI is a common value of a_0 for all sources. Consider any literal χ_m, and let $\mathcal{D}_m$ be the subset of clauses in which χ_m appears. Suppose there is a link group in MPAS formed by link χ_m and those links defined for the clauses in $\mathcal{D}_m$. Conversely, define a link group for $\bar{\chi}_m$ and the corresponding clauses $\bar{\mathcal{D}}_m$.

Let us assume there is a Boolean assignment satisfying all clauses. Consider again χ_m. Suppose it is has value true, meaning that all clauses in χ_m hold true irrespective

[2] To be complete, the reduction has to be polynomial, i.e., the size of the MPAS instance has to be a polynomial of that of 3-SAT. However, this part is straightforward.

of the other value assignments. We schedule link group $\{\chi_m\} \cup \mathcal{D}_m$ once, that is, source χ_m and those in $\mathcal{D}_m$ are emptied in one slot. Conversely, if χ_m is false or equivalently $\bar{\chi}_m$ is true, we schedule the group $\{\bar{\chi}_m\} \cup \mathcal{D}_m$ once. Because there are exactly M Boolean variables, the schedule needs exactly M time slots.[3] As a result, if the 3-SAT instance is satisfiable, the peak AoI is $a_0 + M - 1$. On the other hand, if the 3-SAT instance has no solution, then it is intuitive that at least $a_0 + M + 1$ time slots are needed, leading to a peak AoI higher than $a_0 + M - 1$. This is indeed the case. Thus, finding the optimum to MPAS leads us to the correct yes/no answer to 3-SAT. We refer to [10] for a detailed proof.

Consider a heavily interference-constrained scenario, such that only one link can be scheduled at a time. In other words, the schedule is of type time division multiple access (TDMA). Thus, for any time slot, scheduling amounts to selecting one source for transmission. Consider the first time slot. The packets subject to selection are $U_{11}, \ldots U_{N1}$ with time stamps $\tau_{11}, \ldots, \tau_{N1}$. Without loss of generality, assume that source one has the oldest age at current time slot, that is, $\tau_{10} \leq \tau_{n0}, n \geq 2$, where $\tau_{n0} = t_0 - a_{n0}$. If we select source one, by (9.3) and (9.4), the AoI a_1^1 equals $t_1 - \tau_{10}$. Suppose we defer the transmission of U_{11} to a latter time slot j; then a_1^1 increases to $t_j - \tau_{10}$ as $t_j > t_1$. Thus, it is intuitive that, in order to minimize the peak AoI, selecting link one is optimal. This intuition is indeed correct in general for a TDMA schedule.

THEOREM 2 *With TDMA, it is AoI-optimal to select the link corresponding to the source with the maximal age for every time slot.*

As an immediate corollary of Theorem 2, the TDMA case of MPAS is solved to optimality in polynomial time. The more precise complexity is $O(K \log(K))$, where $K = \sum_{n \in \mathcal{N}} K_n$, due to sorting the time stamps $\tau_{n,i-1}$, $i \in \mathcal{K}_n$, and $\tau_{n0} = t_0 - a_{n0}$.

We can, in fact, make an extension of the preceding observation to the general version of MPAS with link groups. Reusing the example, if for the first time slot we select a link group containing that for source one, the AoI a_1^1 again equals $t_1 - \tau_{10}$; otherwise, the schedule will end up with a higher a_1^1. Thus the same reasoning yields the conclusion that in each time slot, it is optimal to select a group such that the link corresponding to the source with the oldest age at current time can be transmitted. We present the result formally in what follows.

THEOREM 3 *At optimum of MPAS, the schedule has the structure that the link group for every time slot contains the link corresponding to the source with the oldest age.*

Full proofs of both Theorem 2 and Theorem 3 can be found in [10]. The result of Theorem 3 deserves a few remarks. First, the packet having the oldest time stamp must be at the head of its source's queue, because the packets of any source have ascending time stamps. Second, unlike the TDMA case, Theorem 3 does not give us a complete algorithm, as it tells only partial information of optimal group selection, and group selection and which packets are remaining are two intertwined components. However,

[3] The order of the scheduled groups has no effect on the peak AoI, because each source has the same initial age and one packet.

Theorem 3 provides some guideline for algorithm design, as an algorithm for MPAS should reasonably return a schedule satisfying the condition stated in the theorem.

9.3.3 Integer Programming Formulation

In this section we derive a mathematical formulation of MPAS based on integer linear programming (ILP) for two purposes. First, the ILP formulation provides a mathematically formal problem representation. Second, it allows for the computation of the global optimal solution for problem instances of small and moderate sizes, using off-the-shelf optimization tools [11, 12]. The formulation uses four sets of binary optimization variables as defined in what follows, in addition to the AoI variables a_{nj} we have already defined. Moreover, α is used as the auxiliary variable for the overall peak AoI.

$$x_{nij} = \begin{cases} 1 & \text{if packet } U_{ni} \text{ is delivered at } t_j, \\ 0 & \text{otherwise.} \end{cases}$$

$$y_{nj} = \begin{cases} 1 & \text{if link } n \text{ transmits in the } j\text{th time slot,} \\ 0 & \text{otherwise.} \end{cases}$$

$$z_{cj} = \begin{cases} 1 & \text{if group } c \text{ is used in the } j\text{th time slot,} \\ 0 & \text{otherwise.} \end{cases}$$

$$v_{nj} = \begin{cases} 1 & \text{if all packets of link } n \text{ have been delivered} \\ & \text{before/at } t_j, \\ 0 & \text{otherwise.} \end{cases}$$

Let T be the total number of packets, that is, $T = \sum_{n=1}^{N} K_n$. Obviously the schedule length for any MPAS will be at most T. We define $\mathcal{J} \triangleq \{1, 2, \ldots, T\}$, $\mathcal{J}_1 \triangleq \{0, 1, \ldots, T-1\}$, $\mathcal{J}_2 \triangleq \{1, 2, \ldots, T-1\}$, and $\mathcal{K}'_n \triangleq \{1, 2, \ldots, K_n - 1\}$, $\forall n \in \mathcal{N}$. The ILP formulation is given in (9.5).

The objective function defined in (9.5a) is to minimize α that represents the maximum AoI $a_{nj}, \forall n \in \mathcal{N}, j \in \mathcal{J}$, as defined by inequalities (9.5b). The next two sets of constraints, that is, (9.5c) and (9.5d), together achieve the effect of setting AoI a_{nj} as defined in (9.3). To see this, assuming in an optimal solution, no packet from source n is delivered at t_j and not all packets of S_n have been emptied yet, then $y_{nj} = v_{nj} = 0$. The corresponding constraint in (9.5c) reads $a_{nj} \geq a_{n,j-1} + 1$, and (9.5d) is void, as the right-hand side is nonpositive. Since the objective is to minimize the peak AoI and there are no additional constraints on a_{nj}, (9.5c) becomes active, that is, $a_{nj} = a_{n,j-1} + 1$. Next, suppose a packet, for example, U_{ni}, is delivered at t_j and it is not the last one of source n; then $x_{nij} = y_{nj} = 1$, and $v_{nj} = 0$. In this case, (9.5c) becomes void since the right-hand side is at most zero, and (9.5d) can be written as $a_{nj} \geq t_j - \tau_{ni}$, and evidently takes equality at the optimum as analyzed previously. Therefore, the constraint sets (9.5c) and (9.5d) give a linear form of (9.3). Moreover, one can verify that, if $v_{nj} = 1$, that is, all packets of S_n have been delivered, $a_{nj} = 0$ is optimal.

The inequalities in (9.5e) ensure that a schedule uses FCFS for any source. Namely, packet $U_{n,i+1}$ cannot be delivered before U_{ni}, $\forall n \in \mathcal{N}$, $\forall i \in \mathcal{K}'_n$. The constraints in (9.5f) state that when U_{ni} is delivered, the corresponding link must be in transmission. Moreover, by (9.5f), at most one packet from a source can be delivered in a time slot. The value of v_{nj} is constrained by (9.5g) and (9.5h). That is, $v_{nj} = 1$ if and only if the last packet of S_n is delivered. The qualities in (9.5h) also imply that all packets must be delivered by the end of a schedule. By (9.5i), a link can be active only if it is a member of the group that is assigned to the time slot in question. The constraint set (9.5j) states the fact that at most one group is selected in any time slot.

$$\min_{\{x_{nij},\, y_{nj},\, z_{cj},\, v_{nj} \in \{0,1\},\, \alpha,\, a_{nj} \in Z^*\}} \quad \alpha \tag{9.5a}$$

subject to

$$a_{nj} \leq \alpha \quad \forall n \in \mathcal{N},\ \forall j \in \mathcal{J}_1, \tag{9.5b}$$

$$a_{nj} \geq a_{n,j-1} + 1 - (y_{nj} + v_{nj})(a_{n0} + T) \quad \forall n \in \mathcal{N},\ \forall j \in \mathcal{J}_2, \tag{9.5c}$$

$$a_{nj} \geq t_j - \sum_{i \in \mathcal{K}'_n} \tau_{ni} x_{nij} - (1 - y_{nj} + v_{nj}) t_j \quad \forall n \in \mathcal{N},\ \forall j \in \mathcal{J}_2, \tag{9.5d}$$

$$x_{n,i+1,j} \leq \sum_{b=1}^{j} x_{nib} \quad \forall n \in \mathcal{N},\ \forall i \in \mathcal{K}'_n,\ \forall j \in \mathcal{J}, \tag{9.5e}$$

$$y_{nj} = \sum_{i \in \mathcal{K}_n} x_{nij} \quad \forall n \in \mathcal{N},\ \forall j \in \mathcal{J}, \tag{9.5f}$$

$$v_{nj} = \sum_{b=1}^{j} x_{n,K_n,b} \quad \forall n \in \mathcal{N},\ \forall j \in \mathcal{J}, \tag{9.5g}$$

$$v_{nT} = 1 \quad \forall n \in \mathcal{N}, \tag{9.5h}$$

$$y_{nj} \leq \sum_{c \in C_n} z_{cj} \quad \forall n \in \mathcal{N},\ \forall j \in \mathcal{J}, \tag{9.5i}$$

$$\sum_{c \in C} z_{cj} \leq 1 \quad \forall n \in \mathcal{N},\ \forall j \in \mathcal{J}. \tag{9.5j}$$

The reader has probably noticed that the size of (9.5) is significantly larger than that of (9.2) for minimum-length scheduling. The reason is quite obvious. For minimum-length schedule, the order of groups in the schedule is irrelevant, therefore an optimization formulation with given link groups does not need to represent a schedule on a slot-by-slot basis. In contrast, for minimum-AoI scheduling, the order of groups in a schedule is crucial. As a consequence, the optimization formulation has to keep track of the AoI evolution of all sources, time slot by time slot. We have mentioned that for (9.2), column generation leads to an efficient optimization algorithm. For AoI-driven scheduling, to what extent column generation can be used and its performance remain open questions.

9.4 Minimum-Energy Scheduling under Age Constraints

9.4.1 Problem Statement

An important alternative to minimizing the AoI is to minimize the energy consumption of data delivery, subject to a constraint on the average or the peak AoI. In this section we discuss this problem, which we refer to as the minimum-energy scheduling under age constraints (MESA) problem.

Let us denote by A_n the maximum allowed age for source S_n, $\forall n \in \mathcal{N}$, and by $v(n, c)$ the (integer) number of packets that can be transmitted in a time slot on link n if group c is active. Usually $v(n, c)$ is a nondecreasing function of the SINR of link n and is determined by the network configuration and the modulation and coding scheme. Let us now define the binary variable

$$z_{cj} = \begin{cases} 1 & \text{link group } c \text{ is active in time slot } j; \\ 0 & \text{otherwise,} \end{cases} \tag{9.6}$$

and define $\iota_j = \sum_{c \in \mathcal{C}} z_{cj}$, that is, $\iota_j = 1$ if and only if there is a link transmitting during the jth time slot. Intuitively, any transmission schedule that contains an empty time slot can be shortened by skipping the empty time slot, and thus $\iota_j = 1$ during every slot of a transmission schedule. Consequently, the total length of any transmission schedule is at most $\sum_{n \in \mathcal{N}} K_n$. We can leverage this observation for formulating MESA by defining the index set $\mathcal{J} = \{1, 2, \ldots, \sum_{n \in \mathcal{N}} K_n\}$.

$$\min_{\{z_{cj},\ \iota_j \in \{0,1\}\}} \sum_{c \in \mathcal{C},\ j \in \mathcal{J}} P_c z_{cj} \tag{9.7a}$$

subject to (9.3),

$$P_c = \sum_{n \in c} P_n \quad \forall c \in \mathcal{C}, \tag{9.7b}$$

$$\sum_{c \in \mathcal{C},\ j \in \mathcal{J}} v(n, c) z_{cj} \geq K_n \quad \forall n \in \mathcal{N}, \tag{9.7c}$$

$$\sum_{c \in \mathcal{C}} z_{cj} = \iota_j \quad \forall j \in \mathcal{J}, \tag{9.7d}$$

$$\iota_{j-1} \geq \iota_j \quad \forall j \in \mathcal{J} \setminus \{1\}, \tag{9.7e}$$

$$a_{nj} \iota_j \leq A_n \quad \forall n \in \mathcal{N}, \forall j \in \mathcal{J}. \tag{9.7f}$$

The objective function (9.7a) is the total transmission energy in a scheduling solution, computed based on the transmit power P_c of group c, given by (9.7b). The constraints in (9.7c) ensure that all packets are delivered by the end of a schedule. Constraint (9.7d) allows one to identify empty slots and guarantees that at most one group is active in a time slot. The constraint (9.7e) ensures that there are no empty time slots during a transmission schedule. Putting together (9.7d) and (9.7e) implies that $\iota_j = 0$ indicates that all queued packets are delivered and the calculated schedule ends before time slot j. By definition, the maximum peak age is also the maximum age during the schedule, and hence we have the maximum age constraints defined in (9.7f). Note that once all packets are delivered the constraint (9.7f) does not take effect because $\iota_j = 0$. Therefore, we can overestimate the schedule length $|\mathcal{J}|$ in (9.7) without affecting optimality.

9.4.2 Structural Results

MESA too is a combinatorial optimization problem, similar to MPAS. While it is known that minimum-energy scheduling without age constraints is computationally tractable [13], that is, it is solvable in polynomial time, enforcing an age constraint makes that the worst case complexity of MESA is not polynomial.

THEOREM 4 *MESA is NP-hard.*

Here we give a sketch of the proof and provide insights into the MESA problem. We first present a key observation on MESA. For any link $n \in \mathcal{N}$, to empty its queued packets K_n, the energy to be consumed can be written as $\frac{P_n K_n}{\bar{v}_n}$, where $\bar{v}_n$ is the average rate when link n is active. Next, consider a special case of MESA where the links always transmit at rate one (if active), that is, $v(n, c) = 1, \forall n \in \mathcal{N}, \forall c \in \mathcal{C}$. For this case, we have $\bar{v}_n = 1, \forall n \in \mathcal{N}$ and hence the energy consumption of link n is constant. Consequently, for any feasible schedule, the total energy consumption for all links, that is, the objective cost of (9.7), equals $\sum_{n \in \mathcal{N}} P_n K_n$. Therefore, the optimization problem of (9.7) is in fact to determine whether or not there is a feasible solution such that the age constraints in (9.7f) are satisfied. Letting $A_n = A, \forall n \in \mathcal{N}$ makes (9.7) become the decision problem of MPAS, which was shown to be NP-hard in Section 9.3.2. A complete proof of Theorem 4 is provided in [14].

The preceding reasoning leads to two observations. First, it implies that one could hardly expect that the use of simple rate functions, such as binary, would make the problem computationally tractable. Second, for links sharing the same channel, due to their mutual interference, the highest rate of link n is achieved when it transmits alone. Clearly, for any feasible MESA instance, $v(n, \{n\}) > 0$. On the contrary, the lowest rate $v_{n,min}$ of link n (among the groups with this link) is either obtained when all links transmit simultaneously, that is, $\mathcal{N} \in \mathcal{C}$, equalling $v(n, \mathcal{N})$, or at the SINR threshold of RX_n. Therefore, for the average activation rate of link n we have $v(n, \{n\}) \geq \bar{v}_n \geq v_{n,min}$. Considering the fact that to empty the demand of link n, the energy consumption equals $\frac{P_n K_n}{\bar{v}_n}$, we can derive upper and lower bounds of the energy consumption for a MESA instance.

THEOREM 5 *Consider a feasible instance of MESA. The objective value E satisfies* $\sum_{n \in \mathcal{N}} \frac{P_n K_n}{v(n,\{n\})} \leq E \leq \sum_{n \in \mathcal{N}} \frac{P_n K_n}{v_{n,min}}$, *where* $v_{n,min} = v(n, \mathcal{N})$ *if* $\mathcal{N} \in \mathcal{C}$, *otherwise* $v_{n,min} = f(\gamma_n)$ *where* γ_n *is the SINR threshold for link n to be able to transmit and* $f(.)$ *is the rate function.*

Let us now consider the TDMA schedule, which is known to be the optimal solution for minimum-energy scheduling [13] as well as serving as a lower bound of MESA. Note that for minimum-energy scheduling and minimum-energy scheduling with total transmission time constraint the order of activating the selected link groups can be arbitrary, but this is not the case for MESA. Hence, a fundamental question is to understand when a TDMA schedule is optimal for MESA and in which order the links should be activated. In the following theorem we provide a sufficient and necessary condition for when TDMA is optimal for MESA, together with the optimal link activation order.

THEOREM 6 *Define $\mathcal{T}^1$ as the ordered TDMA schedule where the links are activated in the following order: starting from $j = 1$, in the jth time slot, link $n = \text{argmin}_{n \in \mathcal{N}_j}(A_n - a_{n,j-1})$ is active. Here $\mathcal{N}_j$ is the set of links with non-empty demand at t_{j-1}. Ties, if any, can be broken arbitrarily. For any MESA instance, the optimal solution is TDMA if and only if $\mathcal{T}^1$ is feasible.*

The outline of the flow of arguments is as follows. First, sufficiency follows directly since the solution space of MESA is a subset of the corresponding minimum-energy scheduling problem and $\mathcal{T}^1$ is optimal for the latter. For the necessity, we employ an indirect proof. Suppose that there is a feasible TDMA solution with another link activation order. One could then verify that swapping any two links that are not activated following the defined order preserves feasibility, and hence we could obtain $\mathcal{T}^1$ after a finite number of swaps. We refer to [14] for all details of the proof. Following Theorem 6, it is possible to recognize in polynomial time whether TDMA is feasible (i.e., the sufficient and necessary condition), and an optimal solution can be constructed in polynomial time too. Therefore we remark that MESA becomes tractable if TDMA is allowed.

9.4.3 Integer Programming Formulation

In this section we formulate MESA as an ILP to enable the computation of the optimal solution for problem instances of small and moderate sizes, using off-the-shelf optimization tools [11, 12]. For the ILP formulation we define the following binary variables, and we formulate the ILP in (9.8).

$$x_{nij} = \begin{cases} 1 & \text{if packet } U_{ni} \text{ is delivered at } t_j, \\ 0 & \text{otherwise.} \end{cases}$$

$$w_{nij} = \begin{cases} 1 & \text{if } U_{ni} \text{ is the newest one among all packets delivered for link } n \\ & \text{in the } j\text{th time slot,} \\ 0 & \text{otherwise.} \end{cases}$$

$$y_{nj} = \begin{cases} 1 & \text{if link } n \text{ transmits in the } j\text{th time slot,} \\ 0 & \text{otherwise.} \end{cases}$$

$$z_{cj} = \begin{cases} 1 & \text{if group } c \text{ is used in the } j\text{th time slot,} \\ 0 & \text{otherwise.} \end{cases}$$

$$\min_{\{x_{nij},\, w_{nij},\, y_{nj},\, z_{cj} \in \{0,1\},\, a_{nj} \in Z^*\}} \sum_{c \in \mathcal{C},\, j \in \mathcal{J}} P_c z_{cj} \tag{9.8a}$$

subject to

$$P_c = \sum_{n \in c} P_n \quad \forall c \in \mathcal{C}, \tag{9.8b}$$

$$a_{nj} \geq a_{n,j-1} + 1 - y_{nj}(a_{n0} + \sum_{n \in \mathcal{N}} K_n) \quad \forall n \in \mathcal{N}, \forall j \in \mathcal{J}, \tag{9.8c}$$

$$a_{nj} \geq t_j - \sum_{i \in \mathcal{K}'_n} \tau_{ni} w_{nij} - (1 - y_{nj}) t_j \quad \forall n \in \mathcal{N}, \forall j \in \mathcal{J}, \tag{9.8d}$$

$$x_{n,i+1,j} \leq \sum_{b=1}^{j} x_{nib} \quad \forall n \in \mathcal{N}, \forall i \in \mathcal{K}'_n, \forall j \in \mathcal{J}, \tag{9.8e}$$

$$\sum_{i \in \mathcal{K}_n} x_{nij} \leq v(n,c) z_{cj} \quad \forall n \in \mathcal{N}, \forall j \in \mathcal{J}, \tag{9.8f}$$

$$3x_{nij} - x_{n,i,j+1} - 2 \leq 4w_{nij} \leq 3x_{nij} - x_{n,i,j+1} + 1$$
$$\forall n \in \mathcal{N}, \forall i \in \mathcal{K}_n, \forall j \in \mathcal{J} \setminus \{\sum_{n \in \mathcal{N}} K_n\}, \tag{9.8g}$$

$$y_{nj} = \sum_{i \in \mathcal{K}_n} w_{nij} \quad \forall n \in \mathcal{N}, \forall j \in \mathcal{J}, \tag{9.8h}$$

$$y_{nj} \leq \sum_{c \in \mathcal{C}_n} z_{cj} \quad \forall n \in \mathcal{N}, \forall j \in \mathcal{J}, \tag{9.8i}$$

$$\sum_{c \in \mathcal{C}, j \in \mathcal{J}} v(n,c) z_{cj} \geq K_n \quad \forall n \in \mathcal{N}, \tag{9.8j}$$

$$\sum_{c \in \mathcal{C}} z_{cj} - \iota_j \quad \forall j \in \mathcal{J}, \tag{9.8k}$$

$$\iota_{j-1} \geq \iota_j \quad \forall j \in \mathcal{J} \setminus \{1\}, \tag{9.8l}$$

$$a_{nj} \iota_j \leq A_n + (1 - \iota_j)(a_{n0} + \sum_{n \in \mathcal{N}} K_n) \quad \forall n \in \mathcal{N}, \forall j \in \mathcal{J}. \tag{9.8m}$$

Observing the problem formulation (9.7), the key to modeling MESA as an ILP is to linearize the age calculation with given transmission rates. To this end, we introduce the binary variable w_{nij} and represent (9.3) by the constraint sets (9.8c)–(9.8d) following a method similar to that used in Section 9.3.3. The inequalities in (9.8e) guarantee the FCFS order. The constraint sets (9.8f)–(9.8i) model the variables x_{nij}, y_{nj}, z_{cj}, and w_{nij}, with rate v_{nc}. The constraints in (9.8j)–(9.8l) are the same as those defined in (9.7). Finally, the linearization of the age constraints is performed in (9.8m), using the big M method.

9.5 Age of Correlated Information

9.5.1 Motivation and Definition

In the previous sections the assumption was that the AoI changes upon receiving each individual packet. This is based on the underlying assumption that information carried by the packets is independent of each other and thus the receiver could derive new information, for example, status update, upon receiving each packet.

There are, however, many applications where information update requires multiple packets carrying correlated information from one or more information sources. A typical example is an application involving state estimation based on physical models. Consider the example of a wireless camera network, where images from cameras with overlapping field of view (FoV) have to be processed jointly, as doing so can improve the robustness and accuracy of tracking, and enables 3D scene reconstruction [15]. In this case an information update cannot be triggered by a single packet carrying one image nor by packets carrying images captured by a single camera. In addition, to enable timely data delivery and real-time processing of the visual information, it may be necessary to perform computations close to the camera, for example, leveraging the concept of fog computing. In order to minimize the age of the multi-view image data at the fog nodes used for processing, in such a system one needs to consider two coupled resource allocation problems. First, one needs to determine the serving fog node for each camera, which we refer to as the camera-node assignment problem. Second, one needs to solve the scheduling problem for the mutually interfering cameras with overlapping FoVs. We refer to this as camera transmission scheduling. Clearly, the two problems are coupled and require joint optimization for minimizing the age of information.

In view of the preceding, when processing requires information from multiple senders, carried in different packets, such as the case of cameras with overlapping FoVs, age of information should change only when all packets carrying correlated information are received. In this section, we extend the notion of AoI to capture packets carrying correlated data. We refer to the new notion of AoI as the *age of correlated information (AoCI)* and define it as the time difference between the actual time and the generation time of the latest “fully" received correlated set of packets. In Figure 9.4, we illustrate age evolution of a source under this new definition. We remark that the packets carrying correlated information may not be delivered consecutively since they are sent from multiple senders.

9.5.2 Minimizing Age of Correlated Information

We consider a wireless system consisting of a set of sources $\mathcal{S} = \{1, 2, \ldots, S\}$, a set of senders $\mathcal{C} = \{1, 2, \ldots, C\}$, and a set of processing nodes $\mathcal{N} = \{1, 2, \ldots, N\}$. Information from each source is captured by multiple senders and is delivered to the respective processing node via a shared wireless channel. Figure 9.5 illustrates the model for the use case of a wireless camera network.

We denote by $c(s)$ the set of senders that capture correlated information from source s. The senders queue and send the captured data to their respective processing nodes using FCFS. We define the binary variable

$$l_{cn} = \begin{cases} 1 & \text{if information from sender } c \text{ is processed by node } n; \\ 0 & \text{otherwise.} \end{cases} \tag{9.9}$$

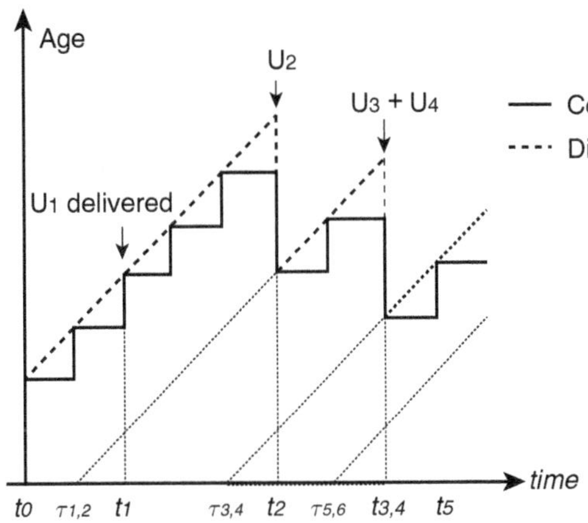

Figure 9.4 Age evolution of correlated information of a source. Packets U_1 and U_2 carrying correlated information were generated at $\tau_{1,2}$ and received by the destination node at t_1 and t_2, respectively. The age increases linearly until t_2, when it decreases to $t_2 - \tau_{1,2}$. After t_2, the age increases linearly until $t_{3,4}$, when packets U_3 and U_4 carrying correlated information and with time stamp $\tau_{3,4}$ were simultaneously received by the destination. The age at $t_{3,4}$ equals $t_{3,4} - \tau_{3,4}$.

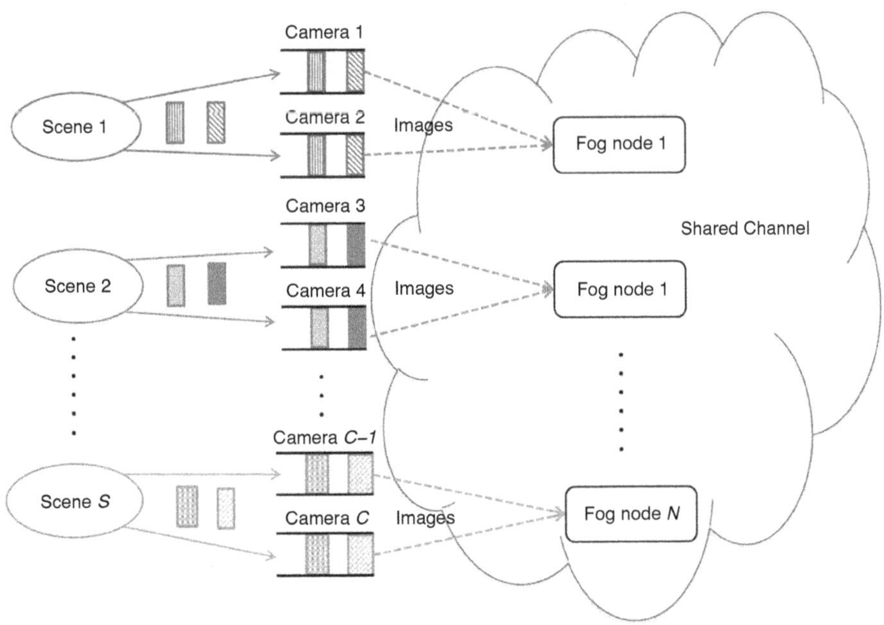

Figure 9.5 Illustration of a camera network, where each source is monitored by at least two cameras, i.e., senders, as indicated by the solid lines, and each camera transmits the captured images to its destination fog node, as shown by the dashed lines.

To allow the processing of correlated information, the senders $c(s)$ that are associated with source s need to transmit their queued packets to the same processing node; hence we have

$$\sum_{n \in N} \prod_{c \in c(s)} l_{cn} = 1, \ \forall s \in \mathcal{S}. \tag{9.10}$$

The processing nodes are responsible for receiving and processing the data from the senders they serve. To capture computing capacity constraints, we denote by M_n the maximum number of senders that can be supported by processing node n, that is,

$$\sum_{c \in C} l_{cn} \leq M_n, \ \forall n \in \mathcal{N}. \tag{9.11}$$

The values of M_n, $\forall n \in \mathcal{N}$, depend on the capacity of each node and could serve the purpose of workload balancing.

We denote by θ a sender group and by Θ the set of all feasible sender groups under a given interference model and network configuration. The transmission rate of a sender–processing node pair in a feasible group is one information packet per time slot. We denote by U_{ci} the ith packet in the queue of sender c, and by K_c the number of queued packets of sender c at the starting time of a schedule cycle. Since the senders that are associated with the same source capture information simultaneously, the ith packet in the queue of each sender $c \in c(s)$ carries time stamp τ_{si}, which indicates the generation time of packet U_{ci}, $\forall c \in c(s)$. We refer to the set of packets $B_{si} = \{U_{ci}, \forall c \in c(s)\}$ as the ith packet block of s. Due to the requirement of correlated information processing, at the processing node n serving $c(s)$, the information of source s will not be updated until all the packets of a packet block are delivered to this processing node. Therefore, at time t_j the age of a source s is defined as

$$a_{sj} = \begin{cases} t_j - \tau_{si} & \text{if all the packets of } B_{si} \text{ have been delivered} \\ & \text{to node } n \text{ exactly by } t_j; \\ a_{s,j-1} + 1 & \text{otherwise.} \end{cases} \tag{9.12}$$

By the definition in (9.12), the peak ages of a source are attained immediately before the last packet of an packet block is delivered to its serving processing node. To express the peak ages, let us denote by t_{ci} the time when packet U_{ci} is delivered, and by $T_{ci} = t_{ci} - t_0$ the number of time slots before U_{ci} is delivered, and observe that T_{ci} is a positive integer in $[1, \sum_{c \in C} K_c]$. Then by defining $\tau_{s0} = t_0 - a_{s0}$, we can express the ith peak age of s as $\alpha_s^i = \max_{c \in c(s)} t_{ci} - \tau_{s,i-1} = t_0 + \max_{c \in c(s)} T_{ci} - \tau_{s,i-1}$.

Now we provide the mathematical formulation of the minimum correlated age scheduling problem (MCAS).

$$\min_{\{T_{ci} \in Z^+,\, l_{cn} \in \{0,1\}\}} \ \max_{s \in \mathcal{S},\, c \in c(s),\, i=1,\ldots,K_c} \alpha_s^i \tag{9.13a}$$

subject to (9.10), (9.11), and

$$\alpha_s^i = t_0 + \max_{c \in c(s)} T_{ci} - \tau_{s,i-1} \ \ \forall s \in \mathcal{S},\ i = 1, \ldots, K_c, \tag{9.13b}$$

$$1 \le T_{c1} < T_{c2} < \cdots < T_{c,K_c} \quad \forall c \in \mathcal{C}, \tag{9.13c}$$

$$\theta_j \in \Theta \quad \theta_j = \{c \in \mathcal{C} : T_{ci} = T_j, i = 1, \ldots, K_c\},\ T_j = 1, \ldots, \sum_{c \in \mathcal{C}} K_c. \tag{9.13d}$$

Note that at the optimum of (9.13), if the scheduling solution uses T time slots and $T < \sum_{c \in \mathcal{C}} K_c$, then θ_j are empty sets for $j = \{T + 1, \ldots, \sum_{c \in \mathcal{C}} K_c\}$. Hence in (9.13d), T can be overestimated without loss of optimality. We remark that the candidate group set Θ is determined by the interference model as well as the sender-processing node assignment. Hence solving MCAS requires joint optimization of the sender to processing node assignment, and of the transmission schedule of the senders to their serving processing nodes.

9.5.3 Structural Results

Akin to MPAS and to MESA, MCAS is a computationally intractable combinatorial optimization problem.

THEOREM 7 *MCAS is NP-hard.*

The result can be proved by a polynomial reduction from the 3-SAT problem, as shown in [16].

Next, we consider MCAS for two baseline scenarios: a wireless network with interference-free channels and with severely interference-limited channels, respectively. In the first scenario, we consider that for each sender there exists at least one processing node such that the SINR exceeds its threshold when all senders are active, thus the network allows all senders to transmit simultaneously. This can be a reasonable model for emerging millimeter wave communications, and is also a good model for a system in which the senders transmit with low power and each sender is deployed close to its serving processing node. The second scenario is the very opposite case in which only one sender is allowed to transmit in a time slot, that is, TDMA. This can be a reasonable model for severely interference-limited networks, for example, when the senders and nodes are densely deployed, and hence senders cause significant interference to each other.

Note that in our problem setup, the serving processing node of a sender is a decision variable. Therefore, for the two preceding baseline scenarios, all senders transmitting simultaneously and TDMA might be infeasible for their corresponding MCAS instances, due to the requirement of multi-view processing and/or due to the processing node capacity constraint. In the following theorems we show MCAS remains hard even for these two scenarios.

THEOREM 8 *The MCAS problem in which all senders transmit simultaneously is NP-hard.*

THEOREM 9 *The MCAS problem in which only TDMA is allowed is NP-hard.*

Both Theorems 8 and 9 can be proved by establishing a polynomial-time reduction from the partition problem, which is known to be NP-complete [9]. For full proofs, please refer to [16]. We now proceed to identifying tractable cases. For the preceding

Table 9.1 Summary of structural results for MCAS.

MCAS case	Arbitrary number of senders per source	Same number of senders per source
Under the physical model	NP-hard	NP-hard
Under the protocol model	NP-hard	NP-hard
With given candidate groups	NP-hard	NP-hard
With all senders transmitting together	NP-hard	tractable
With TDMA	NP-hard	tractable
Only senders with the same source are capable of transmitting together	NP-hard	tractable

two baseline scenarios, we provide sufficient conditions for MCAS instances to be recognizable and solvable in polynomial-time. A thorough tractability analysis can be found in [16]. In Table 9.1, we give a summary of the structural results.

Finally, we provide a general optimality condition of the transmission schedule that applies to all MCAS instances.

THEOREM 10 *Given a transmission schedule, let us denote by Λ_j the set of packets delivered in time slot j, by T the schedule length, and let $v_j = min\{\tau_{s,i-1} : U_{ci} \in \Lambda_j, c \in c(s)\}$. Then for any instance of the MCAS, there exists an optimal schedule in which $v_j, j = 1, 2, \ldots, T$, are nondecreasing.*

Theorem 10 is in fact an extension of Theorem 3 to the MPAS problem, as discussed in 9.3.2. We refer to [16] for a full proof of Theorem 10.

9.5.4 Integer Programming Formulation

We now present an ILP for MCAS under the physical model to solve problem instances of small and moderate sizes, and for effective benchmarking of other, suboptimal algorithms. We define the following variables for the ILP.

$$x_{cij} = \begin{cases} 1 & \text{if packet } U_{ci} \text{ is delivered at } t_j, \\ 0 & \text{otherwise.} \end{cases}$$

$$y_{sij} = \begin{cases} 1 & \text{if any packet } U_{ci}, \forall c \in c(s), \text{ is delivered at } t_j, \\ 0 & \text{otherwise.} \end{cases}$$

$$z_{sj} = \begin{cases} 1 & \text{if all packets of senders } c(s) \text{ have been delivered before/at } t_j, \\ 0 & \text{otherwise.} \end{cases}$$

$$v_{sij} = \begin{cases} 1 & \text{if the last packet of } B_{si} \text{ is delivered at } t_j, \\ 0 & \text{otherwise.} \end{cases}$$

$$\iota_{sn} = \begin{cases} 1 & \text{if the senders } c \in c(s) \text{ are served by node } n, \\ 0 & \text{otherwise.} \end{cases}$$

Letting $\mathcal{J} = \{1, 2, \ldots, T\}$, $\mathcal{K}_c = \{1, 2, \ldots, K_c\}$, and $\mathcal{K}_s = \mathcal{K}_c$, $c \in c(s)$, we formulate the ILP in (9.14). In the objective function we minimize the maximal achievable age for all sources during the schedule cycle. The constraint sets (9.14c) and (9.14d) are linearizations of the age calculation stated in (9.12). By definition, if the last packet of a packet block B_{si} is not delivered at t_j, then $v_{sij} = 0$. For this source, if not all the packets are emptied at the moment, that is, $z_{sj} = 0$, then the corresponding constraint (9.14c) reads $a_{sj} \geq a_{s,j-1} + 1$. If either v_{sij} or z_{sj} (or both) equals one, the right-hand side of (9.14c) is negative because $a_{s0} + T$ is an upper bound of the age of s. In this case, the constraints in (9.14c) take no effect. If the last packet of a packet block B_{si} is delivered at t_j then $v_{sij} = 1$ and (9.14d) is written as $a_{sj} \geq t_j - \tau_{si}$ when $z_{sj} = 0$. If $z_{sj} = 1$, that is, all the packets of s have been delivered, both (9.14c) and (9.14d) become satisfied. Note that there are no constraints on a_{sj} in (9.14e)–(9.14o), and the objective is to minimize the maximal value of a_{sj}. Therefore, at the optimum, for any constraint in (9.14c) and (9.14d) taking effect, equality holds. Hence, the constraint sets (9.14c) and (9.14d) together give the same result as defined in (9.12).

The constraints in (9.14e) ensure the FCFS order of the queues, that is, the packet $U_{c,i+1}$ can be transmitted only if U_{ci} has been delivered. In addition, by (9.14e), one sender delivers at most one packet in each slot $j = 2, \ldots, T$. The transmission rate in the first time slot is governed by (9.14f). The inequalities in (9.14g)–(9.14j) achieve the effect that the auxiliary variables y_{sij}, z_{sj}, and v_{sij} indeed take the desired values. By (9.14g), y_{sij} takes value one if and only if at least one sender $c \in c(s)$ delivers its ith packet at t_j. The value of v_{sij} is defined by (9.14h) and (9.14i). By definition, $v_{sij} = 1$ indicates that all the packets of the ith view of s have been delivered by t_j and there is U_{ci}, $\forall c \in c(s)$, being delivered at t_j. Only when both conditions are fulfilled, the right-hand side of (9.14h), that is, $\sum_{c \in c(s)} \sum_{j'=1}^{j} x_{cij'} + y_{sij}$ achieves its maximal value $|c(s)| + 1$. Otherwise, it is less than this value, resulting in $v_{sij} = 0$. If the two conditions are satisfied, then we must have $v_{sij} = 1$ to fulfill (9.14i). The values of z_{sj} are set by (9.14j), that is, $z_{sj} = 1$ if and only if all packets of source s are delivered by t_j. The constraints in (9.14k) enforce that all packets are delivered to their respective nodes eventually.

$$\min_{\{x_{cij},\, y_{sij},\, z_{sj},\, v_{sij},\, \iota_{sn},\, l_{cn} \in \{0,1\},\, a_{sj} \in Z^*\}} \quad \alpha \tag{9.14a}$$

subject to

$$a_{sj} \leq \alpha \quad \forall s \in \mathcal{S},\ \forall j \in \{0\} \cup \mathcal{J}, \tag{9.14b}$$

$$a_{sj} \geq a_{s,j-1} + 1 - \Big(\sum_{i \in K_s} v_{sij} + z_{sj}\Big)(a_{s0} + T) \quad \forall s \in \mathcal{S},\ \forall j \in \mathcal{J}, \tag{9.14c}$$

$$a_{sj} \geq t_j - \sum_{i \in \mathcal{K}_s} \tau_{si} v_{sij} - \Big(1 - \sum_{i \in \mathcal{K}_s} v_{sij} + z_{sj}\Big) t_j \quad \forall s \in \mathcal{S},\ \forall j \in \mathcal{J}, \tag{9.14d}$$

$$x_{c,i+1,j} \leq \sum_{j'=1}^{j-1} x_{cij'} \quad \forall c \in \mathcal{C},\ \forall i \in \mathcal{K}_c \setminus \{K_c\},\ \forall j \in \mathcal{J} \setminus \{1\}, \tag{9.14e}$$

$$\sum_{i \in \mathcal{K}_c} x_{ci1} \leq 1 \quad \forall c \in \mathcal{C}, \tag{9.14f}$$

$$\sum_{c \in c(s)} x_{cij} \leq |c(s)| y_{sij} \leq |c(s)| \sum_{c \in c(s)} x_{cij} \quad \forall s \in \mathcal{S},\ \forall i \in \mathcal{K}_c,\ \forall j \in \mathcal{J}, \tag{9.14g}$$

$$(|c(s)|+1)v_{sij} \leq \sum_{c\in c(s)}\sum_{j'=1}^{j} x_{cij'} + y_{sij} \quad \forall s \in \mathcal{S},\ \forall i \in \mathcal{K}_c,\ \forall j \in \mathcal{J}, \tag{9.14h}$$

$$(|c(s)|+1)v_{sij} \geq \sum_{c\in c(s)}\sum_{j'=1}^{j} x_{cij'} + y_{sij} - |c(s)| \quad \forall s \in \mathcal{S},\ \forall i \in \mathcal{K}_c,\ \forall j \in \mathcal{J}, \tag{9.14i}$$

$$z_{sj} = \sum_{j'=1}^{j} v_{s,K_c,j'} \quad \forall s \in \mathcal{S},\ \forall j \in \mathcal{J},\ c \in c(s), \tag{9.14j}$$

$$z_{sT} = 1 \quad \forall s \in \mathcal{S}, \tag{9.14k}$$

$$\sum_{n\in\mathcal{N}} \iota_{sn} = 1 \quad \forall s \in \mathcal{S}, \tag{9.14l}$$

$$l_{cn} = \iota_{sn} \quad \forall s \in \mathcal{S},\ \forall c \in c(s),\ \forall n \in \mathcal{N}, \tag{9.14m}$$

$$\sum_{c\in\mathcal{C}} l_{cn} \leq M_n \quad \forall\, n \in \mathcal{N}, \tag{9.14n}$$

$$P_c G_{cn} + (2 - l_{cn} - \sum_{i\in\mathcal{K}_c} x_{cij}) Q_{cn} \geq \gamma_c(\sum_{c'\neq c}(P'_c G_{c'n} \sum_{i\in\mathcal{K}_{c'}} x_{c'ij}) + \sigma_n^2)$$
$$\forall c \in \mathcal{C},\ \forall c \in \mathcal{N},\ \forall j \in \mathcal{J}. \tag{9.14o}$$

The optimization task of sender-node assignment is constrained by (9.14l)–(9.14n). The equalities in (9.14l) ensure that each sender set $c(s)$ is served by a node. By the constraints in (9.14m) and the definition of l_{cn} in (9.9), all senders in $c(s)$ are connected to the same node. In (9.14n) we define the capacity limit for each processing node. The SINR constraints are defined by the inequalities in (9.14o), where Q_{cn} is a positive parameter that is large enough to guarantee that the corresponding constraint is satisfied if sender c does not transmit at this time slot, that is, $\sum_{i\in\mathcal{K}_c} x_{cij} = 0$, or n is not the intended node for c, that is, $l_{cn} = 0$. For this purpose, we set $Q_{cn} = \gamma_c(\sum_{c'\neq c} P'_c G_{c'n} + \sigma_n^2)$. When any constraint of (9.14o) becomes active, that is, $\sum_{i\in\mathcal{K}_c} x_{cij} = l_{cn} = 1$, the corresponding inequality reads $P_c G_{cn} \geq \gamma_c(\sum_{c'\neq c}(P'_c G_{c'n} \sum_{i\in\mathcal{K}_{c'}} x_{c'ij}) + \sigma_n^2)$, of which the left-hand side is the signal strength of c and the right-hand side is the SINR threshold multiplied by the interference and noise at node n in the jth time slot.

9.6 Algorithmic Solutions

Given that the scheduling problems MPAS, MESA, and MCAS are all NP-hard in general, it is necessary to design computationally feasible algorithms even though they may provide suboptimal results. A reasonable candidate would be a greedy-type algorithm that works on a slot-by-slot basis and in each time slot selects the link group that minimizes the objective function with respect to the current slot. This underlying idea leads to the following algorithms for the problems.

- For MPAS, the greedy selection amounts to using a link group containing a link that reduces the maximal AoI among all sources, by transmitting the first packet currently residing at the queue of the link. Moreover, the algorithm shall take in account the age reduction of other sources as well as the schedule length of emptying all queues.

- For MESA, in order to minimize the energy function, a greedy algorithm would select a link group with as few links as possible, because for the problem a TDMA-type solution is optimal if it is feasible (Theorem 6). The algorithm would, however, have to account for the AoI constraints, that is, the algorithm would necessarily select a link group to include links such that the AoI values of the corresponding sources are below the respective upper bounds.
- For MCAS, the solution consists of sender-node assignment in addition to scheduling. Hence we consider a two-step algorithm. The first step deals with source to processing node assignment. This assignment can either be based on maximizing the number of senders that can transmit simultaneously, or based on an AoI-aware approach such that senders containing packets with small time stamps (and thus higher AoI) can transmit together. For scheduling, the algorithm would select a group with priority given to the senders with packets from older sources and carrying smaller time stamps in every time slot.

Based on the preceding design principle, we propose the following optimization algorithms for the three problems.

- For MPAS, starting from the first time slot, at each step we sort $\tau_{n,i-1}$, $\forall n \in \mathcal{N}$, in ascending order, where τ_{ni} denotes the time stamp of the first packet to be served in queue n in the considered time slot. Then the smallest one over the N sources, say $\tau_{l,i-1}$, is selected. We check the candidate group set $\mathcal{C}$ and put all the groups containing link l in a new subset $\mathcal{C}_1$. If there is only one group in $\mathcal{C}_1$, that is, $|\mathcal{C}_1| = 1$, then that group is the one selected for this time slot. If $|\mathcal{C}_1| > 1$, the group with the maximum cardinally in $\mathcal{C}_1$ is chosen. If there are ties remaining, the algorithm uses the second minimal time stamp in each group to break the tie and continues in the same fashion until the tie is broken.
- For MESA, starting from t_0, at the jth time slot the algorithm sorts the non-empty links $\mathcal{N}_j$ in ascending order of $A_n - a_{n,j-1}$. Then the algorithm will first check if there are any "must-do" links, that is, links that if they were not active in this time slot then information from at least one source would expire.[4] If there is only one "must-do" link in this time slot, then evidently the singleton-link group consisting of that link will be the choice. If multiple links have to be active in the jth time slot, then the algorithm will employ a backward construction, as proposed in [14]. If there are no "must-do" links, the algorithm selects the most urgent link, defined as $n = argmin_{n \in \mathcal{N}_j}(A_n - a_{n,j-1})$ and activates the singleton-link group $\{n\}$ in the

[4] There are two possible scenarios: (1) link n has packets in its queue and $A_n - a_{n,j-1} = 0$, then link n must be active at slot j; (2) there exists at least one source, for which the gap between its current age and the corresponding age requirement has decreased to one, and all packets from that source have been delivered. In this case the current schedule needs to be ended in one time slot, so that the age constraint for that source can be met.

jth time slot. We refer to this algorithm as the deadline-first-with-revision (DFR) algorithm.

- For MCAS, we propose the correlated maximum age first (CMAF) algorithm. The algorithm is based on a decomposition of the MCAS problem: it first solves the sender-to-node assignment problem, and for a given assignment it computes a transmission schedule. For the first optimization task, CMAF uses an SINR-based assignment and an age-aware assignment algorithm for computing two sender-node assignments. For both assignments, it executes the following greedy scheduling algorithm: in each time slot, the algorithm first sorts the sources $\mathcal{S}$ in descending order of their current ages, and then it sorts the senders associated with each source, that is, $c(s)$, in ascending order of the time stamps of the packets to be delivered. Ties, if any, are broken in ascending order of source/sender index. The scheduling algorithm adds the sender on the top of the list, that is, the sender associated with the most "aged" source and carrying the packet with the lowest time stamp, to the sender group. It then iterates through the ordered list of senders and adds one sender at a time. In each step, if the SINR threshold is met, then the new added sender is kept; otherwise, it is removed from the group. We refer to [16] for full details of the CMAF algorithm. The algorithm then chooses the better of the solutions among the assignments.

In what follows we show numerical results based on simulations in a wireless network with $N = 30$ links, each associating with an information source. The links are randomly distributed in an area of $1{,}000 \times 1{,}000$ meters. To obtain links with practically meaningful SNR values, the distance between any transmitter and its associated receiver is restricted to be between 3 and 200 meters. The starting time of a schedule is $t_0 = 500$. The initial ages of the 30 sources, that is, a_{n0}, $\forall n \in \mathcal{N}$, are uniformly distributed in $[50, 300]$. For each link, 20 packets are queued at the transmitter at t_0. The time stamps of the packets are integers uniformly distributed in $(t_0 - a_{n0}, t_0)$.

We first evaluate the optimization algorithm for MPAS. The candidate group set $\mathcal{C}$ consists of the 30 single-link groups as well as 200 groups with cardinalities uniformly distributed from 2 to 6. As a baseline solution, we consider that the groups are selected in the descending order of their cardinalities. This baseline solution can be treated as a greedy method of the minimum-length scheduling. For each scenario we run 100 simulations. For each simulations, the peak AoI derived by the proposed optimization algorithm is normalized by that of the baseline solution. In Figure 9.6, we present the empirical cumulative distribution function (CDFs) of the normalized peak AoIs. The results show that with the optimized scheduling solution, the maximum peak age values of 99 instances decrease. The improvement is up to 27 percent, with an average of 7 percent, over the baseline solution.

Next, we evaluate the DFR algorithm for MESA. We assume a bandwidth $B = 2$ kHz, and each packet consists of $\rho = 8$ bits. For all links, the transmit power P_n and the noise variance σ_n are uniformly set to 30 dBm and -100 dBm, respectively. The channel gain follows a distance-based propagation model with a path loss exponent of 4. We adopt the Shannon formula for rate calculation. The number of packets delivered

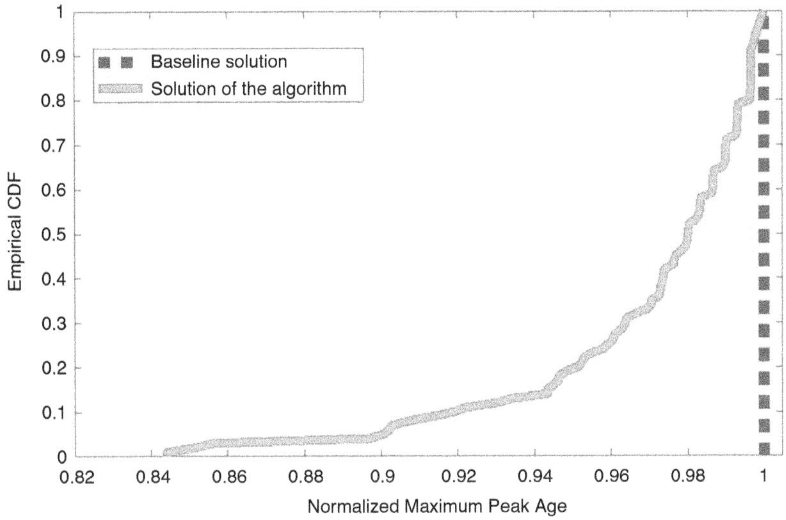

Figure 9.6 Empirical CDF of normalized peak AoI for MPAS.

by link n in a time slot is then given as the minimum of the number of available packets at TX_n and $v(n, g)$, which is calculated as

$$v(n,g) = \lfloor \frac{B}{\rho} \log_2(1 + \text{SINR}(n,g)) \rfloor. \tag{9.15}$$

We set the age constraints $A_n = a_{n0} + x, \forall n \in \mathcal{N}$, where x is a random integer uniformly distributed in $[X, 10 + X]$. For each value of X, we compute the scheduling solutions for 100 MESA instances with the DFR algorithm. Note that it may happen that some instances with low age constraints are infeasible. Hence we collect the results from $X = 11$, that is, $A_n = a_{n0} + x, x \in [11, 25], \forall n \in \mathcal{N}$, for which the DFR obtains feasible solutions for 90 instances out of 100. The energy consumption with the optimized schedule is normalized by its lower bound (see Theorem 5). Figure 9.7 shows the mean of the normalized energy consumption of the 90 instances as a function of the relative age constraint X.

In the inset figure, we compare the resulting energy consumption of DFR with the optimization algorithm for MPAS, both normalized by the lower bound. The figures show that DFR achieves a significant energy reduction ($\sim 70\%$) in comparison to the MPAS algorithm, which aims to minimize the peak AoI. The results also show that the energy consumption decreases with the relative age constraint with a decreasing marginal gain. After $X = 19$, the energy consumption reaches its lower bound. Hence one could roughly estimate that for this dataset TDMA is allowed when $A_n \in [a_{n0} + 19, a_{n0} + 25]$.

Finally, we evaluate the CMAF algorithm for MCAS. The simulation scenario is a wireless camera network monitoring an area of 100×100 meters, resembling an urban surveillance scenario. The area is divided into 16 subareas; each subarea

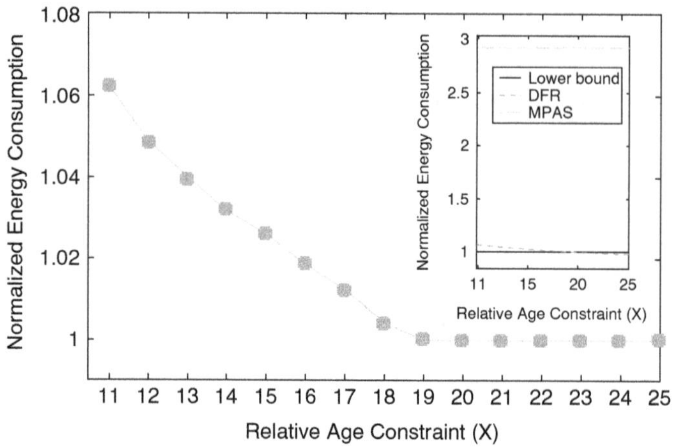

Figure 9.7 Energy consumption with age constraints and Shannon rate function.

occupies 25 × 25 meters and contains one source. The number of cameras (senders) that cover one source is uniformly chosen on $[2, 6]$. For each source s, the cameras $c(s)$ are uniformly distributed in the respective subarea. In the network, we deployed $N \in \{1, 2, 4, 8, 16\}$ fog nodes by splitting the whole area into N equal-sized rectangles and placing one fog node randomly in each rectangle. In Figure 9.8(a) we illustrate the network topology for $N = 16$. The transmit power of the cameras and the noise variance at the fog nodes are uniformly set to 20 dBm and to -100 dBm, respectively. The channel gain follows a distance-based propagation model with a path loss exponent of 4, Rayleigh fading, and log-normal shadowing with standard deviation of 6 dB. The SINR threshold is set to $\gamma = -3$ dB. As a baseline for comparison, we use the location-based greedy (LBG) algorithm, where the cameras are served by the nearest fog node, and a greedy algorithm for minimum-length scheduling [17] is used. For performance comparison as well as for assessing the joint benefit of CMAF and fog computing in peak age reduction, we normalize all results by the maximum peak age achieved by LBG with $N = 1$, which corresponds to a traditional centralized network. Figure 9.8(b) shows the average of the normalized maximum peak age as a function of the number of fog nodes. The averages are computed based on 100 simulations. The results show the number of fog nodes needed for achieving a certain peak AoI is significantly lower when using CMAF, compared to LBG, indicating that significant infrastructure cost savings can be achieved by using CMAF.

9.7 Discussion

Optimizing the transmission scheduling strategy with the objective of minimizing AoI has shown to be a novel and effective approach that enables timely information delivery and thus improves the freshness of information for a wireless system with

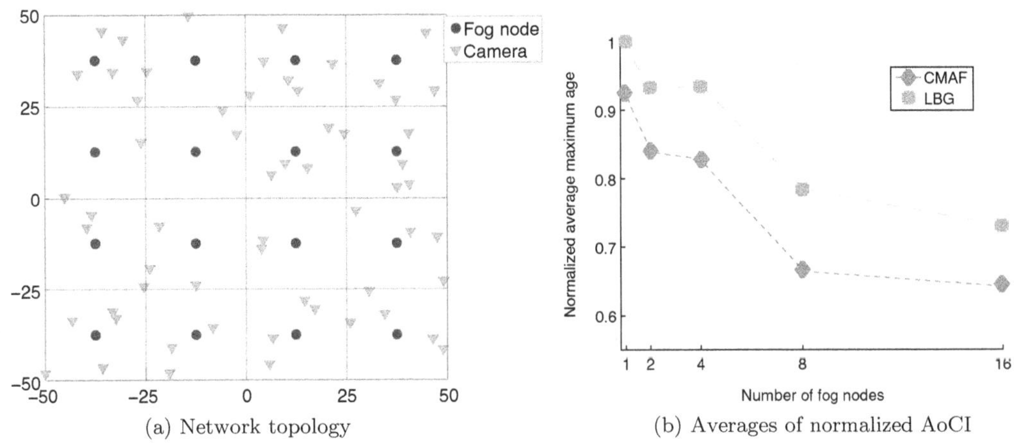

Figure 9.8 MCAS in camera networks with multiple-view image processing.

shared channels. In addition, jointly considering AoI and other optimality goals such as energy efficiency would lead to scheduling solutions that could reduce the environmental impact of communications, and meanwhile meet the expectation of users. The results also indicate that AoI-driven scheduling differs from classic scheduling problems, such as minimum-length scheduling and minimum-energy scheduling, and is of interest to study in its own right. The AoI-driven scheduling problems studied in this chapter are proved to be hard in general, but allow polynomial-time heuristics to compute a scheduling solution that could efficiently improve the network performance.

Moreover, we remark that the study of AoI-driven scheduling is a key step toward designing a complete solution that minimizes AoI in an end-to-end scenario, which involves the joint optimization of status updates scheduling, queue management strategy, transmission scheduling, and data processing. These optimization tasks may take place at different locations, which makes the end-to-end optimization particularly challenging.

9.8 Further Reading

Minimum-length transmission scheduling has been investigated for a long time in wireless networks. Early studies, which mostly assumed the protocol model for interference modeling, have proposed various heuristic algorithms (e.g., [18–20]). The use of column generation has been proposed in [7] and was adopted in a number of subsequent studies [8, 21]. Surveys of minimum-length scheduling and its extensions are provided in [22].

It is worth noting that minimum-length scheduling can be formulated also for continuous time rather than slotted time. For continuous time the problem has been studied extensively in [23, 24].

The wireless link scheduling problem with respective to age of information was first formulated and analyzed in [25], where the optimality goal was to minimize the overall age during a scheduling cycle. In [3], more fundamental insights and numerical results on minimum overall age scheduling were presented. Later on, utilizing the metric of peak AoI introduced in [1], the scheduling problem that aims to minimize the peak AoI, that is, the MPAS problem, is studied in [10]. The study was then extended to age of correlated information, as presented in [16, 26], where the notion of AoCI is defined and the minimum correlated age scheduling problem, that is, MCAS, is studied. In [14], the minimum-energy scheduling problem with age constraints, that is, MESA, is addressed and a heuristic algorithm is proposed.

The most recent developments in AoI-driven or AoI-aware scheduling include the exploration of learning-based algorithms. For example, the authors of [27] employ policy gradients and deep Q-learning methods for scheduling multiple queues with minimum AoI. In [28], reinforcement learning is used for weighted AoI minimization in a UAV-assisted network. In [29] and [30], reinforcement learning is employed in AoI-optimal sampling and scheduling for wireless powered communication systems, respectively. In [31], a deep reinforcement learning based approach is proposed to minimize AoCI for Internet of Things (IoT) systems. Research on AoI and on the potential for using machine learning for scheduling is yet in its early stage. Open issues include the transferability of the learning models, the accuracy of the results, and the sensitivity to adversarial inputs, among others. These exceedingly interesting and relevant, but so far unexplored topics could serve as future research directions in this area.

References

[1] M. Costa, M. Codreanu, and A. Ephremides, "Age of information with packet management," in *2014 IEEE International Symposium on Information Theory*, 2014, pp. 1583–1587.

[2] A. Kosta, N. Pappas, A. Ephremides, and V. Angelakis, "Age of information performance of multiaccess strategies with packet management," *Journal of Communications and Networks*, vol. 21, no. 3, pp. 244–255, 2019.

[3] Q. He, D. Yuan, and A. Ephremides, "Optimal link scheduling for age minimization in wireless systems," *IEEE Transactions on Information Theory*, vol. 64, no. 7, pp. 5381–5394, 2018.

[4] A. Capone, I. Filippini, S. Gualandi, and D. Yuan, "Resource optimization in multi-radio multi-channel wireless mesh networks," in *Mobile Ad Hoc Networking: The Cutting Directions*, 2nd ed., S. Basagni, M. Conti, S. Giordano, and I. Stojmenovic, eds. Wiley and IEEE Press, 2013, pp. 241–274.

[5] P. Gupta and P. R. Kumar, "The capacity of wireless networks," *IEEE Transactions on Information Theory*, vol. 46, no. 2, pp. 388–404, 2000.

[6] Q. He, D. Yuan, and A. Ephremides, "Optimal scheduling for emptying a wireless network: Solution characterization, applications, including deadline constraints," *IEEE Transactions on Information Theory*, vol. 66, no. 3, pp. 1882–1892, 2020.

[7] P. Bjorklund, P. Varbrand, and Di Yuan, "Resource optimization of spatial TDMA in ad hoc radio networks: a column generation approach," in *IEEE INFOCOM 2003. Twenty-second Annual Joint Conference of the IEEE Computer and Communications Societies (IEEE Cat. No.03CH37428)*, vol. 2, 2003, pp. 818–824.

[8] P. Björklund, P. Värbrand, and D. Yuan, "A column generation method for spatial TDMA scheduling in ad hoc networks," *Ad Hoc Networks*, vol. 2, no. 4, pp. 405–418, 2004.

[9] M. R. Garey and D. S. Johnson, *Computers and Intractability: A Guide to the Theory of NP-Completeness*. W. H. Freeman, 1979.

[10] Q. He, D. Yuan, and A. Ephremides, "On optimal link scheduling with min-max peak age of information in wireless systems," in *2016 IEEE International Conference on Communications (ICC)*, 2016, pp. 1–7.

[11] IBM, "IBM CPLEX Optimizer 12.6," www.ibm.com/analytics/cplex-optimizer/, 2015.

[12] GUROBI, "GUROBI Optimizer 6.5," www.gurobi.com/, 2015.

[13] G. D. Nguyen, S. Kompella, C. Kam, J. E. Wieselthier, and A. Ephremides, "Minimum-energy link scheduling for emptying wireless networks," in *2015 13th International Symposium on Modeling and Optimization in Mobile, Ad Hoc, and Wireless Networks (WiOpt)*, 2015, pp. 207–212.

[14] Q. He, G. Dán, and V. Fodor, "On emptying a wireless network with minimum-energy under age constraints," in *2019 IEEE Conference on Computer Communications Workshops (INFOCOM WKSHPS)*, 2019, pp. 668–673.

[15] E. Eriksson, G. Dán, and V. Fodor, "Coordinating distributed algorithms for feature extraction offloading in multi-camera visual sensor networks," *IEEE Transactions on Circuits and Systems for Video Technology*, vol. 28, no. 11, pp. 3288–3299, 2018.

[16] Q. He, G. Dán, and V. Fodor, "Joint assignment and scheduling for minimizing age of correlated information," *IEEE/ACM Transactions on Networking*, vol. 27, no. 5, pp. 1887–1900, 2019.

[17] V. Angelakis, A. Ephremides, Q. He, and D. Yuan, "Minimum-time link scheduling for emptying wireless systems: Solution characterization and algorithmic framework," *IEEE Transactions on Information Theory*, vol. 60, no. 2, pp. 1083–1100, 2014.

[18] R. Nelson and L. Kleinrock, "Spatial TDMA: A collision-free multihop channel access protocol," *IEEE Transactions on Communications*, vol. 33, no. 9, pp. 934–944, 1985.

[19] C. G. Prohazka, "Decoupling link scheduling constraints in multi-hop packet radio networks," *IEEE Transactions on Computers*, vol. 38, no. 3, pp. 455–458, 1989.

[20] S. Ramanathan and E. L. Lloyd, "Scheduling algorithms for multihop radio networks," *IEEE/ACM Transactions on Networking*, vol. 1, no. 2, pp. 166–177, 1993.

[21] S. Kompella, J. E. Wieselthier, A. Ephremides, H. D. Sherali, and G. D. Nguyen, "On optimal SINR-based scheduling in multihop wireless networks," *IEEE/ACM Transactions on Networking*, vol. 18, no. 6, pp. 1713–1724, 2010.

[22] A. Pantelidou and A. Ephremides, "Scheduling in wireless networks," *Foundations and Trends® in Networking*, vol. 4, no. 4, pp. 421–511, 2011. [Online]. Available: http://dx.doi.org/10.1561/1300000030

[23] Q. He, D. Yuan, and A. Ephremides, "Optimal scheduling for emptying a wireless network: Solution characterization, applications, including deadline constraints," *IEEE Transactions on Information Theory*, vol. 66, no. 3, pp. 1882–1892, 2020.

[24] ——, "A general optimality condition of link scheduling for emptying a wireless network," in *2016 IEEE International Symposium on Information Theory (ISIT)*, 2016, pp. 1446–1450.

[25] ——, "Optimizing freshness of information: On minimum age link scheduling in wireless systems," in *2016 14th International Symposium on Modeling and Optimization in Mobile, Ad Hoc, and Wireless Networks (WiOpt)*, 2016, pp. 1–8.

[26] Q. He, G. Dán, and V. Fodor, "Minimizing age of correlated information for wireless camera networks," in *2018 IEEE Conference on Computer Communications Workshops (INFOCOM WKSHPS)*, 2018, pp. 547–552.

[27] H. B. Beytur and E. Uysal, "Age minimization of multiple flows using reinforcement learning," in *2019 International Conference on Computing, Networking and Communications (ICNC)*, 2019, pp. 339–343.

[28] M. A. Abd-Elmagid, A. Ferdowsi, H. S. Dhillon, and W. Saad, "Deep reinforcement learning for minimizing age-of-information in UAV-assisted networks," in *2019 IEEE Global Communications Conference (GLOBECOM)*, 2019, pp. 1–6.

[29] M. A. Abd-Elmagid, H. S. Dhillon, and N. Pappas, "AoI-optimal joint sampling and updating for wireless powered communication systems," *IEEE Transactions on Vehicular Technology*, vol. 69, no. 11, pp. 14 110–14 115, 2020.

[30] ——, "A reinforcement learning framework for optimizing age of information in RF-powered communication systems," *IEEE Transactions on Communications*, vol. 68, no. 8, pp. 4747–4760, 2020.

[31] B. Yin, S. Zhang, and Y. Cheng, "Application-oriented scheduling for optimizing the age of correlated information: A deep reinforcement learning based approach," *IEEE Internet of Things Journal*, pp. 1–1, 2020.

10 Age of Information and Remote Estimation

Tasmeen Zaman Ornee and Yin Sun

Abstract: In this chapter, we discuss the relationship between age of information and signal estimation error in real-time signal sampling and reconstruction. Consider a remote estimation system, where samples of a scalar Gauss–Markov signal are taken at a source node and forwarded to a remote estimator through a channel that is modeled as a queue. The estimator reconstructs an estimate of the real-time signal value from causally received samples. The optimal sampling policy for minimizing the mean square estimation error is presented, in which a new sample is taken once the *instantaneous estimation error* exceeds a predetermined threshold. When the sampler has no knowledge of current and history signal values, the optimal sampling problem reduces to a problem for minimizing a *nonlinear* Age of Information (AoI) metric. In the AoI-optimal sampling policy, a new sample is taken once the *expected estimation error* exceeds a threshold. The threshold can be computed by low-complexity algorithms, and the insights behind these algorithms are provided. These optimal sampling results were established (i) for general service time distributions of the queueing server, (ii) for both stable and unstable scalar Gauss–Markov signals, and (iii) for sampling problems both with and without a sampling rate constraint.[1]

10.1 Introduction

Information is usually of the greatest value when it is fresh. For example, real-time knowledge about the location, orientation, and speed of motor vehicles is imperative in autonomous driving, and access to timely information about the stock price and interest rate movements is essential for developing trading strategies on the stock market. In recent years, the AoI has been adopted as a new metric for quantifying the freshness of information in real-time systems and networks. Consider a sequence of information packets that are sent to a receiver. Let U_t be the generation time of the newest packet that has been delivered to the receiver by time t. AoI, as a function of t, is defined as

$$\Delta(t) = t - U_t. \tag{10.1}$$

[1] This work was supported in part by NSF grant CCF-1813050 and ONR grant N00014-17-1-2417.

Besides the linear AoI $\Delta(t)$, nonlinear functions $p(\Delta(t))$ of the AoI have been recently demonstrated to be useful metrics for information freshness in signal estimation, control, and wireless communications (e.g., the freshness of channel state information); see recent surveys [1], [2], and [3] for more details.

In many real-time systems, the information of interest – for example, the trajectory of UAV mobility, the measurement of temperature sensors, and the price of a stock – is represented by the value of a time-varying signal X_t that may change slowly at some time and vary more dynamically later. Hence, the time difference described by the AoI $\Delta_t = t - U_t$ or its nonlinear functions cannot fully characterize how much the signal value has varied during the same time period, that is, $X_t - X_{U_t}$. Hence, the status-update policy that minimizes the AoI is insufficient for minimizing the signal estimation error.

In this chapter, we discuss some recent results on real-time signal sampling and reconstructions. A problem of sampling a scalar Gauss–Markov signal is considered, where the samples forwarded to a remote estimator through a channel in a first-come, first-served (FCFS) fashion. The considered Gauss–Markov signals are Wiener process, stable and stationary Ornstein–Uhlenbeck (OU) process, and unstable and nonstationary OU process. A well-known example of Gauss–Markov process is Wiener process that is a real-valued continuous-time stochastic process and has independent increments. Another well-known example of stationary Gauss–Markov process is OU process. An OU process X_t is the continuous-time analogue of the well-known first-order autoregressive process, that is, AR(1) process. The OU process is defined as the solution to the stochastic differential equation (SDE) [4], [5],

$$dX_t = \theta(\mu - X_t)dt + \sigma dW_t, \tag{10.2}$$

where $\mu, \theta > 0$, and $\sigma > 0$ are parameters and W_t represents a Wiener process. It is the only nontrivial continuous-time process that is stationary, Gaussian, and Markovian [5]. From (10.2), if $\theta \to 0$ and $\sigma = 1$, the OU process becomes a Wiener process. Therefore, the Wiener process is a special case of the OU process. If the parameter of the OU process $\theta < 0$, then the process becomes unstable and nonstationary. The SDE in (10.2) still holds in such an unstable scenario. Examples of first-order systems that can be described as the Gauss–Markov process include interest rates, currency exchange rates, and commodity prices (with modifications) [6], control systems such as node mobility in the mobile ad hoc networks, robotic swarms, and UAV systems [7], [8], and physical processes such as the transfer of liquids or gases in and out of a tank [9]. Another application is on federated learning where the progression of clients' weights is modeled as an OU process [10].

The samples experience *i.i.d.* random transmission times over the channel, which is caused by random sample size, channel fading, interference, congestions, and so on. For example, UAVs flying close to WiFi access points may suffer from long communication delays and instability issues, because they receive strong interference from the WiFi access points [11]. At any time only one sample can be served by the channel. The samples that are waiting to be sent are stored in a queue at the transmitter. Hence, the channel is modeled as an FCFS queue with *i.i.d.* service times. The service time distributions that we consider are quite general: they are only required to have a finite mean. This queueing model is helpful to analyze the robustness of remote estimation systems with occasionally long transmission times.

The estimator utilizes causally received samples to construct an estimate $\hat{X}_t$ of the real-time signal value X_t. The quality of remote estimation is measured by the time-average mean-squared estimation error, that is,

$$\mathsf{mse} = \limsup_{T\to\infty} \frac{1}{T}\mathbb{E}\left[\int_0^T (X_t - \hat{X}_t)^2 dt\right]. \tag{10.3}$$

The goal is to find the optimal sampling policy that minimizes mse by causally choosing the sampling times. Our results show that if the sampler has no knowledge on the value of the Gauss–Markov signal, the optimal sampling strategy is to minimize the time-average of a nonlinear AoI function; however, by exploiting causal knowledge of the signal values, it is possible to achieve a smaller estimation error. We have also considered the sampling problem that is subject to a maximum sampling rate constraint. In practice, the cost (e.g., energy, CPU cycle, storage) for state updates increases with the average sampling rate. Hence, the optimum trade-off between estimation error and update cost should be found.

Our main results are summarized as follows:

The optimal sampling problem for minimizing the mse under a sampling rate constraint is formulated as a constrained continuous-time Markov decision process (MDP) with an uncountable state space. Because of the curse of dimensionality, such problems are often lack of low-complexity solutions that are arbitrarily accurate. However, this MDP is solved exactly: the optimal sampling policy is proven to be a threshold policy on *instantaneous* estimation error, where the threshold is a nonlinear function $v(\beta)$ of a parameter β. The value of β is equal to the summation of the optimal objective value of the MDP and the optimal Lagrangian dual variable associated with the sampling rate constraint. If there is no sampling rate constraint, the Lagrangian dual variable is zero and hence β is exactly the optimal objective value.

By comparing the optimal sampling policies of the Wiener process and the stable and unstable OU processes, we find that the threshold function $v(\beta)$ changes according to the signal model, whereas the parameter β is determined in the same way for all three signal models.

Further, for a class of signal-agnostic sampling policies, the sampling times are determined without using knowledge of the observed process. The optimal signal-agnostic sampling problem is equivalent to an MDP for minimizing the time-average of a nonlinear age function $p(\Delta_t)$. The age-optimal sampling policy is a threshold policy on *expected* estimation error, where the threshold function is simply $v(\beta) = \beta$ and the parameter β is determined in the same way as previously.

The preceding results hold for (i) general service time distributions of the queueing server, (ii) both stable and unstable scalar Gauss–Markov signals, and (ii) sampling problems both with and without a sampling rate constraint. Numerical results suggest that the optimal sampling policy is better than zero-wait sampling and classic uniform sampling.

The optimal sampling results for the Wiener process and the OU process were proven in [12], [13]. The proofs for the unstable OU process will be provided in a paper that is currently under preparation.

The rest of the chapter is organized as follows: in Section 10.2, we discuss the related work on the age of information and remote estimation. In Section 10.3, the model of remote estimation systems and the formulation of optimal sampling problems are presented. The solutions to signal-aware and signal-agnostic sampling problems are provided in Sections 10.3–10.4, which include both the cases with and without the sampling rate constraint. The numerical results are shown in Section 10.5 and a summary of the chapter is given in Section 10.6.

10.2 Related Work and Contributions

In this section, we will present a brief survey of the related work in the area of age of information and remote estimation.

10.2.1 Age of Information

The results presented in this chapter are significantly related to recent studies on the age of information Δ_t, for example, [1, 14–35]. In [14], the authors provided a simple example about a status updating system, where samples of a Wiener process W_t are forwarded to a remote estimator. The age of the delivered sample is $\Delta_t = t - U_t$ if U_t is the generation time of the latest received sample. Furthermore, the MMSE estimate of W_t is $\hat{W}_t = W_{U(t)}$, and the variance of this estimator is $\mathbb{E}[(W_t - \hat{W}_t)^2] = \Delta_t$. In [15], the authors proposed a sampling policy for a discrete-time source process by incorporating mutual information as a measure for maximizing the information freshness. The results in [15] were further extended for both continuous and discrete-time source processes in [1] where the nonlinear functions of the age had been used to measure data freshness. In [16], sampling and scheduling policy for the multisource system was studied by analyzing the peak-age and peak-average-age. In [17], the authors analyzed status age when the message may take various routes in the network for queueing systems. In [18], the optimal control for information updates traveled from a source to a remote destination was studied and optimal trade-off between the update policy and the age of information was found. The authors also showed that in many cases, the optimal policy is to wait a certain amount before sending the next update. The average age and average peak age have been analyzed for various queueing systems in, for example, [14, 17, 19, 20]. The optimality of the Last-Come, First-Served (LCFS) policy, or more generally the Last-Generated, First-Served (LGFS) policy, was established for various queueing system models in [23–25, 29]. Optimal sampling policies for minimizing nonlinear age functions were developed in, for example, [1, 15, 18, 34]. Age-optimal transmission scheduling of wireless networks were investigated in [21, 22, 26–28, 30, 31, 36, 37]. In [35], a game-theoretic perspective of the age was studied, and the authors proposed a sampling policy by studying the timeliness of the status update where an attacker sabotages the system by jamming the channel and maximizing age-of-information. which does not have a signal model. A broad survey in the area of age of information is presented in [2].

10.2.2 Remote Estimation

The results in this chapter also have a tight connection with the area of remote estimation, for example, [9, 38–43] by adding a queue between the sampler and estimator. In [9], remote state estimation in first-order linear time-invariant (LTI) discrete-time systems was considered with a quadratic cost function and finite time horizon. They showed that a time-dependent threshold-based sampler and Kalman-like estimator are jointly optimal. In [38], the authors investigated the joint optimization of paging and registration policies in cellular networks, which is essentially the same as a joint sampling and estimation optimization problem with an indicator-type cost function and an infinite time horizon. They used majorization theory and Riesz's rearrangement inequality to show that, if the state process is modeled as a symmetric or Gaussian random walk, a threshold-based sampler and a nearest distance estimator are jointly optimal. This is the first study pointing out that the sampler and estimator have different information patterns. In [39], the authors considered a remote estimation problem with an energy-harvesting sensor and a remote estimator, where the sampling decision at the sensor is constrained by the energy level of the battery. They proved that an energy-level-dependent threshold-based sampler and a Kalman-like estimator are jointly optimal. In [40], [41], optimal sampling of Wiener processes was studied, where the transmission time from the sampler to the estimator is zero. Optimal sampling of OU processes was also considered in [40], which is solved by discretizing time and using dynamic programming to solve the discrete-time optimal stopping problems. In [13], optimal sampling of OU processes is obtained analytically. In the optimal sampling policy, sampling is suspended when the server is busy and is reactivated once the server becomes idle. In addition, the threshold was also characterized precisely. The optimal sampling policy for the Wiener process in [12] is a limiting case. Remote estimation of the Wiener process with random two-way delay was considered in [44]. Remote estimation over several different channel models was recently studied in, for example, [42, 43]. In [9, 13, 38–43], the optimal sampling policies were proven to be threshold policies. Because of the queueing model, the optimal sampling policy in [13] has a different structure from those in [9, 38–43]. Specifically, in [45], a jointly optimal sampler, quantizer, and estimator design was found for a class of continuous-time Markov processes under a bit-rate constraint. In [46], the quantization and coding schemes on the estimation performance are studied. A recent survey on remote estimation systems was presented in [47].

10.3 System Model and Problem Formulation

10.3.1 System Model

Let us consider the remote estimation system illustrated in Figure 10.1, where an observer takes samples from a Gauss–Markov process X_t and forwards the samples to

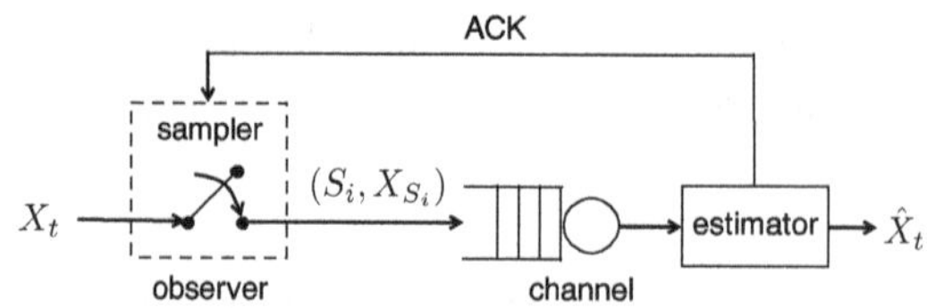

Figure 10.1 System model.

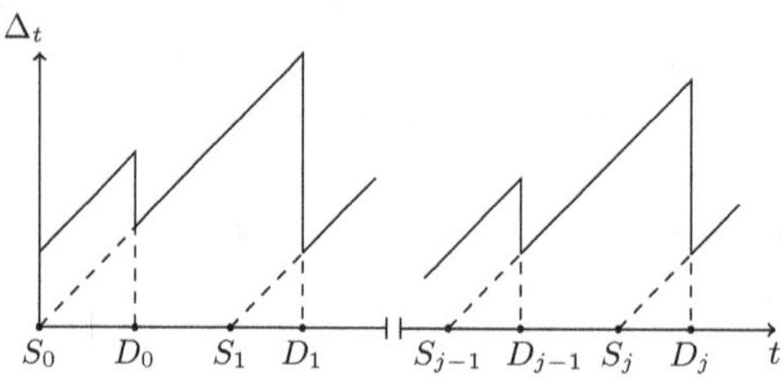

Figure 10.2 Evolution of the age Δ_t over time.

an estimator through a communication channel. The channel is modeled as a single-server FCFS queue with *i.i.d.* service times. The samples experience random service times in the channel due to fading, interference, congestions, and so on. The service times are *i.i.d.* and only one sample can be delivered through the channel at a time.

The system starts to operate at time $t = 0$. The ith sample is generated at time S_i and is delivered to the estimator at time D_i with a service time Y_i, which satisfy $S_i \leq S_{i+1}$, $S_i + Y_i \leq D_i$, $D_i + Y_{i+1} \leq D_{i+1}$, and $0 < \mathbb{E}[Y_i] < \infty$ for all i. Each sample packet (S_i, X_{S_i}) contains the sampling time S_i and the sample value X_{S_i}. Let $U_t = \max\{S_i : D_i \leq t\}$ be the sampling time of the latest received sample at time t. The *age of information* or simply *age*, at time t is defined as [14], [48]

$$\Delta_t = t - U_t = t - \max\{S_i : D_i \leq t\}, \tag{10.4}$$

which is shown in Figure 10.2. Because $D_i \leq D_{i+1}$, Δ_t can be also expressed as

$$\Delta_t = t - S_i, \text{ if } t \in [D_i, D_{i+1}), \ i = 0, 1, 2, \ldots. \tag{10.5}$$

The initial state of the system is assumed to satisfy $S_0 = 0$, $D_0 = Y_0, X_0$, and Δ_0 are finite constants. The parameters μ, θ, and σ in (10.2) are known at both the sampler and estimator.

Let $I_t \in \{0, 1\}$ represent the idle/busy state of the server at time t. We assume that whenever a sample is delivered, an acknowledgement is sent back to the sampler with zero delay. By this, the idle/busy state I_t of the server is known at the sampler. Therefore, the information that is available at the sampler at time t can be expressed as $\{X_s, I_s : 0 \leq s \leq t\}$.

10.3.2 Sampling Policies

In causal sampling policies, each sampling time S_i is chosen by using the up-to-date information available at the sampler. S_i is a *stopping time* with respect to the filtration $\{\mathcal{N}_t^+, t \geq 0\}$ (a nondecreasing and right-continuous family of σ-fields) such that

$$\{S_i \leq t\} \in \mathcal{N}_t^+, \ \forall t \geq 0. \tag{10.6}$$

Let $\pi = (S_1, S_2, ...)$ represent a sampling policy and Π represent the set of *causal* sampling policies.

If the inter-sampling times form a regenerative process [49] and (10.6) is satisfied for each sampling policy $\pi \in \Pi$, we can obtain that S_i is finite almost surely for all i. We assume that the OU process $\{X_t, t \geq 0\}$ and the service times $\{Y_i, i = 1, 2, \ldots\}$ are mutually independent and do not change according to the sampling policy.

A sampling policy $\pi \in \Pi$ is said to be *signal-agnostic* (*signal-aware*), if π is (not necessarily) independent of $\{X_t, t \geq 0\}$. Let $\Pi_{\text{signal-agnostic}} \subset \Pi$ denote the set of signal-agnostic sampling policies, defined as

$$\Pi_{\text{signal-agnostic}} = \{\pi \in \Pi : \pi \text{ is independent of } \{X_t, t \geq 0\}\}. \tag{10.7}$$

10.3.3 MMSE Estimator

According to (10.6), S_i is a finite stopping time. By using [50, Eq. (3)] and the strong Markov property of the Gauss–Markov process [51, Eq. (4.3.27)], X_t is expressed as

$$\begin{aligned} X_t =& X_{S_i} e^{-\theta(t-S_i)} + \mu\left[1 - e^{-\theta(t-S_i)}\right] \\ &+ \frac{\sigma}{\sqrt{2\theta}} e^{-\theta(t-S_i)} W_{e^{2\theta(t-S_i)}-1}, \ \text{if } t \in [S_i, \infty). \end{aligned} \tag{10.8}$$

At any time $t \geq 0$, the estimator uses causally received samples to construct an estimate $\hat{X}_t$ of the real-time signal value X_t. The information available to the estimator consists of two parts: (i) $M_t = \{(S_i, X_{S_i}, D_i) : D_i \leq t\}$, which contains the sampling time S_i, sample value X_{S_i}, and delivery time D_i of the samples that have been delivered by time t and (ii) the fact that no sample has been received after the last delivery time $\max\{D_i : D_i \leq t\}$. Similar to [12, 40, 52], we assume that the estimator neglects the second part of the information. This assumption can be removed by considering a joint sampler and estimator design problem. Specifically, it was shown in [9, 38, 39, 42, 43] that when the sampler and estimator are jointly optimized in discrete-time systems, the optimal estimator has the same expression regardless of whether it is with or without the second part of the information. As pointed out in [38, p. 619], such a structure property of the MMSE estimator can be also established for continuous-time systems. Our goal is to find the closed-form expression of the optimal sampler under this assumption. The remaining task of finding the jointly optimal sampler and estimator design can be done by further using the majorization techniques developed in [9, 38, 39, 42, 43]; see [45] for a recent treatment on this task. Then the minimum mean square error (MMSE) estimator is determined by

$$\hat{X}_t = \mathbb{E}[X_t|M_t] = X_{S_i}e^{-\theta(t-S_i)} + \mu\left[1 - e^{-\theta(t-S_i)}\right],$$
$$\text{if } t \in [D_i, D_{i+1}),\ i = 0, 1, 2, \ldots. \tag{10.9}$$

10.3.4 Problem Formulation

We study the optimal sampling policy that minimizes the mean-squared estimation error subject to an average sampling-rate constraint, which is formulated as the following problem:

$$\mathsf{mse}_{\text{opt}} = \inf_{\pi\in\Pi} \limsup_{T\to\infty} \frac{1}{T}\mathbb{E}\left[\int_0^T (X_t - \hat{X}_t)^2 dt\right] \tag{10.10}$$

$$\text{s.t. } \liminf_{n\to\infty} \frac{1}{n}\mathbb{E}\left[\sum_{i=1}^{n}(S_{i+1} - S_i)\right] \geq \frac{1}{f_{\max}}, \tag{10.11}$$

where $\mathsf{mse}_{\text{opt}}$ is the optimum value of (10.10) and $f_{\max}$ is the maximum allowed sampling rate. When $f_{\max} = \infty$, this problem becomes an unconstrained problem.

Problem (10.10) is a constrained continuous-time MDP with a continuous state space. However, we found an exact solution to this problem.

10.4 Optimal Signal Sampling Policies

The optimal sampling policies for Wiener process, OU process, and unstable OU process are presented in the following theorems, respectively.

10.4.1 Signal-Aware Sampling without a Sampling Rate Constraint

We first consider the unconstrained optimal sampling problem, that is, $f_{\max} = \infty$, such that the sampling rate constraint (10.11) can be removed.

10.4.1.1 Wiener Process

If the signal being sampled, $X_t = W_t$, is a Wiener process, then the optimal sampling policy is provided in the following theorem.

THEOREM 10.1 *If X_t is a Wiener process with $f_{\max} = \infty$, and the Y_i are i.i.d. with $0 < \mathbb{E}[Y_i] < \infty$, then $(S_1(\beta), S_2(\beta), \ldots)$ with a parameter β is an optimal solution to* (10.10), *where*

$$S_{i+1}(\beta) = \inf\left\{t \geq D_i(\beta) : \left|X_t - \hat{X}_t\right| \geq v(\beta)\right\}, \tag{10.12}$$

$D_i(\beta) = S_i(\beta) + Y_i$, $v(\beta)$ is defined by

$$v(\beta) = \sqrt{3(\beta - \mathbb{E}[Y_i])}, \tag{10.13}$$

and β is the unique root of

$$\mathbb{E}\left[\int_{D_i(\beta)}^{D_{i+1}(\beta)} (X_t - \hat{X}_t)^2 dt\right] - \beta\mathbb{E}[D_{i+1}(\beta) - D_i(\beta)] = 0. \tag{10.14}$$

The optimal objective value to (10.15) *is given by*

$$\mathsf{mse}_{\text{opt}} = \frac{\mathbb{E}\left[\int_{D_i(\beta)}^{D_{i+1}(\beta)} (X_t - \hat{X}_t)^2 dt\right]}{\mathbb{E}[D_{i+1}(\beta) - D_i(\beta)]}. \tag{10.15}$$

Furthermore, β is exactly the optimal value to (10.10), *that is, $\beta = \mathsf{mse}_{\text{opt}}$.*

The optimal sampling policy in Theorem 10.1 has a nice structure. Specifically, the $(i+1)$-th sample is taken at the earliest time t satisfying two conditions: (i) The ith sample has already been delivered by time t, that is, $t \geq D_i(\beta)$, and (ii) the estimation error $|X_t - \hat{X}_t|$ is no smaller than a predetermined threshold $v(\beta)$, where $v(\cdot)$ is a nonlinear function defined in (10.13).

10.4.1.2 Ornstein–Uhlenbeck Process (Stable Case, $\theta > 0$)

To present the optimal sampler when X_t is an OU process, we need to introduce another OU process O_t with the initial state $O_t = 0$ and parameter $\mu = 0$. According to (10.8), O_t can be expressed as

$$O_t = \frac{\sigma}{\sqrt{2\theta}} e^{-\theta t} W_{e^{2\theta t}-1}. \tag{10.16}$$

Define

$$\mathsf{mse}_{Y_i} = \mathbb{E}[O_{Y_i}^2] = \frac{\sigma^2}{2\theta}\mathbb{E}[1 - e^{-2\theta Y_i}], \tag{10.17}$$

$$\mathsf{mse}_{\infty} = \mathbb{E}[O_{\infty}^2] = \frac{\sigma^2}{2\theta}. \tag{10.18}$$

In [13], it is shown that mse_{Y_i} and mse_{∞} are the lower and upper bounds of $\mathsf{mse}_{\text{opt}}$, respectively. We will also need to use the following function,

$$G(x) = \frac{e^{x^2}}{x}\int_0^x e^{-t^2} dt = \frac{e^{x^2}}{x}\frac{\sqrt{\pi}}{2}\,\text{erf}(x),\ x \in [0, \infty), \tag{10.19}$$

where, if $x = 0$, $G(x)$ is defined as its right limit, $G(0) = \lim_{x \to 0^+} G(x) = 1$, and $\text{erf}(\cdot)$ is the error function [53], defined as

$$\text{erf}(x) = \frac{2}{\sqrt{\pi}}\int_0^x e^{-t^2} dt. \tag{10.20}$$

In this scenario, the optimal sampler is provided in the following theorem.

THEOREM 10.2 *If X_t is a stable and stationary OU process with $f_{\max} = \infty$ and the Y_i are i.i.d. with $0 < \mathbb{E}[Y_i] < \infty$, then $(S_1(\beta), S_2(\beta), \ldots)$ with a parameter β is an optimal solution to* (10.10), *where*

$$S_{i+1}(\beta) = \inf\left\{t \geq D_i(\beta) : \left|X_t - \hat{X}_t\right| \geq v(\beta)\right\}, \tag{10.21}$$

$D_i(\beta) = S_i(\beta) + Y_i$, $v(\beta)$ is defined by

$$v(\beta) = \frac{\sigma}{\sqrt{\theta}} G^{-1}\left(\frac{\mathsf{mse}_{\infty} - \mathsf{mse}_{Y_i}}{\mathsf{mse}_{\infty} - \beta}\right), \tag{10.22}$$

$G^{-1}(\cdot)$ is the inverse function of $G(\cdot)$ in (10.19), and β is the unique root of

$$\mathbb{E}\left[\int_{D_i(\beta)}^{D_{i+1}(\beta)} (X_t - \hat{X}_t)^2 dt\right] - \beta\mathbb{E}[D_{i+1}(\beta) - D_i(\beta)] = 0. \tag{10.23}$$

The optimal objective value to (10.10) is given by

$$\mathsf{mse}_{\text{opt}} = \frac{\mathbb{E}\left[\int_{D_i(\beta)}^{D_{i+1}(\beta)} (X_t - \hat{X}_t)^2 dt\right]}{\mathbb{E}[D_{i+1}(\beta) - D_i(\beta)]}. \tag{10.24}$$

Similar to Theorem 10.1, $v(\cdot)$ in (10.22) is also a nonlinear function. In [13], it is shown that $\mathsf{mse}_{Y_i} \leq \beta < \mathsf{mse}_\infty$. Further, it is not hard to show that $G(x)$ is strictly increasing on $[0, \infty)$ and $G(0) = 1$. Hence, its inverse function $G^{-1}(\cdot)$ and the threshold $v(\beta)$ are properly defined and $v(\beta) \geq 0$.

10.4.1.3 Ornstein-Uhlenbeck Process (Unstable Case, $\theta < 0$)

Let $\rho = -\theta$; then an unstable OU process with initial state O_0, parameters $\mu = 0$, $\sigma > 0$, and $\theta < 0$ can be expressed as

$$O_t = \frac{\sigma}{\sqrt{2\rho}} e^{\rho t} W_{1-e^{-2\rho t}}. \tag{10.25}$$

Define

$$\mathsf{umse}_{Y_i} = \mathbb{E}[O_{Y_i}^2] = \frac{\sigma^2}{2\rho}\mathbb{E}[e^{2\rho Y_i} - 1], \tag{10.26}$$

$$\mathsf{umse}_\infty = \mathbb{E}[O_\infty^2] \to \infty, \tag{10.27}$$

where umse_{Y_i} and umse_∞ are the lower and upper bounds of $\mathsf{mse}_{\text{opt}}$, respectively. We will also need to use the following function,

$$K(x) = \frac{e^{-x^2}}{x}\int_0^x e^{t^2} dt = \frac{e^{-x^2}}{x}\frac{\sqrt{\pi}}{2}\,\text{erfi}(x),\ x \in [0, \infty), \tag{10.28}$$

where if $x = 0$, $K(x)$ is defined as its right limit $K(0) = \lim_{x \to 0^+} K(x) = 1$, and erfi$(\cdot)$ is the error function [53], defined as

$$\text{erfi}(x) = \frac{2}{\sqrt{\pi}}\int_0^x e^{t^2} dt. \tag{10.29}$$

Note that $K(x)$ is a strictly decreasing function on $x \in [0, \infty)$ and its inverse $K^{-1}(\cdot)$ is properly defined.

The optimal sampler is then provided in the following theorem.

THEOREM 10.3 *If X_t is an unstable and nonstationary OU process with $f_{\max} = \infty$ and the Y_i are i.i.d. with $0 < \mathbb{E}[Y_i] < \infty$, then $(S_1(\beta), S_2(\beta), \ldots)$ with a parameter β is an optimal solution to (10.10), where*

$$S_{i+1}(\beta) = \inf\left\{t \geq D_i(\beta) : \left|X_t - \hat{X}_t\right| \geq v(\beta)\right\}, \tag{10.30}$$

$D_i(\beta) = S_i(\beta) + Y_i$, $\nu(\beta)$ is defined by

$$\nu(\beta) = \frac{\sigma}{\sqrt{\rho}} K^{-1}\left(\frac{\text{umse}_{Y_i} + \frac{\sigma^2}{2\rho}}{\frac{\sigma^2}{2\rho} + \beta} \right), \tag{10.31}$$

$K^{-1}(\cdot)$ is the inverse function of $K(\cdot)$ in (10.28) and β is the unique root of

$$\mathbb{E}\left[\int_{D_i(\beta)}^{D_{i+1}(\beta)} (X_t - \hat{X}_t)^2 dt \right] - \beta \mathbb{E}[D_{i+1}(\beta) - D_i(\beta)] = 0. \tag{10.32}$$

The optimal objective value to (10.10) is given by

$$\text{mse}_{\text{opt}} = \frac{\mathbb{E}\left[\int_{D_i(\beta)}^{D_{i+1}(\beta)} (X_t - \hat{X}_t)^2 dt \right]}{\mathbb{E}[D_{i+1}(\beta) - D_i(\beta)]}. \tag{10.33}$$

The results for unstable OU process will be provided in a paper that is currently under preparation.

10.4.1.4 Low-Complexity Algorithms for Computing β

We now present three algorithms, for example, bisection search, Newton's method, and fixed-point iterations algorithms for computing the root of (10.14), (10.23), and (10.32). Because the $S_i(\beta)$ are stopping times, numerically calculating the expectations in (10.14), (10.23), and (10.32) appears to be a difficult task. Nonetheless, this challenge can be solved by resorting to the following lemma, which is obtained by using Dynkin's formula [54, Theorem 7.4.1] and the optional stopping theorem.

LEMMA 10.4 *In Theorems 10.2 and 10.3, it holds that*

$$\begin{aligned} &\mathbb{E}[D_{i+1}(\beta) - D_i(\beta)] \\ &= \mathbb{E}[\max\{R_1(\nu(\beta)) - R_1(O_{Y_i}), 0\}] + \mathbb{E}[Y_i], \end{aligned} \tag{10.34}$$

$$\begin{aligned} &\mathbb{E}\left[\int_{D_i(\beta)}^{D_{i+1}(\beta)} (X_t - \hat{X}_t)^2 dt \right] \\ &= \mathbb{E}[\max\{R_2(\nu(\beta)) - R_2(O_{Y_i}), 0\}] \\ &\quad + \text{mse}_\infty[\mathbb{E}(Y_i) - \gamma] + \mathbb{E}\left[\max\{\nu^2(\beta), O_{Y_i}^2\} \right] \gamma, \end{aligned} \tag{10.35}$$

where

$$\gamma = \frac{1}{2\theta} \mathbb{E}[1 - e^{-2\theta Y_i}], \tag{10.36}$$

$$R_1(\nu) = \frac{\nu^2}{\sigma^2} {}_2F_2\left(1, 1; \frac{3}{2}, 2; \frac{\theta}{\sigma^2}\nu^2\right), \tag{10.37}$$

$$R_2(\nu) = -\frac{\nu^2}{2\theta} + \frac{\nu^2}{2\theta} {}_2F_2\left(1, 1; \frac{3}{2}, 2; \frac{\theta}{\sigma^2}\nu^2\right). \tag{10.38}$$

In (10.37) and (10.38), we have used the generalized hypergeometric function, which is defined by [55, Eq. 16.2.1],

Algorithm 1 Bisection search method for finding β

given $l = \mathsf{mse}_{Y_i}$, $u = \mathsf{mse}_\infty$, tolerance $\epsilon > 0$.
repeat
  $\beta := (l+u)/2$.
  $o := f_1(\beta) - \beta f_2(\beta)$.
  if $o \geq 0$, $l := \beta$; **else**, $u := \beta$.
until $u - l \leq \epsilon$.
return β.

$$
\begin{aligned}
&{}_pF_q(a_1, a_2, \cdots, a_p; b_1, b_2, \cdots b_q; z)\\
=&\sum_{n=0}^{\infty} \frac{(a_1)_n(a_2)_n \cdots (a_p)_n}{(b_1)_n(b_2)_n \cdots (b_p)_n}\frac{z^n}{n!},
\end{aligned}
\tag{10.39}
$$

where

$$
(a)_0 = 1, \tag{10.40}
$$

$$
(a)_n = a(a+1)(a+2)\cdots(a+n-1),\ n \geq 1. \tag{10.41}
$$

Using Lemma 10.4, the expectations in (10.14), (10.23), and (10.32) can be evaluated by Monte Carlo simulations of one-dimensional random variables O_{Y_i} and Y_i, which is much simpler than directly simulating the entire random process $\{O_t, t \geq 0\}$.

For notational simplicity, we rewrite (10.14), (10.23), and (10.32) as

$$
f(\beta) = f_1(\beta) - \beta f_2(\beta) = 0, \tag{10.42}
$$

where $f_1(\beta) = \mathbb{E}\left[\int_{D_i(\beta)}^{D_{i+1}(\beta)} (X_t - \hat{X}_t)^2 dt\right]$ and $f_2(\beta) = \mathbb{E}[D_{i+1}(\beta) - D_i(\beta)]$. The function $f(\beta)$ has several nice properties, which are asserted in the following lemma.

LEMMA 10.5 *The function $f(\beta)$ has the following properties:*
(i) $f(\beta)$ is concave, continuous, and strictly decreasing on β,
(ii) $f(\mathsf{mse}_{Y_i}) > 0$ and $\lim_{\beta \to \mathsf{mse}_\infty^-} f(\beta) = -\infty$.

The uniqueness of β follows immediately from Lemma 10.5.

Now we are ready to present the associated algorithms for solving β.

Bisection Search

Because $f(\beta)$ is decreasing and has a unique root, one can use a bisection search method to solve (10.14), (10.23), and (10.32), which is illustrated in Algorithm 1. The bisection search method has a globally linear convergence speed.

Newton's Method

To achieve an even faster convergence speed, we can use Newton's method [56],

$$
\beta_{k+1} = \beta_k - \frac{f(\beta_k)}{f'(\beta_k)}, \tag{10.43}
$$

Algorithm 2 Newton's method for finding β

given tolerance $\epsilon > 0$.
Pick initial value $\beta_0 \in [\mathsf{mse}_{\text{opt}}, \mathsf{mse}_\infty)$.
repeat
  $\beta_{k+1} := \beta_k - \frac{f(\beta_k)}{f'(\beta_k)}$.
until $|\frac{f(\beta_k)}{f'(\beta_k)}| \le \epsilon$.
return β_{k+1}.

to solve (10.23), as shown in Algorithm 2. We suggest choosing the initial value β_0 of Newton's method from the set $[\mathsf{mse}_{\text{opt}}, \mathsf{mse}_\infty)$, that is, β_0 is larger than the root $\mathsf{mse}_{\text{opt}}$. Such an initial value β_0 can be found by taking a few bisection search iterations. Because $f(\beta)$ is a concave function, the choice of the initial value $\beta_0 \in [\mathsf{mse}_{\text{opt}}, \mathsf{mse}_\infty)$ ensures that β_k is a decreasing sequence converging to $\mathsf{mse}_{\text{opt}}$ [57]. Moreover, because $R_1(\cdot)$ and $R_2(\cdot)$ are twice continuously differentiable, the function $f(\beta)$ is twice continuously differentiable. Therefore, Newton's method is known to have a locally quadratic convergence speed in the neighborhood of the root $\mathsf{mse}_{\text{opt}}$ [56, Chapter 2].

Fixed-Point Iterations

Newton's method requires to compute the gradient $f'(\beta_k)$, which can be solved by a finite-difference approximation, as in the secant method [56]. In the sequel, we introduce another approximation approach of Newton's method, which is of independent interest. In Theorem 10.2, we have shown that $\mathsf{mse}_{\text{opt}}$ is the optimal solution for $f_1(\beta)/f_2(\beta)$. Hence, the gradient of $f_1(\beta)/f_2(\beta)$ is equal to zero at the optimal solution $\beta = \mathsf{mse}_{\text{opt}}$, which leads to

$$f_1'(\mathsf{mse}_{\text{opt}})f_2(\mathsf{mse}_{\text{opt}}) - f_1(\mathsf{mse}_{\text{opt}})f_2'(\mathsf{mse}_{\text{opt}}) = 0. \tag{10.44}$$

Therefore,

$$\mathsf{mse}_{\text{opt}} = \frac{f_1(\mathsf{mse}_{\text{opt}})}{f_2(\mathsf{mse}_{\text{opt}})} = \frac{f_1'(\mathsf{mse}_{\text{opt}})}{f_2'(\mathsf{mse}_{\text{opt}})}. \tag{10.45}$$

Because $f_1(\beta)$ and $f_2(\beta)$ are smooth functions, when β_k is in the neighborhood of $\mathsf{mse}_{\text{opt}}$, (10.45) implies that $f_1'(\beta_k) - \beta_k f_2'(\beta_k) \approx f_1'(\mathsf{mse}_{\text{opt}}) - \mathsf{mse}_{\text{opt}} f_2'(\mathsf{mse}_{\text{opt}}) = 0$. Substituting this into (10.43) yields

$$\begin{aligned}\beta_{k+1} &= \beta_k - \frac{f_1(\beta_k) - \beta_k f_2(\beta_k)}{f_1'(\beta_k) - f_2(\beta_k) - \beta_k f_2'(\beta_k)} \\ &\approx \beta_k - \frac{f_1(\beta_k) - \beta_k f_2(\beta_k)}{-f_2(\beta_k)} \\ &= \frac{f_1(\beta_k)}{f_2(\beta_k)},\end{aligned} \tag{10.46}$$

Algorithm 3 Fixed-point iterations for finding β

given tolerance $\epsilon > 0$.
Pick initial value $\beta_0 \in [\mathsf{mse}_{\text{opt}}, \mathsf{mse}_{\infty})$.
repeat
$\beta_{k+1} := \frac{f_1(\beta_k)}{f_2(\beta_k)}$.
until $|\beta_{k+1} - \frac{f_1(\beta_k)}{f_2(\beta_k)}| \le \epsilon$.
return β_{k+1}.

which is a fixed-point iterative algorithm that was recently proposed in [58]. Similar to Newton's method, the fixed-point updates in (10.46) converge to $\mathsf{mse}_{\text{opt}}$ if the initial value $\beta_0 \in [\mathsf{mse}_{\text{opt}}, \mathsf{mse}_{\infty})$. Moreover, (10.46) has a locally quadratic convergence speed. See [58] for a proof of these results.

A comparison of these three algorithms is shown in [13]. One can observe that the fixed-point updates and Newton's method converge faster than bisection search.

More Insights

We note that although (10.14), (10.23), and (10.32), and equivalently (10.42), has a unique root $\mathsf{mse}_{\text{opt}}$, the fixed-point equation

$$h(\beta) = \frac{f_1(\beta)}{f_2(\beta)} - \beta = \frac{f_1(\beta) - \beta f_2(\beta)}{f_2(\beta)} = 0 \tag{10.47}$$

has two roots $\mathsf{mse}_{\text{opt}}$ and mse_{∞} if the signal process X_t is a stable and stationary OU process. This is because the denominator $f_2(\beta)$ in (10.47) increases to ∞ as $\beta \to \mathsf{mse}_{\infty}$, which creates an extra root at $\beta = \mathsf{mse}_{\infty}$. See [13, Figure 6] for an illustration of the two roots of $h(\beta)$. It is shown in [13] that the correct root for computing the optimal threshold is $\mathsf{mse}_{\text{opt}}$. Interestingly, Algorithms 1–2 converge to the desired root $\mathsf{mse}_{\text{opt}}$ instead of mse_{∞}. Finally, we remark that these three algorithms can be used to find the optimal threshold in the age-optimal sampling problem studied in, for example, [1, 15].

10.4.2 Signal-Aware Sampling with a Sampling Rate Constraint

When the sampling rate constraint (10.11) is taken into consideration, a solution to (10.10) for the three processes discussed previously can be expressed in the following theorems:

10.4.2.1 Wiener Process

THEOREM 10.6 *If X_t is a Wiener process where Y_i are i.i.d. with $0 < \mathbb{E}[Y_i] < \infty$, then* (10.12)–(10.14) *is an optimal solution to* (10.10). *The value of $\beta \ge 0$ is determined in two cases: β is the unique root of* (10.14) *if the root of* (10.14) *satisfies*

$$\mathbb{E}[D_{i+1}(\beta) - D_i(\beta)] > 1/f_{\max}; \tag{10.48}$$

otherwise, β is the unique root of

$$\mathbb{E}[D_{i+1}(\beta) - D_i(\beta)] = 1/f_{\max}. \tag{10.49}$$

The optimal objective value to (10.10) *is given by*

$$\mathsf{mse}_{\text{opt}} = \frac{\mathbb{E}\left[\int_{D_i(\beta)}^{D_{i+1}(\beta)}(X_t - \hat{X}_t)^2 dt\right]}{\mathbb{E}[D_{i+1}(\beta) - D_i(\beta)]}. \tag{10.50}$$

One can see that Theorem 10.1 is a special case of Theorem 10.6 when $f_{\max} = \infty$.

10.4.2.2 Ornstein–Uhlenbeck Process (Stable Case, $\theta > 0$)

THEOREM 10.7 *If X_t is a stable and stationary OU process where the Y_i are i.i.d. with $0 < \mathbb{E}[Y_i] < \infty$, then* (10.21)–(10.23) *is an optimal solution to* (10.10)*. The value of $\beta \geq 0$ is determined in two cases: β is the unique root of* (10.23) *if the root of* (10.23) *satisfies*

$$\mathbb{E}[D_{i+1}(\beta) - D_i(\beta)] > 1/f_{\max}; \tag{10.51}$$

otherwise, β is the unique root of

$$\mathbb{E}[D_{i+1}(\beta) - D_i(\beta)] = 1/f_{\max}. \tag{10.52}$$

The optimal objective value to (10.10) *is given by*

$$\mathsf{mse}_{\text{opt}} = \frac{\mathbb{E}\left[\int_{D_i(\beta)}^{D_{i+1}(\beta)}(X_t - \hat{X}_t)^2 dt\right]}{\mathbb{E}[D_{i+1}(\beta) - D_i(\beta)]}. \tag{10.53}$$

Theorem 10.2 is also a special case of Theorem 10.7 when $f_{\max} = \infty$.

10.4.2.3 Ornstein–Uhlenbeck Process (Unstable Case, $\theta < 0$)

THEOREM 10.8 *If X_t is a stable and stationary OU process where the Y_i are i.i.d. with $0 < \mathbb{E}[Y_i] < \infty$, then* (10.30)–(10.32) *is an optimal solution to* (10.10)*. The value of $\beta \geq 0$ is determined in two cases: β is the unique root of* (10.32) *if the root of* (10.32) *satisfies*

$$\mathbb{E}[D_{i+1}(\beta) - D_i(\beta)] > 1/f_{\max}; \tag{10.54}$$

otherwise, β is the unique root of

$$\mathbb{E}[D_{i+1}(\beta) - D_i(\beta)] = 1/f_{\max}. \tag{10.55}$$

The optimal objective value to (10.10) *is given by*

$$\mathsf{mse}_{\text{opt}} = \frac{\mathbb{E}\left[\int_{D_i(\beta)}^{D_{i+1}(\beta)}(X_t - \hat{X}_t)^2 dt\right]}{\mathbb{E}[D_{i+1}(\beta) - D_i(\beta)]}. \tag{10.56}$$

Theorem 10.3 is also a special case of Theorem 10.8 when $f_{\max} = \infty$.

In Theorems 10.6, 10.7, and 10.8 the calculation of β falls into two cases. In one case, β can be computed via Algorithms 1–3.

Algorithm 4 Bisection search method for finding β with rate constraint

given $l = \mathsf{mse}_{Y_i}$, $u = \mathsf{mse}_\infty$, tolerance $\epsilon > 0$.
repeat
  $\beta := (l+u)/2$.
  $o := \mathbb{E}[D_{i+1}(\beta) - D_i(\beta)]$.
  if $o \geq 1/f_{\max}$, $u := \beta$; **else**, $l := \beta$.
until $u - l \leq \epsilon$.
return β.

Algorithm 5 Newton's method for finding β with rate constraint

given tolerance $\epsilon > 0$.
Pick initial value $\beta_0 \in [\beta^*, \mathsf{mse}_\infty)$.
repeat
  $\beta_{k+1} := \beta_k - \frac{g(\beta_k)}{g'(\beta_k)}$.
until $|\frac{g(\beta_k)}{g'(\beta_k)}| \leq \epsilon$.
return β_{k+1}.

For this case to occur, the sampling rate constraint (10.11) needs to be inactive at the root of (10.14), (10.23), and (10.32). Because $D_i(\beta) = S_i(\beta) + Y_i$, we can obtain $\mathbb{E}[D_{i+1}(\beta) - D_i(\beta)] = \mathbb{E}[S_{i+1}(\beta) - S_i(\beta)]$ and hence (10.48), (10.51), and (10.54) hold when the sampling rate constraint (10.11) is inactive.

In the other case, β is selected to satisfy the sampling rate constraint (10.11) with equality, as required in (10.49), (10.52), and (10.55). Before we solve (10.49), (10.52), and (10.55), let us first use $f_2(\beta)$ to express (10.49), (10.52), and (10.55) as

$$g(\beta) = \frac{1}{f_{\max}} - f_2(\beta) = 0. \tag{10.57}$$

LEMMA 10.9 *The function $g(\beta)$ has the following properties:*
(i) $g(\beta)$ is continuous and strictly decreasing on β,
(ii) $g(\mathsf{mse}_{Y_i}) \geq 0$ and $\lim_{\beta \to \mathsf{mse}_\infty^-} g(\beta) = -\infty$.

According to Lemma 10.9, (10.49), (10.52), and (10.55) have a unique root in $[\mathsf{mse}_{Y_i}, \mathsf{mse}_\infty)$, which is denoted as β^*. In addition, the numerical results also suggest that $g(\beta)$ should be concave [13].

The root β^* can be solved by using bisection search and Newton's method, which are explained in Algorithms 4–5, respectively. Similar to the discussions in Section 10.4.1.4, the convergence of Algorithm 4 is ensured by Lemma 10.9. Moreover, if $g(\beta)$ is concave and $\beta_0 \in [\beta^*, \mathsf{mse}_\infty)$, β_k in Algorithm 5 is a decreasing sequence converging to the root β^* of (10.49), (10.52), and (10.55) [57].

10.4.3 Signal-Agnostic Sampling

In signal-agnostic sampling policies, the sampling times S_i are determined based only on the service times Y_i, but not on the observed process $\{X_t, t \geq 0\}$, and the following lemma can be introduced.

LEMMA 10.10 *If $\pi \in \Pi_{signal\text{-}agnostic}$, then the mean-squared estimation error of the Wiener process X_t at time t is*

$$p(\Delta_t) = \mathbb{E}\left[(X_t - \hat{X}_t)^2 | \pi, Y_1, Y_2, \ldots\right] = \Delta_t, \tag{10.58}$$

where Δ_t is the age of information at time t. On the other hand, if X_t is an OU process, then the mean-squared estimation error at time t is

$$p(\Delta_t) = \mathbb{E}\left[(X_t - \hat{X}_t)^2 | \pi, Y_1, Y_2, \ldots\right] = \frac{\sigma^2}{2\theta}\left(1 - e^{-2\theta\Delta_t}\right), \tag{10.59}$$

which is a strictly increasing function of the age Δ_t.

According to Lemma 10.10 and Fubini's theorem, for every policy $\pi \in \Pi_{\text{signal-agnostic}}$,

$$\mathbb{E}\left[\int_0^T (X_t - \hat{X}_t)^2 dt\right] = \mathbb{E}\left[\int_0^T p(\Delta_t) dt\right]. \tag{10.60}$$

Hence, minimizing the mean-squared estimation error among signal-agnostic sampling policies can be formulated as the following MDP for minimizing the expected time-average of the nonlinear age function $p(\Delta_t)$:

$$\mathsf{mse}_{\text{age-opt}} = \inf_{\pi \in \Pi_{\text{signal-agnostic}}} \limsup_{T \to \infty} \frac{1}{T}\mathbb{E}\left[\int_0^T p(\Delta_t) dt\right] \tag{10.61}$$

$$\text{s.t.} \quad \liminf_{n \to \infty} \frac{1}{n}\mathbb{E}\left[\sum_{i=1}^{n}(S_{i+1} - S_i)\right] \geq \frac{1}{f_{\max}}, \tag{10.62}$$

where $\mathsf{mse}_{\text{age-opt}}$ is the optimal value of (10.61). By (10.59), $p(\Delta_t)$ and $\mathsf{mse}_{\text{age-opt}}$ are bounded. Because $\Pi_{\text{signal-agnostic}} \subset \Pi$, it follows immediately that $\mathsf{mse}_{\text{opt}} \leq \mathsf{mse}_{\text{age-opt}}$.

Problem (10.61) is one instance of the problems recently solved in Corollary 3 of [1] for general strictly increasing functions $p(\cdot)$.

10.4.3.1 Signal-Agnostic Sampling without a Sampling Rate Constraint

From this, a solution to (10.61) for signal-agnostic sampling without rate constraint is given in the following theorem.

THEOREM 10.11 *If X_t is a Gauss–Markov process with $f_{\max} = \infty$ and the Y_i are i.i.d. with $0 < \mathbb{E}[Y_i] < \infty$, then $(S_1(\beta), S_2(\beta), \ldots)$ where the a parameter β is an optimal solution to* (10.61), *where*

$$S_{i+1}(\beta) = \inf\left\{t \geq D_i(\beta) : \mathbb{E}[(X_{t+Y_{i+1}} - \hat{X}_{t+Y_{i+1}})^2] \geq \beta\right\}, \tag{10.63}$$

$D_i(\beta) = S_i(\beta) + Y_i$, and the optimal threshold β is the root of

$$\mathbb{E}\left[\int_{D_i(\beta)}^{D_{i+1}(\beta)} (X_t - \hat{X}_t)^2 dt\right] - \beta\mathbb{E}[D_{i+1}(\beta) - D_i(\beta)] = 0. \tag{10.64}$$

The optimal objective value to (10.61) *is given by*

$$\mathsf{mse}_{\text{age-opt}} = \frac{\mathbb{E}\left[\int_{D_i(\beta)}^{D_{i+1}(\beta)} (X_t - \hat{X}_t)^2 dt\right]}{\mathbb{E}[D_{i+1}(\beta) - D_i(\beta)]}. \tag{10.65}$$

10.4.3.2 Signal-Agnostic Sampling with a Sampling Rate Constraint

On the other hand, a solution to (10.61) when (10.62) is taken into consideration is given in the following theorem.

THEOREM 10.12 *If X_t is a Gauss–Markov process where the Y_i's are i.i.d. with $0 < \mathbb{E}[Y_i] < \infty$, then* (10.63)–(10.64) *is an optimal solution to* (10.61). *The value of threshold $\beta \geq 0$ is determined in two cases: β is the root of* (10.64), *if the root of* (10.64) *satisfies*

$$\mathbb{E}[D_{i+1}(\beta) - D_i(\beta)] > \frac{1}{f_{\max}}; \tag{10.66}$$

otherwise, β is the root of

$$\mathbb{E}[D_{i+1}(\beta) - D_i(\beta)] = \frac{1}{f_{\max}}. \tag{10.67}$$

The optimal objective value to (10.61) *is given by*

$$\mathsf{mse}_{\text{age-opt}} = \frac{\mathbb{E}\left[\int_{D_i(\beta)}^{D_{i+1}(\beta)} (X_t - \hat{X}_t)^2 dt\right]}{\mathbb{E}[D_{i+1}(\beta) - D_i(\beta)]}. \tag{10.68}$$

Theorem 10.12 follows from Corollary 3 of [1] and Lemma 10.10. Similar to the case of signal-aware sampling, the roots of (10.64) and (10.67) can be solved by using Algorithms 1–5. In fact, Algorithms 1–5 can be used for minimizing general monotonic age penalty functions [1].

10.5 Numerical Results

In this section, we evaluate the estimation error achieved by the following four sampling policies: 1. Uniform sampling, 2. Zero-wait sampling [14, 18], 3. Age-optimal sampling [1], and 4. MSE-optimal sampling.

Let $\mathsf{mse}_{\text{uniform}}$, $\mathsf{mse}_{\text{zero-wait}}$, $\mathsf{mse}_{\text{age-opt}}$, and $\mathsf{mse}_{\text{opt}}$ be the MSEs of uniform sampling, zero-wait sampling, age-optimal sampling, MSE-optimal sampling, respectively. We can obtain

$$\begin{aligned} &\mathsf{mse}_{Y_i} \leq \mathsf{mse}_{\text{opt}} \leq \mathsf{mse}_{\text{age-opt}} \leq \mathsf{mse}_{\text{uniform}} \leq \mathsf{mse}_{\infty}, \\ &\mathsf{mse}_{\text{age-opt}} \leq \mathsf{mse}_{\text{zero-wait}} \leq \mathsf{mse}_{\infty}, \end{aligned} \tag{10.69}$$

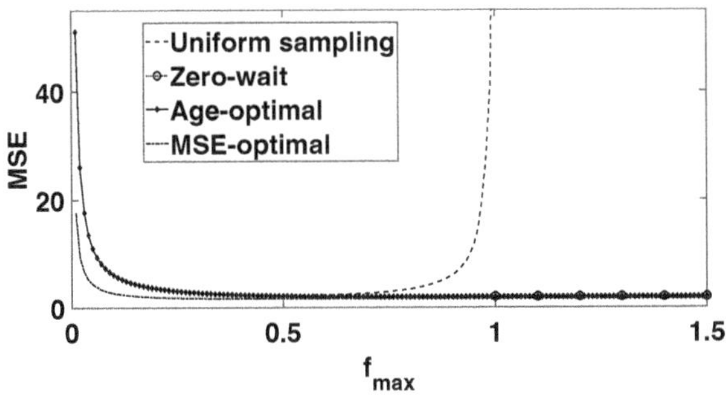

Figure 10.3 MSE vs. $f_{\max}$ trade-off for *i.i.d.* exponential service time for Wiener process with $\mathbb{E}[Y_i] = 1$.

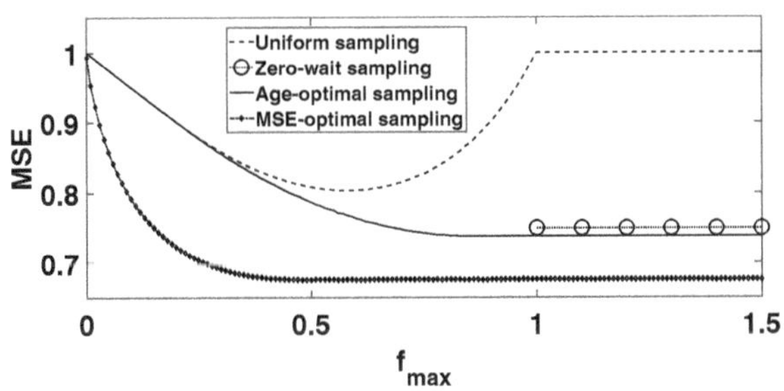

Figure 10.4 MSE vs. $f_{\max}$ trade-off for *i.i.d.* exponential service time with $\mathbb{E}[Y_i] = 1$, where the parameters of the OU process are $\sigma = 1$ and $\theta = 0.5$.

whenever zero-wait sampling is feasible, which fit with our numerical results. The expectations in (10.37) and (10.38) are evaluated by taking the average over 1 million samples. For OU process, the parameters given by $\sigma = 1$, $\theta = 0.5$, and μ can be chosen arbitrarily because it does not affect the estimation error.

Figures 10.3 and 10.4 illustrate the trade-off between the MSE and $f_{\max}$ for *i.i.d.* exponential service times with mean $\mathbb{E}[Y_i] = 1$. Because $\mathbb{E}[Y_i] = 1$, the maximum throughput of the queue is 1. For Figure 10.4, the lower bound mse_{Y_i} is 0.5 and the upper bound mse_∞ is 1. In fact, as Y_i is an exponential random variable with mean 1, $\frac{\sigma^2}{2\theta}(1 - e^{-2\theta Y_i})$ has a uniform distribution on [0, 1]. Hence, $\mathsf{mse}_{Y_i} = 0.5$. For small values of $f_{\max}$, age-optimal sampling is similar to uniform sampling, and hence $\mathsf{mse}_{\text{age-opt}}$ and $\mathsf{mse}_{\text{uniform}}$ are close to each other in the regime. However, as $f_{\max}$ grows, $\mathsf{mse}_{\text{uniform}}$ reaches the upper bound mse_∞ and remains constant for $f_{\max} \geq 1$. This is because the queue length of uniform sampling is large at high sampling frequencies. In particular, when $f_{\max} \geq 1$, the queue length of uniform sampling is infinite. On the other hand, $\mathsf{mse}_{\text{age-opt}}$ and $\mathsf{mse}_{\text{opt}}$ decrease with respect to $f_{\max}$. The reason behind this is that the set of feasible sampling policies satisfying the constraint in (10.10) and (10.61) becomes larger as $f_{\max}$ grows, and hence the optimal values of (10.10) and (10.61) are decreasing in $f_{\max}$. As we expected, $\mathsf{mse}_{\text{zero-wait}}$ is larger than $\mathsf{mse}_{\text{opt}}$ and $\mathsf{mse}_{\text{age-opt}}$. Moreover, all of them are between the lower bound mse_{Y_i} and the upper bound mse_∞.

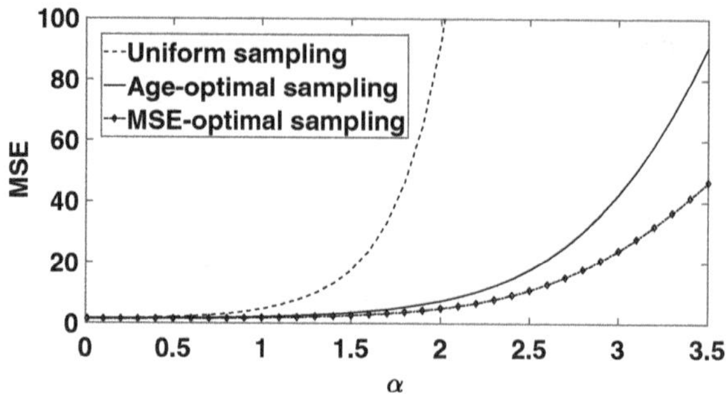

Figure 10.5 MSE vs. the scale parameter α of *i.i.d.* normalized log-normal service time distribution for Wiener process with $\mathbb{E}[Y_i] = 1$ and $f_{\max} = 0.8$. Zero-wait sampling is not feasible here, as $f_{\max} < 1/\mathbb{E}[Y_i]$ and hence is not plotted.

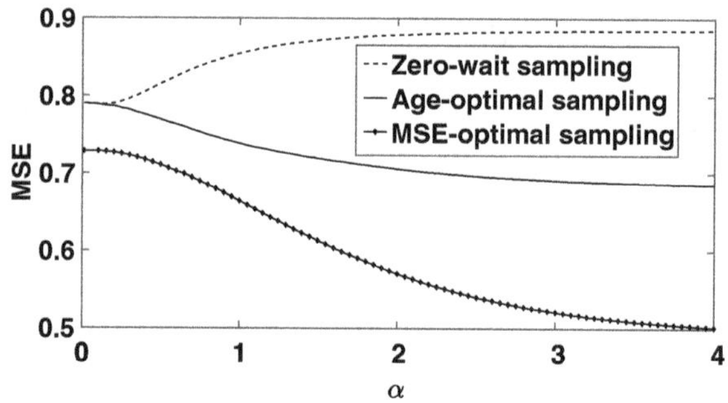

Figure 10.6 MSE vs. the scale parameter α of *i.i.d.* normalized log-normal service time distribution with $\mathbb{E}[Y_i] = 1$ and $f_{\max} = 0.8$, where the parameters of the OU process are $\sigma = 1$ and $\theta = 0.5$. Zero-wait sampling is not feasible here, as $f_{\max} < 1/\mathbb{E}[Y_i]$ and hence is not plotted.

Figures 10.5 and 10.6 depict the MSE of *i.i.d.* normalized log-normal service time for $f_{\max} = 0.8$. Figures 10.7 and 10.8 depict the MSE of *i.i.d.* normalized log-normal service time for $f_{\max} = 1.2$, where $Y_i = e^{\alpha X_i}/\mathbb{E}[e^{\alpha X_i}], \alpha > 0$ is the scale parameter of log-normal distribution, and $(X_1, X_2, \dots)$ are *i.i.d.* Gaussian random variables with zero mean and unit variance. Because $\mathbb{E}[Y_i] = 1$, the maximum throughput of the queue is 1. In Figures 10.5 and 10.6, since $f_{\max} < 1$, zero-wait sampling is not feasible and hence is not plotted. As the scale parameter α grows, the tail of the log-normal distribution becomes heavier.

In Figures 10.5 and 10.7, $\mathsf{mse}_{\text{age-opt}}$ and $\mathsf{mse}_{\text{opt}}$ grow quickly in α. But in Figures 10.6 and 10.8, $\mathsf{mse}_{\text{age-opt}}$ and $\mathsf{mse}_{\text{opt}}$ drop with α. This phenomenon may look surprising at first sight. To understand this phenomenon, let us consider the age penalty function $p(\Delta_t)$ in (10.59) for the OU process. As the scale parameter α grows, the service time tends to become either shorter or much longer than the mean $\mathbb{E}[Y_i]$, rather than being close to $\mathbb{E}[Y_i]$. When Δ_t is short, $p(\Delta_t)$ reduces quickly in Δ_t; meanwhile, when Δ_t is quite long, $p(\Delta_t)$ cannot increase much because it is upper bounded by

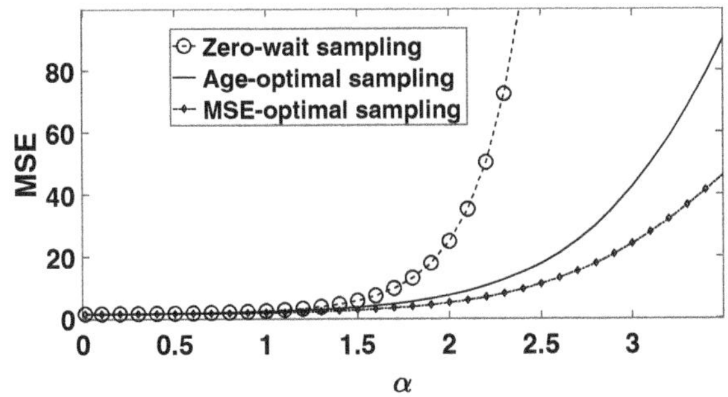

Figure 10.7 MSE vs. the scale parameter α of *i.i.d.* normalized log-normal service time distribution for Wiener process $\mathbb{E}[Y_i] = 1$ and $f_{\max} = 1.2$. Uniform sampling is not feasible here and hence is not plotted.

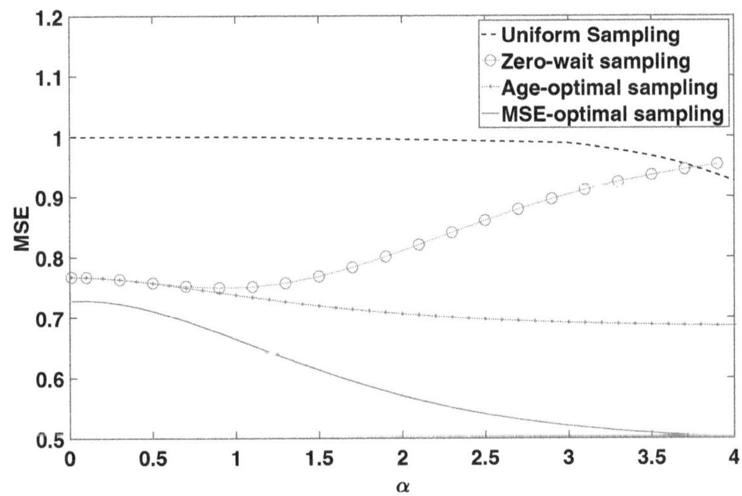

Figure 10.8 MSE vs. the scale parameter α of *i.i.d.* normalized log-normal service time distribution $\mathbb{E}[Y_i] = 1$ and $f_{\max} = 1.2$, where the parameters of the OU process are $\sigma = 1$, $\theta = 0.5$.

mse_∞. Because of these two reasons, the average age penalty $\mathsf{mse}_{\text{age-opt}}$ decreases in α. The dropping of $\mathsf{mse}_{\text{opt}}$ in α can be understood in a similar fashion.

On the other hand, the age penalty function of the Wiener process is $p(\Delta_t) = \Delta_t$, which is quite different from the case considered here. We also observe that in all the four figures, the gap between $\mathsf{mse}_{\text{opt}}$ and $\mathsf{mse}_{\text{age-opt}}$ increases as α grows.

10.6 Summary

In this chapter, we have described the optimal sampling policies for minimizing the estimation error for three scalar Gauss–Markov signal processes. The optimal sampler design problem is solved with and without a sampling rate constraint. A smaller estimation error has been achieved by exploiting causal knowledge of the signal values. On the other hand, when the sampler has no knowledge about the signal process, the optimal sampling policy is to optimize a nonlinear age function. The problem is formulated as a continuous-time (constrained) MDP with a continuous state space. The optimal sampling policy is a threshold policy on instantaneous estimation error and the threshold is found exactly. The optimal sampling policies can be computed by low-complexity algorithms, and the curse of dimensionality is circumvented.

References

[1] Y. Sun and B. Cyr, "Sampling for data freshness optimization: Non-linear age functions," *J. Commun. Netw.*, vol. 21, no. 3, pp. 204–219, 2019.

[2] R. Yates, Y. Sun, D. R. Brown, S. K. Kaul, E. Modiano, and S. Ulukus, "Age of information: An introduction and survey," *ArXiv*, vol. abs/2007.08564, 2020.

[3] M. A. Abd-Elmagid, N. Pappas, and H. S. Dhillon, "On the role of age of information in the internet of things," *IEEE Communications Magazine*, vol. 57, no. 12, pp. 72–77, Dec. 2019.

[4] G. E. Uhlenbeck and L. S. Ornstein, "On the theory of the Brownian motion," *Phys. Rev.*, vol. 36, pp. 823–841, Sept. 1930.

[5] J. L. Doob, "The Brownian movement and stochastic equations," *Annals of Mathematics*, vol. 43, no. 2, pp. 351–369, 1942.

[6] L. Evans, S. Keef, and J. Okunev, "Modelling real interest rates," *Journal of Banking and Finance*, vol. 18, no. 1, pp. 153–165, 1994.

[7] A. Cika, M. Badiu, and J. Coon, "Quantifying link stability in Ad Hoc wireless networks subject to Ornstein-Uhlenbeck mobility," in *IEEE ICC*, 2019.

[8] H. Kim, J. Park, M. Bennis, and S. Kim, "Massive UAV-to-ground communication and its stable movement control: A mean-field approach," in *IEEE SPAWC*, June 2018, pp. 1–5.

[9] G. M. Lipsa and N. C. Martins, "Remote state estimation with communication costs for first-order LTI systems," *IEEE Trans. Auto. Control*, vol. 56, no. 9, pp. 2013–2025, Sept. 2011.

[10] M. Ribero and H. Vikalo, "Communication-efficient federated learning via optimal client sampling," *ArXiv*, vol. abs/2007.15197, 2020.

[11] E. Vinogradov, H. Sallouha, S. D. Bast, M. M. Azari, and S. Pollin, "Tutorial on UAV: A blue sky view on wireless communication," 2019, coRR, abs/1901.02306.

[12] Y. Sun, Y. Polyanskiy, and E. Uysal, "Sampling of the Wiener process for remote estimation over a channel with random delay," *IEEE Trans. Inf. Theory*, vol. 66, no. 2, pp. 1118–1135, Feb. 2020.

[13] T. Z. Ornee and Y. Sun, "Sampling for remote estimation through queues: Age of information and beyond," *CoRR*, vol. abs/1902.03552, 2019.

[14] S. Kaul, R. D. Yates, and M. Gruteser, "Real-time status: How often should one update?" in *IEEE INFOCOM*, 2012.

[15] Y. Sun and B. Cyr, "Information aging through queues: A mutual information perspective," in *IEEE SPAWC Workshop*, 2018.

[16] A. M. Bedewy, Y. Sun, S. Kompella, and N. B. Shroff, "Age-optimal sampling and transmission scheduling in multi-source systems," in *ACM MobiHoc*, 2019.

[17] C. Kam, S. Kompella, G. D. Nguyen, and A. Ephremides, "Effect of message transmission path diversity on status age," *IEEE Trans. Inf. Theory*, vol. 62, no. 3, pp. 1360–1374, Mar. 2016.

[18] Y. Sun, E. Uysal-Biyikoglu, R. D. Yates, C. E. Koksal, and N. B. Shroff, "Update or wait: How to keep your data fresh," *IEEE Trans. Inf. Theory*, vol. 63, no. 11, pp. 7492–7508, Nov. 2017.

[19] C. Kam, S. Kompella, G. D. Nguyen, J. E. Wieselthier, and A. Ephremides, "On the age of information with packet deadlines," *IEEE Trans. Inf. Theory*, vol. 64, no. 9, pp. 6419–6428, Sept. 2018.

[20] R. D. Yates and S. K. Kaul, "The age of information: Real-time status updating by multiple sources," *IEEE Trans. Inf. Theory*, vol. 65, no. 3, pp. 1807–1827, Mar. 2019.

[21] Q. He, D. Yuan, and A. Ephremides, "Optimal link scheduling for age minimization in wireless systems," *IEEE Trans. Inf. Theory*, vol. 64, no. 7, pp. 5381–5394, July 2018.

[22] C. Joo and A. Eryilmaz, "Wireless scheduling for information freshness and synchrony: Drift-based design and heavy-traffic analysis," *IEEE/ACM Trans. Netw.*, vol. 26, no. 6, pp. 2556–2568, Dec 2018.

[23] A. M. Bedewy, Y. Sun, and N. B. Shroff, "Minimizing the age of the information through queues," *IEEE Trans. Inf. Theory*, vol. 65, no. 8, pp. 5215–5232, Aug. 2019.

[24] ——, "The age of information in multihop networks," *IEEE/ACM Trans. Netw.*, vol. 27, no. 3, pp. 1248–1257, June 2019.

[25] Y. Sun, E. Uysal-Biyikoglu, and S. Kompella, "Age-optimal updates of multiple information flows," in *IEEE INFOCOM AoI Workshop*, 2018.

[26] I. Kadota, A. Sinha, and E. Modiano, "Optimizing age of information in wireless networks with throughput constraints," in *IEEE INFOCOM*, April 2018, pp. 1844–1852.

[27] R. Talak, S. Karaman, and E. Modiano, "Optimizing information freshness in wireless networks under general interference constraints," in *ACM MobiHoc*, 2018.

[28] N. Lu, B. Ji, and B. Li, "Age-based scheduling: Improving data freshness for wireless real-time traffic," in *ACM MobiHoc*, 2018.

[29] A. Maatouk, Y. Sun, A. Ephremides, and M. Assaad, "Status updates with priorities: Lexicographic optimality," in *IEEE/IFIP WiOpt*, 2020.

[30] B. Zhou and W. Saad, "Joint status sampling and updating for minimizing age of information in the Internet of things," *IEEE Trans. Commun.*, vol. 67, no. 11, pp. 7468–7482, Nov. 2019.

[31] ——, "Minimum age of information in the Internet of things with non-uniform status packet sizes," *IEEE Trans. Wireless Commun.*, vol. 19, no. 3, pp. 1933–1947, 2020.

[32] A. Kosta, N. Pappas, and V. Angelakis, *Age of Information: A New Concept, Metric, and Tool*. Now Publishers Inc, 2018.

[33] Y. Sun, I. Kadota, R. Talak, and E. Modiano, *Age of Information: A New Metric for Information Freshness*. Morgan & Claypool, 2019.

[34] A. M. Bedewy, Y. Sun, S. Kompella, and N. B. Shroff, "Optimal sampling and scheduling for timely status updates in multi-source networks," 2020, https://arxiv.org/abs/2001.09863.

[35] Y. Xiao and Y. Sun, "A dynamic jamming game for real-time status updates," in *IEEE INFOCOM AoI Workshop*, April 2018, pp. 354–360.

[36] M. A. Abd-Elmagid, H. S. Dhillon, and N. Pappas, "A reinforcement learning framework for optimizing age of information in rf-powered communication systems," *IEEE Transactions on Communications*, vol. 68, no. 8, pp. 4747–4760, Aug. 2020.

[37] ——, "Aoi-optimal joint sampling and updating for wireless powered communication systems," *IEEE Transactions on Vehicular Technology*, vol. 69, no. 11, pp. 14 110–14 115, Nov. 2020.

[38] B. Hajek, K. Mitzel, and S. Yang, "Paging and registration in cellular networks: Jointly optimal policies and an iterative algorithm," *IEEE Trans. Inf. Theory*, vol. 54, no. 2, pp. 608–622, Feb. 2008.

[39] A. Nayyar, T. Başar, D. Teneketzis, and V. V. Veeravalli, "Optimal strategies for communication and remote estimation with an energy harvesting sensor," *IEEE Trans. Auto. Control*, vol. 58, no. 9, pp. 2246–2260, Sept. 2013.

[40] M. Rabi, G. V. Moustakides, and J. S. Baras, "Adaptive sampling for linear state estimation," *SIAM Journal on Control and Optimization*, vol. 50, no. 2, pp. 672–702, 2012.

[41] K. Nar and T. Başar, "Sampling multidimensional Wiener processes," in *IEEE CDC*, Dec. 2014, pp. 3426–3431.

[42] X. Gao, E. Akyol, and T. Başar, "Optimal communication scheduling and remote estimation over an additive noise channel," *Automatica*, vol. 88, pp. 57–69, 2018.

[43] J. Chakravorty and A. Mahajan, "Remote estimation over a packet-drop channel with Markovian state," *IEEE Trans. Auto. Control*, vol. 65, no. 5, pp. 2016–2031, 2020.

[44] C.-H. Tsai and C.-C. Wang, "Unifying AoI minimization and remote estimation: Optimal sensor/controller coordination with random two-way delay," in *IEEE INFOCOM*, 2020.

[45] N. Guo and V. Kostina, "Optimal causal rate-constrained sampling for a class of continuous Markov processes," in *IEEE ISIT*, 2020.

[46] A. Arafa, K. A. Banawan, K. G. Seddik, and H. V. Poor, "Timely estimation using coded quantized samples," *ArXiv*, vol. abs/2004.12982, 2020.

[47] V. Jog, R. J. La, and N. C. Martins, "Channels, learning, queueing and remote estimation systems with a utilization-dependent component," 2019, coRR, abs/1905.04362.

[48] X. Song and J. W. S. Liu, "Performance of multiversion concurrency control algorithms in maintaining temporal consistency," in *Proceedings, Fourteenth Annual International Computer Software and Applications Conference*, Oct. 1990, pp. 132–139.

[49] S. M. Ross, *Applied Probability Models with Optimization Applications*. San Francisco, CA: Holden-Day, 1970.

[50] R. A. Maller, G. Müller, and A. Szimayer, "Ornstein-Uhlenbeck processes and extensions," in *Handbook of Financial Time Series*, T. Mikosch, J.-P. Kreiß, R. A. Davis, and T. G. Andersen, eds. Berlin, Heidelberg: Springer Berlin Heidelberg, 2009, pp. 421–437.

[51] G. Peskir and A. N. Shiryaev, *Optimal Stopping and Free-Boundary Problems*. Basel, Switzerland: Birkhäuswer Verlag, 2006.

[52] T. Soleymani, S. Hirche, and J. S. Baras, "Optimal information control in cyber-physical systems," *IFAC-PapersOnLine*, vol. 49, no. 22, pp. 1–6, 2016.

[53] I. Gradshteyn and I. Ryzhik, *Table of Integrals, Series, and Products*, 7th ed. Academic Press, 2007.

[54] B. Øksendal, *Stochastic Differential Equations: An Introduction with Applications*, 5th ed. Springer-Verlag Berlin Heidelberg, 2000.

[55] F. W. Olver, D. W. Lozier, R. F. Boisvert, and C. W. Clark, *NIST Handbook of Mathematical Functions*. Cambridge University Press, 2010.

[56] J. H. Mathews and K. K. Fink, *Numerical Methods Using Matlab*. Simon & Schuster, Inc., 1998.

[57] M. Spivak, *Calculus*, 4th ed. Publish or Perish, 2008.

[58] C.-H. Tsai and C.-C. Wang, "Age-of-information revisited: Two-way delay and distribution-oblivious online algorithm," in *IEEE ISIT*, 2020.

11 Relation between Value and Age of Information in Feedback Control

Touraj Soleymani, John S. Baras, and Karl H. Johansson

Abstract: In this chapter, we investigate the value of information as a more comprehensive instrument than the age of information for optimally shaping the information flow in a networked control system. In particular, we quantify the value of information based on the variation in a value function and discuss the structural properties of this metric. Through our analysis, we establish the mathematical relation between the value of information and the age of information. We prove that the value of information is, in general, a function of an estimation discrepancy that depends on the age of information and the primitive variables. In addition, we prove that there exists a condition under which the value of information becomes completely expressible in terms of the age of information. Nonetheless, we show that this condition is not achievable without a degradation in the performance of the system.

11.1 Introduction

This chapter is concerned with networked control systems, which are distributed feedback systems where the underlying components, that is, sensors, actuators, and controllers, are connected to each other via communication channels. Such systems have flexible architectures with potential applications in a wide range of areas such as autonomous driving, remote surgery, and space exploration. In these systems, the information that flows through the communication channels plays a key role. Consider, for instance, an unmanned vehicle that should be navigated remotely. For the purpose of the navigation of this vehicle, the sensory information needs to be transmitted from the sensors of the vehicle to a remote controller, and in turn the control commands need to be transmitted from the controller to the actuators of the vehicle. Note that, on the one hand, there exist often different costs and constraints that can hinder the timely updates of the components of a networked control system. On the other hand, it is not hard to see that steering a networked control system based on quite out-of-date information could result in a catastrophic failure. Given this situation, our objective here is to put forward a systematic way for optimally shaping the information flow in a networked control system such that a specific level of performance is attained.

As discussed in various chapters of the present book, the *age of information* [1], which measures the freshness of information at each time, is an appropriate instrument for shaping the information flow in many networked real-time systems. However, we show in this chapter that a more comprehensive instrument than the age of information is required when one deals with networked control systems. This other metric is the *value of information* [2–4], which measures the difference between the benefit and the cost of information at each time.

In our study, we consider a basic networked control system where the channel between the sensor and the controller is costly, and the channel between the controller and the actuator is cost-free. We begin our analysis by making a trade-off that is defined between the packet rate and the regulation cost. The decision makers are an event trigger and the controller. The event trigger decides about the transmission of information from the sensor to the controller at each time, and the controller decides about the control input for the actuator at each time. For the purpose of this study, we design the controller based on the certainty-equivalence principle, and mainly focus on the design of the event trigger. We show that the optimal triggering policy permits a transmission only when the value of information is nonnegative. We quantify the value of information at each time as the variation in a value function with respect to a piece of information that can be communicated to the controller. Through our analysis, we establish the mathematical relation between the value of information and the age of information. We prove that the value of information is, in general, a function of an estimation discrepancy that depends on the age of information and the primitive variables. In addition, we prove that there exists a condition associated with the information structure of the system under which the value of information becomes completely expressible in terms of the age of information. Nonetheless, we show that this condition is not achievable without a degradation in the performance of the system.

There are multiple works that are closely related to our study [5–11]. In these works, optimal triggering policies were characterized in settings that are different from ours. In particular, Lipsa and Martins [5] used majorization theory to study the estimation of a scalar Gauss–Markov process, and proved that the optimal triggering policy is symmetric threshold. Molin and Hirche [6] studied the convergence properties of an iterative algorithm for the estimation of a scalar Markov process with arbitrary noise distribution and found a result coinciding with that in [5]. Chakravorty and Mahajan [7] investigated the estimation of a scalar autoregressive Markov process with symmetric noise distribution based on renewal theory and proved that the optimal triggering policy is symmetric threshold. Rabi et al. [8] formulated the estimation of the scalar Wiener and scalar Ornstein–Uhlenbeck processes as an optimal multiple stopping time problem by discarding the signaling effect and showed that the optimal triggering policy is symmetric threshold. Guo and Kostina [9] addressed the estimation of the scalar Wiener and scalar Ornstein–Uhlenbeck processes in the presence of the signaling effect, and obtained a result that matches with that in [8]. They also looked at the estimation of the scalar Wiener process with fixed communication delay in the presence of the signaling effect in [10] and obtained a similar structural result. Furthermore, Sun et al. [11] studied the estimation of the scalar Wiener process with random communication delay by discarding the signaling effect and showed that the optimal triggering policy is symmetric threshold. In contrast to the preceding works, we here focus on the estimation of a Gauss–Markov process with random processing delay and characterize the optimal triggering policy. Our study builds on the framework we developed previously for the value of information in [2–4, 12].

This chapter is organized in the following way. We formulate the problem in Section 11.2. Then, we present the main results in Section 11.3, and provide a numerical example in Section 11.4. Finally, we conclude the chapter in Section 11.5.

11.2 Problem Statement

In this section, we mathematically formulate the rate-regulation trade-off as a stochastic optimization problem. In our setting, the process under control satisfies the following equations with an n-dimensional state, an m-dimensional input, and an n-dimensional output:

$$x_{k+1} = Ax_k + Bu_k + w_k \tag{11.1}$$

$$y_k = x_{k-\tau_k} \tag{11.2}$$

for $k \in \mathcal{K} = \{0, 1, \ldots, N\}$ with initial condition x_0, where $x_k \in \mathbb{R}^n$ is the state of the process, $A \in \mathbb{R}^{n\times n}$ is the state matrix, $B \in \mathbb{R}^{n\times m}$ is the input matrix, $u_k \in \mathbb{R}^m$ is the control input decided by the controller, $w_k \in \mathbb{R}^n$ is a Gaussian white noise with zero mean and covariance $W \succ 0$, $y_k \in \mathbb{R}^n$ is the output of the process, $\tau_k \in \mathbb{N}_0$ is a random processing delay with known probability distribution, and N is a finite time horizon. It is assumed that x_0 is a Gaussian vector with mean m_0 and covariance M_0, that $\tau_0 = 0$, and that x_0, w_k, and τ_k are mutually independent for all $k \in \mathcal{K}$.

The random processing delay can be due to various sensing constraints. One example of the output model is:

$$y_k = \begin{cases} x_k, & \text{with probability } p_k, \\ x_{k-d}, & \text{otherwise}, \end{cases} \tag{11.3}$$

where d is a fixed delay and p_k is a probability. In (11.3), the output of the process at time k is either the current state or a delayed state. Note that when no sensing constraints exist, we have $y_k = x_k$ for $k \in \mathcal{K}$.

The channel that connects the sensor to the controller is costly, errorless, and with one-step delay. Let $\delta_k \in \{0, 1\}$ be the transmission variable in this channel decided the event trigger. Then, y_k is transmitted over the channel and received by the controller after one-step delay if $\delta_k = 1$. Otherwise, nothing is transmitted and received. It is assumed that, in a successful transmission, the total delay of the received observation can be detected by the controller. For the purpose of this study, we design the controller based on the certainty-equivalence principle without the signaling effect (see [3, 4] for more details). This allows us to concentrate on the design of the event trigger. Let the information sets of the event trigger and the controller, including only causal information, at time k be denoted by $\mathcal{I}_k^e$ and $\mathcal{I}_k^c$, respectively. We say that a triggering policy

π is admissible if $\pi = \{\mathsf{P}(\delta_k|\mathcal{I}_k^e)\}_{k=0}^N$, where $\mathsf{P}(\delta_k|\mathcal{I}_k^e)$ is a Borel measurable transition kernel. We represent the admissible triggering policy set by $\mathcal{P}$. The rate-regulation trade-off between the packet rate and the regulation cost can then be expressed by the following stochastic optimization problem:

$$\underset{\pi \in \mathcal{P}}{\text{minimize}}\ \Phi(\pi), \tag{11.4}$$

where

$$\Phi(\pi) = \mathsf{E}\left[\tfrac{(1-\lambda)}{N+1} \textstyle\sum_{k=0}^N \ell\delta_k + \tfrac{\lambda}{N+1} \textstyle\sum_{k=0}^N \left(\|x_{k+1}\|_Q^2 + \|u_k\|_R^2\right)\right], \tag{11.5}$$

where $\lambda \in (0, 1)$ is the trade-off multiplier, ℓ is a weighting coefficient, $Q \succeq 0$ and $R \succ 0$ are weighting matrices, and $\|.\|$ represents the Euclidean norm. In the following, we seek the optimal triggering policy $\pi^\star$ associated with the problem in (11.4).

11.3 Main Results

First, we provide a formal definition of the age of information, a metric that measures the freshness of information at a component at each time.

DEFINITION 1 The age of information at a component at time k is the time elapsed since the generation of the freshest observation received by that component, that is,

$$\mathrm{AoI}_k = k - t, \tag{11.6}$$

where $t \leq k$ is the time of the freshest received observation.

Let $\zeta_k \in [0, \zeta_{k-1} + 1]$ be the age of information at the event trigger. This implies that

$$\zeta_k = \begin{cases} \tau_k, & \text{if } \tau_k < \zeta_{k-1} + 1, \\ \zeta_{k-1} + 1, & \text{otherwise,} \end{cases} \tag{11.7}$$

for $k \in \mathcal{K}$ with initial condition $\zeta_0 = 0$, given the assumption $\tau_0 = 0$. Moreover, let $\eta_k \in [0, \eta_{k-1} + 1]$ be age of information at the controller. This implies that

$$\eta_k = \begin{cases} \zeta_{k-1} + 1, & \text{if } \delta_{k-1} = 1, \\ \eta_{k-1} + 1, & \text{otherwise,} \end{cases} \tag{11.8}$$

for $k \in \mathcal{K}$ with initial condition $\eta_0 = \infty$, by convention. We say that the generated observation $x_{k-\tau_k}$ at time k is informative if $\tau_k < \zeta_{k-1} + 1$. Otherwise, we say it is obsolete. Since the obsolete observations can safely be discarded, the information set of the event trigger can be determined by the set of the informative observations, the communicated informative observations, and the previous decisions, that is, $\mathcal{I}_k^e = \{x_{t-\zeta_t}, x_{t-\eta_t}, \delta_s, u_s \mid 0 \leq t \leq k, 0 \leq s < k\}$, and the information set of the controller by the set of the communicated informative observations and the previous decisions, that is, $\mathcal{I}_k^c = \{x_{t-\eta_t}, \delta_s, u_s \mid 0 \leq t \leq k, 0 \leq s < k\}$.

Given these information sets, in the next two lemmas, we derive the minimum mean-square-error estimators at the event trigger and at the controller.

LEMMA 11.1 *The conditional mean* $\mathsf{E}[x_k|\mathcal{I}_k^e]$ *is the minimum mean-square-error estimator at the event trigger and satisfies*

$$\mathsf{E}[x_k|\mathcal{I}_k^e] = A^{\zeta_k} x_{k-\zeta_k} + \sum_{t=1}^{\zeta_k} A^{t-1} B u_{k-t} \tag{11.9}$$

for $k \in \mathcal{K}$.

Proof Given $\mathcal{I}_k^e$, it is easy to see that $\mathsf{E}[x_k|\mathcal{I}_k^e]$ minimizes the mean-square error at the event trigger. Moreover, from the definition of ζ_k, $x_{k-\zeta_k}$ represents the freshest observation at the event trigger. Writing x_k in terms of $x_{k-\zeta_k}$ and taking the conditional expectation with respect to $\mathcal{I}_k^e$, we obtain the result. □

LEMMA 11.2 *The conditional mean* $\mathsf{E}[x_k|\mathcal{I}_k^c]$ *is the minimum mean-square-error estimator at the controller and satisfies*

$$\mathsf{E}[x_{k+1}|\mathcal{I}_{k+1}^c] = \begin{cases} A^{\zeta_k+1} x_{k-\zeta_k} + \sum_{t=0}^{\zeta_k} A^t B u_{k-t}, & \text{if } \delta_k = 1, \\ A\,\mathsf{E}[x_k|\mathcal{I}_k^c] + B u_k + \imath_k, & \text{otherwise} \end{cases} \tag{11.10}$$

for $k \in \mathcal{K}$ *with initial condition* $\mathsf{E}[x_0|\mathcal{I}_0^c] = m_0$, *where* $\imath_k = A(\mathsf{E}[x_k|\mathcal{I}_k^c, \delta_k = 0] - \mathsf{E}[x_k|\mathcal{I}_k^c])$ *is the signaling residual.*

Proof Given $\mathcal{I}_k^c$, it is easy to see that $\mathsf{E}[x_k|\mathcal{I}_k^c]$ minimizes the mean-square error at the controller. In addition, writing x_{k+1} in terms of $x_{k-\zeta_k}$ and taking the conditional expectation with respect to $\mathcal{I}_{k+1}^c$, we can write

$$\mathsf{E}[x_{k+1}|\mathcal{I}_{k+1}^c] = A^{\zeta_k+1}\,\mathsf{E}[x_{k-\zeta_k}|\mathcal{I}_{k+1}^c] + \sum_{t=0}^{\zeta_k} A^t B u_{k-t}, \tag{11.11}$$

where we used the fact that $\mathsf{E}[w_{k-t}|\mathcal{I}_{k+1}^c] = 0$ for all $t \in [0, \zeta_k]$ since the freshest observation that the controller might receive at time $k+1$ is $x_{k-\zeta_k}$. Now, note that if $\delta_k = 1$, the controller receives $x_{k-\zeta_k}$ at time $k+1$. In this case, $\mathsf{E}[x_{k-\zeta_k}|\mathcal{I}_{k+1}^c] = x_{k-\zeta_k}$. Inserting this into (11.11), we obtain

$$\mathsf{E}[x_{k+1}|\mathcal{I}_{k+1}^c] = A^{\zeta_k+1} x_{k-\zeta_k} + \sum_{t=0}^{\zeta_k} A^t B u_{k-t}.$$

However, if $\delta_k = 0$, the controller receives nothing at time $k+1$. In this case, $\mathsf{E}[x_{k-\zeta_k}|\mathcal{I}_{k+1}^c] = \mathsf{E}[x_{k-\zeta_k}|\mathcal{I}_k^c, \delta_k = 0]$. Hence, we can write $\mathsf{E}[x_{k-\zeta_k}|\mathcal{I}_k^c, \delta_k = 0] = \mathsf{E}[x_{k-\zeta_k}|\mathcal{I}_k^c] + \tilde{x}_k$ for an appropriate residual $\tilde{x}_k$. Inserting this into (11.11), we obtain

$$\begin{aligned} \mathsf{E}[x_{k+1}|\mathcal{I}_{k+1}^c] &= A^{\zeta_k+1}\,\mathsf{E}[x_{k-\zeta_k}|\mathcal{I}_k^c] + \sum_{t=0}^{\zeta_k} A^t B u_{k-t} + A^{\zeta_k+1}\tilde{x}_k \\ &= A\,\mathsf{E}[x_k|\mathcal{I}_k^c] + B u_k + \imath_k, \end{aligned}$$

where we used the definition of $\mathsf{E}[x_k|\mathcal{I}_k^c]$ and introduced $\imath_k = A^{\zeta_k+1}\tilde{x}_k$. Lastly, we can write $\imath_k$ as

$$\begin{aligned} \imath_k &= A^{\zeta_k+1}\big(\mathsf{E}[x_{k-\zeta_k}|\mathcal{I}_k^c, \delta_k = 0] - \mathsf{E}[x_{k-\zeta_k}|\mathcal{I}_k^c]\big) \\ &= A\big(\mathsf{E}[x_k|\mathcal{I}_k^c, \delta_k = 0] - \mathsf{E}[x_k|\mathcal{I}_k^c]\big). \end{aligned}$$

This completes the proof. □

Now, we can express the control inputs based on the certainty equivalence principle as discussed. This leads to the employment of the linear-quadratic regulator with a state estimate at the controller without the signaling residual. More specifically, we have $u_k = -L_k\hat{x}_k$, where $L_k = (B^T S_{k+1} B + R)^{-1} B^T S_{k+1} A$ is the linear-quadratic regulator gain, $\hat{x}_k$ is a state estimate at the controller satisfying (11.10) with $\iota_k = 0$ for all $k \in \mathcal{K}$, and $S_k \succeq 0$ is a matrix that satisfies the algebraic Riccati equation,

$$S_k = Q + A^T S_{k+1} A - A^T S_{k+1} B (B^T S_{k+1} B + R)^{-1} B^T S_{k+1} A, \tag{11.12}$$

for $k \in \mathcal{K}$ with initial condition $S_{N+1} = Q$.

We continue our analysis by presenting four lemmas that provide some properties associated with the estimation error $e_k = x_k - \hat{x}_k$ and the estimation mismatch $\tilde{e}_k = \check{x}_k - \hat{x}_k$, where $\check{x}_k$ is the minimum mean-square-error state estimate at the event trigger.

LEMMA 11.3 *The estimation error e_k satisfies*

$$e_{k+1} = (1 - \delta_k)(A e_k + w_k) + \delta_k \textstyle\sum_{t=0}^{\zeta_k} A^t w_{k-t} \tag{11.13}$$

for $k \in \mathcal{K}$ with initial condition $e_0 = x_0 - m_0$.

Proof To prove, we need to subtract (11.10) from (11.1), given $\iota_k = 0$ for all $k \in \mathcal{K}$. □

LEMMA 11.4 *The following facts are true:*

$$\mathsf{E}[e_{k+1} | \mathcal{I}_k^e] = (1 - \delta_k) A \tilde{e}_k \tag{11.14}$$

$$\mathsf{cov}[e_{k+1} | \mathcal{I}_k^e] = \textstyle\sum_{t=0}^{\zeta_k} A^t W A^{t^T}. \tag{11.15}$$

Proof By Lemma 11.3, when $\delta_k = 1$, we obtain

$$\mathsf{E}[e_{k+1} | \mathcal{I}_k^e] = 0$$

$$\mathsf{cov}[e_{k+1} | \mathcal{I}_k^e] = \textstyle\sum_{t=0}^{\zeta_k} A^t W A^{t^T}.$$

However, when $\delta_k = 0$, we obtain

$$\mathsf{E}[e_{k+1} | \mathcal{I}_k^e] = A\, \mathsf{E}[e_k | \mathcal{I}_k^e] = A \tilde{e}_k$$

$$\mathsf{cov}[e_{k+1} | \mathcal{I}_k^e] = A\, \mathsf{cov}[e_k | \mathcal{I}_k^e] A^T + W = \textstyle\sum_{t=0}^{\zeta_k} A^t W A^{t^T},$$

where we used the fact that $\mathsf{cov}[e_k | \mathcal{I}_k^e] = \mathsf{cov}[x_k | \mathcal{I}_k^e]$. □

LEMMA 11.5 *The estimation mismatch $\tilde{e}_k$ satisfies*

$$\tilde{e}_{k+1} = (1 - \delta_k) A \tilde{e}_k + \textstyle\sum_{t=\zeta_{k+1}}^{\zeta_k} A^t w_{k-t} \tag{11.16}$$

for $k \in \mathcal{K}$ with initial condition $\tilde{e}_0 = x_0 - m_0$.

Proof By Lemmas 11.1 and 11.2 and from the definitions of ζ_k and η_k, we can write

$$\check{x}_k = A^{\zeta_k} x_{k-\zeta_k} + \textstyle\sum_{t=1}^{\zeta_k} A^{t-1} B u_{k-t} \tag{11.17}$$

$$\hat{x}_k = A^{\eta_k} x_{k-\eta_k} + \textstyle\sum_{t=1}^{\eta_k} A^{t-1} B u_{k-t}. \tag{11.18}$$

Since $\zeta_k \le \eta_k$, we can write $x_{k-\zeta_k}$ in terms of $x_{k-\eta_k}$ as

$$x_{k-\zeta_k} = A^{\eta_k-\zeta_k} x_{k-\eta_k} + \sum_{t=1}^{\eta_k-\zeta_k} (A^{t-1} B u_{k-\zeta_k-t} + A^{t-1} w_{k-\zeta_k-t}). \tag{11.19}$$

Inserting (11.19) in (11.17), we get

$$\begin{aligned}\check{x}_k &= A^{\eta_k} x_{k-\eta_k} + \sum_{t=1}^{\zeta_k} A^{t-1} B u_{k-t} \\ &\quad + \sum_{t=1}^{\eta_k-\zeta_k} (A^{t+\zeta_k-1} B u_{k-\zeta_k-t} + A^{t+\zeta_k-1} w_{k-\zeta_k-t}) \\ &= A^{\eta_k} x_{k-\eta_k} + \sum_{t=1}^{\eta_k} A^{t-1} B u_{k-t} + \sum_{t=\zeta_k+1}^{\eta_k} A^{t-1} w_{k-t}.\end{aligned}$$

Now, using (11.18), we find

$$\tilde{e}_k = \sum_{t=\zeta_k+1}^{\eta_k} A^{t-1} w_{k-t} \tag{11.20}$$

$$\tilde{e}_{k+1} = \sum_{t=\zeta_{k+1}+1}^{\eta_{k+1}} A^{t-1} w_{k+1-t}. \tag{11.21}$$

From (11.8), we know that the dynamics of η_{k+1} depends on δ_k. In particular, when $\delta_k = 1$, we have $\eta_{k+1} = \zeta_k + 1$ and can write

$$\tilde{e}_{k+1} = A^{\zeta_{k+1}} w_{k-\zeta_{k+1}} + \cdots + A^{\zeta_k} w_{k-\zeta_k}. \tag{11.22}$$

However, when $\delta_k = 0$, we have $\eta_{k+1} = \eta_k + 1$, and can write

$$\tilde{e}_{k+1} = A^{\zeta_{k+1}} w_{k-\zeta_{k+1}} + \cdots + A^{\eta_k} w_{k-\eta_k}. \tag{11.23}$$

Hence, putting together (11.22) and (11.23), we get

$$\begin{aligned}\tilde{e}_{k+1} &= (1-\delta_k)(A^{\zeta_k+1} w_{k-\zeta_k-1} + \cdots + A^{\eta_k} w_{k-\eta_k}) \\ &\quad + A^{\zeta_{k+1}} w_{k-\zeta_{k+1}} + \cdots + A^{\zeta_k} w_{k-\zeta_k} \\ &= (1-\delta_k) A \tilde{e}_k + A^{\zeta_{k+1}} w_{k-\zeta_{k+1}} + \cdots + A^{\zeta_k} w_{k-\zeta_k},\end{aligned}$$

where in the second equality we used (11.20) and the fact that $\zeta_k \le \eta_k$. This completes the proof. □

LEMMA 11.6 *Let $f(\tilde{e}_{k+1}) : \mathbb{R}^n \to \mathbb{R}$ be a symmetric function of $\tilde{e}_{k+1}$. Then, $g(\tilde{e}_k, \delta_k) = \mathsf{E}[f(\tilde{e}_{k+1}) | \tilde{e}_k, \delta_k] : \mathbb{R}^n \times \{0,1\} \to \mathbb{R}$ is a symmetric function of $\tilde{e}_k$.*

Proof By Lemma 11.5, we can write

$$\tilde{e}_{k+1} = (1-\delta_k) A \tilde{e}_k + n_k, \tag{11.24}$$

where $n_k = \sum_{t=\zeta_{k+1}}^{\zeta_k} A^t w_{k-t}$ is a Gaussian white noise with zero mean and covariance $N_k = \sum_{t=\zeta_{k+1}}^{\zeta_k} A^t W A^{t^T}$. Define the variable $\bar{n}_k$ as $\bar{n}_k = -n_k$. Then, $\bar{n}_k$ is also a Gaussian variable with zero mean and covariance N_k. Therefore, we have

$$\begin{aligned}g(\tilde{e}_k, \delta_k) &= \mathsf{E}\Big[f(\tilde{e}_{k+1}) \Big| \tilde{e}_k, \delta_k\Big] \\ &= \mathsf{E}\Big[f\big((1-\delta_k) A \tilde{e}_k + n_k\big) \Big| \tilde{e}_k, \delta_k\Big]\end{aligned}$$

$$= \mathsf{E}\Big[f\big(-(1-\delta_k)A\tilde{e}_k - n_k\big)\big|\tilde{e}_k, \delta_k\Big]$$

$$= \int_{\mathbb{R}^n} f\big(-(1-\delta_k)A\tilde{e}_k - n_k\big) c \exp(-\tfrac{1}{2} n_k^T N_k^{-1} n_k)\, dn_k$$

$$= \int_{\mathbb{R}^n} f\big(-(1-\delta_k)A\tilde{e}_k + \bar{n}_k\big) c \exp(-\tfrac{1}{2} \bar{n}_k^T N_k^{-1} \bar{n}_k)\, d\bar{n}_k$$

$$= \mathsf{E}\Big[f\big(-(1-\delta_k)A\tilde{e}_k + n_k\big)\big|\tilde{e}_k, \delta_k\Big]$$

$$= g(-\tilde{e}_k, \delta_k),$$

where c is a constant, the first equality comes from (11.24), and the third equality comes from the hypothesis assumption. This proves the claim. □

Note that, given the state estimate used in the structure of the controller, we proved in Lemmas 11.3 and 11.5 that the estimation error and the estimation mismatch satisfy linear recursive equations, which were then used in the derivations of Lemmas 11.4 and 11.6, respectively. These linear recursive equations, as we will show, lead to a tractable analysis for the characterization of the optimal triggering policy.

In the next lemma, we introduce a loss function that is equivalent to the original loss function, in the sense that optimizing it is equivalent to optimizing the original loss function.

LEMMA 11.7 *Given the adopted certainty-equivalent controller, the following loss function is equivalent to the original loss function $\Phi(\pi)$:*

$$\Psi(\pi) = \mathsf{E}\Big[\textstyle\sum_{k=0}^{N} \theta\delta_k + \|e_k\|_{\Gamma_k}^2\Big], \tag{11.25}$$

where $\theta = \ell(1-\lambda)/\lambda$ is a weighting coefficient and $\Gamma_k = A^T S_{k+1} B(B^T S_{k+1} B + R)^{-1} B^T S_{k+1} A$ for $k \in \mathcal{K}$ is a weighting matrix.

Proof Given (11.1) and (11.12), we can derive the following identities:

$$x_{k+1}^T S_{k+1} x_{k+1} = (Ax_k + Bu_k + w_k)^T S_{k+1}(Ax_k + Bu_k + w_k)$$

$$x_k^T S_k x_k = x_k^T\big(Q + A^T S_{k+1} A - L_k^T(B^T S_{k+1} B + R)L_k\big)x_k$$

$$x_{N+1}^T S_{N+1} x_{N+1} - x_0^T S_0 x_0 = \textstyle\sum_{k=0}^{N}\big(x_{k+1}^T S_{k+1} x_{k+1} - x_k^T S_k x_k\big).$$

Using the preceding identities together with $u_k = -L_k\hat{x}_k$ and applying few algebraic operations, we can obtain $\Psi(\pi)$ as in (11.25) and can see that it is equivalent to $\Phi(\pi)$. □

Based on the loss function $\Psi(\pi)$, we can form the value function $V_k(\mathcal{I}_k^e)$ as

$$V_k(\mathcal{I}_k^e) = \min_{\pi\in\mathcal{P}} \mathsf{E}\Big[\textstyle\sum_{t=k}^{N} \theta\delta_t + \|e_{t+1}\|_{\Gamma_{t+1}}^2 \Big| \mathcal{I}_k^e\Big] \tag{11.26}$$

for $k \in \mathcal{K}$, where $\Gamma_{N+1} = 0$ by convention. Given $V_k(\mathcal{I}_k^e)$, we can formally define the value of information, a metric that measures the difference between the benefit and the cost of the system at each time.

DEFINITION 2 The value of information at time k is the variation in the value function $V_k(\mathcal{I}_k^e)$ with respect to the information $x_{k-\zeta_k}$ that can be communicated to the controller, that is,

$$\mathrm{VoI}_k = V_k(\mathcal{I}_k^e)|_{\delta_k=0} - V_k(\mathcal{I}_k^e)|_{\delta_k=1}, \tag{11.27}$$

where $V_k(\mathcal{I}_k^e)|_{\delta_k}$ denotes $V_k(\mathcal{I}_k^e)$ when δ_k is enforced.

We are now in a position to present our main results. We first have the following theorem in which we obtain the optimal triggering policy under the main information structure, consisting of $\mathcal{I}_k^e$ and $\mathcal{I}_k^c$.

THEOREM 11.1 *The optimal triggering policy $\pi^\star$ under the main information structure is a symmetric threshold policy given by $\delta_k^\star = \mathbb{1}_{\mathrm{VoI}_k \geq 0}$, where VoI_k is the value of information expressed as*

$$\mathrm{VoI}_k = \left\|\tilde{e}_k\right\|^2_{A^T\Gamma_{k+1}A} - \theta + \varrho_k(\tilde{e}_k) \tag{11.28}$$

and $\varrho_k(\tilde{e}_k) = \mathsf{E}[V_{k+1}(\mathcal{I}_{k+1}^e)|\mathcal{I}_k^e, \delta_k = 0] - \mathsf{E}[V_{k+1}(\mathcal{I}_{k+1}^e)|\mathcal{I}_k^e, \delta_k = 1]$.

Proof From the additivity of $V_k(\mathcal{I}_k^e)$, we obtain

$$V_k(\mathcal{I}_k^e) = \min_{\mathsf{P}(\delta_k|\mathcal{I}_k^e)} \mathsf{E}\left[\theta\delta_k + e_{k+1}^T\Gamma_{k+1}e_{k+1} + V_{k+1}(\mathcal{I}_{k+1}^e)\middle|\mathcal{I}_k^e\right]$$

with initial condition $V_{N+1}(\mathcal{I}_{N+1}^e) = 0$. We prove that $V_k(\mathcal{I}_k^e)$ is a symmetric function of $\tilde{e}_k$. We assume that the claim holds at time $k+1$, and shall prove that it also holds at time k. By Lemma 11.4, we find

$$\mathsf{E}\left[e_{k+1}^T\Gamma_{k+1}e_{k+1}\middle|\mathcal{I}_k^e\right] = \mathop{\mathsf{E}}_{\delta_k}\left[(1-\delta_k)\tilde{e}_k^TA^T\Gamma_{k+1}A\tilde{e}_k + \mathrm{tr}(\Gamma_{k+1}\Sigma_k)\middle|\mathcal{I}_k^e\right],$$

where $\Sigma_k = \sum_{t=0}^{\zeta_k} A^tWA^{t^T}$. We can show that

$$V_k(\mathcal{I}_k^e) = \min_{\delta_k}\left\{\theta\delta_k + (1-\delta_k)\tilde{e}_k^TA^T\Gamma_{k+1}A\tilde{e}_k + \mathrm{tr}(\Gamma_{k+1}\Sigma_k) + \mathsf{E}[V_{k+1}(\mathcal{I}_{k+1}^e)|\mathcal{I}_k^e]\right\}. \tag{11.29}$$

The minimizer in (11.29) is obtained as $\delta_k^\star = \mathbb{1}_{\mathrm{VoI}_k\geq 0}$, where

$$\mathrm{VoI}_k = \tilde{e}_k^TA^T\Gamma_{k+1}A\tilde{e}_k - \theta + \varrho_k$$

and $\varrho_k = \mathsf{E}[V_{k+1}(\mathcal{I}_{k+1}^e)|\mathcal{I}_k^e, \delta_k = 0] - \mathsf{E}[V_{k+1}(\mathcal{I}_{k+1}^e)|\mathcal{I}_k^e, \delta_k = 1]$. Besides, by Lemma 11.6, we know that $\mathsf{E}[V_{k+1}(\mathcal{I}_{k+1}^e)|\mathcal{I}_k^e, \delta_k]$ is a symmetric function of $\tilde{e}_k$. Hence, we conclude that $V_k(\mathcal{I}_k^e)$ is a symmetric function of $\tilde{e}_k$. This completes the proof. □

Theorem 11.1 states that under the main information structure, the value of information VoI_k, as characterized in (11.28), depends on the sample path of the process, and is a symmetric function of the estimation mismatch $\tilde{e}_k = \sum_{t=\zeta_k+1}^{\eta_k} A^{t-1}w_{k-t}$. Moreover, it states that the information $x_{k-\zeta_k}$ should be transmitted from the sensor to the controller at time k only if the value of information VoI_k is nonnegative. Note that the value of information VoI_k in this case can be computed with arbitrary accuracy by solving (11.29) recursively and backward in time.

We show in following corollary that given the triggering policy in Theorem 11.1 the minimum mean-square-error state estimate at the controller in fact matches with the state estimate used in the structure of the controller.

COROLLARY 11.1 *Given the triggering policy* $\pi^\star$, *the conditional mean* $\mathsf{E}[x_k|\mathcal{I}_k^c]$ *satisfies (11.10) with* $\iota_k = 0$ *for all* $k \in \mathcal{K}$.

Proof Note that $\hat{x}_0 = \mathsf{E}[x_0|\mathcal{I}_0^c]$ holds regardless of the adopted triggering policy. Assume that $\iota_t = 0$ holds for all $t < k$. Hence, we have $\hat{x}_t = \mathsf{E}[x_t|\mathcal{I}_t^c]$ for all $t \leq k$. We shall prove that the claim also holds at time $t = k$. We can write

$$\mathsf{P}(\tilde{e}_k|\mathcal{I}_k^c, \delta_k = 0) \propto \mathsf{P}(\delta_k = 0|\tilde{e}_k)\,\mathsf{P}(\tilde{e}_k|\mathcal{I}_k^c). \tag{11.30}$$

By the hypothesis assumption, Lemma 11.5, and Theorem 11.1, we see that $\mathsf{P}(\tilde{e}_k|\mathcal{I}_k^c)$ and $\mathsf{P}(\delta_k = 0|\tilde{e}_k)$ are symmetric with respect to $\tilde{e}_k$. Hence, $\mathsf{P}(\tilde{e}_k|\mathcal{I}_k^c, \delta_k = 0)$ is also symmetric with respect to $\tilde{e}_k$. This means that $\mathsf{E}[\tilde{e}_k|\mathcal{I}_k^c, \delta_k = 0] = 0$. Besides, we can write

$$\mathsf{E}\left[e_k|\mathcal{I}_k^c, \delta_k\right] = \mathsf{E}\left[\,\mathsf{E}\left[e_k|\mathcal{I}_k^e\right]|\mathcal{I}_k^c, \delta_k\right] = \mathsf{E}\left[\tilde{e}_k|\mathcal{I}_k^c, \delta_k\right],$$

where in the first equality we used the tower property of conditional expectations and the fact that δ_k is a function of $\mathcal{I}_k^e$. Hence, $\iota_k = A(\mathsf{E}[x_k|\mathcal{I}_k^c, \delta_k = 0] - \mathsf{E}[x_k|\mathcal{I}_k^c]) = A\,\mathsf{E}[e_k|\mathcal{I}_k^c, \delta_k = 0] = A\,\mathsf{E}[\tilde{e}_k|\mathcal{I}_k^c, \delta_k = 0] = 0$. This completes the proof. □

In the rest of this section, we prove that there exists a condition associated with the information structure under which the value of information becomes completely expressible in terms of the age of information. Let the information set of the event trigger be a restricted set that includes only the time stamps of the informative observations, the time stamps of the communicated informative observations, and the previous decisions, that is, $\mathcal{I}_k^r = \{t - \zeta_t, t - \eta_t, \delta_s, u_s \mid 0 \leq t \leq k, 0 \leq s < k\}$, where for clarity we represented the set by $\mathcal{I}_k^r$ instead of $\mathcal{I}_k^e$. A triggering policy π is now admissible if $\pi = \{\mathsf{P}(\delta_k|\mathcal{I}_k^r)\}_{k=0}^N$, where $\mathsf{P}(\delta_k|\mathcal{I}_k^r)$ is a Borel measurable transition kernel. We now have the following theorem in which we obtain the optimal triggering policy under the restricted information structure, consisting of $\mathcal{I}_k^r$ and $\mathcal{I}_k^c$.

THEOREM 11.2 *The optimal triggering policy* $\pi^\star$ *under the restricted information structure is a threshold policy given by* $\delta_k^\star = \mathbb{1}_{\mathrm{VoI}_k \geq 0}$, *where* VoI_k *is the value of information expressed as*

$$\mathrm{VoI}_k = \mathrm{tr}\left(\Gamma_{k+1} \textstyle\sum_{t=\zeta_k+1}^{\eta_k} A^t W A^{t^T}\right) - \theta + \varrho_k(\zeta_k, \eta_k) \tag{11.31}$$

and $\varrho_k(\zeta_k, \eta_k) = \mathsf{E}[V_{k+1}(\mathcal{I}_{k+1}^r)|\mathcal{I}_k^r, \delta_k = 0] - \mathsf{E}[V_{k+1}(\mathcal{I}_{k+1}^r)|\mathcal{I}_k^r, \delta_k = 1]$.

Proof From the additivity of $V_k(\mathcal{I}_k^r)$, we obtain

$$V_k(\mathcal{I}_k^r) = \min_{\mathsf{P}(\delta_k|\mathcal{I}_k^r)} \mathsf{E}\left[\theta\delta_k + e_{k+1}^T \Gamma_{k+1} e_{k+1} + V_{k+1}(\mathcal{I}_{k+1}^r)|\mathcal{I}_k^r\right]$$

with initial condition $V_{N+1}(\mathcal{I}^r_{N+1}) = 0$. We prove that $V_k(\mathcal{I}^r_k)$ is a function of ζ_k and η_k. We assume that the claim holds at time $k+1$ and shall prove that it also holds at time k. Using Lemma 11.3 and applying few operations, we can write

$$\mathsf{E}[e_{k+1}|\mathcal{I}^r_k] = 0$$

$$\mathsf{cov}[e_{k+1}|\mathcal{I}^r_k] = \delta_k \sum_{t=0}^{\zeta_k} A^t W A^{t^T} + (1-\delta_k) \sum_{t=0}^{\eta_k} A^t W A^{t^T}.$$

This implies that

$$\mathsf{E}\left[e^T_{k+1}\Gamma_{k+1}e_{k+1}|\mathcal{I}^r_k\right] = \mathop{\mathsf{E}}_{\delta_k}\Big[\delta_k \operatorname{tr}\left(\Gamma_{k+1}\sum_{t=0}^{\zeta_k} A^t W A^{t^T}\right) + (1-\delta_k)\operatorname{tr}\left(\Gamma_{k+1}\sum_{t=0}^{\eta_k} A^t W A^{t^T}\right)|\mathcal{I}^r_k\Big].$$

Hence, we can show that

$$V_k(\mathcal{I}^r_k) = \min_{\delta_k}\Big\{\theta\delta_k + \delta_k \operatorname{tr}\left(\Gamma_{k+1}\sum_{t=0}^{\zeta_k} A^t W A^{t^T}\right) + (1-\delta_k)\operatorname{tr}\left(\Gamma_{k+1}\sum_{t=0}^{\eta_k} A^t W A^{t^T}\right) + \mathsf{E}[V_{k+1}(\mathcal{I}^r_{k+1})|\mathcal{I}^r_k]\Big\}. \quad (11.32)$$

The minimizer in (11.32) is obtained as $\delta^\star_k = \mathbb{1}_{\mathrm{VoI}_k \geq 0}$, where

$$\mathrm{VoI}_k = \operatorname{tr}\left(\Gamma_{k+1}\sum_{t=0}^{\eta_k} A^t W A^{t^T}\right) - \operatorname{tr}\left(\Gamma_{k+1}\sum_{t=0}^{\zeta_k} A^t W A^{t^T}\right) - \theta + \varrho_k$$

and $\varrho_k = \mathsf{E}[V_{k+1}(\mathcal{I}^r_{k+1})|\mathcal{I}^r_k, \delta_k = 0] - \mathsf{E}[V_{k+1}(\mathcal{I}^r_{k+1})|\mathcal{I}^r_k, \delta_k = 1]$. Note that ζ_{k+1} and η_{k+1} are functions of τ_{k+1}, ζ_k, and η_k. Since τ_{k+1} is an independent random variable with known distribution, we conclude that $V_k(\mathcal{I}^r_k)$ is a function of ζ_k and η_k. This completes the proof. □

Theorem 11.2 states that, under the restricted information structure, the value of information VoI_k, as characterized in (11.31), becomes independent of the sample path of the process and can be expressed in terms of the age of information at the event trigger ζ_k and that at the controller η_k. Note that the value of information VoI_k in this case can be computed with arbitrary accuracy by solving (11.32) recursively and backward in time. Finally, we remark that using the triggering policy in Theorem 11.2 instead of the triggering policy in Theorem 11.1 leads to a degradation in the performance of the system because $\mathcal{I}^r_k$ contains less information than $\mathcal{I}^e_k$.

As before, given the triggering policy in Theorem 11.2 the minimum mean-square-error state estimate at the controller matches with the state estimate used in the structure of the controller

11.4 Numerical Example

Consider a scalar stochastic process defined by the model in (11.1) with state coefficient $A = 1.1$, input coefficient $B = 1$, noise variance $W = 1$, and mean and variance of the initial condition $m_0 = 0$ and $M_0 = 1$. Let the time horizon be $N = 200$, and the weighting coefficients be $\theta = 10$, $Q = 1$, and $R = 0.1$. Moreover, let the output of the

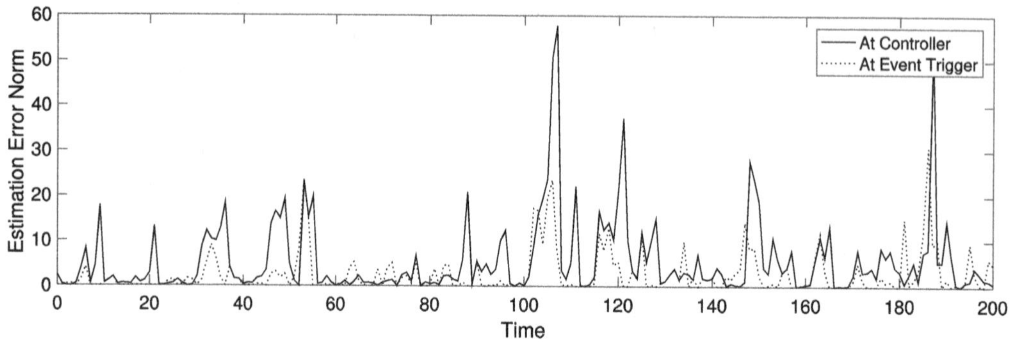

Figure 11.1 The estimation error norm trajectories at the controller and at the event trigger.

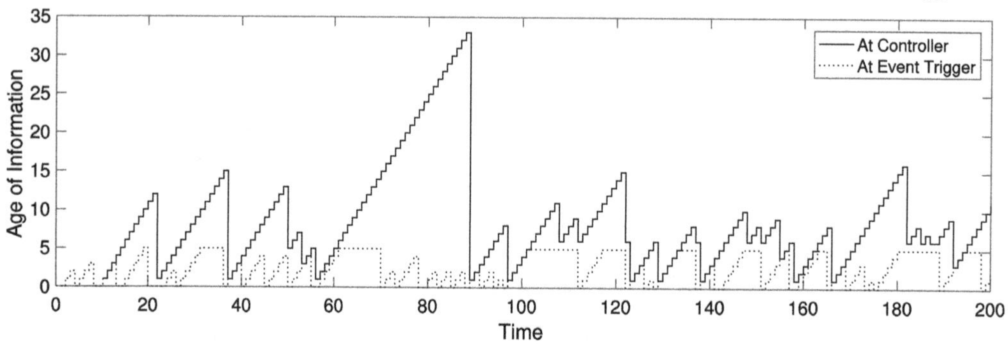

Figure 11.2 The age of information trajectories at the controller and at the event trigger.

process be given by the model in (11.3) with delay $d = 5$ and probability $p_k = 0.2$. For this system, we implemented the triggering policy characterized by Theorem 11.1. For a test run of the system, Figure 11.1 shows the estimation error norm trajectories at the controller and at the event trigger, and Figure 11.2 shows the age of information trajectories at the controller and at the event trigger. Furthermore, Figure 11.3 depicts the value of information and transmission event trajectories.

We recall that the optimal triggering policy in Theorem 11.1 permits a transmission only when the value of information is nonnegative. The value of information itself is a function of the estimation mismatch. This implies that the value of information does not necessarily increase with the age of information at the controller or that at the event trigger. This key point can be observed in our numerical example. Consider the time interval $k \in [56, 87]$. In this interval, the age of information at the controller increases continually. However, the value of information remains small, and no transmission occurs in this interval. This situation continues until time $k = 88$ at which the value of information becomes nonnegative, due to the dramatic increment of the estimation mismatch. As a result, at this time a transmission occurs.

11.5 Conclusion

In this chapter, we studied the value of information as a more comprehensive instrument than the age of information for optimally shaping the information flow

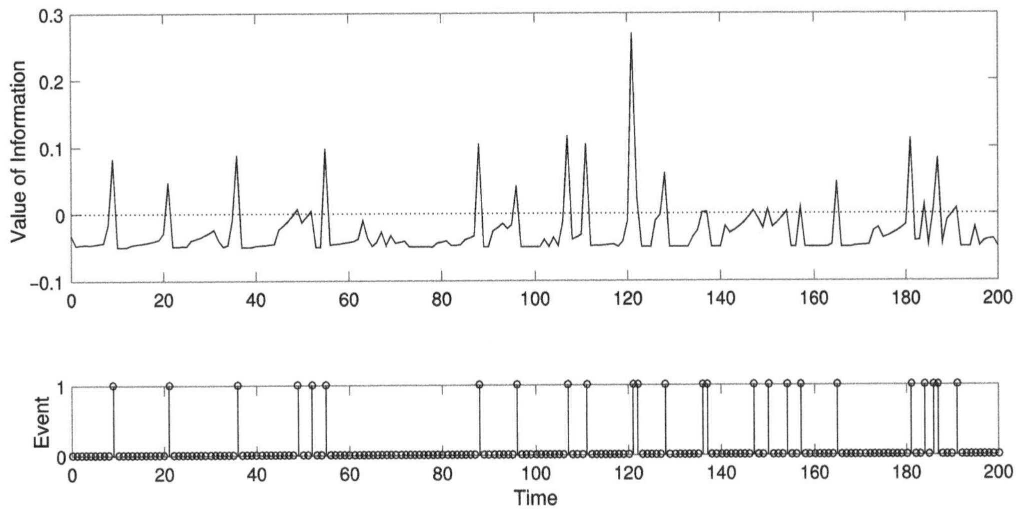

Figure 11.3 The value of information (scaled by $1/N$) and transmission event trajectories.

in a networked control system. Two information structures were considered: the main one and a restricted one. We proved that, under the main information structure, the value of information is a symmetric function of the estimation mismatch. Moreover, we proved that, under the restricted information structure, the value of information becomes completely expressible in terms of the age of information. Accordingly, we characterized two optimal triggering policies of the form $\delta_k^\star = \mathbb{1}_{\mathrm{VoI}_k \geq 0}$. One policy is dependent on the sample path of the process and achieves the best performance, while the other one is independent of the sample of the process and achieves a lower performance.

Acknowledgment

This work was partially funded by the Knut and Alice Wallenberg Foundation, by the Swedish Strategic Research Foundation, and by the Swedish Research Council.

References

[1] R. D. Yates, Y. Sun, D. R. Brown, S. K. Kaul, E. Modiano, and S. Ulukus, "Age of information: An introduction and survey," *IEEE Journal on Selected Areas in Communications*, vol. 39, no. 5, pp. 1183–1210, 2021.

[2] T. Soleymani, "Value of information analysis in feedback control," Ph.D. dissertation, Technical University of Munich, 2019.

[3] T. Soleymani, J. S. Baras, and S. Hirche, "Value of information in feedback control: Quantification," *IEEE Trans. on Automatic Control*, vol. 67, no. 7, pp. 3730–3737, 2022.

[4] T. Soleymani, J. S. Baras, S. Hirche, and K. H. Johansson, "Value of information in feedback control: Global optimality," *arXiv preprint arXiv:2103.14012*, 2021.

[5] G. M. Lipsa and N. C. Martins, "Remote state estimation with communication costs for first-order LTI systems," *IEEE Trans. on Automatic Control*, vol. 56, no. 9, pp. 2013–2025, 2011.

[6] A. Molin and S. Hirche, "Event-triggered state estimation: An iterative algorithm and optimality properties," *IEEE Trans. on Automatic Control*, vol. 62, no. 11, pp. 5939–5946, 2017.

[7] J. Chakravorty and A. Mahajan, "Fundamental limits of remote estimation of autoregressive Markov processes under communication constraints," *IEEE Trans. on Automatic Control*, vol. 62, no. 3, pp. 1109–1124, 2016.

[8] M. Rabi, G. V. Moustakides, and J. S. Baras, "Adaptive sampling for linear state estimation," *SIAM Journal on Control and Optimization*, vol. 50, no. 2, pp. 672–702, 2012.

[9] N. Guo and V. Kostina, "Optimal causal rate-constrained sampling for a class of continuous Markov processes," *IEEE Trans. on Information Theory*, vol. 67, no. 12, pp. 7876–7890, 2021.

[10] ——, "Optimal causal rate-constrained sampling of the Wiener process," *IEEE Trans. on Automatic Control*, vol. 67, no. 4, pp. 1776–1791, 2022.

[11] Y. Sun, Y. Polyanskiy, and E. Uysal, "Sampling of the Wiener process for remote estimation over a channel with random delay," *IEEE Trans. on Information Theory*, vol. 66, no. 2, pp. 1118–1135, 2019.

[12] T. Soleymani, S. Hirche, and J. S. Baras, "Optimal self-driven sampling for estimation based on value of information," in *Proc. Int. Workshop on Discrete Event Systems*, 2016, pp. 183–188.

12 Age of Information in Practice

Elif Uysal, Onur Kaya, Sajjad Baghaee, and Hasan Burhan Beytur[1]

Abstract: While Age of Information (AoI) has gained importance as a metric characterizing the freshness of information in information-update systems and time-critical applications, most previous studies on AoI have been theoretical. In this chapter, we compile a set of recent works reporting AoI measurements in real-life networks and experimental testbeds, and investigating practical issues such as synchronization, the role of various transport layer protocols, congestion control mechanisms, application of machine learning for adaptation to network conditions, and device-related bottlenecks such as limited processing power.

12.1 Introduction

Theoretical studies have by now established AoI as a Key Performance Indicator (KPI) that characterizes information freshness in status update systems and applications that are time-critical. Several principles of age optimization have been developed for network models under simplifying assumptions, when delay and service time statistics are known. However, in practice, it may be difficult, if at all possible, to extract these statistics and optimize for them in real-life network implementations due to complex interactions among different networking layers. Moreover, there are a number of practical system issues that have been largely ignored in theoretical problem formulations exploring the fundamental behaviour of age in networks.

There have been a relatively small number of implementation studies on AoI to date [1–9]. The goal of this chapter is to compile and discuss some results on practical aspects of AoI from this burgeoning literature. We will review experimental and emulation-based AoI measurement results for data flows that are transmitted over wireless/wired links, using different transport layer protocols. Practical issues such as synchronization and the suitability of various congestion control mechanisms and application layer approaches to control AoI will be covered. The impact of device-related bottlenecks on age, such as limited processing power or age constraints that may affect simple IoT nodes will be demonstrated.

[1] This work was supported by TUBITAK Grant 117E215 and by Huawei. We thank Canberk Sonmez and Egemen Sert for their work, which inspired the writing of this chapter.

The first emulation study of AoI in wireless links was reported in [1], and the first real-life implementation measuring the variation of AoI over TCP/IP links served by WiFi, LTE, 3G, 2G, and Ethernet was demonstrated in [2]. The experimental AoI measurement results reported in [2] exhibit a non-monotone (specifically, "U-shaped") AoI versus arrival rate characteristic for TCP/IP connections served by Wifi, Ethernet, and LTE links. This measured characteristic is in line with theoretical results for FCFS systems with Poisson or Gamma distributed arrivals [10].

According to the U-shaped relation between AoI and sampling rate in the real-life TCP/IP connection in [2], the age first falls sharply as the update arrival rate increases. Then, it stays relatively flat. After a certain arrival rate, though, due to building congestion in queues, a sharp increase in age is observed. This indicates that there is an acute need for age-optimal service policies in practical networks, for several reasons: Firstly, the increasing importance of machine type communications, including autonomous systems, remote monitoring, and control, implies an increase in the type of services where freshness of data rather than throughput is the main performance criterion. Unlike conventional data transmission, where the goal is to transmit the entirety of a stream of data as reliably as possible, in status-update type services only the timely data matter. Increasing the rate at which samples are injected into the network starts acting contrary to the purpose of freshness after a certain point. Secondly, the goals of achieving low AoI and high throughput are often aligned, as we shall see in the system implementation examples in the following sections. The principal reason behind this is that a good AoI performance requires a high throughput of data updates, at sufficiently low delay.

Therefore, we believe that AoI optimization should be one of the directions for the future development of network architectures. By now, there is a number of years of theoretical work supporting this vision, and in the rest of this chapter we will elaborate on the initial implementation experience in various layers of the network protocol stack.

One of the challenges in translating theoretical work on the control of AoI to practically viable algorithms is that theoretical work often assumed full knowledge of the statistics of network delay (e.g., [11, 12]). Service or age-control policies proposed in the literature rely heavily on the delay statistics of the network. However, knowledge of these statistics is not easy to obtain, especially as the scale of the network and, hence, the number of connections grow. Basing a decision policy on ill-inferred or invalid statistics can be a suboptimal approach. In such scenarios, AoI-aware sampling, scheduling, and transmission policies can be adaptively generated using machine learning. The idea of introducing a delay to the sampling operation in response to instantaneous age and the delay statistics on the network was proposed in [11]. However, this smart sampling operation was based on a knowledge of the statistics of the network delay. In [3, 12], reinforcement learning was employed to obtain this sampling operation without requiring prior information about the delay statistics. Similarly, in [13–16] age-aware scheduling is studied using RL methods.

The rest of this chapter is organized as follows: Section 12.2 provides the definition of AoI, expressed in terms of time stamps of transmitted and received packets, followed by a brief review of the behaviour of steady-state average AoI under some basic queueing and service disciplines, in order to provide a basis for the interpretation of practical results in the following sections. Section 12.3 describes the measurement of AoI on a simple physical network testbed. Section 12.4 discusses general issues arising while computing age in realistic setups, such as clock bias, and how to mitigate those. Section 12.5 reviews AoI in TCP connections, first in an emulation testbed in subsection 12.5.1, and then a real-world testbed testing TCP/IP connections running on various physical links such as WiFi, Ethernet, LTE, 3G, and 2G in subsection 12.5.2. Section 12.6 contrasts age behavior over UDP and TCP connections. Section 12.7 overviews the application of statistical learning methods to age optimization in practical networks. Section 12.8 reviews recent proposals of application-layer mechanisms for age control over UDP, in particular the ACP protocol. Section 12.9 describes implementation of age-aware scheduling in Wi-Fi uplink or downlink settings. Section 12.10 concludes the chapter by offering a general vision for the future development of this research area.

12.2 Age of Information: Definition, Measurement, and Behaviour in Queueing Systems

Status age of a flow is defined as $\Delta(t) = t - U(t)$, where $U(t)$ is the generation time (i.e., time stamp) of the newest data packet belonging to this flow that has been received by the destination by time t. As a consequence of this linearity, $\Delta(t)$ follows a sawtooth pattern as seen in Figure 12.1. In between updates, age increases linearly in time and drops just after a new update is received.

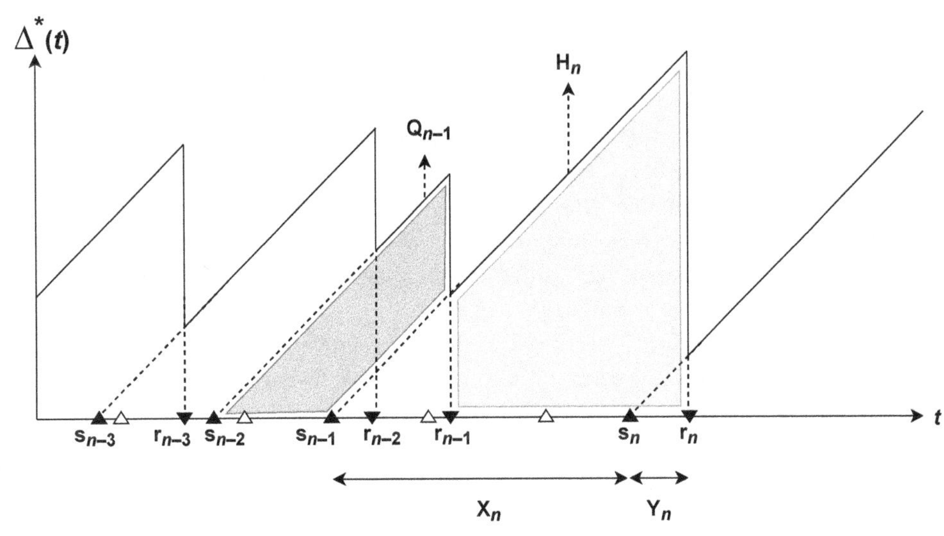

Figure 12.1 Age sample path $\Delta(t)$ for a given realization of packet transmission and reception times r_i and s_i. ©[2020] IEEE. Reprinted, with permission, from [6].

As shown in Figure 12.1, the area under age sample paths can be presented as sum of the trapezoidal areas $\{Q_i\}$, or alternatively, that of the areas $\{H_i\}$.

Let $n(T)$ be the number of received packets by time T. In what follows, we will use n and $n(T)$ interchangeably, depending on whether or not the time dependence needs be explicitly stated in an expression. The area is composed of the area of polygon Q_1, isosceles trapezoids Q_i for $2 \leq i \leq n$, and the triangle of base Y_n:

$$\overline{\Delta} = \lim_{T\to\infty} \frac{Q_1 + \sum_{i=2}^{n(T)} Q_i + Y_{n(T)}^2/2}{T}. \tag{12.1}$$

As $T \to \infty$, it also follows that $n(T) \to \infty$, and that the average age is determined by the total area of the isosceles trapezoids normalized by time. This area depends only on the arrival and departure instants:

$$Q_i = \frac{(2r_i - s_i - s_{i-1})(s_i - s_{i-1})}{2}. \tag{12.2}$$

Note that $(s_i - s_{i-1}) = X_i$, the inter-arrival time between successfully transmitted packets.

The formulation using Q_i facilitates mathematical analysis, however, a formulation using H_i is useful in some scenarios involving practical measurement of AoI. Note that

$$\overline{\Delta} = \lim_{T\to\infty} \frac{1}{T} \sum_{i=1}^{n(T)} H_i \tag{12.3}$$

$$H_i = (r_i - r_{i-1})(r_{i-1} - s_{i-1}) + \frac{(r_i - r_{i-1})^2}{2}, \tag{12.4}$$

where $(r_i - r_{i-1})$ is the inter-departure time between $i-1$th and ith packets and $(r_{i-1} - s_{i-1})$ is equal to Y_{i-1}, which is the system time of $i-1$th packet.

Depending on the location of the age or age penalty computation, it may be more appropriate to measure the age at the receiver or at the transmitter. The location of this computation in scheduling/transmission algorithms that control age often depends on the capabilities of the devices or the essence of the control problem. For example, age of a flow between a simple transmitter and a remote server should be measured at the destination (i.e., the server). However, if the transmitter needs to control the AoI, then it would need to know the AoI values with minimum latency. In such a situation, the AoI is measured at the transmitter.

Note that, as seen in (12.2) and (12.4), the time stamps r_i and s_i should be computed using a common time reference. In practice, however, as the receiver and transmitter are separate machines, they rarely have access to a common clock, which results in a synchronization issue that we will address in Section 12.4.1.

Peak age of information is another metric often investigated in literature:

$$\overline{\Delta_{peak}} = \frac{1}{n} \sum_{i=1}^{n} \Delta(r_i^-). \tag{12.5}$$

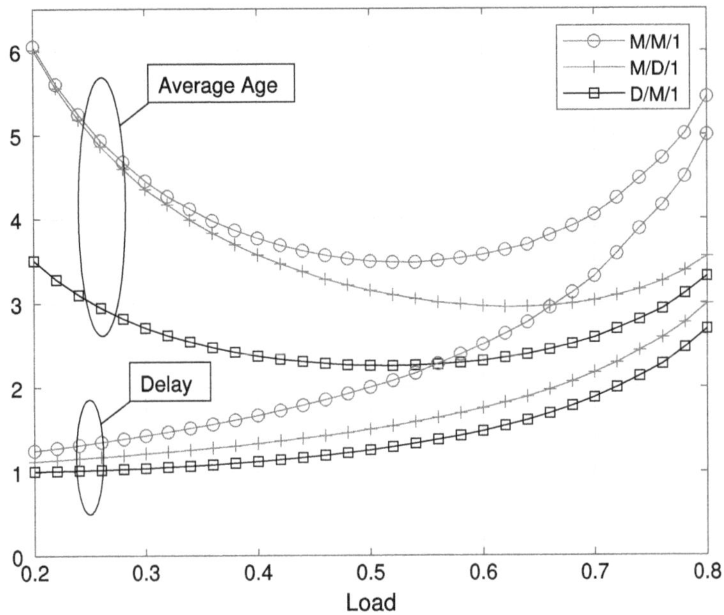

Figure 12.2 Average delay and AoI attained by an M/M/1 queueing system with service rate $\mu = 1$. ©[2018] IEEE. Reprinted, with permission, from [2].

This quantity can also be calculated using the transmission and reception instances of packets:

$$\overline{\Delta_{peak}} = \frac{1}{n}\sum_{i=1}^{n}(r_i - s_{i-1}). \tag{12.6}$$

It should be stressed that minimizing status age is not the same as minimizing delay. It requires, on the other hand, a joint optimization of delay and throughput together [17]. We exhibit this phenomenon on simple queueing systems in Figure 12.2, with Poisson packet arrivals. We plot average delay and AoI as a function of load to the server (the server corresponding to the communication channel in this case). The transmission duration of each packet on the link is modeled as an exponential random variable, independent of all other transmission durations. The load is equivalent to throughput, as long as the queue is operating in the stable regime. This abstraction is an M/M/1 queueing system[2] with a first-come first-served (FCFS) discipline.

[2] An M/M/1 queue is a single-server queueing system abstraction. Packet inter-arrival times are exponentially distributed and i.i.d., with mean $1/\lambda$. Service times are i.i.d. exponential with mean $1/\mu$. The expected delay experienced by packets at steady state is $(1/\mu)/(1-\rho)$, where $\rho = \lambda/\mu$ is the "load," i.e., the steady-state probability that the system is busy. Steady-state delay therefore monotonically increases with the load and blows up as $\rho \to 1$, and the system enters an unstable regime (where delay is not guaranteed to be finite) when $\rho \geq 1$. If discarding packets from the buffer is allowed (as in an M/M/1/k system that holds up to $k-1$ packets in queue), the delay will remain finite at the expense of dropping a fraction $\max(0, \rho - 1)$ of all arriving packets.

The throughput increases linearly with load, ρ, in the range of values plotted, as the system is stable in this range. Note that the AoI decreases with increasing load up to the age optimal operating point $\rho* = 0.53$, as arriving packets get served with no appreciable waiting time in queue, as the queue is often empty up to that point of optimal server utilization. When packets get served with nearly no delay, they contribute to a reduction of age upon service completion. The average number of packets in the system at $\rho = \rho*$ is ≈ 1.13. However, AoI starts increasing due to queueing effects as the load exceeds $\rho*$: this is a natural consequence of the FCFS queueing discipline, which lets newer packets get stale waiting in queue behind older packets. Interestingly, we will see in the experimental results on TCP connections in the Internet in the following sections that the optimality of a load value around 0.53 and a total number of packets in flight being around 1 continue to hold, even though the underlying arrival and service processes are much more complex.

Note from the plots in Figure 12.2 that AoI exhibits different behavior than delay. Increasing update rate, while it may contribute to a larger throughput of updates, does not necessarily improve AoI. In fact, it can even result in an increase in AoI in the FCFS case. Better AoI is achieved in multi-server models (e.g., M/M/2 [18], M/M/∞) due to the increased service rate: the AoI in an M/M/2 system is nearly half that in M/M/1. Note, though, that increasing the update rate in such systems may cause some packets to become obsolete by the time they arrive at the destination, because they have been preceded by newer packets. This makes packet management schemes important [19]. The fraction of packets that are rendered obsolete, causing a waste of resources without a reduction in age, increases with the load ρ. Mitigating this problem has been the subject of previous studies (e.g., [19]) that studied packet management schemes.

The literature on AoI analysis and optimization (e.g., [17, 19–25]) was based on exogenous packet arrivals, prior to [26]. The "generate-at-will" model was introduced in [26–28], where a source can generate and output packets to deliberately control age. The model in [28] will be important as we consider arrival rate control in the application later, on top of various transport layer mechanisms, to obtain favorable age performance in the presence of network bottlenecks. Hence, in what follows we elaborate on that model.

In [28], updates $i = 1, 2, \ldots$ generated by a source were subject to a stochastic process of delays $\{Y_i\}$ in the network. The delay process was arbitrary and not affected by the actions of the transmitter. This modeled a scenario where the packet flow is injected into a large network where the congestion and delays are caused by a multitude of factors, among which the actions of this particular source are negligible. The source is assumed to get instantaneous feedback when an update is delivered, which corresponds to neglecting the delivery time for acknowledgements. The policy for generating updates is determined by the sequence of waiting times $\{Z_i\}$. A possible work-conserving policy is the zero wait (ZW) policy, which achieves maximum link utility and minimum average delay by setting $Z_i = 0$ for all i. This corresponds to submitting a fresh update once the server (i.e., communication link) becomes idle.

[27] investigated the performance of the ZW policy with respect to AoI metric, based on the assumption that it is feasible. Somewhat counterintuitively, ZW is suboptimal for a large class of delay distributions. The results in [27, 28] characterized policies that minimize not only AoI but a general AoI penalty function $g(\Delta(t))$. Often, an optimal update policy sends updates at a rate that is lower than the maximum at which it is allowed to send. This pointed to a very important conclusion: one could control AoI by judiciously adapting the rate of generation of samples (updates) to variable external conditions such as delay. This can be done through the measured or estimated [26] value of the age penalty. Such policies can prevent unnecessary staleness resulting from congestion, queueing, or more generally, inefficient utilization of resources from an overall network perspective.

In the rest of the chapter, we will explore how the conclusions drawn from the analysis of idealized queueing and service models and age control policies from literature extend to real-life networks.

12.3 AoI Measurement on a Simple Physical Network

Age measurement in a real network requires, first and foremost, an accurate timing information at the node where the calculation will be carried out. As the clocks at the client and server nodes are not necessarily synchronized, either an estimation of the timing offset or a system-wide synchronization must be performed prior to the computation of age.

Before discussing synchronization methods in relation to age measurements and the effect of eventual synchronization/timing estimation error on age performance, we first present a toy network model from [5] which circumvents such need for synchronization, so that the effect of sampling rate on age can be observed over a network running a practical transport layer protocol.

The model comprises a sampler-transceiver node in Ankara, Turkey, and an echo server (Figure 12.3) in Istanbul, Turkey. The sampler-transceiver produces and sends UDP packets which contain the time stamp marking the generation time, packet ID, and a dummy payload and transmits the packet to the echo server through the Internet. The echo server echoes the received packet immediately after receiving it. Upon reception and decoding, the receive time stamp of each packet is noted and compared with generation time stamp, and the instantaneous age is calculated. A multi-threaded or multi-process solution, which enables simultaneous sampling, transmission, and reception packets, is employed at the sampler-transceiver.

The two-way connection between the sampler-transceiver node and the echo server can alternatively be viewed as a single end-to-end connection from the transmitter node to the receiver node through the echo server, which may simply be seen as one of intermediate nodes in the network. Since the transmitter and the receiver are the same device, the average AoI can be measured without any synchronization issue. To do so, UDP packets consisting of 1,058 bytes (including header) are transmitted at a constant rate. The average AoI is calculated by plugging the timing information available at the

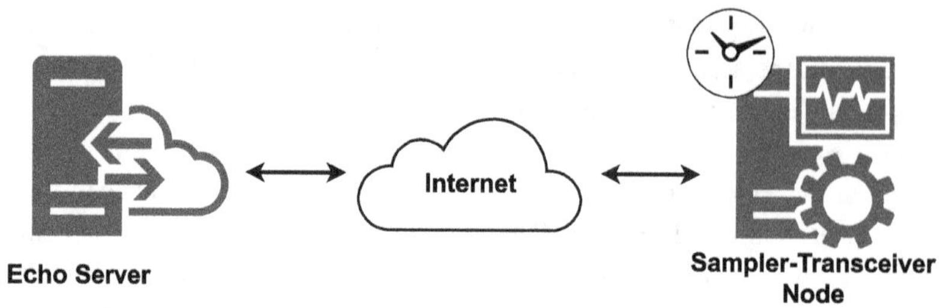

Figure 12.3 An illustration of the physical testbed. ©[2019] IEEE. Reprinted, with permission, from [5].

transceiver in (12.3). The sampling rate is linearly increased, from 1 packet per second up to 370 packets per second, to investigate the age-rate relation, which is plotted in Figure 12.4.

Even though many routers and switches in the Internet infrastructure use FCFS queues, still, it is rather surprising to observe results similar to those obtained in the theoretical analysis of a simple M/M/1 queueing system [29]. Another interesting result is that, unlike the results for M/M/1, M/D/1, and D/M/1 queues shown in Figure 12.2, the average AoI is much less sensitive to sampling rate, over a large range of rate values. For example, the average age remains almost constant as the packet generation rate is increased from 100 pps up to 340 pps. This phenomenon can be explained by (i) the determinism in service time on individual links, (ii) the load being shared by multiple servers due to data being routed on multiple possible paths. It should be noted, however, that packet drops sharply increase as the rate at which samples are injected into the network increases beyond a point.

The experimental results in Figure 12.4 also verify the expected relation between the minimum average age and the round-trip time experienced by the packets in the network. In the testbed, before starting transmission, test packets were used to estimate the number of hops and the average RTT from the transmitter to receiver. It was observed that each packet travels through a total of 18–20 hops from the transmitter to echo server and back, with an average RTT of 12.5 msec. Since, during the transmission of the test packets, the network was not loaded, the average RTT measured was close to the minimum possible RTT, which is essentially the sum of the round trip propagation delay and the total transmission time for one packet along its route. For a deterministic sequence of packet transmissions at rate R_p, the minimum average age of a packet received at the transceiver will be

$$\Delta_{\min} = RTT + \frac{1}{2R_p} \tag{12.7}$$

in the absence of queueing delay. The real-life measurements are in close agreement with this estimate: the age is minimized at 13 msec, around $R_p = 140$ Hz. At this rate, (12.7) predicts an average age estimate of about 16 msec computed using the average RTT, which is expected to be slightly conservative, due to the actual minimum RTT being shorter.

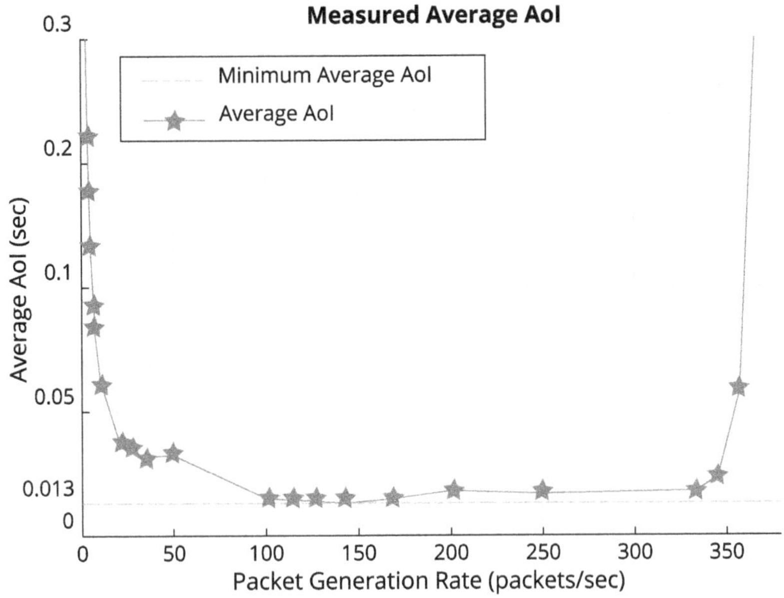

Figure 12.4 AoI vs. sampling rate on the testbed running UDP protocol. ©[2019] IEEE. Reprinted, with permission, from [5].

The synchronization, which was automatic due to the toy network model used in this section that involves an echo server, is not possible in real-life networks where the transmitter and the receiver are not collocated. We will present practical AoI measurement results for such networks in the remainder of this chapter. In order to do so, we need to first discuss possible methods of AoI measurement, synchronization, and effect of synchronization errors on AoI optimization.

12.4 Clock Bias and Synchronization for Age Measurement over Networks

Depending on the application and the network model, the AoI and other AoI-related values may be calculated at the receiver, the transmitter, or centrally. In this section, we will discuss some practical timing-related issues regarding AoI measurement.

12.4.1 Age Measurement via Synchronization

A straightforward approach to AoI measurement is to establish synchronization among the transmitter and receiver clocks. This may be realized in a variety of ways, some of which are listed here:

- **Using a GPS clock**: Using a GPS module is currently one of the most accurate methods for synchronization of distant machines. According to [30], the accuracy of the GPS clock is in the order of tens of nanoseconds. GPS clocks are currently employed in UAVs, base stations, and some software-defined radios [8]. However, this method may be unsuited for networks composed of many low-power devices such as sensor modules and IoT nodes, where it is not feasible to support GPS modules on each node.

- **Pre-synchronization using a common reference**: For some applications, it is possible to physically connect two devices before deploying them in separate locations. In such a case, the receiver and transmitter can be synchronized using a common Real Time Clock (RTC). The disadvantage of this approach is that the synchronization will be lost if one of the devices is reset or powered off.
- **Synchronization using Network Time Protocol (NTP)**: NTP is used to synchronize the clocks of all participating computers in a network. It provides synchronization in the range of 0.1 milliseconds in fast LANs and within the range of a few tens of milliseconds in the intercontinental Internet [31]. The performance of the NTP is affected by jitter on the network, which reduces accuracy.
- **Synchronization via prior signalling:** For purposes of AoI measurement, since only the timing information at the transmitter and the ultimate receiver are required, the synchronization can be directly performed at the application layer by measuring at the transmitter the round trip time based on a simple ping mechanism prior to transmission. The receiver responds to the ping by sending the time stamp of the received packet, and the transmitter can correct its clock assuming that the round trip delay is twice the unidirectional delay in either direction. This assumption is justifiable on the basis that the packets used in this phase are small, hence incurring negligible transmission times in queues, and the RTT times can be thought of as being dominantly due to delays.

Once the synchronization is established, the AoI measurement can take place at the transmitter or the receiver:

- **Measuring AoI at Receiver:** To measure the AoI at the receiver, the transmitter needs to send the generation time stamps of packets. Both UDP and TCP headers carry the time stamp information of the packet, so these can be used in AoI measurement. If the header of the packet is not available, then the time stamp information can be inserted in the payload. For the calculation average AoI, (12.3) is preferable, because of its higher accuracy while averaging in short time-windows.
- **Measuring AoI at Transmitter:** To measure AoI at the transmitter, the time stamps of packets received at the receiver are needed to be fed back to the transmitter. This information can be carried using an ACK for each packet, by inserting it into the payload of the ACKs. The drawbacks of this method are the additional delay in calculation due to need for the extra wait until the ACK is received, and miscalculations that may be caused by ACKs being lost in the return link. Finally, in cases where ACK is not already used as part of the protocol and is transmitted solely for AoI measurement, this method occupies more channel resources than measurement at the receiver. However, in setups where we need to adaptively control the age and

related parameters by adjusting the transmission policy, measuring the age at the transmitter is a more viable option.

12.4.2 Effect of Imperfect Synchronization on Age-Related Parameters

The methods highlighted in the previous section have the drawback that they are all bound to yield some synchronization error, which in turn translates directly to error in age measurements. In this subsection, the effect of the synchronization error on age computation is investigated.

To calculate AoI-related values, such as average peak AoI or average AoI, time stamps from both the receiver and the transmitter are required. However, since the receiver and transmitter are distant from each other, they have their own system clocks, and even after synchronization there will be constant bias, due to the synchronization error. Assuming the drift of each clock is negligible during the time window of interest, the bias between two clocks will remain the same.

Let us denote by s_i' and s_i, the time stamps marking the transmission of the ith packet, at the receiver and the transmitter, respectively. That is, the transmitter knows the actual transmission time, and the receiver knows it up to a synchronization error, that is,

$$s_i' = s_i + B \tag{12.8}$$

for some constant bias (synchronization error) B.

Similarly, let r_i' and r_i denote the time stamps that mark the instant of reception of the packet at the receiver and the transmitter respectively. We have

$$r_i' = r_i + B. \tag{12.9}$$

The receiver returns its perceived receive time stamp r_i' to the transmitter, and the transmitter then calculates the average AoI using Eq. (12.1). Substituting (12.9) in (12.2) and (12.4), we get the erroneous areas of the trapezoids,

$$Q_i' = Q_i + B(s_i - s_{i-1}) \tag{12.10}$$

and

$$H_i' = H_i + B(r_i - r_{i-1}). \tag{12.11}$$

Substituting (12.11) in (12.3), the average age becomes

$$\overline{\Delta}' = \frac{1}{T}\sum_{i=1}^{n(T)}(H_i + B(r_i - r_{i-1})). \tag{12.12}$$

Finally, since the total time elapsed, T, is equal to the sum of inter-arrival times at the receiver, $\sum_{i=1}^{n(T)}(r_i - r_{i-1})$, from (12.12), the average age measurement in the presence of synchronization error yields

$$\overline{\Delta}' = \overline{\Delta} + B. \tag{12.13}$$

That is, the average age is shifted due to synchronization error by the same bias among the clocks of the transmitter and receiver. Similarly, using (12.6), the time synchronization error adds the same constant bias on average peak AoI:

$$\overline{\Delta_{peak}}' = \overline{\Delta_{peak}} + B. \tag{12.14}$$

At this point, an important observation is in order: the effect of error in synchronization, or equivalently, a constant bias among the transmitter and receiver clocks, simply shifts the average AoI and average peak AoI on age axis by the same bias. Therefore, even if there is an unknown bias among clocks is not known, the optimal operating point on the AoI versus sampling rate curve remains unchanged. This opens up the possibility of skipping the synchronization before age calculation altogether, in case the goal is to adjust the transmission rate, provided the age can somehow be estimated. This idea will be revisited in Section 12.4.3.

While a constant bias in the clock measurements has no effect on the value of the optimal sampling rate when the penalty function is a linear function of age, there may be penalty functions that are nonlinear functions of age [27, 28], for which the optimal operating points differ for biased and unbiased age. For an age penalty function, $f(\Delta(t))$, let $F(t) = \int_0^t f(\tau)d\tau$. Then, the time average age penalty is [6]

$$\begin{aligned}\overline{\Delta_{Bias}} = \frac{1}{T}\sum_{i=1}^{n(T)} & F(r_i + B - s_{i-1}) - F(r_{i-1} + B - s_{i-1}) \\ & - F(r_i - s_{i-1}) + F(r_{i-1} - s_{i-1}).\end{aligned} \tag{12.15}$$

The age biases for the commonly used linear ($f(t) = \alpha t$), exponential ($f(t) = e^{\alpha t} - 1$), and logarithmic ($f(t) = \log(\alpha t + 1)$) penalty functions [32, 33], calculated using (12.15), are [6]

$$\overline{\Delta_{Bias,Linear}} = \alpha B; \tag{12.16}$$

$$\overline{\Delta_{Bias,Exp}} = \frac{1}{\alpha\sum_{i=1}^{n(T)}(r_i - r_{i-1})}\left(e^{\alpha(\theta+B)} - e^{\alpha(\beta+B)} - e^{\alpha\theta} + e^{\alpha\beta}\right) \tag{12.17}$$

and

$$\begin{aligned}\overline{\Delta_{Bias,Log}} = \frac{1}{\sum_{i=1}^{n(T)}(r_i - r_{i-1})}\Bigg(\frac{1}{\alpha}\Big(& \log(\alpha\beta + 1) - \log(\alpha\theta + 1) + \log(\alpha(B+\theta) + 1) \\ & - \log(\alpha(B+\beta) + 1)\Big) - \theta\log(\alpha\theta + 1) + \beta\log(\alpha\beta + 1) \\ & + \log(\alpha(B+\theta) + 1)(B+\theta) - \log(\alpha(B+\beta) + 1)(B+\beta)\Bigg),\end{aligned} \tag{12.18}$$

respectively, where $\beta = r_{i-1} - s_{i-1}$ and $\theta = r_i - s_{i-1}$.

Since sampling, remote control, and tracking formulations are known to benefit from nonlinear age penalty functions [32–34], the measurement of correct age to be used in such penalty functions is crucial. Therefore, synchronization is essential in applications requiring nonlinear penalty functions. For cases when a linear relation between age penalty and measured age exists, the bias in age measurement can be tolerated, and the following method for AoI estimation can be used in lieu of synchronization.

12.4.3 Asynchronous Estimation of AoI Using RTT

In cases where the exact value of age is not crucial, the transmitter, upon receiving an ACK from the receiver in response to its transmission of a packet, can directly use the RTT of that packet as an estimate for the age of that packet [4]. Note that this method doesn't require the transmitter and receiver to be synchronized. The idea is that, since the ACK doesn't carry time stamp information, it is shorter. Consequently, the error probabilities and delays faced by the ACKs are lower, and the round trip time can be mostly attributed to the forward link. This method clearly provides an overestimate of average AoI. Yet, it is very suitable for scenarios when the age penalty function is linear, and the bias in age measurement therefore does not affect the optimal operating point. We will review some experimental results that use this method for age estimation in Section 12.6.

12.5 The Effect of the Access Network to AoI in TCP-IP Connections

We begin by reviewing the results of [2], which was a first attempt at studying age behavior on experimental implementations of packet flows in the Internet. The experimental setup was used to measure the relation between AoI and status update generation rates, for an end-to-end TCP flow traversing a physical network. The first hop on the network was varied between WiFi, Ethernet, LTE, 3G, and 2G access links. Additionally, an emulation study was carried out in order to understand the effect of bottlenecks introduced at various links on the relationship between the sample generation rate and end-to-end AoI.

This experimental study was mainly motivated by the following question: is there a non-monotonic convex relationship between AoI and sampling rate in practice, as the one in simple idealized networking setups (e.g., Figure 12.2)? This is a valid question for the following reason: practical communication networks consist of many interacting protocol components at different layers, each of which have its own queueing mechanisms. These tend to be beyond the control of the end-to-end application. In the rest, we elaborate on the results of this study that paved a way for the rest of the works we consider in this chapter.

12.5.1 AoI Measurement over an Emulation Testbed

The emulation component of the work in [2], used an open-source network emulator called CORE [35], which was also employed in [1].

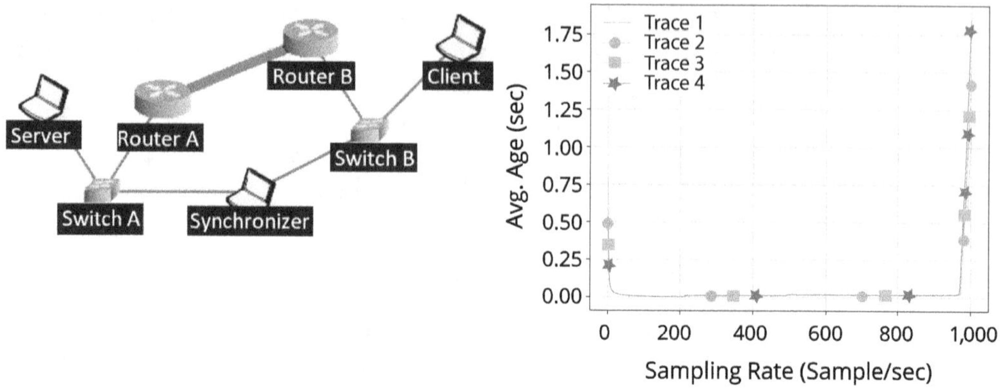

(a) CORE networking topology (b) Age vs. rate

Figure 12.5 Emulation topology and resulting AoI behavior.

The topology in the emulation study is described in Figure 12.5(a). The setup ran the CORE 5.0 emulator on the Fedora 27 Linux distribution on a PC with Intel® Core™ i7-7700HQ CPU @ 2.80GHz × 8, 16 GB RAM, and 1 TB 7200 rpm HDD. The test setup involves three virtual nodes: server, client, and time_sync. The time_sync node was for synchronizing the clocks at the client and the server, for correct computation of the age at the server.

The client generated time-stamped samples at different sampling rates, sending them to the server. Upon receiving each sample, the server computed the status age. Information was exchanged between the server and the client via a link with bandwidth limited to 130 Kbps, marked by the thick line in Figure 12.5(a), by configuring routers accordingly. All links other than this particular one had unlimited bandwidth, allowing instantaneous packet transmission. None of the links had path delay. Therefore, delay occurred due to only transport layer queueing. Each iteration of the experiment started with empty buffers, and the sampling rates were gradually increased, obtaining a sample path for AoI versus rate. Four such sample paths are shown in Figure 12.5(b). All iterations exhibit a sharp U-shape behaviour of age with sampling rate. The starting age and the shape of the curve do not appear to deviate much among iterations, and the AoI nearly reaches zero at moderate data rates, owing to the absence of path delay in this emulation.

12.5.2 AoI Measurements on a Real-World Testbed

A real-world AoI measurement experiment was reported in [2], which evaluated age in TCP/IP connections through 2G, 3G and LTE, WiFi, and Ethernet. The study took a two-step approach: first, the server and client were time-synchronized; then, samples were generated at the client at a variable and increasing rate. The client was a computer located 39°53′29.6″N 32°46′56.6″E. The Internet connection for the client was provided alternatively through WiFi, Ethernet, or a cellular connection. The cellular

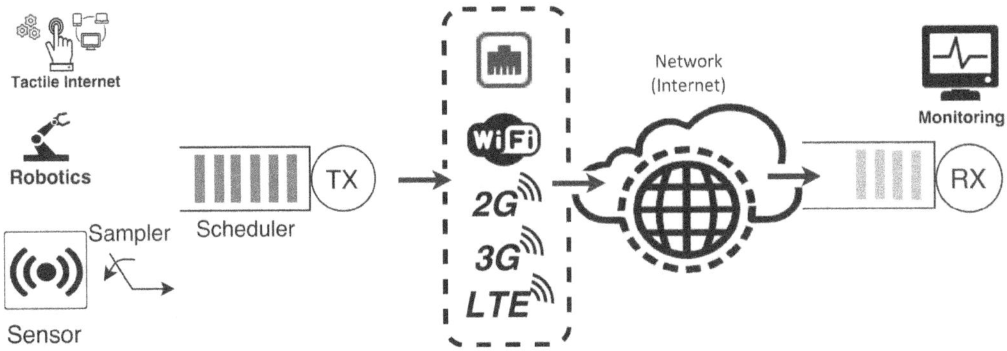

Figure 12.6 A schematic of the physical testbed. ©[2018] IEEE. Reprinted, with permission, from [2].

connection was established on the service provider TURKCELL via USB tethering to a cellular phone. The relatively high bandwidth of the USB 2.0 connection supporting 53 MBps [36] is more than capable of handling data rate resulting from the peak sampling rate. Hence, the effect of the USB buffer on AoI was negligible.

The topology of the physical testbed, involving a remote server in France, is shown in Figure 12.6. The Speed Test application by Ookla was used to measure the upload speeds for each networking scenario: 17.84 Mbps for LTE, 0.47 Mbps for 3G, 0.03 Mbps for 2G, 99.04 Mbps for WiFi, and 93.37 Mbps for Ethernet. Each data sample was approximately 35 kB.

In contrast to the emulation scenario, a time_sync node is not present in this experimental study. The sender and receiver were in distant geographical locations. Synchronization was performed via prior signaling as described in Section 12.4.1. More precisely, continual requests were sent from the client to the server. The server sent its own time stamp in response to each request. The client recorded the time at the beginning of each request, and then again upon receiving the response to the same request. The round trip delay was computed based on this data. Assuming that the delay was symmetric in both directions in the link, the present clock of the server was predicted. Comparing the predicted clock of the server and its own clock, the client calculated an offset value to synchronize clocks. To reduce offset computation errors, offsets of a large number of independent runs were averaged. A reliable Ethernet connection to a local area network, which had lower latency than the cellular connection, was used as a benchmark for synchronization.

The queues at the transport layer operated on a FIFO basis: the head-of-line packet in the queue was popped and served when the transport channel became available. The sampling rate determined how frequently the new data was pushed into the queue.

In Figures 12.7(a)–12.8(c), the results for different first-hop access types are shown. The figures exhibit a non-monotone behavior of age with sampling rate. The curves look similar at lower sampling rates; however, the AoI starts increasing at different values of the sampling rate, due to the variation in the rates supportable in the first-hop

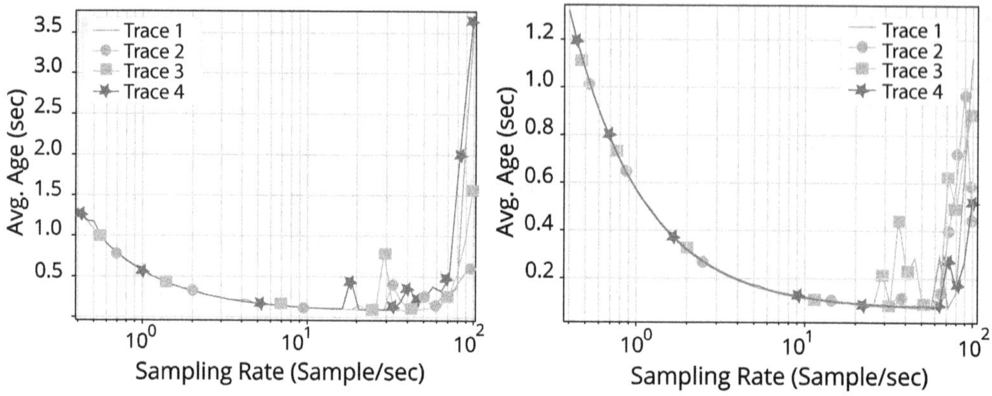

(a) WiFi First-hop

(b) Ethernet First-hop

Figure 12.7 AoI measurements of connections over different access networks. ©[2018] IEEE. Reprinted, with permission, from [2].

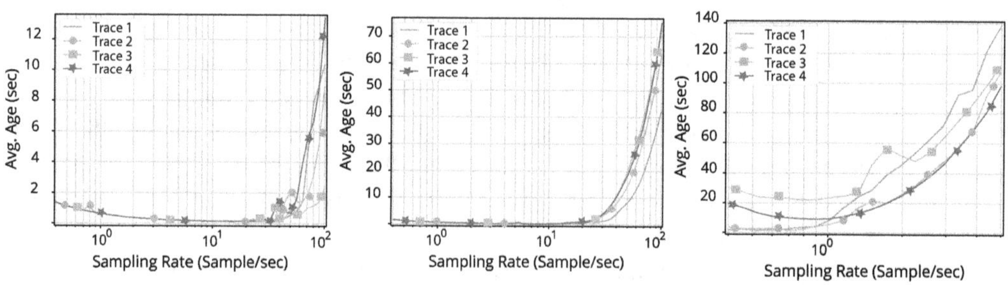

(a) Over LTE network

(b) Over 3G network

(c) Over 2G network

Figure 12.8 AoI measurements on mobile networks. ©[2018] IEEE. Reprinted, with permission, from [2].

links for different access types. In Figure 12.8(c) it is observed that the AoI increases much faster in case of 2G.

12.6 AoI Comparison of UDP versus TCP-IP

TCP and UDP are the two most popularly implemented transport layer protocols in IP networks today. The mechanisms that enable TCP variants to transmit data at high rates without loss, such as adaptive window size, congestion control, and retransmissions make this protocol not only computationally more intense, but also somewhat unsuited to real-time data. In terms of age, the retransmission mechanism of TCP is particularly wasteful, as stale packets do not contribute to an age reduction. Many real-time applications use RTP over UDP. However, do the currently available protocols used for real-time traffic truly cater to age objectives? This question justifies a careful look at the age performance in UDP traffic. The brief answer is, while RTP and more modern application layer modifications such as those used in cloud gaming, for example [37], offer relatively timely data for streaming applications, they do not immediately cater to an age objective: status update type flows ultimately do not need all the data to be necessarily transmitted. When there is a newer packet in the sender buffer, it may deem all older ones obsolete. For this reason, to cater to age objectives, we believe that application-layer age-aware sampling or LCFS queueing needs to be implemented on top of UDP.

As observed in [20, 29, 38], the average age initially decreases with increasing throughput in FCFS systems without any strict buffer management or limitation. Yet, once the communication system struggles to serve the high throughput, the queueing delay becomes significant and the average age starts to increase, thereby resulting in the U-shaped relation among average AoI and throughput. The FCFS buffers in routers, switches, and access points of practical network infrastructures are therefore expected to lead to a similar non-monotone relationship in real-life networks.

The particular shape of this relation between age and throughput is still influenced by the properties of transport layer protocols, CPU capacity, transmit and receive capabilities and so on, as will be demonstrated by the experimental results in the IoT setup, that will be discussed in this section. Achieving a good AoI performance with a low-power IoT node first requires that the CPU can generate packets at a sufficient rate. Secondly, the transceiver modules at sender and receiver nodes must be capable of timely processing of the generated packets. Lack of sufficient processing power in either of these will cause an age bottleneck. For example, if the CPU is not fast enough, the transmission rate cannot be increased sufficiently to reduce the AoI. Low-power IoT devices are especially prone to running into these bottlenecks. With common transport protocols, such as TCP, which already have a high computational overhead, and even with UDP, IoT devices can thus run into problems in terms of AoI, underlining the need for new approaches.

In this section, experimental results from [6], taken in real-life networks, are reviewed. In particular, we first focus on the age behavior of TCP and UDP flows over the Internet. Then, we review age behavior of same type of flows among IoT nodes, this time on a local IoT setup, with the goal of demonstrating the effect of limited computational performance of IoT devices on AoI. In these experiments, RTT information was used to bound the synchronization error as described in Section 12.4.3: before taking measurements, the RTT was estimated by sending several packets and receiving the ACKs. As the ACK packets are small, the estimated RTT was used as an upper bound to the transmission time. Using this as an approximation, the synchronization error was estimated. The remaining age bias was bounded by the RTT. Note that, according to (12.16), even if approximations still contain a constant bias, this does not affect the location of the age-optimal sampling rate operating point, due to the linearity of the penalty function (which in this case is age itself).

12.6.1 UDP vs. TCP-IP over Multi-hop Network Testbed

In the testbed, high-power desktop PCs were used to send TCP or UDP packets through regular Internet/IP infrastructure. Three PCs were located at remote locations,

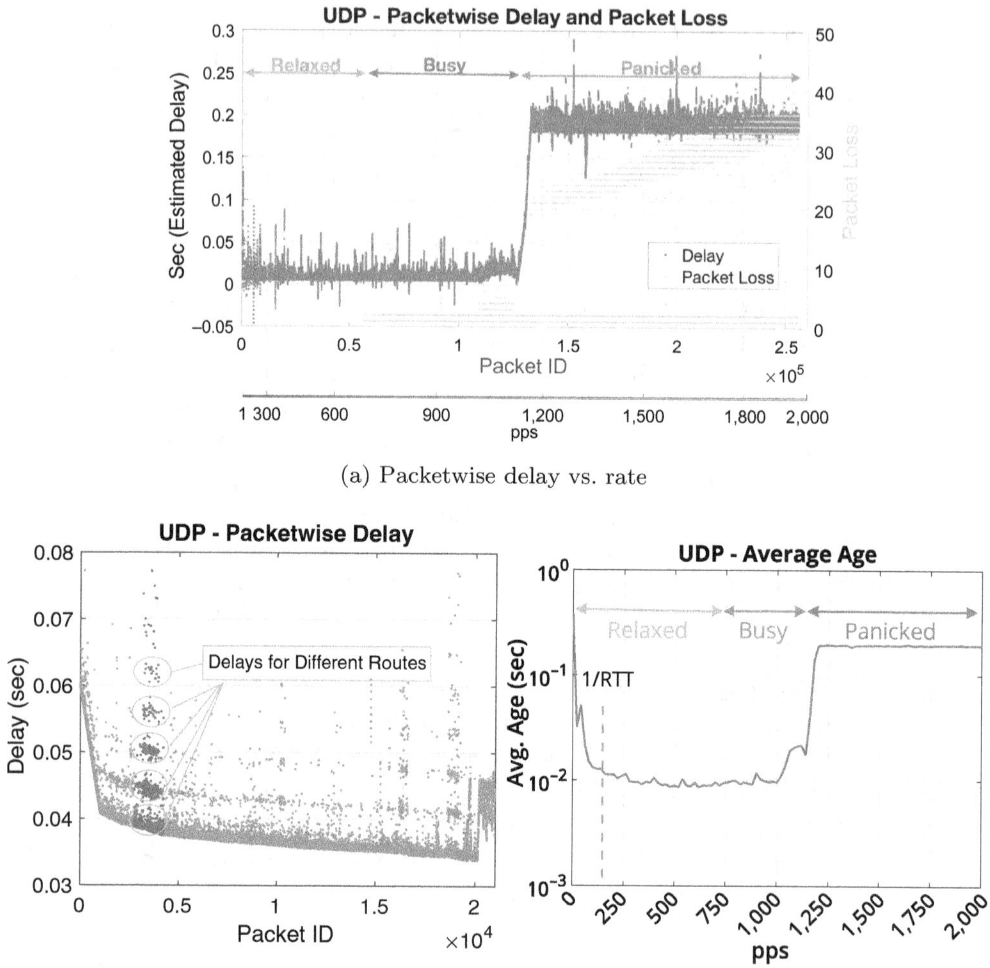

(a) Packetwise delay vs. rate

(b) Delay distribution in the relaxed regime

(c) Average age

Figure 12.9 Age, delay, and packet loss vs. rate for UDP over Internet. ©[2020] IEEE. Reprinted, with permission, from [6].

within a 450 km range, yielding a network with paths with varying delays. One of the PCs was set up as the receiver node, and the other PCs sent packets to it. The path between one of the transmitting PCs and the receiver contained approximately 7 hops and 6 ms RTT, whereas the other contained approximately 12 hops and had 80 ms RTT.

The main cause of AoI increase with UDP was observed to be the high packet loss rate. The delays incurred in transport layer queues were observed to be minor. In Figures 12.9(a)–12.9(d), the results of UDP transmission tests on the Internet testbed just described are shown. Figure 12.9(a) plots delays seen by packets that are transmitted successfully, while Figure 12.9(b) shows the number of lost packets between consecutive successful packets. The horizontal axes in both figures represent the packet IDs

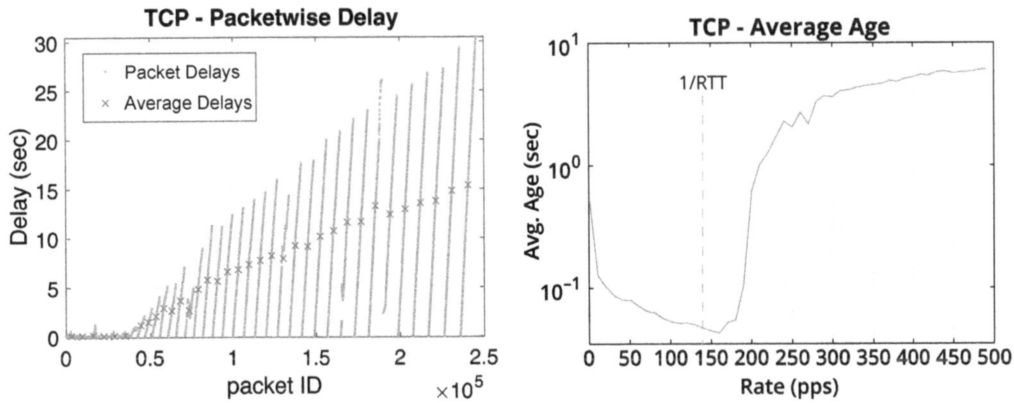

(a) Individual and average packet delays

(b) Average age

Figure 12.10 Age and packetwise delay for TCP over Internet. ©[2020] IEEE. Reprinted, with permission, from [6].

and the increasing sample generation rate. In Figure 12.9(c) delays experienced by all packets are plotted versus increasing rate. The delay distribution suggests the presence of multiple routes, which plays a role in the age remaining constant despite the increasing rate. In Figure 12.9(d) the calculated average age values for the same experiments are shown.

The UDP test results on the Internet testbed reveal three modes of operation, namely relaxed, busy, and panicked. The categorization of these modes is based on the packet loss characteristics. In the relaxed region, number of samples is well below the load that will congest the network. Increasing rate in relaxed mode reduces average age, as more frequent samples provide more timely data. With increasing load, the network gradually gets congested, causing some nodes to drop packets at random, and enters the busy mode. As seen in Figure 12.9(b), the start of the busy region is marked by a step-up in the number of lost packets from 0 to a moderate number between 1 and 3, and throughout the so-called busy region, the number of lost packets generally stays in this range. Despite the occasional packet loss in the busy mode, the delay remains similar to that in the relaxed mode (see Figure 12.9(a)), as intermediate nodes can still handle the high traffic without additional queueing delay and the end-to-end delay is mostly due to propagation delay over the links.

With further increase in rate, we enter the panicked region where both the number of packet losses and the delay increase. In this mode the queueing delays at intermediate nodes become significant. Note that, due to the lack of a retransmission mechanism for failed packets in UDP, the packet delays do not monotonically increase with rate; they saturate at a level that depends on the maximum queueing delay. Hence, with increasing rate, the age also remains constant.

Unlike UDP, in TCP, failed packets are retransmitted, resulting in a steady increase in AoI with increased transmission rates; see Figure 12.10(a). The resulting experimental results, therefore, show a similarity to the U-shaped age-throughput graph of FCFS queues.

Additionally, according to the experiments, TCP flows often observe a sharp increase in average AoI as rate exceeds a certain value, due to congestion. On the other hand, with UDP we do not observe a similar sharp age blowup with increased rate.

12.6.2 UDP vs. TCP-IP over IoT Testbed

The second testbed in [6] was a Wi-Fi network, comprising two NodeMCU ESP32 with Xtensa® LX6 (600 MIPS) IoT devices and a Wi-Fi router operating in 802.11n mode. The two IoT devices were configured to act as a transmitter and receiver respectively, sending TCP/UDP packets through the central router node.

Due to the inability of the nodes to run stand-alone operating systems, the TCP/UDP operations were carried out by Lightweight IP-Stack (LWIP), an open-source software used by different IoT devices.

The results reported in [6] indicate an interesting AoI bottleneck for IoT devices: memory and computational power limitations of these devices, rather than any communication or network constraints, can be the dominant cause of aging. The peak packet generation rate for a simple IoT node (e.g., the Wi-Fi module integrated to a ESP32 has a higher rate than the processing unit on the device) will be significantly lower than what is possible with a regular PC. Hence the device is able to generate packets at a much lower rate than its transceiver is able to send. Moreover, transceiver buffers on those devices are too small to observe the non-monotone age behavior expected from an infinite-buffer FCFS queueing system. Hence, it is not surprising that Figure 12.11 does not indicate a U-shaped AoI behaviour for either UDP or TCP. However, with devices more powerful than those typically employed in a LoRa system, non-monotone age behavior can still be observed due to bandwidth bottlenecks.

As a side note, in many IoT applications, as the nodes are typically small battery-operated devices, energy is a limiting factor that influences the fidelity of operation. Energy limitations have a direct effect on transmission rate and duty cycle, [39–43], as

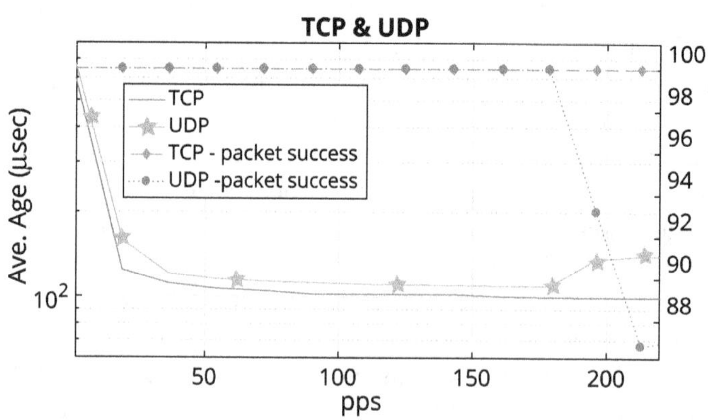

Figure 12.11 Packet loss and average age TCP and UDP measured over a local Wi-Fi network, using IoT devices. ©[2020] IEEE. Reprinted, with permission, from [6].

well as processing capability. We have seen in this section that in simple IoT devices, the processing power tends to be the main limiting factor in age performance, which previous theoretical work has tried to address (e.g., [44]).

12.7 Application of Machine Learning Methods to AoI Optimization in Real-World Networks

AoI optimization methods in many previous studies (e.g., [17, 23, 45–48]) relied on knowledge of the statistics of underlying processes (e.g., network delays, energy arrivals). This reduces the applicability of these approaches to practical settings where these statistics are unknown and result from fairly complicated interactions, deeming the application of machine learning methods promising for AoI optimization in real-life networks. To the best of our knowledge, [3, 49] were the first works that utilized machine learning (specifically, reinforcement learning) for minimizing AoI. [49] formulated the AoI minimization problem on a point-to-point link through the application of HARQ as a constrained Markov Decision Process and solved it by using Value Iteration and SARSA algorithms on discrete state space (ages defined as integers). More recently, [50–54] considered the application of machine learning methods to age optimization in various network settings.

In this section we review [3], which is focused on a real-life implementation of a deep RL-based algorithm for sample rate optimization in a continuous state-space to control the arrival rate of a flow of packets over the Internet. This is similar to the rate control model in earlier TCP-or UDP-related works reviewed in previous sections.

Modeling a flow of packets generated by an application running between a server–client pair over the Internet, [3] introduced a deep reinforcement learning-based approach that can learn to control the rate of generation of updates to minimize the AoI with no prior assumptions about network topology. After evaluating the learning model on an emulated network, it was argued that the method can be scaled up to larger and more realistic networks with unknown delay distribution.

Let us briefly discuss the formulation in [3]. The sender generates update packets at $t_1, t_1, \cdots, t_n$, which are received at times $U(t_1), U(t_2), \cdots, U(t_n)$ (packets can change order in the network.) The AoI optimization problem was modeled as a Markov Decision Process (MDP) with unknown transition probabilities:

- The state at time t is the age at time t: $s_t = \Delta_t$,
- The action space at any time t contains two possible actions (pause or resume): $a_t = \{p, r\}$.

Accordingly, at any time, the next state depends on the current state and the chosen action through the transition probabilities $p(s_{t+1}|s_t, a_t)$. To learn the transition probabilities, RL was utilized. The goal of the RL is to find the action that maximizes the expected cumulative reward $r(s, a)$ over the trajectory distribution $p_\theta(\tau)$. The trajectory was defined as the state–action pair $\tau = (s, a)$. The AoI minimization objective was formulated with a cost function $c(s_t, a_t) = -r(s_t, a_t) = -\Delta_t$, thus the objective of the formulation in [3] is to solve (12.19).

$$\theta^* = \arg\min_{\theta} \left(\limsup_{T\to\infty} \frac{1}{T} \mathbb{E}_{(s_t,a_t)\sim p_\theta(s_t,a_t)}\left[\Sigma_{t=1}^{T} \Delta_t \right] \right) \tag{12.19}$$

To solve (12.19), the well-known deep Q network (DQN) algorithm, introduced in [55], was employed. The goal of this learning algorithm was to estimate the expected reward (Q-value) of possible actions in each state, using neural networks. As it is a cost minimization rather than maximization of a reward, the algorithm was modified such that it minimizes the introduced cost function $c(s_t, a_t) = 1 - \exp\{-\Delta_t\}$.

The algorithm in [55] employed a neural network to map states to the Q-value of each possible action. The Q-values represented the vector of expected costs when each of the actions a was taken in state s, and $\arg\min_a \hat{Q}_\phi(s)$ represented the action, at each state s, that minimized the expected cost $\mathbb{E}[c(s, a)]$. Rather than taking the action with the minimum Q-value at each step with probability 1, the agent was forced to explore by utilizing the ϵ_{greedy} approach [56], where a random action is taken with probability ϵ. Stabilizing approaches such as experience replay [57] and Double DQN [58] were also employed.

The DQN algorithm just described was tested in CORE, an open-source network emulator, using the topology in Figure 12.5(a). In this topology, the path delay on the direct link between routers A and B was selected from a truncated Gaussian distribution with adjustable mean and variance, and the remaining links have zero delay. The routing is planned such that there is only one path between the server and the client, and that involves router A and router B. The algorithm runs on both the server and the client. Synchronization between the server and client is performed using a synchronizer node. The accuracy of this synchronization is guaranteed, as the links between the synchronizer, the server, and the client have no delay. The client assumes one of three modes: (1) generating time-stamped samples at a fixed rate, (2) sending samples to the server, and (3) responding to pause and resume commands from the server. The client stops generating new samples when it receives a pause command and continues upon receiving a resume command.

The case of a link with unlimited bandwidth was tested first. Here, due to the absence of queueing, path delay (1 ± 0.5s) was the sole cause of aging. Therefore, the pause command can only cause an increase in age this case. Having computed age from the time stamps of the received samples, the age is given to a deep RL algorithm, which uses the age value to decide on a command to reduce the age. This really acted as a test of the RL algorithm: the optimal decision is always "resume" in this case, as it is possible to transmit data at any sampling rate without loading the network buffer.

A double DQN was run with two hidden layers of 24 units and experience replay, initially trained for 10,000 iterations. The main network was updated each iteration, whereas the target layer was updated every 100 iterations. The Q-value for the *resume* action is expected to approach $Q(s, a) \approx 1 - \exp\{-\Delta_{min\_age}\}$ according to the boundary conditions applied on the Bellman equation:

$$Q(s, a) = c(s, a) + \gamma \min_{a'} Q(s', a'); \tag{12.20}$$

At the terminal state, $Q^*(s, a) = c(s, a) = 1 - \exp\{-\Delta_{min\_age}\}$. As the delay is fixed at 1 s, the Q-value for the resume action is expected to converge down to

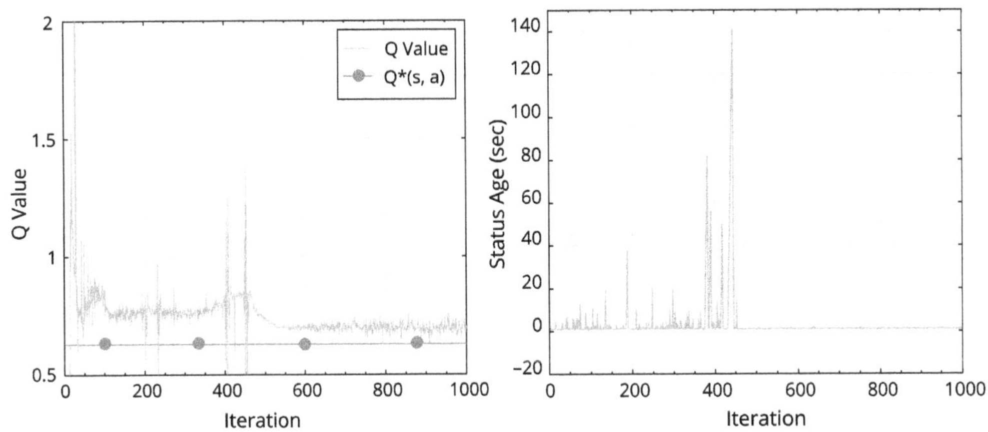

(a) Q-value per 10 iterations (unlimited bandwidth)

(b) Status age per 10 iterations (limited bandwidth)

Figure 12.12 Q-value and status age vs. iteration. ©[2018] IEEE. Reprinted, with permission, from [3].

$Q(s, a) = 1 - \exp\{-1\} \approx 0.63$, which happened after approximately 5,000 iterations. Figure 12.12(a) plots the progress of the instantaneous Q value, as well as its optimal setting. The progress of the status age versus iteration is plotted in Figure 12.12(b).

Based on the experimental results just summarized and the Universal Approximation Theorem [59], the authors argued that the approach can be scaled up to larger networks with general delay distributions.

12.8 Application Layer Age Control Mechanisms over UDP

Age Control Protocol [4] is a transmission control policy designed to work on top of UDP to reduce the Age of Information (AoI) in dynamic networks. The protocol consists of two phases: the initialization phase, and the epochs phase. In the initialization phase, the source (sender) sends a certain number of packets to the monitor (receiver) and calculates RTT. Next, the protocol enters a phase divided into certain periods called epochs. In each epoch, ACP computes the average AoI, an exponentially weighted moving average RTT, exponentially weighted moving average inter-ACK arrival times, and average backlog (i.e., average of the number of packages sent to the receiver but not yet received). At the end of each epoch, the sender's output rate changes according to change in the backlog and average AoI.

There are three different actions that can be chosen at the end of each epoch: additive increase (INC), additive decrease (DEC), and multiplicative decrease (MDEC). If "INC" or "DEC" actions are used, the protocol tries to increase or decrease the average backlog by a step size κ. This step-size parameter is an important determinant of the resulting age dynamics.

It has been shown that over an intercontinental connection, ACP can reduce the median age by about 33% when compared with *Lazy* [4], a simple policy to always keep the total number of packets in flight around 1 (the delay-bandwidth product [60]) by setting the rate of packet transmission to 1/RTT. The logic of Lazy follows the observation made by Kleinrock in [60] to keep the pipe busy, but not too full. Having sufficiently high load to preventing idling, while still low enough to keep the queues stable, is a classic rule of thumb in finding a throughput-optimal operating point in networks. However, doing this control more finely, on the "knife-edge," is especially important with respect to the age metric, as any queueing causes unnecessary aging. Keeping the number of packets in flight around 1 represents this knife-edge: at any time, the network is kept busy transmitting a packet, while there is no packet waiting in queue. It is worth recalling that at the age-optimal operating point of the $M/M/1$ queue, the average number of packets in the system is ≈ 1.13.

In [61], the ACP protocol was implemented on IoT devices, specifically two ESP32 nodes controlled via separate Arduino devices and connected through a Wi-Fi hotspot. The results of the implementation suggested that ACP may run into unnecessary oscillations in a small and relatively static network, and that *Lazy* may be preferable when network conditions are more or less static.

12.9 Link Layer Experimentation of Age-Aware Wireless Access Schedulers

Building on theoretical results on multi-access and broadcast scheduling as in [62, 63], testbed implementations have been reported in [8, 64]. WiFresh RT [8] proposes a novel Wi-Fi uplink implementation using hardware-level programming of schedulers in a network of FPGA-based software-defined radios, as well as an application layer scheme for easier adoption of age adaptation, without lower-layer protocol stack modifications.

For each source, WiFresh RT appends a time stamp to each generated packet and stores it in a Last-Come First-Served (LCFS) queue of size 1, that is, keeping only the freshest packet and discarding older packets. WiFresh RT running on the access point (AP) side coordinates the communication in the network. WiFresh defines two states for the AP: (1) waiting for a data packet, and (2) transmitting a poll packet. The WiFresh App runs over UDP protocol and uses standard Wi-Fi. This App contains all elements of WiFresh RT and some additional features. Fragmentation of large information updates, a simple built-in synchronization algorithm, and support for sources that generate multiple types of information are additional features of the WiFresh App.

The experimental results in [8] indicate that when congestion in the wireless network increases, the AoI degrades sharply, leading to outdated information at the destination. WiFresh is shown to mitigate this problem, achieving near-optimal information freshness in wireless networks of any size, even when the network is overloaded. WiFresh was compared with UDP over Wi-Fi and Wi-Fi Age Control Protocol (ACP) [4] on architectures using a Raspberry Pi wireless base station receiving data from N sources. The results show that ACP improves the average age by a factor of four when compared with Wi-Fi UDP, while WiFresh improves the average age by a factor of forty when compared with Wi-Fi UDP. The reason for the superiority of WiFresh over the other two architectures is that WiFresh has used a combination of a polling multiple-access mechanism, the Max-Weight (MW) policy, and LCFS queues at the Application layer. Unlike ACP, WiFresh and Wi-Fi UDP architectures do not attempt to control the packet generation rate at the sources. Therefore, when the number of sources N increases to the point that the cumulative packet generation rate exceeds the capacity of the network, both Wi-Fi UDP and WiFresh become overloaded. Consequently, the number of backlogged packets at the sources grows rapidly. As Wi-Fi UDP uses the FCFS policy, a large backlog in queues leads to high packet delay, resulting in high average age. In contrast, WiFresh scales gracefully with average age increasing linearly with N.

Results of an experimental implementation of a wireless downlink are reported in [64]. On a multiuser testbed consists of multiple software-defined USRP radios, the authors have implemented Round-Robin, Max-Weight, Whittle's Index, and Greedy policies provided in [63] using Labview and tested them under different node formations and power limitations. In this downlink implementation, at each time frame, the AP generates a status update and sends it to the receiver determined by the scheduling policy. The transmitter uses QPSK modulation and BCH coding. The transmitter and receiver nodes have access to a common clock due to being connected to a common computer. A performance comparison of scheduling policies based on the measured age values is reported to indicate that Max-Weight and Whittle's Index policies outperform Round-Robin and Greedy policies, especially when signal power is strictly limited and the channels are asymmetric, that is, observing different channel gains.

12.10 Conclusions

In this chapter we have provided an overview of implementation studies related to Age of Information as a performance metric and design criterion on communication networks. These studies have each provided a proof of concept of how data freshness can be improved by modifying the design of various protocols and algorithms in the networking stack, without loss of performance with respect to traditional metrics such as throughput and delay. They have each addressed certain practical issues that arise in the process of incorporating AoI into the design of network architectures and algorithms, and have discussed how to handle those.

We believe, based on the theoretical understanding about AoI having gained maturity within the last decade and the results of these pioneering implementation studies, that it is worthwhile to introduce age-aware modifications to the network stack. Perhaps it will be easiest for these principles to penetrate application layer mechanisms, and in fact, most of the rate-control mechanisms that we have reviewed in this chapter have done exactly that.

In conclusion, we believe we are witnessing the first steps of the concept of AoI entering mainstream network protocol design. This may be followed by more general semantics-related metrics, some of which may capture more about the information than its timeliness. One has to be careful, though, in finding a balance between the usefulness of semantics-related metrics and their practical applicability in the protocol stack. While semantics-empowered communication has been a holy grail in communication and information theory for many decades, it has not found its way to implementation mainly because of the requirement of cross-layer operation across the protocol stack. For example, optimizing physical layer coding schemes to directly serve the needs of a control system [65] is potentially more efficient than ignoring the semantics and treating all data equally; however, this would require the transceiver operation to be customized to interact with the application layer, which is difficult from a standardization and industry-wide adoption perspective.

Age is perhaps the simplest of semantics-related metrics, as it depends only on time stamps. While it may be considered simplistic for the same reason, many previous studies have shown that application performance (e.g., location tracking) can be characterized in terms of a function of age, or even if this is not true, age-based sampling still provides a major step toward optimal performance (e.g., [66]). We have seen in the implementation examples reviewed in this chapter that it is possible to obtain significant age improvements by simple modifications such as application layer sample generation rate. We are thus convinced that such revision of protocols has great potential, due to age, as a freshness metric, capturing the needs of IoT and other applications often requiring status-update type data, and constitute an increasingly large portion of Internet traffic.

References

[1] C. Kam, S. Kompella, and A. Ephremides, "Experimental evaluation of the age of information via emulation," in *MILCOM 2015 – 2015 IEEE Military Communications Conference*, October 2015, pp. 1070–1075.

[2] C. Sönmez, S. Baghaee, A. Ergişi, and E. Uysal-Biyikoglu, "Age-of-Information in practice: Status age measured over TCP/IP connections through WiFi, Ethernet and LTE," in *2018 IEEE International Black Sea Conference on Communications and Networking (BlackSeaCom)*, June 2018, pp. 1–5.

[3] E. Sert, C. Sönmez, S. Baghaee, and E. Uysal-Biyikoglu, "Optimizing age of information on real-life TCP/IP connections through reinforcement learning," in *2018 26th Signal Processing and Communications Applications Conference (SIU)*, May 2018, pp. 1–4.

[4] T. Shreedhar, S. K. Kaul, and R. D. Yates, "An age control transport protocol for delivering fresh updates in the Internet-of-Things," in *2019 IEEE 20th International Symposium on " A World of Wireless, Mobile and Multimedia Networks" (WoWMoM)*, 2019, pp. 1–7.

[5] H. B. Beytur, S. Baghaee, and E. Uysal, "Measuring age of information on real-life connections," in *2019 27th Signal Processing and Communications Applications Conference (SIU)*, 2019, pp. 1–4.

[6] ——, "Towards AoI-aware smart IoT systems," in *2020 International Conference on Computing, Networking and Communications (ICNC)*, 2020, pp. 353–357.

[7] I. Kadota, M. S. Rahman, and E. Modiano, "Age of information in wireless networks: From theory to implementation," in *Proceedings of the 26th Annual International Conference on Mobile Computing and Networking*, ser. MobiCom '20. New York, NY, USA: Association for Computing Machinery, 2020.

[8] ——, "WiFresh: Age-of-information from theory to implementation," 2020.

[9] S. Kaul, M. Gruteser, V. Rai, and J. Kenney, "Minimizing age of information in vehicular networks," in *2011 8th Annual IEEE Communications Society Conference on Sensor, Mesh and Ad Hoc Communications and Networks*, 2011, pp. 350–358.

[10] E. Najm and R. Nasser, "Age of information: The gamma awakening," in *2016 IEEE International Symposium on Information Theory (ISIT)*, 2016, pp. 2574–2578.

[11] Y. Sun, E. Uysal-Biyikoglu, R. D. Yates, C. E. Koksal, and N. B. Shroff, "Update or wait: How to keep your data fresh," *IEEE Transactions on Information Theory*, vol. 63, no. 11, pp. 7492–7508, November 2017.

[12] C. Kam, S. Kompella, and A. Ephremides, "Learning to sample a signal through an unknown system for minimum AoI," in *IEEE INFOCOM 2019 – IEEE Conference on Computer Communications Workshops (INFOCOM WKSHPS)*, 2019, pp. 177–182.

[13] H. B. Beytur and E. Uysal, "Age minimization of multiple flows using reinforcement learning," in *2019 International Conference on Computing, Networking and Communications (ICNC)*, Feb. 2019, pp. 339–343.

[14] E. T. Ceran, D. Gündüz, and A. György, "Reinforcement learning to minimize age of information with an energy harvesting sensor with HARQ and sensing cost," in *IEEE INFOCOM 2019 – IEEE Conference on Computer Communications Workshops (INFOCOM WKSHPS)*, 2019, pp. 656–661.

[15] M. A. Abd-Elmagid, H. S. Dhillon, and N. Pappas, "A reinforcement learning framework for optimizing age of information in rf-powered communication systems," *IEEE Transactions on Communications*, vol. 68, no. 8, pp. 4747–4760, 2020.

[16] ——, "Aoi-optimal joint sampling and updating for wireless powered communication systems," *IEEE Transactions on Vehicular Technology*, vol. 69, no. 11, pp. 14 110–14 115, 2020.

[17] S. Kaul, R. Yates, and M. Gruteser, "Real-time status: How often should one update?" in *2012 Proceedings IEEE INFOCOM*, 2012, pp. 2731–2735.

[18] C. Kam, S. Kompella, G. D. Nguyen, and A. Ephremides, "Effect of message transmission path diversity on status age," *IEEE Trans. Inf. Theory*, vol. 62, no. 3, pp. 1360–1374, March 2016.

[19] M. Costa, M. Codreanu, and A. Ephremides, "On the age of information in status update systems with packet management," *IEEE Transactions on Information Theory*, vol. 62, no. 4, pp. 1897–1910, April 2016.

[20] S. K. Kaul, R. D. Yates, and M. Gruteser, "Status updates through queues," in *2012 46th Annual Conference on Information Sciences and Systems (CISS)*, Mar. 2012, pp. 1–6.

[21] R. D. Yates and S. Kaul, "Real-time status updating: Multiple sources," in *IEEE International Symposium on Information Theory (ISIT)*, July 2012, pp. 2666–2670.

[22] C. Kam, S. Kompella, and A. Ephremides, "Age of information under random updates," in *2013 IEEE International Symposium on Information Theory*, July 2013, pp. 66–70.

[23] C. Kam, S. Kompella, and A. Ephremides, "Effect of message transmission diversity on status age," in *2014 IEEE International Symposium on Information Theory*, June 2014, pp. 2411–2415.

[24] M. Costa, M. Codreanu, and A. Ephremides, "Age of information with packet management," in *2014 IEEE International Symposium on Information Theory*, June 2014, pp. 1583–1587.

[25] L. Huang and E. Modiano, "Optimizing age-of-information in a multi-class queueing system," in *2015 IEEE International Symposium on Information Theory (ISIT)*, June 2015, pp. 1681–1685.

[26] B. T. Bacinoglu, E. T. Ceran, and E. Uysal-Biyikoglu, "Age of information under energy replenishment constraints," in *2015 Information Theory and Applications Workshop (ITA)*, February 2015, pp. 25–31.

[27] Y. Sun, E. Uysal-Biyikoglu, R. Yates, C. E. Koksal, and N. B. Shroff, "Update or wait: How to keep your data fresh," in *IEEE INFOCOM 2016 – The 35th Annual IEEE International Conference on Computer Communications*, April 2016, pp. 1–9.

[28] Y. Sun, E. Uysal-Biyikoglu, R. D. Yates, C. E. Koksal, and N. B. Shroff, "Update or wait: How to keep your data fresh," *IEEE Transactions on Information Theory*, vol. 63, no. 11, pp. 7492–7508, November 2017.

[29] S. Kaul, R. Yates, and M. Gruteser, "Real-time status: How often should one update?" in *2012 Proceedings IEEE INFOCOM*, March 2012, pp. 2731–2735.

[30] Official U.S. government information about the Global Positioning System (GPS) and related topics, *GPS Accuracy*, accessed January 10, 2021. [Online]. Available: www.gps.gov/systems/gps/performance/accuracy/

[31] D. L. Mills, *Computer Network Time Synchronization: The Network Time Protocol on Earth and in Space*, 2nd ed. Boca Raton: CRC Press, Inc., 2010.

[32] A. Kosta, N. Pappas, A. Ephremides, and V. Angelakis, "Age and value of information: Non-linear age case," in *2017 IEEE International Symposium on Information Theory (ISIT)*, 2017, pp. 326–330.

[33] M. Klügel, M. H. Mamduhi, S. Hirche, and W. Kellerer, "AoI-Penalty minimization for networked control systems with packet loss," in *IEEE INFOCOM 2019 – IEEE Conference on Computer Communications Workshops (INFOCOM WKSHPS)*, 2019, pp. 189–196.

[34] Y. Sun and B. Cyr, "Sampling for data freshness optimization: Non-linear age functions," *Journal of Communications and Networks*, vol. 21, no. 3, pp. 204–219, 2019.

[35] J. Ahrenholz, C. Danilov, T. R. Henderson, and J. H. Kim, "Core: A real-time network emulator," in *MILCOM 2008 – 2008 IEEE Military Communications Conference*, November 2008, pp. 1–7.

[36] *Universal Serial Bus Specification*, USB-IF, 4 2000, Revision 2.0. [Online]. Available: www.usb.org/document-library/usb-20-specification

[37] A. Alós, F. Morán, P. Carballeira, D. Berjón, and N. García, "Congestion control for cloud gaming over udp based on round-trip video latency," *IEEE Access*, vol. 7, pp. 78 882–78 897, 2019.

[38] Y. Inoue, H. Masuyama, T. Takine, and T. Tanaka, "A general formula for the stationary distribution of the age of information and its application to single-server queues," *IEEE Transactions on Information Theory*, vol. 65, no. 12, pp. 8305–8324, 2019.

[39] H. Erkal, F. M. Ozcelik, and E. Uysal-Biyikoglu, "Optimal offline broadcast scheduling with an energy harvesting transmitter," *EURASIP Journal on Wireless Communications and Networking*, vol. 2013, no. 1, p. 197, 2013.

[40] S. Chamanian, S. Baghaee, H. Ulusan, O. Zorlu, H. Kulah, and E. Uysal-Biyikoglu, "Powering-up wireless sensor nodes utilizing rechargeable batteries and an electromagnetic vibration energy harvesting system," *Energies*, vol. 7, no. 10, pp. 6323–6339, 2014.

[41] S. Baghaee, S. Chamanian, H. Ulusan, and O. Zorlu, "Wirelessenergysim: A discrete event simulator for an energy-neutral operation of iot nodes," in *2018 IEEE International Black Sea Conference on Communications and Networking (BlackSeaCom)*, June 2018, pp. 1–5.

[42] F. M. Ozcelik, G. Uctu, and E. Uysal-Biyikoglu, "Minimization of transmission duration of data packets over an energy harvesting fading channel," *IEEE Communications Letters*, vol. 16, no. 12, pp. 1968–1971, December 2012.

[43] E. T. Ceran, D. Gündüz, and A. György, "Average age of information with hybrid arq under a resource constraint," *IEEE Transactions on Wireless Communications*, vol. 18, no. 3, pp. 1900–1913, 2019.

[44] B. T. Bacinoglu, Y. Sun, E. Uysal, and V. Mutlu, "Optimal status updating with a finite-battery energy harvesting source," *Journal of Communications and Networks*, vol. 21, no. 3, pp. 280–294, 2019.

[45] R. D. Yates and S. K. Kaul, "Status updates over unreliable multiaccess channels," in *2017 IEEE International Symposium on Information Theory (ISIT)*, 2017, pp. 331–335.

[46] Y. Sun, E. Uysal-Biyikoglu, and S. Kompella, "Age-optimal updates of multiple information flows," in *IEEE INFOCOM 2018 – IEEE Conference on Computer Communications Workshops (INFOCOM WKSHPS)*, 2018, pp. 136–141.

[47] B. T. Bacinoglu, E. Uysal-Biyikoglu, and C. E. Koksal, "Finite-horizon energy-efficient scheduling with energy harvesting transmitters over fading channels," *IEEE Transactions on Wireless Communications*, vol. 16, no. 9, pp. 6105–6118, September 2017.

[48] B. T. Bacinoglu, O. Kaya, and E. Uysal-Biyikoglu, "Energy efficient transmission scheduling for channel-adaptive wireless energy transfer," in *2018 IEEE Wireless Communications and Networking Conference (WCNC)*, 2018, pp. 1–6.

[49] E. T. Ceran, D. Gündüz, and A. György, "Average age of information with hybrid ARQ under a resource constraint," in *2018 IEEE Wireless Communications and Networking Conference (WCNC)*, April 2018.

[50] A. Elgabli, H. Khan, M. Krouka, and M. Bennis, "Reinforcement learning based scheduling algorithm for optimizing age of information in ultra reliable low latency networks," in *2019 IEEE Symposium on Computers and Communications (ISCC)*. IEEE, 2019, pp. 1–6.

[51] M. A. Abd-Elmagid, A. Ferdowsi, H. S. Dhillon, and W. Saad, "Deep reinforcement learning for minimizing age-of-information in uav-assisted networks," in *2019 IEEE Global Communications Conference (GLOBECOM)*. IEEE, 2019, pp. 1–6.

[52] M. Ma and V. W. Wong, "A deep reinforcement learning approach for dynamic contents caching in hetnets," in *ICC 2020–2020 IEEE International Conference on Communications (ICC)*. IEEE, 2020, pp. 1–6.

[53] M. Hatami, M. Jahandideh, M. Leinonen, and M. Codreanu, "Age-aware status update control for energy harvesting iot sensors via reinforcement learning," in *2020 IEEE 31st Annual International Symposium on Personal, Indoor and Mobile Radio Communications*. IEEE, 2020, pp. 1–6.

[54] A. Ferdowsi, M. A. Abd-Elmagid, W. Saad, and H. S. Dhillon, "Neural combinatorial deep reinforcement learning for age-optimal joint trajectory and scheduling design in uav-assisted networks," *arXiv preprint arXiv:2006.15863*, 2020.

[55] V. Mnih, K. Kavukcuoglu, D. Silver, A. A. Rusu, J. Veness, M. G. Bellemare, A. Graves, M. Riedmiller, A. K. Fidjeland, G. Ostrovski, S. Petersen, C. Beattie, A. Sadik, I. Antonoglou, H. King, D. Kumaran, D. Wierstra, S. Legg, and D. Hassabis, "Human-level control through deep reinforcement learning," *Nature*, vol. 518, no. 7540, pp. 529–533, February 2015.

[56] R. S. Sutton and A. G. Barto, *Reinforcement learning: An introduction.* Massachusetts Institute of Technology Press, 1998.

[57] L.-J. Lin, "Reinforcement learning for robots using neural networks," Ph.D. dissertation, Carnegie Mellon University, USA, 1992.

[58] H. v. Hasselt, A. Guez, and D. Silver, "Deep reinforcement learning with double q-learning," in *Proceedings of the Thirtieth AAAI Conference on Artificial Intelligence*, ser. AAAI'16. AAAI Press, 2016, p. 2094–2100.

[59] K. Hornik, M. Stinchcombe, and H. White, "Universal approximation of an unknown mapping and its derivatives using multilayer feedforward networks," *Neural Netw.*, vol. 3, no. 5, p. 551–560, October 1990. [Online]. Available: https://doi.org/10.1016/0893-6080(90)90005-6

[60] L. Kleinrock, "Internet congestion control using the power metric: Keep the pipe just full, but no fuller," *Ad Hoc Networks*, vol. 80, pp. 142–157, 2018. [Online]. Available: www.sciencedirect.com/science/article/pii/S1570870518302476

[61] U. Guloglu, "Implementation and evaluation of ACP on an IoT testbed," Middle East Technical University, Technical report, METU, 2021.

[62] I. Kadota, A. Sinha, and E. Modiano, "Scheduling algorithms for optimizing age of information in wireless networks with throughput constraints," *IEEE/ACM Transactions on Networking*, p. 1–14, 2019.

[63] I. Kadota, A. Sinha, E. Uysal-Biyikoglu, R. Singh, and E. Modiano, "Scheduling policies for minimizing age of information in broadcast wireless networks," *IEEE/ACM Transactions on Networking*, vol. 26, no. 6, pp. 2637–2650, 2018.

[64] T. K. Oğuz, E. Uysal, and T. Girici, "Implementation and evaluation of MAC layer age-aware scheduling algorithms in a wireless uplink," Middle East Technical University, Technical report, METU, 2021.

[65] S. Tatikonda, A. Sahai, and S. Mitter, "Control of lqg systems under communication constraints," in *Proceedings of the 37th IEEE Conference on Decision and Control (Cat. No. 98CH36171)*, vol. 1. IEEE, 1998, pp. 1165–1170.

[66] Y. Sun, Y. Polyanskiy, and E. Uysal, "Sampling of the wiener process for remote estimation over a channel with random delay," *IEEE Transactions on Information Theory*, vol. 66, no. 2, pp. 1118–1135, 2019.

13 Reinforcement Learning for Minimizing Age of Information over Wireless Links

Elif Tuğçe Ceran, Deniz Gündüz, and András György

13.1 Introduction

In this chapter, we study the age of information (AoI) when status updates of an underlying process of interest, sampled and recorded by a *source* node, must be transmitted to one or more *destination* nodes over error-prone wireless channels. We consider the practical setting, in which the statistics of the system are not known a priori, and must be learned in an online fashion. This requires designing (RL) algorithms that can adapt their policy dynamically through interactions with the environment. Accordingly, the aim of this chapter is to design and analyze RL algorithms to minimize the average AoI at the destination nodes, taking into account *retransmissions* due to channel errors.

Retransmissions are essential for providing reliability of status updates over error-prone channels, particularly in wireless settings, and are incorporated into almost all wireless communication standards. In the standard automatic repeat request (ARQ) protocol, failed transmissions are repeated until they are successfully received, or a maximum retransmission count is reached. Some of the recent standards including ZigBee (Alliance 2008), Bluetooth *IEEE 802.15.1*, WiFi *IEEE 802.11ac*, and UWB (Ultra-wideband) *IEEE 802.15.4a* (Oppermann, Hamalainen, & Iinatti 2004) use cyclic redundancy check (CRC) together with ARQ. On the other hand, in the hybrid ARQ (HARQ) protocol, the receiver combines information from previous transmission attempts of the same packet in order to increase the success probability of decoding. Recent communication standards including *IEEE 802.16m*, 3GPP LTE, LTE-A (E-UTRA 2013), *IEEE 802.11be*, and Narrow Band IoT (NB-IoT) have adopted HARQ techniques to enhance the system performance, typically through a combination of CRC and forward error correction (FEC) (802.16e 2005 2006). In this chapter, we study both ARQ and HARQ protocols for the minimization of AoI.

Until recently, prior literature in the AoI framework assumed that the perfect statistical information regarding the random processes governing the status-update system is available to the source. However, an increasing number of works are focusing on the practically relevant problem (e.g., sensors embedded in unknown or time-varying environments) and study RL for AoI optimization (Hsu, Modiano, & Duan 2017; Ceran, Gündüz, & György 2018; Ceran, Gündüz, & György 2019;

Sert, Sönmez, Baghaee, & Uysal-Biyikoglu 2018; Beytur & Uysal 2019; Leng & Yener 2019; Abd-Elmagid, Ferdowsi, Dhillon, & Saad 2019; Elgabli, Khan, Krouka, & Bennis 2019; Abd-Elmagid, Dhillon, & Pappas 2020; Hatami, Jahandideh, Leinonen, & Codreanu 2020). An end-to-end IoT application running over the Internet is considered in Sert et al. (2018) without prior assumptions about network topology and a deep RL algorithm is applied. An RL approach to minimize the AoI in an ultrareliable low-latency communication system is considered in Elgabli et al. (2019). The scheduling decisions with multiple receivers over a perfect channel are investigated in Hsu et al. (2017) and Beytur and Uysal (2019), where the goal is to learn data arrival statistics. Q-learning (Sutton & Barto 1998) is used for a generate-at-will model in Hsu et al. (2017), while policy gradients and DQN methods are used for a queue-based multi-flow AoI-optimal scheduling problem in Beytur and Uysal (2019). In Leng and Yener (2019), policy gradients and DQN methods are employed for AoI minimization in a wireless ad hoc network, where nodes exchange status updates with one another over a shared spectrum. Average-cost RL algorithms are proposed in Ceran, Gündüz, and György (2018) and (2019) to learn the decoding error probabilities in a status-update system with HARQ. The work in Ceran, Gündüz, & György (2019) exploits RL methods in order to learn both decoding error probabilities and characteristics.

The rest of the chapter is organized as follows. Section 13.2 provides a brief background on Markov decision processes (MDPs) and RL methods, which will be used to model and solve the AoI minimization problems addressed in this chapter. Section 13.3 investigates a point-to-point status-update system with HARQ under a resource constraint and exploits an average-cost RL algorithm to minimize the average AoI. Section 13.4 extends the results in Section 13.3 to a multiuser status-update system, and presents various RL algorithms with different complexity-performance trade-offs. Section 13.5 considers an energy-harvesting status-update system with HARQ and considers sensing cost at the source node, as well as the transmission cost of the updates. Finally, Section 13.6 concludes the chapter.

13.2 Preliminaries

Reinforcement Learning (RL) is an important area of machine learning where a learning agent learns how to behave in an environment by performing actions and observing the results of its actions in the form of state transitions and costs in order to learn to minimize some notion of cumulative cost Sutton and Barto (1998). In recent years, RL methods have attracted significant attention thanks to groundbreaking achievements in this area of research. Examples include AlphaGo, which incorporates deep RL, beating the world champions at the game of Go (Silver et al. 2016), as well as the Deep Q-Network (DQN) algorithm (Mnih, Kavukcuoglu, Silver, Rusu, Veness, Bellemare, Graves, Riedmiller, Fidjeland, Ostrovski, Petersen, Beattie, Sadik, Antonoglou, King, Kumaran, Wierstra, Legg, & Hassabis 2015) beating humans playing numerous Atari video games. RL methods have also been widely adopted for many wireless

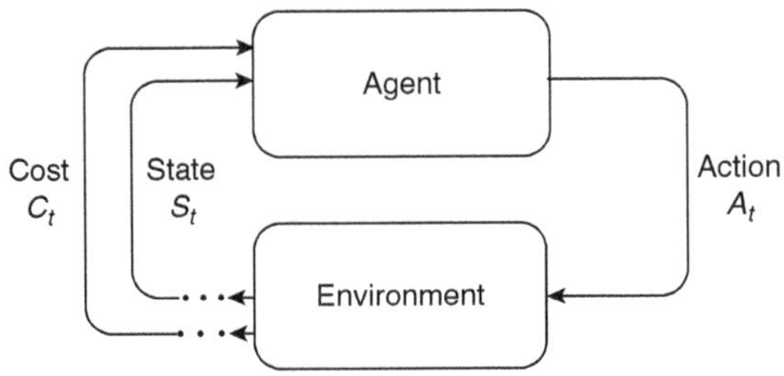

Figure 13.1 Illustrations of the interactions between the agent and the environment in the RL framework.

networking and mobile communication systems and applications (Clancy, Hecker, Stuntebeck, & O'Shea 2007; Somuyiwa, György, & Gündüz 2018; Luong, Hoang, Gong, Niyato, Wang, Liang, & Kim 2019).

In the RL framework, as depicted in Figure 13.1, an agent repeatedly interacts with its environment: At time t the state of the environment is S_t. The agent takes an action A_t, which makes the environment to transition to another state S_{t+1}, and the agent suffers a cost c_t. The agent's goal is to minimize its long-term costs.

This process can be conveniently modeled as a *Markov decision process (MDP)* (Puterman 1994): An MDP is defined with a tuple $\langle \mathcal{S}, \mathcal{A}, \mathrm{P}, c \rangle$, where $\mathcal{S}$ denotes a countable set of states and $\mathcal{A}$ denotes a countable set of actions.[1] The transition kernel $\mathrm{P} : \mathcal{S} \times \mathcal{A} \times \mathcal{S} \to [0, 1]$ defines the transition probabilities: that is, if action $a \in \mathcal{A}$ is taken in state $s \in \mathcal{S}$, the environment transitions to state $s' \in \mathcal{S}$ with probability $\mathrm{P}(s'|s, a)$, independently of previous states and actions. (Note that $\mathrm{P}(\cdot|s, a)$ defines a distribution over $\mathcal{S}$ and hence $\sum_{s' \in \mathcal{S}} \mathrm{P}(s'|s, a) = 1$ for all $s \in \mathcal{S}, a \in \mathcal{A}$.) Thus, if S_t and A_t denote the state and action at time t, then $\mathrm{P}(s'|s, a) = \Pr(S_{t+1} = s'|S_t = s, A_t = a)$, and for any $s_0, \ldots, s_{t+1} \in \mathcal{S}$ and $a_0, \ldots, a_t \in \mathcal{A}$, the state action sequence $S_0, A_0, S_1, A_1, \ldots, S_t, A_t, S_{t+1}$ satisfies $\Pr(S_{t+1} = s_{t+1}|S_t = s_t, A_t = a_t) = \Pr(S_{t+1} = s_{t+1}|S_t = s_t, A_t = a_t, \ldots, S_0 = s_0, A_0 = a_0)$. Finally, the cost suffered by the agent is determined by the state of the environment and the action taken in that state via the cost function $c : \mathcal{S} \times \mathcal{A} \to \mathbb{R}$.

In the MDP formulation it is assumed that in every time step the agent observes the state of the MDP, and it can select its action based on its past observations. Therefore, an agent's strategy can be described by a *policy*, defined as a sequence of decision rules π_t: $(\mathcal{S} \times \mathcal{A})^t \to [0, 1]$, which maps the past states and actions and the current state to a distribution over the actions. That is, after the state-action sequence $s_0, a_0, \ldots, s_{t-1}, a_{t-1}, s_t$, action a_t is selected (in state s_t) with probability $\pi_t(a_t|s_t, a_{t-1}, s_{t-1} \ldots, a_0, s_0)$. We use s_t^π and a_t^π to denote the sequences of states and actions, respectively, induced by policy $\pi = \{\pi_t\}$. A policy $\pi = \{\pi_t\}$ is called *stationary* if the distribution of the next action is independent of the past states and actions

[1] Assuming that $\mathcal{S}$ and $\mathcal{A}$ are countable is not necessary, but simplifies the treatment of MDPs and is sufficient for our applications concerning the age of information. (What is more, we also assume in the rest of the chapter that the action set is finite.)

given the current state, and it is time invariant; that is, with a slight abuse of notation, $\pi_t(a_t|s_t, a_{t-1}, s_{t-1} \ldots, s_0, a_0) = \pi(a_t|s_t)$ for all t and $(s_i, a_i) \in \mathcal{S} \times \mathcal{A}$, $i = 1, \ldots, t$. Finally, a policy is said to be *deterministic* if it chooses an action with probability one; with a slight abuse of notation, we use $\pi(s)$ to denote the action taken with probability one in state s by a stationary deterministic policy.

The goal of the agent is to select a policy that minimizes its expected average cost suffered after starting from state $s_0 \in \mathcal{S}$:

$$J^{\pi}(s_0) \triangleq \limsup_{T\to\infty} \frac{1}{T+1} \mathbb{E}\left[\sum_{t=0}^{T} c(s_t^{\pi}, a_t^{\pi})\Big|s_0\right].$$

A policy π^* achieving the minimum is called optimal. Under general conditions, there exists an optimal policy which is stationary, deterministic, and is independent of the start state s_0 (Puterman 1994).

Oftentimes, in practical problems, the agent has constraints on the actions it can take. For example, in an energy-harvesting system it is not possible to make a transmission if the transmitter's battery does not contain enough energy (Gündüz, Stamatiou, Michelusi, & Zorzi 2014). While such information can be included in the state, it is often simpler to keep the original state space and introduce some extra constraints governing the behavior of the agent. This can be modeled using a *constrained Markov decision process* (CMDP) (Altman 1999), which is an extension of an MDP: A CMDP is defined by the 5-tuple $\langle \mathcal{S}, \mathcal{A}, \mathrm{P}, c, d\rangle$, where $\mathcal{S}, \mathcal{A}, \mathrm{P}$, and c are defined as before, but an additional cost function $d : \mathcal{S} \times \mathcal{A} \to \mathbb{R}$, is introduced to describe the constraints to the system. (In the update systems we consider, this can be the energy cost of a transmission.)

Letting $C^{\pi}(s_0)$ denote the infinite horizon average cost for the constraint, starting from state $s_0 \in \mathcal{S}$, the goal of the agent in a CMDP is to minimize its average cost J^{π} subject to a constraint C_{max} on C^{π}; that is, to find and use a policy π solving the optimization problem:

$$\text{Minimize } J^{\pi}(s_0) \triangleq \limsup_{T\to\infty} \frac{1}{T+1} \mathbb{E}\left[\sum_{t=0}^{T} c(s_t^{\pi}, a_t^{\pi})\Big|s_0\right],$$

$$\text{subject to } C^{\pi}(s_0) \triangleq \limsup_{T\to\infty} \frac{1}{T+1} \mathbb{E}\left[\sum_{t=0}^{T} d(s_t^{\pi}, a_t^{\pi})\Big|s_0\right] \le C_{max}.$$

An optimal policy in an CMDP is a solution of the preceding problem. Under general conditions, an optimal policy is stationary and deterministic except for a single state (Altman 1999; Sennott 1993).[2]

[2] In general, there could be more than one constraint in a CMDP, in which case the optimal policy needs to randomize in more states. In fact, the number of states where randomization is necessary is equal to the number of constraints (Altman 1999; Sennott 1993).

The *differential value function* $h^\pi: \mathcal{S} \to \mathbb{R}$ and the *action-value function* $Q^\pi: \mathcal{S} \times \mathcal{A} \to \mathbb{R}$ of a policy π are defined as

$$h^\pi(s_0) = \limsup_{T\to\infty} \mathbb{E}\left[\sum_{t=0}^{T} c(s_t^\pi, a_t^\pi) - J^\pi(s_t^\pi)\middle| s_0\right];$$

$$Q^\pi(s_0, a_0) = \limsup_{T\to\infty} = \mathbb{E}\left[\sum_{t=0}^{T} c(s_t^\pi, a_t^\pi) - J^\pi(s_t^\pi)\middle| s_0, a_0\right].$$

Under general conditions (Puterman 1994; Bertsekas 2000), h^π and Q^π are the unique solutions (up to an additive constant) of the so-called *Bellman equations*: for all states $s \in \mathcal{S}$ and actions $a \in \mathcal{A}$,

$$Q^\pi(s, a) = c(s, a) - J^\pi(s, a) + \sum_{s' \in \mathcal{S}} P(s'|s, a) h^\pi(s');$$

$$h^\pi(s) = \sum_{a \in \mathcal{A}} \pi(a|s) Q^\pi(s, a).$$

An optimal policy π^* for an MDP satisfies a slightly modified version of these equations, called the *Bellman optimality equations*: for all $s \in \mathcal{S}$,

$$h^{\pi^*}(s) + J^{\pi^*}(s) = \min_a c(s, a) + \sum_{s' \in \mathcal{S}} P(s'|s, a^{\pi^*}) h^{\pi^*}(s').$$

On the other hand, no suboptimal policy can satisfy these equations. We will use Q and J to denote the state-action value function and the differential value function of an optimal policy π^*. It is easy to see that the Bellman optimality equations imply that, in every state s, an optimal policy chooses the greedy action minimizing the action-value function $Q(s, a)$ in a. There exist several algorithms in the literature that are based on the Bellman optimality equations and iteratively improve a policy whenever it violates these optimality conditions.

In the following sections, we model the AoI minimization problem under resource constraints using the MDP formulation defined previously. We study many RL techniques for the AoI minimization problem in different settings and compare their performances under different scenarios when the system characteristics are not known in advance or change with time. We present average-cost RL algorithms to learn transmission policies when the environment determined by the status-update system is not known a priori, including, in particular, the case of unknown decoding error probabilities in a status-update system with HARQ (Ceran, Gündüz, & György 2018; Ceran, Gündüz, & György 2019), and unknown energy-harvesting characteristics of the source node (Ceran, Gündüz, & György 2019).

13.3 RL for Minimizing AoI in Point-to-Point Status-Update Systems

In this section, we consider a point-to-point wireless status-update system. The source monitors an underlying time-varying process and can generate a status update at any

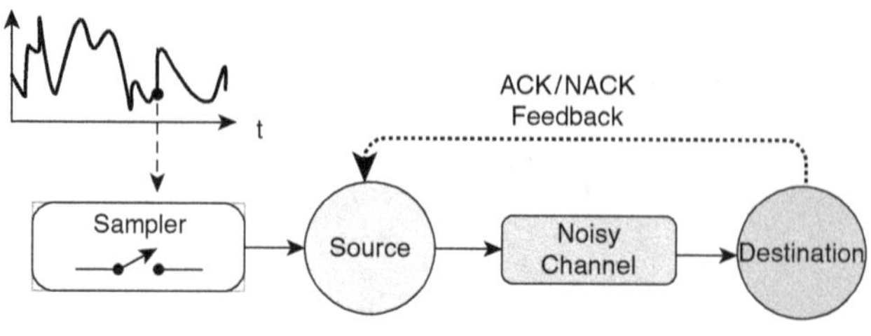

Figure 13.2 System model of a status-update system over an error-prone point-to-point link in the presence of ACK/NACK feedback from the destination.

time slot. The status updates are communicated from the source node to the destination over a time-varying channel (see Figure 13.2). Each transmission attempt of a status update takes constant time, set as one time slot. Throughout the chapter, we will normalize all time durations by the duration of one time slot. We assume that the wireless channel changes randomly from one time slot to the next in an independent and identically distributed (i.i.d.) fashion, and the channel state information is available only at the destination node. We further assume the availability of an error- and delay-free single-bit feedback from the destination to the source node for each transmission attempt. Successful receipt of a status update is acknowledged by an ACK signal, while a NACK signal is sent in case of a failure. In the classical ARQ protocol, a packet is retransmitted after each NACK feedback, until it is successfully decoded (or a maximum number of allowed retransmissions is reached), and the received signal is discarded after each failed transmission attempt. Therefore, the probability of error is the same for all retransmissions. However, in the AoI framework there is no point in retransmitting a failed out-of-date status packet if it has the same error probability as that of a fresh update. Hence, we assume that if the ARQ protocol is adopted, the source always removes failed packets and transmits a fresh status update. If the HARQ protocol is used, the received signals from all previous transmission attempts for the same packet are combined for decoding. Therefore, the probability of error decreases with every retransmission. In general, the error probability of each retransmission attempt depends on the particular combination technique used by the decoder as well as on the channel conditions.

AoI measures the timeliness of the information at the receiver. It is defined as the number of time slots elapsed since the generation of the most up-to-date packet successfully decoded at the receiver. Formally, denoting the latter generation time for any time slot t by $U(t)$, the AoI, denoted by δ_t, is defined as

$$\delta_t \triangleq t - U(t). \tag{13.1}$$

A transmission decision is made at the beginning of each slot. The AoI increases by one when the transmission fails. When it is successfully received, it decreases to one in the case of ARQ, or to the number of retransmissions plus one in the case of HARQ. (Minimum age is set to 1 to reflect that the transmission is one slot long.)

The probability of error after r retransmissions, denoted by $g(r)$, depends on r and the HARQ scheme. We assume that $g(r)$ is nonincreasing in the number of retransmissions r. For simplicity, we assume that $0 < g(0) < 1$, that is, the channel is noisy and there is a possibility that the first transmission is successful. Also, we will denote the maximum number of retransmissions by r_{max}, which may take the value ∞, unless otherwise stated. However, if $g(r) = 0$ for some r (i.e., a packet is always correctly decoded after r retransmissions), we set r_{max} to be the smallest such r. Note that practical HARQ methods only allow a finite number of retransmissions (802.16e 2005 2006).

Let $\delta_t \in \mathbb{Z}^+$ denote the AoI at the beginning of the time slot t, and $r_t \in \{0, \ldots, r_{max}\}$ denote the number of previous transmission attempts. Then the state of the system can be described by $s_t \triangleq (\delta_t, r_t)$. At each time slot, the source node takes one of the three actions, denoted by $a \in \mathcal{A}$, where $\mathcal{A} = \{\text{i}, \text{n}, \text{x}\}$: (i) remain idle ($a = \text{i}$); (ii) transmit a new status update ($a = \text{n}$); or (iii) retransmit the previously failed update ($a = \text{x}$).

If no resource constraint is imposed on the source, remaining idle is clearly suboptimal since it does not contribute to decreasing the AoI. However, continuous transmission is typically not possible in practice due to energy or interference constraints. Accordingly, we impose a constraint on the average number of transmissions, and we require that the long-term average number of transmissions not exceed $C_{max} \in (0, 1]$. (Note that $C_{max} = 1$ corresponds to the case in which transmission is allowed in every slot.)

This leads to the CMDP formulation, defined in Section 13.2: The countable set of states $(\delta, r) \in \mathcal{S}$ and the finite action set $\mathcal{A} = \{\text{i}, \text{n}, \text{x}\}$ have already been defined. P will be explicitly defined in (13.4). The cost function $c\colon \mathcal{S} \times \mathcal{A} \to \mathbb{R}$, is the AoI at the destination and is defined as $c((\delta, r), a) = \delta$ for any $(\delta, r) \in \mathcal{S}$, $a \in \mathcal{A}$, independently of action a. The transmission cost $d\colon \mathcal{S} \times \mathcal{A} \to \mathbb{R}$ is independent of the state and depends only on the action a, where $d = 0$ if $a = \text{i}$, and $d = 1$ otherwise.

Let $J^\pi(s_0)$ and $C^\pi(s_0)$ denote the infinite horizon average age and the average number of transmissions, respectively. The CMDP problem can be stated as follows:

Problem 1

$$\text{Minimize}\, J^\pi(s_0) \triangleq \limsup_{T\to\infty} \frac{1}{T+1}\, \mathbb{E}\left[\sum_{t=0}^{T} \delta_t^\pi \Big| s_0\right], \tag{13.2}$$

$$\text{subject to}\, C^\pi(s_0) \triangleq \limsup_{T\to\infty} \frac{1}{T+1}\, \mathbb{E}\left[\sum_{t=0}^{T} \mathbb{1}[a_t^\pi \neq \text{i}] \Big| s_0\right] \leq C_{max}. \tag{13.3}$$

Without loss of generality, we assume that the sender and the receiver are synchronized, that is, $s_0 = (1, 0)$; and we omit s_0 from the notation for simplicity.

Before formally defining the transition function P, we present a simple observation that simplifies P: Retransmitting a packet immediately after a failed attempt is better than retransmitting it after waiting for some slots. This is true, since waiting increases the age without increasing the success probability.

Proposition 1 For any policy π there exists another policy π' (not necessarily distinct from π) such that $J^{\pi'}(s_0) \le J^{\pi}(s_0)$, $C^{\pi'}(s_0) \le C^{\pi}(s_0)$, and π' takes a retransmission action only following a failed transmission, that is, the probability $Pr(a_{t+1}^{\pi'} = \mathrm{x} | a_t^{\pi'} = \mathrm{i}) = 0$.

P are given as follows (omitting the parenthesis from the state variables (δ, r)):

$$\begin{aligned}
&\mathrm{P}(\delta+1, 0|\delta, r, \mathrm{i}) = 1, \\
&\mathrm{P}(\delta+1, 1|\delta, r, \mathrm{n}) = g(0), \\
&\mathrm{P}(1, 0|\delta, r, \mathrm{n}) = 1 - g(0), \\
&\mathrm{P}(\delta+1, r+1|\delta, r, \mathrm{x}) = g(r), \\
&\mathrm{P}(r+1, 0|\delta, r, \mathrm{x}) = 1 - g(r),
\end{aligned} \tag{13.4}$$

and $\mathrm{P}(\delta', r'|\delta, r, a) = 0$ otherwise. Note that the preceding equations set the retransmission count to 0 after each successful transmission, and it is not allowed to take a retransmission action in states where the transmission count is 0. Also, the property in Proposition 1 is enforced by the first equation in (13.4), that is, $\mathrm{P}(\delta+1, 0|\delta, r, \mathrm{i}) = 1$ (since retransmissions are not allowed in states $(\delta, 0)$). Since the starting state is $(1, 0)$, it also follows that the state set of the CMDP can be described as

$$\mathcal{S} = \{(\delta, r) : r < \min\{\delta, r_{max}+1\}, \delta, r \in \mathbb{N}\}\,. \tag{13.5}$$

13.3.1 Lagrangian Relaxation and the Structure of the Optimal Policy

In this section, we derive the structure of the optimal policy for Problem 1 based on Sennott (1993). A detailed treatment of finite state–finite action CMDPs is considered in Altman (1999), but here we need more general results that apply to countable state spaces. These results require certain technical conditions; roughly speaking, there must exist a deterministic policy that satisfies the transmission constraint while maintaining a finite average AoI, and any "reasonable" policy must induce a positive recurrent Markov chain. The precise formulation of the requirements is given in Ceran, Gündüz, and György (2019), wherein Proposition 2 of Ceran, Gündüz, and György (2019) shows that the conditions of Sennott (1993) are satisfied for Problem 1. Given this result, we follow Sennott (1993) to characterize the optimal policy.

While there exists a stationary and deterministic optimal policy for countable-state finite-action average-cost MDPs (Sennott 1989; Puterman 1994; Bertsekas 2000), this is not necessarily true for CMDPs (Sennott 1993; Altman 1999). To solve the CMDP, we start by rewriting the problem in its Lagrangian form. The average Lagrangian cost of a policy π with Lagrange multiplier $\eta \ge 0$ is defined as

$$L_\eta^\pi = \lim_{T\to\infty} \frac{1}{T+1}\left(\mathbb{E}\left[\sum_{t=0}^{T} \delta_t^\pi\right] + \eta\mathbb{E}\left[\sum_{t=0}^{T} \mathbb{1}[a_t^\pi \neq \mathrm{i}]\right]\right), \tag{13.6}$$

and, for any η, the optimal achievable cost L_η^* is defined as $L_\eta^* \triangleq \min_\pi L_\eta^\pi$. If the constraint on the transmission cost is less than one (i.e., $C_{max} < 1$), then we have $\eta > 0$, which will be assumed throughout the chapter.[3] This formulation is equivalent to an unconstrained countable-state average-cost MDP with overall cost $\delta_t + \eta \mathbb{1}[a_t^\pi \neq \text{i}]$. A policy π is called η-optimal if it achieves L_η^*. Since the assumptions of Proposition 3.2 of Sennott (1993) are satisfied by Proposition 2 of Ceran, Gündüz, and György (2019), the former implies that there exists a *differential cost function* $h_\eta(\delta, r)$ satisfying

$$h_\eta(\delta, r) + L_\eta^* = \min_{a \in \{\text{i,n,x}\}} \left(\delta + \eta \cdot \mathbb{1}[a \neq \text{i}] + \mathbb{E}\left[h_\eta(\delta', r')\right]\right), \tag{13.7}$$

for all states $(\delta, r) \in \mathcal{S}$, where (δ', r') is the next state after taking action a.

We also introduce the *state-action cost function* defined as

$$Q_\eta(\delta, r, a) \triangleq \delta + \eta \cdot \mathbb{1}[a \neq \text{i}] + \mathbb{E}\left[h_\eta(\delta', r')\right] \tag{13.8}$$

for all $(\delta, r) \in \mathcal{S}, a \in \mathcal{A}$. Then, also implied by Proposition 3.2 of Sennott (1993), the optimal deterministic policy for the Lagrangian problem with a given η takes, for any $(\delta, r) \in \mathcal{S}$, the action achieving the minimum in (13.8):

$$\pi_\eta^*(\delta, r) \in \arg\min_{a \in \{\text{i,n,x}\}} Q_\eta(\delta, r, a)\,. \tag{13.9}$$

Focusing on deterministic policies, we can characterize the optimal policies for the CMDP problem. Based on Theorem 2.5 of Sennott (1993), we can prove the following:

THEOREM 13.1 *There exists an optimal stationary policy for the CMDP in Problem 1 that is optimal for the unconstrained problem considered in* (13.6) *for some* $\eta = \eta^*$, *and randomizes in at most one state. This policy can be expressed as a mixture of two deterministic policies* $\pi_{\eta^*,1}^*$ *and* $\pi_{\eta^*,2}^*$ *that differ in at most a single state s, and are both optimal for the Lagrangian problem* (13.6) *with* $\eta = \eta^*$. *More precisely, there exists* $\mu \in [0, 1]$ *such that the mixture policy* $\pi_{\eta^*}^*$, *which selects, in state s,* $\pi_{\eta^*,1}^*(s)$ *with probability* μ *and* $\pi_{\eta^*,2}^*(s)$ *with probability* $1 - \mu$, *and otherwise follows these two policies (which agree in all other states) is optimal for Problem 1, and* (13.3) *is satisfied with equality.*

Proof By Proposition 2 of Ceran, Gündüz, and György (2019), Theorem 2.5, Proposition 3.2, and Lemma 3.9 of Sennott (1993) hold for Problem 1. By Theorem 2.5 of Sennott (1993), there exists an optimal stationary policy that is a mixture of two deterministic policies, $\pi_{\eta^*,1}^*$ and $\pi_{\eta^*,2}^*$, which differ in at most one state and are η^*-optimal by Proposition 3.2 of Sennott (1993) satisfying (13.7) and (13.8). From Lemma 3.9 of Sennott (1993), the mixture policy π_μ^*, for any $\mu \in [0, 1]$, also satisfies (13.7) and (13.8), and is optimal for the unconstrained problem in (13.6) with $\eta = \eta^*$. From the proof of Theorem 2.5 of Sennott (1993), there exists a $\mu \in [0, 1]$ such that $\pi_{\eta^*}^*$ satisfies the constraint in (13.3) with equality. This completes the proof of the theorem. □

[3] If $C_{max} = 1$, a transmission is allowed in every time slot, and we have an infinite state-space MDP with unbounded cost. It follows directly from part (ii) of the theorem in Sennott (1989) (whose conditions can be easily verified for Problem 1) that there exists an optimal stationary policy that satisfies the Bellman equations. In this chapter, we concentrate on the more interesting constrained case.

Some other results in Sennott (1993) will be useful in determining $\pi^*_{\eta^*}$. For any $\eta > 0$, let C_η and J_η denote the average number of transmissions and average AoI, respectively, for the optimal policy π^*_η. Note that these are multivalued functions since there might be more than one optimal policy for a given η. Note also that C_η and J_η can be computed directly by finding the stationary distribution of the chain, or estimated empirically by running the MDP with policy π^*_η. From Lemma 3.4 of Sennott (1993), L^*_η, C_η, and J_η are monotone functions of η: if $\eta_1 < \eta_2$, we have $C_{\eta_1} \geq C_{\eta_2}, J_{\eta_1} \leq J_{\eta_2}$, and $L^*_{\eta_1} \leq L^*_{\eta_2}$. This statement is also intuitive since η effectively represents the cost of a single transmission in (13.7) and (13.8); as η increases, the average number of transmissions of the optimal policy cannot increase, and as a result, the AoI cannot decrease.

To determine the optimal policy, one needs to find η^*, the policies $\pi^*_{\eta^*,1}$ and $\pi^*_{\eta^*,2}$, and the weight μ. In fact, Sennott (1993) shows that η^* is defined as

$$\eta^* \triangleq \inf\{\eta > 0 : C_\eta \leq C_{max}\}, \tag{13.10}$$

where the inequality $C_\eta \leq C_{max}$ is satisfied if it is satisfied for at least one value of (multivalued) C_η. By Lemma 3.12 of Sennott (1993), η^* is finite, and $\eta^* > 0$ if $C_{max} < 1$.

If $C^{\pi^*_{\eta^*,i}} = C_{max}$ for $i=1$ or $i=2$, then it is the optimal policy, that is, $\pi^*_\mu = \pi^*_{\eta^*,i}$ and $\mu=1$ if $i=1$ and 0 if $i=2$. Otherwise, one needs to select μ such that $C^{\pi^*_\mu} = C_{max}$: that is, if $C^{\pi^*_{\eta^*,2}} < C_{max} < C^{\pi^*_{\eta^*,1}}$, then

$$\mu = \frac{C_{max} - C^{\pi^*_{\eta^*,2}}}{C^{\pi^*_{\eta^*,1}} - C^{\pi^*_{\eta^*,2}}}, \tag{13.11}$$

which results in an optimal policy.

In practice, finding both η^* and the policies $\pi^*_{\eta^*,1}$ and $\pi^*_{\eta^*,2}$ is hard. However, given two monotone sequences $\eta_n \uparrow \eta^*$ and $\eta'_n \downarrow \eta^*$, there is a subsequence of η_n (resp., η'_n) such that the corresponding subsequence of the η_n-optimal policies $\pi^*_{\eta_n}$ (η'_n-optimal policies $\pi^*_{\eta'_n}$, resp.) satisfying the Bellman equation (13.7) converge.[4] Then the limit points π and π' are η^*-optimal by Lemma 3.7 (iii) of Sennott (1993) and $C^\pi \geq C_{max} \geq C^{\pi'}$ by the monotonicity of C_η and the same Lemma 3.7. Although there is no guarantee that π and π' only differ in a single point, we can combine them to get an optimal randomized policy using μ defined in (13.11). In this case, Lemma 3.9 of Sennott (1993) implies that the policy that first randomly selects if it should use π or π' (choosing π with probability μ) and then uses the selected policy forever is η^*-optimal. However, since $(1, 0)$ is a positive recurrent state of both policies and they have a single recurrent class by Proposition 3.2 of Sennott (1993), we can do the random selection of between π and π' independently every time the system gets into state $(1, 0)$ without changing the long-term average or expected AoI and transmission cost. (Note that one cannot choose randomly between the two policies in, e.g., every step.) Thus, the resulting randomized policy is η^*-optimal, and

[4] $\pi_n \to \pi$ if for any state s, $\pi_n(s) = \pi(s)$ for n large enough.

since μ is selected in such a way that the total transmission cost is C_{max}, it is also an optimal solution of Problem 1 by Lemma 3.10 of Sennott (1993). Note that to derive two η^*-optimal policies, which provably differ only in a single state, a much more elaborate construction is used in Sennott (1993). However, in practice, the π and π' obtained previously are often the same except for a single state. Furthermore, we can approximate π_1 and π_2 by $\pi^*_{\eta_n}$ and $\pi^*_{\eta'_n}$ for n large enough.

Theorem 13.1 and the succeeding discussion present the general structure of the optimal policy. In Section 13.3.2, for practical implementation, a computationally efficient heuristic algorithm is proposed based on the discussion in this section.

13.3.2 (RVI)

While the state space is countably infinite, since the age can be arbitrarily large, in practice we can approximate the countable state space with a large but finite space by setting an upper bound on the age (which will be denoted by N) and by selecting a finite r_{max}. When we consider the finite state space approximation of the problem, we can employ the *relative value iteration* (RVI) (Puterman 1994) algorithm to solve (13.7) for any given η, and hence find (an approximation of) the optimal policy π^*_η.

Starting with an initialization of $h_0(\delta, r)$, $\forall(\delta, r)$, and setting an arbitrary but fixed reference state (δ^{ref}, r^{ref}), a single iteration for the RVI algorithm is given as follows:

$$Q_{n+1}(\delta, r, a) \leftarrow \delta + \eta \cdot \mathbb{1}[a^\pi \neq \text{i}] + \mathbb{E}\left[h_n(\delta', r')\right], \tag{13.12}$$

$$V_{n+1}(\delta, r) \leftarrow \min_a (Q_{n+1}(\delta, r, a)), \tag{13.13}$$

$$h_{n+1}(\delta, r) \leftarrow V_{n+1}(\delta, r) - V_{n+1}(\delta^{ref}, r^{ref}), \tag{13.14}$$

where $Q_n(\delta, r, a)$, $V_n(\delta, r)$, and $h_n(\delta', r')$ denote the state action value function, value function, and differential value function for iteration n, respectively. Then, h_n converges to h_η, and $\pi^*_n(\delta, r) \triangleq \arg\min_a Q_n(\delta, r, a)$ converges to $\pi^*_\eta(\delta, r)$ (Puterman 1994).

After computing the optimal deterministic policy π^*_η for any given η (more precisely, an arbitrarily close approximation in the finite approximate MDP), we need to find η^* as defined by (13.10). We can use the following heuristic: With the aim of finding a single η value with $C_\eta \approx C_{max}$, we start with an initial parameter η^0 and update η iteratively as $\eta^{m+1} = \eta^m + \alpha_m(C_{\eta^m} - C_{max})$ for a step size parameter α_m[5].

13.3.3 Practical RL Algorithms

We now assume that the source does not have a priori information about the decoding error probabilities, and has to learn them. The literature for average-cost RL is quite limited compared to discounted cost problems (Mahadevan 1996; Sutton and Barto 1998). SARSA (Sutton and Barto 1998) is a well-known RL algorithm, originally proposed for discounted MDPs, that iteratively computes the optimal state-action

[5] α_m is a positive decreasing sequence and satisfies the following conditions: $\sum_m \alpha_m = \infty$ and $\sum_m \alpha_m^2 < \infty$ from the theory of stochastic approximation (Kushner & Yin 1997).

Algorithm 1 Average-cost SARSA with Softmax

Input: Lagrange parameter η /* *error probability* $g(r)$ *is unknown* */
$n \leftarrow 0$ /* *time iteration* */
$\tau \leftarrow 1$ /* *softmax temperature parameter* */
$Q_{\eta}^{N \times M \times 3} \leftarrow 0$
$L_{\eta} \leftarrow 0$ /* *initialization of the gain* */
for n **do**
  OBSERVE current state s_n
  for $a \in \mathcal{A}$ **do** $\pi(a|s_n) = \dfrac{\exp(-Q_{\eta}(s_n, a)/\tau)}{\sum_{a' \in \mathcal{A}} \exp(-Q_{\eta}(s_n, a')/\tau)}$ /* *since it is a minimization problem, use minus Q function in softmax* */
  end for
  SAMPLE a_n from $\pi(a|S_n)$
  OBSERVE next state s_{n+1} and cost $c_n = \delta_n + \eta 1_{\{a_n=1,2\}}$
  for $a \in \mathcal{A}$ **do**
    $\pi(a|s_{n+1}) = \dfrac{\exp(-Q_{\eta}(s_{n+1}, a_{n+1})/\tau)}{\sum_{a'_{n+1} \in \mathcal{A}} \exp(-Q_{\eta}(s_{n+1}, a'_{n+1})/\tau)}$
  end for
  SAMPLE a_{n+1} from $\pi(a_{n+1}|s_{n+1})$
  UPDATE
  $\alpha_n \leftarrow 1/\sqrt{n}$ /* *update parameter* */
  $Q_{\eta}(s_n, a_n) \leftarrow Q_{\eta}(s_n, a_n) + \alpha_n[\delta + \eta \cdot \mathbb{1}[a_n \neq \text{i}] - J_{\eta} + Q_{\eta}(s_{n+1}, a_{n+1}) - Q_{\eta}(s_n, a_n)]$
  $L_{\eta} \leftarrow L_{\eta} + 1/n[\delta + \eta \cdot \mathbb{1}[a_n \neq \text{i}] - J_{\eta}]$ /* *update* L_{η} *at every step* */
  $n \leftarrow n + 1$ /* *increase iteration count* */
end for

value function $Q(s, a)$ and the optimal policy for an MDP based on the action performed by the current policy in a recursive manner. We employ an average-cost version of SARSA with *Boltzmann* (*softmax*) exploration to learn $g(r)$ without degrading the performance significantly. The resulting algorithm is called *average-cost SARSA with softmax*.

Average-cost SARSA with softmax starts with an initial estimate of $Q_{\eta}(s, a)$ and finds the optimal policy by estimating state-action values in a recursive manner. In the nth iteration, after taking action a_n, the source observes the next state s_{n+1} and the instantaneous cost value c_n. Based on this, the estimate of $Q_{\eta}(s, a)$ is updated by weighing the previous estimate and the estimated expected value of the current policy in the next state s_{n+1}. The instantaneous cost c_n is the sum of AoI and the weighted cost of transmission, that is, $\delta_n + \eta \cdot \mathbb{1}[a_n \neq \text{i}]$; hence, it is readily known at the source node. At each time slot, the learning algorithm (see Algorithm 1)

- observes the current state $s_n \in \mathcal{S}$,
- selects and performs an action $a_n \in \mathcal{A}$,
- observes the next state $s_{n+1} \in \mathcal{S}$ and the instantaneous cost c_n,
- updates its estimate of $Q_{\eta}(s_n, a_n)$ using the current estimate of $_{\eta}$ by

$$Q_{\eta}(s_n, a_n) \leftarrow Q_{\eta}(s_n, a_n) + \alpha_n[\delta + \eta \cdot \mathbb{1}[a_n \neq \text{i}] - L_{\eta} + Q_{\eta}(s_{n+1}, a_{n+1}) - Q_{\eta}(s_n, a_n)], \tag{13.15}$$

where α_n is the update parameter (learning rate) in the nth iteration.

- updates its estimate of L_η based on empirical average.

As we have discussed, with the accurate estimate of $Q_\eta(s,a)$ at hand the transmitter can decide for the optimal actions for a given η as in (13.9). However, until the state-action cost function is accurately estimated, the transmitter action selection method should balance the *exploration* of new actions with the *exploitation* of actions known to perform well. In particular, the *Boltzmann* action selection method, which chooses each action probabilistically relative to its expected cost, is used in this chapter. The source assigns a probability to each action for a given state s_n, denoted by $\pi(a|s_n)$:

$$\pi(a|s_n) \triangleq \frac{\exp(-Q_\eta(s_n,a)/\tau)}{\sum_{a' \in \mathcal{A}} \exp(-Q_\eta(s_n,a')/\tau)}, \tag{13.16}$$

where τ is called the temperature parameter such that high τ corresponds to more uniform action selection (exploration) whereas low τ is biased toward the best action (exploitation).

The constrained structure of the average AoI problem requires additional modifications to the algorithm, which is achieved by updating the Lagrange multiplier according to the empirical resource consumption. In each time slot, we keep track of a value η resulting in a transmission cost close to C_{max}, and then find and apply a policy that is optimal (given the observations so far) for the MDP with Lagrangian cost as in Algorithm 1.

13.3.4 Simulation Results

For the numerical simulation, we assume that decoding error reduces exponentially with the number of retransmission, that is, $g(r) \triangleq p_0\lambda^r$ for some $\lambda \in (0,1)$, where p_0 denotes the error probability of the first transmission, and r is the retransmission count (set to 0 for the first transmission). The exact value of λ depends on the particular HARQ protocol and the channel model. Note that ARQ corresponds to the case with $\lambda = 1$ and $r_{max} = 0$. Following the *IEEE 802.16* standard (802.16e 2005 2006), the maximum number of retransmissions is set to $r_{max} = 3$.

Figure 13.3 shows the evolution of the average AoI over time with the average-cost SARSA algorithm. The average AoI achieved by Algorithm 1, denoted by *RL* in the figure, converges to the one obtained from the RVI algorithm, which has a priori knowledge of $g(r)$. We can observe from Figure 13.3 that the performance of SARSA approaches that of RVI in about 10,000 iterations. Figure 13.4 shows the performance of the two algorithms (with again 10,000 iterations in SARSA) as a function of C_{max} in two different setups. We can see that SARSA performs very close to RVI with a gap that is roughly constant for the whole range of C_{max} values. We can also observe that the variance of the average AoI achieved by SARSA is much larger when the number of transmissions is limited, which also limits the algorithm's learning capability.

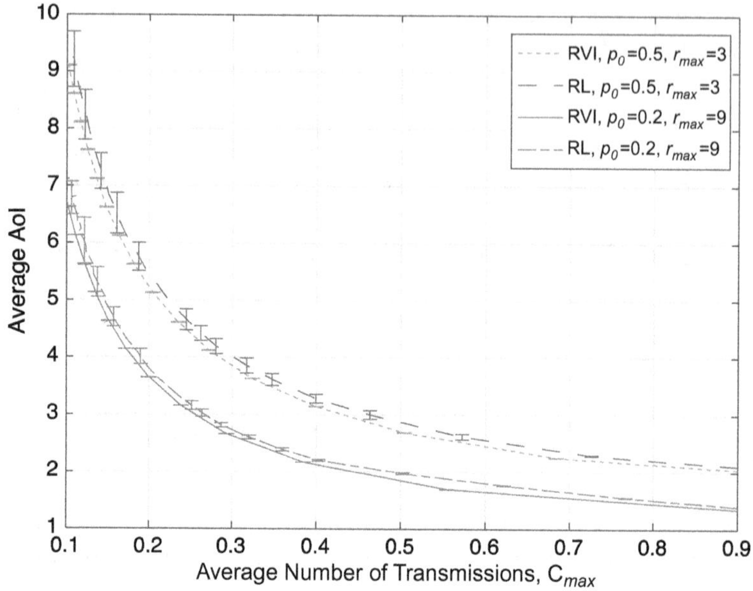

Figure 13.3 Performance of the average-cost SARSA for $r_{max} = 3, p_0 = 0.5, \lambda = 0.5,$ $C_{max} = 0.4$, and $n = 10{,}000$, averaged over 1,000 runs. (Both the mean and the variance are shown.)

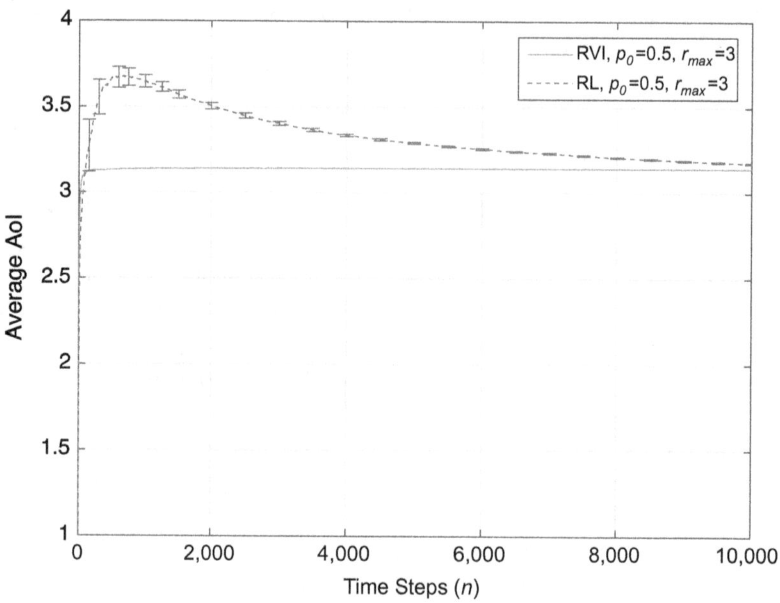

Figure 13.4 Performance of the proposed RL algorithm (average-cost SARSA) and its comparison with the RVI algorithm for $n = 10{,}000$ iterations, and values are averaged over 1,000 runs for different p_0 and r_{max} values when $\lambda = 0.5$. (Both the mean and the variance are shown.)

13.4 RL for Minimizing AoI in Multiuser Status-Update Systems

In this section, we consider a status-update system with M users. The source can transmit the status update to only a single user at each time slot. This can be either because of dedicated orthogonal links to the users, or because the users are interested in distinct processes. As before, each transmission attempt takes one time slot, and the channels change randomly from one time slot to the next in an i.i.d. fashion, with states known only at the corresponding receivers. Successful reception of the status update is acknowledged by an ACK signal (denoted by $K_t = 1$), while a NACK signal is sent in case of a failure (denoted by $K_t = 0$).

To simplify the analysis, we set an upper bound on the age (which will be denoted by $\delta_{max} < \infty$). In practice, the utility of status updates typically becomes zero beyond a certain age, so age can be assumed bounded. Then, if the most up-to-date packet received by the jth user ($j \in [M] \triangleq \{1, \dots, M\}$) before time slot t was generated in slot $U_j(t)$; then the AoI at user j at the beginning of time slot t is defined as $\delta_{j,t} \triangleq \min\{t - U_j(t), \delta_{max}\} \in [N] \triangleq \{1, \dots, \delta_{max}\}$.

At each time slot t, the source node takes an action a_t from the set of actions $\mathcal{A} = \{\mathrm{i}, \mathrm{n}_1, \mathrm{x}_1, \dots, \mathrm{n}_M, \mathrm{x}_M\}$: in particular, the source can (i) remain idle ($a_t = \mathrm{i}$); (ii) generate and transmit a new status update packet to the jth user ($a_t = \mathrm{n}_j, j \in [M]$); or, (iii) retransmit the most recent failed status update to the jth user ($a_t = \mathrm{x}_j, j \in [M]$). We have $|\mathcal{A}| = 2M+1$. For the jth user, the probability of error after r retransmissions, denoted by $g_j(r)$.

Let $\delta^{tx}_{j,t}$ denote the number of time slots elapsed since the generation of the most recently transmitted (whether successfully or not) packet to user j at the transmitter, while $\delta^{rx}_{j,t}$ denote the AoI of the most recently received status update at receiver of the user j. $\delta^{tx}_{j,t}$ resets to 1 if a new status update is generated at time slot $t-1$, and increases by one (up to δ_{max}) otherwise, that is,

$$\delta^{tx}_{j,t+1} = \begin{cases} 1 & \text{if } a_t = \mathrm{n}_j; \\ \min(\delta^{tx}_{j,t} + 1, \delta_{max}) & \text{otherwise.} \end{cases}$$

On the other hand, the AoI at the receiver side evolves as follows:

$$\delta^{rx}_{j,t+1} = \begin{cases} 1 & \text{if } a_t = \mathrm{n}_j \text{ and } K_t = 1; \\ \min(\delta^{tx}_{j,t} + 1, \delta_{max}) & \text{if } a_t = \mathrm{x}_j \text{ and } K_t = 1; \\ \min(\delta^{rx}_{j,t} + 1, \delta_{max}) & \text{otherwise .} \end{cases}$$

Note that once the AoI at the receiver is at least as large as at the transmitter, this relationship holds forever; thus, it is enough to consider cases when $\delta^{rx}_t \geq \delta^{tx}_t$.

Therefore, $\delta^{rx}_{j,t}$ increases by 1 when the source chooses to transmit to another user, or if the transmission fails, while it decreases to 1 or, in the case of HARQ, to $\min(\delta^{tx}_{j,t} + 1, \delta_{max})$, when a status update is successfully decoded. Also, $\delta^{tx}_{j,t}$ increases by 1 if the source chooses not to generate a new packet and transmit it to user j ($a_t \neq \mathrm{n}_j$).

For the jth user, let $r_{j,t} \in [r_{max}] \triangleq \{0, \dots, r_{max}\}$ denote the number of previous transmission attempts of the most recent packet. Thus, the number of retransmissions is zero for a newly sensed and generated status update and increases up to

r_{max} as we keep retransmitting. Then, the state of the system can be described by $s_t \triangleq (\delta^{rx}_{1,t}, \delta^{tx}_{1,t}, r_{1,t}, \ldots, \delta^{rx}_{M,t}, \delta^{tx}_{M,t}, r_{M,t})$, where $s_t \in \mathcal{S} \subset ([\delta_{max}] \times [\delta_{max}] \times [r_{max}])^M$.

Similarly to the previous section, we impose a constraint on the average number of transmissions, denoted by $C_{max} \in (0, 1]$. This leads to CMDP formulation, defined in Section 13.2: The set of states $\mathcal{S}$ and the finite set of actions $\mathcal{A}$ have already been defined. $\mathcal{P}$ can be summarized as follows:

$$\mathcal{P}_{s,s'}(a) = \begin{cases} 1 & \text{if } a = \text{i},\ \delta_i^{rx'} = \min\{\delta_i^{rx}+1, \delta_{max}\},\ \delta_i^{tx'} = \min\{\delta_i^{tx}+1, \delta_{max}\}, \\ & r_i' = r_i,\ \forall i; \\ 1 - g_j(0) & \text{if } a = \text{n}_j,\ \delta_j^{rx'} = 1,\ \delta_j^{tx'} = 1,\ r_j' = 0,\ \delta_i^{rx'} = \min\{\delta_i^{rx}+1, \delta_{max}\}, \\ & \delta_i^{tx'} = \min\{\delta_i^{tx}+1, \delta_{max}\},\ r_i' = r_i,\ \forall i \neq j; \\ g_j(0) & \text{if } a = \text{n}_j,\ \delta_j^{rx'} = \min\{\delta_j^{rx}+1, \delta_{max}\},\ \delta_j^{tx'} = 1,\ r_j' = 1,\ r_i' = r_i; \\ & \delta_i^{rx'} = \min\{\delta_i^{rx}+1, \delta_{max}\},\ \delta_i^{tx'} = \min\{\delta_i^{tx}+1, \delta_{max}\}, \forall i \neq j; \\ 1 - g_j(r_j) & \text{if } a = \text{x}_j,\ \delta_j^{rx'} = \delta_j^{tx}+1,\ \delta_j^{tx'} = \min\{\delta_j^{tx}+1, \delta_{max}\},\ r_j' = 0,\ r_i' = r_i, \\ & \delta_i^{rx'} = \min\{\delta_i^{rx}+1, \delta_{max}\},\ \delta_i^{tx'} = \min\{\delta_i^{tx}+1, \delta_{max}\}, \forall i \neq j; \\ g_j(r_j) & \text{if } a = \text{x}_j,\ \delta_j^{rx'} = \min\{\delta_j^{rx}+1, \delta_{max}\},\ \delta_j^{tx'} = \min\{\delta_j^{tx}+1, \delta_{max}\}, \\ & \delta_i^{rx'} = \min\{\delta_i^{rx}+1, \delta_{max}\},\ \delta_i^{tx'} = \min\{\delta_i^{tx}+1, \delta_{max}\}, \\ & r_j' = \min\{r_j'+1, r_{max}\},\ r_i' = r_i, \forall i \neq j; \\ 0 & \text{otherwise.} \end{cases} \tag{13.17}$$

The instantaneous cost function $c\colon \mathcal{S} \times \mathcal{A} \to \mathbb{R}$ is defined as the weighted sum of AoI for multiple users, independently of a. Formally, $c(s, a) = \Delta \triangleq w_1\delta_1^{rx} + \cdots + w_M\delta_M^{rx}$, where the weight $w_j > 0$ represents priority of user j. The instantaneous transmission cost $d\colon \mathcal{A} \to \mathbb{R}$ is defined as $d(\text{i}) = 0$ and $d(a) = 1$ if $a \neq \text{i}$. We use $s_t^\pi = (\delta_{1,t}^{rx\ \pi}, \delta_{1,t}^{tx\ \pi}, r_{1,t}^\pi, \ldots, \delta_{M,t}^{rx\ \pi}, \delta_{M,t}^{tx\ \pi}, r_{M,t}^\pi)$ and a_t^π to denote the sequences of states and actions, respectively, induced by policy π, while $\Delta_t^\pi \triangleq \sum_{j=1}^M w_j\delta_{j,t}^{rx\pi}$ denotes the instantaneous weighted cost.

The infinite horizon expected weighted average AoI for policy π starting from the initial state $s_0 \in \mathcal{S}$ is defined as

$$J^\pi(s_0) \triangleq \limsup_{T\to\infty} \frac{1}{T+1}\mathbb{E}\left[\sum_{t=0}^{T} \Delta_t^\pi \middle| s_0\right], \tag{13.18}$$

while the average number of transmissions is given by (13.3).

As before, we assume that the source and the users are synchronized at the beginning of the problem, that is, $s_0 = (1, 0, 2, 0, \ldots, M, 0)$; and we omit s_0 from the notation for simplicity.

Problem 2 $\underset{\pi\in\Pi}{\text{Minimize}}\ J^\pi(s_0)$ over $\pi \in \Pi$ such that $C^\pi(s_0) \leq C_{max}$.

Similarly to Section 13.3.1, we can rewrite the problem in its Lagrangian form under policy π with Lagrange multiplier $\eta \geq 0$, denoted by L_η^π,

$$L_\eta^\pi = \lim_{T\to\infty} \frac{1}{T+1}\left(\mathbb{E}\left[\sum_{t=0}^{T} \Delta_t^\pi\right] + \eta\mathbb{E}\left[\sum_{t=0}^{T} \mathbb{1}[a_t^\pi \neq \text{i}]\right]\right) \tag{13.19}$$

and, for any η, the optimal achievable cost is defined as $L_\eta^* \triangleq \inf_\pi L_\eta^\pi$ under policy π_n^*. This formulation is equivalent to an unconstrained finite-state average-cost MDP with the instantaneous overall cost $\Delta_t^\pi + \eta \mathbb{1}[a_t^\pi \neq \mathrm{i}]$.

THEOREM 13.2 *An optimal stationary policy π_n^* that minimizes* (13.19) *exists with constant L_η^* for the unconstrained MDP with Lagrangian parameter η.*

The proof of Theorem 13.2 can be found in Ceran, Gündüz, and György (2020). Also, as in Section 13.3.2, an iterative algorithm to minimize average AoI in multiuser systems can be designed by applying the RVI algorithm (Puterman 1994).

13.4.1 AoI with Standard ARQ Protocol

In this section, we assume that the system adopts the standard ARQ protocol. The action space reduces to $\mathcal{A} = \{\mathrm{i}, \mathrm{n}_1, \ldots, \mathrm{n}_M\}$ and the state space reduces to $(\delta_1^{rx}, \delta_2^{rx}, \ldots, \delta_M^{rx})$ as $r_{j,t} = 0$, $\forall j, t$, and there is no need to store the AoI at the transmitter side. The probability of error of each status update is $p_j \triangleq g_j(0)$ for user j. State transitions in (13.17), the Bellman optimality equations, and the RVI algorithm can all be simplified accordingly. Thanks to these simplifications, we are able to show the structure of the optimal policy and to derive a low-complexity suboptimal policy.

Structure of the Optimal Policy

THEOREM 13.3 *There exists an optimal stationary policy for Problem 2 under standard ARQ that is optimal for the unconstrained problem considered in* (13.6) *for some $\eta = \eta^*$, and randomizes in at most one state. This policy can be expressed as a mixture of two deterministic policies $\pi_{\eta^*,1}^*$ and $\pi_{\eta^*,2}^*$ that differ in at most a single state $\hat{s}$, and are both optimal for the Lagrangian problem* (13.6) *with $\eta = \eta^*$. More precisely, there exist two deterministic policies $\pi_{\eta^*,1}^*$, $\pi_{\eta^*,2}^*$ as described previously and $\mu \in [0, 1]$, such that the mixture policy $\pi_{\eta^*}^*$, which selects, in state $\hat{s}$, $\pi_{\eta^*,1}^*(\hat{s})$ with probability μ and $\pi_{\eta^*,2}^*(\hat{s})$ with probability $1 - \mu$, and otherwise follows these two policies (which agree in all other states) is optimal for Problem 2, and the constraint in (13.3) is satisfied with equality.*

The proof of Theorem 13.3 can be found in Ceran et al. (2020). Some other results in Altman (1999) and Beutler and Ross (1985) will be useful in determining $\pi_{\eta^*}^*$. For any $\eta > 0$, let C_η and J_η denote the average number of transmissions and average AoI, respectively, for the optimal policy π_η^*. Note that C_η and J_η can be computed directly by finding the stationary distribution of the chain, or estimated empirically by running the MDP with policy π_η^*.

A detailed discussion on finding both η^* and the policies $\pi_{\eta^*,1}^*$ and $\pi_{\eta^*,2}^*$ are given in Ceran, Gündüz, and György (2019) and also in Section 13.3.

WI Policy

Although the RVI algorithm (Puterman 1994) provides an optimal solution to Problem 2, its computational complexity is significant for large networks consisting of many users. Instead, we can derive a low-complexity policy for multiuser AoI minimization problem with standard ARQ protocol based on Whittle's approach (Whittle 1988) by modelling the problem as a RMAB (Gittins, Glazebrook, & Weber 2011). The WI policy in Section 13.4.1 results in a possibly suboptimal yet computationally efficient policy, which often performs very well in practice.

MABs (Gittins et al. 2011) constitute a class of RL problems with a single state. In the restless MAB (RMAB) problem (Whittle 1988), each arm is associated with a state that evolves over time, and the reward distribution of the arm depends on its state. (In contrast, in the basic stochastic MAB problems, rewards are i.i.d.) The multiuser AoI minimization problem with ARQ can be formulated as a RMAB with $M+1$ arms: choosing arm j is associated with transmitting to user j, while arm $M+1$ represents the action of staying idle ($a = \mathrm{i}$). RMAB problems are known to be PSPACE-hard in general (Gittins et al. 2011); however, a low-complexity heuristic policy can be found for certain problems by relaxing the constraint that in every round only a single arm can be selected, and instead introducing a bound on the expected number of arms chosen (Whittle 1988). The resulting policy, known as the WI policy, is a suboptimal policy, but it is known to perform close to optimal in many settings (Whittle 1988).

Following Whittle's approach, we decouple the problem into M subproblems each corresponding to a single user, and treat these problems independently. The cost of transmitting to a user (called *subsidy for passivity* (Whittle 1988)) is denoted by C, which will be later used to derive the index policy. Writing the Bellman equation (13.8) for each subproblem, we obtain the optimality equations for the single user AoI minimization problem with the standard ARQ protocol where the action space is $\{\mathrm{i}, \mathrm{n}_j\}$

$$h_C(\delta_j^{rx}) + L_j^* = \min\{Q(\delta_j^{rx}, \mathrm{n}_j), Q(\delta_j^{rx}, \mathrm{i})\}, \tag{13.20}$$

and the optimal policy to each subproblem is given,

$$\begin{aligned} \pi_C^*(\delta_j^{rx}) &\in \operatorname*{arg\,min}_{a\in\{\mathrm{i},\mathrm{n}_j\}} \left\{Q(\delta_j^{rx}, a)\right\}, \text{ where} \\ Q(\delta_j^{rx}, \mathrm{n}_j) &\triangleq w_j\delta_j^{rx} + C + p_j h_C(\delta_j^{rx}+1) + (1-p_j)h_C(1), \\ Q(\delta_j^{rx}, \mathrm{i}) &\triangleq w_j\delta_j^{rx} + h_C(\delta_j^{rx}+1). \end{aligned} \tag{13.21}$$

Given (13.20) and (13.21), let $S_j^{\mathrm{n}_j}(C)$ represent the set of states the optimal action is equal to n_j for a given C, that is, $S_j^{\mathrm{n}_j}(C) = \{s : \pi_C^*(\delta_j^{rx}) = \mathrm{n}_j\}$. Then, we define indexability as follows.

DEFINITION 13.4 An arm is indexable if the set $S_j^{\mathrm{n}_j}(C)$ as a function of C is monotonically decreasing for $C \in \mathbb{R}$, and $\lim_{C\to\infty} S_j^{\mathrm{n}_j}(C) = \varnothing$ and $\lim_{C\to-\infty} S_j^{\mathrm{n}_j}(C) = \mathcal{S}$ (Whittle 1988; Gittins et al. 2011). The problem is indexable if every arm is indexable.

Note that if a problem is indexable as defined in Definition 13.4, $S_j^a(C_1) \subset S_j^a(C_2)$ for $C_1 \geq C_2$, and there exists a C such that both actions are *equally desirable*, that is, $Q(\delta_j^{rx}, \mathrm{i}) = Q(\delta_j^{rx}, \mathrm{n}_j)$ for all δ_j^{rx}. The WI is defined as follows.

DEFINITION 13.5 *The WI for user j at state δ_j^{rx}, denoted by $I_j(\delta_j^{rx})$, is defined as the cost C that makes both actions* n_j *and* i *equally desirable.*

Next, we derive the WI (see Ceran et al. (2020) for derivations):

Proposition 2 Problem 2 with standard ARQ is indexable and the WI for each user j and state δ_j^{rx} can be computed as

$$I_j(\delta_j^{rx}) = \frac{1}{2} w_j \delta_j^{rx}(1-p_j)\left(\delta_j^{rx} + \frac{1+p_j}{1-p_j}\right), \ \forall j \in [M], \tag{13.22}$$

where the WI for the idle action is $I_{M+1} = \eta$.

The WI policy is defined as follows: in state $(\delta_1^{rx}, \delta_2^{rx}, \ldots, \delta_M^{rx})$, compare the highest index with the Lagrange parameter η, and if η is smaller, then the source transmits to the user with the highest index, otherwise the source remains idle. The WI policy, defined below, tends to transmit to the user with a high weight (w_j), low error probability (p_j), and high AoI (δ_j^{rx}). Formally,

$$\pi(\delta_1, \delta_2, \ldots, \delta_M) = \begin{cases} \mathrm{n}_{\arg\max_j (I_j(\delta_j^{rx}))} & \text{if } \max_j I_j(\delta_j^{rx}) \geq \eta, \\ \mathrm{i} & \text{otherwise.} \end{cases} \tag{13.23}$$

The effectiveness of the WI policy is demonstrated numerically in Section 13.4.3.

Lower Bound on the Average AoI under a Resource Constraint

In this section, we derive a closed-form lower bound for the constrained MDP, for which the proof is given in Ceran et al. (2020):

THEOREM 13.6 *For Problem 2 with the standard ARQ protocol, we have $J_{LB} \leq J^{\pi}$, $\forall \pi \in \Pi$, where*

$$J_{LB} = \frac{1}{2C_{max}}\left(\sum_{j=1}^{M}\sqrt{\frac{w_j}{1-p_j}}\right)^2 + \frac{C_{max} w_{j^*} p_{j^*}}{2(1-p_{j^*})} + \frac{1}{2}\sum_{j=1}^{M} w_j, \text{ and } j^* \triangleq \arg\min_j \frac{w_j p_j}{2(1-p_j)}.$$

Previously, Kadota, Uysal-Biyikoglu, Singh, and Modiano (2018) proposed a lower bound on the average AoI for a source node sending time-sensitive information to multiple users through unreliable channels, without any resource constraint (i.e., $C_{max} = 1$). The lower bound in Theorem 13.6 shows the effect of the constraint C_{max}, and even for $C_{max} = 1$, it is tighter than the one provided in Kadota et al. (2018).

13.4.2 Practical RL Algorithms

In Section 13.3, we presented a simple *average-cost SARSA* algorithm to minimize the average AoI for a single user. Due to the large state space of the multiuser network considered in this section, alternative lower-complexity learning algorithms are proposed.

UCRL2 with HARQ

The upper confidence RL (UCRL2) algorithm (Auer, Jaksch, and Ortner 2009) is a well-known RL algorithm for finite state and action MDP problems, with strong theoretical performance guarantees. However, the computational complexity of the algorithm scales quadratically with the size of the state space, which makes it unsuitable for large state spaces. UCRL2 has been initially proposed for generic MDPs with unknown rewards and transition probabilities, which need to be learned for each state–action pair. For the average AoI problem, the rewards are known (i.e., AoIs) while the transition probabilities are unknown. Moreover, the number of parameters to be learned can be reduced to the number of transmission error probabilities to each user; thus, the computational complexity can be reduced significantly.

For a generic tabular MDP, UCRL2 keeps track of the possible MDP models (transition probabilities and expected immediate rewards) in a high-probability sense and finds a policy that has the best performance in the best possible MDP. In our problem, it is enough to optimistically estimate the error probabilities $g_j(r)$ and find an optimal policy for this optimistic MDP. This is possible since the performance corresponding to a fixed sequence of transmission decisions improves if the error probabilities decrease. We will guarantee the average transmission constraint by updating the Lagrange multiplier according to the empirical resource consumption. The details are given in Algorithm 2.

UCRL2 exploits the optimistic MDP characterized by the optimistic estimation of error probabilities within a certain confidence interval, where $\hat{g}_j(r)$ and $\tilde{g}_j(r)$ represent the empirical and the optimistic estimates of the error probability for user j after

Algorithm 2 UCRL2-VI

Input: Confidence parameter $\rho \in (0, 1)$, update parameter α, C_{max}, confidence bound constant U, $|\mathcal{S}|$, $|\mathcal{A}|$

$\eta = 0, t = 1$
Observe initial state s_1
for episodes $k = 1, 2, \dots$ **do** set $t_k \triangleq t$
  for $j \in [M], r \in [r_{max}]$ **do**
    $N_k(j,r) \triangleq |\{\tau < t_k : a_\tau = \mathrm{x}_j, r_{j,\tau} = r\}|, \quad N_k(j,0) \triangleq |\{\tau < t_k : a_\tau = \mathrm{n}_j\}|$
    $E_k(j,r) \triangleq |\{\tau < t_k : a_\tau = \mathrm{x}_j, r_{j,\tau} = r, NACK\}|, E_k(j,0) \triangleq |\{\tau < t_k : a_\tau = \mathrm{n}_j, NACK\}|$
    $\hat{g}_j(r) \triangleq \frac{E_k(j,r)}{\max\{N_k(j,r),1\}}$
  end for
  $C_k \triangleq |\{\tau < t_k : a_\tau \neq \mathrm{i}\}|$
  $\eta \leftarrow \eta + \alpha(C_k/t_k - C_{max})$
  Compute optimistic error probability estimates: $\tilde{g}_j(r) \triangleq \max\left\{0, \hat{g}_j(r) - \sqrt{\frac{U \log(|\mathcal{S}||\mathcal{A}|t_k/\rho)}{max\{1,N_k(j,r)\}}}\right\}$
  Use $\tilde{g}_j(r)$ and VI to find a policy $\tilde{\pi}_k$
  Set $v_k(j,r) \leftarrow 0, \forall j, r$
  while $v_k(j,r) < N_k(j,r)$ **do** /* *run policy* $\tilde{\pi}_k$ */
    Choose action $a_t = \tilde{\pi}_k(s_t)$, and if $a_t \neq \mathrm{i}$, set j_t as target user, otherwise $j_t = 0$
    Obtain cost $\sum_{j=1}^M w_j \delta_j^{rx} + \eta \mathbb{1}[a_t \neq \mathrm{i}]$ and observe s_{t+1}
    Update $v_k(j_t,r) = v_k(j_t,r) + 1$ and set $t \leftarrow t + 1$
  end while
end for

Algorithm 3 UCRL2 for Average AoI with ARQ

Input: Confidence parameter $\rho \in (0,1)$, update parameter α, C_{max}, confidence bound constant U, $|\mathcal{S}|$, $|\mathcal{A}|$

1: $\eta = 0, t = 1$
2: Observe initial state s_1
3: **for** episodes $k = 1, 2, \ldots$ **do** and set $t_k \triangleq t$,
4: $\quad N_k(j) \triangleq |\{\tau < t_k : a_\tau = \mathrm{n}_j\}|$, $E_k(j) \triangleq |\{\tau < t_k : a_\tau = \mathrm{n}_j, NACK\}|$
5: $\quad \hat{p}(j) \triangleq \frac{E_k(j)}{\max\{N_k(j),1\}}$, $C_k \triangleq |\{\tau < t_k : a_\tau \neq \mathrm{i}\}|$,
6: $\quad \eta \leftarrow \eta + \alpha(C_k/t_k - C_{max})$
7: $\quad$ Compute optimistic error probabilities: $\tilde{p}(j) \triangleq \max\{0, \hat{p}(j) - \sqrt{\frac{U\log(|\mathcal{S}||\mathcal{A}|t_k/\rho)}{max\{1,N_k(j)\}}}\}$
8: $\quad$ Use $\tilde{p}(j)$ to find a policy $\tilde{\pi}_k$ and execute policy $\tilde{\pi}_k$
9: $\quad$ **while** $v_k(j) < N_k(j)$ **do**
10: $\qquad$ Choose action $a_t = \tilde{\pi}_k(s_t)$,
11: $\qquad$ Obtain cost $\sum_{j=1}^{M} w_j \delta_j^{rx} + \eta * \mathbb{1}[a_t \neq \mathrm{i}]$ and observe s_{t+1}
12: $\qquad$ Update $v_k(j) = v_k(j) + 1$ and set $t \leftarrow t + 1$;
13: $\quad$ **end while**
14: **end for**

r retransmissions. In each episode, we keep track of a value η resulting in a transmission cost close to C_{max} and then find and apply a policy that is optimal for the optimistic MDP (i.e., the MDP with the smallest total cost from among all plausible ones given the observations) with Lagrangian cost. In contrast to the original UCRL2 algorithm, finding the optimistic MDP in this case is easy (choosing lower estimates of the error probabilities), and we can use standard value iteration (VI) to compute the optimal policy (instead of the more complex extended VI used in UCRL2). Thus, the computational complexity, which is the main drawback of UCRL2 algorithm, reduces significantly for the average AoI problem. UCRL2 is employed for Problem 2 in this chapter since it is an online algorithm (i.e., it does not need any previous training) and it enjoys strong theoretical guarantees for $C_{max} = 1$. The resulting algorithm will be called UCRL2-VI.

A Heuristic Version of the UCRL2 for Standard ARQ

Next, we consider the standard ARQ protocol with unknown error probabilities $p_j = g_j(0)$. The estimation procedure of UCRL2-VI can be immediately simplified, as it only needs to estimate M parameters. In order to reduce the computational complexity, we can replace the costly VI in the algorithm to find the $\tilde{\pi}_k$ with the suboptimal WI policy given in Section 13.4.1. The resulting algorithm, called *UCRL2-Whittle*, selects policy $\tilde{\pi}_k$ in step 16 following the WI policy in Section 13.4.1. The details are given in Algorithm 3, where $\hat{p}(j)$ and $\tilde{p}(j)$ denote the empirical and the optimistic estimate of the error probability for user j.

Average-Cost SARSA with LFA

In Section 13.3.3, the average-cost SARSA algorithm is employed with *Boltzmann* (*softmax*) exploration for the average AoI problem with a single user. When the state-space $\mathcal{S}$ is small and a simulator is available for the system, updates similar

Algorithm 4 Average-cost SARSA with LFA

Input: Lagrange parameter η, update parameters α, β, γ, $\mathcal{A}$

Set $t \leftarrow 1$, $\theta \leftarrow 0$, $J_\eta \leftarrow 0$

for $t = 1, 2, \ldots$ **do**

  Find parameterized policies with Boltzmann exploration: $\pi(a|s_t) = \frac{\exp(-\theta^T \phi(s_t,a))}{\sum_{a' \in \mathcal{A}} \exp(-\theta^T \phi(s_t,a'))}$

  Sample and execute action a_t from $\pi(a|s_t)$

  Observe next state s_{t+1} and cost $\sum_{j=1}^{M} \delta_j^{rx} + \eta * \mathbb{1}[a_t \neq \text{i}]$.

  $\pi(a|s_{t+1}) = \frac{\exp(-\theta^T \phi(s_{t+1},a))}{\sum_{a'_{t+1} \in \mathcal{A}} \exp(-\theta^T \phi(s_{t+1},a'))}$

  Sample a_{t+1} from $\pi(a|s_{t+1})$

  Compute C_η

  Update linear coefficients: $\theta \leftarrow \theta + \alpha_t[\Delta_t + \eta \cdot \mathbb{1}[a_t \neq \text{i}] - J_\eta + \theta^T \phi(s_{t+1}, a_{t+1}) - \theta^T \phi(s_t, a_t)]\phi(s_t, a_t)$

  Update gain: $J_\eta \leftarrow J_\eta + \beta_t[\Delta_t + \eta \cdot \mathbb{1}[a_t \neq \text{i}] - J_\eta]$

  Update Lagrange multiplier: $\eta \leftarrow \eta + \gamma_t(C_\eta - C_{max})$

end for

to Section 13.3.3 can be computed for all state–action pairs. However, this is not possible for large state spaces, or if the Q functions are learned online: that is, to collect data about some states, the system needs to be driven to that state, which may be very costly, severely limiting the set of states for which the updates can be computed. For the problem with multiple users, the cardinality of the state–action space is large and it is difficult to even store a matrix that has the size of the state–action space. Hence, average-cost SARSA with LFA is employed, where a linear function of features can be used to approximate the Q-function in SARSA (Puterman 1994). Average-cost SARSA with LFA is an online algorithm similar to average-cost SARSA and UCRL2 algorithms. It improves the performance of average-cost SARSA by improving the convergence rate significantly for multiuser systems and its application is much simpler than the UCRL2 algorithm.

We approximate the Q function with a linear function Q_θ defined as: $Q_\theta(s, a) \triangleq \theta^T \phi(s, a)$, where $\phi(s, a) \triangleq (\phi_1(s, a), \ldots, \phi_d(s, a))^T$ is a given feature associated with the pair (s, a). In our experiments, we set $\{\phi_i(s, a)\}_{i=1}^{M}$ as the weighted age of each user $(w_j \delta_j^{rx})$ and $\{\phi_i(s, a)\}_{i=M+1}^{2M}$ as the retransmission number of each user (r_j) given an action $a \in \mathcal{A}$ is chosen in state $s \in \mathcal{S}$:

$$Q_\theta(s, a) = \theta_{(0,a)} + \theta_{(1,a)} w_1 \delta_1 + \ldots + \theta_{(M,a)} w_M \delta_M + \theta_{(M+1,a)} r_1 + \ldots + \theta_{(2M,a)} r_M, \tag{13.24}$$

where $\theta_{(0,a)}$ denotes the constant variable. The dimension of θ is $d = (2M + 1)|\mathcal{A}|$. The outline of the algorithm is given in Algorithm 4.

The performance of average-cost SARSA with LFA is demonstrated in Section 13.4.3. We note that linear approximators are not always effective, and the performance can be improved in general by using a nonlinear approximator. However, the performance also depends on the availability of data (i.e., the linear approximator may perform better if the available data set is limited).

Table 13.1 Hyperparameters of DQN algorithm used in the chapter.

Parameter	Value	Parameter	Value	Parameter	Value
Discount factor γ	0.99	Optimizer	Adam	Activation function	ReLU
Minibatch size	32	Loss function	Huber loss	Hidden size	24
Replay memory length	2,000	Exploration coefficient ϵ_0	1	Episode length T	1,000
Learning rate α	10^{-4}	ϵ decay rate β	0.9	ϵ_{min}	0.01

Deep Q-Network (DQN)

A DQN uses a multilayered neural network to estimate $Q(s,a)$; that is, for a given state s, DQN outputs a vector of state–action values, $Q_\theta(s,a)$, where θ denotes the parameters of the network. The neural network is a function from $2M$ inputs to $|\mathcal{A}|$ outputs, which are the estimates of the Q-function $Q_\theta(s,a)$. We apply the DQN algorithm of Mnih et al. (2015) to learn a scheduling policy. We create a fairly simple feed-forward neural network of three layers, one of which is the hidden layer with 24 neurons. We also use *Huber loss* (Huber 1964) and the *Adam* algorithm (Kingma & Ba 2015) to conduct stochastic gradient descent to update the weights of the neural network.

We exploit two important features of DQNs as proposed in Mnih et al. (2015): *experience replay* and a *fixed target network*, both of which provide algorithm stability. For *experience replay*, instead of training the neural network with a single observation $< s, a, s', c(s,a) >$ at the end of each step, many experiences (i.e., (state, action, next state, cost) quadruplets) can be stored in the replay memory for batch training, and a minibatch of observations randomly sampled at each step can be used. The DQN uses two neural networks: a target network and an online network. The *target network*, with parameters θ^-, is the same as the online network except that its parameters are updated with the parameters θ of the online network after every T steps, and θ^- is kept fixed in other time slots. For a minibatch of observations for training, temporal difference estimation error e for a single observation can be calculated as

$$e = Q_\theta(s,a) - (-c(s,a) + \gamma Q_{\theta^-}(s', \arg\max Q_\theta(s',a))). \tag{13.25}$$

Huber loss is defined by the squared error term for small estimation errors, and a linear error term for high estimation errors, allowing less dramatic changes in the value functions and further improving the stability. For a given estimation error e and loss parameter d, the Huber loss function, denoted by $\mathrm{L}^d(e)$, and the average loss over the minibatch, denoted by $\mathcal{B}$, are computed as

$$\mathrm{L}^d(e) = \begin{cases} e^2 & \text{if } e \le d \\ d(|e| - \frac{1}{2}d)) & \text{if } e > d, \end{cases} \quad \text{and} \quad \mathrm{L}_\mathcal{B} = \frac{1}{|\mathcal{B}|} \sum_{<s,a,s',c(s,a)>\in\mathcal{B}} \mathrm{L}^d(e).$$

We apply the ϵ-greedy policy to balance exploration and exploitation. We let ϵ decay gradually from ϵ_0 to ϵ_{min}; in other words, the source explores more at the beginning of training and exploits more at the end. The hyperparameters of the DQN algorithm are tuned experimentally and are given in Table 13.1.

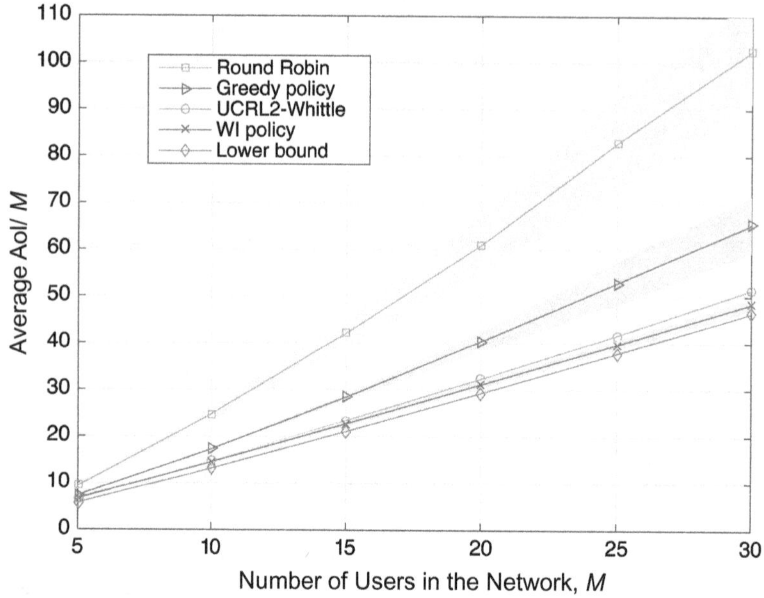

Figure 13.5 Average AoI for networks with different sizes, where $p_j = (j-1)/M$, $C_{max} = 1$, and $w_j = 1$, $\forall j$. The results are obtained after 10^4 time steps and averaged over 100 runs. (Both the mean and the variance are shown.)

13.4.3 Simulation Results

In this section, we provide numerical results for the proposed learning algorithms and compare the achieved average performances. Figure 13.5 illustrates the mean and variance of the average AoI with standard ARQ with respect to the size of the network when there is no constraint on the average number of transmissions (i.e., $C_{max} = 1$) and the performance of the UCRL2-Whittle is compared with the lower bound. (UCRL2-VI is omitted since its performance is very similar to UCLR2-Whittle and has a much higher computational complexity, especially for large M.) The performance of UCRL2-Whittle is close to the lower bound and is very similar to that of the WI policy, which requires a priori knowledge of the error probabilities. Moreover, UCRL2-Whittle outperforms the greedy benchmark policy, which always transmits to the user with the highest age and the round robin policy, which transmits to each user in turns. We can also observe that the variances of the average AoI achieved by benchmark policies are much larger, which also limits their performance.

Figure 13.6 shows the performance of the learning algorithms for the HARQ protocol (the mean and the variance of the average AoI) for a two-user scenario. DQN is trained for 500 episodes with configuration in Table 13.1. It is worth noting that although UCRL2-VI converges to the optimal policy in fewer iterations than average-cost SARSA and average-cost SARSA with LFA, iterations in UCRL2-VI are computationally more demanding since the algorithm uses VI in each epoch. Therefore, UCRL2-VI is not practical for problems with large state spaces, in a case for

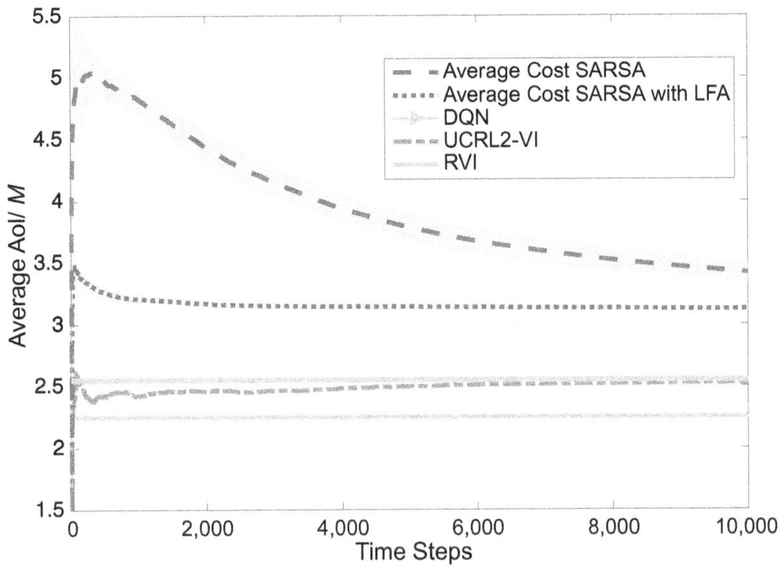

Figure 13.6 Average AoI for a two-user HARQ network with error probabilities $g_1(r_1) = 0.5 \cdot 2^{-r_1}$ and $g_2(r_2) = 0.2 \cdot 2^{-r_2}$, where $C_{max} = 1$ and $w_j = 1$, $\forall j$. The simulation results are averaged over 100 runs. (Both the mean and the variance are shown.)

large M. On the other hand, UCRL2-Whittle can handle a large number of users since it is based on a simple index policy instead of VI. As illustrated in Figure 13.6, LFA significantly improves the performance of average-cost SARSA and DQN with neural network estimator, and UCRL2-Whittle improves the performance of RL even more.

We concluded that the choice of the learning algorithm to be adopted depends on the scenario and system characteristics. It has been shown that average-cost SARSA is not effective, considering the large state space of the multiuser problem. Different state-of-the-art RL methods are presented, including SARSA with LFA, UCRL2, and DQN. The performance of UCRL2-VI algorithm is close to optimal for small networks, that is, consisting of 1–5 users, and enjoys theoretical guaranties. However, UCRL2-VI is not favorable for large networks due to its computational complexity, and UCRL2-Whittle is preferable. On the other hand, UCRL2-Whittle cannot be employed for a general HARQ multiuser system. Similarly, SARSA with LFA has decreased the average AoI significantly for small-size networks with HARQ; however, it is not effective for large networks, and SARSA with LFA lacks stability. A nonlinear approximation with DQN performs for large networks, which is not fully online and requires a training time before running the algorithm.

13.5 AoI in Energy Harvesting Status-Update Systems

Many status-update systems are powered by scavenging energy from renewable sources (e.g., solar cells, wind turbines, or piezoelectric generators). Harvesting energy from ambient sources provides environmentally friendly and ubiquitous operation for remote sensing systems. Therefore, there has been a growing interest in maximizing the timeliness of information in (EH) communication systems (Bacinoglu, Ceran, and Uysal-Biyikoglu 2015; Yates 2015; Abd-Elmagid et al. 2020; Arafa, Yang, Ulukus, and Poor 2020; Stamatakis, Pappas, and Traganitis 2019). In this section, we assume that the source can sense the underlying time-varying process and generate a status update at a certain energy cost.

At the end of each time slot t, a random amount of energy is harvested and stored in a rechargeable battery at the transmitter, denoted by $E_t \in \mathcal{E} \triangleq \{0, 1, \ldots, E_{max}\}$, following a first-order discrete-time Markov model, characterized by the stationary transition probabilities $p_E(e_1|e_2)$, defined as $p_E(e_1|e_2) \triangleq Pr(E_{t+1} = e_2|E_t = e_1)$, $\forall t$ and $\forall e_1, e_2 \in \mathcal{E}$. It is also assumed that $p_E(0|e) > 0$, $\forall e \in \mathcal{E}$. Harvested energy is first stored in a rechargeable battery with a limited capacity of $B_{\max}$ energy units. The energy consumption for status sensing is denoted by $E^{\rm s} \in \mathbb{Z}^+$, while the energy consumption for a transmission attempt is denoted by $E^{\rm tx} \in \mathbb{Z}^+$.

The battery state at the beginning of time slot t, denoted by B_t, can be written as follows:

$$B_{t+1} = \min(B_t + E_t - (E^{\rm s} + E^{\rm tx})\mathbb{1}[A_t = {\rm n}] - E^{\rm tx}\mathbb{1}[A_t = {\rm x}], B_{\max}), \tag{13.26}$$

and the energy causality constraints are given as

$$(E^{\rm s} + E^{\rm tx})\mathbb{1}[A_t = {\rm n}] + E^{\rm tx}\mathbb{1}[A_t = {\rm x}] \leq B_t, \tag{13.27}$$

where the indicator function $\mathbb{1}[C]$ is equal to 1 if event C holds, and zero otherwise. (13.26) implies that the battery overflows if energy is harvested when the battery is full, while (13.27) imposes that the energy consumed by sensing or transmission operations at time slot t is limited by the energy B_t available in the battery at the beginning of that time slot. We set $B_0 = 0$ so that the battery is empty at time $t = 0$.

Let δ_t^{tx} denotes the number of time slots elapsed since the generation of the most recently sensed status update at the transmitter side, while δ_t^{rx} denotes the AoI of the most recently received status update at the receiver side. δ_t^{tx} resets to 1 if a new status update is generated at time slot $t-1$, and increases by one (up to δ_{max}) otherwise, that is,

$$\delta_{t+1}^{tx} = \begin{cases} 1 & \text{if } A_t = {\rm n}; \\ \min(\delta_t^{tx} + 1, \delta_{max}) & \text{otherwise.} \end{cases}$$

On the other hand, the AoI at the receiver side evolves as follows:

$$\delta_{t+1}^{rx} = \begin{cases} \min(\delta_t^{rx} + 1, \delta_{max}) & \text{if } A_t = {\rm i} \text{ or } K_t = 0; \\ 1 & \text{if } A_t = {\rm n} \text{ and } K_t = 1; \\ \min(\delta_t^{tx} + 1, \delta_{max}) & \text{if } A_t = {\rm x} \text{ and } K_t = 1. \end{cases}$$

Note that once the AoI at the receiver is at least as large as at the transmitter, this relationship holds forever; thus, it is enough to consider cases when $\delta_t^{rx} \geq \delta_t^{tx}$.

To determine the success probability of a transmission, we need to keep track of the number of current retransmissions. The number of retransmissions is zero for a newly sensed and generated status update and increases up to r_{max} as we keep retransmitting the same packet. It is easy to see that retransmitting when $\delta^{tx}_{t+1} = \delta_{max}$ is suboptimal, therefore we explicitly exclude this action by setting the retransmission count to 0 in this case. Also, it is suboptimal to generate a new update and retransmit the old one; thus, whenever a new status update is generated, the previous update at the transmitter is dropped and cannot be retransmitted. Thus, the evolution of the retransmission count is given as follows:

$$R_{t+1} = \begin{cases} 0 & \text{if } K_t = 1 \text{ or } \delta^{tx}_{t+1} = \delta_{max}; \\ 1 & \text{if } A_t = \text{n and } K_t = 0; \\ r_t & \text{if } A_t = \text{i and } \delta^{tx}_{t+1} \neq \delta_{max}; \\ \min(r_t + 1, r_{max}) & \text{if } A_t = \text{x}, K_t = 0 \text{ and } \delta^{tx}_{t+1} \neq \delta_{max}. \end{cases}$$

The state of the system is formed by five components $S_t = (E_t, B_t, \delta^{rx}_t, \delta^{tx}_t, r_t)$. In each time slot, the transmitter knows the state and takes action from the set $\mathcal{A} = \{\text{i, n, x}\}$. The goal is to find a policy π that minimizes the expected average AoI at the receiver over an infinite time horizon, which is given by the following:

Problem 3

$$J^* \triangleq \min_{\pi} \lim_{T\to\infty} \frac{1}{T+1}\mathbb{E}\left[\sum_{t=0}^{T} \delta^{rx}_t\right] \quad \text{subject to (13.26) and (13.27).} \tag{13.28}$$

In Section 13.3.3, we have considered status updates with HARQ under an average power constraint. In that case, it is possible to show that it is suboptimal to retransmit a failed update after an idle period. Restricting the actions of the transmitter accordingly, the AoI at the receiver after a successful transmission event is equal to the number of retransmissions of the corresponding update. Therefore, in addition to the AoI at the receiver, we only need to track the retransmission count. However, in the current scenario, retransmissions of a status update can be interrupted due to energy outages, which means that we also need to keep track of the AoI at the transmitter side (hence, we need to have both δ^{rx}_t and δ^{tx}_t in the state of the system).

Problem 3 can be formulated as an average-cost finite-state MDP: The finite set of states $\mathcal{S}$ is defined as $\mathcal{S} = \{s = (e, b, \delta^{rx}, \delta^{tx}, r) : e \in \mathcal{E}, b \in \{0, \ldots, B_{\max}\}, \delta^{rx}, \delta^{tx} \in \{1, \ldots, \delta_{max}\}, r \in \{0, \ldots, r_{max}\}, \delta^{rx} \geq \delta^{tx}\}$, while the finite set of actions $\mathcal{A} = \{\text{i, n, x}\}$ is already defined. Note that action x cannot be taken in states with retransmission count $r = 0$. P is characterized by the EH statistics and channel error probabilities. The cost function, $c: \mathcal{S} \times \mathcal{A} \to \mathbb{Z}$, is the AoI at the receiver and is defined as $c(s, a) = \delta^{rx}$ for any $s \in \mathcal{S}$, $a \in \mathcal{A}$, independent of the action taken, where δ^{rx} is the component of s describing the AoI at the receiver.

It is easy to see that MDP formulated for Problem 3 is a communicating MDP by Proposition 8.3.1 of Puterman (1994),[6] (i.e., for every pair of $(s, s') \in \mathcal{S}$, there

[6] By Proposition 8.3.1 of Puterman (1994), MDP is communicating since there exists a stationary policy that induces a recurrent Markov chain, for example, a state $(0, B_0, \delta_{max}, \delta_{max}, R_0)$ is reachable from all other states considering a policy that never transmits and in a scenario where no energy is harvested.

exists a deterministic policy under which s' is accessible from s). By Theorem 8.3.2 of Puterman (1994), an optimal stationary policy exists with constant gain. In particular, there exists a function, $h\colon \mathcal{S} \to \mathbb{R}$, called the *differential cost function* for all $s = (e, b, \delta^{rx}, \delta^{tx}, r) \in \mathcal{S}$, satisfying the following *Bellman optimality equations* for the average-cost finite-state finite-action MDP (Puterman 1994):

$$h(s) + J^* = \min_{a\in\{\text{i,n,x}\}} \left(\delta^{rx} + \mathbb{E}\left[h(s')|a\right]\right), \tag{13.29}$$

where $s' \triangleq (e', b', \delta^{rx\prime}, \delta^{tx\prime}, r')$ is the next state obtained from $(e, b, \delta^{rx}, \delta^{tx}, r)$ after taking action a, and J^* represents the optimal achievable average AoI under policy π^*.

We also introduce the *state–action cost function*:

$$Q((e, b, \delta^{rx}, \delta^{tx}, r), a) \triangleq \delta^{rx} + \mathbb{E}\left[h(e', b', \delta^{rx\prime}, \delta^{tx\prime}, r')|a\right]. \tag{13.30}$$

Then an optimal policy, for any $(e, b, \delta^{rx}, \delta^{tx}, r) \in \mathcal{S}$, takes the action achieving the minimum in (13.30):

$$\pi^*(e, b, \delta^{rx}, \delta^{tx}, r) \in \arg\min_{a\in\{\text{i,n,x}\}} \left(Q((e, b, \delta^{rx}, \delta^{tx}, r), a)\right). \tag{13.31}$$

An optimal policy solving (13.29), (13.30), and (13.31) can be found by RVI for finite-state finite-action average-cost MDPs from Sections 8.5.5 and 9.5.3 of Puterman (1994): Starting with an arbitrary initialization of $h_0(s)$, $\forall s \in \mathcal{S}$, and setting an arbitrary but fixed reference state $s^{ref} \triangleq (e^{ref}, b^{ref}, \delta^{rx^{ref}}, \delta^{tx^{ref}}, r^{ref})$, a single iteration of the RVI algorithm $\forall (s, a) \in \mathcal{S} \times \mathcal{A}$ is given as follows:

$$Q_{n+1}(s, a) \leftarrow \delta_n^{rx} + \mathbb{E}\left[h_n(s')\right], \tag{13.32}$$

$$V_{n+1}(s) \leftarrow \min_a(Q_{n+1}(s, a)), \tag{13.33}$$

$$h_{n+1}(s) \leftarrow V_{n+1}(s) - V_{n+1}(s^{ref}), \tag{13.34}$$

where $Q_n(s, a)$, $V_n(s)$, and $h_n(s)$ denote the state–action value function, value function, and differential value function at iteration n, respectively. By Theorem 8.5.7 of Puterman (1994), h_n converges to h, and $\pi_n^*(s) \triangleq \arg\min_a Q_n(s, a)$ converges to $\pi^*(s)$.

13.5.1 Structure of the Optimal Policy

Next, we investigate the structure of the optimal policy and show that the solution to Problem 3 is of threshold-type. The proof could be obtained by following the same steps in Section 6.11 of Puterman (1994).

THEOREM 13.7 *There exists an optimal stationary policy $\pi^*(s)$ that is monotone with respect to δ_t^{rx}, that is, $\pi^*(s)$ is of threshold-type.*

Following Theorem 13.7, we present a threshold-based policy that will be termed a *double-threshold policy* in the rest of this section.

$$A_t = \begin{cases} \text{i} & \text{if } \delta_t^{rx} < \mathcal{T}_\text{n}(e, b, \delta^{tx}, r), \\ \text{n} & \text{if } \mathcal{T}_\text{n}(e, b, \delta^{tx}, r) \leq \delta_t^{rx} < \mathcal{T}_\text{x}(e, b, \delta^{tx}, r), \\ \text{x} & \text{if } \delta_t^{rx} \geq \mathcal{T}_\text{x}(e, b, \delta^{tx}, r), \end{cases} \tag{13.35}$$

for some threshold values denoted by $\mathcal{T}_\text{n}(e, b, \delta^{tx}, r)$ and $\mathcal{T}_\text{x}(e, b, \delta^{tx}, r)$, where $\mathcal{T}_\text{n}(e, b, \delta^{tx}, r) \leq \mathcal{T}_\text{x}(e, b, \delta^{tx}, r)$.

We can simplify the problem by making an assumption on the policy space in order to obtain a simpler *single-threshold policy*, which will result in a more efficient learning algorithm: We assume that a packet is retransmitted until it is successfully decoded, provided that there is enough energy in the battery, that is, the transmitter is not allowed to preempt an undecoded packet and transmit a new one.

The solution to the simplified problem is also of threshold-type, that is,

$$A_t = \begin{cases} \text{i} & \text{if } \delta_t^{rx} < \mathcal{T}(e, b, \delta^{tx}, r), \\ \text{n} & \text{if } \delta_t^{rx} \geq \mathcal{T}(e, b, \delta^{tx}, r), \text{ and } r = 0 \\ \text{x} & \text{if } \delta_t^{rx} \geq \mathcal{T}(e, b, \delta^{tx}, r) \text{ and } r \neq 0, \end{cases} \tag{13.36}$$

for some $\mathcal{T}(e, b, \delta^{tx}, r)$.

13.5.2 Practical RL Algorithms

In some scenarios, it can be assumed that the channel and energy arrival statistics remain the same or change very slowly, and the same environment is experienced for a sufficiently long time before the time of deployment. Accordingly, we can assume that the statistics regarding the error probabilities and energy arrivals are available before the time of transmission. In such scenarios, RVI algorithm (Puterman 1994) can be used. However, in most practical scenarios, channel error probabilities for retransmissions and the EH characteristics are not known at the time of deployment, or may change over time. In this section, we assume that the transmitter does not know the system characteristics a-priori, and has to learn them. The work in Ceran, Gündüz, and György (2018) and Ceran, Gündüz, and György (2019) employed learning algorithms for constrained problems with countably infinite state spaces such as average-cost SARSA. While these algorithms can be adopted to the current framework by considering an average transmission constraint of 1, they do not have convergence guarantees. However, Problem 3 constitutes an unconstrained MDP with finite state and action spaces, and there exist RL algorithms for unconstrained MDPs, which enjoy convergence guarantees. Moreover, we have shown the optimality of a threshold-type policy for Problem 3, and RL algorithms that exploit this structure can be considered. Thus, we employ three different RL algorithms and compare their performances in terms of the average AoI as well as the convergence speed. First, we employ a value-based RL algorithm, namely GR-learning, which converges to an optimal policy. Next, we consider a structured policy search algorithm, namely FDPG, which does not necessarily find the optimal policy but performs very well in practice, as demonstrated through simulations in Section 13.5.3. We also note that GR-learning learns from a single trajectory generated during learning steps while FDPG uses Monte-Carlo roll-outs for each policy update. Thus, GR-learning is more amendable for applications in real-time systems. Finally, we employ the DQN algorithm, which implements a nonlinear neural network estimator in order to learn the optimal status update policy.

GR-Learning with Softmax

For Problem (13.28), we employ a modified version of the *GR-learning* algorithm proposed in Gosavi (2004) with *Boltzmann* (*softmax*) exploration. The resulting algorithm is called *GR-learning with softmax* and can find the optimal policy π^* using (13.31) without knowing P, characterized by $g(r)$ and p_E. Notice that, similar to *average-cost SARSA with softmax* in Section 13.3.3, *GR-learning with softmax* starts with an initial estimate of $Q_0(s, a)$ and finds the optimal policy by estimating state–action values in a recursive manner. Update rules for $Q_n(S_n, A_n)$ and J_n in Section 13.3.3 are modified as

$$Q_{n+1}(S_n, A_n) \leftarrow Q_n(S_n, A_n) + \alpha(m(S_n, A_n, n))[\delta_n^{rx} - J_n + Q_n(S_{n+1}, A_{n+1}) - Q_n(S_n, A_n)], \tag{13.37}$$

where $\alpha(m(S_n, A_n, n))$ is the update parameter (learning rate) in the nth iteration, and depends on the function $m(S_n, A_n, n)$, which is the number of times the state–action pair (S_n, A_n) has been visited till the nth iteration,

$$J_{n+1} \leftarrow J_n + \beta(n) \left[\frac{nJ_n + \delta_n^{rx}}{n+1} - J_n \right], \tag{13.38}$$

where $\beta(n)$ is the update parameter in the nth iteration.

The transmitter action selection method should balance the *exploration* of new actions with the *exploitation* of actions known to perform well. We use the *Boltzmann* (*softmax*) action selection method, which chooses each action randomly relative to its expected cost as defined in Section 13.3.3.

According to Theorem 2 of Gosavi (2004), if α, β satisfy $\sum_{m=1}^{\infty} \alpha(m), \sum_{m=1}^{\infty} \beta(m) \to \infty$, $\sum_{m=1}^{\infty} \alpha^2(m), \sum_{m=1}^{\infty} \beta^2(m) < \infty$, $\lim_{x\to\infty} \frac{\beta(m)}{\alpha(m)} \to 0$, *GR*-Learning converges to an optimal policy.

Finite-Difference Policy Gradient (FDPG)

GR-learning in Section 13.5.2 is a value-based RL method, which learns the state–action value function for each state–action pair. In practice, δ_{max} can be large, which might slow down the convergence of GR-learning due to a prohibitively large state space.

In this section, we are going to propose a learning algorithm that exploits the structure of the optimal policy exposed in Theorem 13.7. A monotone policy is shown to be average optimal in the previous section; however, it is not possible to compute the threshold values directly for Problem 3.

Note that $A_t = \text{i}$ if $B_t < E^{\text{tx}}$ ($B_t < E^{\text{tx}} + E^{\text{s}}$) for $r \geq 1$ ($r = 0$); that is, $\mathcal{T}(e, b, \delta^{tx}, r) = \delta_{max} + 1$ if $b < E^{\text{tx}}$ for $r \geq 1$ and $b < E^{\text{tx}} + E^{\text{s}}$ for $r = 0$. This ensures that energy causality constraints in (13.27) hold. Other thresholds will be determined using PG. In order to employ the PG method, we approximate the policy by a parameterized smooth function with parameters $\theta(e, b, \delta^{tx}, r)$ and convert the discrete policy search problem into estimating the optimal values of these continuous parameters, which can be numerically solved by stochastic approximation algorithms (Spall 2003). Continuous stochastic approximation is much more efficient than discrete search algorithms in general.

In particular, with a slight abuse of notation, we let $\pi_\theta(e, b, \delta^{rx}, \delta^{tx}, r)$ denote the probability of taking action $A_t = \text{n}$ ($A_t = \text{x}$) if $r = 0$ ($r \neq 0$) and consider the parameterized sigmoid function:

$$\pi_\theta(e, b, \delta^{rx}, \delta^{tx}, r) \triangleq \frac{1}{1 + e^{-\frac{\delta^{rx} - \theta(e, b, \delta^{tx}, r)}{\tau}}}. \tag{13.39}$$

We note that $\pi_\theta(e, b, \delta^{rx}, \delta^{tx}, r) \to \{0, 1\}$ and $\theta(e, b, \delta^{tx}, r) \to \mathcal{T}(e, b, \delta^{tx}, r)$ as $\tau \to 0$. Therefore, in order to converge to a deterministic policy π, $\tau > 0$ can be taken as a sufficiently small constant, or can be decreased gradually to zero. The total number of parameters to be estimated is $|\mathcal{E}| \times B_{\max} \times \delta_{max} \times r_{max} + 1$ minus the parameters corresponding to $b < E^{\text{tx}}$ ($b < E^{\text{tx}} + E^{\text{s}}$) for $r > 0$ ($r = 0$) due to energy causality constraints, as stated previously.

With a slight abuse of notation, we map the parameters $\theta(e, b, \delta^{tx}, r)$ to a vector $\overline{\theta}$ of size $d \triangleq |\mathcal{E}| \times B_{\max} \times \delta_{max} \times r_{max} + 1$. Starting with some initial estimates of $\overline{\theta}_0$, the parameters can be updated in each iteration n, using the gradients as follows:

$$\overline{\theta}_{n+1} = \overline{\theta}_n - \gamma(n)\, \partial J / \partial \overline{\theta}_n, \tag{13.40}$$

where the step size parameter $\gamma(n)$ is a positive decreasing sequence and satisfies the first two convergence properties given at the end of Section 13.5.2 from the theory of stochastic approximation (Kushner & Yin 1997).

Computing the gradient of the average AoI directly is not possible; however, several methods exist in the literature to estimate the gradient (Spall 2003). In particular, we employ the FDPG (Peters & Schaal 2006) method. In this method, the gradient is estimated by estimating J at slightly perturbed parameter values. First, a random perturbation vector D_n of size d is generated according to a predefined probability distribution, for example, each component of D_n is an independent Bernoulli random variable with parameter $q \in (0, 1)$. The thresholds are perturbed with a small amount $\sigma > 0$ in the directions defined by D_n to obtain $\overline{\theta}_n^{\pm}(e, b, \delta^{tx}, r) \triangleq \overline{\theta}_n(e, b, \delta^{tx}, r) \pm \sigma D_n$. Then, empirical estimates $\hat{J}^{\pm}$ of the average AoI corresponding to the perturbed parameters $\overline{\theta}_n^{\pm}$, obtained from Monte-Carlo rollouts, are used to estimate the gradient:

$$\partial J / \partial \overline{\theta}_n \approx (D_n^{\mathsf{T}} D_n)^{-1} D_n^{\mathsf{T}} \frac{(\hat{J}^{+} - \hat{J}^{-})}{2\sigma}, \tag{13.41}$$

where D_n^{T} denotes the transpose of vector D_n. The pseudo code of the finite difference policy gradient algorithm is given in Algorithm 5.

Algorithm 5 FDPG

$\tau_0 \leftarrow 0.3,$ /* *temperature parameter* */
$\zeta \leftarrow 0.99,$ /* *decaying coefficient for* τ */.
$\overline{\theta}_0 \leftarrow 0$ /* *initialization of* $\overline{\theta}$ */
for $n = \{1, 2, \ldots\}$ **do**
  GENERATE random perturbation vector
    $D_n = binomial(\{0, 1\}, q = 0.5, d)$
  PERTURB parameters $\overline{\theta}_n$
    $\overline{\theta}_n^+ = \overline{\theta}_n + \beta D_n, \overline{\theta}_n^- = \overline{\theta}_n - \beta D_n$
  ESTIMATE $\hat{J}_n^{\pm}$ from Monte-Carlo rollouts using policies $\pi_{\theta_n^{\pm}}$:
  for $t \in \{1, \ldots, T\}$ **do**
    OBSERVE current state S_t and USE policy $\pi_{\theta_n^{\pm}}$
  end for
  ESTIMATE $\hat{J}_n^{\pm}$ from Monte-Carlo rollouts using policy $\pi_{\theta_n^{\pm}}$
    $\hat{J}_n^{\pm} = \frac{1}{T+1} \sum_{t=0}^{T} \delta_t^{rx}$
  COMPUTE estimate of the gradient $\partial J / \partial \overline{\theta}_n$
  UPDATE
    $\overline{\theta}_{n+1} = \overline{\theta}_n - \gamma(n)\, \partial J / \partial \overline{\theta}_n$
    $\tau_{n+1} \leftarrow \zeta \tau_n$ /* *decrease* τ */
end for

Similar steps can be followed to find the thresholds for the double-threshold policy, where $\mathcal{T}(e, b, \delta^{tx}, r)$ and $\theta(e, b, \delta^{tx}, r)$ are replaced by $\mathcal{T}_{\mathrm{n}}(e, b, \delta^{tx}, r)$, $\mathcal{T}_{\mathrm{x}}(e, b, \delta^{tx}, r)$ and $\theta_{\mathrm{n}}(e, b, \delta^{tx}, r)$, $\theta_{\mathrm{x}}(e, b, \delta^{tx}, r)$, respectively.

13.5.3 Simulation Results

In this section, we provide numerical results for all the proposed algorithms and compare the achieved average AoI. In the experiments, the maximum number of retransmissions is set to $r_{max} = 3$, while λ and p_0 are set to 0.5. E^{tx} and E^s are both assumed to be constant and equal to 1 unit of energy unless otherwise stated. δ_{max} is set to 40.

We choose the exact step sizes for the learning algorithms by fine-tuning in order to balance the algorithm stability in the early time steps with nonnegligible step sizes in the later time steps. In particular, we use step size parameters of $\alpha(m), \beta(m), \gamma(m) = y/(m+1)^z$, where $0.5 < z \leq 1$ and $y > 0$ (which satisfy the convergence conditions), and choose y and z such that the oscillations are low and the convergence rate is high. We have observed that a particular choice of parameters gives similar performance results for scenarios addressed in simulation results. DQN algorithm in this section is configured as in Table 13.1 and trained for 500 episodes. The average AoI for DQN is obtained after 10^5 time steps and averaged over 100 runs.

As a baseline, we have also included the performance of a greedy policy, which senses and transmits a new status update whenever there is sufficient energy. It retransmits the last transmitted status update when the energy in the battery is sufficient only for transmission, and it remains idle otherwise.

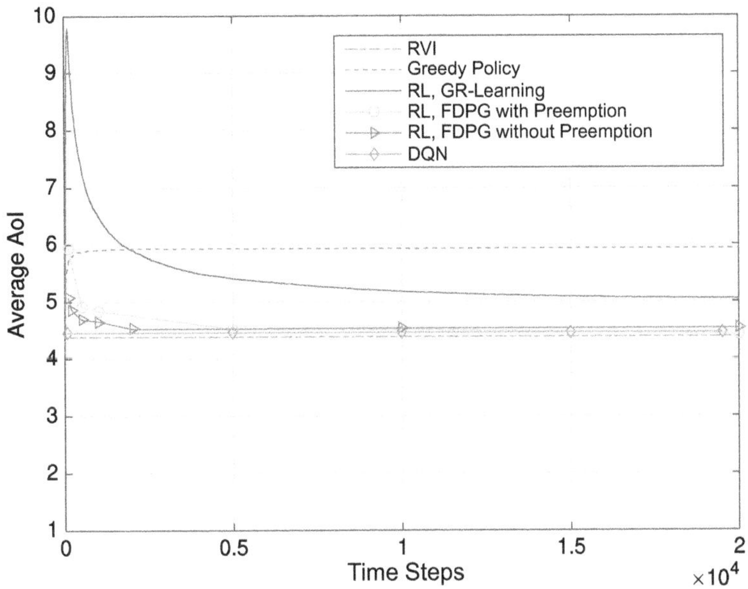

Figure 13.7 The performance of RL algorithms when $B_{\max} = 5, p_E(1,1), p_E(0,0) = 0.7$, and $E^s, E^{tx} = 1$. FDPG with and without preemption represent the double-threshold and the single-threshold policies, respectively.

Next, we investigate the performance when the EH process has temporal correlations. A symmetric two-state Markovian EH process is assumed, such that $\mathcal{E} = \{0,1\}$ and $Pr(E_{t+1} = 1|E_t = 0) = Pr(E_{t+1} = 0|E_t = 1) = 0.3$. That is, if the transmitter is in harvesting state, it is more likely to continue harvesting energy, and vice versa.

Figure 13.7 shows the evolution of the average AoI over time when the average-cost RL algorithms are employed in this scenario. It can be observed again that the average AoI achieved by the FDPG method in Section 13.5.2 performs very close to the one obtained by the RVI algorithm, which has a priori knowledge of $g(r)$ and p_e. GR-learning, on the other hand, outperforms the greedy policy but converges to the optimal policy much more slowly, and the gap between the two RL algorithms is larger compared to the i.i.d. case. Tabular methods in RL, like GR-learning, need to visit each state–action pair infinitely often for RL to converge (Sutton and Barto 1998). GR-learning in the case of temporally correlated EH does not perform as well as in the i.i.d. case since the state space becomes larger with the addition of the EH state. We also observe that the gap between the final performances of single- and double-threshold FDPG solutions is larger compared to the memoryless EH scenario, while that of the single-threshold solutions still converges faster. The DQN algorithm performs better than GR-learning, but it requires a training time before running the simulation and does not have convergence guarantees. Moreover, it still falls short of the final performance of double-threshold FDPG.

Next, we investigate the impact of the burstiness of the EH process, measured by the correlation coefficient between E_t and E_{t+1}. Figure 13.8 illustrates the performance of

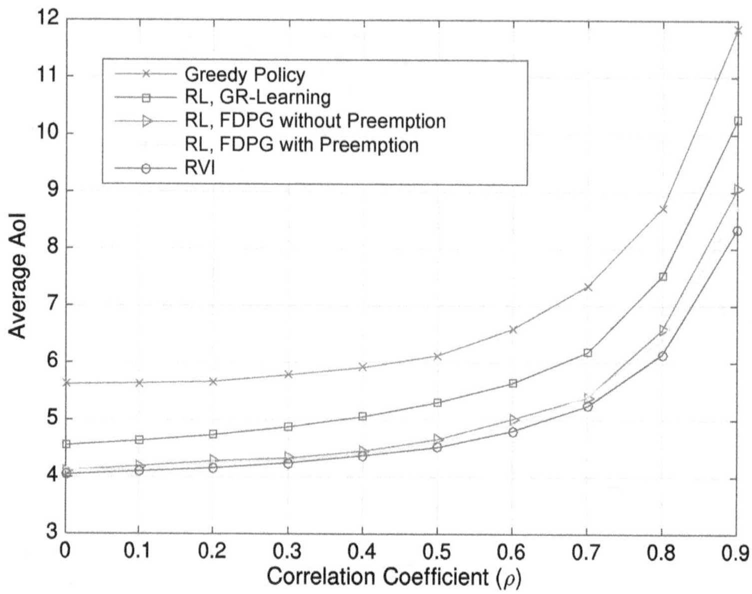

Figure 13.8 The performance of RL algorithms obtained after $2 \cdot 10^4$ time steps and averaged over 1,000 runs for different temporal correlation coefficients.

the proposed RL algorithms for different correlation coefficients, which can be computed easily for the two-state symmetric Markov chain; that is, $\rho \triangleq (2p_E(1,1) - 1)$. Note that $\rho = 0$ corresponds to memoryless EH with $p_e = 1/2$. We note that the average AoI is minimized by transmitting new packets successfully at regular intervals, which has been well investigated in previous works (Bacinoglu et al. 2015; Ceran, Gündüz, and György 2018; Yates 2015). Intuitively, for highly correlated EH, there are either successive transmissions or successive idle time slots, which increases the average AoI. Hence, the AoI is higher for higher values of ρ. Figure 13.8 also shows that both RL algorithms result in much lower average AoI than the greedy policy and FDPG outperforms GR-learning since it benefits from the structural characteristics of a threshold policy. We can also conclude that the single-threshold policy can be preferable in practice, especially in highly dynamic environments, as its performance is very close to that of the double-threshold FDPG, but with faster convergence.

13.6 Conclusions

This chapter investigated communication systems transmitting time-sensitive data over imperfect channels with the average AoI as the performance measure, which quantifies the timeliness of the data available at the receiver. Considering both the classical ARQ and the HARQ protocols, preemptive scheduling policies have been proposed by taking into account retransmissions under a resource constraint. Scenarios in which the system characteristics are not known a priori and must be learned in an online fashion are considered. RL algorithms are employed to balance exploitation and exploration in an unknown environment, so that the source node can quickly learn the environment based on the ACK/NACK feedback and can adapt its scheduling policy accordingly, exploiting its limited resources in an efficient manner. The proposed algorithms and the established results in this chapter are also relevant to different systems concerning the timeliness of information, in addition to systems under average resource constraint or energy replenishment constraints. The proposed methodology and the results regarding the structure of the optimal policies can be used in other wireless communication problems.

References

802.16e 2005, I. S. (2006), "IEEE standard for local and metropolitan area networks part 16: Air interface for fixed and mobile broadband wireless access systems amendment 2 (incorporated into IEEE standard 802.16e-2005 and std 802.16-2004/cor1-2005)."

Abd-Elmagid, M. A., Ferdowsi, A., Dhillon, H. S., & Saad, W. (2019), Deep reinforcement learning for minimizing age-of-information in UAV-assisted networks, *in* "2019 IEEE Global Communications Conference (GLOBECOM)," pp. 1–6.

Abd-Elmagid, M., Dhillon, H., & Pappas, N. (2020), "A reinforcement learning framework for optimizing age-of-information in rf-powered communication systems," *IEEE Transactions on Communications* .

Alliance, Z. (2008), "ZigBee specification (document 053474r17)," *Luettu* **21**.

Altman, E. (1999), *Constrained Markov Decision Processes*, Stochastic modeling, Chapman & Hall/CRC.

Arafa, A., Yang, J., Ulukus, S., & Poor, H. V. (2020), "Age-minimal transmission for energy harvesting sensors with finite batteries: Online policies," *IEEE Transactions on Information Theory* **66**(1), 534–556.

Auer, P., Jaksch, T., & Ortner, R. (2009), Near-optimal regret bounds for reinforcement learning, *in* "Advances in Neural Information Processing Systems 21," Curran Associates, Inc., pp. 89–96.

Bacinoglu, B. T., Ceran, E. T., & Uysal-Biyikoglu, E. (2015), Age of information under energy replenishment constraints, *in* "Inf. Theory and Applications Workshop (ITA)," pp. 25–31.

Bertsekas, D. P. (2000), *Dynamic Programming and Optimal Control*, 2nd ed., Athena Scientific.

Beutler, F. J., & Ross, K. W. (1985), "Optimal policies for controlled Markov chains with a constraint," *Journal of Mathematical Analysis and Applications* **112**(1), 236–252.

Beytur, H. B., & Uysal, E. (2019), Age minimization of multiple flows using reinforcement learning, *in* "International Conference on Computing, Networking and Communications (ICNC)," pp. 339–343.

Ceran, E. T., Gündüz, D., & György, A. (2018), Average age of information with hybrid arq under a resource constraint, *in* "IEEE Wireless Communications and Networking Conference (WCNC)," pp. 1–6.

Ceran, E. T., Gündüz, D., & György, A. (2019), "Average age of information with hybrid ARQ under a resource constraint," *IEEE Transactions on Wireless Communications* **18**, 1900–1913.

Ceran, E. T., Gündüz, D., & György, A. (2018), Reinforcement learning approach to age of information in multi-user networks, *in* "IEEE International Symposium on Personal, Indoor, and Mobile Radio Communications (PIMRC)."

Ceran, E. T., Gündüz, D., & György, A. (2019), Reinforcement learning to minimize age of information with an energy harvesting sensor with HARQ and sensing cost, *in* "IEEE Conf. on Computer Comms. Workshops (INFOCOM WKSHPS)."

Ceran, E. T., Gündüz, D., & György, A. (2020), "A reinforcement learning approach to age of information in multi-user networks with HARQ," *IEEE Journal on Selected Areas in Communications*. arXiv:2102.09774 [cs.IT].

Clancy, C., Hecker, J., Stuntebeck, E., & O'Shea, T. (2007), "Applications of machine learning to cognitive radio networks," *IEEE Wireless Communications* **14**(4), 47–52.

E-UTRA (2013), "LTE PHY; General Description," *3GPP TS 36.201* **21**.

Elgabli, A., Khan, H., Krouka, M., & Bennis, M. (2019), Reinforcement learning based scheduling algorithm for optimizing age of information in ultra reliable low latency networks, *in* "IEEE Symposium on Computers and Communications (ISCC)," pp. 1–6.

Gittins, J., Glazebrook, K. D., & Weber, R. (2011), *Multi-Armed Bandit Allocation Indices*, Wiley-Blackwell, London.

Gosavi, A. (2004), "Reinforcement learning for long-run average cost," *European Journal of Operational Research* **155**, 654 – 674.

Gündüz, D., Stamatiou, K., Michelusi, N., & Zorzi, M. (2014), "Designing intelligent energy harvesting communication systems," *IEEE Communications Magazine* **52**, 210–216.

Hatami, M., Jahandideh, M., Leinonen, M., & Codreanu, M. (2020), Age-aware status update control for energy harvesting iot sensors via reinforcement learning, *in* "2020 IEEE 31st Annual International Symposium on Personal, Indoor and Mobile Radio Communications," pp. 1–6.

Hsu, Y. P., Modiano, E., & Duan, L. (2017), Age of information: Design and analysis of optimal scheduling algorithms, *in* "IEEE International Symposium on Information Theory (ISIT)," pp. 561–565.

Huber, P. J. (1964), "Robust estimation of a location parameter," *The Annals of Math. Statistics* **35**, 73–101.

Kadota, I., Uysal-Biyikoglu, E., Singh, R., & Modiano, E. (2018), "Scheduling policies for minimizing age of information in broadcast wireless networks," *IEEE/ACM Transactions on Networking* **26**, 2637–2650.

Kingma, D. P., & Ba, J. (2015), Adam: A method for stochastic optimization, *in* "3rd International Conference on Learning Representations, ICLR 2015, San Diego, CA, USA, May 7–9, 2015, Conference Track Proceedings."

Kushner, H. J., & Yin, G. G. (1997), *Stochastic Approximation Algorithms and Applications*, New York: Springer-Verlag, Orlando, FL, USA.

Leng, S., & Yener, A. (2019), Age of information minimization for wireless ad hoc networks: A deep reinforcement learning approach, *in* "2019 IEEE Global Communications Conference (GLOBECOM)," pp. 1–6.

Luong, N. C., Hoang, D. T., Gong, S., Niyato, D., Wang, P., Liang, Y., & Kim, D. I. (2019), "Applications of deep reinforcement learning in communications and networking: A survey," *IEEE Communications Surveys Tutorials* **21**(4), 3133–3174.

Mahadevan, S. (1996), "Average reward reinforcement learning: Foundations, algorithms, and empirical results," *Machine Learning* **22**(1), 159–195.

Mnih, V., Kavukcuoglu, K., Silver, D., Rusu, A. A., Veness, J., Bellemare, M. G., Graves, A., Riedmiller, M. A., Fidjeland, A., Ostrovski, G., Petersen, S., Beattie, C., Sadik, A., Antonoglou, I., King, H., Kumaran, D., Wierstra, D., Legg, S., & Hassabis, D. (2015), "Human-level control through deep reinforcement learning," *Nature* **518**(7540), 529–533.

Oppermann, Hamalainen M., & Iinatti, J. (2004), "UWB theory and applications," Wiley.

Peters, J., & Schaal, S. (2006), Policy gradient methods for robotics, *in* "2006 IEEE/RSJ International Conference on Intelligent Robots and Systems," pp. 2219–2225.

Puterman, M. L. (1994), *Markov Decision Processes: Discrete Stochastic Dynamic Programming*, John Wiley & Sons, Hoboken, NJ, USA.

Sennott, L. I. (1989), "Average cost optimal stationary policies in infinite state Markov decision processes with unbounded costs," *Operations Research* **37**(4), 626–633.

Sennott, L. I. (1993), "Constrained average cost Markov decision chains," *Prob. in Eng. and Inf. Sci.* **7**, 69–83.

Sert, E., Sönmez, C., Baghaee, S., & Uysal-Biyikoglu, E. (2018), Optimizing age of information on real-life TCP/IP connections through reinforcement learning, *in* "2018 26th Signal Processing and Communications Applications Conference (SIU)," pp. 1–4.

Silver et al., D. (2016), "Mastering the game of Go with deep neural networks and tree search," *Nature* **529**(7587), 484–489.

Somuyiwa, S. O., György, A., & Gündüz, D. (2018), "A reinforcement-learning approach to proactive caching in wireless networks," *IEEE Journal on Selected Areas in Communications* **36**(6), 1331–1344.

Spall, J. C. (2003), *Introduction to Stochastic Search and Optimization*, John Wiley & Sons, Inc., Hoboken, NJ, USA.

Stamatakis, G., Pappas, N., & Traganitis, A. (2019), Control of status updates for energy harvesting devices that monitor processes with alarms, *in* "2019 IEEE Globecom Workshops (GC Wkshps)," pp. 1–6.

Sutton, R. S., & Barto, A. G. (1998), *Introduction to Reinforcement Learning*, 1st ed., Massachusetts Institute of Technology Press, Cambridge, MA, USA.

Whittle, P. (1988), "Restless bandits: activity allocation in a changing world," *Journal of App. Prob.* **25**, 287–298.

Yates, R. D. (2015), Lazy is timely: Status updates by an energy harvesting source, *in* "IEEE International Symposium on Information Theory (ISIT)," pp. 3008–3012.

14 Information Freshness in Large-Scale Wireless Networks: A Stochastic Geometry Approach

Howard H. Yang and Tony Q. S. Quek

14.1 Introduction

Timeliness is an emerging requirement for a variety of wireless services like unmanned grocery, industrial Internet-of-Things, and vehicular communications, as well as the likes of mobile applications. As a result, this obliges network designers to develop systems that ensure timeliness of information delivery. Recognizing the limitation in conventional latency metrics (e.g., delay or throughput) as not being able to account for such "information lag" in a time-varying environment, there emerges a new metric, namely the *age of information (AoI)*, to quantify the freshness of a typical piece of information. This notion was originally conceived to maintain timely status update in a standard first-come-first-served (FCFS) queue [25]. Soon after that, a host of research has been conducted to investigate different schemes aimed at minimizing the AoI, whereas the results range from controlling the update generating policy [2, 3, 6, 22, 32, 35, 42], deploying last-come-first-served (LCFS) queue [10], to proactively discarding stale packets at the source node [13]. Although these works have extensively explored the minimization of the information age on a single-node basis, many fundamental questions, especially those pertaining to large-scale networks, are not understood satisfactorily. To that end, it spurred a series of studies seeking different approaches, mainly in the form of scheduling protocols, to optimize information freshness in the context of wireless networks [8, 19, 20, 23, 24, 31, 33, 34]. The problem of finding optimal scheduling protocol, despite being NP hard [19], is shown to possess a solution in terms of a greedy algorithm, which schedules the link with the highest age to transmit, in a symmetric network [24], and the optimality of such a maximum age first policy is shown in [31], which provided a general and insightful sample-path proof. Moreover, depending on whether the channel state is perfectly available [34] or not [33], advanced virtual queue and age-based protocols are proposed. Further, the scheduling decision can even be made online using the approximation from Markov decision process [20], which largely boosts the implementational efficiency. However, these models simplify the packet departure process by adopting a Poisson process and do not account for the interference that differs according to the distance between simultaneous transmitters as well as channel gains. As a result, the space–time interactions are yet to be precisely captured.

By nature, the wireless channel is a broadcast medium. Thus, transmitters sharing a common spectrum in space will interact with each other through the interference they cause. To understand the performance of communication links in such networks, stochastic geometry has been introduced as a tool by which one can model node locations as spatial point processes and obtain closed-form expressions for various network statistics, for example, the distribution of interference, the successful transmission probability, and the coverage probability [17]. The power of stochastic geometry has made it a disruptive tool for performance evaluation among various wireless applications, including ad hoc and cellular networks [1], D2D communications [39], MIMO [38], and mmWave systems [7]. While such model has been conventionally relying on the *full buffer* assumption, that is, every link always has a packet to transmit, a line of recent works managed to bring in queueing theory and relax this constraint [15, 40, 41, 43]. The application territory of stochastic geometry is then stretched, allowing one to give a complete treatment for the behavior of wireless links in a network with spatial and temporal dynamics. As a result, the model is further employed to design scheduling policies [41, 43], study the scaling property in IoT networks [15], and analyze the delay performance in cellular networks [40]. Following a similar vein, a recent line of research has been carried out [14, 21, 27, 27, 28, 36] that conflates queueing theory with stochastic geometry to account for the spatial, temporal, and physical level attributes in the analysis of AoI. Particularly, lower and upper bounds for the distribution of average AoI are derived in the context of a Poisson network [21]. Furthermore, via a careful construction of the dominant system, tighter upper bounds for the spatial distribution of mean peak AoI is derived for a large system under both preemptive and non-preemptive queueing disciplines [27]. Additionally, the performance of peak AoI in uplink IoT networks is analytically evaluated under time-triggered and event-triggered traffic profiles [14]. And a distributed algorithm that configures the channel access probabilities at each individual transmitter based on the local observation of the network topology is proposed to minimize the peak AoI [36]. In the context of a cellular-based IoT network, the interplay between throughput and AoI is also investigated [28].

In light of its effectiveness, we demonstrate in this chapter how to leverage a model as in [11] and [12] to develop analytical frameworks for the comprehensive understanding of AoI, followed by the design of a transmission protocol that optimizes information freshness in wireless networks. Particularly, we model the deployment of transmitters and receivers as independent Poisson point processes (PPPs). The temporal dynamic of AoI is modeled as a discrete-time queueing system, in which we consider the arrival of packets at each transmitter to be independent Bernoulli processes. Each transmitter maintains an infinite capacity buffer to store the incoming packets and initiates a transmission attempt at each time slot if the buffer is not empty. Transmissions are successful only if the (SINR) exceeds a predefined threshold, upon which the packet can be removed from the buffer. Because of the shared nature of the spectrum, dynamics over the wireless links are coupled with each other via the interference they caused, which results in memories in the evolution of queues and hence hinders tractable analysis. To that end, we adopt the mean field approximation to decouple such a spatial interaction and assume the queues evolve independently over time. Then, by jointly using tools from stochastic geometry and queueing theory, we derive a closed-form expression for the characterization of average AoI. Additionally, in order to control the cross-network interference level and hence assure the timely delivery of information, a channel access protocol is employed at each node to decide whether the current transmission attempt should be approved or not. Design of the policy exploits local observation, which is encapsulated via the concept of stopping sets [5], to make the transmission decision, aimed at optimizing the some utility function. The analytical results enable us to explore the effect from various network parameters on the AoI and hence devise useful insights for the protocol design. The main contributions are summarized in what follows.

- We develop an analytical framework that captures the interplay between the geographic locations of information source nodes and their temporal traffic dynamics. Due to spatially queueing interactions, dynamics over the wireless links are entangled where a complete analysis of such phenomena is yet available. In that respect, we adopt the mean field approximation to decouple the queueing interactions and derive tractable expressions for various network statistics by taking into account all the key features of a wireless network, including packet arrival rate, small scale fading and large scale path-loss, random network topology, and spatially queueing interactions.
- Using the dominant system, we derive an upper bound for the peak AoI of the network. And based on that we propose a decentralized channel access policy to minimize the information age in a wireless network. The proposed scheme is efficient in the sense that it requires only local information and has very low implementation complexity.
- Numerical results show that although the packet arrival rate directly affects the service process via queueing interaction, our proposed scheme can adapt to the traffic variations and largely reduce the network average AoI. Moreover, the proposed scheme can also adequately adjust according to the change of the ambient environment and thus scales well as the network grows in size.

14.2 System Model

In this section, we set up the network model for the study, including the spatial topology and the traffic profile. We also elaborate on the concept of average AoI of a network.

14.2.1 Network Structure

We model the wireless network as a set of transmitters and their corresponding receivers, all located in the Euclidean plane. The transmitting nodes are scattered according to a homogeneous $\tilde{\Phi}$ of spatial density λ. Each transmitter located at $X_i \in \tilde{\Phi}$ has a dedicated receiver, whose location y_i is at distance r in a random orientation. According to the displacement theorem [4], the location set $\bar{\Phi} = \{y_i\}_{i=0}^{\infty}$ also forms a homogeneous PPP with spatial density λ.

We segment the time into equal-length slots with each being the duration to transmit a single packet. At the beginning of each time slot, every transmitter has the information packets updated according to an independent and identically distributed (i.i.d.) Bernoulli process with parameter ξ. All the incoming packets are stored in a single-server queue with infinite capacity under discipline. In this network, we adopt the protocol at the transmitters to control the radio channel access. Simply put, during each time slot, the transmitters that have non-empty buffers will send out one packet with probability p. The transmission succeeds if the SINR at the corresponding receiver exceeds a predefined threshold, denoted as θ, upon which the receiver will feedback an ACK and the packet can be removed from the buffer. Otherwise, the receiver sends a NACK message and the packet is retransmitted in the next available time slot. We assume the ACK/NACK transmission is instantaneous and error-free, as commonly done in the literature [34]. Note that in this system the delivery of packets incurs a delay of one time slot, namely, packets are transmitted at the beginning of time slots and, if the transmission is successful, they are delivered by the end of the same time slot. In order to investigate the time-domain evolution, we limit the mobility of transceivers by considering a static network, that is, the locations of transmitters and receivers remain unchanged in all the time slots.

We assume that each transmitter uses unit transmission power P_{tx}. The channel is subjected to both Rayleigh fading, which varies independently across time slots, and path-loss that follows power law attenuation. Moreover, the receiver is also subjected to white Gaussian thermal noise with variance σ^2. By applying Slivnyak's theorem [4], it is sufficient to focus on a *typical* receiver located at the origin, with its tagged transmitter at X_0. Thus, when the tagged transmitter sends out a packet during slot t, the corresponding SINR received at the typical node can be written as

$$\text{SINR}_{0,t} = \frac{P_{\text{tx}} H_{00} r^{-\alpha}}{\sum_{j \neq 0} P_{\text{tx}} H_{j0} \zeta_{j,t} \nu_{j,t} \|X_j\|^{-\alpha} + \sigma^2}, \tag{14.1}$$

where α denotes the path-loss exponent, $H_{ji} \sim \exp(1)$ is the channel fading from transmitter j to receiver i, $\zeta_{j,t} \in \{0, 1\}$ is an indicator showing whether the buffer of node j is empty ($\zeta_{j,t} = 0$) or not ($\zeta_{j,t} = 1$), and $\nu_{j,t} \in \{0, 1\}$ represents the scheduling decision of node j, where it is set to 1 upon assuming transmission approval and 0 otherwise.

14.2.2 Age of Information

Without loss of generality, we typify the communication link between the transmitter–receiver pair located at (X_0, y_0). Then, as illustrated in Figure 14.1, the AoI, $A_0(t)$, over the typical link grows linearly in the absence of reception of new packets, and, when a new information packet is successfully received, reduces to the time elapsed since the generation of the delivered packet. To make the statement more precise, we formalize the evolution of $A_0(t)$ via the following expression:

$$A_0(t+1) = \begin{cases} A_0(t) + 1, & \text{if transmission fails or no transmission,} \\ t - G_0(t) + 1, & \text{otherwise,} \end{cases} \tag{14.2}$$

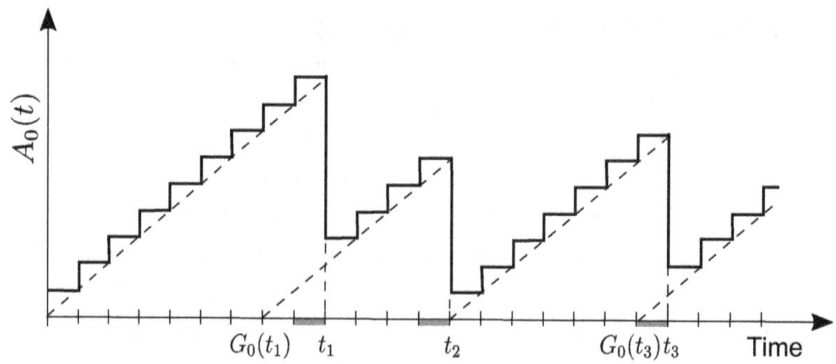

Figure 14.1 An example of the AoI evolution over a typical link. The time instance t_i denotes the moment when the ith packet is successfully delivered, and $G(t_i)$ is the generation moment of the packet to be transmitted in the time slot following $T_0(i)$. The age is set to be $t_i - G(t_i) + 1$.

where $G_0(t)$ is the generation time of the packet delivered over the typical link at time t.

In this paper, we use the *average AoI* as our metric to evaluate the age performance across a wireless network. Formally, the average AoI at one generic link j is defined as:

$$\bar{A}_j = \limsup_{T\to\infty} \frac{\sum_{t=1}^{T} A_j(t)}{T}. \tag{14.3}$$

Following the preceding definition, we can extend this concept to a large-scale network by taking the average of AoI from all the communication links and define the *network average AoI* as follows:

$$\bar{A} = \limsup_{R\to\infty} \frac{\sum_{X_j\in\tilde{\Phi}\cap B(0,R)} \bar{A}_j}{\sum_{X_j\in\tilde{\Phi}} \chi\{X_j\in B(0,R)\}} = \mathbb{E}^0\Big[\limsup_{T\to\infty} \frac{1}{T}\sum_{t=1}^{T} A_0(t)\Big], \tag{14.4}$$

where $B(0,R)$ denotes a disk centered at the origin with radius R; χ_E is the indicator function, which takes value 1 if event E occurs and 0 otherwise; and (a) follows from Campbell's theorem [4]. The notion $E^0[\cdot]$ indicates the expectation is taken with respect to the Palm distribution P^0 of the stationary point process, where under P^0 almost surely there is a node located at the origin [4].

14.2.3 Spatially Interacting Queues

In a wireless network, transmitters sharing the spectrum in space can impact each other's queueing states through the interference they cause. As such, the active state of a generic link j, $\zeta_{j,t}$, is dependent on both the spatial and temporal factors. A pictorial interpretation of this concept is given in Figure 14.2, which illustrates the spatiotemporal interactions among the queues of four wireless transmitter–receiver pairs. From a spatial perspective, we can see that transmitters X_1 and X_2 are located in geographic proximity and hence their transmissions incur strong mutual interference, which slows down the rate of service and eventually prolongs their queue lengths. In sharp contrast,

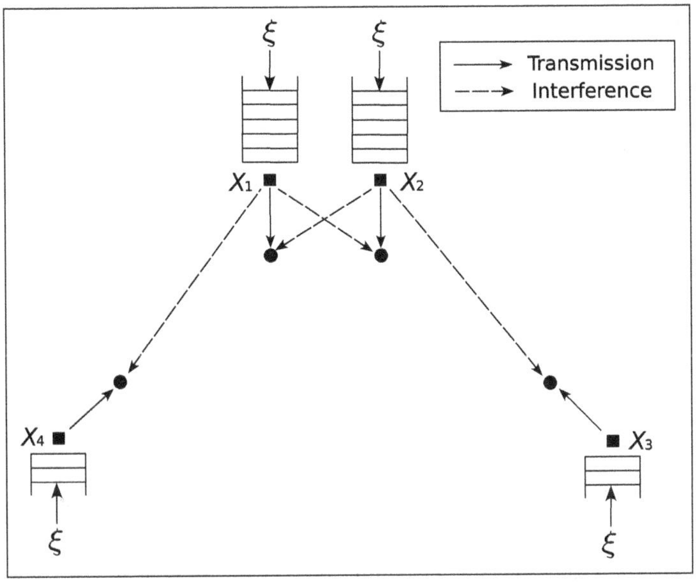

Figure 14.2 Illustration of a wireless network with spatially interacting queues. All the transmitter–receiver pairs are configured with the same distance and packet arrival rates.

transmitters X_3 and X_4 are at relatively long distances to their geographic neighbors. Such advantageous locations benefit the transmissions in these links as they do not suffer severe crosstalk, and hence their buffer lengths are generally much shorter compared to those of transmitters X_1 and X_2. From a temporal perspective, the packet arrival rate also plays a critical role in the process of service and further affects the queue length. Particularly, if packets arrive at a high rate, all the transmitters will be active, which raises up the total interference level, and that can incur many transmission failures, which prolong the active duration of the senders. On the contrary, when packet arrival rates are low, some transmitters may flush their queues and become silent, the reduced interference will also accelerate the depletion of packets at other nodes, which in turn leads to a shortened active period.

As such, in the context of a large-scale network, even if the packet arrivals are homogeneous in time, the spatial interactions result in a large variation of queue status across the nodes because the transmitters located in a crowded area of space will face poor transmission conditions and eventually have longer queue lengths than those situated at far distances from their neighbors. This phenomenon is commonly known as the spatially interacting queues, and the characterization of its behavior is very challenging.

14.3 Analysis

In this section, we derive analytical expressions to characterize the distribution of the conditional transmission success probability, which directly determines the rate of packet depletion, and based on that we calculate the average AoI in large-scale networks.

14.3.1 Preliminaries

Under the depicted system model, when seen from the temporal perspective, dynamics of packet transmissions on any given link can be abstracted as a Geo/G/1 queue where the departure rate is dependent on the communication condition – which is essentially determined by the geographic structure of the network. In order to characterize the departure rate, we condition on the realization of point process $\Phi = \tilde{\Phi} \cup \bar{\Phi}$ and define the conditional transmission success probability of the typical link at a generic time slot t as follows [40]:

$$\mu_{0,t}^{\Phi} = P\big(\text{SINR}_{0,t} > \theta \,|\, \Phi\big). \tag{14.5}$$

Note that $\mu_{0,t}^{\Phi}$ is a random variable that has a distribution equivalent to that of service rate. Another observation of (14.5) is that the quantity $\mu_{0,t}^{\Phi}$ is indexed by the time slot t, which implies that the collection of service rates forms a stochastic process over time. Due to interference, there are temporal correlations in the queue evolution processes across the entire network [30], which complicate the subsequent analysis. It thus necessitates the introduction of the following approximation: *In this network, each queue observes the time-averages of the activity indicators of other queues but evolves independently of their current state.*

This assumption is commonly known as the mean-field approximation which makes the dynamics of packet transmissions at each node conditionally independent, given the positions of all transmitters and receivers in the network. As a result, we can regard the transmissions of packets over the typical link as i.i.d. over time with a success probability $\mu_0^{\Phi} = \lim_{t\to\infty} \mu_{0,t}^{\Phi}$. In consequence, the packet dynamics at the typical transmitter can be abstracted as a Geo/Geo/1 queue, and by leveraging tools from queueing theory, we arrive at a conditional form of the AoI under FCFS.

LEMMA 14.1 *Given $\xi < p\mu_0^{\Phi}$, when conditioned on the point process Φ, the average AoI is given by*

$$\mathbb{E}^0\big[\bar{A}\,|\,\Phi\big] = \frac{1-\xi}{p\mu_0^{\Phi} - \xi} + \frac{\xi}{p\mu_0^{\Phi}} - \frac{\xi}{(p\mu_0^{\Phi})^2} + \frac{1}{\xi} - 1, \tag{14.6}$$

where μ_0^{Φ} denotes the transmission success probability under the considered scenario.

Proof When conditioned on the point process Φ, the transmission process at a typical link can be regarded as a Geo/Geo/1 queue with the rate of arrival and departure being ξ and $p\mu_0^{\Phi}$, respectively, and hence (14.6) follows from leveraging results in [33]. □

14.3.2 Transmission Success Probability and AoI Derivations

According to (14.1), the SINR received at a typical UE comprises a series of random quantities, including the fading, active states, and locations of transmitters, which daunts the task of analysis. To launch an initial step, we first average out the randomness from fading and derive a conditional form of the transmission success probability:

LEMMA 14.2 *Conditioned on the spatial realization Φ, the transmission success probability at the typical link is given by*

$$\mu_0^{\Phi} = e^{-\frac{\theta r^{\alpha}}{\rho}} \prod_{j \neq 0} \left(1 - \frac{p\, a_j^{\Phi}}{1 + D_{j0}}\right), \tag{14.7}$$

where $\rho = P_{\mathrm{tx}}/\sigma^2$, $D_{ij} = \|X_i - y_j\|^{\alpha}/\theta r^{\alpha}$, and $a_j^{\Phi} = \lim_{t\to\infty} P(\zeta_{j,t} = 1|\Phi)$ is the active probability of transmitter j at the steady state.

Proof By conditioning on the spatial realization Φ of all the transceiver locations, the transmission success probability can be derived as follows:

$$\begin{aligned}
\mathbb{P}(\mathrm{SINR}_0 > \theta|\Phi) &= P\left(H_{00} > \sum_{j\neq 0} \frac{H_{j0} \nu_j \zeta_j \theta r^{\alpha}}{\|X_j - y_0\|^{\alpha}} + \frac{\theta r^{\alpha}}{\rho} \,\Big|\, \Phi\right) \\
&= \mathbb{E}\left[e^{-\frac{\theta r^{\alpha}}{\rho}} \prod_{j\neq 0} \exp\left(-\theta r^{\alpha} \frac{\nu_j \zeta_j H_{j0}}{\|X_j - y_0\|^{\alpha}}\right) \Big| \Phi\right] \\
&\overset{(a)}{=} e^{-\frac{\theta r^{\alpha}}{\rho}} \prod_{j\neq 0} \left(1 - p\, a_j^{\Phi} + \frac{p\, a_j^{\Phi}}{1 + 1/D_{j0}}\right),
\end{aligned} \tag{14.8}$$

where (a) follows by the mean-field approximation in Section 14.3.1, with which the active state at every transmitter can be treated as independent from all the each other, and further using the fact that $H_{j0} \sim \exp(1)$. The expression per (14.7) can then be obtained by further simplifying the product factors. □

The result from Lemma 14.2 explicitly reveals that the randomness associated with the is mainly assorted to (i) the random location of the interfering transmitters, and (ii) their corresponding active states. As such, by leveraging the queueing theory, a conditional expression for the active state at each transmitter can be obtained as follows.

LEMMA 14.3 *Given the service rate μ_j^{Φ}, the queue-non-empty probability at a generic transmitter j is given as*

$$a_j^{\Phi} = \begin{cases} 1, & \text{if } p\,\mu_j^{\Phi} \leq \xi, \\ \frac{\xi}{p\mu_j^{\Phi}}, & \text{if } p\,\mu_j^{\Phi} > \xi. \end{cases} \tag{14.9}$$

Proof For a Geo/G/1 queue, the probability of being active in the steady state follows from Little's law [18]. □

With these results in hand, we can now put the pieces together and derive the distribution of the conditional transmission success probability.

THEOREM 14.4 *The cumulative distribution function (CDF) of the conditional transmission success probabilities is given by the following:*

$$F(u) = \frac{1}{2} - \int_0^\infty \mathrm{Im}\left\{u^{-j\omega} \exp\Big(-\frac{j\omega\theta r^\alpha}{\rho} - \lambda\pi\delta\frac{\pi\theta^\delta r^2}{\sin(\pi\delta)}\sum_{k=1}^{\infty}\binom{j\omega}{k}\binom{\delta-1}{k-1} \times \Big[p^k F\Big(\frac{\xi}{p}\Big) + \int_{\xi/p}^1 \frac{\xi^k F(dt)}{t^k}\Big]\Big)\right\}\frac{d\omega}{\pi\omega}, \tag{14.10}$$

in which $j = \sqrt{-1}$, $\delta = 2/\alpha$, *and* $\mathrm{Im}\{\cdot\}$ *denotes the imaginary part of a complex quantity.*

Proof To facilitate the presentation, let us introduce two notations Y_i^Φ and q_i, defined as $Y_i^\Phi = \ln P(\mathrm{SINR}_i > \theta|\Phi)$ and $q_{l,i} = P(\nu_i \times \zeta_i = 1|\|X_i - y_0\| = l)$, respectively. Using Slivnyark's theorem [4], we concentrate on the moment-generating function of Y_0^Φ as follows:

$$\begin{aligned} M_{Y_0^\Phi}(s) &= E\left[\exp\left(sY_0^\Phi\right)\right] = E\left[P\left(\mathrm{SINR}_0 > \theta|\Phi\right)^s\right] \\ &\overset{(a)}{=} E\left[e^{-\frac{s\theta r^\alpha}{\rho}}\prod_{i\neq 0}\left(1-q_i+\frac{q_i}{1+\theta r^\alpha\|X_i\|^\alpha}\right)^s\right] \\ &\overset{(b)}{=} \exp\left(-2\pi\lambda\int_0^\infty\sum_{k=1}^{s}(-1)^{k+1}\frac{E[q_x^k]xdx}{1+x^\alpha/\theta r^\alpha}\right), \end{aligned} \tag{14.11}$$

where (a) follows from the independent evolution of queues according to mean field approximation, and (b) by using the probability-generating functional (PGFL).

The complete expression of (14.11) requires us to further compute $\mathbb{E}[q_x^k]$, which can be written as

$$\begin{aligned} E[q_x^k] &\overset{(a)}{=} E\Big[\Big(p\cdot\min\Big\{\frac{\xi}{p\mu_x^\Phi},1\Big\}\Big)^k\Big] \\ &= E\Big[\Big(p^k\chi_{\{p\mu_x^\Phi<\xi\}}+\Big(\frac{\xi}{\mu_x^\Phi}\Big)^k\chi_{\{p\mu_x^\Phi\geq\xi\}}\Big)\Big], \end{aligned} \tag{14.12}$$

where (a) follows from Little's Law [18].

At this stage, let us temporally assume the CDF $F(u)$ of μ_0^Φ is available. Using Lemma 14.3, the first term of (14.12) can then be derived as

$$E\Big[p^k\chi_{\{p\mu_x^\Phi<\xi\}}\Big] = p^k F(\xi/p). \tag{14.13}$$

Similarly, we have the second term of (14.12) given by

$$E\Big[\Big(\frac{\xi}{\mu_x^\Phi}\Big)^k\chi_{\{p\mu_x^\Phi\geq\xi\}}\Big] = \int_{\xi/p}^1 \frac{\xi^k}{t^k}F(dt). \tag{14.14}$$

Finally, we apply the Gil-Pelaez theorem [16] and derive the CDF of μ_0^Φ as

$$F(u) = P\left(\mu_0^\Phi < u\right) = P\left(Y_0^\Phi < \ln u\right)$$
$$= \frac{1}{2} - \frac{1}{\pi}\int_0^\infty \text{Im}\left\{u^{-j\omega} M_{Y_0^\Phi}(j\omega)\right\} \frac{d\omega}{\omega}. \tag{14.15}$$

To this end, by substituting (14.12), (14.13), and (14.14) into the preceding equation, the result follows. □

Owing to the space–time coupling among the queues, the CDF (14.10) is given in the form of a fixed-point functional equation, which can be solved in an iterative manner [40]. Additionally, approximations via the beta distribution are also available to reduce the computational complexity [40]. At this stage, we are ready to finally derive an expression to quantify the AoI.

THEOREM 14.5 *Under the FCFS discipline, the network average AoI is given as follows:*

$$\bar{A} = \int_{\xi/p}^{1} \left(\frac{1-\xi}{pt-\xi} + \frac{\xi}{pt} - \frac{\xi}{p^2 t^2} \right) F(dt) + \frac{1}{\xi} - 1, \tag{14.16}$$

where $F(\cdot)$ is given in (14.10).

Proof The preceding expressions are attained by deconditioning the random variable μ_0^Φ in (14.6) with respect to its distribution functions given per (14.10). □

The AoI expression given in Theorem 14.5 accounts for all the key features of a wireless network, including the packet arrival rate, channel access probability, deployment density, interference, and interactions among the queues. In Section 14.5, we will explore the effects of different network parameters on the performance of AoI based on this equation. Before that, let us use the analytical framework to construct a location-dependent channel access policy.

14.4 Locally Adaptive Channel Access Policy

In a wireless network, as all transmitters are intertwined through the interference they cause to each other, the queue state of each transmitter varies across the network. It is therefore important to have a link-wise protocol that schedules the appropriate channel access at each node. Inspired by the fact that knowledge from local activities can be utilized to improve the overall network performance, we incorporate such local information in the design of the scheduling policy.

14.4.1 Stopping Sets and Channel Access Policy

For a generic transmitter, note that it can only obtain the information about its geometry vicinity; we thus encapsulate such local knowledge by the notion of *stopping set* $W = W(\tilde{\Phi}, \bar{\Phi})$ [4, 5]. More precisely, the stopping set is a random element taking each realization from the Borel sets in $\mathbb{R}^2$ such that for any observation window A, one can

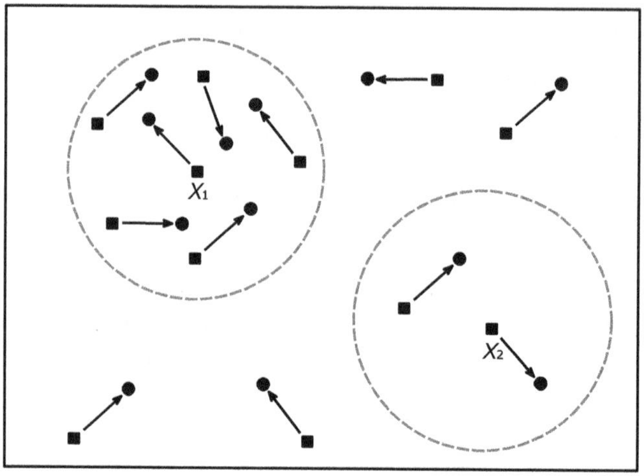

Figure 14.3 Illustration of a Poisson bipolar network with stopping sets being disks with constant radii, where black squares and dots are the transmitters and receivers, respectively, and the circles with dashed lines are two exemplary stopping sets centered at X_1 and X_2.

determine whether $W = W(\tilde{\Phi}, \bar{\Phi}) \subset A$. This concept enables us to model the region in which the information of nodes, including their locations, are known to a typical observer. In particular, depending on the scenarios under consideration, stopping sets can take various forms. For instance, if the transmitters have unified sensing power, the observation region at each node will be a disk with a constant radius and the stopping set takes a deterministic form. When the transmitters are empowered with heterogeneous sensing capabilities, each node may want to obtain information up to the pth nearest neighbor, in which case the observation region varies across different nodes and the stopping set takes a random shape. Aside from disks, the stopping set can have more complicated formats, for example, a hexagon under clustering regulation or different orders of Voronoi cells in the context of cellular networks [37], depending on the specific task under consideration. An illustration of deterministic stopping sets in a Poisson bipolar network is given in Figure 14.3. Note that different transmitters, for example, the nodes located at X_1 and X_2, can have various local observations.

To generalize the concept network-wide, we further introduce a shifting operation, denoted by S_x and performs on $\tilde{\Phi}$ and $\bar{\Phi}$, which translates all the network nodes by the vector $-x$, that is, $S_x\{X_i\} = \{X_i - x\}$ and similarly for the receivers. Extending this operator to a Borel $A \subset \mathbb{R}^2$, we have $S_x(A) = \{a - x : a \in A\}$. Using the shifting operator and the stopping set, we define the set of candidated scheduling policies to be the ones that take into the local knowledge at every source node and output a channel access probability for it. More precisely, given a stopping set $W = W(\tilde{\Phi}, \bar{\Phi})$, the scheduling policy at the typical transmitter can be defined as follows:

$$\gamma_0^{\Phi} = \eta_{\mathrm{W}}(\tilde{\Phi}, \bar{\Phi}) = \eta_{\mathrm{W}}\left(\tilde{\Phi} \cap W, \bar{\Phi} \cap W\right),$$

where $\eta_{\mathrm{W}}(\cdot)$ is a measurable function whose argument is the network geometry $(\tilde{\Phi}, \bar{\Phi})$ and has value in $[0, 1]$. For node i located at X_i, its scheduling policy can be obtained by applying the shifting operator S_{X_i} to (14.4.1), which results in $\gamma_i^{\Phi} = \eta_{\mathrm{W}}(S_{X_i}\tilde{\Phi}, S_{X_i}\bar{\Phi})$. To this end, a feasible scheduling policy is *translation invariant*, which allows us to devise it by only focusing on the typical node. Note that to apply such a policy and evaluate the channel access probability $\eta_{\mathrm{W}}(S_{X_i}\tilde{\Phi}, S_{X_i}\bar{\Phi})$, node i needs to obtain only local knowledge about the other nodes in the stopping set $W_i = W(S_{X_i}\tilde{\Phi}, S_{X_i}\bar{\Phi})$, which can be obtained via, for example, the methods in [5]. In this regard, the scheme can run without a central controller and is thus decentralized.

14.4.2 Locally Adaptive ALOHA Scheme

Based on Lemma 14.1, for a given stopping set $W = W(\tilde{\Phi}, \bar{\Phi})$, the design of an AoI-minimum channel access policy can be written as:

$$\min_{\eta_{\mathrm{W}}} \quad \mathbb{E}_{\Phi}^{0}\left[\frac{1-\xi}{\gamma_0^{\Phi}\mu_0^{\Phi}-\xi} + \frac{\xi}{\gamma_0^{\Phi}\mu_0^{\Phi}} - \frac{\xi}{\left(\gamma_0^{\Phi}\mu_0^{\Phi}\right)^2}\right] + \frac{1}{\xi} - 1 \tag{14.17}$$

$$\text{s.t.} \quad 0 \le \gamma_i^{\Phi} = \eta_{\mathrm{W}}\left(S_{X_i}\tilde{\Phi}, S_{X_i}\bar{\Phi}\right) \le 1, \ \forall i \tag{14.18}$$

$$\xi \le \mathbb{E}_{\Phi}^{0}[\eta_{\mathrm{W}}\mu_0^{\Phi}]. \tag{14.19}$$

It is worthwhile to point out that the design factor η_{W} in (14.17) is not a single parameter but instead a policy, which takes input the state information, that is, the node's location and information observed from the corresponding stopping set, and as an output the channel access probability. As such, the scheduling policy varies from node to node, which is stated in constraint (14.18), because the local knowledge is location dependent. Moreover, the queueing stability shall be guaranteed in the average sense, as shown in (14.19), according to Loynes' theorem [26].

According to the expression of μ_0^{Φ} given in (14.7), we note that the optimization of (14.17) requires at a typical source node the: (i) parameter D_{j0}, which depends on the locations of receivers in $\bar{\Phi} \cap W$, and (ii) the active states of these receivers. However, due to the interaction of queueing dynamics at different nodes, the active state at each transmitter varies over time and is difficult to keep track of. For that reason, we leverage the dominant system [9], in which every transmitter keeps sending out packets in each time slot (if one transmitter has an empty buffer at any given time slot, it sends out a dummy packet) for an approximation. Besides, because transmitters under the dominant system are subject to higher interference levels, the policies devised from a dominant system can also help us prepare for the "worst case" interference condition in the original system when all the transmitters are active. As such, instead of directly solving the original problem (14.17), we minimize the following alternative:

$$\min_{\eta_{\mathrm{W}}} \quad \mathbb{E}_{\Phi}^{0}\left[\frac{1-\xi}{\gamma_0^{\Phi}\hat{\mu}_0^{\Phi}-\xi} + \frac{\xi}{\gamma_0^{\Phi}\hat{\mu}_0^{\Phi}} - \frac{\xi}{\left(\gamma_0^{\Phi}\hat{\mu}_0^{\Phi}\right)^2}\right] + \frac{1}{\xi} - 1 \tag{14.20}$$

$$\text{s.t.} \quad 0 \le \gamma_i^{\Phi} = \eta_{\mathrm{W}}\left(S_{X_i}\tilde{\Phi}, S_{X_i}\bar{\Phi}\right) \le 1, \ \forall i \tag{14.21}$$

$$\xi \le E_{\Phi}^{0}[\eta_{\mathrm{W}}\hat{\mu}_0^{\Phi}], \tag{14.22}$$

where $\hat{\mu}_0^\Phi$ is the transmission success probability at the typical link under the dominant system, given by

$$\hat{\mu}_0^\Phi = e^{-\frac{\theta r^\alpha}{\rho}} \prod_{j \neq 0} \Big(1 - \frac{\gamma_j^\Phi}{1 + D_{j0}}\Big). \tag{14.23}$$

The design of the scheduling policy can now be cast into an optimization problem with explicit terms. And that brings us to the main structural result of this paper.

THEOREM 14.6 *For all given stopping sets $W = W(\tilde{\Phi}, \bar{\Phi})$, the solution to the optimization problem in* (14.20) *is given by the unique solution of the following fixed point equation,*

$$\frac{1}{\eta_W} - \sum_{j \neq 0, y_j \in W} \frac{1}{1 + D_{0j} - \eta_W} - \int_{R^2 \setminus W} \frac{\lambda dz}{1 + \|z\|^\alpha / \theta r^\alpha} = 0, \tag{14.24}$$

if the following condition holds:

$$\sum_{j \neq 0, y_j \in W} \frac{1}{D_{0j}} + \int_{R^2 \setminus W} \frac{\lambda dz}{1 + \|z\|^\alpha / \theta r^\alpha} > 1. \tag{14.25}$$

Otherwise, $\eta_W = 1$.

Proof First of all, we note that under a dominant system, the point process Φ is stationary. As such, by substituting (14.23) into the (14.20) and using the mass transportation theorem [4], we can treat the operator η_W as a variable [5, 29] and take the derivative of (14.20) with respect to η_W and equate it to zero, which yields the following:

$$\frac{1}{\eta_W} - \sum_{j \neq 0, y_j \in W} \frac{1}{1 + D_{0j} - \eta_W} - \int_{R^2 \setminus W} \frac{\lambda\, dy}{1 + y^\alpha / T r^\alpha} = 0. \tag{14.26}$$

If we write the left-hand side (L.H.S.) of the preceding equation as a function $f(\eta_W)$ of η_W, that is,

$$f(\eta_W) = \frac{1}{\eta_W} - \sum_{j \neq 0, y_j \in W} \frac{1}{1 + D_{0j} - \eta_W} - \int_{R^2 \setminus W} \frac{\lambda dz}{1 + \|z\|^\alpha / \theta r^\alpha}, \tag{14.27}$$

it is easy to verify that (*a*) $f(\eta_W)$ monotonically decreases in η_W over [0, 1], and (*b*) $\lim_{\eta_W \to 0} f(\eta_W) = +\infty$. As such, if $f(1) < 0$, that is, the condition (14.25) holds, then according to the Intermediate Value Theorem, the equation in (14.26), or equivalently, (14.27), has a solution and this solution is unique. Otherwise, if (14.25) does not hold, we have the derivative of (14.20) being negative, which indicates that (14.20) monotonically decreases as a function of η_W. Hence, the minimum is achieved at $\eta_W = 1$. □

The preceding theorem gives an explicit way to construct the channel access policy at the typical node, that is, $\gamma_0^\Phi = \eta_W(S_{X_0}\tilde{\Phi}, S_{X_0}\bar{\Phi})$. In view of a generic node i, the corresponding channel access probability can be attained by shifting the origin of the point process Φ to X_i and then applying the same approach as per Theorem 14.6, which gives $\gamma_i^{\Phi} = \eta_W(S_{X_i}\tilde{\Phi}, S_{X_i}\bar{\Phi})$. As such, every source node only need to identify and record the geographic neighbors located inside its observation window, that is, the stopping set W, and solve for the channel access probability via a fixed point equation, which has very low complexity. Essentially, such a channel access policy is a spatial version of ALOHA, where the channel access probability at each transmitter is conceived based on the local topology information. Different from the interference graphs [33], which mute simultaneous transmissions from nodes in proximity locations, the proposed approach allows neighboring transmitters to initiate the radio channel access attempts at the same time while communally controlling the mutual interference with a location-dependent probability and thus achieves better spectrum reuse.

It is also worthwhile to note that condition (14.25) implies the typical transmitter will only opt for an opportunistic channel access when the following holds:

$$\theta > \frac{r^{-\alpha}}{\sum\limits_{j\neq 0, y_j \in W} \|y_j\|^{-\alpha} + \int_{R^2\setminus W} \frac{\lambda dz}{\|z\|^{\alpha}+r^{\alpha}}}$$

$$> \frac{r^{-\alpha}}{\sum\limits_{j\neq 0, y_j \in W} \|y_j\|^{-\alpha} + \int_{R^2\setminus W} \frac{\lambda dz}{\|z\|^{\alpha}}}$$

$$= \frac{\mathbb{E}\big[P_{\text{tx}} H_{00} r^{-\alpha}\big]}{\mathbb{E}\big[\sum_{j\neq 0} P_{\text{tx}} H_{0j} \|X_0 - y_j\|^{-\alpha} | W\big]},$$

namely, the ratio between the average signal power and interference generated by the transmitter is smaller than the decoding threshold. In other words, a given source node will reduce the channel access frequency when its own transmission may cause potential transmission failure to the neighbors. Such an observation also coincides with the intuition that transmitters located close to each other can cause severe mutual interference and hence need to be scheduled for more stringent channel access, while the ones located far away from their neighbors can access the radio channel more frequently.

14.5 Numerical Results

In this section, we use the analytical results to explore the impacts from different network parameters on the performance of AoI. We also evaluate the effectiveness of the proposed channel access policy on reducing AoI in large-scale wireless networks. Unless differently specified, we use the following parameters: $\alpha = 3.8$, $\xi = 0.3$, $\theta = 0$ dB, $P_{\text{tx}} = 17$ dBm, $\sigma^2 = -90$ dBm, $r = 25$ m, and $\lambda = 10^{-4}\text{m}^{-2}$.

Figure 14.4 depicts the network average AoI as a function of the packet arrival rate ξ, under a varying value of channel access probability p. From this figure, we immediately notice the existence of an optimal packet arrival rate that minimizes the AoI. This is mainly attributable to the trade-off between information freshness at the source nodes and the interference level across the entire network where an increase of packet arrival rate leads to fresher updates at the source nodes but also a higher

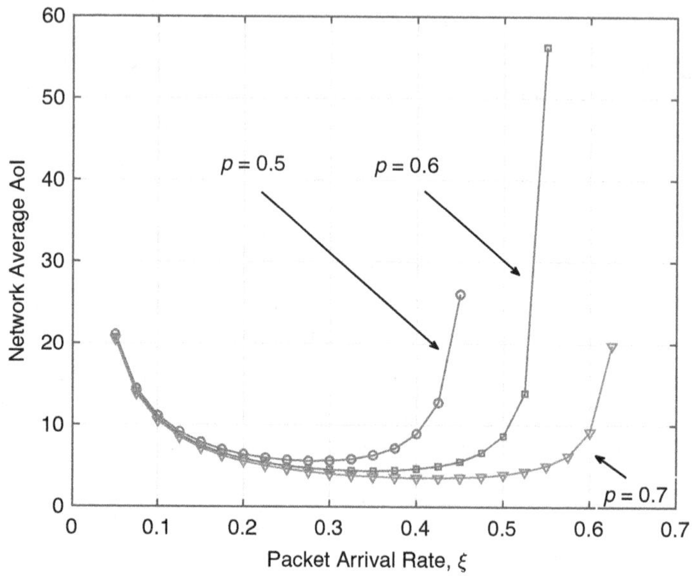

Figure 14.4 Impact of packet arrival rate on the average AoI.

level of interference, and vice versa. Moreover, the figure also shows that the optimal update frequency (resp. optimum AoI) increases (resp. decreases) with the channel access probability, because the network is relatively sparse under this set of parameters and therefore the overall interference power is small. As such, one shall increase the frequency of packet updating to attain fresh information at the receivers.

In Figure 14.5 we plot the AoI as a function of the channel access probability, under different network deployment densities. The figure shows that on the one hand, when the network is sparsely deployed, the average AoI keeps decreasing with the channel access probability. On the other hand, if the infrastructure is rolled out densely, there exists an optimal channel access probability that minimizes the average AoI. This is because when there is an abundant number of wireless links in the network, the excessive interference can devastate the throughput of each link. In this context, it is worthwhile to reduce the channel access probability so as to strike a balance between the radio channel utilization per node and the overall interference level, which attains the minimum AoI by achieving a maximum throughput.

Finally, we compare in Figure 14.6 the proposed scheduling policy with observation from a deterministic stopping set to that with no available local information, that is, $W = \phi$ (in which case, $\eta_W = 1, \forall j \in N$). This figure not only illustrates how network densification affects the information freshness, but also highlights the critical role played by the channel access policy. We hence conclude the observation with the following takeaways:

- The peak AoI always increases with respect to the spatial density, since densifying the network inevitably entails additional interference; thus, the SINR is defected and it further hurts the transmission quality across the network.

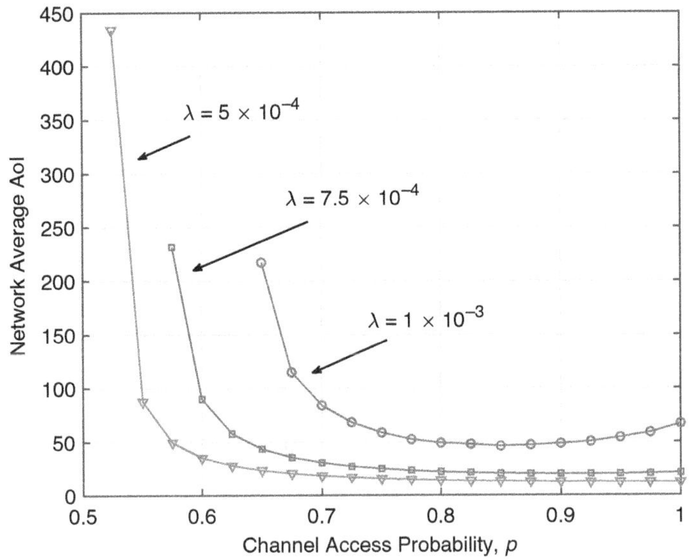

Figure 14.5 Impact of channel access probability on the average AoI.

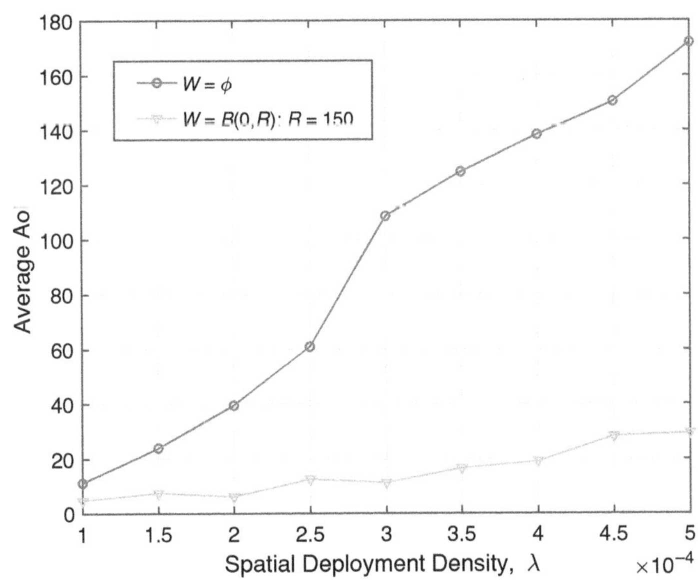

Figure 14.6 Network average AoI vs. spatial density: $r = 25$ m, $\xi = 0.3$, and $R = 150$ m for the deterministic stopping set $W = B(0, R)$.

- By employing the locally adaptive ALOHA (Locally Adaptive ALOHA) at each transmitter, the AoI undergoes a substantial discount and the gain is more pronounced in the dense network scenario. This is because the interference between neighbors becomes more severe when their mutual distance is reduced, and hence adequately scheduling the channel access patterns of transmitters can prevent interference from rising too quickly and maintain the AoI at a low level.

14.6 Conclusion

In this chapter, we have developed a theoretical framework for the understanding of AoI performance in large-scale wireless networks. We have used a general model that accounts for the channel gain and interference, dynamics of status updating, and spatial queueing interactions. Our results have confirmed that the network topology has a direct and sweeping influence on the AoI. Specifically, if the topology infrastructure is densely deployed, there exists an optimal rate of packet arrival that minimizes the average AoI. In addition, ALOHA is instrumental in further reducing the AoI, given the packet arrival rates are high. However, when the network deployment density is low, the average AoI decreases monotonically with the packet arrival rate, and ALOHA cannot contribute to reducing the AoI in this scenario. Building upon the analytical framework, we have also proposed a decentralized protocol that allows every transmitter to make transmission decisions based on the observed local information. Using the concept of stopping sets, we encapsulated the local knowledge from individual nodes and derived tractable expressions to characterize the stochastic behavior of our proposed scheme, as well as quantified its effectiveness in terms of network average AoI. Numerical results showed that our proposed scheme is able to adaptively adjust according to the geographical change of the ambient environment and thus scales well as the network grows in size.

By combining queueing theory with stochastic geometry, the developed framework bridges the gap between the abstract service model – which is widely used in the existing AoI literature – and the physical transmission environment. It thus enables one to devise fundamental insights on the impact from both spatial and temporal aspects of a network on the information freshness. The model can be further applied in the design of scheduling schemes in wireless systems under different queueing disciplines, or with multiple sources as well as multi-hop routings. Investigating up to what extent a nonbinary power control can improve AoI is also regarded as a concrete direction for future work.

References

[1] Andrews, J. G., Baccelli, F., and Ganti, R. K. [Nov. 2011], "A tractable approach to coverage and rate in cellular networks," *IEEE Trans. Commun.* **59**(11), 3122–3134.

[2] Arafa, A., and Ulukus, S. [Aug. 2019], "Timely updates in energy harvesting two-hop networks: Offline and online policies," *IEEE Trans. Commun.* **18**(8), 4017–4030.

[3] Arafa, A., Yang, J., Ulukus, S., and Poor, H. V. [Jan. 2020], "Age-minimal transmission for energy harvesting sensors with finite batteries: Online policies," *IEEE Trans. Inf. Theory* **66**(1), 534–556.

[4] Baccelli, F., and Blaszczyszyn, B. [2009], *Stochastic Geometry and Wireless Networks. Volumn I: Theory*, Now Publishers.

[5] Baccelli, F., Blaszczyszyn, B., and Singh, C. [Apr. 2014], Analysis of a proportionally fair and locally adaptive spatial ALOHA in poisson networks, *in* "Proc. IEEE INFOCOM," Toronto, ON, Canada, pp. 2544–2552.

[6] Bacinoglu, B. T., Sun, Y., Uysal-Bivikoglu, E., and Mutlu, V. [Jun. 2018], Achieving the age-energy tradeoff with a finite-battery energy harvesting source, Vail, CO, pp. 876–880.

[7] Bai, T., and Heath, R. W. [Feb. 2015], "Coverage and rate analysis for millimeter-wave cellular networks," *IEEE Trans. Wireless Commun.* **14**(2), 1100–1114.

[8] Bedewy, A. M., Sun, Y., and Shroff, N. B. [Jun. 2017], Age-optimal information updates in multihop networks, *in* "Proc. IEEE Int. Symp. Inform. Theory," Aachen, Germany, pp. 576–580.

[9] Bordenave, C., McDonald, D., and Proutiere, A. [Sep. 2012], "Asymptotic stability region of slotted aloha," *IEEE Trans. Inf. Theory* **58**(9), 5841–5855.

[10] Chen, K., and Huang, L. [Jul. 2016], Age-of-information in the presence of error, *in* "Proc. IEEE Int. Symp. Inform. Theory," Barcelona, Spain, pp. 2579–2583.

[11] Chisci, G., ElSawy, H., Conti, A., Alouini, M.-S., and Win, M. Z. [Aug. 2017], On the scalability of uncoordinated multiple access for the internet of things, *in* "Int. Symposium on Wireless Commun. Systems (ISWCS)," Bologna, Italy, pp. 402–407.

[12] Chisci, G., Elsawy, H., Conti, A., Alouini, M.-S., and Win, M. Z. [Jun. 2019], "Uncoordinated massive wireless networks: Spatiotemporal models and multiaccess strategies," *IEEE/ACM Trans. Netw.* **27**(3), 918–931.

[13] Costa, M., Codreanu, M., and Ephremides, A. [Feb. 2016], "On the age of information in status update systems with packet management," *IEEE Trans. Inf. Theory* **62**(4), 1897–1910.

[14] Emara, M., ElSawy, H., and Bauch, G. [Aug. 2020], "A spatiotemporal model for peak aoi in uplink IoT networks: Time versus event-triggered traffic," *IEEE Internet of Things Journal* **7**(8), 6762–6777.

[15] Gharbieh, M., ElSawy, H., Bader, A., and Alouini, M.-S. [Aug. 2017], "Spatiotemporal stochastic modeling of IoT enabled cellular networks: Scalability and stability analysis," *IEEE Trans. Commun.* **65**(9), 3585–3600.

[16] Gil-Pelaez, J. [Dec. 1951], "Note on the inversion theorem," *Biometrika* **38**(3–4), 481–482.

[17] Haenggi, M., Andrews, J. G., Baccelli, F., Dousse, O., and Franceschetti, M. [Sept. 2009], "Stochastic geometry and random graphs for the analysis and design of wireless networks," *IEEE J. Sel. Areas in Commun.* **27**(7), 1029–1046.

[18] Harchol-Balter, M. [2013], *Performance modeling and design of computer systems: queueing theory in action*, Cambridge University Press, Cambridge.

[19] He, Q., Yuan, D., and Ephremides, A. [May 2016], Optimizing freshness of information: On minimum age link scheduling in wireless systems, *in* "Proc. Modeling and Optimization in Mobile, Ad Hoc, and Wireless Networks (WiOpt)," Tempe, AZ, pp. 1–8.

[20] Hsu, Y.-P., Modiano, E., and Duan, L. [Jun. 2017], Age of information: Design and analysis of optimal scheduling algorithms, *in* "Proc. IEEE Int. Symp. Inform. Theory," Aachen, Germany, pp. 561–565.

[21] Hu, Y., Zhong, Y., and Zhang, W. [Dec. 2018], Age of information in Poisson networks, *in* "Proc. Int. Conf. Wireless Commun. and Signal Process. (WCSP)," Hangzhou, China, pp. 1–6.

[22] Huang, L., and Modiano, E. [Jun. 2015], Optimizing age-of-information in a multi-class queueing system, *in* "Proc. IEEE Int. Symp. Inform. Theory," Hong Kong, China, pp. 1681–1685.

[23] Kadota, I., Sinha, A., and Modiano, E. [2018], Optimizing age of information in wireless networks with throughput constraints, *in* "Proc. INFOCOM," Honolulu, HI, pp. 1844–1852.

[24] Kadota, I., Uysal-Biyikoglu, E., Singh, R., and Modiano, E. [Sept. 2016], Minimizing the age of information in broadcast wireless networks, *in* "Proc. IEEE Allerton," Monticello, IL, pp. 844–851.

[25] Kaul, S., Yates, R., and Gruteser, M. [Mar. 2012], Real-time status: How often should one update?, *in* "Proc. IEEE INFOCOM," Orlando, FL, pp. 2731–2735.

[26] Loynes, R. M. [1962], The stability of a queue with non-independent inter-arrival and service times, *in* "Math. Proc. Cambridge Philos. Soc.," vol. 58, Cambridge University Press, pp. 497–520.

[27] Mankar, P. D., Abd-Elmagid, M. A., and Dhillon, H. S. [2020], "Stochastic geometry-based analysis of the distribution of peak age of information," *Available as ArXiv:2006.00290*.

[28] Mankar, P. D., Chen, Z., Abd-Elmagid, M. A., Pappas, N., and Dhillon, H. S. [2020], "Throughput and age of information in a cellular-based IoT network," *Available as ArXiv:2005.09547*.

[29] Ramesan, N. S., and Baccelli, F. [Apr. 2019], Powers maximizing proportional fairness among poisson bipoles, *in* "Proc. IEEE INFOCOM," Paris, France, pp. 1666–1674.

[30] Sankararaman, A., Baccelli, F., and Foss, S. [2019], "Interference queueing networks on grids," *Ann. Applied Probability* **29**(5), 2929–2987.

[31] Sun, Y., Uysal-Biyikoglu, E., and Kompella, S. [Apr. 2018], Age-optimal updates of multiple information flows, *in* "Proc. IEEE INFOCOM Workshops," Honolulu, HI, pp. 136–141.

[32] Sun, Y., Uysal-Biyikoglu, E., Yates, R. D., Koksal, C. E., and Shroff, N. B. [Nov. 2017], "Update or wait: How to keep your data fresh?," *IEEE Trans. Inf. Theory* **63**(11), 7492–7508.

[33] Talak, R., Karaman, S., and Modiano, E. [2018*b*], "Optimizing information freshness in wireless networks under general interference constraints," *arXiv preprint arXiv:1803.06467*.

[34] Talak, R., Karaman, S., and Modiano, E. [May 2018*a*], Optimizing age of information in wireless networks with perfect channel state information, *in* "Proc. Modeling and Optimization in Mobile, Ad Hoc, and Wireless Networks (WiOpt)," Shanghai, China, pp. 1–8.

[35] Wu, X., Yang, J., and Wu, J. [Mar. 2018], "Optimal status update for age of information minimization with an energy harvesting source," *IEEE Trans. Green Commun. Netw.* **2**(1), 193–204.

[36] Yang, H. H., Arafa, A., Quek, T. Q. S., and Poor, H. V. [Jun. 2021], "Optimizing information freshness in wireless networks: A stochastic geometry approach," *IEEE Trans. Mobile Comput.* **20**(6), 2269–2280.

[37] Yang, H. H., Geraci, G., Quek, T. Q., and Andrews, J. G. [Jul. 2017], "Cell-edge-aware precoding for downlink massive MIMO cellular networks," *IEEE Trans. Signal Process.* **65**(13), 3344–3358.

[38] Yang, H. H., Geraci, G., and Quek, T. Q. S. [2016], "Energy-efficient design of MIMO heterogeneous networks with wireless backhaul," *IEEE Trans. Wireless Commun.* **5**(7), 4914–4927.

[39] Yang, H. H., Lee, J., and Quek, T. Q. S. [Feb. 2016], "Heterogeneous cellular network with energy harvesting-based D2D communication," *IEEE Trans. Wireless Commun.* **15**(2), 1406–1419.

[40] Yang, H. H., and Quek, T. Q. S. [May 2019], "Spatiotemporal analysis for SINR coverage in small cell networks," *IEEE Trans. Commun.* **67**(8), 5520–5531.

[41] Yang, H. H., Wang, Y., and Quek, T. Q. S. [2018], "Delay analysis of random scheduling and round robin in small cell networks," *IEEE Wireless Commun. Lett.* **7**(6), 978–981.

[42] Yates, R. D. [Jun. 2015], Lazy is timely: Status updates by an energy harvesting source, *in* "2015 IEEE International Symposium on Information Theory (ISIT)," Hong Kong, China, pp. 3008–3012.

[43] Zhong, Y., Quek, T. Q. S., and Ge, X. [Jun. 2017], "Heterogeneous cellular networks with spatio-temporal traffic: Delay analysis and scheduling," *IEEE J. Sel. Areas Commun.* **35**(6), 1373–1386.

15 The Age of Channel State Information

Shahab Farazi, Andrew G. Klein, and D. Richard Brown III

While there are many applications motivating the study of age of information, most of the AoI literature considers "information" in an abstract context. Typically, one or more sensors (or sources) sample one or more underlying time-varying processes and provide updates to one or more monitors. In this chapter, we consider "information" in a more specific context: *channel state information* (CSI) in wireless communication systems. As discussed in more detail in what follows, modern wireless communication systems use CSI to efficiently allocate resources and maintain reliable communication links. Stale CSI can lead to inefficient and unreliable communications. In the setting considered here, the "information" is the state of each time-varying wireless channel in the network. In the simplest possible setting with a single antenna transmitter, a single-antenna receiver, and a flat fading channel, the channel state is simply a complex number representing the magnitude and phase of the wireless link. More complicated examples arise from settings such as wireless mesh networks with multiple transmitter–receiver pairs, frequency-selective fading, multiple antenna transceivers, and/or bidirectional wireless links.

The rest of this chapter is organized as follows. First, we provide several examples of why fresh knowledge of CSI is crucial in modern wireless communication systems. We then develop a general system model for studying the age of channel state information, which we call AoCSI, and discuss how AoCSI is different from the typical abstract setting considered in the AoI literature. Fundamental limits and achievability results are then discussed for two canonical graph structures (complete graphs, i.e., fully connected networks, and cycle graphs, i.e., ring networks) under the assumption that all nodes in the network wish to maintain global tables of CSI. We then conclude with a summary and a discussion of some potential future directions for AoCSI.

15.1 Applications Requiring Fresh Channel State Information

In modern wireless networks, knowledge of CSI is critically important for achieving increased data rates, reduced interference, and improved energy efficiency. There are two sources of error in the CSI at each node in a network: (i) channel estimation error, typically governed by fundamental bounds such as the Cramer–Rao lower bound (CRLB) and (ii) error caused by time-variation and staleness (i.e., the delay from when the time-varying channel was estimated and the current time). While the type (i) error in CSI is very well understood, including the impact of estimation error as well as the design of optimal estimators that minimize this type of error, the type (ii) error has received far less attention and is the focus here. Generally, more advanced communication techniques that maximize network performance are also more demanding in terms of the amount and accuracy of CSI knowledge required throughout the network. We begin with a brief discussion of CSI knowledge in classical wireless networks consisting primarily of point-to-point transmissions, and then discuss wireless techniques that require more comprehensive, global CSI knowledge.

Almost all of today's wireless network standards, including WiFi, require CSI knowledge at each receiver (CSIR) due to the prevalence of coherent modulation formats used for their superior spectral efficiency and data rates [1]. Moreover, the vast majority of wireless networks make use of at least some CSI at the transmitter (CSIT), as well. For example, it is commonplace for wireless networks to employ adaptive modulation and coding, where the transmitter uses CSI knowledge to optimize its transmission rate and error performance [2]. Typically, this is accomplished using a form of CSI called the *channel quality indicator* (CQI), which is obtained through a feedback channel from the receiver to the transmitter, and subsequently used by the transmitter to select an appropriate modulation format and coding scheme. In wireless networks with multiple antennas that employ multiple-input multiple-output (MIMO) techniques, CSIT knowledge between all pairs of transmit and receive antennas allows for coherent transmission techniques like beamforming and spatial multiplexing [3], as well as interference mitigation techniques [4–6]. While the value of CSIT is well established in the literature, there are also many examples of systems where the nodes in the wireless network benefit from having a more comprehensive, global view of the channel states in the network beyond just CSIT.

The design of modern communication systems has expanded beyond centralized point-to-point systems with an access point or base station to include more sophisticated multipoint-to-multipoint systems where a set of nodes may be cooperating in the transmission or reception of wireless packets. Such advanced techniques have been shown to yield significant gains in network throughput, reliability, and power efficiency while minimizing interference. However, these techniques often require knowledge of CSI that goes beyond simple CSIT and CSIR between pairs of communicating nodes, and require the nodes to have knowledge of channels to which they are not directly connected. Examples that serve to motivate the need for more comprehensive CSI knowledge include:

- **Cooperative relaying.** In cooperative relaying, the transmitting node is assisted by one or more other nodes that act as relays, and can be used to increase the diversity in systems with fading channels, particularly when the nodes have limited diversity to counteract that fading (e.g., due to the use of single antenna nodes). This approach can be used to form a virtual MIMO system and allows receiving nodes to exploit the increased diversity of the system, thereby improving throughput and reliability. The receiving nodes decode the combined signal from the relays, as well as the direct signal from the transmitting node. In this scenario, the transmitting node benefits from having knowledge of the CSI between the relays and the destination [7–9]. Moreover, as information is passed through the network, nodes may change roles – acting as a transmitter, relay, or destination at various times – and thus global knowledge of CSI is required in these systems. Relay selection strategies, too, generally require the source to know the CSI between relay and receiver, and in multi-relay systems optimum transmission schemes may require the relays to have knowledge of all CSI between transmitter and relays, as well as relays and destinations [10]. In extensions such as cooperative networks with dynamic relay pairing, stable matchings require global channel state knowledge [11].

- **Distributed beamforming**. Distributed beamforming is related to cooperative relaying in that multiple nodes participate in the transmission. In the most classical form of distributed beamforming, sources simultaneously transmit a common message and control the phase of their transmissions so that the signals combine constructively at the destination. While this approach can be achieved with CSIT [12, 13], more general distributed transmission schemes such as zero-forcing beamforming [14], nullforming [15], and interference alignment [16, 17] require each transmitting node to know all source-destination CSI to be able to compute the desired precoding vector. Similarly, optimum combining in distributed reception systems requires the fusion center to know all source-destination CSI [18, 19].
- **Routing/cross-layer design**. In multi-hop networks, the optimization of end-to-end data rate and scheduling generally requires global CSI [20, 21]. Routing performance in multi-hop networks is also significantly improved if all nodes have comprehensive knowledge of CSI that includes channels to which they are not directly connected [22, 23].

These examples serve to illustrate that a more comprehensive view of CSI in wireless networks allows nodes to dynamically adapt their roles, form efficient cooperative structures, and effectively use the available network resources. Moreover, since CSI is generally time-varying, it is important that each node's estimate of the CSI in the network not become too stale. If CSI becomes stale, techniques that rely on accurate CSI will not perform as intended and network resources will be used inefficiently.

15.2 Related Work

While the age of channel state information has received less attention than more general studies of age, we summarize here some related work. The "channel information age" of a setting with a single transmitter–receiver pair with nonreciprocal channels and a two-state Gilbert–Elliot (good/bad) channel model was studied in [24, 25]. Similar to the studies discussed in more detail in what follows, the receiver estimates the current channel state and disseminates CSI estimates via periodic feedback to the transmitter. A utility function was formulated to study the trade-off between channel estimation errors and the cost of feedback.

Recent interest in massive MIMO cellular communication systems and their reliance on accurate channel state information has led to several studies of the effect of channel aging in this setting [26–28]. In all these studies, the system model is a single base station serving multiple users. Reciprocal channels are assumed and, since there is just one hop between the base station and each user, the base station can estimate the channels from uplink transmissions and there is no need for CSI dissemination. In [26], a Jakes channel model was used to study the combined effects of channel estimation errors and aging. In [27], an autoregressive channel model was considered and uplink and downlink sum rates were analyzed with and without channel prediction. In [28], a scheme was developed to tune the CSI updating frequency of each user based on the temporal correlation of the user's channel. The trade-off between CSI updating frequency and downlink sum throughput was then optimized.

15.3 System Model

We consider a general wireless network setting where the nodes in the network and their connectivity are modeled by a time-invariant directed graph $\mathcal{G} = (\mathcal{V}, \mathcal{E})$, where the vertex set $\mathcal{V}$ represents the wireless nodes and the edge set $\mathcal{E}$ represents the channels between the nodes in the network. Edge $e_{i,j}$ is in set $\mathcal{E}$ when transmissions can be reliably delivered from node i directly to node j in the absence of interference. We assume that nodes have equal transmission range, hence $e_{i,j} \in \mathcal{E} \Leftrightarrow e_{j,i} \in \mathcal{E}$. We denote the number of nodes as $N = |\mathcal{V}|$, the number of edges as $L = |\mathcal{E}|$, and the set of one-hop neighbors of node i as $\mathcal{N}_1(i)$, that is, $j \in \mathcal{N}_1(i) \Leftrightarrow e_{i,j} \in \mathcal{E}$. Finally, we assume that the network is connected, that is, there exists a path between any two distinct vertices $i,j \in \mathcal{V}$.

While the graph is assumed to be time-invariant, each edge $e_{i,j} \in \mathcal{E}$ is associated with a *time-varying channel* denoted as $h_{i,j}(t)\colon \mathbb{R} \mapsto \mathbb{C}$. For ease of exposition, we assume scalar channels. The system model can, however, be straightforwardly extended to higher-dimensional channels to capture the effects of frequency selectivity, multiple transmit antennas, and/or multiple receive antennas. It is important to note that *these channels are the information* in the AoCSI setting. There is no assumption of an exogenous underlying time-varying process that the nodes sample as in the typical AoI setting; rather, the collection of time-varying channels between the nodes is the information and the nodes seek to obtain timely estimates of this information by exchanging wireless messages subject to interference constraints. A different but somewhat similar setting was considered in [29], where the presence/absence of an edge in the graph $\mathcal{G}$ was the underlying information of interest.

A unique feature of the AoCSI setting is that, for reasons previously discussed, every node in the network seeks to obtain fresh CSI and is thus a monitor. This implies that each node in the network may have CSI estimates with different time stamps. Consequently, the age of the CSI estimates at one node may differ from the age of the CSI estimates at another node. Hence, each node maintains a local table of L CSI estimates, and the total number of CSI estimates with corresponding ages in the network is NL. Each node updates its local table of L channel estimates and corresponding ages when it receives a message in two ways: (i) direct channel measurements through standard

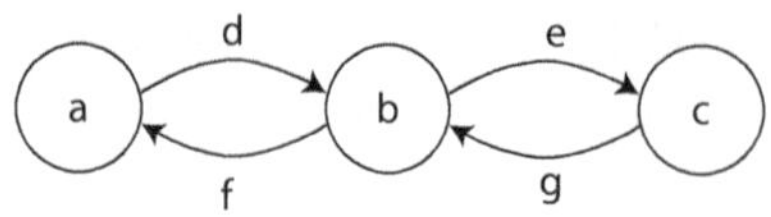

Figure 15.1 Three-node path network example.

physical layer channel estimation techniques and (ii) indirect observations of channels through CSI dissemination from other nodes in the network. CSI dissemination is necessary in this setting since each node can only directly estimate a subset of the channels in the network. As a simple example, consider the three-node path network in Figure 15.1. Node 3 can only directly estimate $h_{2,3}(t)$ when it receives a message from node 2. Node 3 requires CSI dissemination from node 2 to update its estimate of the other channels in the network. Even if the channels are assumed to be reciprocal, that is, $h_{i,j}(t) = h_{j,i}(t)$ for all i, j, node 3 still requires node 2 to disseminate an estimate of the $h_{1,2}(t) = h_{2,1}(t)$ channel since node 3 has no means to directly estimate this channel.

We assume a time-slotted system with a discrete time index $n = 0, 1, \ldots$ such that, during each timeslot, one or more nodes each transmit a packet including M disseminated CSI estimates as well as possibly additional data and overhead. We also assume reliable broadcast transmissions such that a packet transmitted by node i is received by all one-hop neighbors of node i. Figure 15.2 represents the general structure of a packet exchanged among the nodes in the network. All packets are assumed to be received reliably. Each packet contains $M \geq 1$ disseminated CSI estimates, each with a length of one word, as well as $D \geq 0$ words of data and overhead. The total packet length is $P = D+M$ words. Although Figure 15.2 shows a particular packet structure, the position of the overhead, data, and disseminated CSI within any packet is arbitrary.

For a packet transmitted by node i, each node $j \in \mathcal{N}_1(i)$ does two things:

1. It directly estimates the channel $h_{i,j}(t)$, which can be obtained via a known training sequence in the packet, for example, a known preamble embedded in the overhead, and/or through blind channel estimation techniques.
2. It extracts the disseminated CSI in the packet, compares the time stamps to the time stamps in its local CSI table, and then updates any "staler" CSI estimates and corresponding time stamps in its local table.

Note that all disseminated CSI includes time stamps of when each estimate was obtained. This allows each node to determine if the disseminated CSI is fresher than any CSI currently in its table.

Since the system is timeslotted and each timeslot has a duration of P words (or rP seconds if the transmission rate r in words per second is specified), the age dynamics can be analyzed in discrete time. In what follows, we define the age of a CSI estimate in units of words. Note that the notation $\hat{h}_{i,j}^{(k)}[n']$ corresponds to an estimate of the channel $h_{i,j}(t)$ stored in the local table at node k with discrete time stamp n', that is, sampled at time $t = n'rP$.

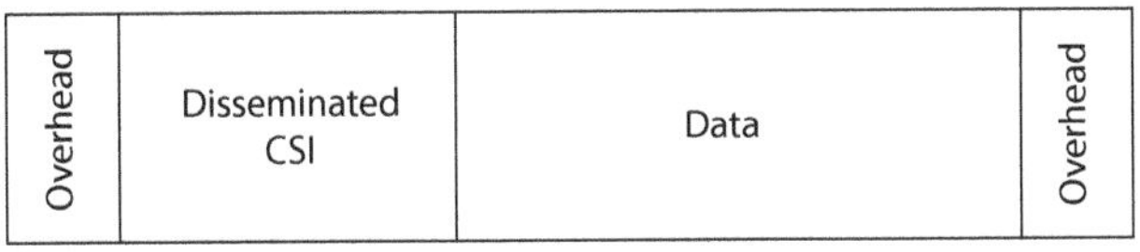

Figure 15.2 Example packet showing overhead, data, and CSI dissemination. The total packet length is $P = D + M$ words.

DEFINITION 15.1 (Age) The age $\Delta_{i,j}^{(k)}[n]$ of the CSI estimate $\hat{h}_{i,j}^{(k)}[n']$ with time stamp n' at time $n \geq n'$ is $(n - n')P$ words.

Directly estimated CSI has an age of zero in the timeslot in which it is estimated. All CSI obtained through dissemination (not directly estimated) has a minimum age of P words. If a given CSI estimate is not updated in a timeslot, either due to direct estimation or dissemination, then its age increases by P words in that timeslot.

Using any ordering of the collection of individual ages $\{\Delta_{i,j}^{(k)}[n]\}$, we can denote a global age vector $\boldsymbol{\Delta}[n] \colon \mathbb{Z} \mapsto \mathbb{Z}^{NL}$. It is not difficult to see that, given a schedule with transmitting node and disseminated channel indices, the age vector obeys a simple time-varying update equation

$$\boldsymbol{\Delta}[n+1] = \boldsymbol{A}[n]\left(\boldsymbol{\Delta}[n] + \mathbf{1}P\right), \tag{15.1}$$

where $\boldsymbol{A}[n] \in \mathbb{Z}^{NL \times NL}$ is a time-varying update matrix with entries equal to either zero or one.

As an explicit example, and assuming channel reciprocity for simplicity so that there are only $L = 2$ channels ($h_{1,2}(t) = h_{2,1}(t)$ and $h_{2,3}(t) = h_{3,2}(t)$) and two corresponding ages at each node in the network, we can define the global age vector for the network in Figure 15.1 as

$$\boldsymbol{\Delta}[n] = \left[\Delta_{1,2}^{(1)}[n], \Delta_{2,3}^{(1)}[n], \Delta_{1,2}^{(2)}[n], \Delta_{2,3}^{(2)}[n], \Delta_{1,2}^{(3)}[n], \Delta_{2,3}^{(3)}[n]\right]^{\top}.$$

Suppose node 2 transmits a packet at time n with disseminated CSI $\hat{h}_{1,2}^{(2)}[n']$ and $n' < n$. This message is received by both node 1 and node 3 since both nodes are in $\mathcal{N}_1(2)$. Node 1 ignores the disseminated CSI since it can directly estimate $h_{2,1}(t) = h_{1,2}(t)$ from the transmitted packet at time n. Node 3 does two things: (i) it directly estimates $h_{2,3}(t) = h_{3,2}(t)$ from the transmitted packet at time n and (ii) it checks the time stamp of its local CSI estimate $\hat{h}_{1,2}^{(3)}[n'']$ and updates this CSI estimate to the disseminated CSI $\hat{h}_{1,2}^{(3)}[n']$ if $n'' < n'$. Assuming the local estimate at node 3 is staler than the disseminated CSI, that is, $n'' < n'$, then we have

$$A[n] = \begin{bmatrix} 0 & 0 & 0 & 0 & 0 & 0 \\ 0 & 1 & 0 & 0 & 0 & 0 \\ 0 & 0 & 1 & 0 & 0 & 0 \\ 0 & 0 & 0 & 1 & 0 & 0 \\ 0 & 0 & 1 & 0 & 0 & 0 \\ 0 & 0 & 0 & 0 & 0 & 0 \end{bmatrix} = \begin{bmatrix} \mathbf{0}^\top \\ \mathbf{e}_2^\top \\ \mathbf{e}_3^\top \\ \mathbf{e}_4^\top \\ \mathbf{e}_3^\top \\ \mathbf{0}^\top \end{bmatrix}$$

where $\mathbf{e}_m$ is the mth standard unit vector. Rows 3–4 of $\mathbf{A}[n]$ reflect the fact that node 2 was transmitting and, as such, no updates are made to its local CSI table (the age of all CSI at node 2 increases by P words). Row 1, being all zeros, reflects the direct estimation $h_{2,1}(t) = h_{1,2}(t)$ at node 1. Direct estimation resets the age to zero. Row 2, on the other hand, is similar to rows 3–4 in that node 1 does not update its estimate of the $h_{2,3}(t) = h_{3,2}(t)$ channel. Row 5 reflects the indirect estimation of $h_{1,2}(t)$ at node 3 from the CSI disseminated by node 2, that is, node 3 updates its local estimate of $h_{1,2}(t)$ with the fresher estimate in the disseminated CSI and now has the same estimate and the same age as node 2. Finally, similar to row 1, row 6 reflects the direct estimation $h_{2,3}(t) = h_{3,2}(t)$ at node 3. This example illustrates some basic properties of $\mathbf{A}[n]$ that will be discussed more generally in Section 15.3.2.

15.3.1 Age Statistics

In this section, we define the maximum and average age statistics used in the remainder of this chapter.

DEFINITION 15.2 (Maximum age at time n) The maximum age at time n is defined as

$$\Delta_{\max}[n] = \max \boldsymbol{\Delta}[n].$$

DEFINITION 15.3 (Average age at time n) The average age at time n is defined as

$$\Delta_{\text{avg}}[n] = \frac{1}{NL}\mathbf{1}^\top \boldsymbol{\Delta}[n].$$

DEFINITION 15.4 (Maximum age) The maximum age $\Delta_{\max}$ is defined as

$$\Delta_{\max} = \max_{n \geq \bar{n}} \Delta_{\max}[n]$$

for $\bar{n}$ sufficiently large such that the effect of any initial age state $\boldsymbol{\Delta}[0]$ can be ignored.

DEFINITION 15.5 (Average age) The average age Δ_{avg} is defined as

$$\Delta_{\text{avg}} = \mathrm{E}\left[\Delta_{\text{avg}}[n]\right],$$

where the expectation is over $n \geq \bar{n}$ for $\bar{n}$ sufficiently large such that the effect of any initial age state $\mathbf{\Delta}[0]$ can be ignored.

15.3.2 General Properties of $A[n]$

From (15.1) and the description of how each node updates its local CSI table through direct estimation and disseminated CSI, we can derive certain basic properties of $\boldsymbol{A}[n]$ that are useful in the analysis of average and maximum ages. To allow for the possibility of multiple simultaneous transmissions, we denote the transmitter set at time n as $\mathcal{T}[n] = \{i_1, \ldots, i_K\}$. We assume (i) there is no overlap in the one-hop neighborhoods of all members of the $\mathcal{T}[n]$ and (ii) the union of the one-hop neighborhoods $\mathcal{N}[n] = \cup_{k=1}^{K} \mathcal{N}_1(i_k)$ of all transmitters in $\mathcal{T}[n]$ has no intersection with $\mathcal{T}[n]$. The former assumption is consistent with satisfying an interference constraint such that no node simultaneously receives multiple packets. The latter assumption is consistent with a node not being able to simultaneously transmit and receive. We denote $\eta[n] = |\mathcal{N}[n]| = \sum_{k=1}^{K} |\mathcal{N}_1(i_k)| \leq N - 1$ as the total number of nodes receiving a packet at time n. It can be shown that

1. Each row of $\boldsymbol{A}[n]$ is either equal to zero or has a single non-zero entry equal to one.
2. Row m of $\boldsymbol{A}[n]$ is equal to zero if the corresponding CSI estimate is directly estimated in timelot n.
3. Exactly $\eta[n]$ rows of $\boldsymbol{A}[n]$ are zero. This is a consequence of the fact that $\eta[n]$ nodes receive a packet at time n and use this packet to directly estimate the channel from the transmitting node.
4. Row m of $\boldsymbol{A}[n]$ is equal to $\boldsymbol{e}_m^\top$, that is, the transposed mth standard unit vector, if the corresponding CSI estimate is not updated.
5. Row m of $\boldsymbol{A}[n]$ is equal to $\boldsymbol{e}_\ell^\top$, that is, the transposed ℓ th standard unit vector, if the corresponding CSI estimate is updated to match the disseminated CSI estimate corresponding to row ℓ. Given a transmitter set $\mathcal{T}[n]$, the M disseminated CSI estimates are received by $\eta[n]$ nodes and it is easy to see that there are at most $M\eta[n]$ such rows in $\boldsymbol{A}[n]$. In other words, there are at most $M\eta[n]$ disseminated CSI estimates in timeslot n.

As discussed in more detail in Section 15.4.1, these properties can be used to characterize the dimension of the nullspace of $\boldsymbol{A}[n]$, which can then be used to develop fundamental bounds on the time it takes to update all CSI estimates in the global CSI vector $\mathbf{\Delta}[n]$ as well as fundamental bounds on the average and maximum ages of CSI in this setting.

We conclude this section by noting that the number of disseminated CSI estimates per packet M is a design parameter with interesting trade-offs that will be explored in the following sections. Using a smaller value of M leads to less CSI dissemination per packet, but also results in shorter packets and more frequent transmissions. Conversely, using a larger value of M leads to more CSI dissemination per packet, but also results in longer packets and less frequent transmissions. Hence, in addition to specifying a schedule to efficiently estimate and disseminate CSI to all nodes in the network, M is a parameter that can also be tuned to optimize the age statistics.

15.4 Fully Connected Networks (K_N)

The fully connected setting with reciprocal channels was considered in [30]. In this setting, there are a total of $L = (N^2 - N)/2$ unique channel gains between all pairs of nodes in the network. Only one node can transmit in each timeslot without collisions, hence $\eta[n] = N - 1$ for all n. When node i transmits a packet at time n, all other nodes $j \neq i$ receive the packet reliably. Out of the L CSI estimates in each node's table, $N - 1$ estimates are directly obtained via channel estimation for $i = k$ or $j = k$, while the remaining $L - N + 1$ estimates are indirectly obtained via disseminated CSI for $i, j \neq k$.

To develop an understanding of how a schedule of transmissions with disseminated CSI can update the CSI tables of all nodes in a fully connected network, consider the $N = 3$ node network shown in Figure 15.3. This graph is also called K_3 in the graph theory literature. The schedule shown here assumes $M = 1$ CSI estimates are disseminated in each timeslot and that there is no data or overhead in each packet ($D = 0$). The following is a description of the schedule shown in Figure 15.3 and resulting CSI estimates at each node in the network:

$n = 0$:
: The first packet is transmitted by node 1. Prior to the first packet transmission, there is no knowledge of CSI anywhere in the network. As such, the first packet ($n = 0$) can only be used for estimation, and not dissemination. In this case, node 2 directly estimates the channel $h_{1,2}[0]$ (computing the estimate $h_{1,2}^{(2)}[0]$), and node 3 directly estimates the channel $h_{1,3}[0]$ (computing the estimate $h_{1,3}^{(3)}[0]$). The age of both of these estimates is zero since they were obtained directly from the current packet.

$n = 1$:
: This packet is transmitted by node 2. Since node 2 now has an estimate of the $(1, 2)$ channel, it disseminates $h_{1,2}^{(2)}[0]$ in its packet. Node 1 directly estimates the channel $h_{1,2}[1]$ (computing the estimate $h_{1,2}^{(1)}[1]$) and node 3 directly estimates the channel $h_{2,3}[1]$ (computing the estimate $h_{2,3}^{(3)}[1]$). Additionally, node 3 extracts the disseminated CSI $h_{1,2}^{(2)}[0]$ since it does not have a prior estimate of the $(1, 2)$ channel. Node 1 does not use the disseminated CSI since it is staler than the direct estimate $h_{1,2}^{(1)}[1]$. To summarize, and as shown in Figure 15.3, after node 2 transmits its packet, node 1

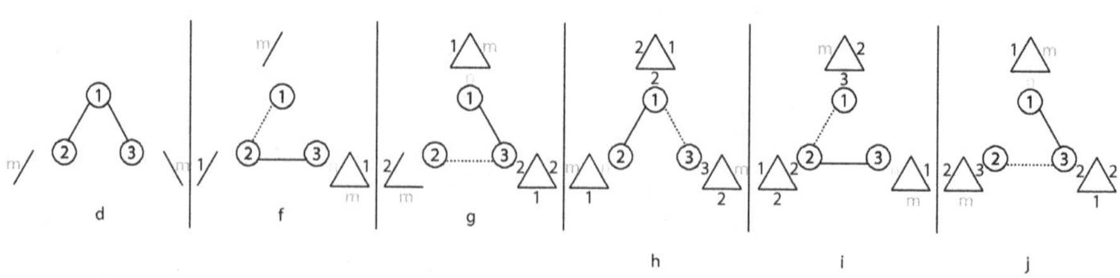

Figure 15.3 Example schedule for three-node fully connected network. Numbers on edges indicate the age of CSI estimates locally at each node. Boldface numbers indicate CSI estimates that have been refreshed through direct estimation. Underlined numbers indicate CSI estimates that have been refreshed through dissemination.

Table 15.1 Example Deterministic schedule for K_3.

Time	Transmitting node	Disseminated channel
$n = 0, 3, 6, \ldots$	1	(1,3)
$n = 1, 4, 7, \ldots$	2	(1,2)
$n = 2, 5, 8, \ldots$	3	(2,3)

has a current estimate of the $(1,2)$ channel, node 2 has a stale estimate of the $(1,2)$ channel, and node 3 has stale estimates of the $(1,2)$ and $(1,3)$ channels as well as a current estimate of the $(2,3)$ channel. Estimates of all channels in the network now exist at node 3, but two more packets are required for node 1 and node 2 to each have complete CSI tables.

$n = 2$:
This packet is transmitted by node 3, and node 3 disseminates $h^{(3)}_{2,3}[1]$. Node 1 directly estimates the channel $h_{1,3}[1]$ (computing the estimate $h^{(1)}_{1,3}[2]$), and node 2 directly estimates the channel $h_{2,3}[2]$ (computing the estimate $h^{(2)}_{2,3}[2]$). Additionally, node 1 extracts the disseminated CSI $h^{(3)}_{2,3}[1]$. Node 1 now has a complete CSI table.

$n = 3, \ldots$:
The nodes repeat this same sequence of steps, which are summarized in the three-round schedule shown in Table 15.1. This periodic 3-round schedule is shown in Figure 15.3, where the time of the most recent information is indicated locally on each of the figures for each node.

For the three-node network example in Figure 15.3 (omitting the time index for notational convenience) the global age vector can be written as

$$\Delta = \left[\Delta^{(1)}_{1,2}, \Delta^{(1)}_{1,3}, \Delta^{(1)}_{2,3}, \Delta^{(2)}_{1,2}, \Delta^{(2)}_{1,3}, \Delta^{(2)}_{2,3}, \Delta^{(3)}_{1,2}, \Delta^{(3)}_{1,3}, \Delta^{(3)}_{2,3}\right]^\top . \tag{15.2}$$

Suppose node 1 disseminates the $(1,3)$ link at time 3ℓ, node 2 disseminates the $(1,2)$ link at time $3\ell + 1$, and node 3 disseminates the $(2,3)$ link at time $3\ell + 2$ for $\ell = 0, 1, 2, \ldots$. Then, using the same notation as our previous three-node path example, we have

$$A[3\ell] = \begin{bmatrix} \boldsymbol{e}_1^\top \\ \boldsymbol{e}_2^\top \\ \boldsymbol{e}_3^\top \\ \mathbf{0}^\top \\ \boldsymbol{e}_2^\top \\ \boldsymbol{e}_6^\top \\ \boldsymbol{e}_7^\top \\ \mathbf{0}^\top \\ \boldsymbol{e}_9^\top \end{bmatrix}, \quad A[3\ell+1] = \begin{bmatrix} \mathbf{0}^\top \\ \boldsymbol{e}_2^\top \\ \boldsymbol{e}_3^\top \\ \boldsymbol{e}_4^\top \\ \boldsymbol{e}_5^\top \\ \boldsymbol{e}_6^\top \\ \boldsymbol{e}_4^\top \\ \boldsymbol{e}_8^\top \\ \mathbf{0}^\top \end{bmatrix}, \quad A[3\ell+2] = \begin{bmatrix} \boldsymbol{e}_1^\top \\ \mathbf{0}^\top \\ \boldsymbol{e}_9^\top \\ \boldsymbol{e}_4^\top \\ \boldsymbol{e}_5^\top \\ \mathbf{0}^\top \\ \boldsymbol{e}_7^\top \\ \boldsymbol{e}_8^\top \\ \boldsymbol{e}_9^\top \end{bmatrix}. \tag{15.3}$$

Rows 1–3 of $\boldsymbol{A}[3\ell]$ reflect the fact that node 1 was transmitting and, as such, no updates are made to its local CSI table (the age of all CSI at node 1 increases by P). Row 4, being all zeros, reflects the direct estimation of the $(1,2)$ link at node 2. Row 5 reflects the indirect estimation of the $(1,3)$ link at node 2 from the CSI disseminated by node 1, that is, node 2 now has the same estimate and the same age of the $(1,3)$ link as node 1. The remaining rows are similar, but it is worth mentioning row 8. In this case, node 3 receives the disseminated CSI of the $(1,3)$ link from node 1 but also directly estimates the $(1,3)$ link. Since the direct estimate is fresher than any disseminated estimate, node 3 ignores the disseminated CSI. This is reflected in the all-zero row 8. Similar observations can be made for $\boldsymbol{A}[3\ell+1]$ and $\boldsymbol{A}[3\ell+2]$.

15.4.1 Fundamental Bounds on AoCSI for K_N with Reciprocal Channels

In this section theoretical lower bounds on the maximum and average AoCSI metrics are reviewed. The maximum age was lower bounded in [30] for the two extreme cases of $M \in \{1, N-1\}$ by

$$\Delta_{\text{max}} \geq \Delta^*_{\text{max}} = \begin{cases} L(D+1), & M=1 \\ (N-1)(D+N-1), & M=N-1 \end{cases}. \tag{15.4}$$

The proof follows from the observation that the age of disseminated estimates is always at least P words. Further, every one of the L channel gains must be indirectly estimated by some nodes in the network. Since L packets are required to be transmitted to disseminate all of the L channel gains, and they each have an age of at least P words, the result is obtained. Similarly the average age was lower bounded in [30] by

$$\Delta_{\text{avg}} \geq \Delta^*_{\text{avg}} = \begin{cases} \frac{(N^3-3N^2+8N-8)}{4N}(D+1), & M=1 \\ \frac{(2N^2-2N-1)}{3N}(D+N-1), & M=N-1 \end{cases}. \tag{15.5}$$

Using the time-varying update equation in (15.1), the work in [31] generalizes the derived lower bounds in [30] to an arbitrary number of CSI estimates disseminated per packet, that is, $M \in \{1, 2, \ldots, N-1\}$. The maximum age is shown to be lower bounded by

$$\Delta_{\text{max}} \geq \Delta^*_{\text{max}} = \left(\Delta^* - 1\right) P, \tag{15.6}$$

where $\Delta^* = \lceil \Delta \rceil$ and

$$\Delta = \frac{NL}{N-1+(N-2)M}. \tag{15.7}$$

Similarly, the average age was shown to be lower bounded by

$$\Delta_{\text{avg}} \geq \Delta^*_{\text{avg}} = \lambda(\Delta^* - 1)P, \tag{15.8}$$

where

$$\lambda \triangleq 1 - \frac{\Delta^*}{2\Delta}. \tag{15.9}$$

Recall from the properties of $\boldsymbol{A}[n]$ in Section 15.3.2 that $M\eta[n]$ CSI estimates are disseminated in timeslot n where $\eta[n]$ is the number of nodes receiving a transmission in timeslot n. In the complete graph case, we have $\eta[n] = N - 1$ for all n. Moreover, due to the fact that some of the disseminated CSI is also directly estimated (and hence ignored, as illustrated in the earlier examples), there will be at most $M\eta[n] - M = M(N-2)$ "useful" disseminated CSI estimates in timeslot n. This observation allows us to characterize the dimension of the nullspace of $\boldsymbol{A}[n]$ for the complete graph case. It can be shown that the dimension of the nullspace of $\boldsymbol{A}[n]$, that is, nullity($\boldsymbol{A}[n]$), satisfies

$$\eta[n] \leq \text{nullity}(\boldsymbol{A}[n]) \leq (N-1) + M(N-2), \tag{15.10}$$

where the $N-1$ term corresponds to the number of rows of $\boldsymbol{A}[n]$ that must be zero due to direct CSI estimation at the $N-1$ nodes receiving a transmission in each timeslot. The inequality is due to the $M(N-2)$ term, which corresponds to the maximum "useful" disseminated CSI. The upper bound in (15.10) is tight, as seen from the example in Figure 15.3 where the nullity of $\boldsymbol{A}[n]$ is 3, and is the key result that facilitates the development of the bounds on the maximum and average age in [31]. Further, observe that from (15.1) and given an initial state at time n_0, the age vector can be written as

$$\boldsymbol{\Delta}[n] = \boldsymbol{\Phi}[n, n_0]\boldsymbol{\Delta}[n_0] + P\sum_{t=n_0}^{n-1} \boldsymbol{\Phi}[n,t]\mathbf{1}, \tag{15.11}$$

where

$$\boldsymbol{\Phi}[n,\tau] = \begin{cases} \boldsymbol{A}[n-1]\boldsymbol{A}[n-2]\cdots\boldsymbol{A}[\tau] & n-\tau > 0 \\ \boldsymbol{I}_{NL} & n-\tau = 0 \\ \text{undefined} & n-\tau < 0. \end{cases}$$

The $\boldsymbol{\Phi}$ matrix has several interesting properties. It was shown that $\mathbf{1}^\top\boldsymbol{\Phi}[n,\tau]\mathbf{1} > 0$ for $0 \leq n-\tau < \Delta^*$. Further, $\mathbf{0} \preceq \boldsymbol{\Phi}[n,t-1]\mathbf{1} \preceq \boldsymbol{\Phi}[n,t]\mathbf{1} \preceq \mathbf{1}$ for all $n \geq t$, where $\preceq$ corresponds to element-wise inequality. Using these properties, the lower bounds in [31] shown previously were obtained.

15.4.2 CSI Dissemination Schedule Design for K_N with Reciprocal Channels

In this section explicit schedules for efficient dissemination of global CSI are reviewed. For $M = 1$, since finding efficient deterministic schedules relies on traversals of K_N, the schedule design was divided into two cases: (1) N is odd, (2) N is even. For odd N, it was shown that a deterministic schedule where the node transmission follows an Eulerian tour and the transmitting node always disseminates its freshest CSI estimate, the lower bound on the maximum age is achieved. Algorithm 1 is a schedule design for $M = 1$ and N odd.

Algorithm 1 ($M = 1$ and N odd).

Node transmission order follows an Eulerian tour composed of $(N-1)/2$ edge-disjoint Hamiltonian cycles of K_N. Let H_0 be the length N zigzag Hamiltonian cycle

$$H_0 = \left\{1, N-1, 2, N-2, .., \frac{N-1}{2}, \frac{N+1}{2}, N\right\}$$

and let σ be the permutation whose disjoint cycle decomposition is

$$\sigma = (1, 2, \ldots, N-1)(N).$$

Let $H_m = \sigma^m(H_0)$, which forms a Hamiltonian cycle decomposition of K_N for $m = 0, 1, \ldots, (N-3)/2$.

Since the schedule repeats with period L, consider only times $0 \leq n \leq L-1$. At time n, node i_n disseminates its estimate of the (i_n, j_n) channel. Let $j_n = i_{n-1}$ so node i_n always disseminates the freshest CSI, that is, an estimate of the channel between itself and the last node that transmitted. The transmitting node at each time n is given by

$$i_{mN:mN+N-1} = \sigma^m(H_0).$$

Recall that Eulerian tours do not exist for N even, hence it is not possible to always disseminate the freshest CSI in this case. Based on permitting dissemination of CSI that was estimated two packets ago, Algorithm 2 is a schedule design for $M = 1$ and N even.

Algorithm 2 ($M = 1$ and N even).

Step 1: Construct Eulerian tour of $K_N - I$ composed of edge-disjoint Hamiltonian cycle decompositions.

Let H_0 be the length N zigzag Hamiltonian cycle

$$H_0 = \left\{1, 2, 3, N, 4, N-1, .., \frac{N}{2}+3, \frac{N}{2}+1, \frac{N}{2}+2\right\}$$

and let σ be the permutation whose disjoint cycle decomposition is

$$\sigma = (1)(2, \ldots, N).$$

Let $H_m = \sigma^m(H_0)$, which forms a Hamiltonian cycle decomposition of $K_N - I$ for $m = 0, 1, \ldots, N/2 - 2$.

Construct an intermediate periodic schedule that disseminates channel estimates corresponding to edges of $K_N - I$, and therefore has period $L - N/2$. The sequence i'_n is given by

$$i'_{mN:mN+N-1} = \sigma^m(H_0),$$

where $i'_{k(L-N/2):k(L-N/2)+L-N/2-1} = i'_{0:L-N/2-1}$ for any k.

Step 2: Insert edges from 1-factor to complete K_N.

For $m = 0, 1, \ldots, N/2 - 2$, the edges

$$(e_m, f_m) = \left(N + 2 - \sigma^{2m+3}\left(\frac{N}{2}\right), \sigma^{2m+3}\left(\frac{N}{2}\right)\right)$$

and $(N/2+1, 1)$ form a 1-factor and do not appear in any of the Hamiltonian cycles from step 1. To complete the graph K_N, insert edges of the 1-factor as follows. For the first Hamiltonian cycle H_0, insert two edges via

$$\begin{aligned}(i,j)_0 &= (i',j')_0 \\ (i,j)_1 &= (N/2+1, 1) \\ (i,j)_{2:N-2} &= (i',j')_{1:N-3} \\ (i,j)_{N-1} &= (e_0, f_0) \\ (i,j)_{N:N+1} &= (i',j')_{N-2:N-1},\end{aligned}$$

where node i_n disseminates its estimate of the (i_n, j_n) channel at time n. For $m = 1, \ldots, N/2 - 2$, insert one edge into each of the other Hamiltonian cycles H_m via

$$\begin{aligned}(i,j)_{m(N+1)+1:m(N+1)+\ell} &= (i',j')_{mN:mN+\ell-1} \\ (i,j)_{m(N+1)+\ell+1} &= (e_m, f_m) \\ (i,j)_{m(N+1)+\ell+2:m(N+1)+N+1} &= (i',j')_{mN+\ell:mN+N-1},\end{aligned}$$

where $\ell = N - 2 - 2m$.

It was shown that there exist deterministic schedules with maximum age $\Delta_{\max} \leq \Delta^*_{\max} + P$ and average age $\Delta_{\text{avg}} \leq \Delta^*_{\text{avg}} + 2P/3$. Moreover, it was shown that the schedules generated by Algorithm 1 and Algorithm 2 achieve these bounds.

For the $M = N - 1$ case, a schedule can be developed based on the round-robin transmitting node selection, where each node disseminates its $N-1$ freshest estimates. Algorithm 3 is a schedule design for $M = N - 1$.

Algorithm 3 ($M = N - 1$).

Node transmission order follows a round-robin schedule, and the schedule has period N. That is, the transmitting node at time n is given by $i_n = n + 1$ for $0 \leq n \leq N - 1$, and each node disseminates its $N - 1$ direct estimates. Due to the periodicity, it follows that $i_{kN:kN+N-1} = i_{0:N-1}$ for any k.

It was shown that there exists a deterministic schedule with maximum age $\Delta_{\max} = \Delta^*_{\max}$ and average age $\Delta_{\text{avg}} = \Delta^*_{\text{avg}}$. Moreover, it was shown that the proposed round-robin schedule in Algorithm 3 achieves these bounds with equality.

As discussed in [31], in general for $M \in \{2, 3, 4, \ldots, N-2\}$, due to many possible combinatorics, it is difficult to determine the best choice of transmitting node in each timeslot and also what choice of CSI estimates should be disseminated among the $\binom{N-1}{M} = \frac{(N-1)!}{(N-1-M)!M!}$ different sets of direct estimates that a node can disseminate from its table. A simple one-step greedy algorithm was developed, which generates CSI dissemination schedules based on maximizing the average age improvement for any number of CSI estimates per packet $M \in \{1,2,\ldots,N-1\}$. This schedule is shown in Algorithm 4.

Algorithm 4 (One-Step Greedy Schedule Design, $M \in \{1,2,\ldots,N-1\}$).
Initialize time, $n \leftarrow 0$. Each of the N nodes shares its table with the rest of the network.

• Compute the average age improvement over all combinations, that is, for each of the N nodes compute the average age improvement over all $\binom{N-1}{M}$ possible M direct estimates. Choose the node and its set of M estimates that maximize the average age improvement:

For $i_{\text{node}} = 1 : N$:

  For $j_{\text{set}} = 1 : \binom{N-1}{M}$:

    compute $\Delta_{\text{avg,improve}}(i_{\text{node}}, j_{\text{set}})$, which denotes the average age improvement in the network, given node i_{node} disseminates its $j_{\text{set}}^{\text{th}}$ set of M direct estimates.

Resolve ties: (i) if $\exists$ a tie between two different nodes, select the node that has least recently transmitted, and (ii) if $\exists$ a tie between two different sets of M direct estimates at the same node, select the set that updates a greater number of staler estimates in the network.

The selected node disseminates its selected set of M direct estimates. All nodes update their local tables with any disseminated CSI that is fresher than the CSI currently in their table.

$n \leftarrow n + 1$. Repeat from •.

Figure 15.4 compares the achievable maximum and average age of the greedy schedule with the lower bounds for $N = 8$ versus the number of CSI estimates per packet M for $D = 2$. Since the greedy schedule is sensitive to the initial age values throughout the network, for each M the schedule is run for 1,000 random initializations and the minimum and maximum ages values are chosen as the best and worst achievable ages of the greedy schedule, respectively. The area between the best and worst achievable age of the greedy schedule is shaded, which represents achievable maximum and average age of the greedy schedule for different initializations. It was shown that for small and large values of D, to minimize the achievable maximum age it is optimal to disseminate $M = 1$ and $M = N - 1$ CSI estimates in each packet, respectively. As Figure 15.4 shows, however, for an intermediate value of D, that is, $D = 2$, the achievable maximum age is minimized when $M = 4$ CSI estimates are disseminated in each packet.

The work in [32] studied opportunistic schedules with random transmit node selection from a fixed probability mass function $\{p_1, p_2, \ldots, p_N\}$, where $p_i > 0$ represents the probability that node i is selected to transmit. For $M \in \{1, N-1\}$, closed-form expressions were derived for the achievable average age.

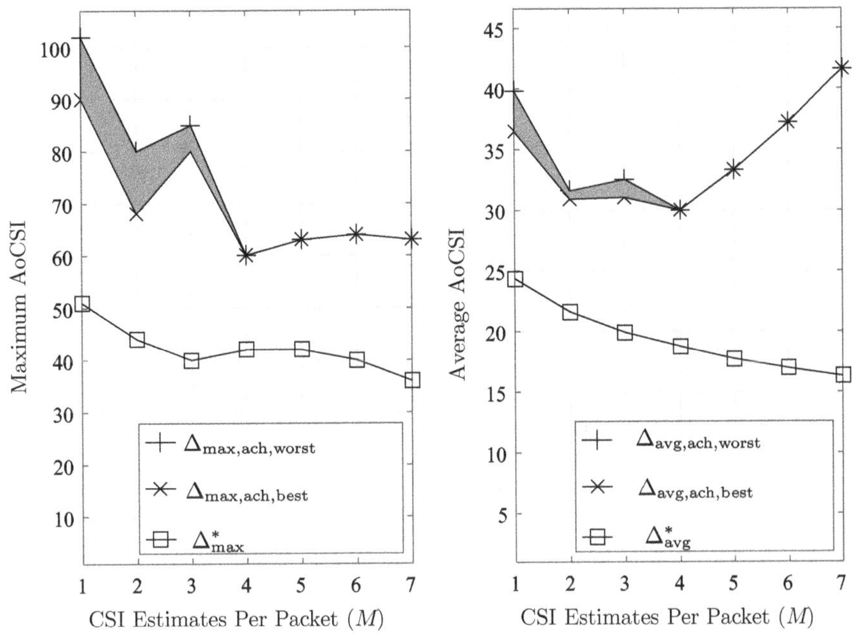

Figure 15.4 Age versus M for $N = 8$ and $D = 2$.

15.4.3 Nonreciprocal Channels

In the case of nonreciprocal channels, where the (i, j) channel gain is not equal to the (j, i) channel gain, the number of CSI estimates in the network doubles [30]. As a result, the maximum and average age double with respect to the reciprocal channels case. For example, for $M = 1$, the work in [30] suggested that the presented schedules can be extended to the nonreciprocal channels case by using the schedules as they are for the first phase, and then in the second phase using a reversed version of the schedule where the sequence of node transmissions arises by traversing the Hamiltonian path in the opposite direction. This leads to doubling the schedule length, and hence the maximum age.

15.5 Ring Networks (C_N)

The age of global CSI in ring networks with reciprocal channels was studied in [33]. In contrast to fully connected networks, where $\mathcal{N}_1(i) = N - 1$ for all i, ring networks have $\mathcal{N}_1(i) = 2$ for all i. The ring network's topology is represented by a cycle graph with $N \geq 3$ vertices and the set $\{(1, 2), (2, 3), \ldots, (N-1, N), (N, 1)\}$ represents the set of all channels in the network. Also, unlike fully connected networks, a nice feature of ring networks is that it is sometimes possible for multiple nodes to transmit simultaneously without collisions. This can reduce the number of timeslots required to update all CSI tables throughout the network compared to the case where only one node transmits at a time.

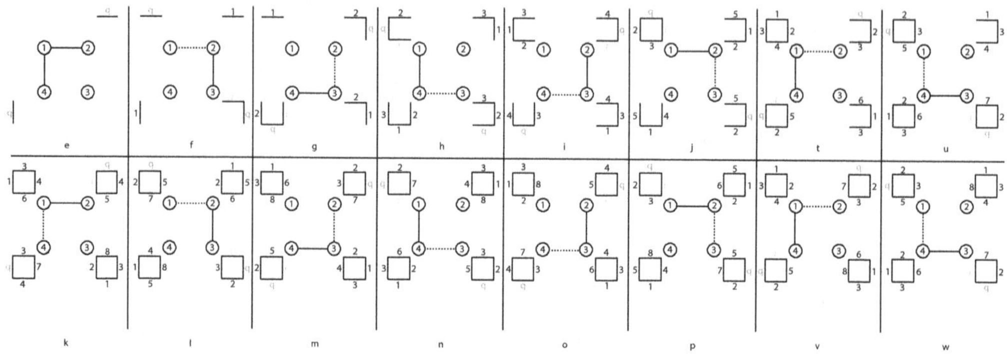

Figure 15.5 Example schedule for four-node ring network with $M = 1$ and one node transmitting in each timeslot. Numbers on edges indicate the age of CSI estimates locally at each node. Boldface numbers indicate CSI estimates that have been refreshed through direct estimation. Underlined numbers indicate CSI estimates that have been refreshed through dissemination.

To develop an understanding of how a schedule of transmissions with disseminated CSI can update the CSI tables of all nodes in a ring network, consider the four-node network and schedule shown in Figure 15.5. This graph is also called C_4 in the graph theory literature. The schedule shown here assumes $M = 1$ CSI estimates are disseminated in each timeslot and that there is no data or overhead in each packet ($D = 0$). Since the schedule is periodic, the age statistics over one period of the schedule are sufficient to calculate the AoCSI. Hence, considering $8 \leq n \leq 15$, the instantaneous maximum age is equal to 8, thus achieving $\Delta_{\max} = 8$. Also, observe that an average age of $\Delta_{\text{avg}} = 3.125$ is achieved. Note that only one node can transmit in each timeslot in this example since two simultaneously transmitting nodes will always cause a collision in a four-node ring network.

15.5.1 Fundamental Bounds on AoCSI for C_N with Reciprocal Channels

Lower bounds on the maximum and average AoCSI in ring networks were developed in [33]. For $M \in \{1, N-1\}$, two extreme scenarios were considered where either (i) one node transmits at a time ($K = 1$) or (ii) the maximum number of nodes transmit at a time without collisions ($K = K_{max} = \lfloor \frac{N}{3} \rfloor$). The maximum age was lower bounded by

$$\Delta_{\max} \geq \Delta^*_{\max} = \begin{cases} \left(\dfrac{2N^2 - 3N - 4}{2}\right) P, & K = 1 \text{ and } M = 1 \\ (2N-4)P, & K = 1 \text{ and } M = N-1 \\ \left(\dfrac{7N-16}{2}\right) P, & K = \lfloor \frac{N}{3} \rfloor \text{ and } M = 1 \\ \left(\dfrac{2N}{3}\right) P, & K = \lfloor \frac{N}{3} \rfloor \text{ and } M = N-1 \end{cases} . \quad (15.12)$$

Table 15.2 Lower bounds on the maximum and average staleness and Algorithm design.

K	M	Δ^*_{max}	Δ^*_{avg}	Algorithm in [33]
1	1	$\left(\frac{2N^2-3N-4}{2}\right)P$	$\left(\frac{2N^2-7N+8}{4}\right)P$	1-a: $N = 2k+1$ and $N \neq 6k-3$ 1-b: $N = 6k-3$ 1-c: $N = 4k+2$ 1-d: $N = 4k$
1	$N-1$	$(2N-4)P$	$\left(\frac{2N^2-5N+4}{2N}\right)P$	2
$\lfloor \frac{N}{3} \rfloor$	1	$\left(\frac{7N-16}{2}\right)P$	$\left(\frac{7N^2-28N+36}{4N}\right)P$	3-a: $N = 6k$ 3-b: $N = 6k-3$ 3-c: $N \neq 3k$ and $N \geq 7$
$\lfloor \frac{N}{3} \rfloor$	$N-1$	$\left(\frac{2N}{3}\right)P$	$\left(\frac{2N^2-N+6}{6N}\right)P$	4

It is interesting to note that all cases except the $K = 1$ and $M = 1$ case have a maximum age that scales linearly with N. The $K = 1$ and $N = 1$ case has a maximum age that scales quadratically with N.

Similarly, the average age was lower bounded by

$$\Delta_{\text{avg}} \geq \Delta^*_{\text{avg}} = \begin{cases} \left(\dfrac{2N^2-7N+8}{4}\right)P, & K = 1 \text{ and } M = 1 \\ \left(\dfrac{2N^2-5N+4}{2N}\right)P, & K = 1 \text{ and } M = N-1 \\ \left(\dfrac{7N^2-28N+36}{4N}\right)P, & K = \lfloor \frac{N}{3} \rfloor \text{ and } M = 1 \\ \left(\dfrac{2N^2-N+6}{6N}\right)P, & K = \lfloor \frac{N}{3} \rfloor \text{ and } M = N-1 \end{cases}. \quad (15.13)$$

Like the maximum age bounds, all cases except the $K = 1$ and $M = 1$ case have an average age that scales linearly with N. (Note the N in the denominator of all cases except the $K = 1$ and $M = 1$ case.) Deriving these bounds involved separately considering the directly and indirectly estimated age parameters throughout the network and separately analyzing the statistics of each of these groups of parameters.

15.5.2 CSI Dissemination Schedule Design for C_N with Reciprocal Channels

As shown in [33], schedule design for ring networks with reciprocal channels (and possible simultaneous transmission) is somewhat more involved than for the fully connected network case. For different choices of parameters K and M, the underlying combinatorics are such that developing schedules achieving near the bounds requires splitting the number of nodes N into different cases. Table 15.2 lists the corresponding algorithm for each case in [33]. In what follows we provide an example of the case when the number of nodes is an integer multiple of six (Algorithm 3-a in [33]). For a thorough description of the other algorithms used for schedule design in ring networks, the reader is referred to [33].

Table 15.3 Schedule for $N = 6,\ K = 2,\ M = 1$.

Timeslot	Tx_1	Disseminated CSI	Tx_2	Disseminated CSI
$n = 0$	1	$(1, 6)$	4	$(3, 4)$
$n = 1$	2	$(1, 6)$	5	$(3, 4)$
$n = 2$	3	$(2, 3)$	6	$(5, 6)$
$n = 3$	4	$(2, 3)$	1	$(5, 6)$
$n = 4$	5	$(4, 5)$	2	$(1, 2)$
$n = 5$	6	$(4, 5)$	3	$(1, 2)$
$n = 6$	5	$(5, 6)$	2	$(2, 3)$
$n = 7$	4	$(5, 6)$	1	$(2, 3)$
$n = 8$	3	$(3, 4)$	6	$(1, 6)$
$n = 9$	2	$(3, 4)$	5	$(1, 6)$
$n = 10$	1	$(1, 2)$	4	$(4, 5)$
$n = 11$	6	$(1, 2)$	3	$(4, 5)$

Algorithm 5 (Algorithm 3-a in [33]) represents the schedule design for $K = K_{max}$, $M = 1$, and $N = 6k$ where $k \in \mathbb{Z}^+$.

Algorithm 5 ($N = 6k, K = K_{max}, M = 1$)

Let $i^{\ell}_{1+m} = \sigma^{1}_{m+\ell\frac{(N-2)}{2}}(\{1, 4, \ldots, N-2\})$ be the simultaneous transmitters' indices for dissemination of the $\sigma^{1}_{\ell\frac{(N-2)}{2}}(\{(N, 1), (3, 4), \ldots, (N-3, N-2)\})$ channels, respectively, during time $1 + m + \ell\frac{(N-2)}{2}$ for $m = \{0, 1, \ldots, \frac{N-4}{2}\}$. Define $H_\ell = \{i^{\ell}_1, i^{\ell}_2, \ldots, i^{\ell}_{\frac{N-2}{2}}\}$ for $\ell = \{0, 1, 2\}$.

Let $i^{\ell}_{1+\frac{3(N-2)}{2}+m} = \sigma^{1}_{-m-\ell\frac{(N-2)}{2}}(\{2, 5, \ldots, N-1\})$ be the simultaneous transmitters' indices for dissemination of the $\sigma^{1}_{-\ell\frac{(N-2)}{2}}(\{(2, 3), (5, 6), \ldots, (N-1, N)\})$ channels, respectively, during time $1 + m + (\ell + 3)\frac{(N-2)}{2}$ for $m = \{0, 1, \ldots, \frac{N-4}{2}\}$. Define $H_{\ell+3} = \{i^{\ell}_{1+\frac{3(N-2)}{2}}, i^{\ell}_{2+\frac{3(N-2)}{2}}, \ldots, i^{\ell}_{2(N-2)}\}$ for $\ell = \{0, 1, 2\}$.

The modulus operator $\sigma^{m}_{k}(i) = 1 + [i + mk \pmod N]$ is used to simplify the notation. The argument i can be a vector, in which case σ operates element-wise. Note that $H = \{H_0, H_1, \ldots, H_5\}$ denotes the node transmission order for this schedule, which has a length of $3(N-2)$ timeslots. The schedule generated by Algorithm 5 achieves maximum age matching the lower bound, that is, $\Delta_{\text{max}} = \Delta^*_{\text{max}}$, and average age of $\Delta_{\text{avg}} \leq \Delta^*_{\text{avg}} + P/2$. This schedule must be repeated with period $3(N-2)$ to maintain its achieved maximum and average ages. Table 15.3 shows one period of the schedule generated by Algorithm 5 for $N = 6$. Observe that, for this example, two nodes (Tx_1 and Tx_2) can transmit simultaneously without collisions in each timeslot.

For $N = 6$, $M = 1$, and $D = 0$, comparing the schedules generated for $K = 1$ and $K = K_{max} = 2$ shows that the latter improves the achieved maximum age by at least 48% and the achieved average age by at least 42%.

15.6 Summary and Future Directions

This chapter considers an application of age of information called AoCSI in which the channel states in a wireless network represent the information of interest and the goal is to maintain fresh estimates of these channel states at each node in the network. Rather than sampling some underlying time-varying process and propagating updates through a queue or graph, the AoCSI setting obtains direct updates of the channels as a by-product of wireless communication through standard physical layer channel estimation techniques. These CSI estimates are then disseminated through the network to provide global snapshots of the CSI to all of the nodes in the network. What makes the AoCSI setting unique is that disseminating some CSI updates and directly sampling/estimating other CSI occur simultaneously. Moreover, as illustrated in this chapter, there are inherent trade-offs on how much CSI should be disseminated in each transmission to minimize the average or maximum age.

While the system model for AoCSI is sufficiently general to allow for general network topologies, only two network topologies have been studied in detail: fully connected networks (complete graphs) and ring networks (cycle graphs). Open problems include the development of fundamental bounds on average and maximum ages and the development of efficient schedules for general network topologies. An important consideration in these settings, first considered for the AoCSI setting with ring networks in [33] and also for a general AoI setting with general network topologies in [34], is the notion of "feasible activation sets." For networks with general topologies, multiple nodes may be able to transmit simultaneously due to frequency reuse, the use of multiple access strategies, or other interference avoidance methods. It is also possible that allowing for collisions (and lost updates) at some nodes in the network may reduce the age at other nodes in the network and result in an overall improvement in average or maximum age. Other open problems include relaxing the requirement to maintain global CSI knowledge at all nodes in the network. For example, there may be settings where nodes value nearby CSI more than distant CSI, which gives rise potentially to a weighted AoCSI optimization problem.

In summary, the initial results on AoCSI discussed in this chapter establish fundamental bounds and efficient schedules for two specific network topologies. These results provide a promising foundation for additional studies in more general network settings.

References

[1] C. R. Johnson Jr., W. A. Sethares, and A. G. Klein, *Software receiver design: Build your own digital communication system in five easy steps*. Cambridge University Press, 2011.

[2] R. A. Berry and R. G. Gallagher, "Communication over fading channels with delay constraints," *IEEE Trans. Inf. Theory*, vol. 48, no. 5, pp. 1135–1149, May 2002.

[3] A. Scaglione, P. Stoica, S. Barbarossa, G. Giannakis, and H. Sampath, "Optimal designs for space-time linear precoders and decoders," *IEEE Trans. Signal Process.*, vol. 50, no. 5, pp. 1051–1064, May 2002.

[4] Q. Spencer, A. Swindlehurst, and M. Haardt, "Zero-forcing methods for downlink spatial multiplexing in multiuser MIMO channels," *IEEE Trans. Signal Process.*, vol. 52, no. 2, pp. 461–471, Feb. 2004.

[5] D. R. Brown III and D. Love, "On the performance of MIMO nullforming with random vector quantization limited feedback," *IEEE Trans. Wireless Commun.*, vol. 13, no. 5, pp. 2884–2893, May 2014.

[6] A. Sendonaris, E. Erkip, and B. Aazhang, "Interference alignment and degrees of freedom of the k-user interference channel," *IEEE Trans. Inf. Theory*, vol. 54, no. 8, pp. 3425–3441, Aug. 2008.

[7] ——, "User cooperation diversity, Part I: System description," *IEEE Trans. Commun.*, vol. 51, no. 11, pp. 1927–1938, Nov. 2003.

[8] J. N. Laneman, D. N. Tse, and G. W. Wornell, "Cooperative diversity in wireless networks: Efficient protocols and outage behavior," *IEEE Trans. Inf. Theory*, vol. 50, no. 12, pp. 3062–3080, Dec. 2004.

[9] Y. Li, Q. Yin, W. Xu, and H.-M. Wang, "On the design of relay selection strategies in regenerative cooperative networks with outdated CSI," *IEEE Trans. Wireless Commun.*, vol. 10, no. 9, pp. 3086–3097, Sep. 2011.

[10] R. Madan, N. Mehta, A. Molisch, and J. Zhang, "Energy-efficient cooperative relaying over fading channels with simple relay selection," *IEEE Trans. Wireless Commun.*, vol. 7, no. 8, pp. 3013–3025, Aug. 2008.

[11] F. Fazel and D. R. Brown III, "On the endogenous formation of energy efficient cooperative wireless networks," in *Proceedings of the 2009 Allerton Conference on Communications, Control and Computing*, Montecello, IL, Sep. 2009, pp. 879–886.

[12] R. Mudumbai, D. R. Brown III, U. Madhow, and H. V. Poor, "Distributed transmit beamforming: Challenges and recent progress," *IEEE Commun. Mag.*, vol. 47, no. 2, pp. 102–110, Feb. 2009.

[13] D. R. Brown III, P. Bidigare, and U. Madhow, "Receiver-coordinated distributed transmit beamforming with kinematic tracking," in *2012 IEEE International Conference on Acoustics, Speech and Signal Processing (ICASSP)*, Mar. 2012, pp. 5209–5212.

[14] D. R. Brown III, P. Bidigare, S. Dasgupta, and U. Madhow, "Receiver-coordinated zero-forcing distributed transmit nullforming," in *2012 IEEE Statistical Signal Processing Workshop (SSP)*, Aug. 2012, pp. 269–272.

[15] D. R. Brown III, U. Madhow, S. Dasgupta, and P. Bidigare, "Receiver-coordinated distributed transmit nullforming with channel state uncertainty," in *Conf. Inf. Sciences and Systems (CISS2012)*, Mar. 2012.

[16] V. Cadambe and S. Jafar, "Interference alignment and spatial degrees of freedom for the k-user interference channel," in *IEEE International Conference on Communications (ICC'08)*, May 2008, pp. 971–975.

[17] ——, "Interference alignment and degrees of freedom of the k-user interference channel," *IEEE Trans. Inf. Theory*, vol. 54, no. 8, pp. 3425–3441, Aug. 2008.

[18] D. R. Brown III, U. Madhow, M. Ni, M. Rebholz, and P. Bidigare, "Distributed reception with hard decision exchanges," *IEEE Trans. Wireless Commun.*, vol. 13, no. 6, pp. 3406–3418, Jun. 2014.

[19] J. Choi, D. Love, D. R. Brown III, and M. Boutin, "Quantized distributed reception for mimo wireless systems using spatial multiplexing," *IEEE Trans. Wireless Commun.*, vol. 14, no. 13, pp. 3537–3548, Jul. 2015.

[20] X. Lin, N. B. Shroff, and R. Srikant, "A tutorial on cross-layer optimization in wireless networks," *IEEE J. Sel. Areas Commun.*, vol. 24, no. 8, pp. 1452–1463, Aug. 2006.

[21] L. Georgiadis, M. J. Neely, and L. Tassiulas, *Resource allocation and cross-layer control in wireless networks*. Now Publishers Inc, 2006.

[22] R. Babaee and N. Beaulieu, "Cross-layer design for multihop wireless relaying networks," *IEEE Trans. Wireless Commun.*, vol. 9, no. 11, pp. 3522–3531, Nov. 2010.

[23] B. Gui, L. Dai, and L. J. Cimini Jr, "Routing strategies in multihop cooperative networks," *IEEE Trans. Wireless Commun.*, vol. 8, no. 2, pp. 843–855, Feb. 2009.

[24] J. Yang, S. C. Draper, and R. Nowak, "Learning the interference graph of a wireless network," *IEEE Trans. Signal Inf. Process. Netw.*, vol. 3, no. 3, pp. 631–646, 2016.

[25] A. G. Klein, S. Farazi, W. He, and D. R. Brown III, "Staleness bounds and efficient protocols for dissemination of global channel state information," *IEEE Trans. Wireless Commun.*, vol. 16, no. 9, pp. 5732–5746, Sep. 2017.

[26] S. Farazi, A. G. Klein, and D. R. Brown III, "Bounds on the age of information for global channel state dissemination in fully-connected networks," in *Proc. Intl. Conf. on Computer Communication and Networks (ICCCN)*, Jul. 2017, pp. 1–7.

[27] S. Farazi, A. G. Klein, and D. R. Brown III, "On the average staleness of global channel state information in wireless networks with random transmit node selection," in *Proc. IEEE Intl. Conf. on Acoustics, Speech and Signal Processing (ICASSP)*, Mar. 2016, pp. 3621–3625.

[28] S. Farazi, D. R. Brown III, and A. G. Klein, "On global channel state estimation and dissemination in ring networks," in *Proc. Asilomar Conf. on Signals, Systems and Computers*, Nov. 2016, pp. 1122–1127.

[29] M. Costa, S. Valentin, and A. Ephremides, "On the age of channel information for a finite-state markov model," in *IEEE International Conference on Communications (ICC)*, 2015, pp. 4101–4106.

[30] ——, "On the age of channel state information for non-reciprocal wireless links," in *IEEE International Symposium on Information Theory (ISIT)*, Jun. 2015, pp. 2356–2360.

[31] K. T. Truong and R. W. Heath, "Effects of channel aging in massive mimo systems," *Journal of Communications and Networks*, vol. 15, no. 4, pp. 338–351, 2013.

[32] C. Kong, C. Zhong, A. K. Papazafeiropoulos, M. Matthaiou, and Z. Zhang, "Sum-rate and power scaling of massive mimo systems with channel aging," *IEEE transactions on communications*, vol. 63, no. 12, pp. 4879–4893, 2015.

[33] R. Deng, Z. Jiang, S. Zhou, and Z. Niu, "How often should csi be updated for massive mimo systems with massive connectivity?" in *GLOBECOM 2017–2017 IEEE Global Communications Conference*. IEEE, 2017, pp. 1–6.

[34] S. Farazi, A. G. Klein, and D. R. Brown, "Fundamental bounds on the age of information in general multi-hop interference networks," in *IEEE INFOCOM 2019-IEEE Conference on Computer Communications Workshops (INFOCOM WKSHPS)*. IEEE, 2019, pp. 96–101.

16 Transmission Preemption for Information Freshness Optimization

Songtao Feng, Boyu Wang, Chenghao Deng, and Jing Yang

Enabled by the proliferation of ubiquitous sensing devices and pervasive wireless data connectivity, real-time status monitoring has become a reality in large-scale cyber-physical systems, such as power grids, manufacturing facilities, and smart transportation systems. However, the unprecedented high-dimensionality and generation rate of the sensing data also impose critical challenges on its timely delivery. In order to measure and ensure the freshness of the information available to the monitor, novel metrics, such as Age of Information, have been introduced and analyzed in various status updating systems. Such metrics exhibit properties that are fundamentally different from traditional network performance metrics, such as throughput and delay, and imply a paradigm shift in the design and analysis of communication protocols for fresh information delivery.

While researchers have examined information freshness in various communication systems from many different aspects, in this chapter, we focus on optimal service preemption policies in point-to-point status updating systems for information freshness optimization. We consider two different information freshness metrics, namely, Age of Information (AoI) and Age of Synchronization (AoS).

16.1 Service Preemption for Timely Information Delivery

The difference between information freshness metrics, such as AoI, and traditional network performance metrics, such as throughput, is manifested in a setting where the average transmission time of an update is long compared with the average interarrival time of the updates. This may lead to a scenario where new updates arrive at the transmitter during the transmission of an "old" update. While preempting the transmission of the old update and starting to transmit the new one degrades the throughput performance, it may actually improve the information freshness at the destination. Whether the source should preempt the current transmission in general requires sophisticated analysis.

16.1.1 Summary of Related Work

The impact of service preemption on information freshness has intrigued researchers recently. A great amount of the effort has been made to analyze the time-average AoI in various queueing models under fixed service preemption principles.

For single-source system updating, analysis has been performed under various assumptions on the update generation and service processes. In [1], it analyzes the AoI for $M/M/1$ Last-Come-First-Serve (LCFS) queues with and without service preemption. Explicit expressions for the time-average AoI have been derived. In [2], it studies the peak age of information (PAoI) in $M/M/1$ queues with random packet losses. Analytical results under the LCFS and retransmission schemes with and without service preemption have been obtained. Those results show that when preemption is allowed in service, the age performance may be improved. When the service time is non-memoryless gamma distributed, reference [3] analyzes AoI and PAoI for the LCFS queues under the preemptive and non-preemptive schemes, and shows that LCFS with preemption may not outperform LCFS without preemption. More recently, [4] generalizes the analysis to $G/G/1/1$ systems when preemption in service is allowed. It derives an exact average AoI expression for $G/G/1/1$ systems, as well as upper bounds for certain update inter-arrival and service time distributions. The probability that the age of an update gets above a threshold in the $M/G/1$ LCFS system with preemption is identified in [5]. Reference [6] investigates an energy harvesting status updating system where the transmitter is powered by energy harvested from the environment. It assumes that the inter-arrival time of the updates and that of the energy units, as well as the service time are all exponential. The analysis shows that preempting packets in service is not always age-optimal, while the best policy depends on the system parameters. In [7], it considers the average AoI for a symbol erasure channel under the $M/G/1/1$ framework where hybrid ARQ (HARQ) schemes are adopted. Numerical results indicate that HARQ without preemption outperforms its counterpart with preemption. Besides, in a similar noisy wireless channel setting, [8] analyzes the PAoI under an HARQ scheme and shows that service preemption is more advantageous when the signal-to-noise ratio is high.

When data streams from multiple sources are considered, [9] studies a multi-stream $M/G/1/1$ queue with preemption, where the transmitter always preempts the packet being served when a new update is generated. It computes a closed-form expression for the average AoI and PAoI of each stream using the detour flow graph method. Reference [10] analyzes the age performance in a multiple-source $M/M/1$ system with preemption by utilizing the stochastic hybrid system method. The same approach has been utilized to analyze a similar setting in [11], where sources are assigned different priorities, and the transmission of an update can be preempted by updates of equal or higher priority. Simulation indicates that the age of the least-priority source performs worse when preemption in service is allowed. Reference [12] further studies a similar setting where each stream may have a size-one buffer to store the preempted update and resume the transmission afterward. Results indicate that the extra buffer may not always improve the time-average AoI of certain streams. In [13], it considers a multi-source probabilistically preemptive $M/PH/1/1$ queue with packet errors, where the update arrival of a source m may preempt the transmission of an update from another source n with certain probability $p_{m,n}$. The exact distributions of the AoI and PAoI for each of the information sources are derived using the theory of Markov fluid queues and sample path arguments. Numerical results indicate that by selecting the optimal preemption probabilities, the AoI may be improved substantially compared with conventional non-preemptive, self-preemptive, or globally preemptive policies.

Besides the just mentioned works that analytically characterize the AoI or PAoI for given preemption policies, there also exist works that aim to explicitly identify the optimal service preemption policies. It is shown in [14] that within a class of stationary Markov policies where the decision only depends on the AoI at the monitor, policies such as always dropping the new update (no preemption) or the old update (preemption) are optimal for certain service time distributions. Reference [15] considers a multi-flow system where the packet generation and arrival times are synchronized across the flows. It shows that for single-server systems with i.i.d. exponential service times, the preemptive Maximum Age First (MAF) Last-Generated First-Serve (LGFS) policy, where the flow with the maximum age is served the first, minimizes the age penalty process among all causal policies in a stochastic ordering sense. Reference [16] investigates a single-source multi-server system and shows that if the service times are i.i.d. exponentially distributed, the preemptive LGFS policy is proven to be age-optimal in a stochastic ordering sense, and if the service times are i.i.d. and satisfy a New-Better-than-Used (NBU) distributional property, the non-preemptive LGFS policy is shown to be within a constant gap from the optimum age performance. In [17] it considers a multi-flow single-server system where individual flows have different priorities. It shows that a policy named preemptive priority MAF LGFS is lexicographic age-optimal. Reference [18] focuses on deterministic policies, in which the source may preempt the processing of an update if its age grows above a fixed cut-off time. It shows that replacing such updates by fresher ones can enhance the overall AoI.

Different from the continuous-time queueing models previously discussed, in this chapter, we focus on discrete-time systems where the service preemption decision is made at the beginning of each time slot. Such modeling naturally fits the Markov Decision Processes (MDP) framework, where the optimal decision depends on the instantaneous state of the system. We note that similar MDP formulations have been utilized to obtain age-optimal policies numerically in other status updating systems [19–22]. In most of those works, it assumes that each update takes exactly one time slot to transmit. The service preemption problem studied in this chapter is considerably more complicated. This is because under the model considered in this chapter, each update may take more than one slot to transmit. Therefore, the optimal decision needs to take the portion of an update that has been delivered into consideration, which increases the dimension of the system state and makes the analysis extremely challenging.

16.1.2 Chapter Outline

Over the next three sections, we will discuss works on transmission preemption in point-to-point status updating systems with nonideal communication channels [23–25]. Specifically, in Section 16.2 we consider AoI optimization under link capacity constraint, where updates arrive randomly at the source. In Section 16.2, we focus on the symbol erasure channel where rateless codes are adopted to encode the updates. While updates are generated at will, the source needs to decide whether to preempt the current transmission based on erasure patterns it has encountered. In Section 16.4 we consider a setting similar to that in Section 16.2, except that the objective is to optimize AoS instead of AoI.

The main takeaway from the findings of these works is summarized as follows. *To optimize information freshness, strict preemptive or non-preemptive policies are often nonoptimal. Rather, the optimal policy should judiciously decide when to preempt based on factors such as the instantaneous AoI at the destination, the age of the update being transmitted, as well as its remaining transmission time. Moreover, the optimal policy usually exhibits certain threshold structures on those factors.*

16.2 AoI-Optimal Policy under Link Capacity Constraint

In this section, we discuss the results reported in [23] in greater detail. We consider a point-to-point status updating system where updates are generated randomly at the source according to an i.i.d. Bernoulli process $\{a_t\}_t$ with parameter p. The updates are sent to the destination through a communication link. We consider the scenario where the size of each update is large compared with the link capacity, so that it takes multiple time slots to transmit. We assume that at most one update can be transmitted during each time slot, and there is *no buffer* at the source to store any updates other than the one being transmitted. Therefore, once an update arrives at the source, it needs to decide whether to transmit it and drop the one being transmitted if there is any, or to drop the new arrival.

A status update policy is denoted as π, which consists of a sequence of transmission decisions $\{w_t\}$. We let $w_t \in \{0, 1\}$. Specifically, when $a_t = 1$, w_t can take both values 1 and 0: If $w_t = 1$, the source will start transmitting the new arrival in time slot t and drop the unfinished update if there is one. We term this as *switch*. Otherwise, if $w_t = 0$, the source will drop the new arrival and continue transmitting the unfinished update if there is one, or be idle otherwise. We term this as *skip*. When $a_t = 0$, we can show that dropping the update being transmitted is suboptimal. Thus, we restrict to the policies under which w_t can only take value 0, that is, to continue transmitting the unfinished update, if there is one, or to idle.

Let S_n, $n = 1, 2, \ldots$, be the time slot when an update is completely transmitted to the destination. $\{S_n\}_n$ divide the time axis into epochs, where the length of the nth update, denoted as X_n, equals $S_n - S_{n-1}$. Without loss of generality, we assume $S_0 = 0$. We note that since the update arrivals will either be dropped or transmitted immediately, the AoI after a completed transmission is reset to the transmission time of the delivered update, denoted as d_n. An example sample path of the AoI evolution under a given status update policy is shown in Figure 16.1. As illustrated, some updates are skipped when they arrive, while others are transmitted partially or completely.

We focus on a set of *online* policies Π, under which the decision w_t is based on the decision history $\{w_i\}_{i=1}^{t-1}$, the update arrival profile $\{a_i\}_{i=1}^{t}$, as well as the statistics of the update arrivals (i.e., arrival rate p and the distribution of update sizes in this

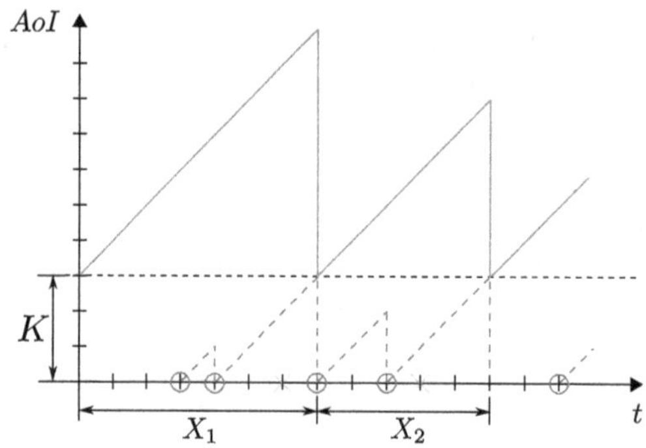

Figure 16.1 Example of AoI evolution with uniform update size. The required transmission time of each update equals 3. Circles represent transmitted updates, crosses represent skipped ones, and the dashed zigzag curve indicates the age of the update being transmitted.

scenario). Let $R(T)$ be the total AoI over time period $[0, T]$. The objective is to choose an online policy to minimize the long-term average AoI

$$\min_{\pi \in \Pi} \limsup_{T \to \infty} \mathbb{E}\left[\frac{R(T)}{T}\right], \tag{16.1}$$

where the expectation is taken over all possible update arrival patterns and possibly random decisions.

16.2.1 Updates of Uniform Sizes

We start with the scenario where the updates are of the same size, and the required transmission time is equal to K time slots. Here K is an integer greater than or equal to two.

Consider the nth epoch, that is, the duration between time slots $S_{n-1}+1$ and S_n under any online policy in Π. Let $a_{n,k}$ be the time slot when the kth update arrival after S_{n-1} occurs, and let $x_{n,k} := a_{n,k} - S_{n-1}$. Denote the update arrival profile in epoch n as $\mathbf{x}_n := (x_{n,1}, x_{n,2}, \ldots)$. Then, we introduce the following definition.

DEFINITION 16.1 (Uniformly Bounded Policy) *Under an online policy $\pi \in \Pi$, if there exists a function $g(\mathbf{x})$ such that for any $\mathbf{x}_n = \mathbf{x}$ the length of the corresponding epoch X_n is upper bounded by $g(\mathbf{x})$, $\mathbb{E}[g^2(\mathbf{x}_n)] < \infty$, then this policy is a uniformly bounded policy.*

Denote the subset of uniformly bounded policies as Π'. Then, we have the following theorem.

THEOREM 16.2 ([23]) *The time-average AoI under any uniformly bounded policy $\pi \in \Pi'$ can always be improved by a renewal policy, under which $\{S_n\}_{n=1}^{\infty}$ form a renewal process, and the decision $\{w_t\}$ over the nth renewal epoch only depends on $\mathbf{x}_n$ causally.*

Theorem 16.2 is mainly due to the memoryless property of the update arrival process. Intuitively, if we are able to obtain a policy to achieve the minimum average AoI over one epoch among all possible epochs, with all possible history and external randomness, the memoryless update arrival process ensures that this policy can be applied over all epochs irrespective of history and external randomness. This policy is a renewal policy and only causally depends on $\mathbf{x}_n$.

Based on Theorem 16.2, in the following we will focus on renewal policies that depend on $\mathbf{x}_n$ only. This will reduce the complexity of the problem vastly, as it basically states that history prior to the latest successful update does not really matter. We then introduce the definition of sequential switching policy.

DEFINITION 16.3 (Sequential Switching Policy) *A sequential switching (SS) policy is a renewal policy such that the source is allowed to switch to an update arriving at time slot t* ***only if*** *it has switched to all updates that arrive before t in the same epoch.*

Under the SS policy, once a source skips a new update generated at time t, it will skip all of the upcoming updates generated afterward until it finishes the one being transmitted at time t. In general, an SS policy does not necessarily exhibit threshold structure, as it does impose any threshold on when the source should skip or switch. One main result of [23] is to show that the optimal policy is an SS policy. Moreover, the optimal SS policy exhibits a threshold structure.

LEMMA 16.4 *The optimal renewal policy in Π' is an SS policy.*

LEMMA 16.5 *Consider a renewal epoch under the optimal SS policy in Π'. If the source switches to an update at the ith time slot in that epoch, then, there exists a threshold τ_i, $\tau_i \geq i$, which depends on i only, such that if the next update arrives before or at the τ_ith time slot in that epoch, the source will switch to the new arrival; otherwise, it will skip any new updates until the end of the epoch.*

Lemmas 16.4 and 16.5 can be proved through contradiction; that is, if the optimal policy does not exhibit the SS and threshold structures, respectively, we can always construct another renewal policy to achieve lower average AoI. Technically, we construct an alternative policy under which the first moment of the length of each renewal interval stays the same while strictly reducing the second moment. Note that the average AoI within each renewal interval is the ratio between the second moment and the first moment. Thus, the alternative policy renders a lower expected average AoI.

Based on the two lemmas, the structure of the optimal renewal policy is characterized in the following theorem:

THEOREM 16.6 *Consider a renewal epoch under the optimal SS policy. There exists a sequence of thresholds $\tau_1 \geq \tau_2 \geq \cdots \geq \tau_\gamma$, such that if the source switches to an update at the ith time slot in that epoch, $1 \leq i \leq \gamma$, and the next update arrives before or at the τ_i-th time slot in the epoch, the source will switch to the new arrival; otherwise, if the next update arrives after the τ_ith time slot or the update being transmitted*

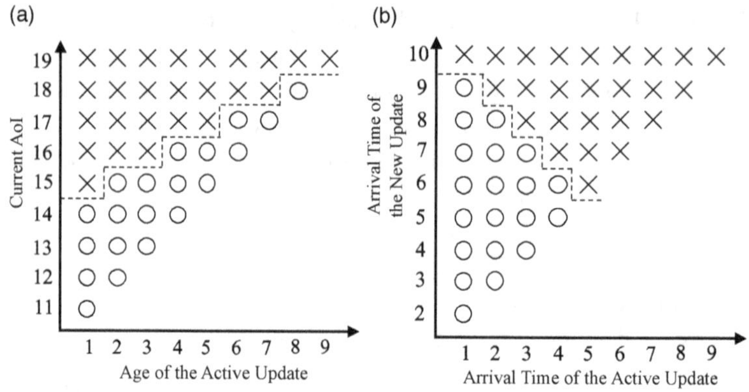

Figure 16.2 The optimal policy when $p = 0.07$, $K = 10$. Circles represent *switch*, while crosses represent *skip*.

arrives after the γth slot in the epoch, the source will skip all upcoming arrivals until the end of the epoch.

The monotonicity of the switch thresholds depicted in Theorem 16.6 can be proved through contradiction, that is, for any SS policy that does not exhibit monotonic threshold structure, we can construct another randomized SS policy to strictly improve its AoI.

Theorem 16.6 indicates that the optimal decision of the source in the nth epoch only depends on the arrival time of the update being transmitted and the arrival time of the new update, both relative to S_{n-1}, the end of the previous renewal epoch.

Motivated by the Markovian structure of the optimal policy in Theorem 16.6, the problem can be cast as a three-dimensional MDP and numerically search for the optimal thresholds $\tau_1, \tau_2, \ldots, \tau_\gamma$, and γ. The state of the MDP can be captured by a 3-tuple $\mathbf{s}_t := (\delta_t, d_t, a_t)$, where δ_t and d_t are the AoI in the system and the age of the unfinished update at the beginning of time slot t.

Figure 16.2(a) shows the optimal action for states $(\delta, d, 1)$ when the transmission time of updates is set as $K = 10$ and the update generation rate is set as $p = 0.07$. We observe the monotonicity of the thresholds in both δ and d. Moreover, in Figure 16.2(b) we depict the optimal action for each pair of the arrival time of the update being transmitted and that of the new arrival in a renewal epoch. We see that the thresholds $\tau_1 = 9$, $\tau_2 = 8$, $\tau_3 = 7$, and $\tau_4 = 6$ and they are monotonically decreasing, as stated in Theorem 16.6.

16.2.2 Updates of Nonuniform Sizes

In this section we consider the scenario where the updates are of different sizes. We assume the required transmission time of each update is an i.i.d. random variable following a known probability mass function (PMF) f_b over a bounded support in $\mathbb{N}$. Compared with the uniform update size case, this problem becomes more challenging. This is because each update may take a different number of time slots to transmit, and the AoI will be reset to different values when an update is delivered successfully. Therefore, the optimal policy depends on the arrival times of the updates, as well as their sizes. In order to make the problem tractable, we restrict it to Markovian polices, where the decision at any time slot t depends on the AoI at the destination, denoted as δ_t; the remaining transmission time of the update being transmitted, denoted as r_t; the total required transmission time of the update being transmitted, denoted as c_t; and the total required transmission time of the new arrival, denoted as b_t. The system state at time t can thus be denoted as

$$\mathbf{s}_t := (\delta_t, r_t, c_t, b_t). \tag{16.2}$$

We note that $0 \leq r_t \leq c_t - 1$. Denote the set of all valid states as $\mathcal{S}$.

Then, if $w_t = 0$, the source either continues its transmission and skips the new update, or stays idle. Thus, the state at time $t+1$ can be specified as follows:

$$\delta_{t+1} = \begin{cases} \delta_t + 1, & \text{if } r_t \neq 1, \\ c_t, & \text{if } r_t = 1, \end{cases} \tag{16.3}$$

$$r_{t+1} = \begin{cases} r_t - 1, & \text{if } r_t > 1, \\ 0, & \text{if } r_t = 0, 1, \end{cases} \tag{16.4}$$

$$c_{t+1} = \begin{cases} c_t, & \text{if } r_t > 1, \\ 0, & \text{if } r_t = 0, 1. \end{cases} \tag{16.5}$$

If $b_t \neq 0$ and $w_t = 1$, that is, the source switches to the new arrival, $\delta_{t+1} = \delta_t + 1$, $r_{t+1} = b_t - 1$, $c_{t+1} = b_t$.

Let $C(\mathbf{s}_t; w_t)$ be the immediate cost under state $\mathbf{s}_t$ with action w_t. To minimize the time-average AoI, we let $C(\mathbf{s}_t; w_t) = \delta_t$.

Assume $\mathbf{s} := (\delta, r, c, b)$. Let $\mathbf{s}'$ be the next state after taking action at state $\mathbf{s}$. In order to obtain some structural properties of the optimal policy, we introduce an infinite horizon α-discounted MDP as follows:

$$V^\alpha(\mathbf{s}) = \min_{w \in \{0,1\}} C(\mathbf{s}; w) + \alpha \mathbb{E}[V^\alpha(\mathbf{s}') | \mathbf{s}, w], \tag{16.6}$$

where $0 < \alpha < 1$. It has been shown that the optimal policy to minimize the average AoI can be obtained by solving (16.6) when $\alpha \to 1$ [26]. We start with the following value iteration formulation:

$$V^\alpha_{n+1}(\mathbf{s}) = \min_{w \in \{0,1\}} C(\mathbf{s}; w) + \alpha \mathbb{E}[V^\alpha_n(\mathbf{s}') | \mathbf{s}, w], \tag{16.7}$$

where $V^\alpha_0(\mathbf{s}) = 0$ for any state $\mathbf{s} \in \mathcal{S}$.

Denote the corresponding state-action value function at the nth iteration as $Q^\alpha_n(\mathbf{s}; w)$. We have the following observation:

- $b \neq 0$:

$$Q_k(\mathbf{s}; 0) = \begin{cases} \delta + \alpha \mathbb{E}[V^\alpha_k(\delta + 1, (r-1)^+, c, \bar{b})], & \text{if } r \neq 1, \\ \delta + \alpha \mathbb{E}[V^\alpha_k(c, 0, 0, \bar{b})], & \text{if } r = 1, \end{cases} \tag{16.8}$$

$$Q_k(\mathbf{s}; 1) = \delta + \alpha \mathbb{E}[V^\alpha_k(\delta + 1, b - 1, b, \bar{b})], \tag{16.9}$$

where the expectation is taken with respect to $\bar{b}$, and $(x)^+ = \max\{x, 0\}$. We adopt $(r-1)^+$ in (16.8) to indicate that if $r = c = 0$, and $w = 0$, the system will remain idle.

- $b = 0$: $V^{\alpha}_{k+1}(\mathbf{s}) = Q_k(\mathbf{s}; 0)$, where $Q_k(\mathbf{s}; 0)$ follows the same form of (16.8).

Based on the value iteration defined in (16.7), one can show the properties of the state value function $V^{\alpha}_n(\mathbf{s})$ as follows.

LEMMA 16.7 *For any state $\mathbf{s} \in \mathcal{S}$, at every iteration, the following properties hold:*

(1) $V^{\alpha}_n(\delta, r, c, b)$ is monotonically increasing in δ;
(2) $V^{\alpha}_n(\delta, r, c, b)$ is monotonically increasing in c for $c > 0$;
(3) $V^{\alpha}_n(\delta, r, c, b)$ is monotonically increasing in b for $b > 0$;
(4) $\mathbb{E}[V^{\alpha}_n(\delta, r_1, c_1, \bar{b})] \leq \mathbb{E}[V^{\alpha}_n(\delta, r_2, c_2, \bar{b})]$ if and only if $Q_{n-1}(\delta, r_1, c_1, .; 0) \leq Q_{n-1}(\delta, l_2, c_2, .; 0)$.

Lemma 16.7 can be proved through value iteration, and the detailed proof can be found in [23]. Based on these properties, we have the following theorem.

THEOREM 16.8 *Denote $\mathbf{s} = (\delta, r, c, b)$ and assume $b > 0$. The optimal policy for the α-discounted MDP has the following structure:*

(1) If the optimal action for $\mathbf{s}$ is to switch, then the optimal action for any state (δ, r, c, b'), $0 < b' < b$, is to switch as well;
(2) If $0 < b \leq l$, then the optimal action for $\mathbf{s}$ is to switch to the new update;
(3) If the optimal action for $\mathbf{s}$ is to switch, then for any $c' > c$, the optimal action for state $\mathbf{s}' := (\delta, r, c', b)$ is to switch as well;
(4) If $r = c = 0$, then the optimal action for $\mathbf{s}$ is to switch to the new update;
(5) If the optimal action for $\mathbf{s}$ is to skip, then the optimal action for state $(\delta + 1, r, c, b)$ is to skip as well.

Property (1) indicates the threshold structure on the size of the new update arrival, that is, the source prefers to switch to a new update if its size is small, and will skip it if its size is large. Property (2) is a consequence of property (1). Property (3) shows that there exists a threshold on the size of the update being transmitted, that is, the source prefers to drop updates with larger sizes and switch to new updates. Property (4) says that the source should immediately start transmitting the new update arrival if it has been idle, which is consistent with the uniform update size case. Property (5) essentially indicates the threshold structure on the instantaneous AoI at the destination: the source prefers to skip new updates when the AoI is large, as it is in more urgent need to complete the current transmission and reset the AoI to a smaller value.

We point out that all of the structural properties of the optimal policy derived for the α-discounted problem hold when $\alpha \to 1$ [26]. Thus, the optimal policy for the time-average problem also exhibits similar structures.

Figure 16.3 illustrates the optimal policy at different states in a system where the transmission time of each update randomly takes value between 5 and 8 with equal probability, and the update arrival rate p is fixed as 0.14. We note that the optimal

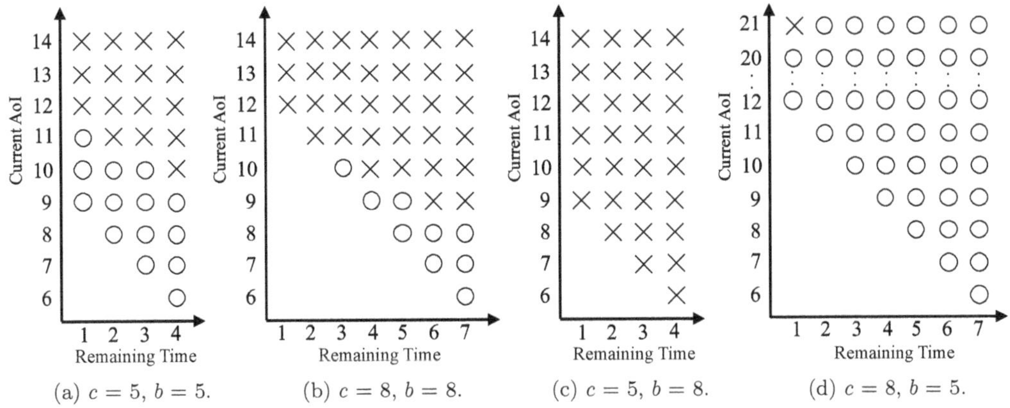

Figure 16.3 The optimal policy when $p = 0.14$. Circles represent *switch*, while crosses represent *skip*.

policy exhibits the threshold structure described in Theorem 16.8. We also note that in Figure 16.3(c), the source always prefers to skip when $c = 5, b = 8$. This can be explained by the intuition that switching to an update with longer transmission time would lead to larger AoI when p is not very small.

Figure 16.4 compares the time-average AoI under the optimal policy with two baseline policies, that is, the non-preemptive policy (termed as *Always Skip*), and the preemptive policy under which the source will always switch to new updates upon their arrivals (termed as *Always Switch*). As observed, the performance differences between those three policies are negligible when p is close to 0. This is because when p is small, the source won't receive a new update before it finishes transmitting the current update with high probability, thus the source behaves almost identically under all three policies. As p increases, the performances of the optimal policy and the Always Skip policy are still very close to each other, while the Always Switch policy renders the highest AoI.

16.3 AoI-Optimal Transmission over Symbol Erasure Channels

In this section, we present the results reported in [24]. In contrast to the deterministic communication link between the source and the destination assumed in Section 16.2, in this work, we consider a random symbol erasure channel for update delivery. Specifically, we assume each update consists of K information symbols and is encoded using rateless codes. Each transmitted encoded symbol will be erased according to an independent and identically distributed (i.i.d.) Bernoulli process with parameter ε, and an update can be successfully decoded if K encoded symbols are received. We consider an ideal scenario where instant feedback is available to the source after the transmission of each symbol. Upon receiving feedback, the source has the choice to preempt the transmission of the old update and start transmitting a new one, or to continue with the transmission of the previous update if it is not delivered yet.

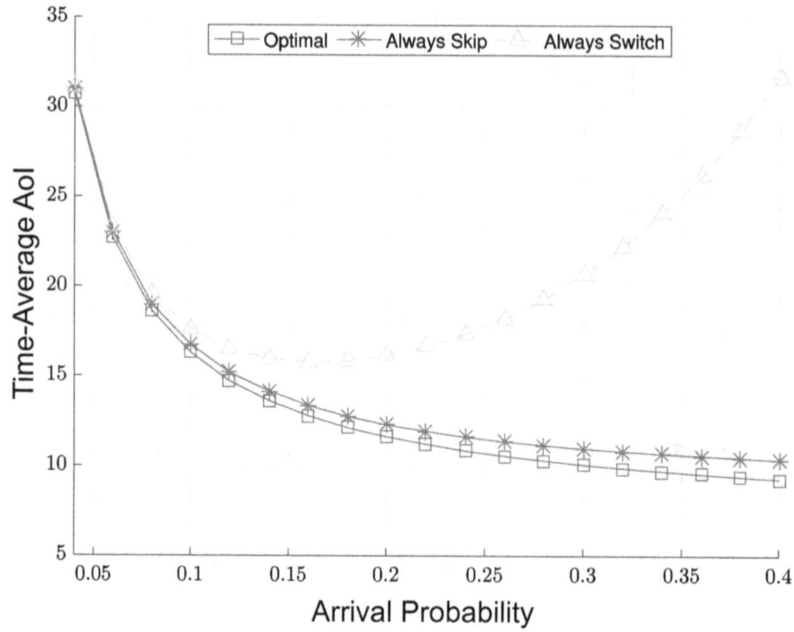

Figure 16.4 Average AoI comparison. $b \in \{5, 8\}$.

Denote the decision of the source at the beginning of time slot t as $w_t \in \{0, 1\}$, where $w_t = 1$ represents transmitting a new update and $w_t = 0$ represents transmitting the unfinished update (if there is one) or being idle. We point out that in this scenario, updates are generated at will, that is, when $w_t = 1$, a new update will be generated. As illustrated in Figure 16.5, whenever $w_t = 1$, the source starts sending a new update, whose age will continuously grow (as indicated by the dashed curve), and the AoI at the destination will grow simultaneously as well (as indicated by the solid curve). Once the update is successfully decoded, the AoI at the destination will be reset to the age of the delivered update. Due to the random erasures happening during the transmission, each update may take a different time to get delivered.

We focus on a set of online policies Π, under which the information available for determining w_t includes the decision history $\{w_i\}_{i=1}^{t-1}$, the update-to-date feedback information, as well as the system parameters (i.e., K, ε in this scenario). The goal is to choose an online policy $\{w_t\}_t \in \Pi$ such that that the long-term average AoI is minimized.

16.3.1 MDP formulation

To make the optimization problem tractable, we restrict it to the set of Markovian policies under which the decision only depends on current state, and formulate the problem as an MDP.

Let δ_t, d_t, and ℓ_t denote the AoI at the destination, the age of the unfinished update, and the number of successfully delivered symbols of the unfinished update, respectively. Then, the state of the MDP at time t is denoted as

$$\mathbf{s}_t := (\delta_t, d_t, \ell_t). \tag{16.10}$$

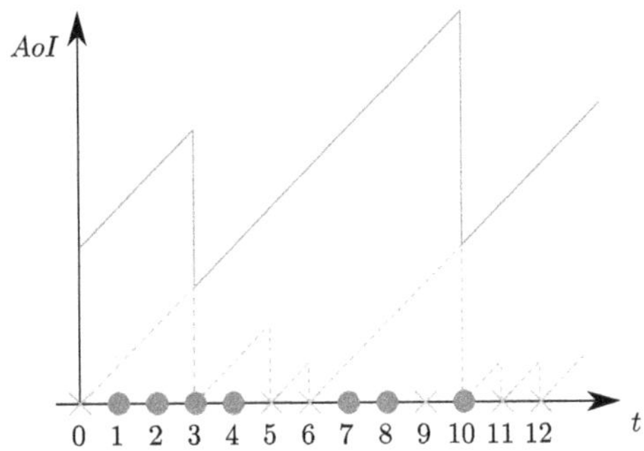

Figure 16.5 Example of AoI evolution when $K = 3$. Circles and crosses represent successful transmissions and erasures, respectively. The dashed curve indicates the age of the update being transmitted.

Then, the set of all valid states $\mathcal{S} = \{(\delta, d, \ell) : \delta \geq d + K, d \geq \ell, 0 \leq \ell < K\}$.

When the system state at time t is $\mathbf{s} = (\delta, d, \ell)$, the state at time $t+1$ is specified as follows:

- $w_t = 1$:

$$\mathbf{s}_{t+1} = \begin{cases} (\delta+1, 1, 1), & \text{with prob. } 1-\varepsilon, \\ (\delta+1, 1, 0), & \text{with prob. } \varepsilon. \end{cases} \tag{16.11}$$

- $w_t = 0$ and $\ell < K - 1$:

$$\mathbf{s}_{t+1} = \begin{cases} (\delta+1, d+1, \ell+1), & \text{with prob. } 1-\varepsilon, \\ (\delta+1, d+1, \ell), & \text{with prob. } \varepsilon, \end{cases} \tag{16.12}$$

- $w_t = 0$ and $\ell = K - 1$:

$$\mathbf{s}_{t+1} = \begin{cases} (d+1, 0, 0), & \text{with prob. } 1-\varepsilon, \\ (\delta+1, d+1, k-1), & \text{with prob. } \varepsilon. \end{cases} \tag{16.13}$$

Then, the state value function $V_n^\alpha(\mathbf{s})$ and the state-action value function $Q_n^\alpha(\mathbf{s}; w)$ can be defined in a way similar to that in Section 16.2.

16.3.2 Threshold Structure of the Optimal Policy

To obtain structural properties of the optimal policy, we establish the properties of the state value function $V_n^\alpha(\mathbf{s})$ first.

LEMMA 16.9 *For any $\mathbf{s} \in \mathcal{S}$, the following properties hold at every iteration:*

(1) $V_n^{\alpha}(\delta, d, \ell)$ *is monotonically increasing in* δ;
(2) $V_n^{\alpha}(\delta, d, \ell)$ *is monotonically increasing in* d;
(3) $V_n^{\alpha}(\delta, d, \ell) \leq V_n^{\alpha}(\delta, 0, 0)$;
(4) $V_n^{\alpha}(\delta, d, k-1) \geq V_n^{\alpha}(d, 0, 0)$;
(5) $V_n^{\alpha}(\delta, d, \ell)$ *is monotonically decreasing in* ℓ.

Lemma 16.9 can be proved based on value iteration and induction. Interested readers may refer to [24] for more details. Properties (1), (2), and (5) indicate that lower age at the destination or of the update being transmitted, or more symbols of the unfinished update successfully delivered, is more desirable. Those properties coincide with the intuition that to minimize the long-term average AoI, the instantaneous age at the destination should remain low, which requires updates to be delivered with short delay. Further, property (3) implies that $(\delta, 0, 0)$ is the least favorable state among all states sharing the same instantaneous AoI at the destination (i.e., δ). Intuitively, at state (δ, d, ℓ), it is always permissible to discard the current update and start to transmit a new one, and the system evolves in the same way afterward as that emanating from $(\delta, 0, 0)$. Therefore, $V_n^{\alpha}(\delta, d, \ell) \leq V_n^{\alpha}(\delta, 0, 0)$. Property (4) is based on a similar observation.

The next corollary is an immediate result of property (3) in Lemma 16.9.

COROLLARY 1 *Whenever the source successfully delivers an update, it should start transmitting a new update immediately.*

Corollary 1 indicates that zero-wait policy is optimal in this setting. This is in contrast to the result in [27], showing that zero-wait policy is not age-optimal in general. This is mainly due to the assumption that under our setup, the source is allowed to preempt the transmission of an update at any time, while in [27], the service is non-preemptive.

The next corollary is based on properties (2) and (3) in Lemma 16.9.

COROLLARY 2 *Whenever the source starts transmitting a new update and the first symbol is erased, it should discard it and start transmitting a new update immediately.*

Lemma 16.9 implies the following threshold structure of the optimal policy.

THEOREM 16.10 *The optimal MDP policy in* Π *must satisfy the following properties:*

(1) If the source decides to discard the unfinished update and start transmitting a new update at state (δ, d, l), *it should make the same decision at state* $(\delta, d+1, l)$;
(2) If the source decides to keep sending the unfinished update at state (δ, d, l), *it should make the same decision at state* $(\delta, d, l+1)$;
(3) If the source decides to keep sending the unfinished update at state (δ, d, l), *it should make the same decision at state* $(\delta+1, d+1, l+1)$.

Applying property (3) in Theorem 16.10 recursively and combining with Corollary 2, and properties (1) and (2) in Theorem 16.10, we characterize the structure of the optimal policy in the following theorem.

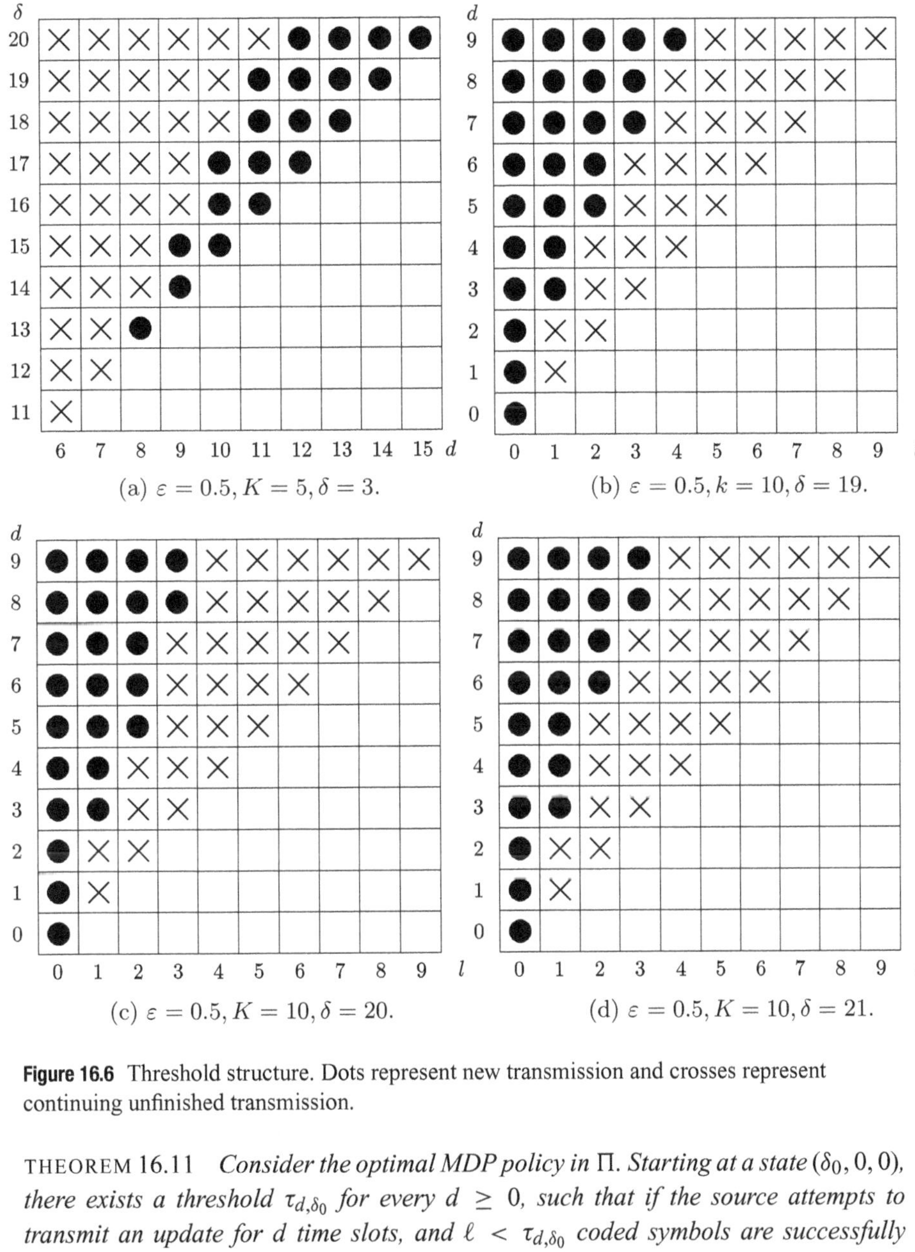

Figure 16.6 Threshold structure. Dots represent new transmission and crosses represent continuing unfinished transmission.

THEOREM 16.11 *Consider the optimal MDP policy in* Π. *Starting at a state* $(\delta_0, 0, 0)$, *there exists a threshold* τ_{d,δ_0} *for every* $d \geq 0$, *such that if the source attempts to transmit an update for d time slots, and* $\ell < \tau_{d,\delta_0}$ *coded symbols are successfully delivered, the source should quit the unfinished update and start transmitting a new update at state* $(\delta_0 + d, d, \ell)$; *otherwise, it will keep transmitting until the update is successfully decoded. Besides,* τ_{d,δ_0} *monotonically increases in d.*

Figure 16.6 shows the structure of the optimal policy obtained through dynamic programming, where we fix $K = 5$ and $\varepsilon = 0.5$. The threshold structure on δ and ℓ as well as the monotonicity of the thresholds corroborate the theoretical results in Theorem 16.11.

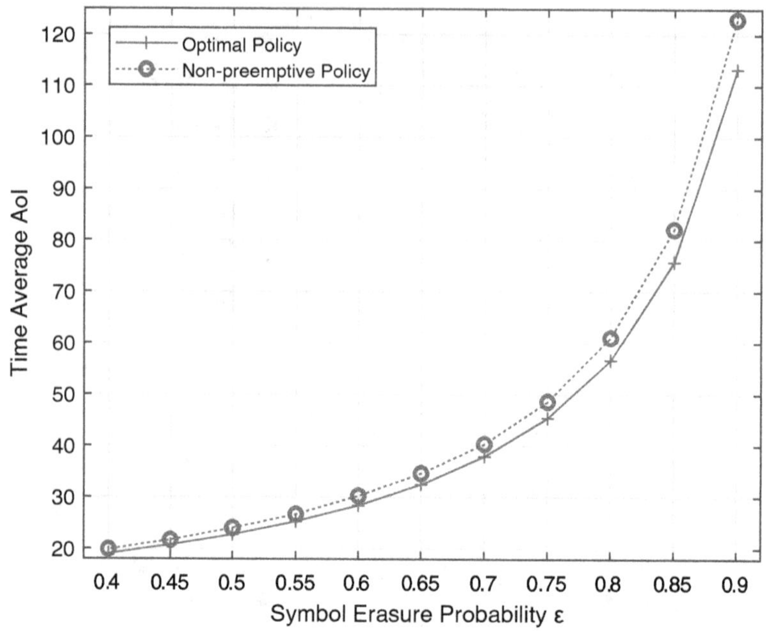

Figure 16.7 Performance comparison with $K = 8$.

The performances of the optimal policy and the non-preemptive policy are compared in Figure 16.7. Under the baseline non-preemptive policy, the source keeps transmitting an update until it is delivered. The update size is fixed as $K = 8$ and the symbol erasure probability ε varies in $(0.4, 0.9)$. Note that the optimal policy outperforms the non-preemptive policy and the performance gap decreases as $\varepsilon \to 0$. The intuition behind this is, when ε decreases, the probability for an update to experience many symbol erasures during transmission is low, thus, the source has less chance to discard an unfinished update, that is, the policy is actually non-preemptive for most cases.

16.4 AoS-Optimal Policy under Link Capacity Constraint

In this section we briefly describe the results reported in [25]. We consider the scenario where the status of the monitored signal only changes at discrete time points. The objective is to let the destination be synchronized with the source in a timely manner once a status change happens. What complicates the problem is that the transmission takes multiple time slots due to the link-rate constraint. Similar to the system model presented in Section 16.2, the transmitter has to decide to switch or to skip a new update when the status of the monitored signal changes and it has not completed the transmission of the previous one yet. We adopt a metric called "Age of Synchronization" (AoS) to measure the "dissatisfaction" of the destination when it is desynchronized with the source.

AoS is first introduced in [28] and refers to the duration since the destination became *desynchronized* with the source. This metric of information freshness is motivated by the observation that in certain applications, such as caching systems [28], the status of the source only changes sporadically in time; as long as the destination is updated with the latest status change at the source, the information is considered "fresh."

To simplify the analysis, we restrict it to the case where the updates are of the same size, and it takes K time slots to transmit one update to the destination. We assume the transmitter can transmit only one update at any time slot, and there is no buffer at the transmitter. Let $w_t \in \{0, 1\}$ be a binary decision variable, where $w_t = 0$ corresponds to *skip* and $w_t = 1$ corresponds to *switch*.

We label the updates in the order of their generation times and use T_m to denote the generation time of the mth update. Denote $D(t)$ as the index of the latest update received by the monitor at the beginning of the tth time slot. Then the age of synchronization is defined as

$$\text{AoS}(t) := (t - T_{D(t)+1})^+, \tag{16.14}$$

where $T_{D(t)+1}$ refers to the time when the source generates a new update after $D(t)$ and the destination becomes desynchronized, and $(x)^+ = \max(x, 0)$.

In order to capture the state of the system, we introduce the AoS at the transmitter as well. Specifically, let $A(t)$ be the index of the latest update the transmitter transmits. Then the AoS at the transmitter is denoted as $(t - T_{A(t)+1})^+$. If no new update is generated over duration $(T_{A(t)}, t]$, $T_{A(t)+1}$ would be greater than t. As a result, we have $(t - T_{A(t)+1})^+ = 0$. If the transmitter *switches* to a new update once it is generated, the AoS at the transmitter is zero; otherwise, if it *skips* a new update, the transmitter becomes desynchronized with the source, and its AoS starts growing.

Let S_i be the time slot that the destination has been updated successfully for the ith time, where $i = 0, 1, 2, \cdots$. Without loss of generality, we assume $S_0 = 0$. S_is partition the time axis into epochs. Depending on the evolution of the monitored process, two scenarios may happen when the destination receives an update. For the first scenario, the update arriving at the monitor is the latest update generated by the monitored process, which is regarded as a "fresh" update. Therefore, the monitor is synchronized to the observed process successfully and $\text{AoS}(S_i) = 0$. It will not increase until the monitored process changes. On the other hand, if the monitored process changes during the transmission of the latest received update, the monitor will not be synchronized with the monitored process, thus $\text{AoS}(S_i) > 0$. For both scenarios, $\text{AoS}(t)$ will be increased by 1 for each time slot of desynchronization. Finally, when a new update is delivered to the monitor, this epoch ends and a new one begins. An example of AoS evolution is shown in Figure 16.8.

Intuitively, it is suboptimal to terminate an unfinished update if no new update arrives at the source. We focus on the set of Markovian policies satisfying this property and cast the problem as a four-dimensional MDP, as detailed in what follows.

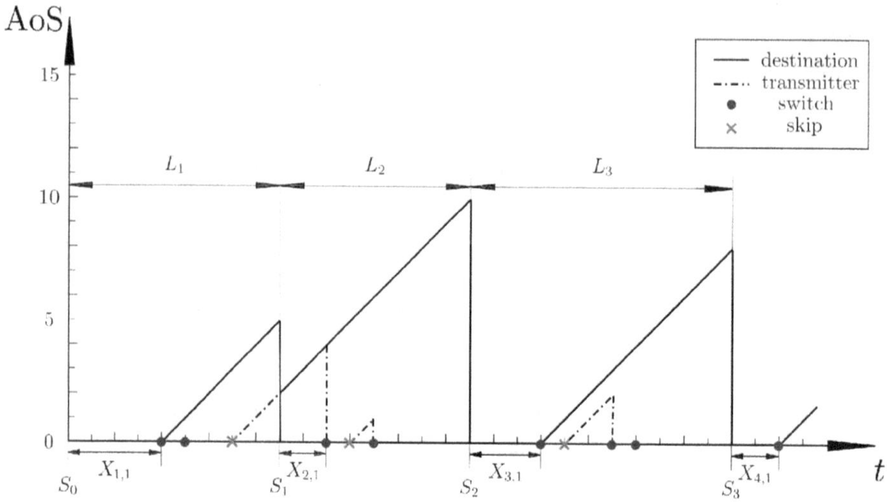

Figure 16.8 Evolution of the AoS at the transmitter and at the destination with $K = 4$. Once an update is received at the destination, the AoS at the destination is reset to the AoS at the transmitter.

16.4.1 MDP Formulation

With a bit of abuse of the notation, in this section, we denote the AoS at the destination and at the transmitter at the beginning of time slot t as δ_t, d_t, respectively, and the remaining transmission time of the unfinished update as r_t. Let $a_t \in \{0, 1\}$ denote whether a new update is generated at time t. Then, the state of the MDP at the beginning of time slot t is denoted as

$$\mathbf{s}_t := (\delta_t, d_t, r_t, a_t). \tag{16.15}$$

Note that $\delta_t = d_t$ if and only if the system is idle, $r_t = 0$. Meanwhile, if the system is busy, $r_t > 0$, we must have $d_t < K - r_t \leq \delta_t$. We collectively denote the set valid states as

$$\mathcal{S} = \{(\delta, d, r, a) : \delta = d \text{ if } r = 0; d < K - r \leq \delta \text{ if } r > 0\}. \tag{16.16}$$

At the beginning of time t, if $a_t = 1$, when $w_t = 0$, the transmitter skips the new update, and when $w_t = 1$, the transmitter begins to send the new update. If $a_t = 0$, we must have $w_t = 0$.

We note that a_{t+1} evolves according to an independent Bernoulli random variable with parameter p. Then, based on a_t and w_t, we divide the states into two categories:

- $a_t = 0$ or $w_t = 0$: When there is no new update generated at the source or a new update is generated but dropped, the transmitter will continue its previous operation, that is, either transmitting an unfinished update or being idle. Thus, we have

$$\begin{aligned} \delta_{t+1} &= \begin{cases} \delta_t + 1, & r_t \neq 1 \text{ and } (\delta_t > 0 \text{ or } a_t = 1), \\ d_t + 1, & r_t = 1 \text{ and } d_t > 0, \\ 0, & \text{otherwise.} \end{cases} \\ d_{t+1} &= \begin{cases} d_t + 1, & d_t > 0 \text{ or } a_t = 1, \\ 0, & \text{otherwise.} \end{cases} \\ r_{t+1} &= (r_t - 1)^+. \end{aligned} \tag{16.17}$$

- $a_t = 1, w_t = 1$: If there is a new update generated at the source and the transmitter decides to switch, the transmitter will be refreshed and the destination becomes desynchronized. Thus,

$$\delta_{t+1} = \delta_t + 1, \quad d_{t+1} = 0, \quad r_{t+1} = K - 1. \tag{16.18}$$

Then, similar to previous sections, the value function $V_n^\alpha(\mathbf{s})$ and the state–action value function $Q_n^\alpha(\mathbf{s}; w)$ can be defined accordingly.

16.4.2 Threshold Structure of the Optimal Policy

We establish the properties of the state-value function $V_n^\alpha(\mathbf{s})$ as follows.

LEMMA 16.12 *For any $\mathbf{s} \in \mathcal{S}$, at every iteration n, the following properties hold:*

(1) $V_n^\alpha(\delta, d, r, a)$ is monotonically increasing in δ for any busy state;
(2) $V_n^\alpha(\delta, \delta, 0, a)$ is monotonically increasing in δ for any idle state;
(3) $V_n^\alpha(\delta, d, r, a)$ is nondecreasing in d for any busy state;
(4) $V_n^\alpha(\delta, d, r, a) \leq V_n^\alpha(\delta, \delta, 0, a)$ for any busy state with $r > 0$;
(5) $V_n^\alpha(d, d, 0, a) \leq V_n^\alpha(\delta, d, 1, a)$ for any state with $r = 1$;
(6) $V_n^\alpha(\delta, d, r, a)$ is nondecreasing in r for any state with $r > 0$.

The properties can be proved sequentially and we omit the detail here for brevity. Properties (1), (3), and (6) indicate that the value function has higher value with larger δ, d, or r, which is consistent with our intuition that states with larger AoS at the destination or at the transmitter, or longer remaining transmission time, are less preferable. Properties (4) and (5) can also be explained intuitively as follows. With the same AoS at the destination, the busy state is always more favorable than the idle state, as active transmission is helpful to a successful update delivery. Therefore, property (4) follows. On the other hand, property (6) suggests that with the same AoS at the transmitter, the state requiring one more slot to update the destination is "worse" than the state that the destination has just been synchronized. This is due to the fact that the AoS at the destination in the former state is not lower than that in the latter state, and this relationship holds for any upcoming state after transition with the same action taken at the transmitter.

Motivated by the monotonicity of the state value function $V_n^\alpha(\mathbf{s})$ in δ, d, and r, we show that the state–action value function $Q_n^\alpha(\mathbf{s}; w)$ must preserve the following properties.

LEMMA 16.13 *Under the optimal policy, the following properties hold:*

(1) *If* $Q_n^\alpha(\delta, d, r, 1; 1) \leq Q_n^\alpha(\delta, d, r, 1; 0)$*, then for any state* $(\delta, d, r', 1) \in \mathcal{S}$ *with* $r' > r$*, it must have* $Q_n^\alpha(\delta, d, r', 1; 1) \leq Q_n^\alpha(\delta, d, r', 1; 0)$*;*
(2) *If* $Q_n^\alpha(\delta, d, r, 1; 0) \leq Q_n^\alpha(\delta, d, r, 1; 1)$*, then for any state* $(\delta', d, r, 1) \in \mathcal{S}$ *with* $\delta' > \delta$*, it must have* $Q_n^\alpha(\delta', d, r, 1; 0) \leq Q_n^\alpha(\delta', d, r, 1; 1)$*.*

Based on the property (1), there exists a threshold on r such that the transmitter will switch to a new update only when the remaining transmission time for the update being transmitted is above the threshold. Similarly, property (2) suggests that the transmitter should skip a new update only if the current AoS at the destination is above the threshold. We characterize the structural properties of the optimal policy in the following theorem.

THEOREM 16.14 *Under the optimal policy, for any fixed AoS at destination* δ *and remaining transmission time* r*, there exists a threshold* $\tau_{\delta,r}$*, such that when* $d \geq \tau_{\delta,r}$*, the optimal action is to transmit the new update, that is,* $w^*(\delta, d, r, 1) = 1$ *and when* $d < \tau_{\delta,r}$*, the optimal action is to continue the transmitter's previous action, that is,* $w^*(\delta, d, r, 1) = 0$*. Especially,* $\tau_{\delta,r} = K$ *if the optimal policy for all states with* δ *and* r *is to skip. Besides, for any fixed* δ*, the set of thresholds is decreasing in* r*, that is,* $\tau_{\delta,1} \geq \tau_{\delta,2} \geq \cdots \geq \tau_{\delta,K-1-d}$*. Similarly, for any fixed* r*, the set of threshold is increasing in* δ*, that is,* $\tau_{K-r,r} \leq \tau_{K-r+1,r} \leq \cdots \leq \tau_{\delta,r} \leq \cdots$*.*

Figure 16.9 shows the thresholds on d with fixed δ and r. For any state $(\delta, d, r, 1)$, the optimal action is to switch if d is above the bar located at the corresponding δ and r, otherwise the optimal action is to skip. Note that the thresholds are monotonically increasing in δ and decreasing in r, as summarized in Theorem 16.14. Besides, the optimal policy is inclined to skip as the update arrival rate p increases.

We plot the average age performance for the AoS-optimal policy, the AoI-optimal policy, as well as the always skip and always switch policies in Figure 16.10. Note that when $p \to 1$, the AoS-optimal policy and the always skip policy tend to be identical, since the optimal policy behaves the same as the always skip policy with high probability. When p is small, the AoS-optimal policy and the always switch policy tend to be the same, as the chance that a new update is generated when the transmitter is busy is small. We observe that the AoI-optimal policy is close to the AoS-optimal policy when $p \to 1$, while there is a constant gap when p is small. Thus, AoI minimization is not efficient for AoS minimization. The most interesting distinction between AoS and AoI lies in the different trending curves as the update generation rate p increases: the minimum AoI monotonically decreases as p increases, while the minimum AoS exhibits the opposite trend. This is due to the fact that the AoI only depends on the age of the freshest information at the destination without considering the underlying status evolution, and any information ages linearly in time since its generation. Correspondingly, lower generation rate increases the duration between two successful updates at the destination, leading to higher AoI. On the other hand, AoS depends on the change

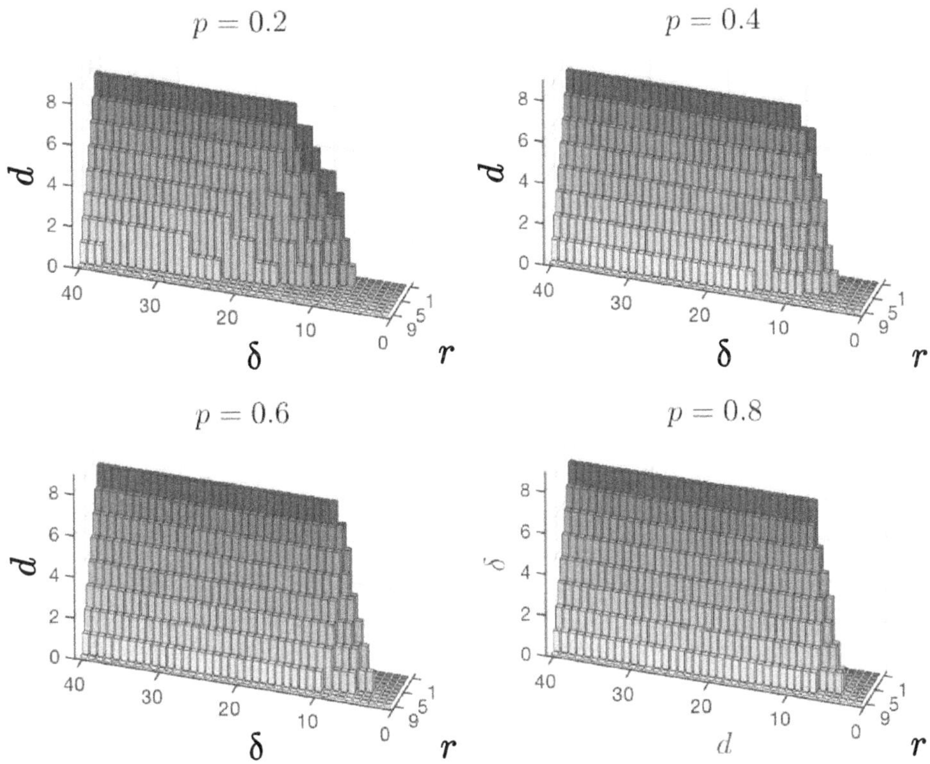

Figure 16.9 Thresholds on d with different update arrival rate p.

of the system status, and lower generation rate implies that each update can stay fresh for a long time, which leads to low AoS.

16.5 Conclusion and Outlook

In this chapter, we investigate service preemption policies for single-link status updating systems for the optimization of information freshness. Two information freshness metrics are considered: Age of Information and Age of Synchronization. For AoI optimization, we investigate two settings. Under the first setting, updates are generated randomly in time, and the required transmission time for each update is either fixed (Section 16.2.1) or random (Section 16.2.2). For the second setting (Section 16.3), updates are generated at will, and the delivery time of an update depends on the realization of the symbol erasure channel. For AoS optimization (Section 16.4), we consider the setting where the updates of the same size are generated randomly. For all cases, we identify threshold structures on the optimal policies and numerically identify them through structured value iteration under the MDP framework. Our results

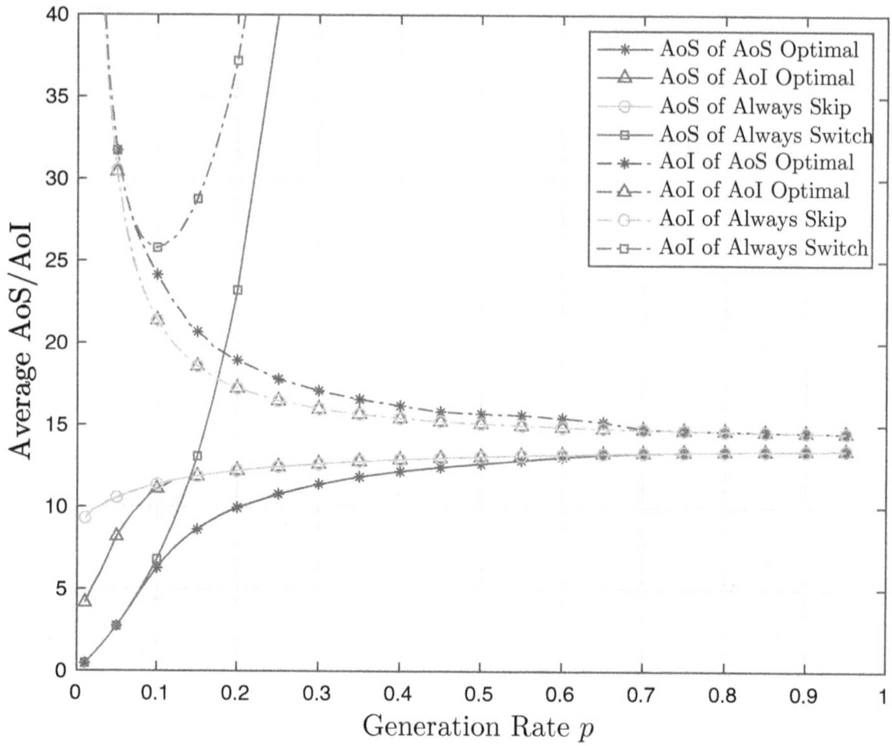

Figure 16.10 Average AoS and AoI with different generation rate p.

indicate that obtaining the optimal preemption policy for information freshness optimization is, in general, very challenging, as it depends on various factors of the system, such as the update arrival process, the sizes of updates, and the characteristics of the communication channel.

While several problems have been discussed in detail in this chapter, there are various open problems that could be tackled in this line of research. We highlight two of them next.

(1) *Obtain practical and analytically tractable suboptimal policies.* As noted previously, obtaining the optimal preemption policy is, in general, quite complicated and analytically intractable. As a result, the results presented in this chapter rely on MDP and value iteration to identify the optimal policy numerically. It is thus of practical interest to design analytically tractable policies that perform within some confidence interval away from the optimal solutions and are relatively simple to implement in practice.

(2) *Investigate age-optimal preemption protocols for communication networks.* Another interesting aspect is to extend the single-link system into more complicated communication models, such as broadcast channels, multiple-access channels, and even multi-hop networks. We envision that the problem would become even more

challenging in such scenarios. Thus, instead of striving for the optimal solutions, searching for analytically tractable policies with performance guarantees would be of more practical interest.

References

[1] S. K. Kaul, R. D. Yates, and M. Gruteser, "Status updates through queues," in *2012 46th Annual Conference on Information Sciences and Systems (CISS)*, 2012, pp. 1–6.

[2] K. Chen and L. Huang, "Age-of-information in the presence of error," in *IEEE International Symposium on Information Theory (ISIT)*, 2016, pp. 2579–2583.

[3] E. Najm and R. Nasser, "Age of information: The gamma awakening," in *IEEE International Symposium on Information Theory (ISIT)*, 2016, pp. 2574–2578.

[4] A. Soysal and S. Ulukus, "Age of information in G/G/1/1 systems," in *53rd Asilomar Conference on Signals, Systems, and Computers*, 2019, pp. 2022–2027.

[5] A. Franco, B. Landfeldt, and U. Körner, "Extended analysis of age of information threshold violations," *Computer Communications*, vol. 161, pp. 191–201, 2020.

[6] S. Farazi, A. G. Klein, and D. R. Brown, "Age of information in energy harvesting status update systems: When to preempt in service?" in *IEEE International Symposium on Information Theory (ISIT)*, 2018, pp. 2436–2440.

[7] E. Najm, R. Yates, and E. Soljanin, "Status updates through M/G/1/1 queues with HARQ," in *IEEE International Symposium on Information Theory (ISIT)*, 2017, pp. 131–135.

[8] S. Asvadi, S. Fardi, and F. Ashtiani, "Analysis of peak age of information in blocking and preemptive queueing policies in a HARQ-based wireless link," *IEEE Wireless Communications Letters*, vol. 9, no. 9, pp. 1338–1341, 2020.

[9] E. Najm and E. Telatar, "Status updates in a multi-stream M/G/1/1 preemptive queue," in *IEEE Conference on Computer Communications Workshops (INFOCOM WKSHPS)*, 2018, pp. 124–129.

[10] R. D. Yates and S. K. Kaul, "The age of information: Real-time status updating by multiple sources," *IEEE Transactions on Information Theory*, vol. 65, no. 3, pp. 1807–1827, 2019.

[11] S. K. Kaul and R. D. Yates, "Age of information: Updates with priority," in *2018 IEEE International Symposium on Information Theory (ISIT)*, 2018, pp. 2644–2648.

[12] A. Maatouk, M. Assaad, and A. Ephremides, "Age of information with prioritized streams: When to buffer preempted packets?" in *IEEE International Symposium on Information Theory (ISIT)*, 2019, pp. 325–329.

[13] O. Dogan and N. Akar, "The multi-source preemptive M/PH/1/1 queue with packet errors: Exact distribution of the age of information and its peak," *ArXiv e-prints*, 2020.

[14] V. Kavitha, E. Altman, and I. Saha, "Controlling packet drops to improve freshness of information," 2018.

[15] Y. Sun, E. Uysal-Biyikoglu, and S. Kompella, "Age-optimal updates of multiple information flows," in *IEEE INFOCOM – Workshop on Age of Information*, April 2018, pp. 136–141.

[16] A. M. Bedewy, Y. Sun, and N. B. Shroff, "Minimizing the age of information through queues," *IEEE Transactions on Information Theory*, pp. 1–1, 2019.

[17] A. Maatouk, Y. Sun, A. Ephremides, and M. Assaad, "Status updates with priorities: Lexicographic optimality," 2020.

[18] A. Arafa, R. D. Yates, and H. V. Poor, "Timely cloud computing: Preemption and waiting," in *57th Annual Allerton Conference on Communication, Control, and Computing (Allerton)*, 2019, pp. 528–535.

[19] B. Zhou and W. Saad, "Joint status sampling and updating for minimizing age of information in the internet of things," *IEEE Transactions on Communications*, vol. 67, no. 11, pp. 7468–7482, 2019.

[20] M. A. Abd-Elmagid, H. S. Dhillon, and N. Pappas, "A reinforcement learning framework for optimizing age of information in rf-powered communication systems," *IEEE Transactions on Communications*, vol. 68, no. 8, pp. 4747–4760, 2020.

[21] ——, "Aoi-optimal joint sampling and updating for wireless powered communication systems," *IEEE Transactions on Vehicular Technology*, vol. 69, no. 11, pp. 14 110–14 115, 2020.

[22] G. Stamatakis, N. Pappas, and A. Traganitis, "Optimal policies for status update generation in an iot device with heterogeneous traffic," *IEEE Internet of Things Journal*, vol. 7, no. 6, pp. 5315–5328, 2020.

[23] B. Wang, S. Feng, and J. Yang, "When to preempt? Age of information minimization under link capacity constraint," *Journal of Communications and Networks*, vol. 21, no. 3, pp. 220–232, 2019.

[24] S. Feng and J. Yang, "Age-optimal transmission of rateless codes in an erasure channel," in *IEEE International Conference on Communications (ICC)*, 2019, pp. 1–6.

[25] C. Deng, J. Yang, and C. Pan, "Timely synchronization with sporadic status changes," in *2020 IEEE International Conference on Communications (ICC)*, 2020, pp. 1–6.

[26] D. P. Bertsekas, *Dynamic Programming and Optimal Control*, 3rd ed. Belmont, MA, USA: Athena Scientific, 2005, vol. I.

[27] Y. Sun, E. Uysal-Biyikoglu, R. D. Yates, C. E. Koksal, and N. B. Shroff, "Update or wait: How to keep your data fresh," *IEEE Transactions on Information Theory*, vol. 63, no. 11, pp. 7492–7508, 2017.

[28] J. Zhong, R. D. Yates, and E. Soljanin, "Two freshness metrics for local cache refresh," in *IEEE ISIT*, Jun. 2018, pp. 1924–1928.

17 Economics of Fresh Data Trading

Meng Zhang, Ahmed Arafa, Jianwei Huang, and H. Vincent Poor

The ever-increasing number of real-time applications has driven the need for timely data. Frequent generation of fresh data, however, may require significant operational cost, introducing a fundamental economic challenge to design proper incentives for fresh data generation and transmission. This dilemma motivates the study of economics of fresh data markets, which provide proper incentives for the fresh data providers and economic regulations of such markets. Keeping data fresh for real-time applications is not only a pure engineering issue but also a complicated economic issue, as it affects the interests of massive users and providers in a wide range of real-time applications, including the Internet of Things (IoT), real-time cloud computing, and financial markets. In addition, destinations' valuations of fresh data are *time-interdependent*, making such an economic issue more challenging, compared with classical settings.

In this chapter we study the economic issues of fresh data trading markets, where the data freshness is captured by Age-of-Information (AoI). In our model a destination user requests and pays for fresh data updates from a source provider. In this work the destination incurs an age-related cost, modeled as a general increasing function of the AoI. To understand the economic viability and profitability of fresh data markets, we consider a pricing mechanism to maximize the source's profit, while the destination chooses a data update schedule to trade off its payments to the source and its age-related cost. The problem is exacerbated when the source has incomplete information regarding the destination's age-related cost, which requires one to exploit *(economic) mechanism design* to induce truthful information. This chapter attempts to build such a fresh data trading framework that centers around the following two key questions: (a) How should a source choose the pricing scheme to maximize its profit in a fresh data market under complete market information? (b) Under incomplete information, how should a source design an optimal mechanism to maximize its profit while ensuring the destination's truthful report of its age-related cost information?

17.1 Background

There has been an explosive growth in data-driven businesses that recognize the necessity to leverage the power of big data, by exploiting, for example, machine learning techniques. While machine learning techniques require high-quality training data,

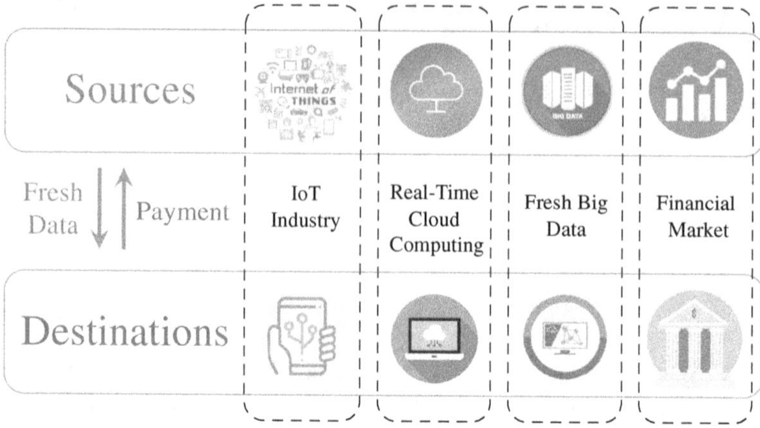

Figure 17.1 Examples of potential fresh data markets.

this problem is further exacerbated, as increasingly many real-time businesses are now largely relying on *fresh data*. Such real-time businesses built on real-time systems range from IoT systems [1], multimedia, cloud-computing services, real-time data analytics, to even financial markets, as shown in Figure 17.1. More specifically, examples of real-time applications demanding timely data updates include monitoring, data analytics, control systems, and phasor data updates in power grid stabilization systems; examples of real-time datasets include real-time map and traffic data, for example, the Google Maps Platform [2]. The systems involving these applications and datasets put high emphasis on the data freshness, and we adopt the *Age-of-Information (AoI)* metric to measure the freshness of data [1, 3].

Despite the increasing significance of fresh data, keeping data fresh relies on frequent data generation, processing, and transmission, which can lead to significant operational costs for the data sources (providers). Such operational costs make *proper incentives* of an essential role in the fresh data trading interaction between data sources and data destinations (users). One such approach is to design pricing schemes, which provide incentives for the sources to update the data and prohibit the destinations from requesting data updates unnecessarily often.

Furthermore, in addition to enabling necessary fresh data trading, *pricing design* is also one of the core techniques of *revenue management*. It facilitates the understanding of the economic viability of real-time businesses and to achieve data sources' profit maximization. The pricing for fresh data, however, is under-explored, as not all existing pricing schemes for communication systems account for data timeliness. To understand the economic viability and profitability of pricing schemes in fresh data markets, this chapter aims at answering the following question:

Question 1 How should the source choose a pricing scheme to maximize its profit in fresh data trading?

We study the following two types of pricing schemes by exploiting two different dimensions, namely *time and quantity*, described in the following:

- *Time-dependent pricing scheme*: The source of fresh data prices each data update based on the time at which the update is requested. Due to the nature of the AoI, the destination's desire for updates increases as time (since the most recent update) goes by, which makes it natural to explore this time sensitivity. This pricing scheme is also motivated by practical pricing schemes for mobile networks (in which users are not age-sensitive) [4, 5, 6, 7].
- *Quantity-based pricing scheme*: The price for each update depends on the number of updates requested so far (but does not depend on the timing of the updates) [8]. The source may reward the destination by reducing prices for each additional request to incentivize more fresh data updates. Such a pricing scheme is motivated by practical pricing schemes for data services (e.g., for data analytics services [9] and cloud computing services [2, 10]). For instance, the storage provider RimuHosting charges a smaller price for each additional gigabyte of storage purchased [10].

The nature of data freshness poses the twofold challenge of designing the above pricing schemes. First, the destination's valuation is *time-interdependent*. Specifically, the demands for fresh data over time are coupled due to the nature of AoI, which differs from the existing settings (e.g., [4–7]).[1] That is, the desire for an update at each time instance depends on the time elapsed since the latest update. Hence, the source's pricing scheme choice needs to take such interdependence into consideration. Second, the flexibility in different pricing choices renders the optimization over (infinitely) many dimensions.

Another crucial economic challenge, in addition to providing incentives, arises in the presence of *market information asymmetry*. Specifically, destinations in practice may have private (market) information regarding their age-related costs. Therefore, they may manipulate the outcomes (e.g., their payments and the update policies) by misreporting such private information to their own advantages. In the economics literature, a standard approach to designing markets with asymmetric information is to leverage the *optimal mechanism design* [11]. To this end, we study the optimal mechanism design problem to address the following question:

Question 2 How should a source design an optimal mechanism to maximize its profit while ensuring to incentivize the destination to report its truthful age-related cost information?

The focus of this chapter is to lay a foundation of the economic analysis (including both pricing design and mechanism design) of the AoI-based fresh data trading through a heuristic one-source one-destination system model. The main results and contributions of this chapter include the following:

- *Pricing Scheme Design:* In Section 17.4, we study the time-dependent pricing scheme and show in this case that the game equilibrium leads to only one data

[1] Note that time dependence in pricing scheme design implies the price may change over time, whereas time interdependence here emphasizes the impact of different decisions across time on the destination's valuation.

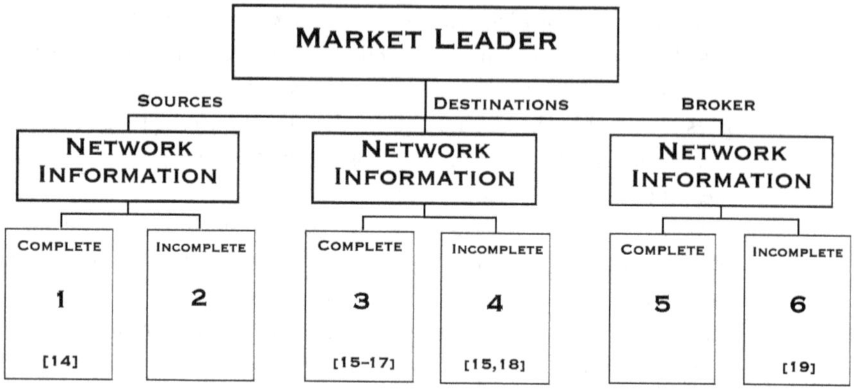

Figure 17.2 Taxonomy of fresh data trading.

update, which does not yield the maximum profit to the source. This motivates us to consider a quantity-based pricing scheme, in which the price of each update depends on how many updates have been previously requested. We show that among all possible pricing schemes, the quantity-based pricing scheme performs the best: it maximizes the source's profit and minimizes the social cost of the system, defined as the aggregate source's operational cost and the destination's age-related cost.

- *Mechanism Design:* In Section 17.5, we design an optimal (economic) mechanism following Myerson's seminal work to cope with incomplete information. Our proposed optimal mechanism maximizes the source's profit, while ensuring that the destinations truthfully report their private information and will voluntarily participate in the mechanism. In addition, our analysis shows the existence of a quantity-based pricing scheme that is equivalent to the optimal mechanism.

17.2 Taxonomy and Brief Literature Review

In this section, we aim at building a clear taxonomy by classifying the design space of fresh data trading in the context of AoI as shown in Figure 17.2, depending on which entities are market leaders (the market participants who have dominant market power) and whether the market leaders have complete information. Each scenario leads to a different formulation. Scenario 3 is relatively well understood. Scenarios 1 and 4 are partially understood. The rest of the scenarios have received little or no attention in the literature.

Next, we briefly review the related literature. In Scenarios 1–2, sources are market leaders (e.g., fresh data providers). In this case, sources are often real-time datasets owners (e.g., real-time Google Map data owned by Google) or real-time data service providers (such as real-time cloud computing services [12, 13]). In [14] we studied the source's pricing scheme design under both a predictable deadline and a predictable deadline.

In Scenarios 3–4, related AoI studies focused on the cases in which destinations are market leaders [15–18]. For instance, destinations are crowdsourcing platforms who aim at incentivizing workers (e.g., mobile users) to participatorily generate and transmit their real-time sensory data. In [15], Hao et al. studied a repeated game between two AoI-aware platforms. Specifically, they formulated competition among crowdsourcing platforms as a noncooperative game and showed that they want to over-sample to reduce their own AoIs. The authors proposed a trigger mechanism that incentivizes platform cooperation to approach the social optimum under both complete information scenarios and incomplete information scenarios. In [16], Wang et al. proposed dynamic pricing for the provider to offer age-dependent monetary returns and encourage users to sample information at different rates over time. This dynamic pricing design problem needs to balance the monetary payments to users and the AoI evolution over time, and is challenging to solve, especially under the incomplete information about users' arrivals and their private sampling costs. This study finally conducted the steady-state analysis of the optimized approximate dynamic pricing scheme and showed that such a pricing scheme can be further simplified to an ϵ-optimal version. Li et al. in [17] considered AoI-based efficient reward mechanisms for a crowd-learning platform under complete information. Such mechanisms incentivize mobile users to send real-time information and achieve the optimal AoI performance asymptotically. Zhang et al. in [18] studied mechanism design to tackle with the incomplete information issue for multisource fresh data markets, which achieves the minimal overall cost (including age-related cost and payment to the sources) and ensures sources' truthful reports of their private sampling cost information.

Scenarios 5–6 consider a third possibility: a broker, as a centralized controller, serves to be a market leader to match and coordinate sources and destinations. This framework was only considered in a recent study that designed a marketplace for real-time matching to efficiently buy and sell data for machine learning tasks, but without using age-of-information [19]. This line of work on fresh data trading is important but has been largely overlooked in the literature.

Overall, the existing understanding of many scenarios is limited. In the following we will discuss our own contributions in Scenarios 1–2.

17.3 System Model

In this section, we introduce the system model of a single-source single-destination information update system.

17.3.1 System Overview

17.3.1.1 Single-Source Single-Destination System

We consider an information update system, in which one source node generates data packets and sends them to one destination through a channel. For instance, Google (the source) generates real-time Google Map data (the data packets) and sends them to Uber (the destination).

We note that the single-source single-destination model has been widely considered in the AoI literature (e.g., [22–26]). The insights derived from this model allow us to potentially extend the results to multi-destination scenarios.[2]

17.3.1.2 Data Updates and Age-of-Information

We consider a fixed time period of $\mathcal{T} = [0, T]$, during which the source sends its updates to the destination. We consider a generate-at-will model (as in, e.g., [24–27]), in which the source is able to generate and send a new update when requested by the destination. Updates reach the destination instantly, with negligible transmission time (as in, e.g., [25, 27]).[3]

We denote by $S_k \in \mathcal{T}$ the time of the kth update is generated and submitted. The set of all update time instances is $\mathcal{S} \triangleq \{S_k\}_{1 \le k \le K}$, where K is the number of total updates, that is, $|\mathcal{S}| = K$ with $|\cdot|$ denoting the cardinality of a set. The set $\mathcal{S}$ (and hence the value of K) is the destination's decision. Let x_k denote the kth update inter-arrival time, which is the time elapsed between the generation of $(k-1)$th update and kth update, that is, x_k is

$$x_k \triangleq S_k - S_{k-1}, \ \forall k \in [K+1]. \tag{17.1}$$

In (17.1), we define $S_0 = 0$ and $S_{K+1} = T$, and let $[N] \triangleq \{1, 2, ..., N\}$ denote the set of all integers from 1 up to N. The following definition characterizes the freshness of data:

DEFINITION 17.1 (Age-of-Information (AoI)) The age-of-information $\Delta_t(\mathcal{S})$ at time t is [3]

$$\Delta_t(\mathcal{S}) = t - U_t, \tag{17.2}$$

where U_t is the time stamp of the most recently received update before time t, that is, $U_t = \max_{S_k \le t}\{S_k\}$.

17.3.1.3 Destination's General AoI Cost

The destination experiences an (instantaneous) AoI cost $f(\Delta_t)$ characterizing its desire for the new data update (or dissatisfaction with stale data). We assume that $f(\Delta_t)$ is a general *increasing* function in Δ_t. For instance, a convex AoI cost implies the destination gets more desperate when its data grows stale, an example of which is

$$f(\Delta_t) = \Delta_t^{\kappa}, \ \kappa \ge 1, \tag{17.3}$$

which exists in the online learning in real-time applications such as online advertisement placement and online Web ranking [28, 29].

[2] The system constraints (e.g., congestion and interference constraints) in a multi-destination model can make the joint scheduling and pricing scheme design much more challenging, which will be left for future work.

[3] This assumption is practical when inter-update times are on a scale that is order of magnitudes larger than the transmission times of the updates themselves.

17.3.1.4 Source's Operational Cost

We denote the source's operational cost by $C(K)$, which is modeled as an increasing convex function in the number of updates K, with $C(0) = 0$. This can represent sampling costs in case the source is an IoT service provider, the computing resource consumption in case the source is a cloud computing service provider.[4]

17.3.2 Stackelberg Games

We use the Stackelberg game [31] to model the interactions between the source and the destination. In particular, the Stackelberg game consists of a leader and followers. In our case, the source is the market leader and the destination is the follower, and they make decisions sequentially. Prior works [15–18] have studied the cases in which the destinations are the leader and the sources are the follower.

We categorize the interaction scenarios into both *complete information* and *incomplete information* scenarios, depending on the source's knowledge of the destination's AoI cost function:

- *Complete information:* The source knows the destination's AoI cost function $f(\Delta_t)$. Though the complete information is a strong assumption, it serves as the benchmark of practical designs and provides important insights for the incomplete information analysis. In this case, the source designs a pricing scheme to maximize its profit.
- *Incomplete information:* The source only knows statistical information (of some parameters) regarding the destination's AoI cost function $f(\Delta_t)$, assuming that $f(\Delta_t)$ is parameterized. Such statistical information can be obtained through long-term observations and requires the source to design proper economic mechanisms to ensure the truthfulness of the destination while achieving profit maximization.

We will specify the games and analyze the pricing scheme design and the mechanism design problems in Sections 17.4 and 17.5, respectively.

17.4 Pricing Scheme Design under Complete Information

In this section, we discuss our results reported in [14] in greater detail. We first formulate the pricing scheme design problem when the source is aware of the destination's AoI cost $f(\Delta_t)$. We will then separately consider two special cases of the pricing Π by exploiting different dimensions: time-dependent pricing Π_t and quantity-based pricing Π_q. We will show the existence of the optimal Π_q schemes that can maximize the source's profit among all possible pricing schemes.

[4] The convexity satisfies, e.g., the sublinear speedup assumption: the consumed computing resources multiplied by the completion time for each task is increasingly higher under a shorter completion time [30].

17.4.1 Problem Formulation and Methodology

The source designs the pricing scheme, denoted by Π, for sending the data updates. A pricing scheme may exploit two dimensions: *time* and *quantity*. Specifically, we consider a *time-dependent* pricing scheme $\Pi_t \triangleq \{p(t)\}_{t\in\mathcal{T}}$, in which the price for each update $p(t)$ depends on t, that is, when each update is requested; we also consider a *quantity-based* pricing scheme $\Pi_q \triangleq \{p_k\}_{k\in\mathbb{N}}$, in which the price for each update p_k varies.[5] We will specify the destination's total payment $P(\mathcal{S},\Pi)$, which depends on the destination's update policy $\mathcal{S}$ and the source's pricing scheme Π. We first introduce the following definitions.

DEFINITION 17.2 (Aggregate AoI Cost) The destination's aggregate AoI cost $H(\mathcal{S})$ is

$$H(\mathcal{S}) \triangleq \int_0^T f(\Delta_t(\mathcal{S}))dt. \tag{17.4}$$

DEFINITION 17.3 (Cumulative AoI Cost) The destination's cumulative AoI cost $F(x)$ for each inter-arrival time x (between two updates) is

$$F(x) \triangleq \int_0^x f(\Delta_t)d\Delta_t. \tag{17.5}$$

Let $\mathcal{S}^*(\Pi)$ be the destination's optimal update policy in response to the pricing scheme Π chosen by the source, which will be defined soon. The following definition of the Stackelberg Game captures the interaction between the source and the destination:

GAME 1 (Source-Destination Interaction Game) *The interaction game between the source and the destination involves the following two stages:*

- *In Stage I, the source decides on the pricing scheme Π at the beginning of the period, in order to maximize its profit, given by the following:*

$$\textit{\textbf{Source}}:\ \max_{\Pi}\ P(\mathcal{S}^*(\Pi),\Pi) - C(|\mathcal{S}^*(\Pi)|), \tag{17.6a}$$

$$s.t.\ \ \Pi \in \{\Pi : p_k(t) \ge 0,\ \forall t \in \mathcal{T}, k \in \mathbb{N}\}. \tag{17.6b}$$

- *In Stage II, given the source's decided pricing scheme Π, the destination decides on its update policy to minimize its overall cost (aggregate AoI cost plus payment):*

$$\textit{\textbf{Destination}}:\ \mathcal{S}^*(\Pi) \triangleq \arg\min_{\mathcal{S}\in\Phi}\ H(\mathcal{S}) + P(\mathcal{S},\Pi), \tag{17.7}$$

where Φ is the set of all feasible $\mathcal{S}$ satisfying $S_k \ge S_{k-1}$ for all $1 \le k \le |\mathcal{S}|$.

[5] As mentioned, these pricing schemes are motivated by (i) the time-sensitive demand for an update due to the nature of AoI, and (ii) the wide consideration of time-dependent and quantity-based pricing schemes in practice.

Stage I
The source determines the pricing scheme Π.
$\Downarrow$
Stage II
The destination determines its update policy $\mathcal{S}$.

Figure 17.3 Two-stage Stackelberg game under complete information.

17.4.2 Time-Dependent Pricing Scheme

We first consider a *time-dependent* pricing scheme $\Pi_t = \{p(t)\}_{t\in\mathcal{T}}$, in which the price $p(t)$ for each update depends on the time at which each update k is requested (i.e., S_k) and does not depend on the number of updates so far. Hence, the payment under a *time-dependent* pricing scheme Π_t is

$$P(\mathcal{S}, \Pi_t) \triangleq \sum_{k=1}^{K} p(S_k). \tag{17.8}$$

We derive the *(Stackelberg subgame perfect)* equilibrium price-update profile $(\Pi_t^*, \mathcal{S}^*(\Pi_t^*))$ by backward induction. First, given any pricing scheme Π_t in Stage I, we characterize the destination's update policy $\mathcal{S}^*(\Pi_t)$ that minimizes its overall cost in Stage II. Then in Stage I, by characterizing the equilibrium pricing structure, we convert the continuous function optimization into a vector one, based on which we characterize the source's optimal pricing scheme Π_t^*.

17.4.2.1 Destination's Update Policy in Stage II

We analyze the destination's update policy under arbitrary Π_t within the fixed time period $[0, T]$. Recall that K is the total number of updates, and x_k defined in (17.1) is the kth inter-arrival time.

Given the pricing scheme Π_t, we can simplify the destination's overall cost minimization problem in (17.29) as

$$\min_{K\in\mathbb{N}\cup\{0\}, x\in\mathbb{R}_{++}^{K+1}} \quad \sum_{k=1}^{K+1} F(x_k) + \sum_{k=1}^{K} p\left(\sum_{j\leq k} x_j\right), \tag{17.9a}$$

$$\text{s.t.} \quad \sum_{k=1}^{K+1} x_k = T, \tag{17.9b}$$

where $x = \{x_k\}_{k\in[K+1]}$ and $\mathbb{R}_{++}^{K+1}$ is the space of $(K+1)$-dimensional positive vectors (i.e., the value of every entry is positive).

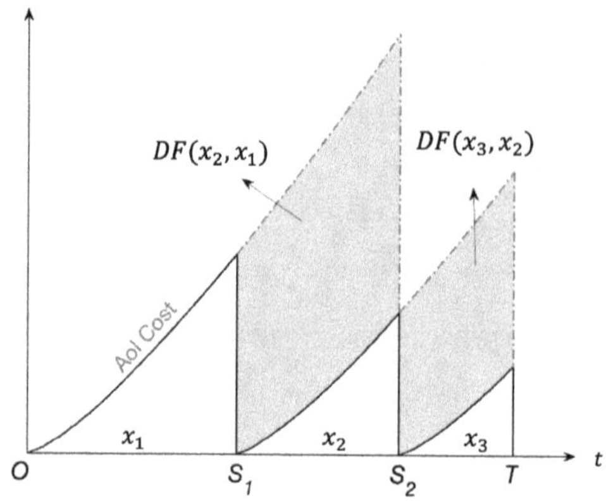

Figure 17.4 An illustrative example of the differential aggregate AoI cost function and Lemma 17.5.

To understand how the destination evaluates fresh data, we introduce the following definition:

DEFINITION 17.4 (Differential Aggregate AoI Cost) The differential aggregate AoI cost function is

$$DF(x,y) \triangleq \int_0^x [f(t+y) - f(t)]dt. \tag{17.10}$$

As illustrated in Figure 17.4, for each update k, $DF(x_{k+1}, x_k)$ is the aggregate AoI cost increase if the destination changes its update policy from $\mathcal{S}$ to $\mathcal{S}\backslash\{S_k\}$ (i.e., removing the update at S_k). We now derive the optimal time-dependent pricing based on (17.10) in the following lemma:

LEMMA 17.5 *Any equilibrium price-update tuple* $(\Pi_t^*, K^*, \boldsymbol{x}^*)$ *should satisfy*

$$p^*\left(\sum_{j=1}^{k} x_j^*\right) = DF(x_{k+1}^*, x_k^*), \quad \forall k \in [K^*+1]. \tag{17.11}$$

Intuitively, the differential aggregate AoI cost equals the destination's *maximal willingness to pay* for each data update. Note that, given that the optimal time-dependent pricing scheme satisfies (17.11), there might exist multiple optimal update policies as the optimal solutions to problem (17.29). This may lead to a multivalued source's profit and thus an ill-defined problem (17.6). To ensure the uniqueness of the received profit for the source without affecting the optimality to the source's pricing problem, one can impose infinitely large prices to ensure that the destination does not update at any time instance other than $\sum_{j=1}^{k} x_j^*$ for all $k \in [K^*+1]$. We refer readers to [14] for the complete proof.

17.4.2.2 Source's Time-Dependent Pricing Design in Stage I

Based on Lemma 17.5, we can reformulate the time-dependent pricing scheme as follows. In particular, the decision variables in problem (17.12) correspond to the inter-arrival time interval vector $\boldsymbol{x}$ instead of the continuous-time pricing function $p(t)$. By converting a functional optimization problem into a finite-dimensional vector optimization problem, we simplify the problem as follows.

Proposition 1 The time-dependent pricing problem in (17.6) is equivalent to the following problem:

$$\max_{K\in\mathbb{N}\cup\{0\},\boldsymbol{x}\in\mathbb{R}_{++}^{K+1}} \quad \sum_{k=1}^{K} DF(x_{k+1},x_k) - C(K), \tag{17.12a}$$

$$\text{s.t.} \quad \sum_{k=1}^{K+1} x_k = T. \tag{17.12b}$$

Proof Sketch: By getting the optimal solution (K^o, x^o) to problem (17.12), we can obtain a time-dependent pricing from (17.11) and impose infinitely large prices at any time instance other than $\sum_{j=1}^{k} x_j^*$ for all $k \in [K^* + 1]$. Under such a pricing scheme, we show that the destination would update at (K^o, x^o).

We refer the readers to [14] for more details. □

To rule out trivial cases with no update at the equilibrium, we adopt the following assumption throughout this chapter:

Assumption 1 The source's operational cost function $C(K)$ satisfies

$$C(1) \leq DF\left(\frac{T}{2}, \frac{T}{2}\right). \tag{17.13}$$

Proposition 2 When Assumption 1 holds and the AoI function $f(x)$ is convex, then there will be only one update (i.e., $K^* = 1$) under any equilibrium time-dependent pricing scheme.

We can prove Proposition 2 by induction, showing that for an arbitrary time-dependent pricing scheme yielding more than $K > 1$ updates (K-update pricing), there always exists a pricing scheme with a single-update equilibrium that is more profitable. Based on the preceding technique, we can show that the previous argument works for any increasing convex AoI cost function.

EXAMPLE 1 *Consider a linear AoI cost $f(\Delta_t) = \Delta_t$ and an arbitrary update policy $(K, \boldsymbol{x})$, as shown in Figure 17.5, for any time-dependent pricing scheme that induces only $K \geq 2$ updates. We will prove by induction that there exists a time-dependent pricing inducing $K - 1$ updates that is more profitable.*

- ***Base case:*** *When there are $K = 2$ updates, as shown in Figure 17.5, the source's profit (the objective value in* (17.12a)*) is $x_1x_2 + x_2x_3 - C(2)$. Consider another update policy $(1, x_1', x_2')$, where $x_1' = x_2$ and $x_2' = x_3 + x_1$. The objective value in* (17.12a) *becomes $x_2(x_1 + x_3) - C(1)$. Comparing these two values, we see that $(1, x_1', x_2')$ is strictly more profitable than (K, x).*

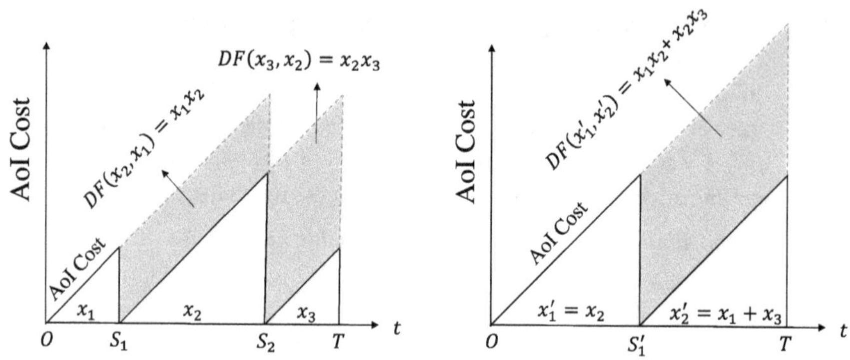

Figure 17.5 Illustrations of Example 1 with a linear cost function ($f(\Delta_t) = \Delta_t$), $\forall t$. Combining the first interval with the third interval maintains the payment.

- ***Induction step:*** *Let $K \geq n$ and suppose the statement that, for an arbitrary K-update pricing, there exists a more profitable $(K-1)$-update pricing is true for $K = n$. The objective value in* (17.12a) *is* $\sum_{k=1}^{K} x_k x_{k+1} - C(K)$. *Consider another update policy* $(K' = K-1, x')$, *where* $x_1' = x_2$, $x_2' = x_3 + x_1$, *and* $x_k' = x_{k+1}$ *for all other k. The objective value in* (17.12a) *becomes* $(x_1+x_3)(x_2+x_4) + \sum_{k=4}^{K} x_k x_{k+1} - C(K-1)$, *which is strictly larger than P. It is then readily verified that* (K', x') *is strictly more profitable than* (K, x). *Based on induction, we can show that we can find a* $(K'-1)$-*update policy that is more profitable than the* (K', x') *policy. This eventually leads to the conclusion that a single update policy is the most profitable.*

Based on the above technique, we can show that the above argument works for any increasing convex AoI cost function. We refer the readers to [14] for more details.

From Proposition 2, it is readily verified that the optimal time-dependent pricing scheme is:

COROLLARY 1 *Under a convex AoI function $f(x)$, there exists an optimal time-dependent pricing scheme Π_t^* such that*[6]

$$p^*(t) = DF\left(\frac{T}{2}, \frac{T}{2}\right), \ \forall t \in \mathcal{T}, \tag{17.14}$$

where the equilibrium update takes place at $S_1^* = T/2$.

Corollary 1 suggests that there exists an optimal time-dependent pricing scheme that is in fact *time-invariant*. That is, although our original intention is to exploit the time sensitivity/flexibility of the destination through the time-dependent pricing, it turns out not to be very effective. This motivates us to consider a quantity-based pricing scheme next.

[6] There actually exist multiple optimal pricing schemes; the only difference among all optimal pricing schemes is the prices for time instances other than $T/2$, which can be arbitrarily larger than $DF(T/2, T/2)$.

We remark that the preceding analysis in Propositions 1 and 2 relies on the convex AoI cost function assumption. The analysis here for a general AoI cost function is difficult due to the resulted non-convexity of the problem in (17.12). Nevertheless, we will show that the optimal quantity-based pricing scheme is optimal among all pricing schemes under the general AoI cost functions.

17.4.3 Quantity-Based Pricing Scheme

In this subsection, we focus on a quantity-based pricing scheme $\Pi_q = \{p_k\}_{k\in\mathbb{N}}$, that is, the price p_k may differ across each update k. Specifically, p_k represents the price for the kth update. The payment to the source is then given by

$$P(\mathcal{S}, \Pi_q) \triangleq \sum_{k=1}^{K} p_k. \tag{17.15}$$

The source determines the quantity-based pricing scheme Π_q in Stage I. Based on Π_q, the destination in Stage II chooses its update policy $(K, \boldsymbol{x})$. We derive the *(Stackelberg)* price-update equilibrium using the bilevel optimization framework [34]. Specifically, the bilevel optimization embeds the optimality condition of the destination's problem (17.29) in Stage II into the source's problem (17.6) in Stage I. We first characterize the conditions of the destination's update policy $(K^*(\Pi_q^*), \boldsymbol{x}^*(\Pi_q^*))$ that minimizes its overall cost in Stage II, based on which we characterize the source's optimal pricing Π_q^* in Stage I.

17.4.3.1 Destination's Update Policy in Stage II

Given the quantity-based pricing scheme Π_q, the destination solves the following overall cost minimization problem:

$$\min_{K\in\mathbb{N}\cup\{0\}, x\in\mathbb{R}_{++}^{K+1}} \quad \sum_{k=1}^{K+1} F(x_k) + \sum_{k=1}^{K} p_k, \tag{17.16a}$$

$$\text{s.t.} \quad \sum_{k=1}^{K+1} x_k = T. \tag{17.16b}$$

If we fix the value of K in (17.16), then problem (17.16) is convex with respect to $\boldsymbol{x}$. Such convexity allows to exploit the Karush–Kuhn–Tucker (KKT) conditions in $\boldsymbol{x}$ to analyze the destination's optimal update policy in the following lemma:

LEMMA 17.6 *Under any given quantity-based pricing scheme Π_q in Stage I, the destination's optimal update policy $(K^*(\Pi_q^*), x^*(\Pi_q^*))$ satisfies*

$$x_k^*(\Pi_q^*) = \frac{T}{K^*(\Pi_q^*)+1}, \quad \forall k \in [K^*(\Pi_q^*)+1]. \tag{17.17}$$

Intuitively, the KKT conditions of the problem in (17.16) equalize $f(x_k)$ for all k and hence lead to the equal-spacing optimal update policy in (17.17). We refer the readers to [14] for the complete proof.

17.4.3.2 Source's Quantity-Based Pricing in Stage I

Instead of solving $(K^*(\Pi_q^*), \boldsymbol{x}^*(\Pi_q))$ explicitly in Stage II, we apply the bilevel optimization to solving the optimal quantity-based pricing Π_q^* in Stage I, which leads to the price-update equilibrium of our entire two-stage game [34]. Substituting (17.17) into the source's pricing in (17.6) yields the following bilevel problem:

$$\textbf{Bilevel:} \max_{\Pi_q, K, x} \sum_{k=1}^{K} p_k - C(K), \tag{17.18a}$$

$$\text{s.t. } K \in \arg \min_{K' \in \mathbb{N} \cup \{0\}} \Upsilon(K', \Pi_q), \tag{17.18b}$$

$$x_k = \frac{T}{K+1}, \ \forall k \in [K+1], \tag{17.18c}$$

where $\Upsilon(K', \Pi_q) \triangleq (K'+1)F\left(\frac{T}{K'+1}\right) + \sum_{k=1}^{K'} p_k$ is the overall cost given the equalized inter-arrival time intervals.

We are now ready to present the optimal solution to the bilevel optimization in (17.18):

Proposition 3 The equilibrium update count $K^*(\Pi_q^*)$ and the optimal quantity-based pricing scheme Π_q^* satisfy

$$\sum_{k=1}^{K^*(\Pi_q^*)} p_k^* = F(T) - (K^*(\Pi_q^*) + 1)F\left(\frac{T}{K^*(\Pi_q^*) + 1}\right), \tag{17.19}$$

$$\sum_{k=1}^{K'} p_k^* \geq F(T) - (K' + 1)F\left(\frac{T}{K' + 1}\right), \forall K' \in \mathbb{N} \backslash \{K^*(\Pi_q^*)\}. \tag{17.20}$$

Proof Sketch: $F(T) - (K^*(\Pi_q^*) + 1)F\left(T/(K^*(\Pi_q^*) + 1)\right)$ is the aggregate AoI cost difference between the no-update scheme and the optimal update policy. Inequality (17.20) together with (17.19) will ensure that constraint (17.18c) holds. That is, if (17.20) is not satisfied or $\sum_{k=1}^{K^*(\Pi_q^*)} p_K^*(\Pi_q^*) > F(T) - (K^*(\Pi_q^*) + 1) F\left(T/(K^*(\Pi_q^*) + 1)\right)$, then $K^*(\Pi_q^*)$ would violate constraint (17.18c). We refer the readers to [14] for more details. On the other hand, if $\sum_{k=1}^{K^*(\Pi_q^*)} p_K^*(\Pi_q^*) < F(T) - (K^*(\Pi_q^*) + 1)F\left(T/(K^*(\Pi_q^*) + 1)\right)$, then the source can always properly increase p_1^* until (17.19) is satisfied. Such an increase does not violate constraint (17.18c) but improves the source's profit, contradicting with the optimality of Π_q^*.

We refer the readers to [14] for more details. □

We will present an illustrative example of the optimal quantity-based pricing in Section 17.4.5.

Substituting the pricing structure in (17.19) into (17.18), we can obtain $K^*(\Pi_q^*)$ through solving the following problem:

$$\max_{K \in \mathbb{N} \cup \{0\}} \ -(K+1)F\left(\frac{T}{K+1}\right) - C(K). \tag{17.21}$$

To solve the problem in (17.21), we first relax the constraint $K \in \mathbb{N} \cup \{0\}$ into $K \in \mathbb{R}_+$, and then recover the integer solution by rounding. We start with relaxing the integer constraint $K \in \mathbb{N} \cup \{0\}$ constraint in (17.21) into $K \geq 0$, which leads to a convex problem.[7]

After obtaining $K^*(\Pi_q^*)$, we can construct an equilibrium pricing scheme based on Proposition 3. An example of optimal quantity-based pricing is

$$p_k^* = \begin{cases} F(T) - 2F(\frac{T}{2}) + \epsilon & \text{if } k = 1, \\ F(T) - (k+1)F(\frac{T}{k+1}) - \sum_{j=1}^{k-1} p_j^* + \epsilon, & \text{if } 1 < k < K^*(\Pi_q^*), \\ F(T) - (K^*(\Pi_q^*)+1)F(\frac{T}{K^*(\Pi_q^*)+1}) - \sum_{j=1}^{K^*(\Pi_q^*)-1} p_j^*, & \text{if } k \geq K^*(\Pi_q^*), \end{cases} \tag{17.22}$$

where $\epsilon > 0$ is infinitesimal to ensure (17.20). We present an illustrative example of (17.22) in [14].

We will next show that the optimal quantity-based pricing scheme is in fact profit-maximizing among all possible pricing schemes.

17.4.4 Social Cost Minimization and Surplus Extraction

To evaluate the performances of the proposed pricing schemes, we consider an achievable upper bound of the source's profit for any pricing scheme. Note that the outcome attaining such an upper bound of the profit collides with the achievement of another system-level goal, namely the *social optimum*:

DEFINITION 17.7 (Social Optimum) A social optimum update policy $\mathcal{S}$ solves the following social cost minimization problem:

$$\min_{\mathcal{S} \in \Phi} \; C(|\mathcal{S}|) + H(\mathcal{S}). \tag{17.23}$$

That is, the socially optimal update policy minimizes the source's operational cost $C(|\mathcal{S}|)$ and the destination's aggregate AoI cost $H(\mathcal{S})$ combined. We further introduce the following definition:

DEFINITION 17.8 (Surplus Extraction) A pricing scheme Π is surplus-extracting if it satisfies

$$P(\mathcal{S}^o(\Pi), \Pi) = F(T) - H(\mathcal{S}^o(\Pi)), \tag{17.24}$$

where $\mathcal{S}^o(\Pi)$ is the social optimal update policy, that is, it solves (17.23).

That is, the surplus extracting pricing leads to a payment equal to the destination's overall AoI cost reduction, that is, the overall AoI cost with no updates $F(T)$ minus the overall AoI cost under a socially optimal update policy $H(\mathcal{S}^o(\Pi))$. We are now ready to show the optimality of a surplus-extracting pricing:

[7] To see the convexity of $(K+1)F(T/(K+1))$, note that $(K+1)F(T/(K+1))$ is the *perspective* function of $F(T)$. The perspective function of $F(T)$ is convex since $F(T)$ is convex.

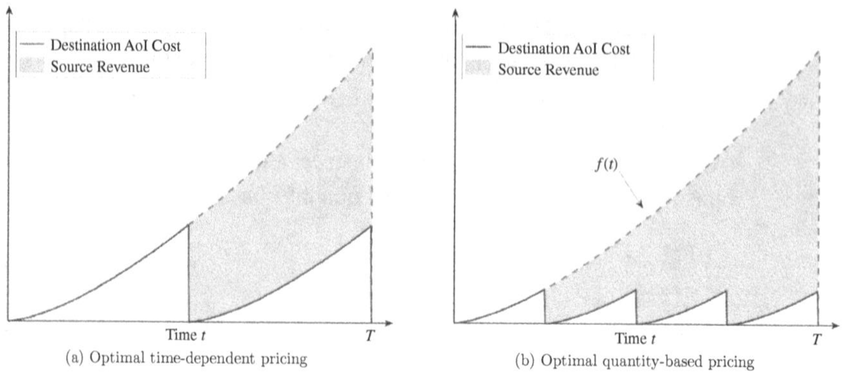

Figure 17.6 Performance comparison in terms of the AoI cost and the revenue under a convex AoI cost.

LEMMA 17.9 *A surplus-extracting pricing scheme maximizes the source's profit among all possible pricing schemes, that is, it corresponds to the optimal solution of the problem in* (17.6).

In later analysis we will show that the optimal quantity-based pricing scheme is surplus-extracting for the predictable deadline case. However, the time-dependent pricing in general is not.

THEOREM 17.10 (Surplus Extraction) *The optimal quantity-based pricing* Π_q^* *is surplus extracting, that is, it achieves the maximum source profit among all possible pricing schemes.*

Proof The optimal quantity-based pricing involves solving the problem in (17.21), which is equivalent to achieving the social optimum in Definition 17.7. Therefore, the optimal quantity-based pricing is surplus extracting by Definition 17.8 and maximizes the source profit among all possible pricing schemes from Lemma 17.9. □

Theorem 17.10 implies that the optimal quantity-based pricing scheme is already one of the optimal pricing schemes. Hence, even without exploiting the time flexibility explicitly, it is still possible to obtain the optimal pricing structure, which again implies that utilizing time flexibility may not be necessary under complete information.

17.4.5 Summary

To summarize our key results in this section, we graphically compare the AoI costs and the revenues under the studied pricing schemes in Figure 17.6 under a convex AoI cost. As in Figure 17.6(a), the optimal time-dependent pricing scheme generates a revenue for the source equal to the differential aggregate AoI cost (Lemma 17.5) and induces a unique update at $T/2$ (Proposition 2). We present the results regarding the optimal quantity-based pricing in Figure 17.6(b). The generated revenue equals the difference of the aggregate AoI costs under a no-update policy and

Stage I
The source announces its designed mechanism $\mathcal{M} = (\mathcal{S}^\star, \mathcal{P}^\star)$.
$\Downarrow$
Stage II
The destination determines a report of its type $\tilde{\theta} \in \Theta$.

Figure 17.7 Two-stage Stackelberg game with incomplete information.

the social optimum update policy. Finally, since the optimal quantity-based pricing is surplus-extracting (Theorem 17.10), it can achieve a higher profit than the optimal time-dependent pricing scheme.

17.5 Mechanism Design under Incomplete Information

In this section, we extend our results in [14] to the incomplete information scenario. Specifically, we will formulate the mechanism design problem in such a case where the destination's AoI cost is only partially known.

We assume that the destination's AoI cost is parameterized by a scalar: the destination's *type* $\theta \in \Theta \triangleq [\underline{\theta}, \bar{\theta}]$. That is, we can rewrite the destination's AoI cost as

$$f(\Delta_t|\theta) = \theta\tilde{f}(\Delta_t), \tag{17.25}$$

where $\tilde{f}(\Delta_t)$ is the *normalized AoI cost,* which we assume to be a known function. It characterizes the shape of the AoI cost; type θ characterizes the destination's sensitivity of the AoI cost compared to its payment.

We define the destination's cumulative AoI cost as

$$F(x|\theta) \triangleq \theta \int_0^x \tilde{f}(\Delta_t) d\Delta_t = \theta\tilde{F}(x), \tag{17.26}$$

where $\tilde{F}(x)$ is the *normalized* cumulative AoI cost, given by $\tilde{F}(x) \triangleq \int_0^x \tilde{f}(\Delta_t) d\Delta_t$. We further define destination's aggregate AoI cost function as

$$H\left(\mathcal{S}\middle|\theta\right) \triangleq \int_0^T f(\Delta_t(\mathcal{S})|\theta)dt, \tag{17.27}$$

for some update policy $\mathcal{S}$.

The source only knows the statistical information regarding θ, including the probability density function (PDF) $\gamma(\theta)$ and the cumulative distribution function (CDF) $\Gamma(\theta)$. We will then design an economic mechanism to maximize the source's expected profit while satisfying some economic properties to be further elaborated.

17.5.1 Problem Formulation and Methodology

Figure 17.7 depicts the interaction between the destination and the source: the source in Stage I designs an (economic) mechanism for acquiring each destination's report of its age sensitivity and data updates; the source in Stage II reports its type $\tilde{\theta}$. A mechanism, denoted by $\mathcal{M}$, takes the destination's report (a potential misreport) of its type $\tilde{\theta}$ as the input of the update policy and for determining the payment from the destination to the source. Mathematically, a mechanism $\mathcal{M} = (\mathcal{S}^\star, P^\star)$ is a tuple of a payment rule $\mathcal{P}^\star$ and an update policy $\mathcal{S}^\star$, which are chosen functions in $\tilde{\theta}$.

The source needs to make sure that the destination should not receive a negative payoff; otherwise, it may choose not to participate in the mechanism. Such a requirement is captured by the following *individual rationality (IR)* constraint:

$$\text{IR}: \quad \min_{\tilde{\theta}\in\Theta} H\left(\mathcal{S}^\star\left(\tilde{\theta}\right)\middle|\theta\right) + P^\star\left(\tilde{\theta}\right) \le \theta\tilde{F}(T),\ \forall\theta\in\Theta. \tag{17.28}$$

The following definition of the Stackelberg game captures the interaction between the source and the destination:

GAME 2 (Source-Destination Interaction Game with Incomplete Information) *With incomplete information, the interaction between the source and the destination involves the following two stages:*

- In Stage I, the source decides on a mechanism $\mathcal{M}$ at the beginning of the period, in order to maximize its profit, given by:

$$\begin{aligned}\textbf{Source}: \max_{\mathcal{M}} \quad & \mathbb{E}_\theta\left[P^\star\left(\tilde{\theta}^\star(\mathcal{M}|\theta)\right) - C\left(\left|\mathcal{S}^\star\left(\tilde{\theta}^\star(\mathcal{M}|\theta)\right)\right|\right)\right]\\ s.t. \quad & (17.28),\end{aligned}$$

  where $\tilde{\theta}^\star(\mathcal{M}|\theta)$ is the destination's optimal reported type to be defined.
- In Stage II, given the source's designed mechanism $\mathcal{M}$, the destination decides on its report of its type $\tilde{\theta}$ to minimize its overall cost (aggregate AoI cost plus payment):

$$\textbf{Destination}: \tilde{\theta}^\star(\mathcal{M}|\theta) \in \arg\min_{\tilde{\theta}\in\Theta} H\left(\mathcal{S}^\star\left(\tilde{\theta}\right)\middle|\theta\right) + P^\star\left(\tilde{\theta}\right). \tag{17.29}$$

As in Game 2, the destination may have incentive to misreport its private information $\tilde{\theta}$. However, the following *revelation principle* [11] allows us to focus on a specific form of mechanisms:

THEOREM 17.11 (Revelation Principle [11]) *For any mechanism $\mathcal{M}$, there exists an incentive compatible (i.e., truthful) mechanism $\tilde{\mathcal{M}} = (\tilde{\mathcal{S}}^\star, \tilde{P}^\star)$ such that, for all $\theta\in\Theta$,*

$$\mathbb{E}_\theta\left[\tilde{P}^\star(\theta) - C\left(\left|\tilde{\mathcal{S}}^\star(\theta)\right|\right)\right] = \mathbb{E}_\theta\left[P^\star\left(\tilde{\theta}^\star(\mathcal{M}|\theta)\right) - C\left(\left|\mathcal{S}^\star\left(\tilde{\theta}^\star(\mathcal{M}|\theta)\right)\right|\right)\right], \tag{17.30}$$

$$\theta \in \arg\min_{\tilde{\theta}\in\Theta} \quad H\left(\tilde{\mathcal{S}}^\star\left(\tilde{\theta}\right)\middle|\theta\right) + \tilde{P}^\star\left(\tilde{\theta}\right). \tag{17.31}$$

The revelation principle allows us to restrict our attention to incentive compatible mechanisms only, that is, those that satisfy (17.31). Hence, we can replace all $\tilde{\theta}^\star(\mathcal{M}|\theta)$

by θ to restrict our attention to incentive compatible mechanisms, and impose the following *incentive compatibility* (IC) constraint:

$$\text{IC}: \quad \theta \in \arg\min_{\tilde{\theta}\in\Theta} H\left(\mathcal{S}^{\star}\left(\tilde{\theta}\right)\middle|\theta\right) + P^{\star}\left(\tilde{\theta}\right). \tag{17.32}$$

In a nutshell, the source seeks to find a mechanism $\mathcal{M}$ to maximize its profit by solving the following *mechanism optimization problem*:

$$\max_{\mathcal{M}} \quad \mathbb{E}_{\theta}[P^{\star}(\theta) - C(|\mathcal{S}^{\star}(\theta)|)] \tag{17.33a}$$

$$\text{s.t.} \quad (17.28), (17.32). \tag{17.33b}$$

This is potentially a challenging optimization problem as the space of all mechanisms is infinite-dimensional. Furthermore, the IC constraint in (17.32) *is nontrivial.* We next consider the following lemma to simplify our analysis:

LEMMA 17.12 *The update policy $\mathcal{S}^{\star}(\cdot) = \{S_k^{\star}(\cdot)\}$ of the optimal mechanism to the problem in* (17.33) *should satisfy that, there exists a $K^{\star}(\cdot)$ such that*

$$S_k^{\star}(\theta) = \frac{kT}{K^{\star}(\theta)+1}, \ \forall\theta \in \Theta, \ \forall k \in [K^{\star}(\theta)+1]. \tag{17.34}$$

Proof Sketch: Intuitively, by exploiting the KKT conditions and the Jensen's inequality, we can show that for any arbitrary optimal mechanism, there always exists an equal-spacing mechanism satisfying (17.34) that leads to at least the same profit as the optimal mechanism. □

Lemma 17.12 implies that the update policy of the optimal mechanism should be equal-spacing. Therefore, based on Lemma (17.12), we can focus on optimizing $\mathcal{M} = (K^{\star}, P^{\star})$ and then recover the optimal update policy $\mathcal{S}^{\star}$ based on (17.34).

17.5.2 Characterizations of Incentive Compatibility

The challenge of the optimization problem in (17.33) mainly lies in the characterization of the IC constraint in (17.32). To overcome such a challenge, Myerson's seminar work in [11] developed a general framework to characterize the incentive compatible mechanisms. In the following we derive the necessary and sufficient conditions of a mechanism satisfying (17.32) based on [11]:

THEOREM 17.13 *A mechanism $\mathcal{M} = (K^{\star}, P^{\star})$ is incentive-compatible if and only if there exists a set of thresholds $\boldsymbol{\theta}_{th} = \{\theta_0, \theta_1, ..., \theta_{K_{\max}}\}$ satisfying*

$$\theta_0 = \underline{\theta}, \theta_{K_{\max}} = \bar{\theta}, \theta_{k-1} \le \theta_k, \forall k \in \{1 \le k \le K_{\max}\} \tag{17.35}$$

such that, for all $k \in \{1 \le k \le K_{\max}\}$,

$$K^{\star}(\theta) = k, \ \textit{if}\, \theta \in [\theta_k, \theta_{k+1}), \tag{17.36}$$

$$P^{\star}(\theta) = \sum_{k=1}^{K^{\star}(\theta)} \theta_k \left[k\tilde{F}\left(\frac{T}{k}\right) - (k+1)\tilde{F}\left(\frac{T}{k+1}\right)\right]. \tag{17.37}$$

We present the complete proof in Section 17.8.1. To understand Theorem 17.13, we note that the described payment rule in (17.37) implies that every incentive-compatible mechanism is equivalent to a quantity-based pricing scheme $\Pi_q^\star = \{p_k^\star\}_{k\in\mathbb{N}}$ such that

$$p_k^\star = \theta_k \left[k\tilde{F}\left(\frac{T}{k}\right) - (k+1)\tilde{F}\left(\frac{T}{k+1}\right)\right], \ \forall k \in \mathbb{N}, \tag{17.38}$$

for some threshold vector $\boldsymbol{\theta}_{\rm th}$ satisfying (17.35). The equivalence is because IC ensures that, under the quantity-based pricing in (17.38), the destination would choose the update policy equivalent to (17.36). To understand why (17.38) ensures the destination to report its true type, the (equivalent) price of update k in (17.38) is set, assuming the destination's type being θ_k. Therefore, for a destination with type $\theta \in [\theta_k, \theta_{k+1})$, the payment for the first kth updates is discounted compared to the optimal quantity-based pricing under the complete information case in Proposition 3. Therefore, it has the incentive to purchase them. The destination does not have the incentive to update more as the more updates are designed for the destination with a larger type value.

The significance of Theorem 17.13 is twofold. First, it shows that all incentive-compatible mechanisms can be fully characterized by a vector $\boldsymbol{\theta}_{\rm th}$, which can thus reduce the decision space of the mechanism optimization design problem in (17.33) from the function space into the vector space. Second, it also implies *the existence of a quantity-based pricing corresponding to an optimal mechanism (satisfying* (17.28) *and* (17.32)*).*

By Theorem 17.13, the source's expected profit under an incentive-compatible mechanism is

$$J(\boldsymbol{\theta}_{\rm th}) \triangleq \sum_{k=1}^{K_{\max}} \theta_{k-1}\left((k-1)\tilde{F}\left(\frac{T}{k-1}\right) - k\tilde{F}\left(\frac{T}{k}\right)\right)\int_{\theta_{k-1}}^{\bar{\theta}} \gamma(\theta')d\theta' \\ - \sum_{k=1}^{K_{\max}} C(k)\int_{\theta_{k-1}}^{\theta_k} \gamma(\theta')d\theta'. \tag{17.39}$$

17.5.3 Mechanism Optimization Problem

Based on the characterization of the IC constraint in Theorem 17.13, we can now transform the source's mechanism optimization problem in (17.33) into the following reduced form:

$$\max_{K_{\max}\in\mathbb{N},\ \boldsymbol{\theta}_{\rm th}\in\mathbb{R}^{K_{\max}}} \quad J(\boldsymbol{\theta}_{\rm th}) \tag{17.40a}$$

$$\text{s.t.} \quad (17.35). \tag{17.40b}$$

To solve the above problem and derive insightful results, we introduce the commonly used definition of the *virtual value*, which is first introduced in [11]:

DEFINITION 17.14 (Virtual Value) The virtual value is

$$\phi(\theta) \triangleq \theta - \frac{1-\Gamma(\theta)}{\gamma(\theta)}, \ \forall \theta \in \Theta. \tag{17.41}$$

The virtual value in Definition 17.14 is a function that measures the surplus that can be extracted from the destination under incomplete information. To solve the problem in (17.40) we take the derivative of the objective in (17.40), which yields

$$\frac{\partial J(\boldsymbol{\theta}_{\text{th}})}{\partial \theta_k} = -\left(k\tilde{F}\left(\frac{T}{k}\right) - (k+1)\tilde{F}\left(\frac{T}{k+1}\right)\right)\gamma(\theta_k)\phi(\theta_k) \\ + \gamma(\theta_k)[C(k+1) - C(k)],\ \forall k \in \mathbb{N}. \tag{17.42}$$

We can then obtain the closed-form solution to the problem in (17.40) under a mild assumption, as in the following proposition:

Proposition 4 If $\phi(\theta)$ is nondecreasing, the optimal thresholds $\boldsymbol{\theta}^{\star}_{\text{th}}$ to the problem in (17.40) satisfies that

$$\phi(\theta_k^{\star}) = \left[\frac{C(k+1) - C(k)}{k\tilde{F}\left(\frac{T}{k}\right) - (k+1)\tilde{F}\left(\frac{T}{k+1}\right)}\right]_{\phi(\underline{\theta})}^{\phi(\bar{\theta})},\ \forall k \in \{1 \leq k \leq K_{\max}\}, \tag{17.43}$$

where $[\cdot]_a^b = \max\{\min\{\cdot, b\}, a\}$.

To see the feasibility of (17.43), if $\phi(\theta)$ is nondecreasing, then $\theta_k^{\star}$ is increasing in k, which thus satisfies the constraint in (17.40b).

The condition of the virtual value $\phi(\theta)$ being nondecreasing is known as the *regularity condition* in [11]. Such a nondecreasing virtual value $\phi(\theta)$ is in fact satisfied for a wide range of distributions, including the uniform distributions and truncated exponential distributions.

17.6 Numerical Results

In this section, we perform simulation studies to compare the proposed pricing schemes under complete information and the optimal mechanism under incomplete information. We then evaluate the significance of the performance gain of the profit-maximizing pricing under complete information and the impact of incomplete information.

17.6.1 Simulation Setup

We consider a convex power AoI cost function:

$$f(\theta, \Delta_t) = \theta \Delta_t^{\kappa}, \tag{17.44}$$

where the coefficient $\kappa \geq 1$ is termed the *destination's age sensitivity*. Such an AoI cost function is useful for online learning due to the recent emergence of real-time applications such as advertisement placement and online web ranking [23, 28, 29]. Hence, the cumulative AoI cost function $F(t)$ is

$$F(t) = \frac{\theta t^{\kappa+1}}{\kappa + 1}. \tag{17.45}$$

The source has a linear operational cost, that is,

$$C(K) = c \cdot K, \tag{17.46}$$

where c is the source's *operational cost coefficient.*

Let κ follow a normal distribution $\mathcal{N}(1.5, 0.2)$ truncated into the interval $[1, 2]$, let c follow a normal distribution $\mathcal{N}(50, 20)$ truncated into the interval $[0, 100]$, and let θ follow a uniform distribution $\mathcal{U}(1, 2)$. Our simulation results take an average of 100,000 experiments.

17.6.2 Performance Comparison

We compare the performances of three proposed schemes, including the optimal time-dependent pricing, the optimal quantity-based pricing, and the optimal mechanism, together with a no-update benchmark. We will show that the profit-maximizing pricing scheme, the optimal quantity-based pricing, can lead to significant profit gains compared against the no-update benchmark. In Figure 17.8(a), we first compare the four schemes in terms of the aggregate AoI and the aggregate AoI cost. The no-update scheme incurs a significantly larger aggregate AoI than all three proposed schemes. Moreover, the optimal quantity-based pricing incurs an aggregate AoI that is only 59% of that of the optimal time-dependent pricing scheme. An interesting observation is that the optimal mechanism leads to an even smaller aggregate AoI, reducing another 40% percent of that of the optimal quantity-based pricing. The reason lies in the smaller price for each update under the optimal mechanism, compared to that of the optimal quantity-based pricing scheme, and hence the destination is incentivized to update more. In terms of the aggregate AoI cost, we observe a similar trend.

In Figure 17.8(b) we compare the four schemes in terms of the social cost and the source's profit. We observe that the optimal quantity-based pricing is 27% more profitable than the optimal time-dependent pricing. In addition, the optimal time-dependent pricing incurs only 34% of the social cost of the no-update scheme. The optimal quantity-based pricing further reduces the social cost and incurs only 46% of that of the optimal time-dependent pricing. On the other hand, the optimal mechanism receives a smaller profit, even compared to the time-dependent pricing under complete information. This implies that incomplete information regarding the AoI cost may reduce the source's profit, while its impact on the social cost is relatively smaller. Therefore, *the optimal quantity-based pricing and the optimal mechanism can significantly outperform the benchmark in terms of the aggregate AoI cost and the social cost.*

17.7 Conclusions

In this chapter, we have outlined a framework to understand and tackle the economic problems of fresh data trading. We have considered a system with one data source (seller) and one destination (buyer), and studied their economic interactions under

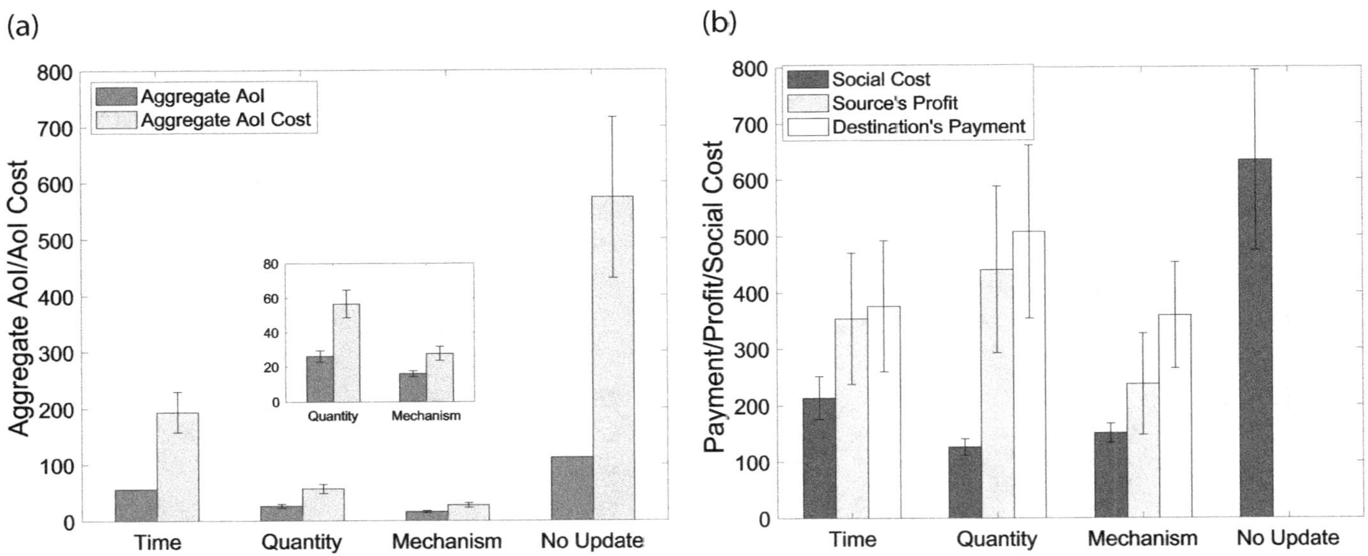

Figure 17.8 Performance comparison in terms of (a) the discounted AoI and the discounted AoI cost, and (b) discounted profit, payment, and social cost. The error bars represent the standard deviations.

both complete information and incomplete information. We have presented two key methodologies, pricing scheme design and mechanism design, for complete information scenario and incomplete information scenario, respectively. Our analysis has revealed the following key insights: (i) pricing scheme design by exploiting time dimension may not be effective under complete information; (ii) a quantity-based pricing may be the optimal pricing structure among all possible pricing schemes; (iii) the optimal mechanism under incomplete information is also equivalent to the optimal quantity-based pricing scheme.

There are some interesting future directions:

- **Pricing scheme and mechanism design for the multisource multi-destination markets.** In this case, a centralized controller serves as a broker to match sources and destinations and price fresh data. Such a problem is challenging, especially when we need to take practical system constraints (such as the general interference constraint in [35]) into account.
- **Prior-free mechanism design for fresh data markets.** Another interesting aspect is to extend our designed mechanisms to the case where the statistical information (of the age-related cost in this chapter) is also unknown. This motivates the design of prior-free mechanisms for fresh data markets.

We hope that the proposed framework and insights can pave the way for these future directions and facilitate the development of future fresh data markets.

17.8 Proofs

17.8.1 Proof of Theorem 17.13

For notational simplicity we denote the payoff of a type-θ destination reporting $\tilde{\theta}$ by

$$G\left(\tilde{\theta}\middle|\theta\right) = H\left(S^{\star}\left(\tilde{\theta}\right)\middle|\theta\right) + P^{\star}\left(\tilde{\theta}\right), \ \forall \theta, \tilde{\theta} \in \Theta. \tag{17.47}$$

By the definition of the IC constraint in (17.32), under an incentive-compatible mechanism, we must have, for any two types $t_1, t_2 \in \Theta$,

$$G(t_1|t_1) \leq G(t_2|t_1), \tag{17.48a}$$
$$G(t_2|t_2) \leq G(t_1|t_2). \tag{17.48b}$$

From (17.48) we have, for all $t_1, t_2 \in \Theta$,

$$(t_1 - t_2)\left[(K^{\star}(t_1)+1)\tilde{F}\left(\frac{T}{K^{\star}(t_1)+1}\right) - (K^{\star}(t_2)+1)\tilde{F}\left(\frac{T}{K^{\star}(t_2)+1}\right)\right] \leq 0. \tag{17.49}$$

Since $K\tilde{F}(\frac{T}{K})$ is decreasing in K, (17.49) requires that $K^{\star}(t_1) \geq K^{\star}(t_2)$ whenever $t_1 \geq t_2$. Therefore, $K^{\star}(\theta)$ must be nondecreasing in $\theta \in \Theta$.

In the following we will first prove the sufficiency of (17.37) and then the necessity of (17.37). A type-θ destination reporting $\tilde{\theta}$ will receive a payoff of

$$G\left(\tilde{\theta}\middle|\theta\right) = \left(K^*\left(\tilde{\theta}\right)+1\right)\theta\tilde{F}\left(\frac{T}{K^*(\tilde{\theta})+1}\right) + \sum_{k=1}^{K^*(\tilde{\theta})}\theta_k\left[(k-1)\tilde{F}\left(\frac{T}{k-1}\right) - k\tilde{F}\left(\frac{T}{k}\right)\right]. \tag{17.50}$$

It follows that $G(\theta|\theta) - G(\tilde{\theta}|\theta)$ satisfies that, for any $\theta \leq \tilde{\theta}$, we have

$$\begin{aligned} G(\theta|\theta) - G(\tilde{\theta}|\theta) &= \theta\left[(K^\star(\theta)+1)\tilde{F}\left(\frac{T}{K^\star(\theta)+1}\right) - (K^\star(\tilde{\theta})+1)\tilde{F}\left(\frac{T}{K^\star(\tilde{\theta})+1}\right)\right] \\ &\quad - \sum_{k=K^\star(\theta)}^{K^\star(\tilde{\theta})}\theta_k\left[(k-1)\tilde{F}\left(\frac{T}{k-1}\right) - k\tilde{F}\left(\frac{T}{k}\right)\right] \\ &\leq \theta\left[(K^\star(\theta)+1)\tilde{F}\left(\frac{T}{K^\star(\tilde{\theta})+1}\right) - (K^\star(\theta)+1)\tilde{F}\left(\frac{T}{K^\star(\tilde{\theta})+1}\right)\right] \\ &\quad - \sum_{k=K^\star(\theta)+1}^{K^\star(\tilde{\theta})}\theta\left[(k-1)\tilde{F}\left(\frac{T}{k-1}\right) - k\tilde{F}\left(\frac{T}{k}\right)\right] \\ &= 0. \end{aligned} \tag{17.51}$$

In a similar fashion, we can also show that $G(\theta|\theta) - G\left(\tilde{\theta}|\theta\right) \leq 0$ for any $\theta \geq \tilde{\theta}$. Having proven the sufficiency of (17.37), we will prove its necessity next. First, for every threshold k, it follows that

$$\lim_{\tilde{\theta}\to\theta_k^-} G\left(\tilde{\theta}\middle|\theta_k\right) = \lim_{\tilde{\theta}\to\theta_k^+} G\left(\tilde{\theta}\middle|\theta_k\right). \tag{17.52}$$

That is, each type-θ_k destination should receive exactly the overall cost. Otherwise, there exists a type-$(\tilde{\theta}_k - \epsilon)$ (or a type-$(\tilde{\theta}_k + \epsilon)$) destination that can always improve its payoff by reporting $(\tilde{\theta}_k + \epsilon)$ (or $(\tilde{\theta}_k - \epsilon)$). Therefore, from (17.52) we must have, for all thresholds θ_k,

$$\lim_{\tilde{\theta}\to\theta_k^+} P\left(\tilde{\theta}\right) - \lim_{\tilde{\theta}\to\theta_k^-} P\left(\tilde{\theta}\right) = \theta_k\left[k\tilde{F}\left(\frac{T}{k}\right) - (k+1)\tilde{F}\left(\frac{T}{k+1}\right)\right]. \tag{17.53}$$

In addition, we must have

$$P(\theta) = P(\theta'), \quad \theta, \theta' \in [\theta_k, \theta_k'), \forall k. \tag{17.54}$$

To prove this, suppose that there exist $\theta, \theta' \in [\theta_k, \theta_k')$ such that $P(\theta) < P(\theta')$. In this case, it follows that a type-θ' destination can always reduce its overall cost by reporting θ, as it reduces the payment while leading to the same total number of updates. Hence, we have proven the necessity of (17.37).

References

[1] M. A. Abd-Elmagid, N. Pappas, and H. S. Dhillon, "On the role of Age of Information in the Internet of Things," *IEEE Commun. Mag.*, vol. 57, no. 12, pp. 72–77, Dec. 2019.

[2] Accessed at https://cloud.google.com/maps-platform.

[3] S. Kaul, R. D. Yates, and M. Gruteser, "Real-time status: How often should one update?" in *Proc. IEEE INFOCOM*, 2012.

[4] C. Joe-Wong, S. Ha, and M. Chiang, "Time-dependent broadband pricing: Feasibility and benefits," in *Proc. ICDCS*, 2011.

[5] P. Hande, M. Chiang, R. Calderbank, and J. Zhang, "Pricing under constraints in access networks: Revenue maximization and congestion management," in *Proc. IEEE INFOCOM*, 2010.

[6] L. Zhang, W. Wu, and D. Wang, "Time dependent pricing in wireless data networks: Flat-rate vs. usage-based schemes," in *Proc. IEEE INFOCOM*, 2014.

[7] S. Ha, S. Sen, C. Joe-Wong, Y. Im, and M. Chiang, "TUBE: Time-dependent pricing for mobile data," *ACM SIGCOMM Comput. Commun. Rev.*, vol. 42, no. 4, pp. 247–258, 2012.

[8] R. L. Phillips, "Pricing and revenue optimization." *Stanford University Press,* 2005.

[9] Accessed at https://cloud.google.com/apigee-api-management.

[10] Accessed at https://rimuhosting.com.

[11] R. B. Myerson, "Incentive compatibility and the bargaining problem," *Econometrica*, vol. 47, no. 1, pp. 61–73, Jan. 1979.

[12] Z. Huang, S. M. Weinberg, L. Zheng, C. Joe-Wong, and M. Chiang, "Discovering valuations and enforcing truthfulness in a deadline-aware scheduler," in *Proc. IEEE INFOCOM*, 2007.

[13] Z. Huang, S. M. Weinberg, L. Zheng, C. Joe-Wong, and M. Chiang, "RUSH: A robust scheduler to manage uncertain completion-times in shared clouds," in *Proc. IEEE ICDCS*, 2016.

[14] M. Zhang, A. Arafa, J. Huang, and H. V. Poor, "Pricing fresh data," *IEEE J. Sel. Areas Commun.*, vol. 39, no. 5, pp. 1211–1225, May 2021.

[15] S. Hao and L. Duan, "Regulating competition in age of information under network externalities," *IEEE J. Sel. Areas Commun.*, vol. 38, no. 4, pp. 697–710, April 2020.

[16] X. Wang and L. Duan, "Dynamic pricing and mean field analysis for controlling age of information," Accessed at https://arxiv.org/abs/2004.10050.

[17] B. Li, and J. Liu, "Can we achieve fresh information with selfish users in mobile crowd-learning?" in *Proc. WiOpt*, 2019.

[18] M. Zhang, A. Arafa, E. Wei, and R. A. Berry, "Optimal and quantized mechanism design for fresh data acquisition," *IEEE J. Sel. Areas Commun.*, vol. 39, no. 5, pp. 1226–1239, May 2021.

[19] A. Agarwal, M. Dahleh, and T. Sarkar, "A marketplace for data: An algorithmic solution," in *Proc. ACM Econ. Comput.*, pp. 701–726, 2019.

[20] A. Sundararajan, "Nonlinear pricing of information goods," *Manage. Sci.*, vol. 50, no. 12, pp. 1660–1673, Dec. 2004.

[21] H. Shen and T. Basar, "Optimal nonlinear pricing for a monopolistic network service provider with complete and incomplete Information," *IEEE J. Sel. Areas Commun.*, vol. 25, no. 6, pp. 1216–1223, Aug. 2007.

[22] C. Kam, S. Kompella, and A. Ephremides, "Age of information under random updates," in *Proc. IEEE ISIT*, 2013.

[23] Y. Sun, E. Uysal-Biyikoglu, R. D. Yates, C. E. Koksal, and N. B. Shroff, "Update or wait: How to keep your data fresh," *IEEE Trans. Inf. Theory*, vol. 63, no. 11, pp. 7492–7508, Nov. 2017.

[24] R. D. Yates, "Lazy is timely: Status updates by an energy harvesting source," in *Proc. IEEE ISIT*, 2015.

[25] A. Arafa, J. Yang, S. Ulukus, and H. V. Poor, "Age-minimal transmission for energy harvesting sensors with finite batteries: Online policies," *IEEE Trans. Inf. Theory*, vol. 66, no. 1, pp. 534–556, Jan. 2020.

[26] A. Arafa and S. Ulukus, "Timely updates in energy harvesting two-hop networks: Offline and online policies," *IEEE Trans. Wireless Commun.*, vol. 18, no. 8, pp. 4017–4030, Aug. 2019.

[27] X. Wu, J. Yang, and J. Wu, "Optimal status update for age of information minimization with an energy harvesting source," *IEEE Trans. on Green Commun. Netw.*, vol. 2, no.1, pp. 193–204, March 2018.

[28] S. Shalev-Shwartz, "Online learning and online convex optimization," *Found. Trends Mach. Learn.*, vol. 4, no. 2, pp. 107–194, 2012.

[29] X. He et al., "Practical lessons from predicting clicks on ads at Facebook," in *Proc. 8th Int. Workshop Data Mining Online Advertising*, 2014.

[30] G. Manimaran and C. S. R. Murthy, "An efficient dynamic scheduling algorithm for multiprocessor real-time systems," *IEEE Trans. Parall. Distr.*, vol. 9, no. 3, pp. 312–319, March 1998.

[31] R. B. Myerson, *Game theory*, Harvard University Press, 2013.

[32] S. Sen, C. Joe-Wong, S. Ha, and M. Chiang, "A survey of smart data pricing: Past proposals, current plans, and future trends," *ACM Comput. Surv.*, vol. 46, no. 2, Nov. 2013.

[33] S. Sen, C. Joe-Wong, S. Ha, and M. Chiang, "Incentivizing time-shifting of data: A survey of time-dependent pricing for Internet access," *IEEE Commun. Mag.*, vol. 50, no. 11, pp. 91–99, Nov. 2012.

[34] B. Colson, P. Marcotte, and G. Savard, "An overview of bilevel optimization," *Ann. Oper. Res.*, vol. 153, no. 1, pp. 235–256, 2007.

[35] R. Talak, S. Karaman, and E. Modiano, "Optimizing information freshness in wireless networks under general interference constraints," *IEEE/ACM Trans. Netw.*, vol. 28, no. 1, pp. 15–28, Feb. 2019.

18 UAV-Assisted Status Updates

Juan Liu, Xijun Wang, Bo Bai, and Huaiyu Dai

18.1 Abstract

With the rapid development of unmanned aerial vehicles (UAVs), extensive attentions have been paid to UAV-aided data collection in wireless sensor networks. However, it is very challenging to maintain the information freshness of the sensor nodes (SNs) subject to the UAV's limited energy capacity and/or the large network scale. This chapter introduces two modes of data collection: single and continuous data collection with the aid of UAV, respectively. In the former case, the UAVs are dispatched to gather sensing data from each SN just once according to a preplanned data collection strategy. To keep information fresh, a multistage approach is proposed to find a set of data collection points at which the UAVs hover to collect and the age-optimal flight trajectory of each UAV. In the later case, the UAVs perform data collection continuously and make real-time decisions on the uploading SN and flight direction at each step. A deep reinforcement learning (DRL) framework incorporating deep Q-network (DQN) algorithm is proposed to find the age-optimal data collection solution subject to the maximum flight velocity and energy capacity of each UAV. Numerical results are presented to show the effectiveness of the proposed methods in different scenarios.

18.2 Introduction

18.2.1 UAV-Enabled Wireless Sensor Networks

Recently, UAVs have become enablers of many important applications and have been widely applied in wireless sensor networks (WSNs) [1]. The UAV-enabled WSN is regarded as a special mobile WSN [2], where the UAVs can move quickly, operate flexibly, and provide effective and fair coverage for ground sensor nodes (SNs). In particular, they can provide timely aerial data services such as search and rescue [3, 4]. The fully controllable mobility of UAVs enables them to move very close to the ground devices and establish low-altitude air-to-ground communication links [5]. Accordingly, the transmission energy of sensor devices is reduced, and the lifetime of the WSN is prolonged. However, the UAV mobility inevitably leads to changes of network topology and hence reduces reliability in data collection. Hence, it is very important to manage mobility and allocate system resources efficiently in UAV-aided communication [6–12]. This chapter investigates efficient UAV-aided data collection and focuses on keeping information freshness in different data collection scenarios.

18.2.2 AoI-Oriented Data Collection: Challenges and Related Works

The freshness is an important metric, referred to as the age of information (AoI) or status age [13], which is very crucial in time-sensitive applications [14–17]. AoI is defined as the amount of time elapsed since the instant at which the freshest delivered update takes place [14]. In UAV-enabled WSNs, the major challenge lies in the design of AoI-oriented data collection strategies subject to the limited energy capacity of each UAV and/or the large network scale. Specifically, the mobility of UAV changes its relative distances with ground SNs, leading to unreliable data transmissions and deteriorated AoI performance. AoI-oriented trajectory planning schemes should be carefully designed subject to the UAVs' maximum flight velocity and limited energy capacity. With the increase of network scale, one single UAV with limited energy capacity may not fulfill the data collection task with guaranteed information freshness. Instead, multiple UAVs should take part in data collection collaborately to meet the energy constraint and AoI requirement. Accordingly, it is a big challenge to schedule the multi-UAV aided data collection and energy replenishment to provide sustainable and timely data service.

Recently, many related works on AoI-oriented UAV communications and trajectory planning have been conducted [18–23]. In [18] the authors studied the age-optimal information gathering and dissemination problems on graphs, where a mobile agent flies along the randomized trajectory that is designed to minimize peak and average AoIs. With the aid of convex programming method, [19] optimized the UAV's flight trajectory, energy, and service time allocation for packet transmissions in order to minimize the average peak AoI for a source-destination pair. In our recent works [20, 21], we proposed a joint SN association and trajectory planning strategy to strike balance between the SNs' uploading time and the UAV's flight time, and thus to minimize the SNs' average and maximal AoIs. In [22], the authors studied the UAV path planning and data acquisition problem while considering the data acquisition mode selection, the node's energy consumption, and age evolution of collected information. The authors of [23] studied the route scheduling of a UAV for fresh data collection over a finite time horizon and developed a graph-labeling-based algorithm to achieve a trade-off between complexity and solution quality.

With the development of the artificial intelligence (AI) technology, increasing interests are aroused in applying learning algorithms in UAV trajectory planning [24–28]. In [24] the authors studied the UAV trajectory planning problem for data collection with the objective of minimizing the number of expired data packets and solved it using the Q-learning-based method. [25] designed AoI-based UAV trajectory planning algorithms using the deep Q-Network (DQN)-based approach in Internet of Things (IoT) networks without considering energy consumption at the IoT devices or UAV. To minimize the normalized weighted sum of AoI values, the UAV's flight trajectory and scheduling of status updates were jointly optimized using the DQN-based approach in [26, 27]. In [28], the authors studied the optimal multi-UAV navigation policy using the DQN-based approach in IoT networks where the data freshness and energy efficiency were improved by optimizing the UAVs' trajectories.

The above works mainly investigated the age-optimal SN scheduling and trajectory planning in single-UAV scenarios without considering the impacts of the sampling and queueing processes of the ground SNs and the trajectories of the energy-constrained UAVs on their AoI values. However, AoI is highly affected by the arriving instants of status update packets at each SN and the packet-relaying mechanism used by the UAVs. Hence, the sampling and queueing models of the SNs and the energy consumption model of the UAVs should be considered in designing age-optimal SN scheduling and UAV trajectory planning.

18.2.3 The AoI-Oriented Data Collection Framework

Motivated by the above works, we consider two modes of UAV-aided data collection: single and continuous data collection, respectively. Then, offline and online strategies are designed using optimization and learning algorithms to improve the AoI performance, respectively.

In the first mode, the UAVs are dispatched to gather sensing data from each SN just once according to a preplanned data collection strategy. A two-stage data collection framework is proposed to find a suboptimal solution to the AoI minimization problem in the single-UAV scenario. Firstly, an Affinity Propagation (AP)-based algorithm with weight δ is applied to find a set of data collection points (CPs) at which the UAV hovers to collect and establish the SN-CP association. Secondly, an age-optimal flight trajectory is found using dynamic programming (DP) or genetic algorithm (GA). The two stages are iteratively performed until the weight δ is optimized to achieve the minimum AoI. This framework can be extended to the multi-UAV scenario, taking into account the energy constraint at each UAV. This framework adapts to the case where the UAVs' data collection operations are prescheduled based on offline computations.

In the second mode, the UAVs perform the data collection task continuously and make real-time decisions on the uploading SN and the UAVs' flight directions at each step. A Markov decision process (MDP) is formulated to minimize the long-term cost, which is the weighted sum of the SNs' average AoI and energy consumption at the UAVs. Considering the SNs' different sampling and queueing processes and AoI values, the varying energy, and location of each UAV, the MDP problem could have large state and action spaces. Hence, it is very challenging to solve the MDP problem with traditional methods such as value iteration. To deal with this issue, this chapter presents a deep reinforcement learning (DRL) framework in which efficient learning algorithms, for example, DQN, can be used to find the age-optimal data collection scheme. This learning-based framework is suitable to large-scale networks or the case without prior network information. Numerical results are presented to show the effectiveness of the proposed method in the modes of single and continuous UAV-aided data collection.

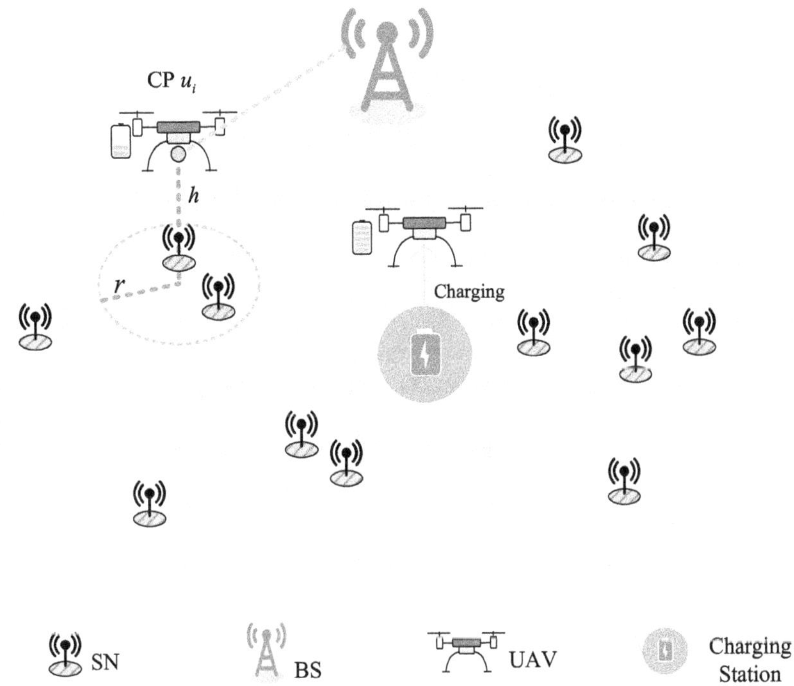

Figure 18.1 The UAV-assisted WSN model.

18.2.4 Outline

Section 18.3 presents the UAV-enabled WSN. Two typical UAV-aided data collection modes are introduced in Section 18.4. Accordingly, two-stage and DRL-based frameworks are designed to find AoI-optimal data collection strategies in Section 18.5 and Section 18.6, respectively. Numerical results are presented in Section 18.7. Section 18.8 presents concluding remarks and potential future works.

18.3 UAV-Aided Data Collection in WSNs

18.3.1 Network Description

Consider a UAV-enabled WSN that consists of one BS v_0 and M sensor nodes (SNs), denoted by $\mathcal{V}=\{v_1, v_2, \cdots, v_M\}$, as shown in Figure 18.1. The location of each ground node including the BS is expressed as $\mathbf{s}_m=(x_m, y_m, 0) \in \mathbb{R}^3$ $(m=0,1,\cdots,M)$ in a three-dimensional Cartesian coordinate system. Several multi-rotor UAVs denoted by $\mathcal{A}=\{a_1, a_2, \cdots, a_K\}$ are dispatched to collect information. The UAVs operate in either flying or hovering. The flight trajectory of UAV a_k consists of the points visited by itself, and its coordinate can be denoted as $\mathbf{u}_k(t)=(X_k(t), Y_k(t), H_k(t)) \in \mathbb{R}^3$, where t is the time index. The SN v_m is said to be in the coverage of UAV a_k, if their horizontal distance

$d^{hor}_{m,k} = \parallel \mathbf{s}_m - \mathbf{u}_k(t) \parallel = \sqrt{(x_m - X_k(t))^2 + (y_m - Y_k(t))^2}$ is less than or equal to the UAV's coverage radius r. Then, each UAV a_k is able to establish communication links with the SNs in its coverage and collect their sensed data. During the data collection period, each UAV is allowed to get energy replenishment in the WSN. For example, the UAV can fly back directly to the BS or some charging station for energy recharging, when the residual energy reaches some threshold, as shown in Figure 18.1.

18.3.2 Sampling Model: Passive Sampling and Proactive Sampling

In the WSN, the SNs are responsible for monitoring the environment and sensing the information of interest. The information sampling models can be divided into two categories: passive sampling and proactive sampling. With passive sampling, the information sampling at the each SN is event-driven. The sampling process is activated at any time when the SN receives a corresponding signal, following the generate-at-will rule [16]. With proactive sampling, each SN senses the environment and generates samples at some rate according to its own working mode. For example, the interval of two samples follows an exponential distribution if the sampling process of one SN is a Poisson process.

18.3.3 Channel Model

There might exist both line-of-sight (LoS) and non–line-of-sight (NLoS) links between the UAVs and the ground SNs. In addition to the free space path loss, LoS and NLoS links have different excessive path losses [29, 30]. At time t, the channel gain between UAV a_k and SN v_m is given by

$$h_{m,k}(t) = \begin{cases} \beta_0[d_{m,k}(t)]^{-\epsilon}, & \text{LoS link,} \\ \varrho\beta_0[d_{m,k}(t)]^{-\epsilon}, & \text{NLoS link,} \end{cases} \tag{18.1}$$

where β_0 denotes the channel gain at the reference distance of 1 meter, ϵ is the path loss exponent, $\varrho \in (0, 1)$ is the additional attenuation factor due to the NLoS effects, and $d_{m,k}(t) = \sqrt{(x_m - X_k(t))^2 + (y_m - Y_k(t))^2 + (H_k(t))^2}$ is the distance between UAV a_k and SN v_m. The LoS link exists with a probability given by [31]:

$$P^{LoS}_{m,k}(t) = \frac{1}{1 + a\exp\left(-b\left[\frac{180}{\pi}\arcsin\left(\frac{H_k(t)}{d_{m,k}(t)}\right) - a\right]\right)}, \tag{18.2}$$

where a and b are constants related to the environment. The probability of NLoS is thus $P^{NLoS}_{m,k}(t) = 1 - P^{LoS}_{m,k}(t)$. When the UAV hovers over this SN, the distance is $d_{m,k}(t) = H_k(t)$ and the LoS probability is approximately equal to one.

When SN v_m uploads data to UAV a_k, the data rate can be expressed as $R_{m,k}(t) = W\log_2\left(1 + \frac{h_{m,k}(t)P_m(t)}{\sigma^2}\right)$, where $P_m(t)$ is the transmit power of SN v_m, and W and σ^2 denote the system bandwidth and noise power at the UAV receiver, respectively. Similarly, when UAV a_k transmits data to the BS at power $P_u(t)$, the data rate is evaluated as $R_{k,BS}(t) = W \log_2 \left(1 + \frac{h_{k,BS}(t) P_u(t)}{\sigma^2}\right)$, where $h_{k,BS}(t)$ is the gain of the UAV-BS channel.

18.3.4 Propulsion Power Model

For a rotary-wing UAV, the propulsion power consumption mainly depends on the UAV's flight velocity and acceleration. Since the acceleration/deceleration time is very small, the power consumed for acceleration can be ignored. When the UAV flies at velocity V, the propulsion power includes the lade profile power, induced power, and parasite power [30], given by

$$F(V) = P_0 \left(1 + \frac{3V^2}{U_{tip}^2}\right) + P_1 \sqrt{\sqrt{1 + \frac{V^4}{4V_0^4}} - \frac{V^2}{2V_0^2}} + \frac{1}{2} d\rho s A_r V^3, \tag{18.3}$$

where P_0 and P_1 are the blade profile and induced powers of the UAV in the hovering status, respectively; V_0 is the mean rotor induced velocity of the UAV in hovering; U_{tip} is the tip speed of the rotor blade; ρ is the air density; and d, s, and A_r are the fuselage drag ratio, rotor solidity, and the rotor disc area, respectively. When the flight velocity is $V = 0$, the UAV is hovering and its power consumption is expressed as $F(0) = P_0 + P_1$.

18.4 Two Typical UAV-Aided Data Collection Modes

18.4.1 Mode 1: A Preplanned Single Data Collection

In this mode, the UAV-aided data collection task is preplanned and performed only once each time. Specifically, K UAVs are dispatched from the BS to fly and hover to collect information from the M SNs following a prescheduled flight trajectory. The sensing data collected by each UAV is offloaded to the BS for data processing when the UAV flies back to the BS.

To keep information fresh, the SNs sample the environment using the passive sample mode and upload data to the UAVs. The sampling time and communication overhead of each SN are assumed to be negligible. Each of the UAVs can fly over some SN and hover to collect its sensing data. Alternatively, it hovers to collect data from multiple SNs in its coverage. The points at which the UAVs hover to collect data are called data collection points (CPs). The design of UAV-aided data collection boils down to the coupled SN-CP association and trajectory planning for each UAV. Let $\mathcal{C} = \{c_1, c_2, \cdots, c_L\}$ denote the set of CPs, where L is the number of CPs. Similarly, the set of CPs visited by UAV a_k is denoted by $\mathcal{C}_k \subseteq \mathcal{C}$ and the number of CPs in $\mathcal{C}_k$ is denoted by $L_k = |\mathcal{C}_k|$, respectively. The horizontal location of each CP c_l is denoted by $\mathbf{s}_l^c = (x_l^c, y_l^c) \in \mathbb{R}^2$ $(l = 1, \cdots, L)$, and $\mathbf{s} = [\mathbf{s}_1^c, \mathbf{s}_2^c, \cdots, \mathbf{s}_L^c]$ is the vector of the CPs' positions.

Flight Trajectory

To save flight time and energy, each UAV can fly directly from one CP to another one. The flight trajectory of each UAV a_k can be represented by the sequence of the CPs visited by this UAV. It is expressed as a vector $\mathbf{c}_k = [c_{\pi_k(1)}, \cdots, c_{\pi_k(L_k)}]$, where π_k is a permutation of the indices of all the CPs and $c_{\pi_k(j)} \in \mathcal{C}$ is the jth CP in the trajectory of UAV a_k. Hence, the flight trajectory can be expressed as $\mathbf{u}_k = [c_{\pi_k(0)}, \mathbf{c}_k, c_{\pi_k(L_k+1)}]$, where $c_{\pi_k(0)} = c_{\pi_k(L_k+1)} = v_0$. Thus, $\mathbf{u} = [\mathbf{u}_1, \mathbf{u}_2, \cdots, \mathbf{u}_K]$ represents the trajectories of all the UAVs. Without loss of generality, each UAV a_k flies at a fixed flight velocity V_k and an appropriate altitude H_k. The UAV's flight time between any two CPs c_i and c_j is calculated as $\tau_{i,j} = V_k^{-1}||\mathbf{s}_i^c - \mathbf{s}_j^c||$. The time index t can be omitted, since the flight trajectory of each UAV with respect to time t can be uniquely described by the visiting sequence of the CPs and the flight time between any two CPs.

In the single-UAV scenario, the UAV flies across all the L CPs and collects data from the SNs in the coverage of each CP. In the multi-UAV scenario, there exist associations between the K UAVs and L CPs. A binary parameter $\eta_{l,k} \in \{0, 1\}$ is used to represent the association between CP c_l and UAV a_k. That is, $\eta_{l,k} = 1$ means CP c_l is visited by UAV a_k, and otherwise $\eta_{l,k} = 0$. Since each CP is just visited by one UAV, $\sum_{k=1}^{K} \eta_{l,k} = 1$ holds for any CP c_l. Let $\eta = [\eta_{l,k}]$ be the CP-UAV association vector.

SN-CP Association

Each SN is associated with just one single CP even when it is covered by multiple CPs. For ease of exposition, we use $\zeta_{m,l} \in \{0, 1\}$ to denote a binary indicator specifying whether SN v_m is visited by one UAV over CP c_l. In particular, $\zeta_{m,l} = 1$ means that SN v_m is associated with CP c_l and otherwise $\zeta_{m,l} = 0$. The SN-CP association can be represented by the vector $\zeta = [\zeta_{m,l}]$ that satisfies $\sum_{l=1}^{L} \zeta_{m,l} = 1$ for each SN v_m. All the SNs are visited by the UAVs through the CPs, that is, $\sum_{m=1}^{M} \sum_{l=1}^{L} \sum_{k=1}^{K} \zeta_{m,l} \cdot \eta_{l,k} = M$.

The SNs are classified into L $(1 \leq L \leq M)$ clusters, and each cluster of SNs is associated with one CP and visited by one of the UAVs. In a special case, each SN belongs to one cluster, and a total of M clusters are created, that is, $L = M$. At each CP the associated SNs can be scheduled to upload their sensed data to the visiting UAV by some multiple access scheme. For example, when a simple TDMA transmission scheme is applied, the SNs associated with each CP upload data to the UAV sequentially. The set of SNs associated with the lth CP of the trajectory $\mathbf{u}_k$ is denoted by $\mathcal{V}_{\pi_k(l)}$. Let $M_{\pi_k(l)}$ denote the number of SNs in the set $\mathcal{V}_{\pi_k(l)}$. If SN v_m is the jth one that uploads to the UAV a_k over the lth CP (or CP $c_{\pi_k(l)}$), it is also labeled as SN $v_{\pi_k(l)}^j \in \mathcal{V}$ $(j = 1, \cdots, M_{\pi_k(l)})$, and the SN's uploading time is rewritten as $T_{\pi_k(l)}^j$. Here, when $v_m = v_{\pi_k(l)}^j$ is the jth uploading SN at CP $c_{\pi_k(l)}$, the uploading time $T_{\pi_k(l)}^j$ can be evaluated as $T_{m,\pi_k(l)} = \mathbb{E}\{L_m/R_{m,\pi_k(l)}\}$, where L_m is the length of a data packet at SN v_m, and $\mathbb{E}$ means the expectation over the LoS and NLoS links. Also, $\mathbf{v}_k$ is used to denote the vector of all the sequentially ordered SNs collected by UAV a_k, and $\mathbf{v} = [\mathbf{v}_1, \mathbf{v}_2, \cdots, \mathbf{v}_K]$ is the vector of the SNs' uploading sequence.

Performance Metrics: AoI and Energy Consumption

With the passive sampling model, each SN samples the sensing data at the time when it receives a signal from the visiting UAV. Hence, the AoI of this SN depends on the data uploading time and the UAV's flight time afterward. We use $\Gamma_m(t)$ to track the age of information of SN v_m at time t. When TDMA is adopted and SN v_m happens to be the jth uploading SN of CP $c_{\pi_k(l)}$, the AoI of this SN can be expressed as [21]:

$$\Gamma_m = \Gamma^j_{\pi_k(l)} = \sum_{i=j}^{M_{\pi_k(l)}} T^i_{\pi_k(l)} + \sum_{n=l+1}^{L_k} T_{\pi_k(n)} + \sum_{l=j}^{L_k} \tau_{\pi_k(l),\pi_k(l+1)} + T^k_o, \forall m, \tag{18.4}$$

where $T_{\pi_k(l)}$ is the overall hovering time of UAV a_k at CP $c_{\pi_k(l)}$, $\tau_{\pi_k(l),\pi_k(l+1)}$ is the flight time of UAV a_k from CP $c_{\pi_k(l)}$ to CP $c_{\pi_k(l+1)}$, and T^k_o is the time of offloading data from UAV a_k to the BS. In (18.4), the hovering time $T_{\pi_k(l)}$ can be evaluated by the total amount of uploading time of the SNs covered by this CP $c_{\pi_k(l)}$, that is, $T_{\pi_k(l)} = \sum_{j=1}^{M_{\pi_k(l)}} T^j_{\pi_k(l)}$. $M_k = \sum_{l=1}^{L_k} M_{\pi_k(l)}$ is the total number of SNs collected by UAV a_k. In the same way, the offloading time is calculated as $T^k_o = \frac{\sum_{l=1}^{L_k} \sum_{v_m \in \mathcal{V}_{\pi_k(l)}} L_m}{R_{k,BS}}$, where $R_{k,BS}$ is the offloading data rate when the UAV hovers over and offloads to the BS.

For each UAV, the total amount of energy consumption is a function of the trajectory $\mathbf{u}_k$ and can be calculated as

$$e_k(\mathbf{u}_k) = F(V_k) \sum_{l=0}^{L_k} \tau_{\pi_k(l),\pi_k(l+1)} + F(0) \cdot \sum_{l=1}^{L_k} T_{\pi_k(l)} + P_u \cdot T^k_o, \tag{18.5}$$

where the three terms mean the propulsion energy, the hovering energy, and the transmission energy, respectively.

18.4.2 Mode 2: Continuous Real-Time Data Collections

In this mode, the SNs sample the environment with proactive sampling and generate the samples of sensing data continuously. To maintain information freshness, the UAVs should make real-time data collection decisions on the flight directions and uploading SNs. For simplicity, the decision space is supposed to be finite, and the target region is divided into J equally-sized square grids in the time-slotted WSN. The length and width of a grid are denoted by L_x and L_y, respectively. The location of the center of grid i is denoted by $\mathbf{w}_i = (w_{x,i}, w_{y,i}) \in \mathbb{R}^2$ $(i = 1, 2, \cdots, J)$ and the set containing the center coordinates of all the grids is denoted by $\mathcal{W} = \{\mathbf{w}_1, \cdots, \mathbf{w}_J\}$. The duration of each slot is equal to T_s seconds. Without causing confusion, the discrete slot index n is used instead of the time index t in this scenario.

UAV Aided Relaying

The UAV-aided data collection consists of two stages. In the first stage, the target SN v_m transmits one update packet of L_m bits to the UAV a_k through a LoS or NLoS link when the UAV hovers over some grid center $\mathbf{w}_i$. In the second stage, the UAV forwards the update packet to the BS. From (18.1), the NLoS path loss is typically larger than the

LoS path loss due to the obstacles in the propagation path. Accordingly, the data rate of a NLoS link $R^{nlos}_{m,k}(n)$ is smaller than that of a LoS link $R^{los}_{m,k}(n)$, and hence it takes more time to upload on the NLoS link. The mobile UAV should schedule its transmit power to meet the transmission time requirement, that is, $\frac{L_m}{R^{nlos}_{m,k}(n)} + \frac{L_m}{R^{nlos}_{k,BS}(n)} = T_{tx}$, where T_{tx} is the overall transmission time. In other words, the UAV a_k relays the update packet of SN v_m at a transmit power,

$$P^k_u(n) = \frac{\sigma^2[d_{k,BS}(n)]^\epsilon}{\varrho\beta_0}\left[2^{W^{-1}\left(\frac{T_{tx}}{L_m} - \frac{1}{R^{nlos}_{m,k}(n)}\right)^{-1}} - 1\right], \tag{18.6}$$

which depends on the SN-UAV distance $d_{m,k}(n)$ and the UAV-BS distance $d_{k,BS}(n)$ that change with the UAV's trajectory $\mathbf{u}_k(n)$. After packet delivery it directly flies from the current grid to one of its adjacent grids due to the UAV's maximum flight speed limitation. Accordingly, one slot can be divided into two subslots: one for the UAV-aided data gathering, and the other for the UAV's flight. The subslot length for flight is denoted by T_{fl}, which satisfies $T_{tx} + T_{fl} = T_s$. If no SN exists in the current grid or the SN's buffer is empty, no packet delivery takes place. The UAV directly flies to one of the neighboring grids based on its decision.

Sampling and Queueing System

Consider two proactive sampling scenarios: fixed sampling and random sampling. By fixed sampling, the sampling interval of each SN is fixed and periodic. In random sampling, the SN's sampling interval is random and follows some distribution, for example, a Poisson process[32].

The UAV-aided data collection system can be modeled as an equivalent queueing system. We use $o_m(n)$ and $z_m(n)$ to denote the update packet arrival and service processes of SN v_m in the nth timeslot, respectively. Specifically, $o_m(n) = 1$ means that one update packet is generated and stored in the buffer of SN v_m, and otherwise $o_m(n) = 0$. Assume that the SNs follow the simple sample-and-replace policy. When an update packet is newly generated, the "old" packet waiting in the buffer has to be discarded and replaced by this new one, since it carries fresher information. Therefore, each update packet is queued in the buffer until it is collected by one UAV or replaced by a new packet. The service process is $z_m(n) = 1$, if an update packet of SN v_m located in grid i is collected by the visiting UAV a_k, and otherwise $z_m(n) = 0$. The queue length of each SN v_m, denoted by $q_m(n)$, is updated as

$$q_m(n) = \min\{q_m(n-1) + o_m(n), 1\} - z_m(n), \tag{18.7}$$

where $\min\{q_m(n-1)+o_m(n), 1\}$ means at most one update packet is ready for delivery.

To capture the impact of sampling and queueing on the AoI, we use $U_m(n)$ to track down the lifetime of the update packet queued at SN v_m at the end of slot n, given by

$$U_m(n) = \begin{cases} 0, & \text{if} z_m(n) = 1 \\ 1, & \text{if} z_m(n) = 0 \& o_m(n) = 1 \\ U_m(n-1) + 1, & \text{otherwise.} \end{cases} \tag{18.8}$$

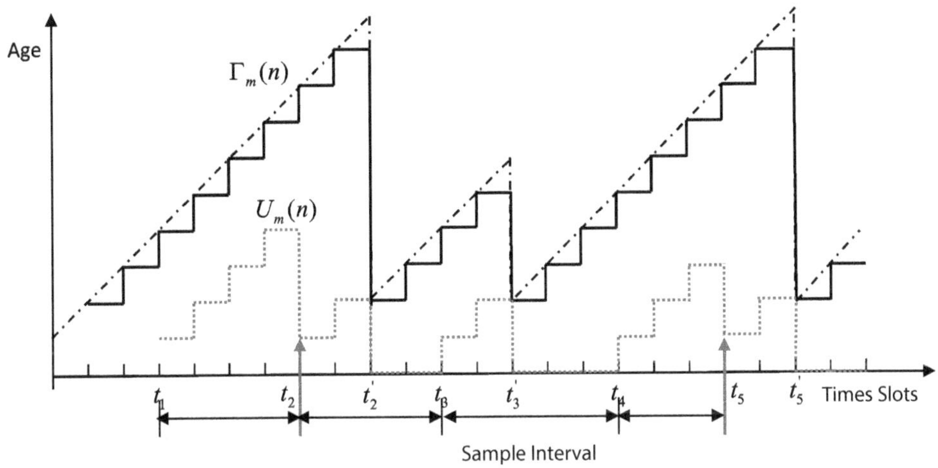

Figure 18.2 Illustration of the AoI of SN v_m and the lifetime of its update packet.

The buffer of SN v_m becomes empty and $U_m(n) = 0$ after a packet delivery, that is, $z_m(n) = 1$. Otherwise, when $o_m(n) = 1$, a new update packet arrives at SN v_m and its lifetime is set to one. In the other cases, $U_m(n) = U_m(n-1) + 1$, indicating that the lifetime of the update packet in the queue linearly increases by one. The lifetime $U_m(n)$ is also illustrated in Figure 18.2.

Performance Metrics: AoI and Energy Consumption

Denote by $\Gamma_m(n) \in \mathcal{I}_m$ the AoI of SN v_m in slot n, where $\mathcal{I}_m \triangleq \{1, 2, 3, \cdots, I_{max}\}$ is a set containing all possible AoI values of SN v_m. Here, I_{max} denotes the maximum value that $\Gamma_m(n)$ can take, which can be an arbitrarily large integer. As shown in Figure 18.2, the evolution of $\Gamma_m(n)$ can be expressed as

$$\Gamma_m(n+1) = \begin{cases} U_m(n), & \text{if } z_m(n) = 1, \\ \Gamma_m(n) + 1, & \text{otherwise,} \end{cases} \tag{18.9}$$

where the AoI of SN v_m increases linearly with time and is reset to the lifetime of the SN's update packet when this packet is delivered.

In each time slot, each UAV a_k could take three actions: flight, hovering, and transmission. Accordingly, the UAV's propulsion energy for flight depends on the flight velocity $V_k(n)$ and the flight time T_{fl}, given by $e_{fl}^k(n) = F(V_k(n)) \cdot T_{fl}$, where $V_k(n) = \frac{\|\mathbf{u}_k(n+1) - \mathbf{u}_k(n)\|}{T_{fl}}$. The transmit power is given by (18.6), and the maximum transmission time is $\frac{L_m}{R_{k,BS}^{nlos}(n)}$. Hence, the UAV's energy consumption for packet transmission can be approximately evaluated as $e_{tx}^k(n) = P_u^k(n) \frac{L_m}{R_{k,BS}^{nlos}(n)}$. The UAV's hovering energy is $e_{hv}^k(n) = F(0) \cdot T_{tx} = (P_0 + P_1) T_{tx}$, which depends on the hovering (or transmission) time T_{tx}.

18.5 A Multistage Framework for Single Data Collection

For the single-UAV scenario, an AoI minimization problem is first formulated, and a two-stage approach is proposed to find the AoI-optimal data collection solution. Then, this framework is extended to the multi-UAV scenario.

18.5.1 Problem Formulation

In the single-UAV scenario, one UAV is able to finish the data collection task without considering the energy constraint. The index k can be omitted with one UAV, that is, $K = 1$, when no confusion is caused. Since the SN first visited by the UAV has the "oldest" information, the SNs' maximum AoI given by (18.4) can be rewritten as

$$\Gamma_{max}(\mathbf{s}, \zeta, \mathbf{u}) = \sum_{m=1}^{M}\sum_{l=1}^{L} T_{m,l}\zeta_{m,l} + \sum_{l=1}^{L} \tau_{\pi(l),\pi(l+1)} + T_o, \tag{18.10}$$

where the first term is the aggregate uploading time of all the SNs. Hence, the maximum AoI of the SNs is a function of the CPs' positions $\mathbf{s}$, the SN-CP association ζ, and the UAV's flight trajectory $\mathbf{u}$. Similarly, the SNs' average AoI can be derived as

$$\begin{aligned}\Gamma_{ave}(\mathbf{s}, \zeta, \mathbf{v}, \mathbf{u}) =& \sum_{l=1}^{L}\sum_{j=1}^{M_{\pi(l)}} \frac{j}{M} T^{j}_{\pi(l)} \\ &+ \sum_{l=2}^{L}\sum_{n=1}^{l-1} \frac{M_{\pi(n)}}{M}\left(\tau_{\pi(l-1),\pi(l)} + T_{\pi(l)}\right) + \tau_{\pi(L),\pi(L+1)} + T_o,\end{aligned} \tag{18.11}$$

where the first term is the weighted uploading time, and the last three terms capture the weighted time since the UAV departs from the first CP. From (18.10) and (18.11), the uploading sequence of the SNs greatly affects the average AoI but the maximum AoI.

The objective is to design two age-optimal SN-association and trajectory planning strategies. One is to minimize the age of the "oldest" sensing information among the SNs. The other is to minimize the average age of the sensing information. Accordingly, two combinatorial optimization problems are formulated as follows:

$$\begin{aligned}\mathcal{P}_1 : \quad & \min_{\mathbf{s},\zeta,\mathbf{u}} \quad \Gamma_{max}(\mathbf{s}, \zeta, \mathbf{u}) \\ s.t. \quad & \begin{cases} \sum_{k=1}^{K} \eta_{l,k} = 1, & (a) \\ \sum_{l=1}^{L} \zeta_{m,l} = 1, & (b) \\ \mathbf{s}^c_l \in \mathbb{R}^2, \zeta_{m,l} \in \{0,1\}, \mathbf{c} \in \sigma(\mathcal{C}), \forall l, m & (c) \end{cases}\end{aligned} \tag{18.12}$$

$$\begin{aligned}\mathcal{P}_2 : \quad & \min_{\mathbf{s},\zeta,\mathbf{v},\mathbf{u}} \quad \Gamma_{ave}(\mathbf{s}, \zeta, \mathbf{v}, \mathbf{u}) \\ s.t. \quad & \begin{cases} \sum_{k=1}^{K} \eta_{l,k} = 1, & (a) \\ \sum_{l=1}^{L} \zeta_{m,l} = 1, & (b) \\ \sum_{l=1}^{L} \zeta_{m,l} = 1 \mathbf{s}^c_l \in \mathbb{R}^2, \zeta_{m,l} \in \{0,1\}, \mathbf{c} \in \sigma(\mathcal{C}), \forall l, m & (c), \end{cases}\end{aligned} \tag{18.13}$$

where $\sigma(\mathcal{C})$ is the set including all the permutations of the set $\mathcal{C}$, and the flight trajectory $\mathbf{u} = [v_0, \mathbf{c}, v_0]$ depends on the permutation $\mathbf{c} \in \sigma(\mathcal{C})$. The optimal solutions to Problems $\mathcal{P}_1$ and $\mathcal{P}_2$ are denoted by $(\mathbf{s}^*_{max}, \zeta^*_{max}, \mathbf{u}^*_{max})$ and $(\mathbf{s}^*_{ave}, \zeta^*_{ave}, \mathbf{v}^*_{ave}, \mathbf{u}^*_{ave})$, respectively, referred to as the optimal maximum-age-of-information (max-AoI-optimal) and average-age-of-information (ave-AoI-optimal) strategies.

Notice that the SNs' average AoI can be further reduced when the SNs associated with each CP u_k sequentially upload sensing data in descending order of their offloading times: $T^1_{\pi(l)} \geq \cdots \geq T^j_{\pi(l)} \geq \cdots \geq T^{M_{\pi(l)}}_{\pi(l)}$ given any SN-CP association ζ and the CPs' positions $\mathbf{s}$. It is very difficult to solve Problems $\mathcal{P}_1$ and $\mathcal{P}_2$, which are NP-hard. In the sequel, a two-stage graph-based approach is proposed to solve these two problems in an efficient and unified way.

18.5.2 A Two-Stage Graph-Based Approach

In the first stage, the AP-based clustering method is used to find the CPs' positions $\mathbf{s}$ and establish the SN-CP association ζ. In the second stage, DP or other intelligent methods are applied to find the AoI-optimal flight trajectory $\mathbf{u}$.

AP-Based SN-CP Association

To reduce the search space, the AP clustering algorithm is adopted to set up the SN-CP association based on a restricted set of candidate CPs. The AP algorithm is designed to partition all the SNs into L clusters and identify some of them as cluster heads. It is applied to minimize the weighted sum of the SNs' uploading time and the number of clusters. Let δ denote the weighting factor. Here, the SN-CP association parameter $\zeta_{m,j} \in \{0, 1\}$ $(m,j = 1, \ldots, M)$ is used to indicate the co-cluster relationship between SN v_m and SN v_j. $\zeta_{m,j} = 1$ means whether SN v_m chooses SN v_j as its cluster head. Let $\mathcal{N}_m$ denote the set of the neighboring nodes of SN v_m with their distances less than or equal to the coverage radius r, that is, $\mathcal{N}_m = \{v_j | \|\mathbf{s}_j - \mathbf{s}_m\| \leq r, j \neq m, v_j \in \mathcal{V}\}$. And $\mathcal{N}^+_m$ represents SN v_m and its neighboring nodes, that is, $\mathcal{N}^+_m = \{v_m\} \cup \mathcal{N}_m$.

Based on the factor graph model and message passing mechanism, there are two kinds of iterative messages passed between adjacent SNs: (1) $\alpha_{m,j}$, a message sent by SN v_j to its neighbor v_m indicating to what degree it is suitable to serve as the cluster head; (2) $\gamma_{m,j}$, a message sent by SN v_m to its neighbor v_j indicating to what degree SN v_j is suitable to be its cluster head. Iteratively, the two messages are updated as follows:

$$\alpha_{m,j} = \begin{cases} \sum\limits_{i \in \mathcal{N}_j} \max\{\gamma_{i,j}, 0\}, & m = j, \\ \min\left\{0, \gamma_{j,j} + \sum\limits_{v_i \in \mathcal{N}_j \setminus \{v_m\}} \max\{\gamma_{i,j}, 0\}\right\}, & m \neq j, \end{cases} \tag{18.14}$$

$$\gamma_{m,j} = \zeta'_{m,j} - \max_{v_i \in \mathcal{N}^+_m \setminus \{v_j\}} \{\zeta'_{m,i} + \alpha_{m,i}\}. \tag{18.15}$$

Here, $\zeta'_{m,j}$ is defined as $\zeta'_{m,j} = -\zeta_{m,j} - \delta$ for $m = j$, and $\zeta'_{m,j} = -\zeta_{m,j}$ for $m \neq j$, respectively. Upon convergence, SN v_m is identified as a cluster head when its belief

$\gamma_{m,m} + \alpha_{m,m}$ is larger than zero. Thus, the set of cluster heads is found with their indices given by $\mathcal{L} = \{m|\gamma_{m,m} + \alpha_{m,m} > 0\}$. And each SN v_m is co-clustered with its cluster head v_j that satisfies $j = \arg\min_{j \in \mathcal{M}} \zeta'_{m,j}$. In this way, all the M SNs are partitioned into $L = |\mathcal{L}|$ clusters. Accordingly, the SN-CP association $\zeta = [\zeta_{m,l}]$ can be established. Based on this clustering result, a set of CPs with their locations $\mathbf{s}$ can be further obtained by solving the 1-center problem for each cluster of SNs, that is, $\mathbf{s}_l^c = \arg\min_{\mathbf{w}} \max_{\{v_m|\zeta_{m,l}=1\}} ||\mathbf{s}_m - \mathbf{w}||$.

Age-Optimal Trajectory Planning

From (18.10) and (18.11), the max-AoI-optimal and ave-AoI-optimal flight trajectories are shown to be the shortest Hamiltonian paths [21].

- The max-AoI-optimal trajectory is a shortest Hamiltonian path, where the distance between any two nodes c_i and c_j is measured by the flight time $\tau_{i,j}$ ($\forall c_i, c_j \in \mathcal{C} \cup \{v_0\}$).
- The ave-AoI-optimal trajectory is a stage-weighted shortest Hamiltonian path, where the distance between any two nodes c_i and c_j is measured by the sum of the flight time $\tau_{i,j}$ and the transmission time T_j at node c_j, and the stage weight is equal to the number of SNs visited from the first CP, that is, $\sum_{n=1}^{l} \frac{M_{\pi(n)}}{M} = 1 - \sum_{n=l+1}^{L+1} \frac{M_{\pi(n)}}{M}$, if node c_i is chosen as the lth CP.

Then, the dynamic programming (DP) method [33] can be applied to find the two AoI-optimal trajectories. However, its computational complexity is intolerable when the number of CPs is large. Hence, with the increase of the network scale, heuristic algorithms such as genetic algorithm (GA) can be applied to find the max-AoI-optimal and ave-AoI-optimal flight trajectories.

Algorithm Design

According to the preceding discussions, a two-stage algorithm is proposed to find the two AoI-optimal solutions approximately. The main procedure consists of two steps: (1) Run the AP-based clustering algorithm with parameter δ to find the positions of the CPs and the SN-CP association; (2) Apply the DP/GA-based trajectory planning algorithm to find an age-optimal trajectory through all the CPs. These two steps are performed alternately until the clustering result remains unchanged. Furthermore, the parameter δ can be adjusted to control the cluster size so as to achieve a good balance between the SNs' uploading time and the UAV's flight time. Hence, the parameter δ significantly affects the age-optimal clustering and trajectory planning result.

The Multi-UAV Scenario

When one single UAV cannot fulfill the data collection task due to limited energy capacity, the BS has to send multiple UAVs to gather the sensing data from the SNs collaboratively. In this scenario, all the CPs are visited by multiple UAVs exclusively. Accordingly, the multistage data collection framework can be decomposed as follows.

Firstly, the positions of the CPs are determined and the SN-CP association is established based on graph theory; secondly, the CPs are classified into clusters based on the some clustering method, and each cluster of CPs is visited by one UAV; thirdly, the max-AoI-optimal or ave-AoI-optimal flight trajectory is found subject to the limited energy capacity of each UAV. The key is to figure out how many UAVs should be applied in data collection.

18.6 A DRL-Based Framework for Continuous Data Collection

An MDP problem is formulated to model the continuous real-time data collection problem. Then, a DRL-based approach is proposed for the single-UAV scenario. The multi-UAV scenario is discussed afterward.

18.6.1 MDP Formulation

At first, the state space, action space, and long-term expected cost of the MDP problem are described. The system state in each time slot n is defined as $S_n \triangleq (\boldsymbol{\Gamma}(n), \boldsymbol{U}(n), \boldsymbol{u}^{hor}(n), e(n))$, which consists of the following four parts:

- The AoI values of the M SNs $\boldsymbol{\Gamma}(n) = (\Gamma_1(n), \cdots, \Gamma_M(n)) \in \prod_{m=1}^{M} \mathcal{I}_m$;
- The lifetimes of the SNs' update packets $\boldsymbol{U}(n) = (U_1(n), \cdots, U_M(n)) \in \prod_{m=1}^{M} \mathcal{U}_m$, where $\mathcal{U}_m = \{0\} \cup \mathcal{T}_m$;
- The UAV's horizontal location $\boldsymbol{u}^{hor}(n) \in \mathcal{W}$;
- The UAV's residual energy $e(n) \in \mathcal{E} = [0, E_{max}]$, where E_{max} is the battery capacity of the UAV.

Hence, the state space can be characterized by $\mathcal{S} = \prod_{m=1}^{M} \mathcal{I}_m \times \prod_{m=1}^{M} \mathcal{U}_m \times \mathcal{W} \times \mathcal{E}$.

At the beginning of each time slot n, the UAV selects the next flight direction and uploads SN. Accordingly, the system action is expressed as $a_n = [a_h(n), a_v(n), \omega(n)] \in \mathcal{A}_c = \{-L_r, \cdots, L_r\}^2 \times \{0, 1, \cdots, M\}$. Here, $a_h(n)$ and $a_v(n)$ denote the number of grids the UAV flies horizontally and vertically, respectively, and L_r is the largest number of grids the UAV can fly horizontally or vertically due to the maximum velocity limit. No SN is selected to upload when $\omega(n) = 0$, and SN v_m will upload for $\omega(n) = m \in \{1, \cdots, M\}$.

The AoI of each SN and the lifetime of its update packet are updated according to (18.9) and (18.8), respectively. If some SN v_m uploads its update packet to the UAV at $\mathbf{u}(n)$, its AoI is reset to $U_m(n)$ and increases by one if no packet delivery takes place. The dynamics of the UAV's horizontal location $\boldsymbol{u}^{hor}(n)$ is expressed as

$$\mathbf{u}^{hor}(n+1) = \mathbf{u}^{hor}(n) + [a_h(n)L_x, a_v(n)L_y]. \tag{18.16}$$

When the UAV reaches one of the boundary grids, some special process is required to prevent the UAV from flying outside the region. If the UAV has sufficient energy, that is, $e(n-1) > E_{th}$, the selected SN v_m gets collected with $z_m(n) = 1$, where E_{th} is the energy threshold. In this case, the UAV's energy consumption includes the transmission energy, hovering energy, and propulsion energy for moving forward. Otherwise, if no SN is selected, that is, $\omega(n) = 0$ or there is no update packet, that is, $U_m(n) = 0$, no SN is served with $\mathbf{z}(n) = \mathbf{0}$, and the UAV flies directly to the next target grid. The UAV's energy consumption involves only propulsion energy. If the UAV's residual energy reaches the threshold E_{th}, the UAV has to fly directly to the BS for energy replenishment, which could deplete all the residual energy. The energy consumption in slot n can be regarded as $e_c(n) = e(n-1)$. Hence, the UAV's energy consumption in slot n is calculated as

$$e_c(n) = \begin{cases} e_{tx}(n) + e_{hv}(n) + e_{f1}(n), & \text{if}\, e(n-1) > E_{th} \& z_m(n) = 1, \\ e_{f2}(n), & \text{if}\, e(n-1) > E_{th} \& \mathbf{z}(n) = \mathbf{0}, \\ e(n-1), & \text{otherwise}, \end{cases} \tag{18.17}$$

where $e_{f1}(n)$ and $e_{f2}(n)$ are the propulsion energy consumptions with different flight velocities for one packet delivery and no packet delivery, respectively. The evolution of the UAV's residual energy is given by $e(n) = e(n-1) - e_c(n)$. After recharging, the UAV has an initial amount of E_{max} (Joule) energy.

When action a_n is taken at state S_n, the cost function $C(S_n, a_n)$ is modeled as a weighted sum of the SNs' average AoI and the UAV's energy consumption, as given by

$$r_n = C(S_n, a_n) = \frac{1}{M} \sum_{m=1}^{M} \Gamma_m(n) + \delta_1 \cdot e_c(n), \tag{18.18}$$

where δ_1 is the weighting factor. Then, an MDP problem is formulated to find an optimal policy Π so as to minimize the long-term expected cost,

$$\bar{C}_{\gamma,\Pi} = \frac{1}{N} \mathbb{E}_{\Pi}[\sum_{n=0}^{N} \gamma^n C(S_n, a_n) | S_0], \tag{18.19}$$

subject to the UAV's residual energy in each slot, where $\mathbb{E}_{\Pi}$ means taking expectation on the policy Π, S_0 is the initial system state, and $\gamma \in [0, 1]$ is the discount rate.

18.6.2 A DRL-Based Approach

Our goal is to find the optimal policy that minimizes the long-term expected cost, that is, the cumulative future return. To this end, we can use the model-free Q-learning algorithm to estimate the action-value function $Q(S, a)$ of taking an action a at a given state S when following the optimal policy Π^*. In fact, Q-learning estimates the optimal $Q^*(S, a)$ iteratively from the sequence of samples. In each time slot, the estimated Q-function is updated as $Q(S_n, a_n) = (1-\alpha)Q(S_n, a_n) + \alpha[r_n + \min_{a'} Q(S_{n+1}, a')]$, where α is the learning rate. However, when the state and action spaces become large, it is highly challenging to compute and update the action-value function $Q(S, a)$ for each state-action pair (S, a). To overcome the curse of dimensionality, we adopt DQN as a remedy, which combines Q-learning with deep neural networks, to solve the UAV-aided data collection problem.

In DQN, one deep neural network (DNN) is used as the approximator of the Q-function. To improve stability of DNN, the authors of [34] designed two effective techniques: experience replay and target network. Thus, DQN contains two neural networks of the same structure with different parameters $\boldsymbol{\theta}$ and $\boldsymbol{\theta}^{'}$, respectively. One is the evaluation network, which accepts the input current state S_n and outputs the evaluation of the value $Q(S_n, a_n; \boldsymbol{\theta})$. The other one is the target network, which accepts the next state S_{n+1} as input and outputs the evaluation of the value $Q(S_{n+1}, a_{n+1}; \boldsymbol{\theta}^{'})$. The neural network can be trained by minimizing the loss function: $J(\boldsymbol{\theta}) = \mathbb{E}[(\psi_n - Q(S_n, a_n; \boldsymbol{\theta}))^2]$, where ψ_n is the target value estimated by the target network as $\psi_n = C(S_n, a_n) + \gamma \cdot \max\limits_{a_{n+1}} Q(S_{n+1}, a_{n+1}; \boldsymbol{\theta}^{'})$.

The main procedure of the DQN-based algorithm is described as follows.

Step 1: Initialize $Q(S_n, a_n; \boldsymbol{\theta}) = Q(S_n, a_n; \boldsymbol{\theta}^{'}) = 0$ with random weights $\boldsymbol{\theta}^{'} = \boldsymbol{\theta}$, and start an episode with an initial state S_0.

Step 2: The agent selects an action a_n randomly with probability ε_1 and action $a_n = \arg\min\limits_{a_n} Q(S_n, a_n)$ with probability $1 - \varepsilon_1$, respectively; it executes action a_n, updates the AoI $\boldsymbol{\Gamma}(n)$, energy consumption $e_c(n)$, and the immediate cost $C(S_n, a_n)$ by (18.9), (18.17), and (18.18), respectively; the system transits to the next state S_{n+1}.

Step 3: Store transition (S_n, a_n, r_n, S_{n+1}) in replay memory with random replacement.

Step 4: Sample a mini-batch of transitions (S_n, a_n, r_n, S_{n+1}) from replay memory; calculate the target network value ψ_n; update the weights of the evaluation network as $\boldsymbol{\theta} = \boldsymbol{\theta} + \nabla_{\boldsymbol{\theta}} J(\boldsymbol{\theta})$, and update the weights of the target network as $\boldsymbol{\theta}^{'} = \boldsymbol{\theta}$ every N_s steps.

Step 5: Update the UAV's residual energy $e(n) = e(n) - e_c(n)$.

Step 6: Repeat the above steps till the end of this episode, that is, the UAV's residual energy $e(n)$ is less than the energy threshold E_{th}.

18.6.3 The Multi-UAV Scenario

The MDP formulation and DRL solution framework can be extended to the multi-UAV scenario. However, there still exist two major challenges. One is to avoid crashes and severe interference between the UAVs by efficient trajectory planning. The other is to deal with a higher-dimensional problem with the enlarged state and action spaces when more UAVs are used and more constraints are considered. Therefore, more intelligent DRL algorithms should be adopted to speed up convergence in the training period.

18.7 Numerical Results

In this section, numerical results are presented to show the AoI performances of the UAV-aided data collection methods in various scenarios. Assume that the BS is located at the origin $(0, 0)$ and the M SNs are randomly located in a UAV-enabled WSN. In the preplanned single data collection mode with one UAV, the SNs' minimum average

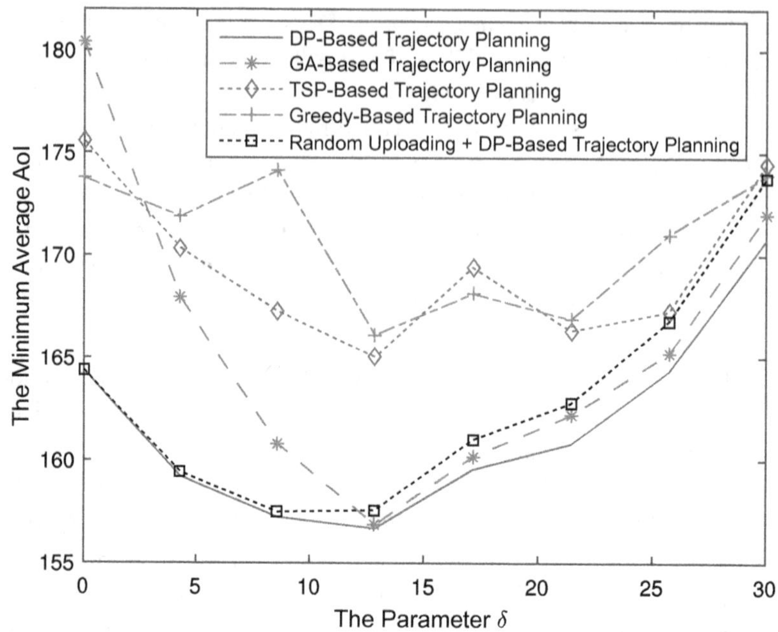

Figure 18.3 The SNs' minimum average AoI versus the parameter δ.

AoI curves are plotted in Figure 18.3, when different trajectory planning algorithms are applied jointly with the AP-based SN-CP association algorithm. The experiment parameters are set as follows: $H = 50$ m, $V = 20$ m/s, $r = 1000$ m, $W = 10^7$ Hz, $P_m = 0.1$ W, $a = 9.61, b = 0.16$, $\beta_0 = -60$ dB, $\varrho = 0.7$, $\sigma^2 = -110$ dBm, $L_m = 10^6$ bits. As shown in Figure 18.3, the SNs' average AoIs decrease or fluctuate to the minimum values and then increase with the increase of δ. This is due to the fact that the SNs' overall uploading time monotonously increases while the UAV's flight time monotonously decreases when δ is increased. Considering these two opposite effects, it might not be optimal for the UAV to collect data from the SNs one by one when $\delta = 0$. Among the four trajectory planning algorithms, the DP-based algorithm achieves the optimal AoI performance for any δ, since it finds the ave-AoI-optimal trajectory by comparing all the candidate flight paths of the UAV. Through intelligent search the GA-based algorithm performs quite well, especially when δ is relatively large. In contrast, the TSP-based algorithm does not perform well, which means that the TSP trajectories are not age-optimal ones. The greedy algorithm, which visits the nearest unvisited CP and collects data from the associated SNs, performs the worst in most cases. As can be seen, the SNs' random uploading causes an enlarged average AoI, especially when $\delta \geq 10$, which means the uploading sequence of SNs does affect the AoI performance.

In the preplanned single data collection scenario with multiple UAVs, the SNs' maximal AoI performance subject to the energy capacity of each UAV is demonstrated in Figure 18.4. In this experiment the related parameters are listed as follows:

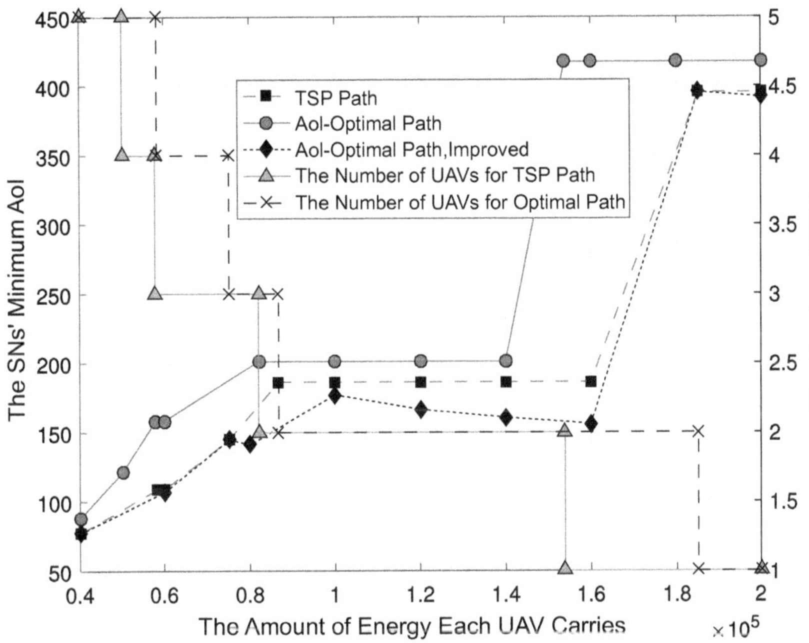

Figure 18.4 The SNs' maximal AoI versus the amount of energy carried by each UAV.

$H = 100$ m, $V_k = 20$ m/s, $\epsilon = 2.3$, $P_0 = 79.0$ W, $P_1 = 88.6$ W, $U_{tip} = 120$ m/s, $V_0 = 4.3$ m/s, $\rho = 1.225$ kg/m^3, $d = 0.6$, $s = 0.05$, $A_r = 0.503$ m^2, $\sigma^2 = -100$ dBm, $L_m = 2 * 10^6$ bits, $W = 10^6$ Hz, $P_m = 1$ W. The SNs are randomly located in a bounded area of size 2,000 m × 2,000 m, and the coverage radius r of each SN is defined as 40 m. When the max-AoI-optimal and TSP trajectory planning methods are applied, the SNs' maximal AoI increases monotonically, while the required number of UAVs decreases with the increase of the energy capacity at each UAV. It is easily understood that fewer UAVs can be applied in the data collection task if each UAV carries more energy. At the same time, each UAV has to visit and collect data from more ground SNs before flying back to the BS, which causes a larger AoI. Compared to the TSP trajectory, the max-AoI-optimal trajectory always achieves a smaller AoI while more UAVs might be required. The SNs' maximal AoI can be reduced by increasing the UAVs' flight velocity when more energy is carried by each UAV.

In the continuous real-time data collection scenario, numerical results of the DQN and Sarsa algorithm are presented for performance comparison. Combining the ideas of Monte Carlo and DP, Sarsa is a simple on-policy temporal-difference learning method that can gradually converge to a close-to-optimal solution [35]. In implementation of DQN, a fully connected feedforward neural network with one layer of 300 neurons is applied and the ReLu activation function is adopted. The experiment parameters are set as follows: $M = 3$, $\mathbf{s}_1 = [320, 280]$, $\mathbf{s}_2 = [360, 120]$, $\mathbf{s}_3 = [80, 40]$, $J = 90$, $L_x = L_y = 40$, $T_s = 1$ s, $T_{tx} = 0.3$ s, $T_{fl} = 0.7$ s, $a = 10, b = 0.6$, $\beta_0 = -60$ dB, $\varrho = 0.2$, $E_{max} = 1.2 * 10^5$ J, $E_{th} = 8 * 10^3$ J, $W = 5 * 10^6$ Hz,

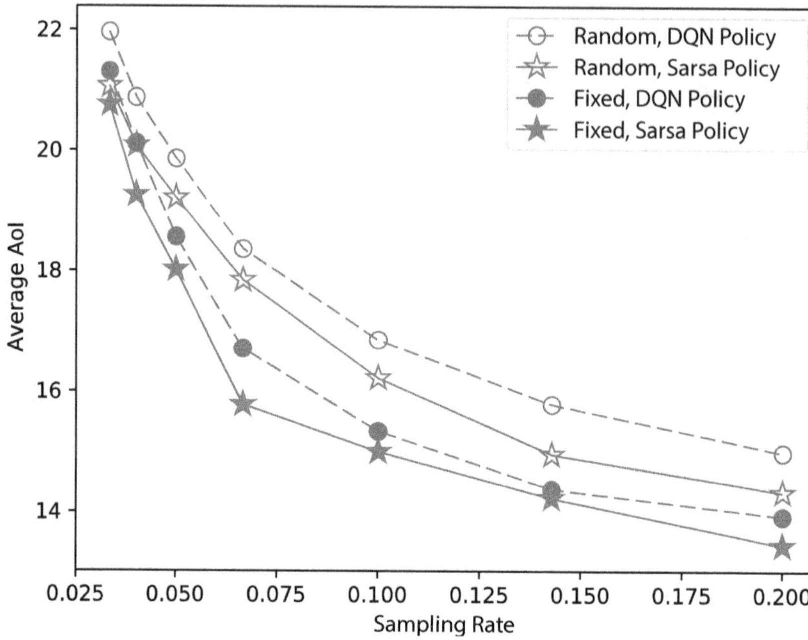

Figure 18.5 The SNs' average AoI versus sampling rate.

$P_0 = 99.66$ W, $P_1 = 120.16$ W, $\delta = 0.002$. The learning rate is set as 0.008. The Sarsa and DQN polices, achieve a lower average AoI when applying fixed sampling instead of random sampling at the same rate. Compared with the fixed sampling model, the sampling intervals in random sampling may become very large or small from time to time, which is more difficult to be predicted. Hence, the random sampling mechanism causes a lot of trouble in predicting the information arrival instants, resulting in a larger average AoI. For both Sarsa and DQN policies, a higher sampling rate leads to a smaller average AoI since the SNs sample more frequently and deliver more fresh information to the BS with the aid of the UAV. When random or fixed sampling is applied, the DQN policy achieves a comparable AoI performance to the Sarsa policy that approximately converges to the optimal solution.

18.8 Summary and Future Work

This chapter proposed different ways to deal with the age-optimal data collection problem in UAV-enabled WSNs. Offline and online algorithms were designed to improve the AoI performance in single and continuous data collection scenarios, respectively. In the former case, a two-stage data collection framework was proposed to find the max-AoI-optimal or ave-AoI-optimal trajectory along multiple CPs based on optimization methods. In the later case, a DRL-based framework was proposed to make real-time decisions on which SN to collect and which direction to fly at each step. Numerical results showed that the proposed data collection methods can achieve a very significant improvement in terms of the maximum or average AoI in different scenarios compared to baseline algorithms.

The proposed frameworks can be applied to many age-optimal data collection scenarios. However, the following issues should be considered to provide more practical solutions.

1. When the network scale is relatively large, it is difficult to apply the multistage framework since the computation complexity could be too high for practical use. Instead, a learning-based framework may be used in the age-optimal trajectory planning stage, which is usually computation-intensive.
2. Similarly, more efficient online learning algorithms should be considered to make quick decisions, when more UAVs are applied to large-scale WSNs subject to the constraints on flight, communication, and energy capacity.
3. To support sustainable data service, efficient and safe energy replenishment strategies should be carefully designed. It brings more challenges to the design of the UAV trajectory planning and communications, when different energy charging methods, such as wireless transfer and energy harvesting, are considered.

References

[1] R. Z. Y. Zeng and T. J. Lim, "Wireless communications with unmanned aerial vehicles: Opportunities and challenges," *IEEE Communications Magazine*, vol. 54, no. 5, pp. 36–42, May 2016.

[2] G. Xing, T. Wang, Z. Xie, and W. Jia, "Rendezvous planning in wireless sensor networks with mobile elements," *IEEE Transactions on Mobile Computing*, vol. 7, no. 12, pp. 1430–1443, Dec. 2008.

[3] E. Y. S. Hayat and R. Muzaffar, "Survey on unmanned aerial vehicle networks for civil applications: A communications viewpoint," *IEEE Communications Surveys & Tutorials*, vol. 18, no. 4, pp. 2624–2661, 2016.

[4] N. H. Motlagh, M. Bagaa, and T. Taleb, "UAV-based IoT platform: A crowd surveillance use case," *IEEE Communications Magazine*, vol. 55, no. 2, pp. 128–134, Feb. 2017.

[5] D. Yang, Q. Wu, Y. Zeng, and R. Zhang, "Energy tradeoff in ground-to-UAV communication via trajectory design," *IEEE Transactions on Vehicular Technology*, vol. 67, no. 7, pp. 6721–6726, Jul. 2018.

[6] Y. Zeng and R. Zhang, "Energy-efficient UAV communication with trajectory optimization," *IEEE Transactions on Wireless Communications*, vol. 16, no. 6, pp. 3747–3760, Jun. 2017.

[7] A. E. A. A. Abdulla, Z. M. Fadlullah, H. Nishiyama, N. Kato, F. Ono, and R. Miura, "An optimal data collection technique for improved utility in UAS-aided networks," in *Proc. IEEE INFOCOM*, Toronto, Canada, May 2014, pp. 736–744.

[8] S. Say, H. Inata, J. Liu, and S. Shimamoto, "Priority-based data gathering framework in UAV-assisted wireless sensor networks," *IEEE Sensors Journal*, vol. 16, no. 14, pp. 5785–5794, Jul. 2016.

[9] C. Zhan, Y. Zeng, and R. Zhang, "Energy-efficient data collection in UAV enabled wireless sensor network," *IEEE Wireless Commun. Letters*, vol. 7, no. 3, Jun. 2018.

[10] C. Liu, Z. Wei, Z. Guo, X. Yuan, and Z. Feng, "Performance analysis of uavs assisted data collection in wireless sensor network," in *Proc. IEEE VTC Spring*, Porto, Portugal, Jun. 2018.

[11] M. Mozaffari, W. Saad, M. Bennis, and M. Debbah, "Mobile unmanned aerial vehicles (UAVs) for energy-efficient internet of things communications," *IEEE Transactions on Wireless Communications*, vol. 16, no. 11, pp. 7574–7589, Nov. 2017.

[12] J. Gong, T. Chang, C. Shen, and X. Chen, "Flight time minimization of UAV for data collection over wireless sensor networks," *IEEE Journal on Selected Areas in Communications*, vol. 36, no. 9, pp. 1942–1954, Sep. 2018.

[13] S. Kaul, M. Gruteser, V. Rai, and J. Kenney, "Minimizing age of information in vehicular networks," in *Proc. IEEE SECON*, Salt Lake City, USA, Jun. 2011.

[14] S. Kaul, R. Yates, and M. Gruteser, "Real-time status: How often should one update?" in *Proc. INFOCOM*, Orlando, FL, USA, Mar. 2012, pp. 2731–2735.

[15] C. Kam, S. Kompella, and A. Ephremides, "Age of information under random updates," in *Proc. IEEE ISIT*, Jul. 2013, pp. 66–70.

[16] Y. Sun, E. Uysal-Biyikoglu, R. Yates, C. E. Koksal, and N. B. Shroff, "Update or wait: How to keep your data fresh," *IEEE Transactions on Information Theory*, vol. 63, no. 11, pp. 7492–7508, Nov. 2017.

[17] R. Talak, S. Karaman, and E. Modiano, "Optimizing information freshness in wireless networks under general interference constraints," in *Proc. ACM MobiHoc*, Jun. 2018.

[18] V. Tripathi, R. Talak, and E. Modiano, "Age optimal information gathering and dissemination on graphs," in *Proc. IEEE INFOCOM*, Paris, France, Apr. 2019, pp. 2422–2430.

[19] M. A. Abd-Elmagid and H. S. Dhillon, "Average peak age-of-information minimization in UAV-assisted iot networks," *IEEE Transactions on Vehicular Technology*, vol. 68, no. 2, pp. 2003–2008, 2018.

[20] J. Liu, X. Wang, B. Bai, and H. Dai, "Age-optimal trajectory planning for UAV-assisted data collection," in *Proc. IEEE INFOCOM WKSHPS*, Honolulu, HI, Apr. 2018.

[21] P. Tong, J. Liu, X. Wang, B. Bai, and H. Dai, "UAV-enabled age-optimal data collection in wireless sensor networks," in *Proc. IEEE ICC Workshop*, Shanghai, China, May 2019.

[22] Z. Jia, X. Qin, Z. Wang, and B. Liu, "Age-based path planning and data acquisition in UAV-assisted iot networks," in *Proc. IEEE ICC Workshops*, Shanghai, China, May. 2019.

[23] G. Ahani, D. Yuan, and Y. Zhao, "Age-optimal UAV scheduling for data collection with battery recharging," *accessed at https://arxiv.org/abs/2005.00252v1*, May 2020.

[24] W. Li, L. Wang, and A. Fei, "Minimizing packet expiration loss with path planning in UAV-assisted data sensing," *IEEE Wireless Communications Letters*, vol. 8, no. 6, pp. 1520–1523, Dec. 2019.

[25] C. Zhou, H. He, P. Yang, F. Lyu, W. Wu, N. Cheng, and X. Shen, "Deep RL-based trajectory planning for AoI minimization in UAV-assisted IoT," in *Proc. WCSP*, Xi'an, China, Oct. 2019, pp. 1–6.

[26] M. A. Abd-Elmagid, A. Ferdowsi, H. S. Dhillon, and W. Saad, "Deep reinforcement learning for minimizing age-of-information in UAV-assisted networks," in *Proc. IEEE GLOBECOM*, Waikoloa, HI, USA, Dec. 2019.

[27] A. Ferdowsi, M. A. Abd-Elmagid, W. Saad, and H. S. Dhillon, "Neural combinatorial deep reinforcement learning for age-optimal joint trajectory and scheduling design in UAV-assisted networks," *accessed at https://arxiv.org/abs/2006.15863*, Jun. 2020.

[28] S. F. Abedin, M. S. Munir, N. H. Tran, Z. Han, and C. S. Hong, "Data freshness and energy-efficient UAV navigation optimization: A deep reinforcement learning approach," *IEEE Transactions on Intelligent Transportation Systems*, 2020.

[29] M. Mozaffari, W. Saad, M. Bennis, and M. Debbah, "Unmanned aerial vehicle with underlaid device-to-device communications: Performance and tradeoffs," *IEEE Transactions on Wireless Communications*, vol. 15, no. 6, pp. 3949–3963, Jun. 2016.

[30] Y. Zeng, J. Xu, and R. Zhang, "Energy minimization for wireless communication with rotary-wing uav," *IEEE Transactions on Wireless Communications*, vol. 18, no. 4, pp. 2329–2345, Apr. 2019.

[31] A. Hourani, S. Kandeepan, and A. Jamalipour, "Modeling air-to-ground path loss for low altitude platforms in urban environments," in *Proc. IEEE GLOBECOM*, Austin, TX, USA, Dec. 2014.

[32] P. Tong, J. Liu, X. Wang, B. Bai, and H. Dai, "UAV-enabled age-optimal data collection in wireless sensor networks," in *Proc. IEEE ICC Workshops*, Shanghai, China, May, 2019.

[33] D. P. Bertsekas, *Dynamic programming and optimal control.* Boston, MA: Athena Scientific, 2000.

[34] V. Mnih, K. Kavukcuoglu, D. Silver, A. A. Rusu, J. Veness, M. G. Bellemare, A. Graves, M. Riedmiller, A. K. Fidjeland, G. Ostrovski et al., "Human-level control through deep reinforcement learning," *Nature*, vol. 518, no. 7540, p. 529, 2015.

[35] R. S. Sutton and A. G. Barto, *Reinforcement learning: An introduction.* Massachusetts Institute of Technology Press, 2018.

Index

For EU product safety concerns, contact us at Calle de José Abascal, 56–1°,
28003 Madrid, Spain or eugpsr@cambridge.org.

www.ingramcontent.com/pod-product-compliance
Lightning Source LLC
LaVergne TN
LVHW060055220926
848484LV00006BA/1306
* 9 7 8 1 1 0 8 8 3 7 8 7 3 *